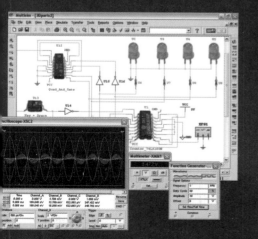

DIGITAL FUNDAMENTALS

Ninth Edition

Thomas L. Floyd

PEARSON

Prentice Hall

Upper Saddle River, New Jersey
Columbus, Ohio

Library of Congress Cataloging-in-Publication Data

Floyd, Thomas L.
   Digital fundamentals / Thomas L. Floyd.—9th ed.
     p. cm.
   Includes index.
   ISBN 0-13-194609-9
    1. Digital electronics. 2. Logic circuits. I. Title.

TK7868.D5F53 2006
621.381—dc22

2005047607

Acquisitions Editor: Kate Linsner
Production Editor: Rex Davidson
Design Coordinator: Diane Ernsberger
Editorial Assistant: Lara Dimmick
Cover Designer: Jim Hunter
Cover art: Getty One
Production Manager: Matt Ottenweller
Marketing Manager: Ben Leonard
Illustrations: Jane Lopez

This book was set in Times Roman by Carlisle Communications, Ltd. and was printed and bound by Courier Kendallville, Inc. The cover was printed by Coral Graphic Services, Inc.

Chapter opening photos and special feature photos by Getty Images.

Multisim is a registered trademark of Electronics Workbench.
Altera Quartus II and other names of Altera products, product features, and services are trademarks and/or service marks of Altera Corporation in the United States and other countries.
Xilinx ISE is a registered trademark of Xilinx, Inc.

Pearson Education Ltd.
Pearson Education Singapore Pte. Ltd.
Pearson Education Canada, Ltd.
Pearson Education—Japan

Pearson Education Australia Pty. Limited
Pearson Education North Asia Ltd.
Pearson Educación de Mexico, S.A. de C.V.
Pearson Education Malaysia Pte. Ltd.

10 9 8 7 6 5 4 3 2
ISBN: 0-13-194609-9

Preface

Welcome to *Digital Fundamentals, Ninth Edition.* A strong foundation in the core fundamentals of digital technology is vital to anyone pursuing a career in this exciting, fast-paced industry. This text is carefully organized to include up-to-date coverage of topics that can be covered in their entirety, used in a condensed format, or omitted altogether, depending upon the course emphasis.

The topics in this text are covered in the same clear, straightforward, and well-illustrated format that has been so successful in the previous editions of *Digital Fundamentals.* Many topics have been strengthened or enhanced and numerous improvements can be found throughout the book.

You will probably find more topics than you can cover in a single course. This range of topics provides the flexibility to accommodate a variety of program requirements. For example, some of the design-oriented or system application topics may not be appropriate in some courses. Other programs may not cover programmable logic, while some may not have time to include topics such as computers, microprocessors, or digital signal processing. Also, in some courses there may be no need to go into the details of "inside-the-chip" circuitry. These and other topics can be omitted or lightly covered without affecting the coverage of the fundamental topics. A background in transistor circuits is not a prerequisite for this textbook although coverage of integrated circuit technology (inside-the-chip circuits) is included in a "floating chapter," which is optional.

Following this Preface is a color-coded table of contents to indicate a variety of approaches for meeting most unique course requirements. The text has a modular organization that allows inclusion or omission of various topics without impacting the other topics that are covered in your course. Because programmable logic continues to grow in importance, an entire chapter (Chapter 11) is devoted to the topic, including PALs, GALs, CPLDs, and FPGAs; specific Altera and Xilinx devices are introduced. Also a generic introduction to programmable logic software is provided and boundary scan logic is covered.

New in This Edition

- The Hamming error detecting and correcting code

- Carry look-ahead adders

- A brief introduction to VHDL

- Expanded and improved coverage of test instruments

- An expanded and reorganized coverage of programmable logic

- Improved troubleshooting coverage

- New approach to Digital System Applications

Features

- Full-color format

- Margin notes provide information in a very condensed form.

- Key terms are listed in each chapter opener. Within the chapter, the key terms are in boldface color. Each key term is defined at the end of the chapter, as well as at the end of the book in the comprehensive glossary along with other glossary terms that are indicated by black boldface in the text.

- Chapter 14 is designed as a "floating chapter" to provide optional coverage of IC technology (inside-the-chip circuitry) at any point in your course.

- Overview and objectives in each chapter opener

- Introduction and objectives at the beginning of each section within a chapter

- Review questions and exercises at the end of each section in a chapter

- A Related Problem in each worked example

- Computer Notes interspersed throughout to provide interesting information about computer technology as it relates to the text coverage

- Hands-On Tips interspersed throughout to provide useful and practical information

- The Digital System Application is a feature at the end of many chapters that provides interesting and practical applications of logic fundamentals.

- Chapter summaries at the end of each chapter

- Multiple-choice self-test at the end of each chapter

- Extensive sectionalized problem sets at the end of each chapter include basic, troubleshooting, system application, and special design problems.

- The use and application of test instruments, including the oscilloscope, logic analyzer, function generator, and DMM, are covered.

- Chapter 12 provides an introduction to computers.

- Chapter 13 introduces digital signal processing, including analog-to-digital and digital-to-analog conversion.

- Concepts of programmable logic are introduced beginning in Chapter 1.

- Specific fixed-function IC devices are introduced throughout.

- Chapter 11 provides a coverage of PALs, GALs, CPLDs and FPGAs as well as a generic coverage of PLD programming.

- Selected circuit diagrams in the text, identified by the special icon shown here, are rendered in Multisim® 2001 and Multisim® 7, and these circuit files are provided on the enclosed CD-ROM. These files (also available on the Companion Website at **www.prenhall.com/floyd**) are provided at no extra cost to the consumer and are for use by anyone who chooses to use Multisim software. Multisim is widely regarded as an excellent simulation tool for classroom and laboratory learning. However, successful use of this textbook is not dependent upon use of the circuit files.

- Boundary scan logic associated with programmable devices is introduced in Chapter 11.

- In addition to boundary scan, troubleshooting coverage includes methods for testing programmable logic, such as traditional, bed-of-nails, and flying probe. Boundary scan and these other methods are important in manufacturing and industry.

- For those who wish to include ABEL programming, an introduction is provided on the Companion Website at **www.prenhall.com/floyd.**

Accompanying Student Resources

■ *Experiments in Digital Fundamentals*, a laboratory manual by David M. Buchla. Solutions for this manual are available in the Instructor's Resource Manual.

■ Two CD-ROMs included with each copy of the text:
   Circuit files in Multisim for use with Multisim software
   Texas Instruments digital devices data sheets

Instructor Resources

■ *PowerPoint*® *slides.* These presentations feature Lecture Notes and figures from the text. (On CD-ROM and online.)

■ *Companion Website.* (**www.prenhall.com/floyd**). For the instructor, this website offers the ability to post your syllabus online with our Syllabus Manager™. This is a great solution for classes taught online, that are self-paced, or in any computer-assisted manner.

■ *Instructor's Resource Manual.* Includes worked-out solutions to chapter problems, solutions to Digital System Applications, a summary of Multisim simulation results, and worked-out lab results for the lab manual by David M. Buchla. (Print and online.)

■ *Test Item File.* This edition of the Test Item File features over 900 questions.

■ *TestGen.*® This is an electronic version of the Test Item File, enabling instructors to customize tests for the classroom.

To access supplementary materials online, instructors need to request an instructor access code. Go to **www.prenhall.com**, click the **Instructor Resource Center** link, and then click **Register Today** for an instructor access code. Within 48 hours after registering you will receive a confirming e-mail including an instructor access code. Once you have received your code, go to the website and log on for full instructions on downloading the materials you wish to use.

Illustration of Chapter Features

Chapter Opener   Each chapter begins with a two-page spread, as shown in Figure P–1. The left page includes a list of the sections in the chapter and a list of chapter objectives. A typical right page includes an overview of the chapter, a list of specific devices introduced in the chapter (each new device is indicated by an IC logo at the point where it is introduced), a brief Digital System Application preview, a list of key terms, and a website reference for chapter study aids.

Section Opener   Each of the sections in a chapter begins with a brief introduction that includes a general overview and section objectives. An illustration is shown in Figure P–2.

Section Review   Each section ends with a review consisting of questions or exercises that emphasize the main concepts presented in the section. This feature is shown in Figure P–2. Answers to the Section Reviews are at the end of the chapter.

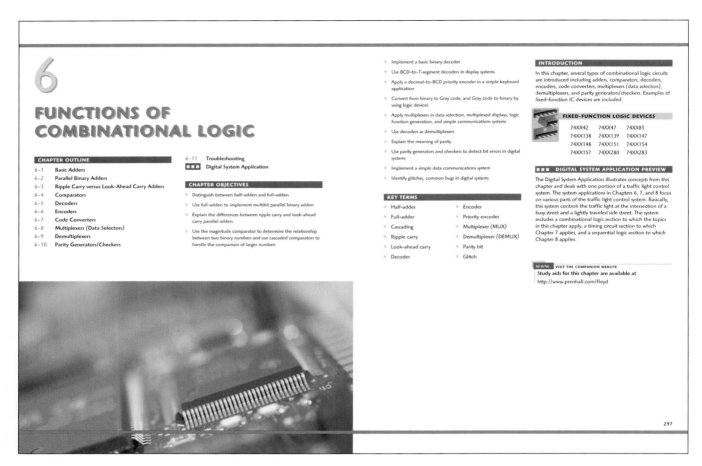

▲ FIGURE P–1

Chapter opener.

Worked Examples and Related Problems   An abundance of worked examples help to illustrate and clarify basic concepts or specific procedures. Each example ends with a Related Problem that reinforces or expands on the example by requiring the student to work through a problem similar to the example. A typical worked example with a Related Problem is shown in Figure P–3.

Troubleshooting Section   Many chapters include a troubleshooting section that relates to the topics covered in the chapter and that emphasizes troubleshooting techniques and the use of test instruments. A portion of a typical troubleshooting section is illustrated in Figure P–4.

Digital System Application   Appearing at the end of many chapters, this feature presents a practical application of the concepts covered in the chapter. This feature presents a "real-world" system in which analysis, troubleshooting, and design elements are implemented using procedures covered in the chapter. Some Digital System Applications are limited to a single chapter and others extend over two or more chapters. Specific Digital System Applications are as follows:

- Tablet counting and control system: Chapter 1

- Digital display: Chapters 4 and 11.

- Storage tank control system: Chapter 5

▶ FIGURE P–2

Section opener and section review.

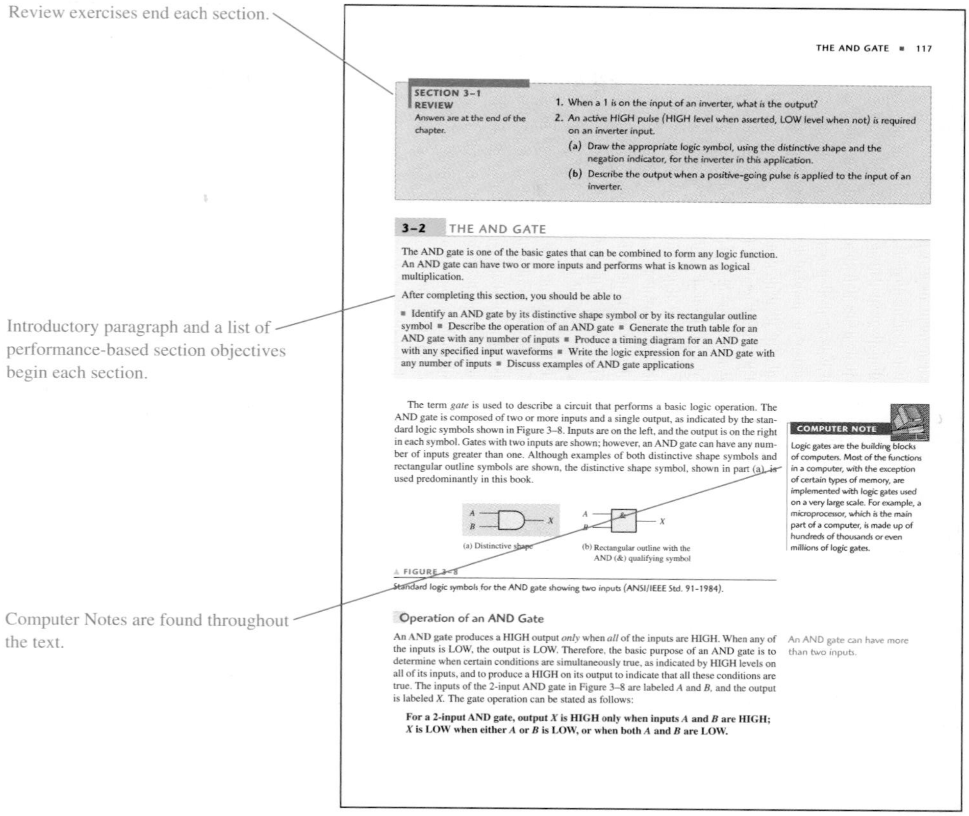

Review exercises end each section.

SECTION 3–1
REVIEW
Answers are at the end of the chapter.

1. When a 1 is on the input of an inverter, what is the output?
2. An active HIGH pulse (HIGH level when asserted, LOW level when not) is required on an inverter input.
   (a) Draw the appropriate logic symbol, using the distinctive shape and the negation indicator, for the inverter in this application.
   (b) Describe the output when a positive-going pulse is applied to the input of an inverter.

Introductory paragraph and a list of performance-based section objectives begin each section.

3–2   THE AND GATE

The AND gate is one of the basic gates that can be combined to form any logic function. An AND gate can have two or more inputs and performs what is known as logical multiplication.

After completing this section, you should be able to

■ Identify an AND gate by its distinctive shape symbol or by its rectangular outline symbol ■ Describe the operation of an AND gate ■ Generate the truth table for an AND gate with any number of inputs ■ Produce a timing diagram for an AND gate with any specified input waveforms ■ Write the logic expression for an AND gate with any number of inputs ■ Discuss examples of AND gate applications

The term *gate* is used to describe a circuit that performs a basic logic operation. The AND gate is composed of two or more inputs and a single output, as indicated by the standard logic symbols shown in Figure 3–8. Inputs are on the left, and the output is on the right in each symbol. Gates with two inputs are shown; however, an AND gate can have any number of inputs greater than one. Although examples of both distinctive shape symbols and rectangular outline symbols are shown, the distinctive shape symbol, shown in part (a), is used predominantly in this book.

Computer Notes are found throughout the text.

COMPUTER NOTE

Logic gates are the building blocks of computers. Most of the functions in a computer, with the exception of certain types of memory, are implemented with logic gates used on a very large scale. For example, a microprocessor, which is the main part of a computer, is made up of hundreds of thousands or even millions of logic gates.

(a) Distinctive shape   (b) Rectangular outline with the AND (&) qualifying symbol

▲ FIGURE 3–8
Standard logic symbols for the AND gate showing two inputs (ANSI/IEEE Std. 91-1984).

Operation of an AND Gate

An AND gate produces a HIGH output *only* when *all* of the inputs are HIGH. When any of the inputs is LOW, the output is LOW. Therefore, the basic purpose of an AND gate is to determine when certain conditions are simultaneously true, as indicated by HIGH levels on all of its inputs, and to produce a HIGH on its output to indicate that all these conditions are true. The inputs of the 2-input AND gate in Figure 3–8 are labeled A and B, and the output is labeled X. The gate operation can be stated as follows:

An AND gate can have more than two inputs.

For a 2-input AND gate, output X is HIGH only when inputs A and B are HIGH; X is LOW when either A or B is LOW, or when both A and B are LOW.

▶ FIGURE P–3

An example and related problem.

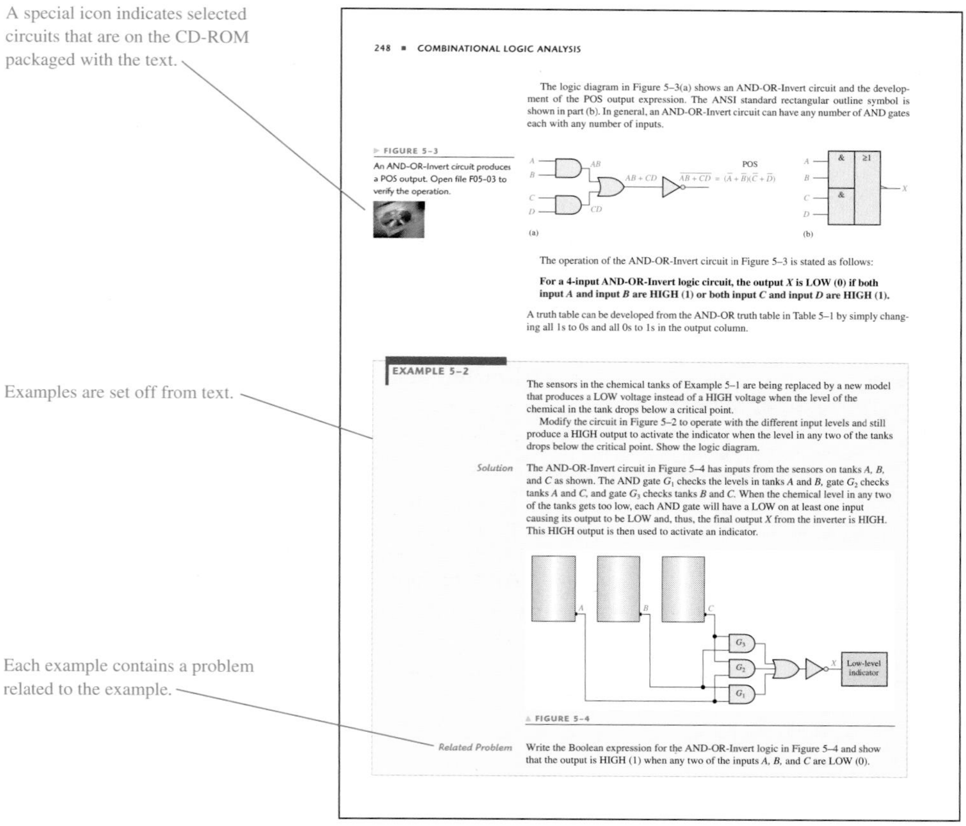

A special icon indicates selected circuits that are on the CD-ROM packaged with the text.

The logic diagram in Figure 5–3(a) shows an AND-OR-Invert circuit and the development of the POS output expression. The ANSI standard rectangular outline symbol is shown in part (b). In general, an AND-OR-Invert circuit can have any number of AND gates each with any number of inputs.

▶ FIGURE 5–3
An AND-OR-Invert circuit produces a POS output. Open file F05-03 to verify the operation.

$$AB + CD \qquad \overline{AB + CD} = (\overline{A} + \overline{B})(\overline{C} + \overline{D})$$

(a)

(b)

The operation of the AND-OR-Invert circuit in Figure 5–3 is stated as follows:

For a 4-input AND-OR-Invert logic circuit, the output X is LOW (0) if both input A and input B are HIGH (1) or both input C and input D are HIGH (1).

A truth table can be developed from the AND-OR truth table in Table 5–1 by simply changing all 1s to 0s and all 0s to 1s in the output column.

Examples are set off from text.

EXAMPLE 5–2

The sensors in the chemical tanks of Example 5–1 are being replaced by a new model that produces a LOW voltage instead of a HIGH voltage when the level of the chemical in the tank drops below a critical point.

Modify the circuit in Figure 5–2 to operate with the different input levels and still produce a HIGH output to activate the indicator when the level in any two of the tanks drops below the critical point. Show the logic diagram.

Solution   The AND-OR-Invert circuit in Figure 5–4 has inputs from the sensors on tanks A, B, and C as shown. The AND gate G_1 checks the levels in tanks A and B, gate G_2 checks tanks A and C, and gate G_3 checks tanks B and C. When the chemical level in any two of the tanks gets too low, each AND gate will have a LOW on at least one input causing its output to be LOW and, thus, the final output X from the inverter is HIGH. This HIGH output is then used to activate an indicator.

Each example contains a problem related to the example.

▲ FIGURE 5–4

Related Problem   Write the Boolean expression for the AND-OR-Invert logic in Figure 5–4 and show that the output is HIGH (1) when any two of the inputs A, B, and C are LOW (0).

160 ■ LOGIC GATES

3–9 TROUBLESHOOTING

Troubleshooting is the process of recognizing, isolating, and correcting a fault or failure in a circuit or system. To be an effective troubleshooter, you must understand how the circuit or system is supposed to work and be able to recognize incorrect performance. For example, to determine whether or not a certain logic gate is faulty, you must know what the output should be for given inputs.

After completing this section, you should be able to

■ Test for internally open inputs and outputs in IC gates ■ Recognize the effects of a shorted IC input or output ■ Test for external faults on a PC board ■ Troubleshoot a simple frequency counter using an oscilloscope

Internal Failures of IC Logic Gates

Opens and shorts are the most common types of internal gate failures. These can occur on the inputs or on the output of a gate inside the IC package. *Before attempting any troubleshooting, check for proper dc supply voltage and ground.*

Effects of an Internally Open Input An internal open is the result of an open component on the chip or a break in the tiny wire connecting the IC chip to the package pin. An open input prevents a signal on that input from getting to the output of the gate, as illustrated in Figure 3–67(a) for the case of a 2-input NAND gate. An open TTL input acts effectively as a HIGH level, so pulses applied to the good input get through to the NAND gate output as shown in Figure 3–67(b).

(a) Application of pulses to the open input will produce no pulses on the output.

(b) Application of pulses to the good input will produce output pulses for TTL NAND and AND gates because an open input typically acts as a HIGH. It is uncertain for CMOS.

FIGURE 3–67
The effect of an open input on a NAND gate.

Conditions for Testing Gates When testing a NAND gate or an AND gate, always make sure that the inputs that are not being pulsed are HIGH to enable the gate. When checking a NOR gate or an OR gate, always make sure that the inputs that are not being pulsed are LOW. When checking an XOR or XNOR gate, the level of the nonpulsed input does not matter because the pulses on the other input will force the inputs to alternate between the same level and opposite levels.

Troubleshooting an Open Input Troubleshooting this type of failure is easily accomplished with an oscilloscope and function generator, as demonstrated in Figure 3–68 for the case of a 2-input NAND gate package. When measuring digital signals with a scope, always use dc coupling.

TROUBLESHOOTING ■ 161

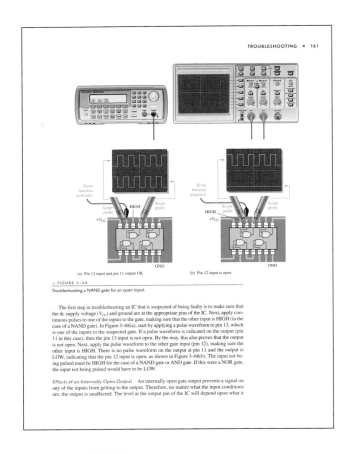

(a) Pin 13 input and pin 11 output OK.

(b) Pin 12 input is open.

FIGURE 3–68
Troubleshooting a NAND gate for an open input.

The first step in troubleshooting an IC that is suspected of being faulty is to make sure that the dc supply voltage (V_{CC}) and ground are at the appropriate pins of the IC. Next, apply continuous pulses to one of the inputs to the gate, making sure that the other input is HIGH (in the case of a NAND gate). In Figure 3–68(a), start by applying a pulse waveform to pin 13, which is one of the inputs to the suspected gate. If a pulse waveform is indicated on the output (pin 11 in this case), then the pin 13 input is not open. By the way, this also proves that the output is not open. Next, apply the pulse waveform to the other gate input (pin 12), making sure the other input is HIGH. There is no pulse waveform on the output at pin 11 and the output is LOW, indicating that the pin 12 input is open, as shown in Figure 3–68(b). The input not being pulsed must be HIGH for the case of a NAND gate or AND gate. If this were a NOR gate, the input not being pulsed would have to be LOW.

Effects of an Internally Open Output An internally open gate output prevents a signal on any of the inputs from getting to the output. Therefore, no matter what the input conditions are, the output is unaffected. The level at the output pin of the IC will depend upon what it

▲ FIGURE P–4

Representative pages from a portion of a typical Troubleshooting section.

■ Traffic light control system: Chapters 6, 7, and 8

■ Security system: Chapters 9 and 10

Digital System Applications may be treated as optional because omitting them will not affect any other material in the text. Figure P–5 shows a portion of a Digital System Application feature.

Chapter End   The following study aids end each chapter:

■ Summary

■ Key term glossary

■ Self-test

■ Problem set that includes some or all of the following categories: Basic, Troubleshooting, Digital System Application, Design, and Multisim Troubleshooting Practice

■ Answers to Section Reviews

■ Answers to Related Problems for Examples

■ Answers to Self-Test

Book End

■ Appendices: Code conversion and table of powers of two (Appendix A) and traffic light interface circuits (Appendix B)

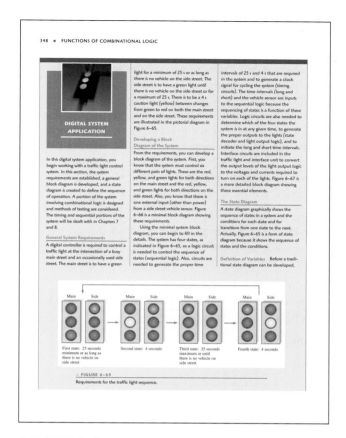 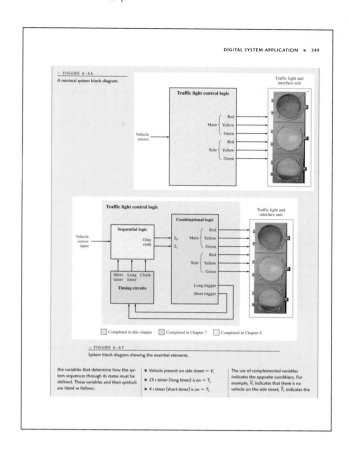

▲ FIGURE P–5

Representative pages from a typical Digital System Application.

- Answers to odd-numbered problems
- Comprehensive glossary
- Index

To the Student

Digital technology is hot! Most everything has already gone digital or will in the near future. For example, cell phones and other types of wireless communication, television, radio, process controls, automotive electronics, consumer electronics, global navigation, military systems, to name only a few applications, depend heavily on digital electronics.

A strong foundation in the fundamentals of digital technology will prepare you for the highly skilled and high-paying jobs of the future. The single most important thing you can do is to understand the core fundamentals. From there you can go anywhere.

In addition, programmable logic is becoming extremely important in today's technology and that topic is introduced in this book. Of course, efficient troubleshooting is a skill that is also widely sought. Troubleshooting and testing methods from traditional testing to manufacturing techniques, such as bed-of-nails, flying probe, and boundary scan, are covered in this book. These are examples of the skills you can acquire with a serious effort to learn the concepts presented.

The CD-ROMs Two CDs are included with this book. One contains Texas Instruments data sheets for digital integrated circuits. The other contains circuit files in Multisim for use with Multisim software Versions 2001 or 7. (These Version 2001 and Version 7 circuit files—as well as those for use with Multisim 8—also appear on the Companion Website at **www.prenhall.com/floyd**.)

User's Guide for Instructors

Generally, time limitation or program emphasis determine the topics to be covered in a course. It is not uncommon to omit or condense topics or to alter the sequence of certain topics in order to customize the material for a particular course. The author recognizes this and has designed this textbook specifically to provide great flexibility in topic coverage.

Using a modular approach, certain topics are organized in separate sections or features such that if they are omitted, the rest of the coverage is not affected. Also, if these topics are included, they flow seamlessly with the rest of the coverage. The book is organized around a core of fundamental topics that are, for the most part, essential in any digital course. Around this core, there are other topics that can be included or omitted depending on the course emphasis or other factors. Figure P-6 illustrates this modular concept.

■ *Core Fundamentals*   The fundamental topics of digital logic should, for the most part, be covered in all programs. Linked to the core are several "satellite" topics that may be considered for omission or inclusion, depending on your course goals. Any block surrounding the core can be omitted without affecting the core fundamentals.

■ *Programmable Logic*   Although it is an important topic, programmable logic can be omitted, but it is recommended that you cover this topic if at all possible. You can cover as little or as much as you consider practical for your program.

■ *Troubleshooting*   Troubleshooting sections appear in many chapters.

■ *Digital System Applications*   System applications appear in many chapters.

■ *Integrated Circuit Technologies*   Some or all of the topics in Chapter 14 can be covered at selected points if you wish to discuss details of the circuitry that make up digital integrated circuits.

■ *Special Topics*   These topics are *Introduction to Computers* and *Digital Signal Processing* in Chapters 12 and 13, respectively. These are special topics and may not be essential to your digital course.

Also, within each block in Figure P-6 you can choose to omit or de-emphasize some topics because of time constraints or other priorities. For example, in the core fundamentals, error correction codes, carry look-ahead adders, sequential logic design, and other selected topics could be omitted.

Customizing the Table of Contents   You can take any one of several paths through *Digital Fundamentals, Ninth Edition,* depending on the goals of your particular program. Whether you choose a minimal coverage of only core fundamentals, a full-blown coverage of all the topics, or anything in between, this book can be adapted to your needs. The

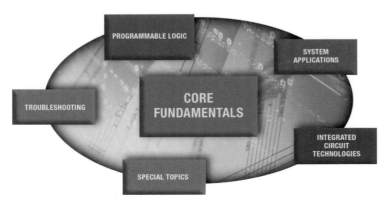

▲ **FIGURE P–6**

Table of Contents following this preface is color coded to match the blocks in Figure P-6. This allows you to identify topics for omission or inclusion for customizing your course.

Several options for use of *Digital Fundamentals, Ninth Edition* are shown below in terms of topics color coded to Figure P-6. Other options are possible, too, including partial coverage of some topics.

Option 1 ■

Option 2 ■ ■

Option 3 ■ ■ ■

Option 4 ■ ■ ■ ■

Option 5 ■ ■ ■ ■ ■

Acknowledgments

This innovative text has been realized by the efforts and the skills of many people. I think that we have accomplished what we set out to do, which was to produce a textbook second to none. At Prentice Hall, Kate Linsner and Rex Davidson have contributed a great amount of time, talent, and effort to move this project through its many phases in order to produce the book as you see it. Lois Porter has done a fantastic job of editing the manuscript. She has unraveled the mysteries of this author's markups and often nearly illegible notes and, from that tangled mess, extracted an unbelievably organized and superbly edited manuscript. Also, Jane Lopez has done another beautiful job with the graphics. Another individual who contributed significantly to this book is Gary Snyder, who has provided all of the Multisim circuit files (in Multisim Versions 2001, 7, and 8, all of which appear on the Companion Website at **www.prenhall.com/floyd**). I extend my thanks and appreciation to all of these people and others who were indirectly involved in the project.

In the revision of this and all textbooks, I depend on expert input from many users as well as non-users. I want to offer my sincere thanks to the following reviewers, who submitted many valuable suggestions and provided lots of constructive criticism: Bo Barry, University of North Carolina–Charlotte; Chuck McGlumphy, Belmont Technical College; and Amy Ray, Mitchell Community College.

My appreciation goes to David Buchla for his efforts to make sure that the lab manual is closely coordinated with the text and for his valuable input. I would also like to mention Muhammed Arif Shabir for his suggestion concerning shift registers.

I thank all of the members of the Prentice Hall sales force whose efforts have helped make my books available to a large number of users throughout the world. In addition, I am grateful to all of you who have adopted this text for your classes or for your own use. Without you we would not be in business. I hope that you find this book to be a valuable learning tool and reference for students.

Tom Floyd

Contents

DIGITAL
FUNDAMENTALS

1

DIGITAL CONCEPTS

CHAPTER OBJECTIVES

- Explain the basic differences between digital and analog quantities

- Show how voltage levels are used to represent digital quantities

- Describe various parameters of a pulse waveform such as rise time, fall time, pulse width, frequency, period, and duty cycle

- Explain the basic logic operations of NOT, AND, and OR

- Describe the logic functions of the comparator, adder, code converter, encoder, decoder, multiplexer, demultiplexer, counter, and register

- Identify fixed-function digital integrated circuits according to their complexity and the type of circuit packaging
- Identify pin numbers on integrated circuit packages
- Describe programmable logic, discuss the various types, and describe how PLDs are programmed
- Recognize various instruments and understand how they are used in measurement and troubleshooting digital circuits and systems
- Show how a complete digital system is formed by combining the basic functions in a practical application

KEY TERMS

Key terms are in order of appearance in the chapter.

Analog	Output
Digital	Gate
Binary	NOT
Bit	Inverter
Pulse	AND
Clock	OR
Timing diagram	Integrated circuit (IC)
Data	SPLD
Serial	CPLD
Parallel	FPGA
Logic	Compiler
Input	Troubleshooting

INTRODUCTION

The term *digital* is derived from the way computers perform operations, by counting digits. For many years, applications of digital electronics were confined to computer systems. Today, digital technology is applied in a wide range of areas in addition to computers. Such applications as television, communications systems, radar, navigation and guidance systems, military systems, medical instrumentation, industrial process control, and consumer electronics use digital techniques. Over the years digital technology has progressed from vacuum-tube circuits to discrete transistors to complex integrated circuits, some of which contain millions of transistors.

This chapter introduces you to digital electronics and provides a broad overview of many important concepts, components, and tools.

■■■ DIGITAL SYSTEM APPLICATION PREVIEW

The last feature in many chapters of this textbook uses a system application to bring together the principal topics covered in the chapter. Each system is designed to fit the particular chapter to illustrate how the theory and devices can be used. Throughout the book, five different systems are introduced, some covering two or more chapters.

All of the systems are simplified to make them manageable in the context of the chapter material. Although they are based on actual system requirements, they are designed to accommodate the topical coverage of the chapter and are not intended to necessarily represent the most efficient or ultimate approach in a given application.

This chapter introduces the first system, which is an industrial process control system for counting and controlling items for packaging on a conveyor line. It is designed to incorporate all of the logic functions that are introduced in this chapter so that you can see how they are used and how they work together to achieve a useful objective.

WWW. VISIT THE COMPANION WEBSITE
Study aids for this chapter are available at
http://www.prenhall.com/floyd

1-1 DIGITAL AND ANALOG QUANTITIES

Electronic circuits can be divided into two broad categories, digital and analog. Digital electronics involves quantities with discrete values, and analog electronics involves quantities with continuous values. Although you will be studying digital fundamentals in this book, you should also know something about analog because many applications require both; and interfacing between analog and digital is important.

After completing this section, you should be able to

▪ Define *analog* ▪ Define *digital* ▪ Explain the difference between digital and analog quantities ▪ State the advantages of digital over analog ▪ Give examples of how digital and analog quantities are used in electronics

An **analog*** quantity is one having continuous values. A **digital** quantity is one having a discrete set of values. Most things that can be measured quantitatively occur in nature in analog form. For example, the air temperature changes over a continuous range of values. During a given day, the temperature does not go from, say, 70° to 71° instantaneously; it takes on all the infinite values in between. If you graphed the temperature on a typical summer day, you would have a smooth, continuous curve similar to the curve in Figure 1–1. Other examples of analog quantities are time, pressure, distance, and sound.

▷ **FIGURE 1–1**

Graph of an analog quantity (temperature versus time).

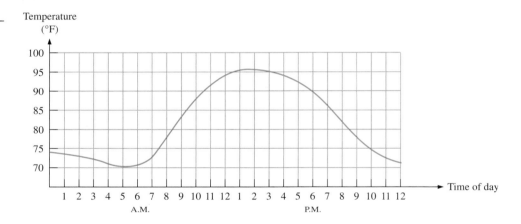

Rather than graphing the temperature on a continuous basis, suppose you just take a temperature reading every hour. Now you have sampled values representing the temperature at discrete points in time (every hour) over a 24-hour period, as indicated in Figure 1–2. You have effectively converted an analog quantity to a form that can now be digitized by representing each sampled value by a digital code. It is important to realize that Figure 1–2 itself is not the digital representation of the analog quantity.

The Digital Advantage Digital representation has certain advantages over analog representation in electronics applications. For one thing, digital data can be processed and transmitted more efficiently and reliably than analog data. Also, digital data has a great advantage when storage is necessary. For example, music when converted to digital form can be stored more compactly and reproduced with greater accuracy and clarity than is possible when it is in analog form. Noise (unwanted voltage fluctuations) does not affect digital data nearly as much as it does analog signals.

***All bold terms are important and are defined in the end-of-book glossary. The blue bold terms are key terms and are included in a Key Term glossary at the end of each chapter.**

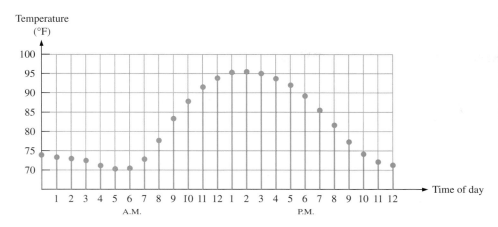

◀ FIGURE 1–2

Sampled–value representation (quantization) of the analog quantity in Figure 1–1. Each value represented by a dot can be digitized by representing it as a digital code that consists of a series of 1s and 0s.

An Analog Electronic System

A public address system, used to amplify sound so that it can be heard by a large audience, is one simple example of an application of analog electronics. The basic diagram in Figure 1–3 illustrates that sound waves, which are analog in nature, are picked up by a microphone and converted to a small analog voltage called the audio signal. This voltage varies continuously as the volume and frequency of the sound changes and is applied to the input of a linear amplifier. The output of the amplifier, which is an increased reproduction of input voltage, goes to the speaker(s). The speaker changes the amplified audio signal back to sound waves that have a much greater volume than the original sound waves picked up by the microphone.

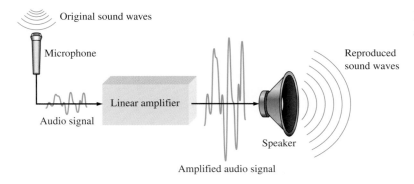

◀ FIGURE 1–3

A basic audio public address system.

A System Using Digital and Analog Methods

The compact disk (CD) player is an example of a system in which both digital and analog circuits are used. The simplified block diagram in Figure 1–4 illustrates the basic principle. Music in digital form is stored on the compact disk. A laser diode optical system picks up the digital data from the rotating disk and transfers it to the **digital-to-analog converter (DAC).**

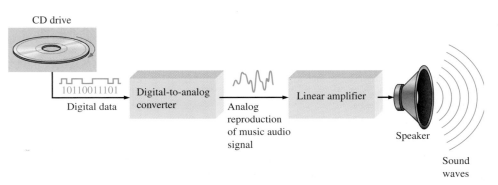

◀ FIGURE 1–4

Basic block diagram of a CD player. Only one channel is shown.

The DAC changes the digital data into an analog signal that is an electrical reproduction of the original music. This signal is amplified and sent to the speaker for you to enjoy. When the music was originally recorded on the CD, a process, essentially the reverse of the one described here, using an **analog-to-digital converter (ADC)** was used.

**SECTION 1–1
REVIEW**

Answers are at the end of the chapter.

1. Define *analog*.
2. Define *digital*.
3. Explain the difference between a digital quantity and an analog quantity.
4. Give an example of a system that is analog and one that is a combination of both digital and analog. Name a system that is entirely digital.

1–2 BINARY DIGITS, LOGIC LEVELS, AND DIGITAL WAVEFORMS

Digital electronics involves circuits and systems in which there are only two possible states. These states are represented by two different voltage levels: A HIGH and a LOW. The two states can also be represented by current levels, bits and bumps on a CD or DVD, etc. In digital systems such as computers, combinations of the two states, called *codes*, are used to represent numbers, symbols, alphabetic characters, and other types of information. The two-state number system is called *binary*, and its two digits are 0 and 1. A binary digit is called a *bit*.

After completing this section, you should be able to

■ Define *binary* ■ Define *bit* ■ Name the bits in a binary system ■ Explain how voltage levels are used to represent bits ■ Explain how voltage levels are interpreted by a digital circuit ■ Describe the general characteristics of a pulse ■ Determine the amplitude, rise time, fall time, and width of a pulse ■ Identify and describe the characteristics of a digital waveform ■ Determine the amplitude, period, frequency, and duty cycle of a digital waveform ■ Explain what a timing diagram is and state its purpose ■ Explain serial and parallel data transfer and state the advantage and disadvantage of each

Binary Digits

Each of the two digits in the **binary** system, 1 and 0, is called a **bit,** which is a contraction of the words *binary digit*. In digital circuits, two different voltage levels are used to represent the two bits. Generally, 1 is represented by the higher voltage, which we will refer to as a HIGH, and a 0 is represented by the lower voltage level, which we will refer to as a LOW. This is called **positive logic** and will be used throughout the book.

HIGH = 1   and   LOW = 0

Another system in which a 1 is represented by a LOW and a 0 is represented by a HIGH is called *negative logic*.

Groups of bits (combinations of 1s and 0s), called *codes*, are used to represent numbers, letters, symbols, instructions, and anything else required in a given application.

Logic Levels

The voltages used to represent a 1 and a 0 are called *logic levels*. Ideally, one voltage level represents a HIGH and another voltage level represents a LOW. In a practical digital circuit, however, a HIGH can be any voltage between a specified minimum value and a specified maximum value. Likewise, a LOW can be any voltage between a specified minimum

COMPUTER NOTE

The concept of a digital computer can be traced back to Charles Babbage, who developed a crude mechanical computation device in the 1830s. John Atanasoff was the first to apply electronic processing to digital computing in 1939. In 1946, an electronic digital computer called ENIAC was implemented with vacuum–tube circuits. Even though it took up an entire room, ENIAC didn't have the computing power of your handheld calculator.

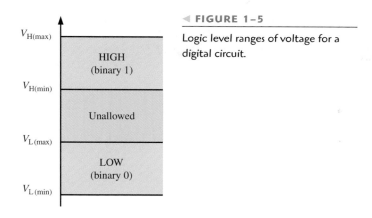

Logic level ranges of voltage for a digital circuit.

and a specified maximum. There can be no overlap between the accepted range of HIGH levels and the accepted range of LOW levels.

Figure 1–5 illustrates the general range of LOWs and HIGHs for a digital circuit. The variable $V_{H(max)}$ represents the maximum HIGH voltage value, and $V_{H(min)}$ represents the minimum HIGH voltage value. The maximum LOW voltage value is represented by $V_{L(max)}$, and the minimum LOW voltage value is represented by $V_{L(min)}$. The voltage values between $V_{L(max)}$ and $V_{H(min)}$ are unacceptable for proper operation. A voltage in the unallowed range can appear as either a HIGH or a LOW to a given circuit and is therefore not an acceptable value. For example, the HIGH values for a certain type of digital circuit called CMOS may range from 2 V to 3.3 V and the LOW values may range from 0 V to 0.8 V. So, for example, if a voltage of 2.5 V is applied, the circuit will accept it as a HIGH or binary 1. If a voltage of 0.5 V is applied, the circuit will accept it as a LOW or binary 0. For this type of circuit, voltages between 0.8 V and 2 V are unacceptable.

Digital Waveforms

Digital waveforms consist of voltage levels that are changing back and forth between the HIGH and LOW levels or states. Figure 1–6(a) shows that a single positive-going **pulse** is generated when the voltage (or current) goes from its normally LOW level to its HIGH level and then back to its LOW level. The negative-going pulse in Figure 1–6(b) is generated when the voltage goes from its normally HIGH level to its LOW level and back to its HIGH level. A digital waveform is made up of a series of pulses.

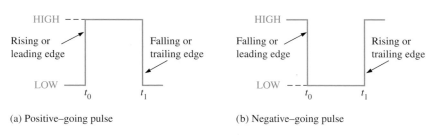

(a) Positive–going pulse   (b) Negative–going pulse

Ideal pulses.

The Pulse   As indicated in Figure 1–6, a pulse has two edges: a **leading edge** that occurs first at time t_0 and a **trailing edge** that occurs last at time t_1. For a positive-going pulse, the leading edge is a rising edge, and the trailing edge is a falling edge. The pulses in Figure 1–6 are ideal because the rising and falling edges are assumed to change in zero time (instantaneously). In practice, these transitions never occur instantaneously, although for most digital work you can assume ideal pulses.

Figure 1–7 shows a nonideal pulse. In reality, all pulses exhibit some or all of these characteristics. The overshoot and ringing are sometimes produced by stray inductive and

Nonideal pulse characteristics.

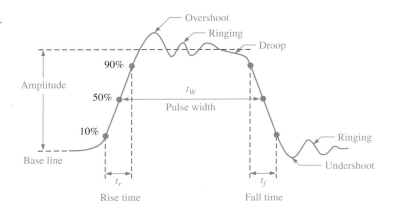

capacitive effects. The droop can be caused by stray capacitive and circuit resistance, forming an *RC* circuit with a low time constant.

The time required for a pulse to go from its LOW level to its HIGH level is called the **rise time** (t_r), and the time required for the transition from the HIGH level to the LOW level is called the **fall time** (t_f). In practice, it is common to measure rise time from 10% of the pulse **amplitude** (height from baseline) to 90% of the pulse amplitude and to measure the fall time from 90% to 10% of the pulse amplitude, as indicated in Figure 1–7. The bottom 10% and the top 10% of the pulse are not included in the rise and fall times because of the nonlinearities in the waveform in these areas. The **pulse width** (t_W) is a measure of the duration of the pulse and is often defined as the time interval between the 50% points on the rising and falling edges, as indicated in Figure 1–7.

Waveform Characteristics   Most waveforms encountered in digital systems are composed of series of pulses, sometimes called *pulse trains,* and can be classified as either periodic or nonperiodic. A **periodic** pulse waveform is one that repeats itself at a fixed interval, called a **period** (*T*). The **frequency** (*f*) is the rate at which it repeats itself and is measured in hertz (Hz). A nonperiodic pulse waveform, of course, does not repeat itself at fixed intervals and may be composed of pulses of randomly differing pulse widths and/or randomly differing time intervals between the pulses. An example of each type is shown in Figure 1–8.

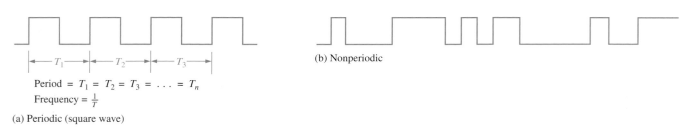

(b) Nonperiodic

Period = $T_1 = T_2 = T_3 = \ldots = T_n$

Frequency = $\frac{1}{T}$

(a) Periodic (square wave)

Examples of digital waveforms.

The frequency (*f*) of a pulse (digital) waveform is the reciprocal of the period. The relationship between frequency and period is expressed as follows:

Equation 1–1

$$f = \frac{1}{T}$$

Equation 1–2

$$T = \frac{1}{f}$$

An important characteristic of a periodic digital waveform is its **duty cycle,** which is the ratio of the pulse width (t_W) to the period (T). It can be expressed as a percentage.

$$\text{Duty cycle} = \left(\frac{t_W}{T}\right)100\%$$

Equation 1–3

EXAMPLE 1–1

A portion of a periodic digital waveform is shown in Figure 1–9. The measurements are in milliseconds. Determine the following:

(a) period   **(b)** frequency   **(c)** duty cycle

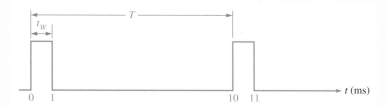

▲ FIGURE 1–9

Solution   **(a)** The period is measured from the edge of one pulse to the corresponding edge of the next pulse. In this case T is measured from leading edge to leading edge, as indicated. T equals **10 ms.**

(b) $f = \dfrac{1}{T} = \dfrac{1}{10\text{ ms}} = \textbf{100 Hz}$

(c) $\text{Duty cycle} = \left(\dfrac{t_W}{T}\right)100\% = \left(\dfrac{1\text{ ms}}{10\text{ ms}}\right)100\% = \textbf{10}\%$

Related Problem*   A periodic digital waveform has a pulse width of 25 μs and a period of 150 μs. Determine the frequency and the duty cycle.

---

*Answers are at the end of the chapter.

A Digital Waveform Carries Binary Information

Binary information that is handled by digital systems appears as waveforms that represent sequences of bits. When the waveform is HIGH, a binary 1 is present; when the waveform is LOW, a binary 0 is present. Each bit in a sequence occupies a defined time interval called a **bit time.**

COMPUTER NOTE

The speed at which a computer can operate depends on the type of microprocessor used in the system. The speed specification, for example 3.5 GHz, of a computer is the maximum clock frequency at which the microprocessor can run.

The Clock   In digital systems, all waveforms are synchronized with a basic timing waveform called the **clock.** The clock is a periodic waveform in which each interval between pulses (the period) equals the time for one bit.

An example of a clock waveform is shown in Figure 1–10. Notice that, in this case, each change in level of waveform *A* occurs at the leading edge of the clock waveform. In other cases, level changes occur at the trailing edge of the clock. During each bit time of the clock, waveform *A* is either HIGH or LOW. These HIGHs and LOWs represent a sequence of bits as indicated. A group of several bits can be used as a piece of binary information, such as a number or a letter. The clock waveform itself does not carry information.

▶ FIGURE 1–10

Example of a clock waveform synchronized with a waveform representation of a sequence of bits.

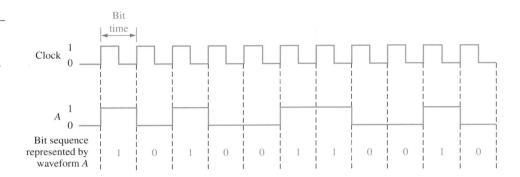

Timing Diagrams A **timing diagram** is a graph of digital waveforms showing the actual time relationship of two or more waveforms and how each waveform changes in relation to the others. By looking at a timing diagram, you can determine the states (HIGH or LOW) of all the waveforms at any specified point in time and the exact time that a waveform changes state relative to the other waveforms. Figure 1–11 is an example of a timing diagram made up of four waveforms. From this timing diagram you can see, for example, that the three waveforms *A, B,* and *C* are HIGH only during bit time 7 and they all change back LOW at the end of bit time 7 (shaded area).

▶ FIGURE 1–11

Example of a timing diagram.

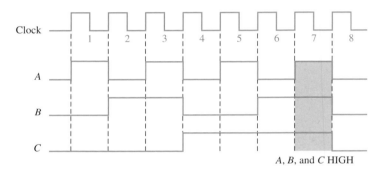

Data Transfer

Data refers to groups of bits that convey some type of information. Binary data, which are represented by digital waveforms, must be transferred from one circuit to another within a digital system or from one system to another in order to accomplish a given purpose. For example, numbers stored in binary form in the memory of a computer must be transferred to the computer's central processing unit in order to be added. The sum of the addition must then be transferred to a monitor for display and/or transferred back to the memory. In computer systems, as illustrated in Figure 1–12, binary data are transferred in two ways: serial and parallel.

When bits are transferred in **serial** form from one point to another, they are sent one bit at a time along a single line, as illustrated in Figure 1–12(a) for the case of a computer-to-modem transfer. During the time interval from t_0 to t_1, the first bit is transferred. During the time interval from t_1 to t_2, the second bit is transferred, and so on. To transfer eight bits in series, it takes eight time intervals.

When bits are transferred in **parallel** form, all the bits in a group are sent out on separate lines at the same time. There is one line for each bit, as shown in Figure 1–12(b) for the example of eight bits being transferred from a computer to a printer. To transfer eight bits in parallel, it takes one time interval compared to eight time intervals for the serial transfer.

To summarize, an advantage of serial transfer of binary data is that a minimum of only one line is required. In parallel transfer, a number of lines equal to the number of

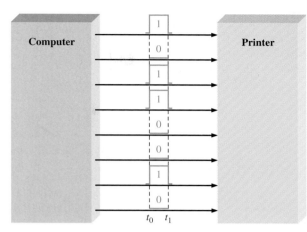

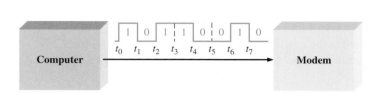

(a) Serial transfer of 8 bits of binary data from computer to modem. Interval t_0 to t_1 is first.

(b) Parallel transfer of 8 bits of binary data from computer to printer. The beginning time is t_0.

▲ **FIGURE 1–12**

Illustration of serial and parallel transfer of binary data. Only the data lines are shown.

bits to be transferred at one time is required. A disadvantage of serial transfer is that it takes longer to transfer a given number of bits than with parallel transfer. For example, if one bit can be transferred in 1 μs, then it takes 8 μs to serially transfer eight bits but only 1 μs to parallel transfer eight bits. A disadvantage of parallel transfer is that it takes more lines than serial transfer.

EXAMPLE 1–2

(a) Determine the total time required to serially transfer the eight bits contained in waveform A of Figure 1–13, and indicate the sequence of bits. The left-most bit is the first to be transferred. The 100 kHz clock is used as reference.

(b) What is the total time to transfer the same eight bits in parallel?

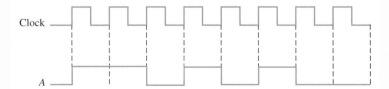

Clock

A

▲ **FIGURE 1–13**

Solution (a) Since the frequency of the clock is 100 kHz, the period is

$$T = \frac{1}{f} = \frac{1}{100 \text{ kHz}} = 10 \ \mu\text{s}$$

It takes 10 μs to transfer each bit in the waveform. The total transfer time for 8 bits is

$$8 \times 10 \ \mu\text{s} = \mathbf{80 \ \mu s}$$

To determine the sequence of bits, examine the waveform in Figure 1–13 during each bit time. If waveform A is HIGH during the bit time, a 1 is transferred. If waveform *A* is LOW during the bit time, a 0 is transferred. The bit sequence is illustrated in Figure 1–14. The left-most bit is the first to be transferred.

▲ FIGURE 1–14

(b) A parallel transfer would take **10 μs** for all eight bits.

Related Problem If binary data are transferred at the rate of 10 million bits per second (10 Mbits/s), how long will it take to parallel transfer 16 bits on 16 lines? How long will it take to serially transfer 16 bits?

SECTION 1–2 REVIEW

1. Define *binary*.
2. What does *bit* mean?
3. What are the bits in a binary system?
4. How are the rise time and fall time of a pulse measured?
5. Knowing the period of a waveform, how do you find the frequency?
6. Explain what a clock waveform is.
7. What is the purpose of a timing diagram?
8. What is the main advantage of parallel transfer over serial transfer of binary data?

1–3 BASIC LOGIC OPERATIONS

In its basic form, logic is the realm of human reasoning that tells you a certain proposition (declarative statement) is true if certain conditions are true. Propositions can be classified as true or false. Many situations and processes that you encounter in your daily life can be expressed in the form of propositional, or logic, functions. Since such functions are true/false or yes/no statements, digital circuits with their two-state characteristics are applicable.

After completing this section, you should be able to

■ List three basic logic operations ■ Define the NOT operation ■ Define the AND operation ■ Define the OR operation

Several propositions, when combined, form propositional, or logic, functions. For example, the propositional statement "The light is on" will be true if "The bulb is not burned out" is true and if "The switch is on" is true. Therefore, this logical statement can be made: *The light is on only if the bulb is not burned out and the switch is on.* In this example the first statement is true only if the last two statements are true. The first statement ("The light is on") is then the basic proposition, and the other two statements are the conditions on which the proposition depends.

In the 1850s, the Irish logician and mathematician George Boole developed a mathematical system for formulating logic statements with symbols so that problems can be written and solved in a manner similar to ordinary algebra. Boolean algebra, as it is known today, is applied in the design and analysis of digital systems and will be covered in detail in Chapter 4.

The term **logic** is applied to digital circuits used to implement logic functions. Several kinds of digital logic **circuits** are the basic elements that form the building blocks for such complex digital systems as the computer. We will now look at these elements and discuss their functions in a very general way. Later chapters will cover these circuits in detail.

Three basic logic operations (NOT, AND, and OR) are indicated by standard distinctive shape symbols in Figure 1–15. Other standard symbols for these logic operations will be introduced in Chapter 3. The lines connected to each symbol are the **inputs** and **outputs.** The inputs are on the left of each symbol and the output is on the right. A circuit that performs a specified logic operation (AND, OR) is called a logic **gate.** AND and OR gates can have any number of inputs, as indicated by the dashes in the figure.

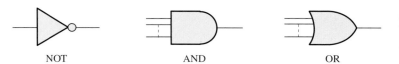

NOT   AND   OR

◀ FIGURE 1–15

The basic logic operations and symbols.

In logic operations, the true/false conditions mentioned earlier are represented by a HIGH (true) and a LOW (false). Each of the three basic logic operations produces a unique response to a given set of conditions.

NOT

The **NOT** operation changes one logic level to the opposite logic level, as indicated in Figure 1–16. When the input is HIGH (1), the output is LOW (0). When the input is LOW, the output is HIGH. In either case, the output is *not* the same as the input. The NOT operation is implemented by a logic circuit known as an **inverter.**

◀ FIGURE 1–16

The NOT operation.

AND

The **AND** operation produces a HIGH output only when all the inputs are HIGH, as indicated in Figure 1–17 for the case of two inputs. When one input is HIGH *and* the other input is HIGH, the output is HIGH. When any or all inputs are LOW, the output is LOW. The AND operation is implemented by a logic circuit known as an *AND gate.*

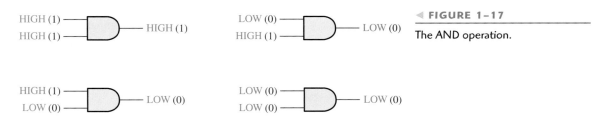

◀ FIGURE 1–17

The AND operation.

OR

The **OR** operation produces a HIGH output when one or more inputs are HIGH, as indicated in Figure 1–18 for the case of two inputs. When one input is HIGH *or* the other input is HIGH *or* both inputs are HIGH, the output is HIGH. When both inputs are LOW, the output is LOW. The OR operation is implemented by a logic circuit known as an *OR gate.*

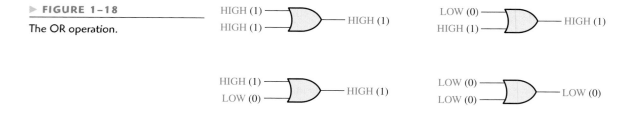

1. When does the NOT operation produce a HIGH output?
2. When does the AND operation produce a HIGH output?
3. When does the OR operation produce a HIGH output?
4. What is an inverter?
5. What is a logic gate?

1–4 OVERVIEW OF BASIC LOGIC FUNCTIONS

The three basic logic elements AND, OR, and NOT can be combined to form more complex logic circuits that perform many useful operations and that are used to build complete digital systems. Some of the common logic functions are comparison, arithmetic, code conversion, encoding, decoding, data selection, storage, and counting. This section provides a general overview of these important functions so that you can begin to see how they form the building blocks of digital systems such as computers. Each of the basic logic functions will be covered in detail in later chapters.

After completing this section, you should be able to

■ Identify nine basic types of logic functions ■ Describe a basic magnitude comparator ■ List the four arithmetic functions ■ Describe a basic adder ■ Describe a basic encoder ■ Describe a basic decoder ■ Define multiplexing and demultiplexing ■ State how data storage is accomplished ■ Describe the function of a basic counter

The Comparison Function

Magnitude comparison is performed by a logic circuit called a **comparator,** covered in Chapter 6. A comparator compares two quantities and indicates whether or not they are equal. For example, suppose you have two numbers and wish to know if they are equal or not equal and, if not equal, which is greater. The comparison function is represented in Figure 1–19. One number in binary form (represented by logic levels) is applied to input A,

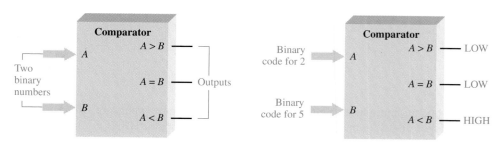

(a) Basic magnitude comparator

(b) Example: A is less than B ($2 < 5$) as indicated by the HIGH output ($A < B$)

and the other number in binary form (represented by logic levels) is applied to input B. The outputs indicate the relationship of the two numbers by producing a HIGH level on the proper output line. Suppose that a binary representation of the number 2 is applied to input A and a binary representation of the number 5 is applied to input B. (We discuss the binary representation of numbers and symbols in Chapter 2.) A HIGH level will appear on the $A < B$ (A is less than B) output, indicating the relationship between the two numbers (2 is less than 5). The wide arrows represent a group of parallel lines on which the bits are transferred.

The Arithmetic Functions

Addition   Addition is performed by a logic circuit called an **adder,** covered in Chapter 6. An adder adds two binary numbers (on inputs A and B with a carry input C_{in}) and generates a sum (Σ) and a carry output (C_{out}), as shown in Figure 1–20(a). Figure 1–20(b) illustrates the addition of 3 and 9. You know that the sum is 12; the adder indicates this result by producing 2 on the sum output and 1 on the carry output. Assume that the carry input in this example is 0.

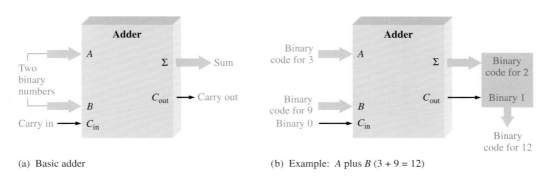

(a) Basic adder (b) Example: *A* plus *B* (3 + 9 = 12)

▲ FIGURE 1–20

The addition function.

Subtraction   Subtraction is also performed by a logic circuit. A **subtracter** requires three inputs: the two numbers that are to be subtracted and a borrow input. The two outputs are the difference and the borrow output. When, for instance, 5 is subtracted from 8 with no borrow input, the difference is 3 with no borrow output. You will see in Chapter 2 how subtraction can actually be performed by an adder because subtraction is simply a special case of addition.

Multiplication   Multiplication is performed by a logic circuit called a *multiplier.* Numbers are always multiplied two at a time, so two inputs are required. The output of the multiplier is the product. Because multiplication is simply a series of additions with shifts in the positions of the partial products, it can be performed by using an adder in conjunction with other circuits.

Division   Division can be performed with a series of subtractions, comparisons, and shifts, and thus it can also be done using an adder in conjunction with other circuits. Two inputs to the divider are required, and the outputs generated are the quotient and the remainder.

The Code Conversion Function

A **code** is a set of bits arranged in a unique pattern and used to represent specified information. A code converter changes one form of coded information into another coded form. Examples are conversion between binary and other codes such as the binary coded decimal

(BCD) and the Gray code. Various types of codes are covered in Chapter 2, and code converters are covered in Chapter 6.

The Encoding Function

The encoding function is performed by a logic circuit called an **encoder,** covered in Chapter 6. The encoder converts information, such as a decimal number or an alphabetic character, into some coded form. For example, one certain type of encoder converts each of the decimal digits, 0 through 9, to a binary code. A HIGH level on the input corresponding to a specific decimal digit produces logic levels that represent the proper binary code on the output lines.

Figure 1–21 is a simple illustration of an encoder used to convert (encode) a calculator keystroke into a binary code that can be processed by the calculator circuits.

▶ FIGURE 1–21

An encoder used to encode a calculator keystroke into a binary code for storage or for calculation.

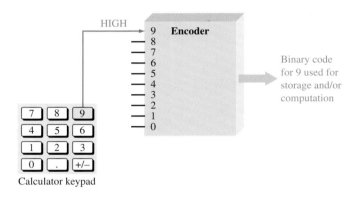

The Decoding Function

The decoding function is performed by a logic circuit called a **decoder,** covered in Chapter 6. The decoder converts coded information, such as a binary number, into a noncoded form, such as a decimal form. For example, one particular type of decoder converts a 4-bit binary code into the appropriate decimal digit.

Figure 1–22 is a simple illustration of one type of decoder that is used to activate a 7-segment display. Each of the seven segments of the display is connected to an output line from the decoder. When a particular binary code appears on the decoder inputs, the appropriate output lines are activated and light the proper segments to display the decimal digit corresponding to the binary code.

▶ FIGURE 1–22

A decoder used to convert a special binary code into a 7-segment decimal readout.

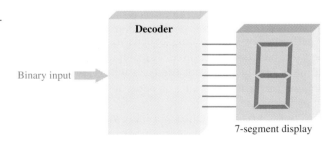

The Data Selection Function

Two types of circuits that select data are the multiplexer and the demultiplexer. The **multiplexer,** or mux for short, is a logic circuit that switches digital data from several input lines onto a single output line in a specified time sequence. Functionally, a multiplexer can be represented by an electronic switch operation that sequentially connects each of the input lines to the output line. The **demultiplexer** (demux) is a logic circuit that switches digital data from one input line to several output lines in a specified time sequence. Essentially, the demux is a mux in reverse.

Multiplexing and demultiplexing are used when data from several sources are to be transmitted over one line to a distant location and redistributed to several destinations. Figure 1–23 illustrates this type of application where digital data from three sources are sent out along a single line to three terminals at another location.

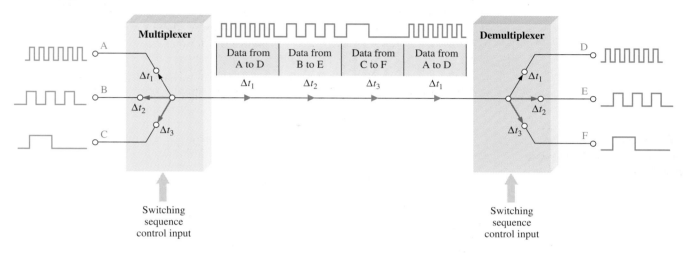

▲ **FIGURE 1–23**

Illustration of a basic multiplexing/demultiplexing application.

In Figure 1–23, data from input A are connected to the output line during time interval Δt_1 and transmitted to the demultiplexer that connects them to output D. Then, during interval Δt_2, the multiplexer switches to input B and the demultiplexer switches to output E. During interval Δt_3, the multiplexer switches to input C and the demultiplexer switches to output F.

To summarize, during the first time interval, input A data go to output D. During the second time interval, input B data go to output E. During the third time interval, input C data go to output F. After this, the sequence repeats. Because the time is divided up among several sources and destinations where each has its turn to send and receive data, this process is called *time division multiplexing* (TDM).

The Storage Function

Storage is a function that is required in most digital systems, and its purpose is to retain binary data for a period of time. Some storage devices are used for short-term storage and some are used for long-term storage. A storage device can "memorize" a bit or a group of bits and retain the information as long as necessary. Common types of storage devices are flip-flops, registers, semiconductor memories, magnetic disks, magnetic tape, and optical disks (CDs).

Flip-flops   A **flip-flop** is a bistable (two stable states) logic circuit that can store only one bit at a time, either a 1 or a 0. The output of a flip-flop indicates which bit it is storing. A HIGH output indicates that a 1 is stored and a LOW output indicates that a 0 is stored. Flip-flops are implemented with logic gates and are covered in Chapter 7.

Registers   A **register** is formed by combining several flip-flops so that groups of bits can be stored. For example, an 8-bit register is constructed from eight flip-flops. In addition to storing bits, registers can be used to shift the bits from one position to another within the register or out of the register to another circuit; therefore, these devices are known as *shift registers*. Shift registers are covered in Chapter 9.

COMPUTER NOTE

The internal computer memories, RAM and ROM, as well as the smaller caches are semiconductor memories. The registers in a microprocessor are constructéd of semiconductor flip-flops. Magnetic disk memories are used in the internal hard drive, the floppy drive, and for the CD-ROM.

The two basic types of shift registers are serial and parallel. The bits are stored in a serial shift register one at a time, as illustrated in Figure 1–24. A good analogy to the serial shift register is loading passengers onto a bus single file through the door. They also exit the bus single file.

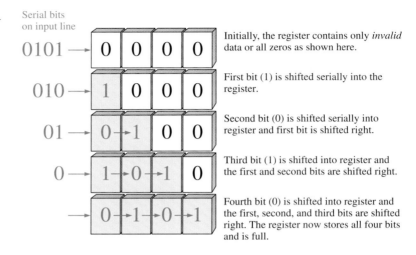

Serial bits on input line

Initially, the register contains only *invalid* data or all zeros as shown here.

First bit (1) is shifted serially into the register.

Second bit (0) is shifted serially into register and first bit is shifted right.

Third bit (1) is shifted into register and the first and second bits are shifted right.

Fourth bit (0) is shifted into register and the first, second, and third bits are shifted right. The register now stores all four bits and is full.

The bits are stored in a parallel register simultaneously from parallel lines, as shown in Figure 1–25. For this case, a good analogy is loading passengers on a roller coaster where they enter all of the cars in parallel.

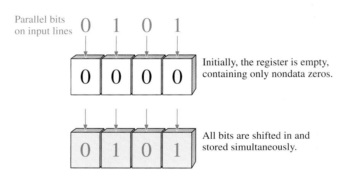

Parallel bits on input lines

Initially, the register is empty, containing only nondata zeros.

All bits are shifted in and stored simultaneously.

Semiconductor Memories   Semiconductor memories are devices typically used for storing large numbers of bits. In one type of memory, called the *read-only memory* or ROM, the binary data are permanently or semipermanently stored and cannot be readily changed. In the *random-access memory* or RAM, the binary data are temporarily stored and can be easily changed. Memories are covered in Chapter 10.

Magnetic Memories   Magnetic disk memories are used for mass storage of binary data. Examples are the so-called floppy disks used in computers and the computer's internal hard disk. Magneto-optical disks use laser beams to store and retrieve data. Magnetic tape is still used in memory applications and for backing up data from other storage devices.

The Counting Function

The counting function is important in digital systems. There are many types of digital **counters,** but their basic purpose is to count events represented by changing levels or pulses. To count, the counter must "remember" the present number so that it can go to the next proper number in sequence. Therefore, storage capability is an important characteristic of all counters, and flip-flops are generally used to implement them. Figure 1–26 illustrates the basic idea of counter operation. Counters are covered in Chapter 8.

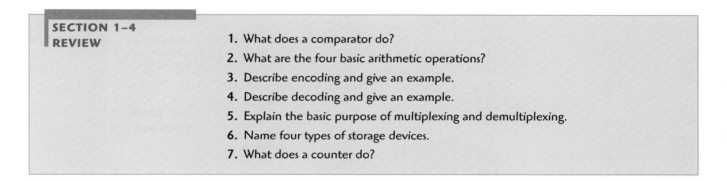

▲ FIGURE 1–26

Illustration of basic counter operation.

SECTION 1–4 REVIEW

1. What does a comparator do?
2. What are the four basic arithmetic operations?
3. Describe encoding and give an example.
4. Describe decoding and give an example.
5. Explain the basic purpose of multiplexing and demultiplexing.
6. Name four types of storage devices.
7. What does a counter do?

1–5 FIXED-FUNCTION INTEGRATED CIRCUITS

All the logic elements and functions that have been discussed are generally available in integrated circuit (IC) form. Digital systems have incorporated ICs for many years because of their small size, high reliability, low cost, and low power consumption. It is important to be able to recognize the IC packages and to know how the pin connections are numbered, as well as to be familiar with the way in which circuit complexities and circuit technologies determine the various IC classifications.

After completing this section, you should be able to

■ Recognize the difference between through-hole devices and surface-mount fixed-function devices ■ Identify dual in-line packages (DIP) ■ Identify small-outline integrated circuit packages (SOIC) ■ Identify plastic leaded chip carrier packages (PLCC) ■ Identify leadless ceramic chip carrier packages (LCCC) ■ Determine pin numbers on various types of IC packages ■ Explain the complexity classifications for fixed-function ICs

A monolithic **integrated circuit (IC)** is an electronic circuit that is constructed entirely on a single small chip of silicon. All the components that make up the circuit—transistors, diodes, resistors, and capacitors—are an integral part of that single chip. Fixed-function logic and programmable logic are two broad categories of digital ICs. In fixed-function logic, the logic functions are set by the manufacturer and cannot be altered.

Figure 1–27 shows a cutaway view of one type of fixed-function IC package with the circuit chip shown within the package. Points on the chip are connected to the package pins to allow input and output connections to the outside world.

▶ **FIGURE 1–27**

Cutaway view of one type of fixed-function IC package showing the chip mounted inside, with connections to input and output pins.

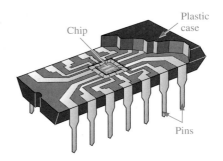

IC Packages

Integrated circuit (IC) packages are classified according to the way they are mounted on printed circuit (PC) boards as either through-hole mounted or surface mounted. The through-hole type packages have pins (leads) that are inserted through holes in the PC board and can be soldered to conductors on the opposite side. The most common type of through-hole package is the dual in-line package (**DIP**) shown in Figure 1–28(a).

▶ **FIGURE 1–28**

Examples of through-hole and surface-mounted devices. The DIP is larger than the SOIC with the same number of leads. This particular DIP is approximately 0.785 in. long, and the SOIC is approximately 0.385 in. long.

(a) Dual in-line package (DIP)

(b) Small-outline IC (SOIC)

Another type of IC package uses surface-mount technology (**SMT**). Surface mounting is a space-saving alternative to through-hole mounting. The holes through the PC board are unnecessary for SMT. The pins of surface-mounted packages are soldered directly to conductors on one side of the board, leaving the other side free for additional circuits. Also, for a circuit with the same number of pins, a surface-mounted package is much smaller than a dual in-line package because the pins are placed closer together. An example of a surface-mounted package is the small-outline integrated circuit (SOIC) shown in Figure 1–28(b).

Three common types of SMT packages are the **SOIC** (small-outline IC), the **PLCC** (plastic leaded chip carrier), and the **LCCC** (leadless ceramic chip carrier). These types of packages are available in various sizes depending on the number of leads (more leads are required for more complex circuits). Examples of each type are shown in Figure 1–29. As you can see, the leads of the SOIC are formed into a "gull-wing" shape. The leads of the PLCC are turned under the package in a J-type shape. Instead of leads, the LCCC has metal contacts molded into its ceramic body. Other variations of SMT packages include **SSOP** (shrink small-outline package), **TSSOP** (thin shrink small-outline package), and **TVSOP** (thin very small-outline package).

Pin Numbering

All IC packages have a standard format for numbering the pins (leads). The dual in-line packages (DIPs) and the small-outline IC packages (SOICs) have the numbering arrangement illustrated in Figure 1–30(a) for a 16-pin package. Looking at the top of the package,

End view   End view   End view

(a) SOIC with "gull-wing" leads   (b) PLCC with J-type leads   (c) LCCC with no leads (contacts are part of case)

◀ FIGURE 1–29

Examples of SMT package configurations.

pin 1 is indicated by an identifier that can be either a small dot, a notch, or a beveled edge. The dot is always next to pin 1. Also, with the notch oriented upward, pin 1 is always the top left pin, as indicated. Starting with pin 1, the pin numbers increase as you go down, then across and up. The highest pin number is always to the right of the notch or opposite the dot.

The PLCC and LCCC packages have leads arranged on all four sides. Pin 1 is indicated by a dot or other index mark and is located at the center of one set of leads. The pin numbers increase going counterclockwise as viewed from the top of the package. The highest pin number is always to the right of pin 1. Figure 1–30(b) illustrates this format for a 20-pin PLCC package.

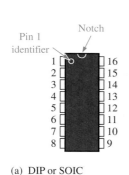
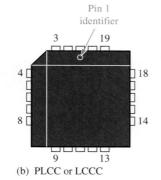

(a) DIP or SOIC   (b) PLCC or LCCC

◀ FIGURE 1–30

Pin numbering for two standard types of IC packages. Top views are shown.

Complexity Classifications for Fixed-Function ICs

Fixed-function digital ICs are classified according to their complexity. They are listed here from the least complex to the most complex. The complexity figures stated here for SSI, MSI, LSI, VLSI, and ULSI are generally accepted, but definitions may vary from one source to another.

- **Small-scale integration (SSI)** describes fixed-function ICs that have up to ten equivalent gate circuits on a single chip, and they include basic gates and flip-flops.

- **Medium-scale integration (MSI)** describes integrated circuits that have from 10 to 100 equivalent gates on a chip. They include logic functions such as encoders, decoders, counters, registers, multiplexers, arithmetic circuits, small memories, and others.

- **Large-scale integration (LSI)** is a classification of ICs with complexities of from more than 100 to 10,000 equivalent gates per chip, including memories.

- **Very large-scale integration (VLSI)** describes integrated circuits with complexities of from more than 10,000 to 100,000 equivalent gates per chip.

■ **Ultra large-scale integration (ULSI)** describes very large memories, larger **microprocessors,** and larger single-chip computers. Complexities of more than 100,000 equivalent gates per chip are classified as ULSI.

Integrated Circuit Technologies

The types of transistors with which all integrated circuits are implemented are either MOSFETs (metal-oxide semiconductor field-effect transistors) or bipolar junction transistors. A circuit technology that uses MOSFETs is CMOS (complementary MOS). A type of fixed-function digital circuit technology that uses bipolar junction transistors is TTL (transistor-transistor logic). BiCMOS uses a combination of both CMOS and TTL.

All gates and other functions can be implemented with either type of circuit technology. SSI and MSI circuits are generally available in both CMOS and TTL. LSI, VLSI, and ULSI are generally implemented with CMOS or NMOS because it requires less area on a chip and consumes less power. There is more on these integrated technologies in Chapter 3. In addition, Chapter 14 provides a complete circuit-level coverage.

Handling Precautions for CMOS   Because of their particular structure, CMOS devices are very sensitive to static charge and can be damaged by electrostatic discharge (ESD) if not handled properly. The following precautions should be taken when you work with CMOS devices:

■ CMOS devices should be shipped and stored in conductive foam.

■ All instruments and metal benches used in testing should be connected to earth ground.

■ The handler's wrist should be connected to earth ground with a length of wire and high-value series resistor.

■ Do not remove a CMOS device (or any device for that matter) from a circuit while the dc power is on.

■ Do not connect ac or signal voltages to a CMOS device while the dc power supply is off.

SECTION 1–5 REVIEW

1. What is an integrated circuit?
2. Define the terms DIP, SMT, SOIC, SSI, MSI, LSI, VLSI and ULSI.
3. Generally, in what classification does a fixed-function IC with the following number of equivalent gates fall?

   (a) 10   (b) 75   (c) 500   (d) 15,000   (e) 200,000

1–6   INTRODUCTION TO PROGRAMMABLE LOGIC

Programmable logic requires both hardware and software. Programmable logic devices can be programmed to perform specified logic functions by the manufacturer or by the user. One advantage of programmable logic over fixed-function logic is that the devices use much less board space for an equivalent amount of logic. Another advantage is that, with programmable logic, designs can be readily changed without rewiring or replacing components. Also, a logic design can generally be implemented faster and with less cost with programmable logic than with fixed-function ICs.

After completing this section, you should be able to

■ State the major types of programmable logic and discuss the differences ■ Discuss methods of programming ■ List the major programming languages used for programmable logic ■ Discuss the programmable logic design process

Types of Programmable Logic Devices

Many types of programmable logic are available, ranging from small devices that can replace a few fixed-function devices to complex high-density devices that can replace thousands of fixed-function devices. Two major categories of user-programmable logic are **PLD** (programmable logic device) and **FPGA** (field programmable gate array), as indicated in Figure 1–31. PLDs are either SPLDs (simple PLDs) or CPLDs (complex PLDs).

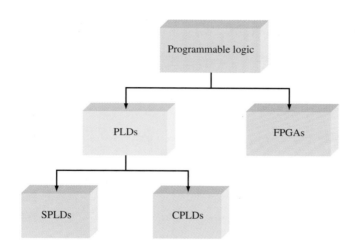

◀ **FIGURE 1–31**

Programmable logic.

Simple Programmable Logic Device (SPLD) The SPLD was the original PLD and is still available for small-scale applications. Generally, an **SPLD** can replace up to ten fixed-function ICs and their interconnections, depending on the type of functions and the specific SPLD. Most SPLDs are in one of two categories: PAL and GAL. A **PAL** (programmable array logic) is a device that can be programmed one time. It consists of a programmable array of AND gates and a fixed array of OR gates, as shown in Figure 1–32(a). A **GAL** (generic array logic) is a

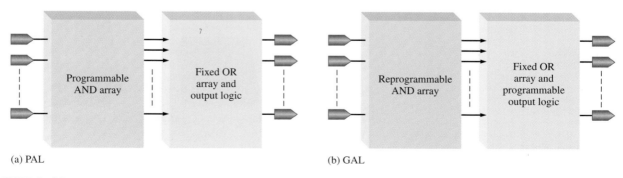

▲ **FIGURE 1–32**

Block diagrams of simple programmable logic devices (SPLDs).

device that is basically a PAL that can be reprogrammed many times. It consists of a reprogrammable array of AND gates and a fixed array of OR gates with programmable ouputs, as shown in Figure 1–32(b). A typical SPLD package is shown in Figure 1–33 and generally has from 24 to 28 pins.

▶ **FIGURE 1–33**

Typical SPLD package.

Complex Programmable Logic Device (CPLD) As technology progressed and the amount of circuitry that could be put on a chip (chip density) increased, manufacturers were able to put more than one SPLD on a single chip and the CPLD was born. Essentially, the **CPLD** is a device containing multiple SPLDs and can replace many fixed-function ICs. Figure 1–34 shows a basic CPLD block diagram with four logic array blocks (LABs) and a programmable interconnection array (PIA). Depending on the specific CPLD, there can be from two to sixty-four LABs. Each logic array block is roughly equivalent to one SPLD.

▶ **FIGURE 1–34**

General block diagram of a CPLD.

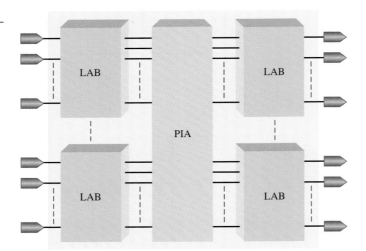

Generally, CPLDs can be used to implement any of the logic functions discussed earlier, for example, decoders, encoders, multiplexers, demultiplexers, and adders. They are available in a variety of configurations, typically ranging from 44 to 160 pin packages. Examples of CPLD packages are shown in Figure 1–35.

▶ **FIGURE 1–35**

Typical CPLD packages.

(a) 84-pin PLCC package   (b) 128-pin PQFP package

Field Programmable Gate Array (FPGA) An **FPGA** is generally more complex and has a much higher density than a CPLD, although their applications can sometimes overlap. As mentioned, the SPLD and the CPLD are closely related because the CPLD basically contains a number of SPLDs. The FPGA, however, has a different internal structure (architec-

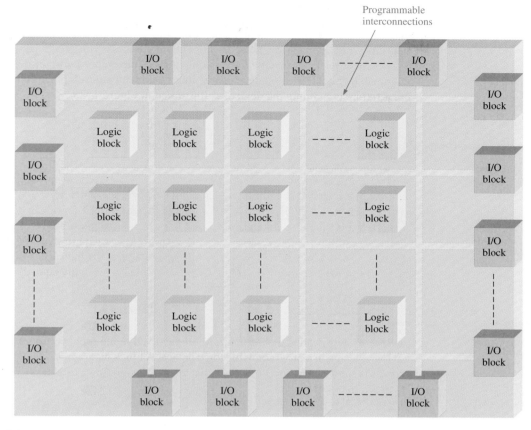

▲ **FIGURE 1–36**

Basic structure of an FPGA.

ture), as illustrated in Figure 1–36. The three basic elements in an FPGA are the logic block, the programmable interconnections, and the input/output (I/O) blocks.

The logic blocks in an FPGA are not as complex as the logic array blocks (LABs) in a CPLD, but generally there are many more of them. When the logic blocks are relatively simple, the FPGA architecture is called *fine-grained*. When the logic blocks are larger and more complex, the architecture is called *coarse-grained*. The I/O blocks are on the outer edges of the structure and provide individually selectable input, output, or bidirectional access to the outside world. The distributed programmable interconnection matrix provides for interconnection of the logic blocks and connection to inputs and outputs. Large FPGAs can have tens of thousands of logic blocks in addition to memory and other resources. A typical FPGA ball-grid array package is shown in Figure 1–37. These types of packages can have over 1000 input and output pins.

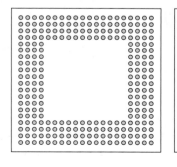

◀ **FIGURE 1–37**

A typical ball–grid array package configuration.

The Programming Process

An SPLD, CPLD, or FPGA can be thought of as a "blank slate" on which you implement a specified circuit or system design using a certain process. This process requires a software development package installed on a computer to implement a circuit design in the programmable chip. The computer must be interfaced with a development board or programming fixture containing the device, as illustrated in Figure 1–38.

Basic configuration for programming a PLD or FPGA.

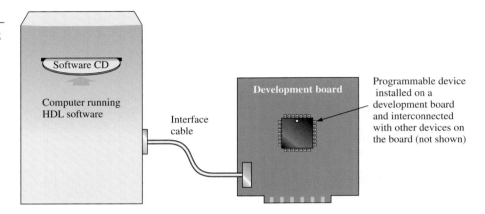

Several steps, called the *design flow,* are involved in the process of implementing a digital logic design in a programmable logic device. A block diagram of a typical programming process is shown in Figure 1–39. As indicated, the design flow has access to a design library.

Basic programmable logic design flow block diagram.

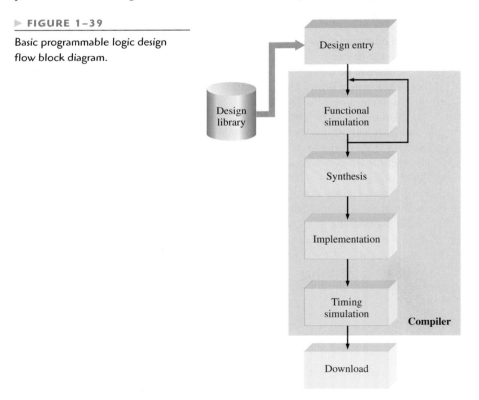

Design Entry   This is the first programming step. The circuit or system design must be entered into the design application software using text-based entry, graphic entry (schematic capture), or state diagram description. Design entry is device independent. Text-based entry is accomplished with a hardware description language (HDL) such as VHDL, Verilog,

AHDL, or ABEL. Graphic (schematic) entry allows prestored logic functions from a library to be selected, placed on the screen, and then interconnected to create a logic design. State-diagram entry requires specification of both the states through which a sequential logic circuit progresses and the conditions that produce each state change.

Once a design has been entered, it is compiled. A **compiler** is a program that controls the design flow process and translates source code into object code in a format that can be logically tested or downloaded to a target device. The source code is created during design entry, and the object code is the final code that actually causes the design to be implemented in the programmable device.

Functional Simulation   The entered and compiled design is simulated by software to confirm that the logic circuit functions as expected. The simulation will verify that correct outputs are produced for a specified set of inputs. A device-independent software tool for doing this is generally called a *waveform editor*. Any flaws demonstrated by the simulation would be corrected by going back to design entry and making appropriate changes.

Synthesis   **Synthesis** is where the design is translated into a netlist, which has a standard form and is device independent.

Implementation   **Implementation** is where the logic structures described by the netlist are mapped into the actual structure of the specific device being programmed. The implementation process is called *fitting* or *place and route* and results in an output called a bitstream, which is device dependent.

Timing Simulation   This step comes after the design is mapped into the specific device. The timing simulation is basically used to confirm that there are no design flaws or timing problems due to propagation delays.

Download   Once a bitstream has been generated for a specific programmable device, it has to be downloaded to the device to implement the software design in hardware. Some programmable devices have to be installed in a special piece of equipment called a *device programmer* or on a development board. Other types of devices can be programmed while in a system—called in-system programming (ISP)—using a standard JTAG (Joint Test Action Group) interface. Some devices are volatile, which means they lose their contents when reset or when power is turned off. In this case, the bitstream data must be stored in a memory and reloaded into the device after each reset or power-off. Also, the contents of an ISP device can be manipulated or upgraded while it is operating in a system. This is called "on-the-fly" reconfiguration.

**SECTION 1–6
REVIEW**

1. List three major categories of programmable logic devices and specify their acronyms.
2. How does a CPLD differ from an SPLD?
3. Name the steps in the programming process.
4. Briefly explain each step named in question 3.

1–7   TEST AND MEASUREMENT INSTRUMENTS

Troubleshooting is the process of systematically isolating, identifying, and correcting a fault in a circuit or system. A variety of instruments are available for use in troubleshooting and testing. Some common types of instruments are introduced and discussed in this section.

After completing this section, you should be able to

■ Distinguish between an analog and a digital oscilloscope ■ Recognize common oscilloscope controls ■ Determine amplitude, period, frequency, and duty cycle of a pulse waveform with an oscilloscope ■ Discuss the logic analyzer and some common formats ■ Describe the purpose of the dc power supply, function generator, and digital multimeter (DMM)

The Oscilloscope

The oscilloscope (scope for short) is one of the most widely used instruments for general testing and troubleshooting. The scope is basically a graph-displaying device that traces the graph of a measured electrical signal on its screen. In most applications, the graph shows how signals change over time. The vertical axis of the display screen represents voltage, and the horizontal axis represents time. Amplitude, period, and frequency of a signal can be measured using the oscilloscope. Also, the pulse width, duty cycle, rise time, and fall time of a pulse waveform can be determined. Most scopes can display at least two signals on the screen at one time, enabling their time relationship to be observed. A typical oscilloscope is shown in Figure 1–40.

▶ FIGURE 1–40

A typical dual-channel oscilloscope. Used with permission from Tektronix, Inc.

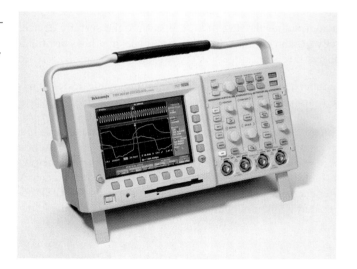

Two basic types of oscilloscopes, analog and digital, can be used to view digital wave-forms. As shown in Figure 1–41(a), the analog scope works by applying the measured waveform directly to control the up and down motion of the electron beam in the cathode-ray tube (CRT) as it sweeps across the display screen. As a result, the beam traces out the waveform pattern on the screen. As shown in Figure 1–41(b), the digital scope converts the measured waveform to digital information by a sampling process in an analog-to-digital converter (ADC). The digital information is then used to reconstruct the waveform on the screen.

The digital scope is more widely used than the analog scope. However, either type can be used in many applications, each has characteristics that make it more suitable for certain situations. An analog scope displays waveforms as they occur in "real time." Digital scopes are useful for measuring transient pulses that may occur randomly or only once. Also, because information about the measured waveform can be stored in a digital scope, it may be viewed at some later time, printed out, or thoroughly analyzed by a computer or other means.

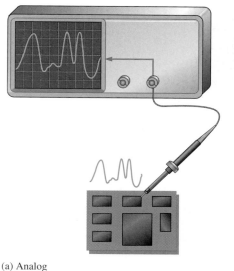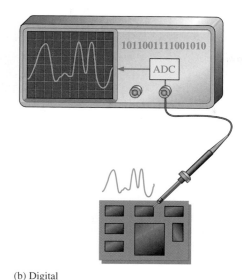

(a) Analog　　　　　　　　　(b) Digital

Comparison of analog and digital oscilloscopes.

Basic Operation of Analog Oscilloscopes　To measure a voltage, a **probe** must be connected from the scope to the point in a circuit at which the voltage is present. Generally, a ×10 probe is used that reduces (attenuates) the signal amplitude by ten. The signal goes through the probe into the vertical circuits where it is either further attenuated or amplified, depending on the actual amplitude and on where you set the vertical control of the scope. The vertical circuits then drive the vertical deflection plates of the CRT. Also, the signal goes to the trigger circuits that trigger the horizontal circuits to initiate repetitive horizontal sweeps of the electron beam across the screen using a sawtooth waveform. There are many sweeps per second so that the beam appears to form a solid line across the screen in the shape of the waveform. This basic operation is illustrated in Figure 1–42.

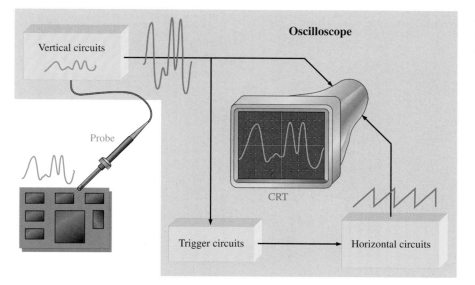

Block diagram of an analog oscilloscope.

Basic Operation of Digital Oscilloscopes　Some parts of a digital scope are similar to the analog scope. However, the digital scope is more complex than an analog scope and typically has an LCD screen rather than a CRT. Rather than displaying a waveform as it occurs, the digital scope first acquires the measured analog waveform and converts it to a digital format using an analog-to-digital converter (ADC). The digital data is stored and processed.

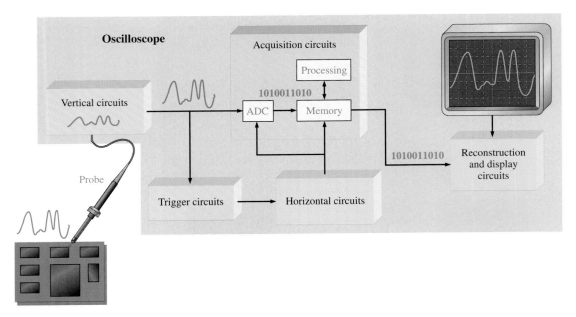

▲ **FIGURE 1–43**

Block diagram of a digital oscilloscope.

The data then goes to the reconstruction and display circuits for display in its original analog form. Figure 1–43 shows a basic block diagram for a digital oscilloscope.

Oscilloscope Controls   A front panel view of a typical dual-channel oscilloscope is shown in Figure 1–44. Instruments vary depending on model and manufacturer, but most have certain common features. For example, the two vertical sections contain a Position control, a channel menu button, and a V/div control. The horizontal section contains a sec/div control.

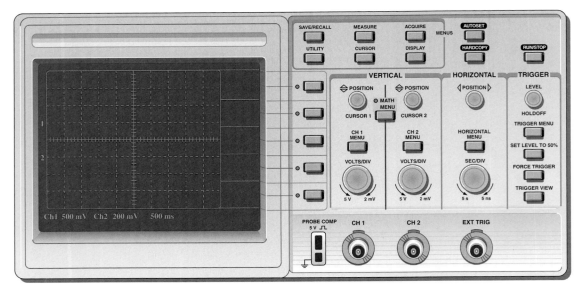

▲ **FIGURE 1–44**

A typical dual-channel oscilloscope. Numbers below screen indicate the values for each division on the vertical (voltage) and horizontal (time) scales and can be varied using the vertical and horizontal controls on the scope.

Some of the main oscilloscope controls are now discussed. Refer to the user manual for complete details of your particular scope.

Vertical Controls   In the vertical section of the scope in Figure 1–44, there are identical controls for each of the two channels (CH1 and CH2). The Position control lets you move a displayed waveform up or down vertically on the screen. The Menu button provides for the selection of several items that appear on the screen, such as the coupling modes (ac, dc, or ground), coarse or fine adjustment for the V/div, probe attenuation, and other parameters. The V/div control adjusts the number of volts represented by each vertical division on the screen. The V/div setting for each channel is displayed on the bottom of the screen. The Math Menu button provides a selection of operations that can be performed on the input waveforms, such as subtraction, addition, or inversion.

Horizontal Controls   In the horizontal section, the controls apply to both channels. The Position control lets you move a displayed waveform left or right horizontally on the screen. The Menu button provides for the selection of several items that appear on the screen such as the main time base, expanded view of a portion of a waveform, and other parameters. The sec/div control adjusts the time represented by each horizontal division or main time base. The sec/div setting is displayed at the bottom of the screen.

Trigger Controls   In the Trigger control section, the Level control determines the point on the triggering waveform where triggering occurs to initiate the sweep to display input waveforms. The Menu button provides for the selection of several items that appear on the screen, including edge or slope triggering, trigger source, trigger mode, and other parameters. There is also an input for an external trigger signal.

Triggering stabilizes a waveform on the screen or properly triggers on a pulse that occurs only one time or randomly. Also, it allows you to observe time delays between two waveforms. Figure 1–45 compares a triggered to an untriggered signal. The untriggered signal tends to drift across the screen, producing what appears to be multiple waveforms.

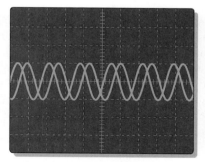

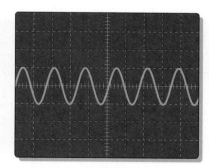

(a) Untriggered waveform display   (b) Triggered waveform display

◀ **FIGURE 1–45**

Comparison of an untriggered and a triggered waveform on an oscilloscope.

Coupling a Signal into the Scope   Coupling is the method used to connect a signal voltage to be measured into the oscilloscope. DC and AC coupling are usually selected from the Vertical menu on a scope. DC coupling allows a waveform including its dc component to be displayed. AC coupling blocks the dc component of a signal so that you see the waveform centered at 0 V. The Ground mode allows you to connect the channel input to ground to see where the 0 V reference is on the screen. Figure 1–46 illustrates the result of DC and AC coupling using a pulse waveform that has a dc component.

The voltage probe, shown in Figure 1–47, is essential for connecting a signal to the scope. Since all instruments tend to affect the circuit being measured due to loading, most scope probes provide a high series resistance to minimize loading effects. Probes that

▶ FIGURE 1–46

Displays of the same waveform having a dc component.

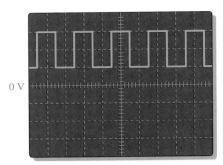

(a) DC coupled waveform

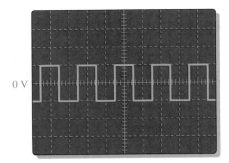

(b) AC coupled waveform

▶ FIGURE 1–47

An oscilloscope voltage probe. Used with permission from Tektronix, Inc.

have a series resistance ten times larger than the input resistance of the scope are called ×10 probes. Probes with no series resistance are called ×1 probes. The oscilloscope adjusts its calibration for the attenuation of the type of probe being used. For most measurements, the ×10 probe should be used. However, if you are measuring very small signals, a ×1 may be the best choice.

The probe has an adjustment that allows you to compensate for the input capacitance of the scope. Most scopes have a probe compensation output that provides a calibrated square wave for probe compensation. Before making a measurement, you should make sure that the probe is properly compensated to eliminate any distortion introduced. Typically, there is a screw or other means of adjusting compensation on a probe. Figure 1–48 shows scope waveforms for three probe conditions: properly compensated, undercompensated, and overcompensated. If the waveform appears either over- or undercompensated, adjust the probe until the properly compensated square wave is achieved.

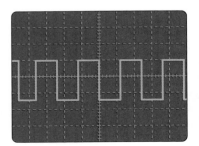

Properly compensated

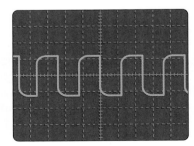

Undercompensated

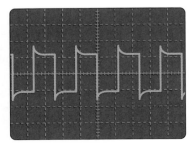

Overcompensated

▲ FIGURE 1–48

Probe compensation conditions.

EXAMPLE 1–3

Based on the readouts, determine the amplitude and the period of the pulse waveform on the screen of a digital oscilloscope as shown in Figure 1–49. Also, calculate the frequency.

▶ FIGURE 1–49

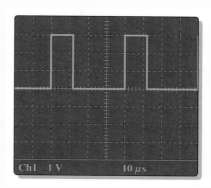

Ch1 1 V   10 μs

Solution   The V/div setting is 1 V. The pulses are three divisions high. Since each division represents 1 V, the pulse amplitude is

$$\text{Amplitude} = (3 \text{ div})(1 \text{ V/div}) = \textbf{3 V}$$

The sec/div setting is 10 μs. A full cycle of the waveform (from beginning of one pulse to the beginning of the next) covers four divisions; therefore, the period is

$$\text{Period} = (4 \text{ div})(10 \text{ μs/div}) = \textbf{40 μs}$$

The frequency is calculated as

$$f = \frac{1}{T} = \frac{1}{40 \text{ μs}} = \textbf{25 kHz}$$

Related Problem   For a V/div setting of 4 V and sec/div setting of 2 ms, determine the amplitude and period of the pulse shown on the screen in Figure 1–49.

The Logic Analyzer

Logic analyzers are used for measurements of multiple digital signals and measurement situations with difficult trigger requirements. Basically, the logic analyzer came about as a result of microprocessors in which troubleshooting or debugging required many more inputs than an oscilloscope offered. Many oscilloscopes have two input channels and some are available with four. Logic analyzers are available with from 34 to 136 input channels. Generally, an oscilloscope is used either when amplitude, frequency, and other timing parameters of a few signals at a time or when parameters such an rise and fall times, overshoot, and delay times need to be measured. The logic analyzer is used when the logic levels of a large number of signals need to be determined and for the correlation of simultaneous signals based on their timing relationships. A typical logic analyzer is shown in Figure 1–50, and a simplified block diagram is in Figure 1–51.

Data Acquisition   The large number of signals that can be acquired at one time is a major factor that distinguishes a logic analyzer from an oscilloscope. Generally, the two types of data acquisition in a logic analyzer are the timing acquisition and the state acquisition. Timing acquisition is used primarily when the timing relationships among the various signals

▶ **FIGURE 1–50**

Typical logic analyzer. Used with permission from Tektronix, Inc.

▶ **FIGURE 1–51**

Simplified block diagram of a logic analyzer.

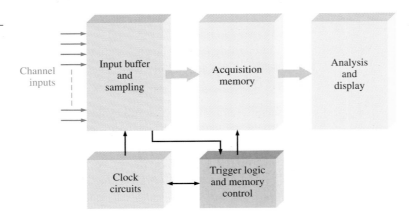

need to be determined. State acquisition is used when you need to view the sequence of states as they appear in a system under test.

It is often helpful to have correlated timing and state data, and most logic analyzers can simultaneously acquire that data. For example, a problem may initially be detected as an invalid state. However, the invalid condition may be caused by a timing violation in the system under test. Without both types of information available at the same time, isolating the problem could be very difficult.

Channel Count and Memory Depth   Logic analyzers contain a real-time acquisition memory in which sampled data from all the channels are stored as they occur. Two features that are of primary importance are the channel count and the memory depth. The acquisition memory can be thought of as having a width equal to the number of channels and a depth that is the number of bits that can be captured by each channel during a certain time interval.

Channel count determines the number of signals that can be acquired simultaneously. In certain types of systems, a large number of signals are present, such as on the data bus in a microprocessor-based system. The depth of the acquisition memory determines the amount of data from a given channel that you can view at any given time.

Analysis and Display   Once data has been sampled and stored in the acquisition memory, it can typically be used in several different display and analysis modes. The waveform display is much like the display on an oscilloscope where you can view the time relationship of multiple signals. The listing display indicates the state of the system under test by showing the values of the input waveforms (1s and 0s) at various points in time (sample points). Typically, this data can be displayed in hexadecimal or other formats. Figure 1–52 shows simplified versions of these two display modes. The listing display samples correspond to the sampled points shown in red on the waveform display. You will study binary and hexadecimal (hex) numbers in the next chapter.

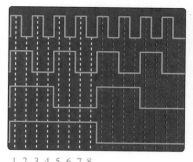

1 2 3 4 5 6 7 8

(a) Waveform display

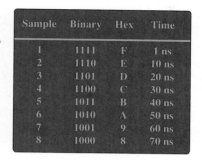

Sample	Binary	Hex	Time
1	1111	F	1 ns
2	1110	E	10 ns
3	1101	D	20 ns
4	1100	C	30 ns
5	1011	B	40 ns
6	1010	A	50 ns
7	1001	9	60 ns
8	1000	8	70 ns

(b) Listing display

◀ FIGURE 1–52

Two logic analyzer display modes.

Two more modes that are useful in computer and microprocessor-based system testing are the instruction trace and the source code debug. The instruction trace determines and displays instructions that occur, for example, on the data bus in a microprocessor-based system. In this mode the op-codes and the mnemonics (English-like names) of instructions are generally displayed as well as their corresponding memory address. Many logic analyzers also include a source code debug mode, which essentially allows you to see what is actually going on in the system under test when a program instruction is executed.

Probes   Three basic types of probes are used with logic analyzers. One is a multichannel compression probe that can be attached to points on a circuit board, as shown in Figure 1–53. Another type of multichannel probe, similar to the one shown, plugs into dedicated sockets mounted on a circuit board. A third type is a single-channel clip-on probe.

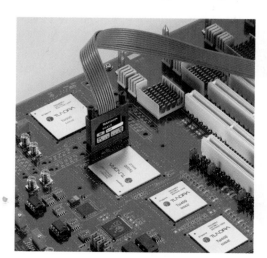

◀ FIGURE 1–53

A typical multichannel logic analyzer probe. Used with permission from Tektronix, Inc.

Signal Generators

Logic Signal Source   These instruments are also known as pulse generators and pattern generators. They are specifically designed to generate digital signals with precise edge placement and amplitudes and to produce the streams of 1s and 0s needed to test computer buses, microprocessors, and other digital systems.

Arbitrary Waveform Generators and Function Generators The arbitrary waveform generator can be used to generate standard signals like sine waves, triangular waves, and pulses as well as signals with various shapes and characteristics. Waveforms can be defined by mathematical or graphical input. A typical arbitrary waveform generator is shown in Figure 1–54(a).

The function generator provides pulse waveforms as well as sine waves and triangular waves. Most function generators have logic-compatible outputs to provide the proper level and drive for inputs to digital circuits. Typical function generators are shown in Figure 1–54(b).

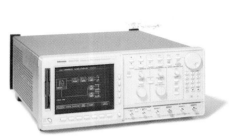

(a) An arbitrary waveform generator.

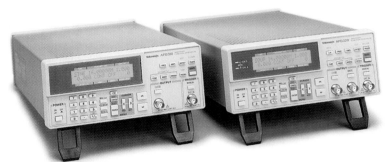

(b) Examples of function generators.

▲ FIGURE 1–54

Typical signal generators. Used with permission from Tektronix, Inc.

The Logic Probe and Logic Pulser The logic probe is a convenient, inexpensive hand-held tool that provides a means of troubleshooting a digital circuit by sensing various conditions at a point in a circuit, as illustrated in Figure 1–55. The probe can detect high-level voltage, low-level voltage, single pulses, repetitive pulses, and opens on a PC

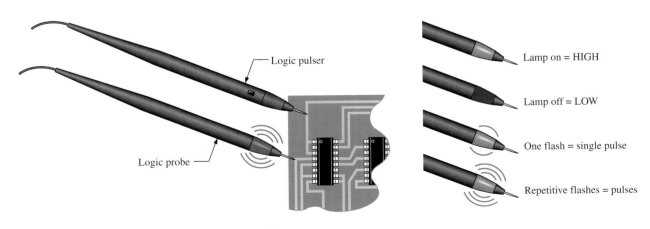

▲ FIGURE 1–55

Illustration of how a logic pulser and a logic probe can be used to apply a pulse to a given point and check for resulting pulse activity at another part of the circuit.

board. The probe lamp indicates the condition that exists at a certain point, as indicated in the figure.

The logic pulser produces a repetitive pulse waveform that can be applied to any point in a circuit. You can apply pulses at one point in a circuit with the pulser and check another point for resulting pulses with a logic probe.

Other Instruments

The DC Power Supply   This instrument is an indispensable instrument on any test bench. The power supply converts ac power from the standard wall outlet into regulated dc voltage. All digital circuits require dc voltage. Many logic circuits require +5 V or +3.3 V to operate. The power supply is used to power circuits during design, development, and troubleshooting when in-system power is not available. Typical test bench dc power supplies are shown in Figure 1–56.

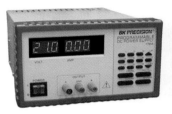  

◀ **FIGURE 1–56**

Typical dc power supplies. Courtesy of B+K Precision.®

The Digital Multimeter (DMM)   The DMM is used for measuring dc and ac voltage and resistance. Figure 1–57 shows typical test bench and handheld DMMs.

◀ **FIGURE 1–57**

Typical DMMs. Courtesy of B+K Precision.®

SECTION 1–7 REVIEW

1. What is the main difference between a digital and an analog oscilloscope?

2. Name two main differences between a logic analyzer and an oscilloscope?

3. What does the V/div control on an oscilloscope do?

4. What does the sec/div control on an oscilloscope do?

5. What is the purpose of a function generator?

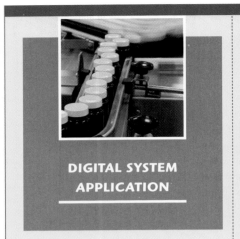

DIGITAL SYSTEM APPLICATION

In this digital system application (DSA), a simplified system application of the logic elements and functions that were discussed in Section 1–4 is presented. It is important that you understand how various digital functions can operate together as a total system to perform a specified task. It is also important to begin to think in terms of system-level operation because, in practice, a large part of your work will involve systems rather than individual functions. Of course, to understand systems, you must first understand the basic elements and functions that make up a system.

This DSA introduces you to the system concept. The example shows you how logic functions can work together to perform a higher-level task and gets you started thinking at the system level. The specific system used here to illustrate the system concept serves as an instructional model and is not necessarily the approach that would be used in practice, although it could be. In modern industrial applications like the one discussed here, instruments known as programmable controllers are often used.

About the System

Let's imagine that a factory uses the process control system shown in the simplified block diagram of Figure 1–58 for automatically counting and bottling tablets. The tablets are fed into a large funnel-like hopper. The narrow neck of the funnel allows only one tablet at a time to fall into a bottle on the conveyor belt below.

The digital system controls the number of tablets going into each bottle and displays a continually updated total near the conveyor line as well as at a remote location in another part of the plant. This system utilizes all the basic logic functions that were introduced in Section 1–4, and its only purpose is to show you how these functions may be combined to achieve a desired result.

The general operation is as follows. An optical sensor at the bottom of the funnel neck detects each tablet that passes and produces an electrical pulse. This pulse goes to the counter and advances it by one count; thus, at any time during the filling of a bottle, the counter contains the binary representation of the number of tablets in the bottle. The binary count is transferred from the counter on parallel lines to the B input of the comparator (comp). A preset binary number equal to the number of tablets that are to go into each bottle is placed on the A input of the comparator. The preset number comes from the keypad and the associated circuits, which include the encoder, register A, and code converter A. When the desired number of tablets is entered on the keypad, it is encoded and then stored by parallel register A until a change in the quantity of tablets per bottle is required.

Suppose, for example, that each bottle is to hold fifty tablets. When the number in the counter reaches 50, the $A = B$ output of the comparator goes HIGH, indicating that the bottle is full.

The HIGH output of the comparator immediately closes the valve in the neck of the funnel to stop the flow of tablets, and at the same time it activates the conveyor to move the next bottle into place under the funnel. When the next bottle is positioned properly under the neck of the funnel, the conveyor control circuit produces a pulse that resets the counter to zero. The $A = B$ output of comparator goes back LOW, opening the funnel valve to restart the flow of tablets.

In the display portion of the system, the number in the counter is transferred in parallel to the A input of the adder. The B input of the adder comes from parallel register B that holds the total number of tablets bottled, up through the last bottle filled. For example, if ten bottles have been filled and each bottle holds fifty tablets, register B contains the binary representation for 500. Then, when the next bottle has been filled, the binary number for 50 appears on the A input of the adder, and the binary number for 500 is on the B input. The adder produces a new sum of 550, which is stored in register B, replacing the previous sum of 500.

The binary number in register B is transferred in parallel to the code converter and decoder, which changes it from binary form to decimal form for display on a readout near the conveyor line. The binary number in the register is also transferred to a multiplexer (mux) so that it can be converted from parallel to series form and transmitted along a single line to a remote location some distance away. It is more economical to run a single line than to run several parallel lines when significant distances are involved, and speed of data transmission is not a factor in this application. At the remote location, the serial data are demultiplexed and sent to register C. From there the data are then decoded for display on the remote readout.

Keep in mind that this system is purely an instructional model and does not necessarily represent the ultimate or most efficient way to implement this hypothetical process. Although there are certainly other approaches, this particular approach has been used in order to illustrate an application of the logic functions that were introduced in Section 1–4 and that will be covered in detail in future chapters. It shows you an application of the various functional devices at the system level and how they can be connected to accomplish a specific objective.

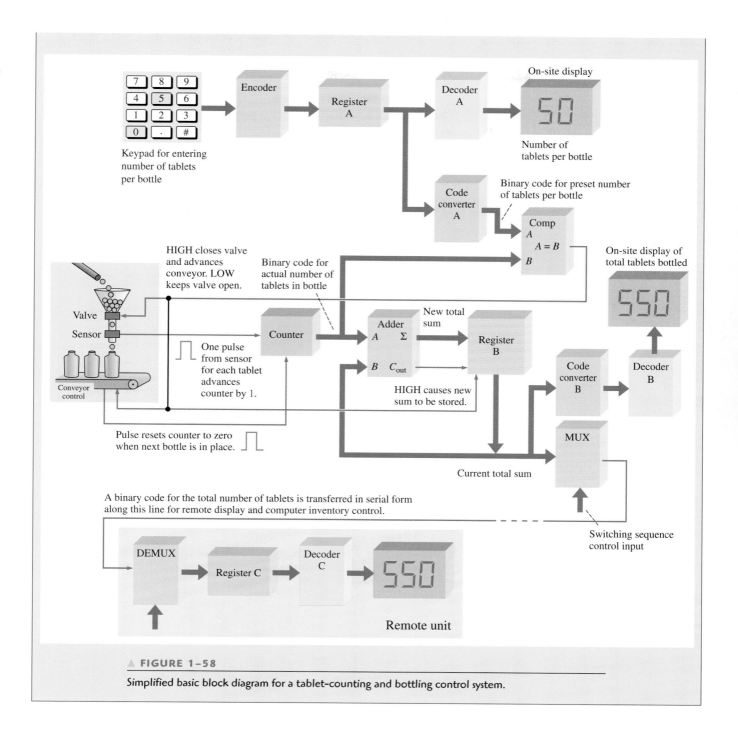

▲ **FIGURE 1–58**

Simplified basic block diagram for a tablet-counting and bottling control system.

SUMMARY

- An analog quantity has continuous values.
- A digital quantity has a discrete set of values.
- A binary digit is called a bit.
- A pulse is characterized by rise time, fall time, pulse width, and amplitude.

- The frequency of a periodic waveform is the reciprocal of the period. The formulas relating frequency and period are

$$f = \frac{1}{T} \text{ and } T = \frac{1}{f}$$

- The duty cycle of a pulse waveform is the ratio of the pulse width to the period, expressed by the following formula as a percentage:

$$\text{Duty cycle} = \left(\frac{t_W}{T}\right)100\%$$

- A timing diagram is an arrangement of two or more waveforms showing their relationship with respect to time.
- Three basic logic operations are NOT, AND, and OR. The standard symbols for these are given in Figure 1–59.

▶ FIGURE 1–59

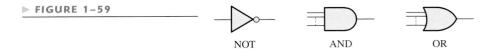

NOT   AND   OR

- The basic logic functions are comparison, arithmetic, code conversion, decoding, encoding, data selection, storage, and counting.
- The two broad physical categories of IC packages are through-hole mounted and surface mounted.
- The categories of ICs in terms of circuit complexity are SSI (small-scale integration), MSI (medium-scale integration), LSI, VLSI, and ULSI (large-scale, very large-scale, and ultra large-scale integration).
- Two types of SPLDs (simple programmable logic devices) are PAL (programmable array logic) and GAL (generic array logic).
- The CPLD (complex programmable logic device) contains multiple SPLDs with programmable interconnections.
- The FPGA (field programmable gate array) has a different internal structure than the CPLD and is generally used for more complex circuits and systems.
- Common instruments used in testing and troubleshooting digital circuits are the oscilloscope, logic analyzer, waveform generator, function generator, dc power supply, digital multimeter, logic probe, and logic pulser.

KEY TERMS

Key terms and other bold terms in the chapter are defined in the end-of-book glossary.

Analog Being continuous or having continuous values.

AND A basic logic operation in which a true (HIGH) output occurs only when all the input conditions are true (HIGH).

Binary Having two values or states; describes a number system that has a base of two and utilizes 1 and 0 as its digits.

Bit A binary digit, which can be either a 1 or a 0.

Clock The basic timing signal in a digital system; a periodic waveform in which each interval between pulses equals the time for one bit.

Compiler A program that controls the design flow process and translates source code into object code in a format that can be logically tested or downloaded to a target device.

CPLD A complex programmable logic device that consists basically of multiple SPLD arrays with programmable interconnections.

Data Information in numeric, alphabetic, or other form.

Digital Related to digits or discrete quantities; having a set of discrete values.

FPGA Field programmable gate array.

Gate A logic circuit that performs a specified logic operation such as AND or OR.

Input The signal or line going into a circuit.

Integrated circuit (IC) A type of circuit in which all of the components are integrated on a single chip of semiconductive material of extremely small size.

Inverter A NOT circuit; a circuit that changes a HIGH to a LOW or vice versa.

Logic In digital electronics, the decision-making capability of gate circuits, in which a HIGH represents a true statement and a LOW represents a false one.

NOT A basic logic operation that performs inversions.

OR A basic logic operation in which a true (HIGH) output occurs when one or more of the input conditions are true (HIGH).

Output The signal or line coming out of a circuit.

Parallel In digital systems, data occurring simultaneously on several lines; the transfer or processing of several bits simultaneously.

Pulse A sudden change from one level to another, followed after a time, called the pulse width, by a sudden change back to the original level.

Serial Having one element following another, as in a serial transfer of bits; occurring in sequence rather than simultaneously.

SPLD Simple programmable logic device.

Timing diagram A graph of digital waveforms showing the time relationship of two or more waveforms.

Troubleshooting The technique or process of systematically identifying, isolating, and correcting a fault in a circuit or system.

SELF-TEST

Answers are at the end of the chapter.

1. A quantity having continuous values is
   - (a) a digital quantity
   - (b) an analog quantity
   - (c) a binary quantity
   - (d) a natural quantity

2. The term *bit* means
   - (a) a small amount of data
   - (b) a 1 or a 0
   - (c) binary digit
   - (d) both answers (b) and (c)

3. The time interval on the leading edge of a pulse between 10% and 90% of the amplitude is the
   - (a) rise time   (b) fall time   (c) pulse width   (d) period

4. A pulse in a certain waveform occurs every 10 ms. The frequency is
   - (a) 1 kHz   (b) 1 Hz   (c) 100 Hz   (d) 10 Hz

5. In a certain digital waveform, the period is twice the pulse width. The duty cycle is
   - (a) 100%   (b) 200%   (c) 50%

6. An inverter
   - (a) performs the NOT operation
   - (b) changes a HIGH to a LOW
   - (c) changes a LOW to a HIGH
   - (d) does all of the above

7. The output of an AND gate is HIGH when
   - (a) any input is HIGH
   - (b) all inputs are HIGH
   - (c) no inputs are HIGH
   - (d) both answers (a) and (b)

8. The output of an OR gate is HIGH when
   - (a) any input is HIGH
   - (b) all inputs are HIGH
   - (c) no inputs are HIGH
   - (d) both answers (a) and (b)

9. The device used to convert a binary number to a 7-segment display format is the
   - (a) multiplexer   (b) encoder   (c) decoder   (d) register

10. An example of a data storage device is

   (a) the logic gate   (b) the flip-flop   (c) the comparator

   (d) the register   (e) both answers (b) and (d)

11. A fixed-function IC package containing four AND gates is an example of

   (a) MSI   (b) SMT   (c) SOIC   (d) SSI

12. An LSI device has a circuit complexity of from

   (a) 10 to 100 equivalent gates   (b) more than 100 to 10,000 equivalent gates

   (c) 2000 to 5000 equivalent gates   (d) more than 10,000 to 100,000 equivalent gates

13. VHDL is a

   (a) logic device   (b) PLD programming language

   (c) computer language   (d) very high density logic

14. A CPLD is a

   (a) controlled program logic device   (b) complex programmable logic driver

   (c) complex programmable logic device   (d) central processing logic device

15. An FPGA is a

   (a) field programmable gate array   (b) fast programmable gate array

   (c) field programmable generic array   (d) flash process gate application

PROBLEMS

Answers to odd-numbered problems are at the end of the book.

SECTION 1–1   Digital and Analog Quantities

1. Name two advantages of digital data as compared to analog data.

2. Name an analog quantity other than temperature and sound.

SECTION 1–2   Binary Digits, Logic Levels, and Digital Waveforms

3. Define the sequence of bits (1s and 0s) represented by each of the following sequences of levels:

   (a) HIGH, HIGH, LOW, HIGH, LOW, LOW, LOW, HIGH

   (b) LOW, LOW, LOW, HIGH, LOW, HIGH, LOW, HIGH, LOW

4. List the sequence of levels (HIGH and LOW) that represent each of the following bit sequences:

   (a) 1 0 1 1 1 0 1   (b) 1 1 1 0 1 0 0 1

5. For the pulse shown in Figure 1–60, graphically determine the following:

   (a) rise time   (b) fall time   (c) pulse width   (d) amplitude

▶ FIGURE 1–60

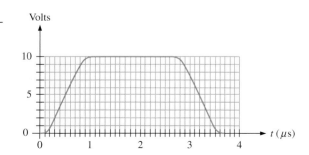

6. Determine the period of the digital waveform in Figure 1–61.

7. What is the frequency of the waveform in Figure 1–61?

8. Is the pulse waveform in Figure 1–61 periodic or nonperiodic?

9. Determine the duty cycle of the waveform in Figure 1–61.

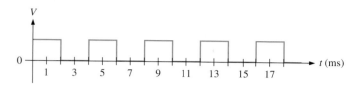

▲ FIGURE 1–61

10. Determine the bit sequence represented by the waveform in Figure 1–62. A bit time is 1 μs this case.

11. What is the total serial transfer time for the eight bits in Figure 1–62? What is the total parallel transfer time?

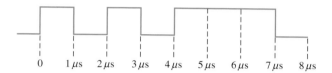

▲ FIGURE 1–62

SECTION 1–3   Basic Logic Operations

12. A logic circuit requires HIGHs on all its inputs to make the output HIGH. What type of logic circuit is it?

13. A basic 2-input logic circuit has a HIGH on one input and a LOW on the other input, and the output is LOW. Identify the circuit.

14. A basic 2-input logic circuit has a HIGH on one input and a LOW on the other input, and the output is HIGH. What type of logic circuit is it?

SECTION 1–4   Overview of Basic Logic Functions

15. Name the logic function of each block in Figure 1–63 based on your observation of the inputs and outputs.

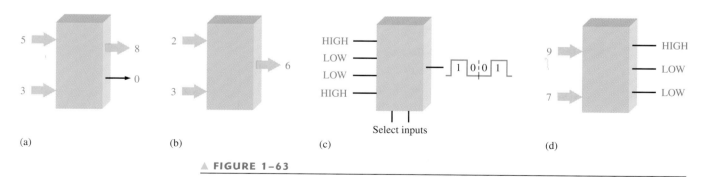

▲ FIGURE 1–63

16. A pulse waveform with a frequency of 10 kHz is applied to the input of a counter. During 100 ms, how many pulses are counted?

17. Consider a register that can store eight bits. Assume that it has been reset so that it contains zeros in all positions. If you transfer four alternating bits (0101) serially into the register, beginning with a 1 and shifting to the right, what will the total content of the register be as soon as the fourth bit is stored?

SECTION 1–5   Fixed-Function Integrated Circuits

18. A fixed-function digital IC chip has a complexity of 200 equivalent gates. How is it classified?

19. Explain the main difference between the DIP and SMT packages.

20. Label the pin numbers on the packages in Figure 1–64. Top views are shown.

▶ FIGURE 1–64

 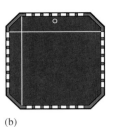

(a)   (b)

SECTION 1–6   Introduction to Programmable Logic

21. Which of the following acronyms do not describe programmable logic?

PAL, GAL, SPLD, ABEL, CPLD, CUPL, FPGA

22. What do each of the following stand for?

(a) SPLD   **(b)** CPLD   **(c)** HDL   **(d)** FPGA   **(e)** GAL

23. Define each of the following PLD programming terms:

(a) design entry   **(b)** simulation   **(c)** compilation   **(d)** download

24. Describe the process of place-and-route.

SECTION 1–7   Test and Measurement Instruments

25. A pulse is displayed on the screen of an oscilloscope, and you measure the base line as 1 V and the top of the pulse as 8 V. What is the amplitude?

26. A logic probe is applied to a contact point on an IC that is operating in a system. The lamp on the probe flashes repeatedly. What does this indicate?

SECTION 1–8   Digital System Application

27. Define the term *system*.

28. In the system depicted in Figure 1–58, why are the multiplexer and demultiplexer necessary?

29. What action can be taken to change the number of tablets per bottle in the system of Figure 1–58?

ANSWERS

SECTION REVIEWS

SECTION 1–1   Digital and Analog Quantities

1. *Analog* means continuous.

2. *Digital* means discrete.

3. A digital quantity has a discrete set of values and an analog quantity has continuous values.

4. A public address system is analog. A CD player is analog and digital. A computer is all digital.

SECTION 1–2   Binary Digits, Logic Levels, and Digital Waveforms

1. Binary means having two states or values.

2. A bit is a binary digit.

3. The bits are 1 and 0.

4. Rise time: from 10% to 90% of amplitude. Fall time: from 90% to 10% of amplitude.

5. Frequency is the reciprocal of the period.

6. A clock waveform is a basic timing waveform from which other waveforms are derived.

7. A timing diagram shows the time relationship of two or more waveforms.

8. Parallel transfer is faster than serial transfer.

SECTION 1–3 **Basic Logic Operations**

1. When the input is LOW
2. When all inputs are HIGH
3. When any or all inputs are HIGH
4. An inverter is a NOT circuit.
5. A logic gate is a circuit that performs a logic operation (AND, OR).

SECTION 1–4 **Overview of Basic Logic Functions**

1. A comparator compares the magnitudes of two input numbers.
2. Add, subtract, multiply, and divide
3. Encoding is changing a familiar form such as decimal to a coded form such as binary.
4. Decoding is changing a code to a familiar form such as binary to decimal.
5. Multiplexing puts data from many sources onto one line. Demultiplexing takes data from one line and distributes it to many destinations.
6. Flip-flops, registers, semiconductor memories, magnetic disks
7. A counter counts events with a sequence of binary states.

SECTION 1–5 **Fixed-Function Integrated Circuits**

1. An IC is an electronic circuit with all components integrated on a single silicon chip.
2. DIP—dual in-line package; SMT—surface-mount technology; SOIC—small-outline integrated circuit; SSI—small-scale integration; MSI—medium-scale integration; LSI—large-scale integration; VLSI—very large-scale integration; ULSI—ultra large-scale integration
3. **(a)** SSI **(b)** MSI **(c)** LSI **(d)** VLSI **(e)** ULSI

SECTION 1–6 **Introduction to Programmable Logic**

1. Simple programmable logic device (SPLD), complex programmable logic device (CPLD), and field programmable gate array (FPGA)
2. A CPLD is made up of multiple SPLDs.
3. Design entry, functional simulation, synthesis, implementation, timing simulation, and download
4. *Design entry:* The logic design is entered using development software. *Functional simulation:* The design is software simulated to make sure it works logically. *Synthesis:* The design is translated into a netlist. *Implementation:* The logic developed by the netlist is mapped into the programmable device. *Timing simulation:* The design is software simulated to confirm that there are no timing problems. *Download:* The design is placed into the programmable device.

SECTION 1–7 **Test and Measurement Instruments**

1. The analog scope applies the measured waveform directly to the display circuits. The digital scope first converts the measured signal to digital form.
2. The logic analyzer has more channels than the oscillosope and has more than one data display format.
3. The V/div control sets the voltage for each division on the screen.
4. The sec/div control sets the time for each division on the screen.
5. The function generator produces various types of waveforms.

RELATED PROBLEMS FOR EXAMPLES

1–1 $f = 6.67$ kHz; Duty cycle $= 16.7\%$

1–2 Parallel transfer: 100 ns; Serial transfer: $1.6 \ \mu s$

1–3 Amplitude $= 12$ V; $T = 8$ ms

SELF-TEST

1. (b) **2.** (d) **3.** (a) **4.** (c) **5.** (c) **6.** (d) **7.** (b) **8.** (d)

9. (c) **10.** (e) **11.** (d) **12.** (d) **13.** (b) **14.** (c) **15.** (a)

2

NUMBER SYSTEMS, OPERATIONS, AND CODES

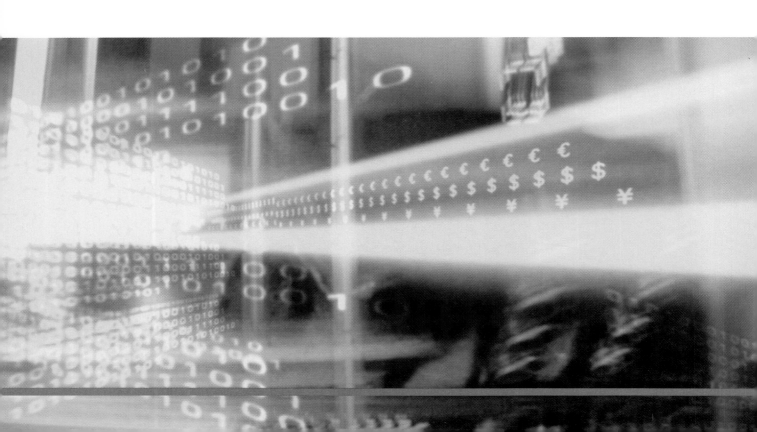

- Review the decimal number system

- Count in the binary number system

- Convert from decimal to binary and from binary to decimal

- Apply arithmetic operations to binary numbers

- Determine the 1's and 2's complements of a binary number

- Express signed binary numbers in sign-magnitude, 1's complement, 2's complement, and floating-point format

- Carry out arithmetic operations with signed binary numbers

- Convert between the binary and hexadecimal number systems

- Add numbers in hexadecimal form

- Convert between the binary and octal number systems

- Express decimal numbers in binary coded decimal (BCD) form

- Add BCD numbers

- Convert between the binary system and the Gray code

- Interpret the American Standard Code for Information Interchange (ASCII)

- Explain how to detect and correct code errors

INTRODUCTION

The binary number system and digital codes are fundamental to computers and to digital electronics in general. In this chapter, the binary number system and its relationship to other number systems such as decimal, hexadecimal, and octal is presented. Arithmetic operations with binary numbers are covered to provide a basis for understanding how computers and many other types of digital systems work. Also, digital codes such as binary coded decimal (BCD), the Gray code, and the ASCII are covered. The parity method for detecting errors in codes is introduced and a method for correcting errors is described. The tutorials on the use of the calculator in certain operations are based on the TI-86 graphics calculator and the TI-36X calculator. The procedures shown may vary on other types.

www. **VISIT THE COMPANION WEBSITE**
Study aids for this chapter are available at
http://www.prenhall.com/floyd

KEY TERMS

- LSB
- MSB
- Byte
- Floating-point number
- Hexadecimal
- Octal
- BCD
- Alphanumeric
- ASCII
- Parity
- Hamming code

2–1 DECIMAL NUMBERS

You are familiar with the decimal number system because you use decimal numbers every day. Although decimal numbers are commonplace, their weighted structure is often not understood. In this section, the structure of decimal numbers is reviewed. This review will help you more easily understand the structure of the binary number system, which is important in computers and digital electronics.

After completing this section, you should be able to

■ Explain why the decimal number system is a weighted system ■ Explain how powers of ten are used in the decimal system ■ Determine the weight of each digit in a decimal number

The decimal number system has ten digits.

In the **decimal** number system each of the ten digits, 0 through 9, represents a certain quantity. As you know, the ten symbols (**digits**) do not limit you to expressing only ten different quantities because you use the various digits in appropriate positions within a number to indicate the magnitude of the quantity. You can express quantities up through nine before running out of digits; if you wish to express a quantity greater than nine, you use two or more digits, and the position of each digit within the number tells you the magnitude it represents. If, for example, you wish to express the quantity twenty-three, you use (by their respective positions in the number) the digit 2 to represent the quantity twenty and the digit 3 to represent the quantity three, as illustrated below.

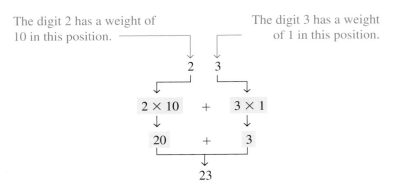

The decimal number system has a base of 10.

The position of each digit in a decimal number indicates the magnitude of the quantity represented and can be assigned a **weight.** The weights for whole numbers are positive powers of ten that increase from right to left, beginning with $10^0 = 1$.

$$\ldots 10^5\ 10^4\ 10^3\ 10^2\ 10^1\ 10^0$$

For fractional numbers, the weights are negative powers of ten that decrease from left to right beginning with 10^{-1}.

The value of a digit is determined by its position in the number.

$$10^2\ 10^1\ 10^0.10^{-1}\ 10^{-2}\ 10^{-3}\ldots$$

$$\uparrow\!\!\!\!\!\!\!\text{——— Decimal point}$$

The value of a decimal number is the sum of the digits after each digit has been multiplied by its weight, as Examples 2–1 and 2–2 illustrate.

EXAMPLE 2–1

Express the decimal number 47 as a sum of the values of each digit.

Solution   The digit 4 has a weight of 10, which is 10^1, as indicated by its position. The digit 7 has a weight of 1, which is 10^0, as indicated by its position.

$$47 = (4 \times 10^1) + (7 \times 10^0)$$
$$= (4 \times 10) + (7 \times 1) = \mathbf{40 + 7}$$

Related Problem*   Determine the value of each digit in 939.

_____

*Answers are at the end of the chapter.

EXAMPLE 2–2

Express the decimal number 568.23 as a sum of the values of each digit.

Solution   The whole number digit 5 has a weight of 100, which is 10^2, the digit 6 has a weight of 10, which is 10^1, the digit 8 has a weight of 1, which is 10^0, the fractional digit 2 has a weight of 0.1, which is 10^{-1}, and the fractional digit 3 has a weight of 0.01, which is 10^{-2}.

$$568.23 = (5 \times 10^2) + (6 \times 10^1) + (8 \times 10^0) + (2 \times 10^{-1}) + (3 \times 10^{-2})$$
$$= (5 \times 100) + (6 \times 10) + (8 \times 1) + (2 \times 0.1) + (3 \times 0.01)$$
$$= \quad 500 \quad + \quad 60 \quad + \quad 8 \quad + \quad 0.2 \quad + \quad 0.03$$

Related Problem   Determine the value of each digit in 67.924.

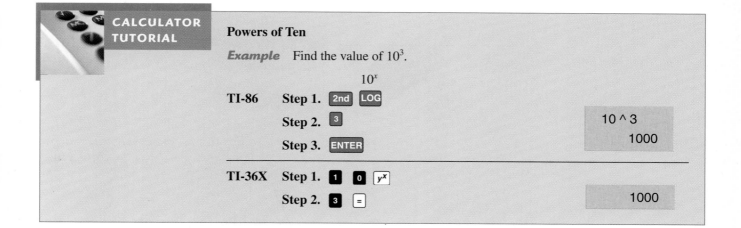

CALCULATOR TUTORIAL

Powers of Ten

Example   Find the value of 10^3.

$$10^x$$

TI-86   Step 1. [2nd] [LOG]

Step 2. [3]   10 ^ 3
                                                        1000
Step 3. [ENTER]

_____

TI-36X   Step 1. [1] [0] [y^x]

Step 2. [3] [=]   1000

SECTION 2–1 REVIEW

Answers are at the end of the chapter.

1. What weight does the digit 7 have in each of the following numbers?
   (a) 1370   (b) 6725   (c) 7051   (d) 58.72
2. Express each of the following decimal numbers as a sum of the products obtained by multiplying each digit by its appropriate weight:
   (a) 51   (b) 137   (c) 1492   (d) 106.58

2–2 BINARY NUMBERS

The binary number system is another way to represent quantities. It is less complicated than the decimal system because it has only two digits. The decimal system with its ten digits is a base-ten system; the binary system with its two digits is a base-two system. The two binary digits (bits) are 1 and 0. The position of a 1 or 0 in a binary number indicates its weight, or value within the number, just as the position of a decimal digit determines the value of that digit. The weights in a binary number are based on powers of two.

After completing this section, you should be able to

■ Count in binary ■ Determine the largest decimal number that can be represented by a given number of bits ■ Convert a binary number to a decimal number

Counting in Binary

The binary number system has two digits (bits).

To learn to count in the binary system, first look at how you count in the decimal system. You start at zero and count up to nine before you run out of digits. You then start another digit position (to the left) and continue counting 10 through 99. At this point you have exhausted all two-digit combinations, so a third digit position is needed to count from 100 through 999.

A comparable situation occurs when you count in binary, except that you have only two digits, called *bits*. Begin counting: 0, 1. At this point you have used both digits, so include another digit position and continue: 10, 11. You have now exhausted all combinations of two digits, so a third position is required. With three digit positions you can continue to count: 100, 101, 110, and 111. Now you need a fourth digit position to continue, and so on. A binary count of zero through fifteen is shown in Table 2–1. Notice the patterns with which the 1s and 0s alternate in each column.

The binary number system has a base of 2.

▶ TABLE 2–1

DECIMAL NUMBER	BINARY NUMBER			
0	0	0	0	0
1	0	0	0	1
2	0	0	1	0
3	0	0	1	1
4	0	1	0	0
5	0	1	0	1
6	0	1	1	0
7	0	1	1	1
8	1	0	0	0
9	1	0	0	1
10	1	0	1	0
11	1	0	1	1
12	1	1	0	0
13	1	1	0	1
14	1	1	1	0
15	1	1	1	1

As you have seen in Table 2–1, four bits are required to count from zero to 15. In general, with n bits you can count up to a number equal to $2^n - 1$.

The value of a bit is determined by its position in the number.

Largest decimal number = $2^n - 1$

For example, with five bits ($n = 5$) you can count from zero to thirty-one.

$2^5 - 1 = 32 - 1 = 31$

With six bits ($n = 6$) you can count from zero to sixty-three.

$2^6 - 1 = 64 - 1 = 63$

A table of powers of two is given in Appendix A.

CALCULATOR TUTORIAL

Powers of Two

Example   Find the value of 2^5.

| TI-86 | Step 1. | 2 ^ |
| | Step 2. | 5 ENTER |

2 ^ 5

32

| TI-36X | Step 1. | 2 y^x |
| | Step 2. | 5 = |

32

An Application

Learning to count in binary will help you to basically understand how digital circuits can be used to count events. This can be anything from counting items on an assembly line to counting operations in a computer. Let's take a simple example of counting tennis balls going into a box from a conveyor belt. Assume that nine balls are to go into each box.

The counter in Figure 2–1 counts the pulses from a sensor that detects the passing of a ball and produces a sequence of logic levels (digital waveforms) on each of its four parallel outputs. Each set of logic levels represents a 4-bit binary number (HIGH = 1 and LOW = 0), as indicated. As the decoder receives these waveforms, it decodes each set of four bits and converts it to the corresponding decimal number in the 7-segment display. When the counter gets to the binary state of 1001, it has counted nine tennis balls, the display shows decimal 9, and a new box is moved under the conveyor. Then the counter goes back to its zero state (0000), and the process starts over. (The number 9 was used only in the interest of single-digit simplicity.)

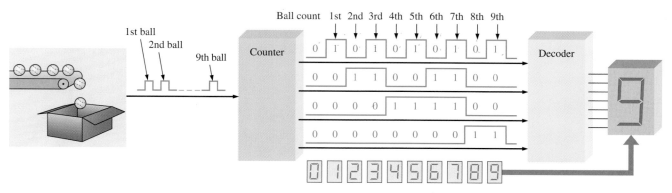

▲ FIGURE 2–1

Illustration of a simple binary counting application.

The Weighting Structure of Binary Numbers

The weight or value of a bit increases from right to left in a binary number.

A binary number is a weighted number. The right-most bit is the **LSB** (least significant bit) in a binary whole number and has a weight of $2^0 = 1$. The weights increase from right to left by a power of two for each bit. The left-most bit is the **MSB** (most significant bit); its weight depends on the size of the binary number.

Fractional numbers can also be represented in binary by placing bits to the right of the binary point, just as fractional decimal digits are placed to the right of the decimal point. The left-most bit is the MSB in a binary fractional number and has a weight of $2^{-1} = 0.5$. The fractional weights decrease from left to right by a negative power of two for each bit.

The weight structure of a binary number is

$$2^{n-1} \ldots 2^3 \, 2^2 \, 2^1 \, 2^0 \, . \, 2^{-1} \, 2^{-2} \, \ldots \, 2^{-n}$$

$\uparrow$ — Binary point

where n is the number of bits from the binary point. Thus, all the bits to the left of the binary point have weights that are positive powers of two, as previously discussed for whole numbers. All bits to the right of the binary point have weights that are negative powers of two, or fractional weights.

The powers of two and their equivalent decimal weights for an 8-bit binary whole number and a 6-bit binary fractional number are shown in Table 2–2. Notice that the weight doubles for each positive power of two and that the weight is halved for each negative power of two. You can easily extend the table by doubling the weight of the most significant positive power of two and halving the weight of the least significant negative power of two; for example, $2^9 = 512$ and $2^{-7} = 0.0078125$.

COMPUTER NOTE

Computers use binary numbers to select memory locations. Each location is assigned a unique number called an *address*. Some Pentium microprocessors, for example, have 32 address lines which can select 2^{32} (4,294,967,296) unique locations.

▼ TABLE 2–2

Binary weights.

POSITIVE POWERS OF TWO (WHOLE NUMBERS)									NEGATIVE POWERS OF TWO (FRACTIONAL NUMBER)					
2^8	2^7	2^6	2^5	2^4	2^3	2^2	2^1	2^0	2^{-1}	2^{-2}	2^{-3}	2^{-4}	2^{-5}	2^{-6}
256	128	64	32	16	8	4	2	1	1/2	1/4	1/8	1/16	1/32	1/64
									0.5	0.25	0.125	0.0625	0.03125	0.015625

Binary-to-Decimal Conversion

Add the weights of all 1s in a binary number to get the decimal value.

The decimal value of any binary number can be found by adding the weights of all bits that are 1 and discarding the weights of all bits that are 0.

EXAMPLE 2–3

Convert the binary whole number 1101101 to decimal.

Solution Determine the weight of each bit that is a 1, and then find the sum of the weights to get the decimal number.

$$\text{Weight: } 2^6 \, 2^5 \, 2^4 \, 2^3 \, 2^2 \, 2^1 \, 2^0$$
$$\text{Binary number: } 1 \; 1 \; 0 \; 1 \; 1 \; 0 \; 1$$
$$1101101 = 2^6 + 2^5 + 2^3 + 2^2 + 2^0$$
$$= 64 + 32 + 8 + 4 + 1 = \mathbf{109}$$

Related Problem Convert the binary number 10010001 to decimal.

EXAMPLE 2–4

Convert the fractional binary number 0.1011 to decimal.

Solution Determine the weight of each bit that is a 1, and then sum the weights to get the decimal fraction.

$$
\begin{aligned}
\text{Weight:} &\quad 2^{-1}\ \ 2^{-2}\ \ 2^{-3}\ \ 2^{-4} \\
\text{Binary number:}\ 0\,. &\quad 1\quad\ \ 0\quad\ \ 1\quad\ \ 1
\end{aligned}
$$

$$
\begin{aligned}
0.1011 &= 2^{-1} + 2^{-3} + 2^{-4} \\
&= 0.5 + 0.125 + 0.0625 = \mathbf{0.6875}
\end{aligned}
$$

Related Problem Convert the binary number 10.111 to decimal.

**SECTION 2–2
REVIEW**

1. What is the largest decimal number that can be represented in binary with eight bits?
2. Determine the weight of the 1 in the binary number 10000.
3. Convert the binary number 10111101.011 to decimal.

2–3 DECIMAL-TO-BINARY CONVERSION

In Section 2–2 you learned how to convert a binary number to the equivalent decimal number. Now you will learn two ways of converting from a decimal number to a binary number.

After completing this section, you should be able to

■ Convert a decimal number to binary using the sum-of-weights method ■ Convert a decimal whole number to binary using the repeated division-by-2 method ■ Convert a decimal fraction to binary using the repeated multiplication-by-2 method

Sum-of-Weights Method

One way to find the binary number that is equivalent to a given decimal number is to determine the set of binary weights whose sum is equal to the decimal number. An easy way to remember binary weights is that the lowest is 1, which is 2^0, and that by doubling any weight, you get the next higher weight; thus, a list of seven binary weights would be 64, 32, 16, 8, 4, 2, 1 as you learned in the last section. The decimal number 9, for example, can be expressed as the sum of binary weights as follows:

To get the binary number for a given decimal number, find the binary weights that add up to the decimal number.

$$9 = 8 + 1 \quad \text{or} \quad 9 = 2^3 + 2^0$$

Placing 1s in the appropriate weight positions, 2^3 and 2^0, and 0s in the 2^2 and 2^1 positions determines the binary number for decimal 9.

$$
\begin{array}{cccc}
2^3 & 2^2 & 2^1 & 2^0 \\
1 & 0 & 0 & 1
\end{array}
\quad \text{Binary number for decimal 9}
$$

EXAMPLE 2–5

Convert the following decimal numbers to binary:

(a) 12   (b) 25   (c) 58   (d) 82

Solution

(a) $12 = 8 + 4 = 2^3 + 2^2$ ⟶ **1100**

(b) $25 = 16 + 8 + 1 = 2^4 + 2^3 + 2^0$ ⟶ **11001**

(c) $58 = 32 + 16 + 8 + 2 = 2^5 + 2^4 + 2^3 + 2^1$ ⟶ **111010**

(d) $82 = 64 + 16 + 2 = 2^6 + 2^4 + 2^1$ ⟶ **1010010**

Related Problem   Convert the decimal number 125 to binary.

Repeated Division-by-2 Method

To get the binary number for a given decimal number, divide the decimal number by 2 until the quotient is 0. Remainders form the binary number.

A systematic method of converting whole numbers from decimal to binary is the *repeated division-by-2* process. For example, to convert the decimal number 12 to binary, begin by dividing 12 by 2. Then divide each resulting quotient by 2 until there is a 0 whole-number quotient. The **remainders** generated by each division form the binary number. The first remainder to be produced is the LSB (least significant bit) in the binary number, and the last remainder to be produced is the MSB (most significant bit). This procedure is shown in the following steps for converting the decimal number 12 to binary.

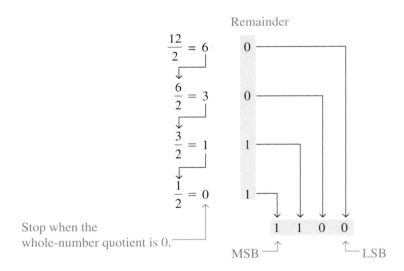

EXAMPLE 2–6

Convert the following decimal numbers to binary:

(a) 19   (b) 45

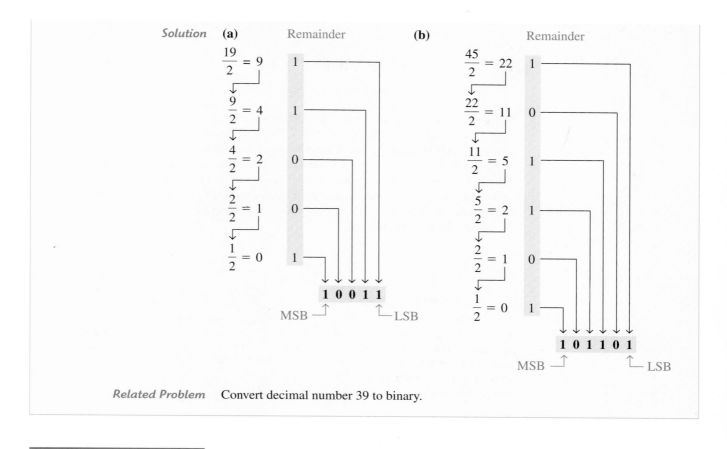

Solution

(a) Remainder

$$\frac{19}{2} = 9$$ 1

$$\frac{9}{2} = 4$$ 1

$$\frac{4}{2} = 2$$ 0

$$\frac{2}{2} = 1$$ 0

$$\frac{1}{2} = 0$$ 1

1 0 0 1 1

MSB ⌐   ⌐ LSB

(b) Remainder

$$\frac{45}{2} = 22$$ 1

$$\frac{22}{2} = 11$$ 0

$$\frac{11}{2} = 5$$ 1

$$\frac{5}{2} = 2$$ 1

$$\frac{2}{2} = 1$$ 0

$$\frac{1}{2} = 0$$ 1

1 0 1 1 0 1

MSB ⌐   ⌐ LSB

Related Problem   Convert decimal number 39 to binary.

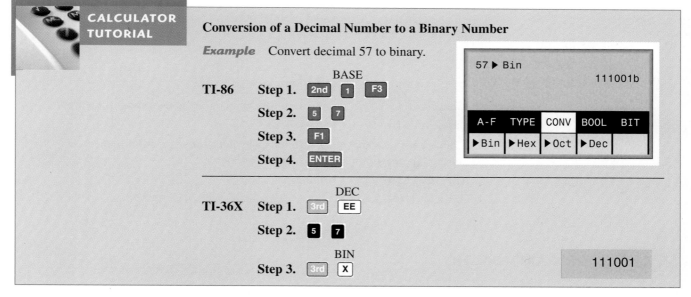

CALCULATOR TUTORIAL

Conversion of a Decimal Number to a Binary Number

Example   Convert decimal 57 to binary.

                     BASE

TI-86   Step 1.   [2nd] [1] [F3]

           Step 2.   [5] [7]

           Step 3.   [F1]

           Step 4.   [ENTER]

```
57 ▶ Bin
                    111001b
```
A-F	TYPE	CONV	BOOL	BIT
▶Bin	▶Hex	▶Oct	▶Dec	

                     DEC

TI-36X   Step 1.   [3rd] [EE]

            Step 2.   [5] [7]

                     BIN

            Step 3.   [3rd] [X]

                                   111001

Converting Decimal Fractions to Binary

Examples 2–5 and 2–6 demonstrated whole-number conversions. Now let's look at fractional conversions. An easy way to remember fractional binary weights is that the most significant weight is 0.5, which is 2^{-1}, and that by halving any weight, you get the next lower weight; thus a list of four fractional binary weights would be 0.5, 0.25, 0.125, 0.0625.

Sum-of-Weights   The sum-of-weights method can be applied to fractional decimal numbers, as shown in the following example:

$$0.625 = 0.5 + 0.125 = 2^{-1} + 2^{-3} = 0.101$$

There is a 1 in the 2^{-1} position, a 0 in the 2^{-2} position, and a 1 in the 2^{-3} position.

Repeated Multiplication by 2   As you have seen, decimal whole numbers can be converted to binary by repeated division by 2. Decimal fractions can be converted to binary by repeated multiplication by 2. For example, to convert the decimal fraction 0.3125 to binary, begin by multiplying 0.3125 by 2 and then multiplying each resulting fractional part of the product by 2 until the fractional product is zero or until the desired number of decimal places is reached. The carry digits, or **carries,** generated by the multiplications produce the binary number. The first carry produced is the MSB, and the last carry is the LSB. This procedure is illustrated as follows:

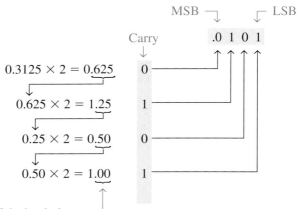

Continue to the desired number of decimal places
or stop when the fractional part is all zeros.

**SECTION 2–3
REVIEW**

1. Convert each decimal number to binary by using the sum-of-weights method:
   (a) 23   (b) 57   (c) 45.5
2. Convert each decimal number to binary by using the repeated division-by-2 method (repeated multiplication-by-2 for fractions):
   (a) 14   (b) 21   (c) 0.375

2–4   BINARY ARITHMETIC

Binary arithmetic is essential in all digital computers and in many other types of digital systems. To understand digital systems, you must know the basics of binary addition, subtraction, multiplication, and division. This section provides an introduction that will be expanded in later sections.

After completing this section, you should be able to

■ Add binary numbers ■ Subtract binary numbers ■ Multiply binary numbers
■ Divide binary numbers

Binary Addition

The four basic rules for adding binary digits (bits) are as follows:

$0 + 0 = 0$ Sum of 0 with a carry of 0
$0 + 1 = 1$ Sum of 1 with a carry of 0
$1 + 0 = 1$ Sum of 1 with a carry of 0
$1 + 1 = 10$ Sum of 0 with a carry of 1

Remember, in binary $1 + 1 = 10$, not 2.

Notice that the first three rules result in a single bit and in the fourth rule the addition of two 1s yields a binary two (10). When binary numbers are added, the last condition creates a sum of 0 in a given column and a carry of 1 over to the next column to the left, as illustrated in the following addition of $11 + 1$:

Carry   Carry

```
  1 ←   1 ←
  0      1     1
+ 0      0     1
  1      0     0
```

$11 = 3$
$1 = 1$
$4 = 0100$

In the right column, $1 + 1 = 0$ with a carry of 1 to the next column to the left. In the middle column, $1 + 1 + 0 = 0$ with a carry of 1 to the next column to the left. In the left column, $1 + 0 + 0 = 1$.

When there is a carry of 1, you have a situation in which three bits are being added (a bit in each of the two numbers and a carry bit). This situation is illustrated as follows:

Carry bits ⟶

$1 + 0 + 0 = 01$ Sum of 1 with a carry of 0
$1 + 1 + 0 = 10$ Sum of 0 with a carry of 1
$1 + 0 + 1 = 10$ Sum of 0 with a carry of 1
$1 + 1 + 1 = 11$ Sum of 1 with a carry of 1

EXAMPLE 2–7

Add the following binary numbers:

(a) $11 + 11$ **(b)** $100 + 10$ **(c)** $111 + 11$ **(d)** $110 + 100$

Solution   The equivalent decimal addition is also shown for reference.

(a)		(b)		(c)		(d)	
11	3	100	4	111	7	110	6
+11	+3	+10	+2	+ 11	+3	+100	+4
110	6	**110**	6	**1010**	10	**1010**	10

Related Problem   Add 1111 and 1100.

Binary Subtraction

The four basic rules for subtracting bits are as follows:

$0 - 0 = 0$
$1 - 1 = 0$
$1 - 0 = 1$
$10 - 1 = 1$ $0 - 1$ with a borrow of 1

Remember in binary $10 - 1 = 1$, not 9.

When subtracting numbers, you sometimes have to borrow from the next column to the left. A borrow is required in binary only when you try to subtract a 1 from a 0. In this case, when a 1 is borrowed from the next column to the left, a 10 is created in the column being subtracted, and the last of the four basic rules just listed must be applied. Examples 2–8 and 2–9 illustrate binary subtraction; the equivalent decimal subtractions are also shown.

EXAMPLE 2–8

Perform the following binary subtractions:

(a) $11 - 01$ **(b)** $11 - 10$

Solution

$$\textbf{(a)} \quad \begin{array}{r} 11 \\ -01 \\ \hline \mathbf{10} \end{array} \qquad \begin{array}{r} 3 \\ -1 \\ \hline 2 \end{array} \qquad \textbf{(b)} \quad \begin{array}{r} 11 \\ -10 \\ \hline \mathbf{01} \end{array} \qquad \begin{array}{r} 3 \\ -2 \\ \hline 1 \end{array}$$

No borrows were required in this example. The binary number 01 is the same as 1.

Related Problem   Subtract 100 from 111.

EXAMPLE 2–9

Subtract 011 from 101.

Solution

$$\begin{array}{r} 101 \\ -011 \\ \hline \mathbf{010} \end{array} \qquad \begin{array}{r} 5 \\ -3 \\ \hline 2 \end{array}$$

Let's examine exactly what was done to subtract the two binary numbers since a borrow is required. Begin with the right column.

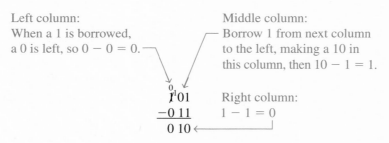

Left column:
When a 1 is borrowed,
a 0 is left, so $0 - 0 = 0$.

Middle column:
Borrow 1 from next column to the left, making a 10 in this column, then $10 - 1 = 1$.

Right column:
$1 - 1 = 0$

Related Problem   Subtract 101 from 110.

Binary Multiplication

Binary multiplication of two bits is the same as multiplication of the decimal digits 0 and 1.

The four basic rules for multiplying bits are as follows:

$$0 \times 0 = 0$$
$$0 \times 1 = 0$$
$$1 \times 0 = 0$$
$$1 \times 1 = 1$$

Multiplication is performed with binary numbers in the same manner as with decimal numbers. It involves forming partial products, shifting each successive partial product left one place, and then adding all the partial products. Example 2–10 illustrates the procedure; the equivalent decimal multiplications are shown for reference.

EXAMPLE 2–10

Perform the following binary multiplications:

(a) 11×11 **(b)** 101×111

Solution   **(a)**

$$
\begin{array}{r}
11 \\
\times\,11 \\
\hline
11 \\
+11 \\
\hline
\mathbf{1001}
\end{array}
\qquad
\begin{array}{r}
3 \\
\times\,3 \\
\hline
9
\end{array}
$$

Partial products $\{$

(b)

$$
\begin{array}{r}
111 \\
\times\,101 \\
\hline
111 \\
000 \\
+111 \\
\hline
\mathbf{100011}
\end{array}
\qquad
\begin{array}{r}
7 \\
\times\,5 \\
\hline
35
\end{array}
$$

Partial products $\{$

Related Problem   Multiply 1101×1010.

Binary Division

Division in binary follows the same procedure as division in decimal, as Example 2–11 illustrates. The equivalent decimal divisions are also given.

> A calculator can be used to perform arithmetic operations with binary numbers as long as the capacity of the calculator is not exceeded.

EXAMPLE 2–11

Perform the following binary divisions:

(a) $110 \div 11$ **(b)** $110 \div 10$

Solution   **(a)**

$$
\begin{array}{r}
\mathbf{10} \\
11\overline{)110} \\
\underline{11} \\
000
\end{array}
\qquad
\begin{array}{r}
2 \\
3\overline{)6} \\
\underline{6} \\
0
\end{array}
$$

(b)

$$
\begin{array}{r}
\mathbf{11} \\
10\overline{)110} \\
\underline{10} \\
10 \\
\underline{10} \\
00
\end{array}
\qquad
\begin{array}{r}
3 \\
2\overline{)6} \\
\underline{6} \\
0
\end{array}
$$

Related Problem   Divide 1100 by 100.

SECTION 2–4 REVIEW

1. Perform the following binary additions:
   (a) $1101 + 1010$ (b) $10111 + 01101$
2. Perform the following binary subtractions:
   (a) $1101 - 0100$ (b) $1001 - 0111$
3. Perform the indicated binary operations:
   (a) 110×111 (b) $1100 \div 011$

2–5   1'S AND 2'S COMPLEMENTS OF BINARY NUMBERS

The 1's complement and the 2's complement of a binary number are important because they permit the representation of negative numbers. The method of 2's complement arithmetic is commonly used in computers to handle negative numbers.

After completing this section, you should be able to

■ Convert a binary number to its 1's complement ■ Convert a binary number to its 2's complement using either of two methods

Finding the 1's Complement

Change each bit in a number to get the 1's complement.

The 1's **complement** of a binary number is found by changing all 1s to 0s and all 0s to 1s, as illustrated below:

$$1\ 0\ 1\ 1\ 0\ 0\ 1\ 0 \qquad \text{Binary number}$$
$$\downarrow \downarrow \downarrow \downarrow \downarrow \downarrow \downarrow \downarrow$$
$$0\ 1\ 0\ 0\ 1\ 1\ 0\ 1 \qquad \text{1's complement}$$

The simplest way to obtain the 1's complement of a binary number with a digital circuit is to use parallel inverters (NOT circuits), as shown in Figure 2–2 for an 8-bit binary number.

▶ **FIGURE 2–2**

Example of inverters used to obtain the 1's complement of a binary number.

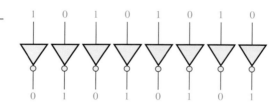

Finding the 2's Complement

Add 1 to the 1's complement to get the 2's complement.

The 2's complement of a binary number is found by adding 1 to the LSB of the 1's complement.

2's complement = (1's complement) + 1

EXAMPLE 2–12

Find the 2's complement of 10110010.

Solution

$$
\begin{array}{ll}
10110010 & \text{Binary number} \\
01001101 & \text{1's complement} \\
+\qquad 1 & \text{Add 1} \\
\hline
\mathbf{01001110} & \text{2's complement}
\end{array}
$$

Related Problem   Determine the 2's complement of 11001011.

An alternative method of finding the 2's complement of a binary number is as follows:

1. Start at the right with the LSB and write the bits as they are up to and including the first 1.

2. Take the 1's complements of the remaining bits.

Change all bits to the left of the least significant 1 to get 2's complement.

EXAMPLE 2–13

Find the 2's complement of 10111000 using the alternative method.

Solution

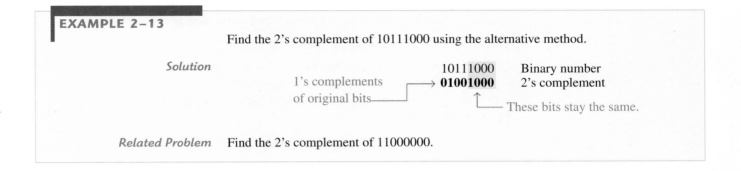

1's complements of original bits ⟶ 10111000 Binary number
01001000 2's complement

└──── These bits stay the same.

Related Problem Find the 2's complement of 11000000.

The 2's complement of a negative binary number can be realized using inverters and an adder, as indicated in Figure 2–3. This illustrates how an 8-bit number can be converted to its 2's complement by first inverting each bit (taking the 1's complement) and then adding 1 to the 1's complement with an adder circuit.

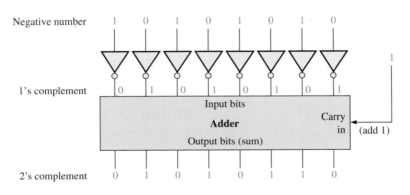

◀ **FIGURE 2–3**

Example of obtaining the 2's complement of a negative binary number.

To convert from a 1's or 2's complement back to the true (uncomplemented) binary form, use the same two procedures described previously. To go from the 1's complement back to true binary, reverse all the bits. To go from the 2's complement form back to true binary, take the 1's complement of the 2's complement number and add 1 to the least significant bit.

SECTION 2–5 REVIEW

1. Determine the 1's complement of each binary number:
   (a) 00011010   (b) 11110111   (c) 10001101

2. Determine the 2's complement of each binary number:
   (a) 00010110   (b) 11111100   (c) 10010001

2–6 SIGNED NUMBERS

Digital systems, such as the computer, must be able to handle both positive and negative numbers. A signed binary number consists of both sign and magnitude information. The sign indicates whether a number is positive or negative, and the magnitude is the value of the number. There are three forms in which signed integer (whole) numbers can be represented in binary: sign-magnitude, 1's complement, and 2's complement. Of these, the 2's complement is the most important and the sign-magnitude is the least used. Noninteger and very large or small numbers can be expressed in floating-point format.

After completing this section, you should be able to

■ Express positive and negative numbers in sign-magnitude ■ Express positive and negative numbers in 1's complement ■ Express positive and negative numbers in 2's complement ■ Determine the decimal value of signed binary numbers ■ Express a binary number in floating-point format

The Sign Bit

The left-most bit in a signed binary number is the **sign bit,** which tells you whether the number is positive or negative.

A 0 sign bit indicates a positive number, and a 1 sign bit indicates a negative number.

Sign-Magnitude Form

When a signed binary number is represented in sign-magnitude, the left-most bit is the sign bit and the remaining bits are the magnitude bits. The magnitude bits are in true (uncomplemented) binary for both positive and negative numbers. For example, the decimal number +25 is expressed as an 8-bit signed binary number using the sign-magnitude form as

$$\underset{\text{Sign bit} \underset{}{\longrightarrow}}{\underline{0}}\underset{\underset{}{\longleftarrow}\text{Magnitude bits}}{0011001}$$

The decimal number −25 is expressed as

    10011001

Notice that the only difference between +25 and −25 is the sign bit because the magnitude bits are in true binary for both positive and negative numbers.

In the sign-magnitude form, a negative number has the same magnitude bits as the corresponding positive number but the sign bit is a 1 rather than a zero.

1's Complement Form

Positive numbers in 1's complement form are represented the same way as the positive sign-magnitude numbers. Negative numbers, however, are the 1's complements of the corresponding positive numbers. For example, using eight bits, the decimal number −25 is expressed as the 1's complement of +25 (00011001) as

    11100110

In the 1's complement form, a negative number is the 1's complement of the corresponding positive number.

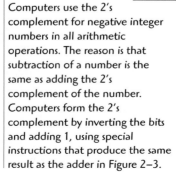

COMPUTER NOTE

Computers use the 2's complement for negative integer numbers in all arithmetic operations. The reason is that subtraction of a number is the same as adding the 2's complement of the number. Computers form the 2's complement by inverting the bits and adding 1, using special instructions that produce the same result as the adder in Figure 2–3.

2's Complement Form

Positive numbers in 2's complement form are represented the same way as in the sign-magnitude and 1's complement forms. Negative numbers are the 2's complements of the corresponding positive numbers. Again, using eight bits, let's take decimal number -25 and express it as the 2's complement of $+25$ (00011001).

> 11100111

In the 2's complement form, a negative number is the 2's complement of the corresponding positive number.

EXAMPLE 2–14

Express the decimal number -39 as an 8-bit number in the sign-magnitude, 1's complement, and 2's complement forms.

Solution   First, write the 8-bit number for $+39$.

$$00100111$$

In the *sign-magnitude form*, -39 is produced by changing the sign bit to a 1 and leaving the magnitude bits as they are. The number is

10100111

In the *1's complement form*, -39 is produced by taking the 1's complement of $+39$ (00100111).

11011000

In the *2's complement form*, -39 is produced by taking the 2's complement of $+39$ (00100111) as follows:

$$
\begin{array}{ll}
11011000 & \text{1's complement} \\
+\qquad 1 & \\
\hline
\mathbf{11011001} & \text{2's complement}
\end{array}
$$

Related Problem   Express $+19$ and -19 in sign-magnitude, 1's complement, and 2's complement.

The Decimal Value of Signed Numbers

Sign–magnitude   Decimal values of positive and negative numbers in the sign-magnitude form are determined by summing the weights in all the magnitude bit positions where there are 1s and ignoring those positions where there are zeros. The sign is determined by examination of the sign bit.

EXAMPLE 2–15

Determine the decimal value of this signed binary number expressed in sign-magnitude: 10010101.

Solution   The seven magnitude bits and their powers-of-two weights are as follows:

$$
\begin{array}{ccccccc}
2^6 & 2^5 & 2^4 & 2^3 & 2^2 & 2^1 & 2^0 \\
0 & 0 & 1 & 0 & 1 & 0 & 1
\end{array}
$$

Summing the weights where there are 1s,

$$16 + 4 + 1 = 21$$

The sign bit is 1; therefore, the decimal number is **−21.**

Related Problem Determine the decimal value of the sign-magnitude number 01110111.

1's Complement Decimal values of positive numbers in the 1's complement form are determined by summing the weights in all bit positions where there are 1s and ignoring those positions where there are zeros. Decimal values of negative numbers are determined by assigning a negative value to the weight of the sign bit, summing all the weights where there are 1s, and adding 1 to the result.

EXAMPLE 2–16

Determine the decimal values of the signed binary numbers expressed in 1's complement:

(a) 00010111   **(b)** 11101000

Solution **(a)** The bits and their powers-of-two weights for the positive number are as follows:

$$-2^7 \quad 2^6 \quad 2^5 \quad 2^4 \quad 2^3 \quad 2^2 \quad 2^1 \quad 2^0$$
$$0 \quad\ \ 0 \quad\ \ 0 \quad\ \ 1 \quad\ \ 0 \quad\ \ 1 \quad\ \ 1 \quad\ \ 1$$

Summing the weights where there are 1s,

$$16 + 4 + 2 + 1 = \textbf{+23}$$

(b) The bits and their powers-of-two weights for the negative number are as follows. Notice that the negative sign bit has a weight of -2^7 or -128.

$$-2^7 \quad 2^6 \quad 2^5 \quad 2^4 \quad 2^3 \quad 2^2 \quad 2^1 \quad 2^0$$
$$1 \quad\ \ 1 \quad\ \ 1 \quad\ \ 0 \quad\ \ 1 \quad\ \ 0 \quad\ \ 0 \quad\ \ 0$$

Summing the weights where there are 1s,

$$-128 + 64 + 32 + 8 = -24$$

Adding 1 to the result, the final decimal number is

$$-24 + 1 = \textbf{−23}$$

Related Problem Determine the decimal value of the 1's complement number 11101011.

2's Complement Decimal values of positive and negative numbers in the 2's complement form are determined by summing the weights in all bit positions where there are 1s and ignoring those positions where there are zeros. The weight of the sign bit in a negative number is given a negative value.

EXAMPLE 2–17

Determine the decimal values of the signed binary numbers expressed in 2's complement:

(a) 01010110   **(b)** 10101010

Solution **(a)** The bits and their powers-of-two weights for the positive number are as follows:

$$-2^7 \quad 2^6 \quad 2^5 \quad 2^4 \quad 2^3 \quad 2^2 \quad 2^1 \quad 2^0$$
$$0 \quad 1 \quad 0 \quad 1 \quad 0 \quad 1 \quad 1 \quad 0$$

Summing the weights where there are 1s,

$$64 + 16 + 4 + 2 = \textbf{+86}$$

(b) The bits and their powers-of-two weights for the negative number are as follows. Notice that the negative sign bit has a weight of $-2^7 = -128$.

$$-2^7 \quad 2^6 \quad 2^5 \quad 2^4 \quad 2^3 \quad 2^2 \quad 2^1 \quad 2^0$$
$$1 \quad 0 \quad 1 \quad 0 \quad 1 \quad 0 \quad 1 \quad 0$$

Summing the weights where there are 1s,

$$-128 + 32 + 8 + 2 = \textbf{-86}$$

Related Problem Determine the decimal value of the 2's complement number 11010111.

From these examples, you can see why the 2's complement form is preferred for representing signed integer numbers: To convert to decimal, it simply requires a summation of weights regardless of whether the number is positive or negative. The 1's complement system requires adding 1 to the summation of weights for negative numbers but not for positive numbers. Also, the 1's complement form is generally not used because two representations of zero (00000000 or 11111111) are possible.

Range of Signed Integer Numbers That Can Be Represented

We have used 8-bit numbers for illustration because the 8-bit grouping is common in most computers and has been given the special name **byte**. With one byte or eight bits, you can represent 256 different numbers. With two bytes or sixteen bits, you can represent 65,536 different numbers. With four bytes or 32 bits, you can represent 4.295×10^9 different numbers. The formula for finding the number of different combinations of n bits is

The range of magnitude of a binary number depends on the number of bits (n).

Total combinations $= 2^n$

For 2's complement signed numbers, the range of values for n-bit numbers is

Range $= -(2^{n-1})$ to $+(2^{n-1} - 1)$

where in each case there is one sign bit and $n - 1$ magnitude bits. For example, with four bits you can represent numbers in 2's complement ranging from $-(2^3) = -8$ to $2^3 - 1 = +7$. Similarly, with eight bits you can go from -128 to $+127$, with sixteen bits you can go from $-32,768$ to $+32,767$, and so on.

Floating-Point Numbers

To represent very large **integer** (whole) numbers, many bits are required. There is also a problem when numbers with both integer and fractional parts, such as 23.5618, need to be represented. The floating-point number system, based on scientific notation, is capable of representing very large and very small numbers without an increase in the number of bits and also for representing numbers that have both integer and fractional components.

A **floating-point number** (also known as a *real number*) consists of two parts plus a sign. The **mantissa** is the part of a floating-point number that represents the magnitude of the number. The **exponent** is the part of a floating-point number that represents the number of places that the decimal point (or binary point) is to be moved.

A decimal example will be helpful in understanding the basic concept of floating-point numbers. Let's consider a decimal number which, in integer form, is 241,506,800. The mantissa is .2415068 and the exponent is 9. When the integer is expressed as a floating-point number, it is normalized by moving the decimal point to the left of all the digits so that the mantissa is a fractional number and the exponent is the power of ten. The floating-point number is written as

$$0.2415068 \times 10^9$$

For binary floating-point numbers, the format is defined by ANSI/IEEE Standard 754-1985 in three forms: *single-precision, double-precision,* and *extended-precision.* These all have the same basic formats except for the number of bits. Single-precision floating-point numbers have 32 bits, double-precision numbers have 64 bits, and extended-precision numbers have 80 bits. We will restrict our discussion to the single-precision floating-point format.

Single-Precision Floating-Point Binary Numbers   In the standard format for a single-precision binary number, the sign bit (S) is the left-most bit, the exponent (E) includes the next eight bits, and the mantissa or fractional part (F) includes the remaining 23 bits, as shown next.

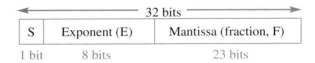

In the mantissa or fractional part, the binary point is understood to be to the left of the 23 bits. Effectively, there are 24 bits in the mantissa because in any binary number the left-most (most significant) bit is always a 1. Therefore, this 1 is understood to be there although it does not occupy an actual bit position.

The eight bits in the exponent represent a *biased exponent,* which is obtained by adding 127 to the actual exponent. The purpose of the bias is to allow very large or very small numbers without requiring a separate sign bit for the exponents. The biased exponent allows a range of actual exponent values from -126 to $+128$.

To illustrate how a binary number is expressed in floating-point format, let's use 1011010010001 as an example. First, it can be expressed as 1 plus a fractional binary number by moving the binary point 12 places to the left and then multiplying by the appropriate power of two.

$$1011010010001 = 1.011010010001 \times 2^{12}$$

Assuming that this is a positive number, the sign bit (S) is 0. The exponent, 12, is expressed as a biased exponent by adding it to 127 (12 + 127 = 139). The biased exponent (E) is expressed as the binary number 10001011. The mantissa is the fractional part (F) of the binary number, .011010010001. Because there is always a 1 to the left of the binary point in the power-of-two expression, it is not included in the mantissa. The complete floating-point number is

S	E	F
0	10001011	01101001000100000000000

Next, let's see how to evaluate a binary number that is already in floating-point format. The general approach to determining the value of a floating-point number is expressed by the following formula:

$$\text{Number} = (-1)^S(1 + F)(2^{E-127})$$

To illustrate, let's consider the following floating-point binary number:

S	E	F
1	10010001	10001110001000000000000

The sign bit is 1. The biased exponent is $10010001 = 145$. Applying the formula, we get

$$\text{Number} = (-1)^1(1.10001110001)(2^{145-127})$$
$$= (-1)(1.10001110001)(2^{18}) = -1100011100010000000$$

This floating-point binary number is equivalent to $-407,688$ in decimal. Since the exponent can be any number between -126 and $+128$, extremely large and small numbers can be expressed. A 32-bit floating-point number can replace a binary integer number having 129 bits. Because the exponent determines the position of the binary point, numbers containing both integer and fractional parts can be represented.

There are two exceptions to the format for floating-point numbers: The number 0.0 is represented by all 0s, and infinity is represented by all 1s in the exponent and all 0s in the mantissa.

EXAMPLE 2–18

Convert the decimal number 3.248×10^4 to a single-precision floating-point binary number.

Solution   Convert the decimal number to binary.

$$3.248 \times 10^4 = 32480 = 111111011100000_2 = 1.11111011100000 \times 2^{14}$$

The MSB will not occupy a bit position because it is always a 1. Therefore, the mantissa is the fractional 23-bit binary number 11111011100000000000000 and the biased exponent is

$$14 + 127 = 141 = 10001101_2$$

The complete floating-point number is

0	10001101	11111011100000000000000

Related Problem   Determine the binary value of the following floating-point binary number:

0 10011000 10000100010100110000000

SECTION 2–6 REVIEW

1. Express the decimal number $+9$ as an 8-bit binary number in the sign-magnitude system.

2. Express the decimal number -33 as an 8-bit binary number in the 1's complement system.

3. Express the decimal number -46 as an 8-bit binary number in the 2's complement system.

4. List the three parts of a signed, floating-point number.

2-7 ARITHMETIC OPERATIONS WITH SIGNED NUMBERS

In the last section, you learned how signed numbers are represented in three different forms. In this section, you will learn how signed numbers are added, subtracted, multiplied, and divided. Because the 2's complement form for representing signed numbers is the most widely used in computers and microprocessor-based systems, the coverage in this section is limited to 2's complement arithmetic. The processes covered can be extended to the other forms if necessary.

After completing this section, you should be able to

■ Add signed binary numbers ■ Explain how computers add strings of numbers ■ Define *overflow* ■ Subtract signed binary numbers ■ Multiply signed binary numbers using the direct addition method ■ Multiply signed binary numbers using the partial products method ■ Divide signed binary numbers

Addition

The two numbers in an addition are the **addend** and the **augend.** The result is the **sum.** There are four cases that can occur when two signed binary numbers are added.

1. Both numbers positive

2. Positive number with magnitude larger than negative number

3. Negative number with magnitude larger than positive number

4. Both numbers negative

Let's take one case at a time using 8-bit signed numbers as examples. The equivalent decimal numbers are shown for reference.

Addition of two positive numbers yields a positive number.

Both numbers positive:

$$
\begin{array}{r}
00000111 \\
+\ 00000100 \\
\hline
00001011
\end{array}
\qquad
\begin{array}{r}
7 \\
+\ 4 \\
\hline
11
\end{array}
$$

The sum is positive and is therefore in true (uncomplemented) binary.

Addition of a positive number and a smaller negative number yields a positive number.

Positive number with magnitude larger than negative number:

$$
\begin{array}{r}
00001111 \\
+\ 11111010 \\
\hline
\text{Discard carry} \longrightarrow\ 1\ \ 00001001
\end{array}
\qquad
\begin{array}{r}
15 \\
+\ -6 \\
\hline
9
\end{array}
$$

The final carry bit is discarded. The sum is positive and therefore in true (uncomplemented) binary.

Addition of a positive number and a larger negative number or two negative numbers yields a negative number in 2's complement.

Negative number with magnitude larger than positive number:

$$
\begin{array}{r}
00010000 \\
+\ 11101000 \\
\hline
11111000
\end{array}
\qquad
\begin{array}{r}
16 \\
+\ -24 \\
\hline
-8
\end{array}
$$

The sum is negative and therefore in 2's complement form.

Both numbers negative:

$$
\begin{array}{r}
11111011 \\
+\ 11110111 \\
\hline
\text{Discard carry} \longrightarrow\ 1\ \ 11110010
\end{array}
\qquad
\begin{array}{r}
-5 \\
+\ -9 \\
\hline
-14
\end{array}
$$

The final carry bit is discarded. The sum is negative and therefore in 2's complement form.

In a computer, the negative numbers are stored in 2's complement form so, as you can see, the addition process is very simple: *Add the two numbers and discard any final carry bit.*

Overflow Condition　When two numbers are added and the number of bits required to represent the sum exceeds the number of bits in the two numbers, an **overflow** results as indicated by an incorrect sign bit. An overflow can occur only when both numbers are positive or both numbers are negative. The following 8-bit example will illustrate this condition.

$$
\begin{array}{ll}
01111101 & 125 \\
+\ 00111010 & +\ \ 58 \\
\hline
10110111 & 183 \\
\end{array}
$$

Sign incorrect ——————
Magnitude incorrect ——————

In this example the sum of 183 requires eight magnitude bits. Since there are seven magnitude bits in the numbers (one bit is the sign), there is a carry into the sign bit which produces the overflow indication.

Numbers Are Added Two at a Time　Now let's look at the addition of a string of numbers, added two at a time. This can be accomplished by adding the first two numbers, then adding the third number to the sum of the first two, then adding the fourth number to this result, and so on. This is how computers add strings of numbers. The addition of numbers taken two at a time is illustrated in Example 2–19.

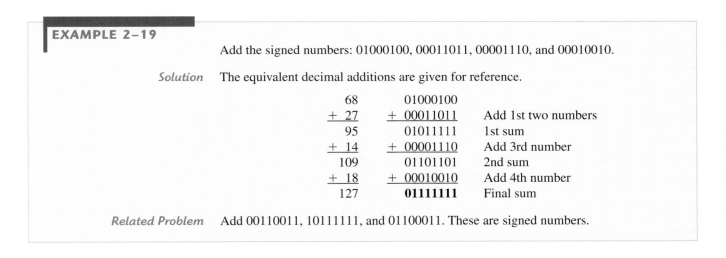

EXAMPLE 2–19

Add the signed numbers: 01000100, 00011011, 00001110, and 00010010.

Solution　The equivalent decimal additions are given for reference.

$$
\begin{array}{lll}
68 & 01000100 & \\
+\ 27 & +\ 00011011 & \text{Add 1st two numbers} \\
\hline
95 & 01011111 & \text{1st sum} \\
+\ 14 & +\ 00001110 & \text{Add 3rd number} \\
\hline
109 & 01101101 & \text{2nd sum} \\
+\ 18 & +\ 00010010 & \text{Add 4th number} \\
\hline
127 & \mathbf{01111111} & \text{Final sum} \\
\end{array}
$$

Related Problem　Add 00110011, 10111111, and 01100011. These are signed numbers.

Subtraction

Subtraction is a special case of addition. For example, subtracting $+6$ (the **subtrahend**) from $+9$ (the **minuend**) is equivalent to adding -6 to $+9$. Basically, *the subtraction operation changes the sign of the subtrahend and adds it to the minuend.* The result of a subtraction is called the **difference.**

Subtraction is addition with the sign of the subtrahend changed.

The sign of a positive or negative binary number is changed by taking its 2's complement.

For example, when you take the 2's complement of the positive number 00000100 $(+4)$, you get 11111100, which is -4 as the following sum-of-weights evaluation shows:

$$-128 + 64 + 32 + 16 + 8 + 4 = -4$$

As another example, when you take the 2's complement of the negative number 11101101 (-19), you get 00010011, which is $+19$ as the following sum-of-weights evaluation shows:

$$16 + 2 + 1 = 19$$

Since subtraction is simply an addition with the sign of the subtrahend changed, the process is stated as follows:

To subtract two signed numbers, take the 2's complement of the subtrahend and add. Discard any final carry bit.

Example 2–20 illustrates the subtraction process.

EXAMPLE 2–20

Perform each of the following subtractions of the signed numbers:

(a) $00001000 - 00000011$ **(b)** $00001100 - 11110111$

(c) $11100111 - 00010011$ **(d)** $10001000 - 11100010$

Solution Like in other examples, the equivalent decimal subtractions are given for reference.

(a) In this case, $8 - 3 = 8 + (-3) = 5$.

```
                         00001000      Minuend (+8)
                       + 11111101      2's complement of subtrahend (−3)
    Discard carry ────→ 1 00000101     Difference (+5)
```

(b) In this case, $12 - (-9) = 12 + 9 = 21$.

```
              00001100      Minuend (+12)
            + 00001001      2's complement of subtrahend (+9)
              00010101      Difference (+21)
```

(c) In this case, $-25 - (+19) = -25 + (-19) = -44$.

```
                         11100111      Minuend (−25)
                       + 11101101      2's complement of subtrahend (−19)
    Discard carry ────→ 1 11010100     Difference (−44)
```

(d) In this case, $-120 - (-30) = -120 + 30 = -90$.

```
              10001000      Minuend (−120)
            + 00011110      2's complement of subtrahend (+30)
              10100110      Difference (−90)
```

Related Problem Subtract 01000111 from 01011000.

Multiplication

The numbers in a multiplication are the **multiplicand**, the **multiplier**, and the **product**. These are illustrated in the following decimal multiplication:

```
     8     Multiplicand
   × 3     Multiplier
    24     Product
```

Multiplication is equivalent to adding a number to itself a number of times equal to the multiplier.

The multiplication operation in most computers is accomplished using addition. As you have already seen, subtraction is done with an adder; now let's see how multiplication is done.

Direct addition and *partial products* are two basic methods for performing multiplication using addition. In the direct addition method, you add the multiplicand a number of times equal to the multiplier. In the previous decimal example (3×8), three multiplicands are added: $8 + 8 + 8 = 24$. The disadvantage of this approach is that it becomes very lengthy if the multiplier is a large number. For example, to multiply 75×350, you must add 350 to itself 75 times. Incidentally, this is why the term *times* is used to mean multiply.

When two binary numbers are multiplied, both numbers must be in true (uncomplemented) form. The direct addition method is illustrated in Example 2–21 adding two binary numbers at a time.

EXAMPLE 2–21

Multiply the signed binary numbers: 01001101 (multiplicand) and 00000100 (multiplier) using the direct addition method.

Solution Since both numbers are positive, they are in true form, and the product will be positive. The decimal value of the multiplier is 4, so the multiplicand is added to itself four times as follows:

01001101	1st time
+ 01001101	2nd time
10011010	Partial sum
+ 01001101	3rd time
11100111	Partial sum
+ 01001101	4th time
100110100	Product

Since the sign bit of the multiplicand is 0, it has no effect on the outcome. All of the bits in the product are magnitude bits.

Related Problem Multiply 01100001 by 00000110 using the direct addition method.

The partial products method is perhaps the more common one because it reflects the way you multiply longhand. The multiplicand is multiplied by each multiplier digit beginning with the least significant digit. The result of the multiplication of the multiplicand by a multiplier digit is called a *partial product*. Each successive partial product is moved (shifted) one place to the left and when all the partial products have been produced, they are added to get the final product. Here is a decimal example.

239	Multiplicand
× 123	Multiplier
717	1st partial product (3×239)
478	2nd partial product (2×239)
+ 239	3rd partial product (1×239)
29,397	Final product

The sign of the product of a multiplication depends on the signs of the multiplicand and the multiplier according to the following two rules:

■ **If the signs are the same, the product is positive.**
■ **If the signs are different, the product is negative.**

The basic steps in the partial products method of binary multiplication are as follows:

Step 1. Determine if the signs of the multiplicand and multiplier are the same or different. This determines what the sign of the product will be.

Step 2. Change any negative number to true (uncomplemented) form. Because most computers store negative numbers in 2's complement, a 2's complement operation is required to get the negative number into true form.

Step 3. Starting with the least significant multiplier bit, generate the partial products. When the multiplier bit is 1, the partial product is the same as the multiplicand. When the multiplier bit is 0, the partial product is zero. Shift each successive partial product one bit to the left.

Step 4. Add each successive partial product to the sum of the previous partial products to get the final product.

Step 5. If the sign bit that was determined in step 1 is negative, take the 2's complement of the product. If positive, leave the product in true form. Attach the sign bit to the product.

EXAMPLE 2–22

Multiply the signed binary numbers: 01010011 (multiplicand) and 11000101 (multiplier).

Solution **Step 1:** The sign bit of the multiplicand is 0 and the sign bit of the multiplier is 1. The sign bit of the product will be 1 (negative).

Step 2: Take the 2's complement of the multiplier to put it in true form.

$$11000101 \longrightarrow 00111011$$

Steps 3 and 4: The multiplication proceeds as follows. Notice that only the magnitude bits are used in these steps.

1010011	Multiplicand
× 0111011	Multiplier
1010011	1st partial product
+ 1010011	2nd partial product
11111001	Sum of 1st and 2nd
+ 0000000	3rd partial product
011111001	Sum
+ 1010011	4th partial product
1110010001	Sum
+ 1010011	5th partial product
100011000001	Sum
+ 1010011	6th partial product
1001100100001	Sum
+ 0000000	7th partial product
1001100100001	Final product

Step 5: Since the sign of the product is a 1 as determined in step 1, take the 2's complement of the product.

$$1001100100001 \longrightarrow 0110011011111$$

Attach the sign bit

$$\longrightarrow \mathbf{1\ 0110011011111}$$

Related Problem Verify the multiplication is correct by converting to decimal numbers and performing the multiplication.

Division

The numbers in a division are the **dividend,** the **divisor,** and the **quotient.** These are illustrated in the following standard division format.

$$\frac{\text{dividend}}{\text{divisor}} = \text{quotient}$$

The division operation in computers is accomplished using subtraction. Since subtraction is done with an adder, division can also be accomplished with an adder.

The result of a division is called the *quotient;* the quotient is the number of times that the divisor will go into the dividend. This means that the divisor can be subtracted from the dividend a number of times equal to the quotient, as illustrated by dividing 21 by 7.

21	Dividend
− 7	1st subtraction of divisor
14	1st partial remainder
− 7	2nd subtraction of divisor
7	2nd partial remainder
− 7	3rd subtraction of divisor
0	Zero remainder

In this simple example, the divisor was subtracted from the dividend three times before a remainder of zero was obtained. Therefore, the quotient is 3.

The sign of the quotient depends on the signs of the dividend and the divisor according to the following two rules:

■ **If the signs are the same, the quotient is positive.**
■ **If the signs are different, the quotient is negative.**

When two binary numbers are divided, both numbers must be in true (uncomplemented) form. The basic steps in a division process are as follows:

Step 1. Determine if the signs of the dividend and divisor are the same or different. This determines what the sign of the quotient will be. Initialize the quotient to zero.

Step 2. Subtract the divisor from the dividend using 2's complement addition to get the first partial remainder and add 1 to the quotient. If this partial remainder is positive, go to step 3. If the partial remainder is zero or negative, the division is complete.

Step 3. Subtract the divisor from the partial remainder and add 1 to the quotient. If the result is positive, repeat for the next partial remainder. If the result is zero or negative, the division is complete.

Continue to subtract the divisor from the dividend and the partial remainders until there is a zero or a negative result. Count the number of times that the divisor is subtracted and you have the quotient. Example 2–23 illustrates these steps using 8-bit signed binary numbers.

EXAMPLE 2–23

Divide 01100100 by 00011001.

Solution **Step 1:** The signs of both numbers are positive, so the quotient will be positive. The quotient is initially zero: 00000000.

Step 2: Subtract the divisor from the dividend using 2's complement addition (remember that final carries are discarded).

$$
\begin{array}{ll}
01100100 & \text{Dividend} \\
+\ 11100111 & \text{2's complement of divisor} \\
\hline
01001011 & \text{Positive 1st partial remainder}
\end{array}
$$

Add 1 to quotient: 00000000 + 00000001 = 00000001.

Step 3: Subtract the divisor from the 1st partial remainder using 2's complement addition.

$$
\begin{array}{ll}
01001011 & \text{1st partial remainder} \\
+\ 11100111 & \text{2's complement of divisor} \\
\hline
00110010 & \text{Positive 2nd partial remainder}
\end{array}
$$

Step 4: Subtract the divisor from the 2nd partial remainder using 2's complement addition.

$$
\begin{array}{ll}
00110010 & \text{2nd partial remainder} \\
+\ 11100111 & \text{2's complement of divisor} \\
\hline
00011001 & \text{Positive 3rd partial remainder}
\end{array}
$$

Add 1 to quotient: 00000010 + 00000001 = 00000011.

Step 5: Subtract the divisor from the 3rd partial remainder using 2's complement addition.

$$
\begin{array}{ll}
00011001 & \text{3rd partial remainder} \\
+\ 11100111 & \text{2's complement of divisor} \\
\hline
00000000 & \text{Zero remainder}
\end{array}
$$

Add 1 to quotient: 00000011 + 00000001 = **00000100** (final quotient). The process is complete.

Related Problem Verify that the process is correct by converting to decimal numbers and performing the division.

**SECTION 2–7
REVIEW**

1. List the four cases when numbers are added.
2. Add 00100001 and 10111100.
3. Subtract 00110010 from 01110111.
4. What is the sign of the product when two negative numbers are multiplied?
5. Multiply 01111111 by 00000101.
6. What is the sign of the quotient when a positive number is divided by a negative number?
7. Divide 00110000 by 00001100.

2–8 HEXADECIMAL NUMBERS

The hexadecimal number system has sixteen characters; it is used primarily as a compact way of displaying or writing binary numbers because it is very easy to convert between binary and hexadecimal. As you are probably aware, long binary numbers are difficult to read and write because it is easy to drop or transpose a bit. Since computers and microprocessors understand only 1s and 0s, it is necessary to use these digits when you program in "machine language." Imagine writing a sixteen bit instruction for a microprocessor system in 1s and 0s. It is much more efficient to use hexadecimal or octal; octal numbers are covered in Section 2–9. Hexadecimal is widely used in computer and microprocessor applications.

After completing this section, you should be able to

■ List the hexadecimal characters ■ Count in hexadecimal ■ Convert from binary to hexadecimal ■ Convert from hexadecimal to binary ■ Convert from hexadecimal to decimal ■ Convert from decimal to hexadecimal ■ Add hexadecimal numbers ■ Determine the 2's complement of a hexadecimal number ■ Subtract hexadecimal numbers

The **hexadecimal** number system has a base of sixteen; that is, it is composed of 16 **numeric** and alphabetic **characters.** Most digital systems process binary data in groups that are multiples of four bits, making the hexadecimal number very convenient because each hexadecimal digit represents a 4-bit binary number (as listed in Table 2–3).

The hexadecimal number system consists of digits 0–9 and letters A–F.

◄ TABLE 2–3

DECIMAL	BINARY	HEXADECIMAL
0	0000	0
1	0001	1
2	0010	2
3	0011	3
4	0100	4
5	0101	5
6	0110	6
7	0111	7
8	1000	8
9	1001	9
10	1010	A
11	1011	B
12	1100	C
13	1101	D
14	1110	E
15	1111	F

Ten numeric digits and six alphabetic characters make up the hexadecimal number system. The use of letters A, B, C, D, E, and F to represent numbers may seem strange at first, but keep in mind that any number system is only a set of sequential symbols. If you understand what quantities these symbols represent, then the form of the symbols themselves is less important once you get accustomed to using them. We will use the subscript 16 to designate hexadecimal numbers to avoid confusion with decimal numbers. Sometimes you may see an "h" following a hexadecimal number.

Counting in Hexadecimal

How do you count in hexadecimal once you get to F? Simply start over with another column and continue as follows:

10, 11, 12, 13, 14, 15, 16, 17, 18, 19, 1A, 1B, 1C, 1D, 1E, 1F, 20, 21, 22, 23, 24, 25, 26, 27, 28, 29, 2A, 2B, 2C, 2D, 2E, 2F, 30, 31, . . .

With two hexadecimal digits, you can count up to FF_{16}, which is decimal 255. To count beyond this, three hexadecimal digits are needed. For instance, 100_{16} is decimal 256, 101_{16} is decimal 257, and so forth. The maximum 3-digit hexadecimal number is FFF_{16}, or decimal 4095. The maximum 4-digit hexadecimal number is $FFFF_{16}$, which is decimal 65,535.

Binary-to-Hexadecimal Conversion

Converting a binary number to hexadecimal is a straightforward procedure. Simply break the binary number into 4-bit groups, starting at the right-most bit and replace each 4-bit group with the equivalent hexadecimal symbol.

EXAMPLE 2–24

Convert the following binary numbers to hexadecimal:

(a) 1100101001010111   (b) 111111000101101001

Solution (a) 1100 1010 0101 0111

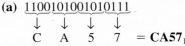

C   A   5   7 $= \textbf{CA57}_{16}$

(b) 0011 1111 0001 0110 1001

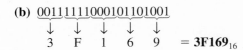

3   F   1   6   9 $= \textbf{3F169}_{16}$

Two zeros have been added in part (b) to complete a 4-bit group at the left.

Related Problem Convert the binary number 1001111011110011100 to hexadecimal.

Hexadecimal-to-Binary Conversion

Hexadecimal is a convenient way to represent binary numbers.

To convert from a hexadecimal number to a binary number, reverse the process and replace each hexadecimal symbol with the appropriate four bits.

EXAMPLE 2–25

Determine the binary numbers for the following hexadecimal numbers:

(a) $10A4_{16}$ (b) $CF8E_{16}$ (c) 9742_{16}

Solution (a) 1   0   A   4   (b) C   F   8   E   (c) 9   7   4   2

1000010100100   **1100111110001110**   **1001011101000010**

In part (a), the MSB is understood to have three zeros preceding it, thus forming a 4-bit group.

Related Problem Convert the hexadecimal number 6BD3 to binary.

It should be clear that it is much easier to deal with a hexadecimal number than with the equivalent binary number. Since conversion is so easy, the hexadecimal system is widely used for representing binary numbers in programming, printouts, and displays.

Conversion between hexadecimal and binary is direct and easy.

Hexadecimal-to-Decimal Conversion

One way to find the decimal equivalent of a hexadecimal number is to first convert the hexadecimal number to binary and then convert from binary to decimal.

EXAMPLE 2–26

Convert the following hexadecimal numbers to decimal:

(a) $1C_{16}$ (b) $A85_{16}$

Solution   Remember, convert the hexadecimal number to binary first, then to decimal.

(a)   1   C
   ↓   ↓
   $\overline{0001\,1100} = 2^4 + 2^3 + 2^2 = 16 + 8 + 4 = \textbf{28}_{10}$

(b)   A   8   5
   ↓   ↓   ↓
   $\overline{1010\,1000\,0101} = 2^{11} + 2^9 + 2^7 + 2^2 + 2^0 = 2048 + 512 + 128 + 4 + 1 = \textbf{2693}_{10}$

Related Problem   Convert the hexadecimal number 6BD to decimal.

Another way to convert a hexadecimal number to its decimal equivalent is to multiply the decimal value of each hexadecimal digit by its weight and then take the sum of these products. The weights of a hexadecimal number are increasing powers of 16 (from right to left). For a 4-digit hexadecimal number, the weights are

16^3 16^2 16^1 16^0
4096   256   16   1

EXAMPLE 2–27

Convert the following hexadecimal numbers to decimal:

(a) $E5_{16}$ (b) $B2F8_{16}$

Solution   Recall from Table 2–3 that letters A through F represent decimal numbers 10 through 15, respectively.

(a) $E5_{16} = (E \times 16) + (5 \times 1) = (14 \times 16) + (5 \times 1) = 224 + 5 = \textbf{229}_{10}$

(b) $B2F8_{16} = (B \times 4096) + (2 \times 256) + (F \times 16) + (8 \times 1)$
$= (11 \times 4096) + (2 \times 256) + (15 \times 16) + (8 \times 1)$
$= 45{,}056 + 512 + 240 + 8 = \textbf{45,816}_{10}$

Related Problem   Convert $60A_{16}$ to decimal.

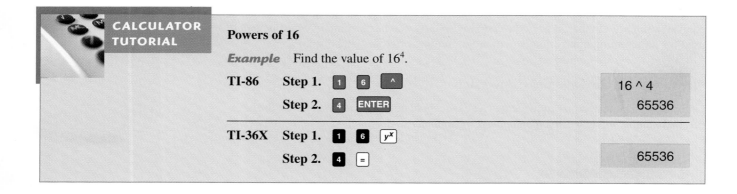

Decimal-to-Hexadecimal Conversion

Repeated division of a decimal number by 16 will produce the equivalent hexadecimal number, formed by the remainders of the divisions. The first remainder produced is the least significant digit (LSD). Each successive division by 16 yields a remainder that becomes a digit in the equivalent hexadecimal number. This procedure is similar to repeated division by 2 for decimal-to-binary conversion that was covered in Section 2–3. Example 2–28 illustrates the procedure. Note that when a quotient has a fractional part, the fractional part is multiplied by the divisor to get the remainder.

EXAMPLE 2–28

Convert the decimal number 650 to hexadecimal by repeated division by 16.

Solution

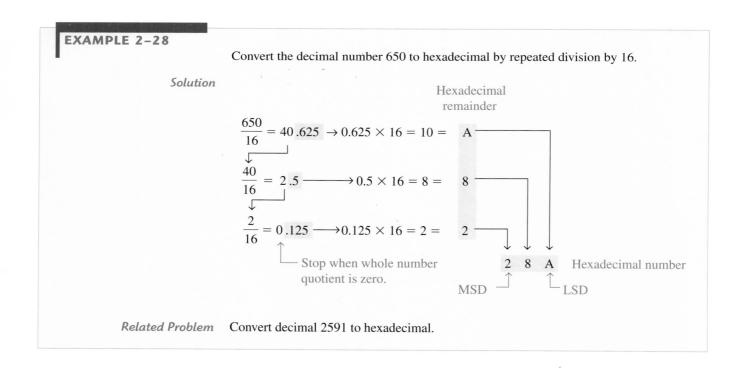

Related Problem Convert decimal 2591 to hexadecimal.

Hexadecimal Addition

Addition can be done directly with hexadecimal numbers by remembering that the hexadecimal digits 0 through 9 are equivalent to decimal digits 0 through 9 and that hexadecimal digits A through F are equivalent to decimal numbers 10 through 15. When adding two hexadecimal numbers, use the following rules. (Decimal numbers are indicated by a subscript 10.)

1. In any given column of an addition problem, think of the two hexadecimal digits in terms of their decimal values. For instance, $5_{16} = 5_{10}$ and $C_{16} = 12_{10}$.

2. If the sum of these two digits is 15_{10} or less, bring down the corresponding hexadecimal digit.

3. If the sum of these two digits is greater than 15_{10}, bring down the amount of the sum that exceeds 16_{10} and carry a 1 to the next column.

A calculator can be used to perform arithmetic operations with hexadecimal numbers.

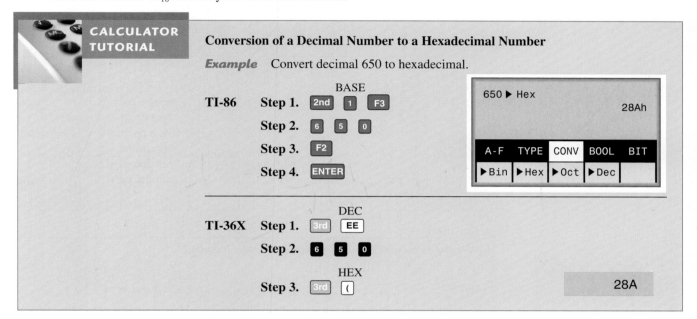

CALCULATOR TUTORIAL

Conversion of a Decimal Number to a Hexadecimal Number

Example   Convert decimal 650 to hexadecimal.

BASE

TI-86   Step 1.   [2nd] [1] [F3]

Step 2.   [6] [5] [0]

Step 3.   [F2]

Step 4.   [ENTER]

650 ▶ Hex
                            28Ah

| A-F | TYPE | CONV | BOOL | BIT |
| ▶Bin | ▶Hex | ▶Oct | ▶Dec | |

DEC

TI-36X   Step 1.   [3rd] [EE]

Step 2.   [6] [5] [0]

HEX

Step 3.   [3rd] [( ]

28A

EXAMPLE 2–29

Add the following hexadecimal numbers:

(a) $23_{16} + 16_{16}$ **(b)** $58_{16} + 22_{16}$ **(c)** $2B_{16} + 84_{16}$ **(d)** $DF_{16} + AC_{16}$

Solution

(a) $\begin{array}{r} 23_{16} \\ +16_{16} \\ \hline \mathbf{39}_{16} \end{array}$ right column: $3_{16} + 6_{16} = 3_{10} + 6_{10} = 9_{10} = 9_{16}$
left column: $2_{16} + 1_{16} = 2_{10} + 1_{10} = 3_{10} = 3_{16}$

(b) $\begin{array}{r} 58_{16} \\ +22_{16} \\ \hline \mathbf{7A}_{16} \end{array}$ right column: $8_{16} + 2_{16} = 8_{10} + 2_{10} = 10_{10} = A_{16}$
left column: $5_{16} + 2_{16} = 5_{10} + 2_{10} = 7_{10} = 7_{16}$

(c) $\begin{array}{r} 2B_{16} \\ +84_{16} \\ \hline \mathbf{AF}_{16} \end{array}$ right column: $B_{16} + 4_{16} = 11_{10} + 4_{10} = 15_{10} = F_{16}$
left column: $2_{16} + 8_{16} = 2_{10} + 8_{10} = 10_{10} = A_{16}$

(d) $\begin{array}{r} DF_{16} \\ +AC_{16} \\ \hline \mathbf{18B}_{16} \end{array}$ right column: $F_{16} + C_{16} = 15_{10} + 12_{10} = 27_{10}$
$27_{10} - 16_{10} = 11_{10} = B_{16}$ with a 1 carry
left column: $D_{16} + A_{16} + 1_{16} = 13_{10} + 10_{10} + 1_{10} = 24_{10}$
$24_{10} - 16_{10} = 8_{10} = 8_{16}$ with a 1 carry

Related Problem   Add $4C_{16}$ and $3A_{16}$.

Hexadecimal Subtraction

As you have learned, the 2's complement allows you to subtract by adding binary numbers. Since a hexadecimal number can be used to represent a binary number, it can also be used to represent the 2's complement of a binary number.

There are three ways to get the 2's complement of a hexadecimal number. Method 1 is the most common and easiest to use. Methods 2 and 3 are alternate methods.

Method 1.   Convert the hexadecimal number to binary. Take the 2's complement of the binary number. Convert the result to hexadecimal. This is illustrated in Figure 2–4.

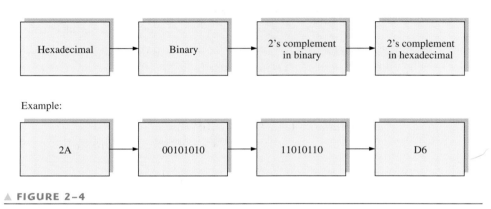

▲ **FIGURE 2–4**

Getting the 2's complement of a hexadecimal number, Method 1.

Method 2.   Subtract the hexadecimal number from the maximum hexadecimal number and add 1. This is illustrated in Figure 2–5.

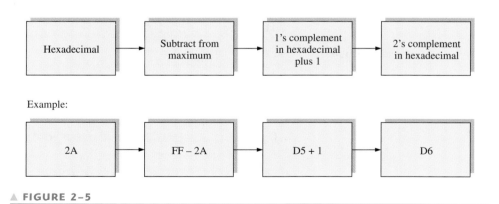

▲ **FIGURE 2–5**

Getting the 2's complement of a hexadecimal number, Method 2.

Method 3.   Write the sequence of single hexadecimal digits. Write the sequence in reverse below the forward sequence. The 1's complement of each hex digit is the digit directly below it. Add 1 to the resulting number to get the 2's complement. This is illustrated in Figure 2–6.

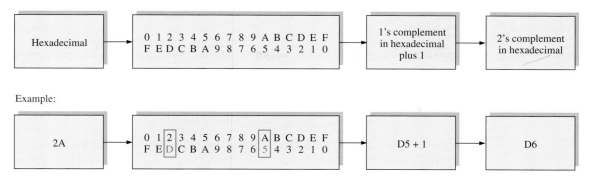

▲ FIGURE 2–6

Getting the 2's complement of a hexadecimal number, Method 3.

EXAMPLE 2–30

Subtract the following hexadecimal numbers:

(a) $84_{16} - 2A_{16}$ (b) $C3_{16} - 0B_{16}$

Solution   (a) $2A_{16} = 00101010$

2's complement of $2A_{16} = 11010110 = D6_{16}$ (using Method 1)

$$\begin{array}{r} 84_{16} \\ + \; D6_{16} \\ \hline \cancel{1}5A_{16} \end{array}$$ Add
Drop carry, as in 2's complement addition

The difference is **$5A_{16}$**.

(b) $0B_{16} = 00001011$

2's complement of $0B_{16} = 11110101 = F5_{16}$ (using Method 1)

$$\begin{array}{r} C3_{16} \\ + \; F5_{16} \\ \hline \cancel{1}B8_{16} \end{array}$$ Add
Drop carry

The difference is **$B8_{16}$**.

Related Problem   Subtract 173_{16} from BCD_{16}.

**SECTION 2–8
REVIEW**

1. Convert the following binary numbers to hexadecimal:
   (a) 10110011   (b) 110011101000
2. Convert the following hexadecimal numbers to binary:
   (a) 57_{16} (b) $3A5_{16}$ (c) $F80B_{16}$
3. Convert $9B30_{16}$ to decimal.
4. Convert the decimal number 573 to hexadecimal.
5. Add the following hexadecimal numbers directly:
   (a) $18_{16} + 34_{16}$ (b) $3F_{16} + 2A_{16}$
6. Subtract the following hexadecimal numbers:
   (a) $75_{16} - 21_{16}$ (b) $94_{16} - 5C_{16}$

2–9 OCTAL NUMBERS

Like the hexadecimal number system, the octal number system provides a convenient way to express binary numbers and codes. However, it is used less frequently than hexadecimal in conjunction with computers and microprocessors to express binary quantities for input and output purposes.

After completing this section, you should be able to

■ Write the digits of the octal number system ■ Convert from octal to decimal ■ Convert from decimal to octal ■ Convert from octal to binary ■ Convert from binary to octal

The octal number system is composed of eight digits, which are

0, 1, 2, 3, 4, 5, 6, 7

To count above 7, begin another column and start over:

10, 11, 12, 13, 14, 15, 16, 17, 20, 21, . . .

The octal number system has a base of 8.

Counting in octal is similar to counting in decimal, except that the digits 8 and 9 are not used. To distinguish octal numbers from decimal numbers or hexadecimal numbers, we will use the subscript 8 to indicate an octal number. For instance, 15_8 in octal is equivalent to 13_{10} in decimal and D in hexadecimal. Sometimes you may see an "o" or a "Q" following an octal number.

Octal-to-Decimal Conversion

Since the octal number system has a base of eight, each successive digit position is an increasing power of eight, beginning in the right-most column with 8^0. The evaluation of an octal number in terms of its decimal equivalent is accomplished by multiplying each digit by its weight and summing the products, as illustrated here for 2374_8.

$$\text{Weight:} \quad 8^3 \; 8^2 \; 8^1 \; 8^0$$
$$\text{Octal number:} \quad 2 \; 3 \; 7 \; 4$$

$$
\begin{aligned}
2374_8 &= (2 \times 8^3) \; + (3 \times 8^2) + (7 \times 8^1) + (4 \times 8^0) \\
&= (2 \times 512) + (3 \times 64) + (7 \times 8) \; + (4 \times 1) \\
&= \quad 1024 \quad + \quad 192 \quad + \quad 56 \quad + \quad 4 \quad = 1276_{10}
\end{aligned}
$$

Decimal-to-Octal Conversion

A method of converting a decimal number to an octal number is the repeated division-by-8 method, which is similar to the method used in the conversion of decimal numbers to binary or to hexadecimal. To show how it works, let's convert the decimal number 359 to octal. Each successive division by 8 yields a remainder that becomes a digit in the equivalent octal number. The first remainder generated is the least significant digit (LSD).

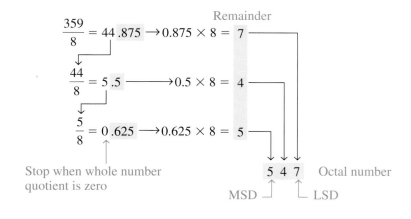

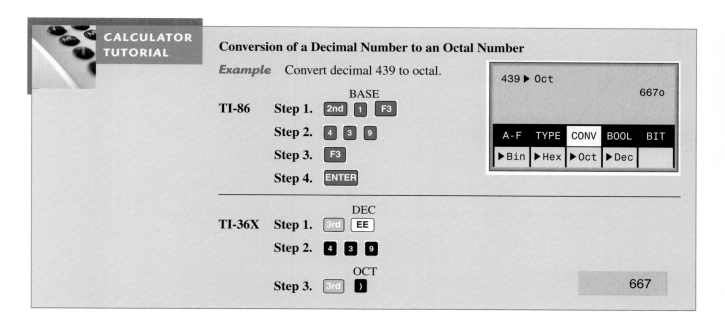

CALCULATOR TUTORIAL

Conversion of a Decimal Number to an Octal Number

Example Convert decimal 439 to octal.

BASE

TI-86 Step 1. [2nd] [1] [F3]

Step 2. [4] [3] [9]

Step 3. [F3]

Step 4. [ENTER]

439 ▶ Oct

667o

| A-F | TYPE | CONV | BOOL | BIT |

▶Bin ▶Hex ▶Oct ▶Dec

DEC

TI-36X Step 1. [3rd] [EE]

Step 2. [4] [3] [9]

OCT

Step 3. [3rd] [)]

667

Octal-to-Binary Conversion

Because each octal digit can be represented by a 3-bit binary number, it is very easy to convert from octal to binary. Each octal digit is represented by three bits as shown in Table 2–4.

Octal is a convenient way to represent binary numbers, but it is not as commonly used as hexadecimal.

▼ TABLE 2–4

Octal/binary conversion.

OCTAL DIGIT	0	1	2	3	4	5	6	7
BINARY	000	001	010	011	100	101	110	111

To convert an octal number to a binary number, simply replace each octal digit with the appropriate three bits.

EXAMPLE 2–31

Convert each of the following octal numbers to binary:

(a) 13_8 (b) 25_8 (c) 140_8 (d) 7526_8

Solution

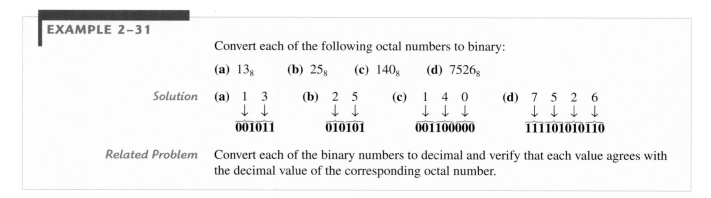

Related Problem Convert each of the binary numbers to decimal and verify that each value agrees with the decimal value of the corresponding octal number.

Binary-to-Octal Conversion

Conversion of a binary number to an octal number is the reverse of the octal-to-binary conversion. The procedure is as follows: Start with the right-most group of three bits and, moving from right to left, convert each 3-bit group to the equivalent octal digit. If there are not three bits available for the left-most group, add either one or two zeros to make a complete group. These leading zeros do not affect the value of the binary number.

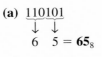

EXAMPLE 2–32

Convert each of the following binary numbers to octal:

(a) 110101   (b) 101111001   (c) 100110011010   (d) 11010000100

Solution

(a) 110101
$\quad\downarrow\quad\downarrow$
$\quad$6$\quad$5 = **65**$_8$

(b) 101111001
$\quad\downarrow\quad\downarrow\quad\downarrow$
$\quad$5$\quad$7$\quad$1 = **571**$_8$

(c) 100110011010
$\quad\downarrow\quad\downarrow\quad\downarrow\quad\downarrow$
$\quad$4$\quad$6$\quad$3$\quad$2 = **4632**$_8$

(d) 011010000100
$\quad\downarrow\quad\downarrow\quad\downarrow\quad\downarrow$
$\quad$3$\quad$2$\quad$0$\quad$4 = **3204**$_8$

Related Problem

Convert the binary number 1010101000111110010 to octal.

SECTION 2–9 REVIEW

1. Convert the following octal numbers to decimal:
   (a) 73$_8$ (b) 125$_8$
2. Convert the following decimal numbers to octal:
   (a) 98$_{10}$ (b) 163$_{10}$
3. Convert the following octal numbers to binary:
   (a) 46$_8$ (b) 723$_8$ (c) 5624$_8$
4. Convert the following binary numbers to octal:
   (a) 110101111   (b) 1001100010   (c) 10111111001

2–10 BINARY CODED DECIMAL (BCD)

Binary coded decimal (BCD) is a way to express each of the decimal digits with a binary code. There are only ten code groups in the BCD system, so it is very easy to convert between decimal and BCD. Because we like to read and write in decimal, the BCD code provides an excellent interface to binary systems. Examples of such interfaces are keypad inputs and digital readouts.

After completing this section, you should be able to

■ Convert each decimal digit to BCD ■ Express decimal numbers in BCD ■ Convert from BCD to decimal ■ Add BCD numbers

The 8421 Code

In BCD, 4 bits represent each decimal digit.

The 8421 code is a type of **BCD** (binary coded decimal) code. Binary coded decimal means that each decimal digit, 0 through 9, is represented by a binary code of four bits. The designation 8421 indicates the binary weights of the four bits ($2^3, 2^2, 2^1, 2^0$). The ease of conversion between 8421 code numbers and the familiar decimal numbers is the main advantage of this code. All you have to remember are the ten binary combinations that represent the ten decimal digits as shown in Table 2–5. The 8421 code is the predominant BCD code, and when we refer to BCD, we always mean the 8421 code unless otherwise stated.

DECIMAL DIGIT	0	1	2	3	4	5	6	7	8	9
BCD	0000	0001	0010	0011	0100	0101	0110	0111	1000	1001

◀ TABLE 2–5

Decimal/BCD conversion.

Invalid Codes   You should realize that, with four bits, sixteen numbers (0000 through 1111) can be represented but that, in the 8421 code, only ten of these are used. The six code combinations that are not used—1010, 1011, 1100, 1101, 1110, and 1111—are invalid in the 8421 BCD code.

To express any decimal number in BCD, simply replace each decimal digit with the appropriate 4-bit code, as shown by Example 2–33.

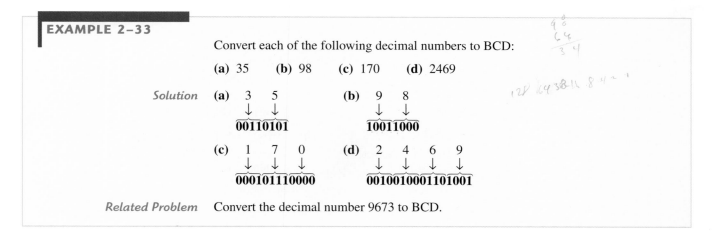

EXAMPLE 2–33

Convert each of the following decimal numbers to BCD:

(a) 35   (b) 98   (c) 170   (d) 2469

Solution   (a) 3   5   (b) 9   8
                  ↓   ↓   ↓   ↓
                 00110101   10011000

              (c) 1   7   0   (d) 2   4   6   9
                  ↓   ↓   ↓   ↓   ↓   ↓   ↓
                 000101110000   0010010001101001

Related Problem   Convert the decimal number 9673 to BCD.

It is equally easy to determine a decimal number from a BCD number. Start at the rightmost bit and break the code into groups of four bits. Then write the decimal digit represented by each 4-bit group.

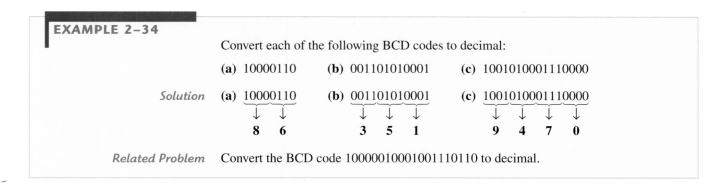

EXAMPLE 2–34

Convert each of the following BCD codes to decimal:

(a) 10000110   (b) 001101010001   (c) 1001010001110000

Solution   (a) 10000110   (b) 001101010001   (c) 1001010001110000
                   ↓   ↓   ↓   ↓   ↓   ↓   ↓   ↓   ↓
                   8   6   3   5   1   9   4   7   0

Related Problem   Convert the BCD code 10000010001001110110 to decimal.

BCD Addition

BCD is a numerical code and can be used in arithmetic operations. Addition is the most important operation because the other three operations (subtraction, multiplication, and division) can be accomplished by the use of addition. Here is how to add two BCD numbers:

Step 1.   Add the two BCD numbers, using the rules for binary addition in Section 2–4.

Step 2.   If a 4-bit sum is equal to or less than 9, it is a valid BCD number.

Step 3. If a 4-bit sum is greater than 9, or if a carry out of the 4-bit group is generated, it is an invalid result. Add 6 (0110) to the 4-bit sum in order to skip the six invalid states and return the code to 8421. If a carry results when 6 is added, simply add the carry to the next 4-bit group.

Example 2–35 illustrates BCD additions in which the sum in each 4-bit column is equal to or less than 9, and the 4-bit sums are therefore valid BCD numbers. Example 2–36 illustrates the procedure in the case of invalid sums (greater than 9 or a carry).

EXAMPLE 2–35

Add the following BCD numbers:

(a) 0011 + 0100   **(b)** 00100011 + 00010101

(c) 10000110 + 00010011   **(d)** 010001010000 + 010000010111

Solution   The decimal number additions are shown for comparison.

(a)	0011		3	**(b)**	0010	0011		23
	+ 0100		+ 4		+ 0001	0101		+ 15
	0111		7		**0011**	**1000**		38

(c)	1000	0110		86	**(d)**	0100	0101	0000	450
	+ 0001	0011		+ 13		+ 0100	0001	0111	+ 417
	1001	**1001**		99		**1000**	**0110**	**0111**	867

Note that in each case the sum in any 4-bit column does not exceed 9, and the results are valid BCD numbers.

Related Problem   Add the BCD numbers: 1001000001000011 + 0000100100100101.

EXAMPLE 2–36

Add the following BCD numbers

(a) 1001 + 0100   **(b)** 1001 + 1001

(c) 00010110 + 00010101   **(d)** 01100111 + 01010011

Solution   The decimal number additions are shown for comparison.

(a)
```
          1001                                          9
        + 0100                                        + 4
          1101      Invalid BCD number (>9)            13
        + 0110      Add 6
   0001   0011      Valid BCD number
    ↓      ↓
    1      3
```

(b)
```
          1001                                          9
        + 1001                                        + 9
     1    0010      Invalid because of carry           18
        + 0110      Add 6
   0001   1000      Valid BCD number
    ↓      ↓
    1      8
```

(c)

	0001	0110		16
	+ 0001	0101		+ 15
	0010	1011	Right group is invalid (>9), left group is valid.	31
		+ 0110	Add 6 to invalid code. Add carry, 0001, to next group.	
	0011	**0001**	Valid BCD number	
	↓	↓		
	3	1		

(d)

	0110	0111		67
	+ 0101	0011		+ 53
	1011	1010	Both groups are invalid (>9)	120
	+ 0110	+ 0110	Add 6 to both groups	
0001	**0010**	**0000**	Valid BCD number	
↓	↓	↓		
1	2	0		

Related Problem Add the BCD numbers: 01001000 + 00110100.

SECTION 2–10 REVIEW

1. What is the binary weight of each 1 in the following BCD numbers?
   (a) 0010   (b) 1000   (c) 0001   (d) 0100
2. Convert the following decimal numbers to BCD:
   (a) 6   (b) 15   (c) 273   (d) 849
3. What decimal numbers are represented by each BCD code?
   (a) 10001001   (b) 001001111000   (c) 000101010111
4. In BCD addition, when is a 4-bit sum invalid?

2–11 DIGITAL CODES

Many specialized codes are used in digital systems. You have just learned about the BCD code; now let's look at a few others. Some codes are strictly numeric, like BCD, and others are alphanumeric; that is, they are used to represent numbers, letters, symbols, and instructions. The codes introduced in this section are the Gray code and the ASCII code.

After completing this section, you should be able to

■ Explain the advantage of the Gray code ■ Convert between Gray code and binary
■ Use the ASCII code

The Gray Code

The **Gray code** is unweighted and is not an arithmetic code; that is, there are no specific weights assigned to the bit positions. The important feature of the Gray code is that *it exhibits only a single bit change from one code word to the next in sequence.* This property is important in many applications, such as shaft position encoders, where error susceptibility increases with the number of bit changes between adjacent numbers in a sequence.

The single bit change characteristic of the Gray code minimizes the chance for error.

Table 2–6 is a listing of the 4-bit Gray code for decimal numbers 0 through 15. Binary numbers are shown in the table for reference. Like binary numbers, *the Gray code can have any number of bits*. Notice the single-bit change between successive Gray code words. For instance, in going from decimal 3 to decimal 4, the Gray code changes from 0010 to 0110, while the binary code changes from 0011 to 0100, a change of three bits. The only bit change is in the third bit from the right in the Gray code; the others remain the same.

▶ **TABLE 2–6**

Four-bit Gray code.

DECIMAL	BINARY	GRAY CODE	DECIMAL	BINARY	GRAY CODE
0	0000	0000	8	1000	1100
1	0001	0001	9	1001	1101
2	0010	0011	10	1010	1111
3	0011	0010	11	1011	1110
4	0100	0110	12	1100	1010
5	0101	0111	13	1101	1011
6	0110	0101	14	1110	1001
7	0111	0100	15	1111	1000

Binary-to-Gray Code Conversion Conversion between binary code and Gray code is sometimes useful. The following rules explain how to convert from a binary number to a Gray code word:

1. The most significant bit (left-most) in the Gray code is the same as the corresponding MSB in the binary number.

2. Going from left to right, add each adjacent pair of binary code bits to get the next Gray code bit. Discard carries.

For example, the conversion of the binary number 10110 to Gray code is as follows:

$$
\begin{array}{ccccc}
1 - + \rightarrow & 0 - + \rightarrow & 1 - + \rightarrow & 1 - + \rightarrow & 0 \quad \text{Binary} \\
\downarrow & \downarrow & \downarrow & \downarrow & \downarrow \\
1 & 1 & 1 & 0 & 1 \quad \text{Gray}
\end{array}
$$

The Gray code is 11101.

Gray-to-Binary Conversion To convert from Gray code to binary, use a similar method; however, there are some differences. The following rules apply:

1. The most significant bit (left-most) in the binary code is the same as the corresponding bit in the Gray code.

2. Add each binary code bit generated to the Gray code bit in the next adjacent position. Discard carries.

For example, the conversion of the Gray code word 11011 to binary is as follows:

$$
\begin{array}{ccccc}
1 & 1 & 0 & 1 & 1 \quad \text{Gray} \\
\downarrow \; + \nearrow & \downarrow \; + \nearrow & \downarrow \; + \nearrow & \downarrow \; + \nearrow & \downarrow \\
1 & 0 & 0 & 1 & 0 \quad \text{Binary}
\end{array}
$$

The binary number is 10010.

EXAMPLE 2–37

(a) Convert the binary number 11000110 to Gray code.

(b) Convert the Gray code 10101111 to binary.

Solution (a) Binary to Gray code:

$$1 \xrightarrow{+} 1 \xrightarrow{+} 0 \xrightarrow{+} 0 \xrightarrow{+} 0 \xrightarrow{+} 1 \xrightarrow{+} 1 \xrightarrow{+} 0$$
$$\downarrow \quad \downarrow \quad \downarrow \quad \downarrow \quad \downarrow \quad \downarrow \quad \downarrow \quad \downarrow$$
$$\mathbf{1} \quad \mathbf{0} \quad \mathbf{1} \quad \mathbf{0} \quad \mathbf{0} \quad \mathbf{1} \quad \mathbf{0} \quad \mathbf{1}$$

(b) Gray code to binary:

$$1 \quad 0 \quad 1 \quad 0 \quad 1 \quad 1 \quad 1 \quad 1$$
$$\downarrow {\nearrow} + \downarrow {\nearrow} + \downarrow {\nearrow} + \downarrow {\nearrow} + \downarrow {\nearrow} + \downarrow {\nearrow} + \downarrow {\nearrow} + \downarrow$$
$$\mathbf{1} \quad \mathbf{1} \quad \mathbf{0} \quad \mathbf{0} \quad \mathbf{1} \quad \mathbf{0} \quad \mathbf{1} \quad \mathbf{0}$$

Related Problem (a) Convert binary 101101 to Gray code. (b) Convert Gray code 100111 to binary.

An Application

A simplified diagram of a 3-bit shaft position encoder mechanism is shown in Figure 2–7. Basically, there are three concentric conductive rings that are segmented into eight sectors. The more sectors there are, the more accurately the position can be represented, but we are using only eight for purposes of illustration. Each sector of each ring is fixed at either a high-level or a low-level voltage to represent 1s and 0s. A 1 is indicated by a color sector and a 0 by a white sector. As the rings rotate with the shaft, they make contact with a brush arrangement that is in a fixed position and to which output lines are connected. As the shaft rotates counterclockwise through 360°, the eight sectors move past the three brushes producing a 3-bit binary output that indicates the shaft position.

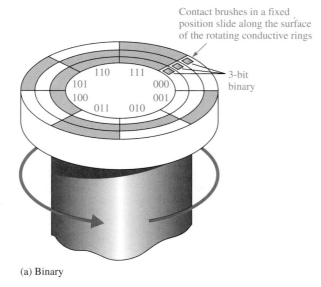
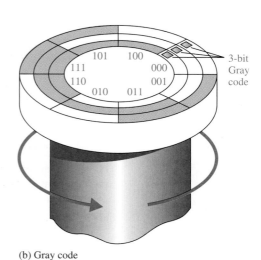

Contact brushes in a fixed position slide along the surface of the rotating conductive rings

(a) Binary

(b) Gray code

▲ **FIGURE 2–7**

A simplified illustration of how the Gray code solves the error problem in shaft position encoders.

In Figure 2–7(a), the sectors are arranged in a straight binary pattern, so that the brushes go from 000 to 001 to 010 to 011, and so on. When the brushes are on color sectors, they output a 1 and when on white sectors, they output a 0. If one brush is slightly ahead of the others during the transition from one sector to the next, an erroneous output can occur. Consider what happens when the brushes are on the 111 sector and about to enter the 000 sector. If the MSB brush is slightly ahead, the position would be incorrectly indicated by a transitional 011 instead of a 111 or a 000. In this type of application, it is virtually impossible to maintain precise mechanical alignment of all the brushes; therefore, some error will always occur at many of the transitions between sectors.

The Gray code is used to eliminate the error problem which is inherent in the binary code. As shown in Figure 2–7(b), the Gray code assures that only one bit will change between adjacent sectors. This means that even though the brushes may not be in precise alignment, there will never be a transitional error. For example, let's again consider what happens when the brushes are on the 111 sector and about to move into the next sector, 101. The only two possible outputs during the transition are 111 and 101, no matter how the brushes are aligned. A similar situation occurs at the transitions between each of the other sectors.

Alphanumeric Codes

In order to communicate, you need not only numbers, but also letters and other symbols. In the strictest sense, **alphanumeric** codes are codes that represent numbers and alphabetic characters (letters). Most such codes, however, also represent other characters such as symbols and various instructions necessary for conveying information.

At a minimum, an alphanumeric code must represent 10 decimal digits and 26 letters of the alphabet, for a total of 36 items. This number requires six bits in each code combination because five bits are insufficient ($2^5 = 32$). There are 64 total combinations of six bits, so there are 28 unused code combinations. Obviously, in many applications, symbols other than just numbers and letters are necessary to communicate completely. You need spaces, periods, colons, semicolons, question marks, etc. You also need instructions to tell the receiving system what to do with the information. With codes that are six bits long, you can handle decimal numbers, the alphabet, and 28 other symbols. This should give you an idea of the requirements for a basic alphanumeric code. The ASCII is the most common alphanumeric code and is covered next.

ASCII

COMPUTER NOTE

A computer keyboard has a dedicated microprocessor that constantly scans keyboard circuits to detect when a key has been pressed and released. A unique scan code is produced by computer software representing that particular key. The scan code is then converted to an alphanumeric code (ASCII) for use by the computer.

ASCII is the abbreviation for American Standard Code for Information Interchange. Pronounced "askee," ASCII is a universally accepted alphanumeric code used in most computers and other electronic equipment. Most computer keyboards are standardized with the ASCII. When you enter a letter, a number, or control command, the corresponding ASCII code goes into the computer.

ASCII has 128 characters and symbols represented by a 7-bit binary code. Actually, ASCII can be considered an 8-bit code with the MSB always 0. This 8-bit code is 00 through 7F in hexadecimal. The first thirty-two ASCII characters are nongraphic commands that are never printed or displayed and are used only for control purposes. Examples of the control characters are "null," "line feed," "start of text," and "escape." The other characters are graphic symbols that can be printed or displayed and include the letters of the alphabet (lowercase and uppercase), the ten decimal digits, punctuation signs and other commonly used symbols.

Table 2–7 is a listing of the ASCII code showing the decimal, hexadecimal, and binary representations for each character and symbol. The left section of the table lists the names of the 32 control characters (00 through 1F hexadecimal). The graphic symbols are listed in the rest of the table (20 through 7F hexadecimal).

American Standard Code for Information Interchange (ASCII).

CONTROL CHARACTERS

NAME	DEC	BINARY	HEX
NUL	0	0000000	00
SOH	1	0000001	01
STX	2	0000010	02
ETX	3	0000011	03
EOT	4	0000100	04
ENQ	5	0000101	05
ACK	6	0000110	06
BEL	7	0000111	07
BS	8	0001000	08
HT	9	0001001	09
LF	10	0001010	0A
VT	11	0001011	0B
FF	12	0001100	0C
CR	13	0001101	0D
SO	14	0001110	0E
SI	15	0001111	0F
DLE	16	0010000	10
DC1	17	0010001	11
DC2	18	0010010	12
DC3	19	0010011	13
DC4	20	0010100	14
NAK	21	0010101	15
SYN	22	0010110	16
ETB	23	0010111	17
CAN	24	0011000	18
EM	25	0011001	19
SUB	26	0011010	1A
ESC	27	0011011	1B
FS	28	0011100	1C
GS	29	0011101	1D
RS	30	0011110	1E
US	31	0011111	1F

GRAPHIC SYMBOLS

SYMBOL	DEC	BINARY	HEX
space	32	0100000	20
!	33	0100001	21
"	34	0100010	22
#	35	0100011	23
$	36	0100100	24
%	37	0100101	25
&	38	0100110	26
'	39	0100111	27
(	40	0101000	28
)	41	0101001	29
*	42	0101010	2A
+	43	0101011	2B
,	44	0101100	2C
–	45	0101101	2D
.	46	0101110	2E
/	47	0101111	2F
0	48	0110000	30
1	49	0110001	31
2	50	0110010	32
3	51	0110011	33
4	52	0110100	34
5	53	0110101	35
6	54	0110110	36
7	55	0110111	37
8	56	0111000	38
9	57	0111001	39
:	58	0111010	3A
;	59	0111011	3B
<	60	0111100	3C
=	61	0111101	3D
>	62	0111110	3E
?	63	0111111	3F
@	64	1000000	40
A	65	1000001	41
B	66	1000010	42
C	67	1000011	43
D	68	1000100	44
E	69	1000101	45
F	70	1000110	46
G	71	1000111	47
H	72	1001000	48
I	73	1001001	49
J	74	1001010	4A
K	75	1001011	4B
L	76	1001100	4C
M	77	1001101	4D
N	78	1001110	4E
O	79	1001111	4F
P	80	1010000	50
Q	81	1010001	51
R	82	1010010	52
S	83	1010011	53
T	84	1010100	54
U	85	1010101	55
V	86	1010110	56
W	87	1010111	57
X	88	1011000	58
Y	89	1011001	59
Z	90	1011010	5A
[	91	1011011	5B
\	92	1011100	5C
]	93	1011101	5D
^	94	1011110	5E
_	95	1011111	5F
`	96	1100000	60
a	97	1100001	61
b	98	1100010	62
c	99	1100011	63
d	100	1100100	64
e	101	1100101	65
f	102	1100110	66
g	103	1100111	67
h	104	1101000	68
i	105	1101001	69
j	106	1101010	6A
k	107	1101011	6B
l	108	1101100	6C
m	109	1101101	6D
n	110	1101110	6E
o	111	1101111	6F
p	112	1110000	70
q	113	1110001	71
r	114	1110010	72
s	115	1110011	73
t	116	1110100	74
u	117	1110101	75
v	118	1110110	76
w	119	1110111	77
x	120	1111000	78
y	121	1111001	79
z	122	1111010	7A
{	123	1111011	7B
		124	1111100	7C
}	125	1111101	7D
~	126	1111110	7E
Del	127	1111111	7F

EXAMPLE 2–38

Determine the binary ASCII codes that are entered from the computer's keyboard when the following BASIC program statement is typed in. Also express each code in hexadecimal.

20 PRINT "A=";X

Solution   The ASCII code for each symbol is found in Table 2–7.

Symbol	Binary	Hexadecimal
2	0110010	32_{16}
0	0110000	30_{16}
Space	0100000	20_{16}
P	1010000	50_{16}
R	1010010	52_{16}
I	1001001	49_{16}
N	1001110	$4E_{16}$
T	1010100	54_{16}
Space	0100000	20_{16}
"	0100010	22_{16}
A	1000001	41_{16}
=	0111101	$3D_{16}$
"	0100010	22_{16}
;	0111011	$3B_{16}$
X	1011000	58_{16}

Related Problem   Determine the sequence of ASCII codes required for the following program statement and express them in hexadecimal:

80 INPUT Y

The ASCII Control Characters   The first thirty-two codes in the ASCII table (Table 2–7) represent the control characters. These are used to allow devices such as a computer and printer to communicate with each other when passing information and data. Table 2–8 lists the control characters and the control key function that allows them to be entered directly from an ASCII keyboard by pressing the control key (CTRL) and the corresponding symbol. A brief description of each control character is also given.

Extended ASCII Characters

In addition to the 128 standard ASCII characters, there are an additional 128 characters that were adopted by IBM for use in their PCs (personal computers). Because of the popularity of the PC, these particular extended ASCII characters are also used in applications other than PCs and have become essentially an unofficial standard.

The extended ASCII characters are represented by an 8-bit code series from hexadecimal 80 to hexadecimal FF.

NAME	DECIMAL	HEX	KEY	DESCRIPTION
NUL	0	00	CTRL @	null character
SOH	1	01	CTRL A	start of header
STX	2	02	CTRL B	start of text
ETX	3	03	CTRL C	end of text
EOT	4	04	CTRL D	end of transmission
ENQ	5	05	CTRL E	enquire
ACK	6	06	CTRL F	acknowledge
BEL	7	07	CTRL G	bell
BS	8	08	CTRL H	backspace
HT	9	09	CTRL I	horizontal tab
LF	10	0A	CTRL J	line feed
VT	11	0B	CTRL K	vertical tab
FF	12	0C	CTRL L	form feed (new page)
CR	13	0D	CTRL M	carriage return
SO	14	0E	CTRL N	shift out
SI	15	0F	CTRL O	shift in
DLE	16	10	CTRL P	data link escape
DC1	17	11	CTRL Q	device control 1
DC2	18	12	CTRL R	device control 2
DC3	19	13	CTRL S	device control 3
DC4	20	14	CTRL T	device control 4
NAK	21	15	CTRL U	negative acknowledge
SYN	22	16	CTRL V	synchronize
ETB	23	17	CTRL W	end of transmission block
CAN	24	18	CTRL X	cancel
EM	25	19	CTRL Y	end of medium
SUB	26	1A	CTRL Z	substitute
ESC	27	1B	CTRL [	escape
FS	28	1C	CTRL /	file separator
GS	29	1D	CTRL]	group separator
RS	30	1E	CTRL ^	record separator
US	31	1F	CTRL _	unit separator

◀ **TABLE 2–8**

ASCII control characters.

The extended ASCII contains characters in the following general categories:

1. Foreign (non-English) alphabetic characters
2. Foreign currency symbols
3. Greek letters
4. Mathematical symbols
5. Drawing characters
6. Bar graphing characters
7. Shading characters

Table 2–9 is a list of the extended ASCII character set with the decimal and hexadecimal representations.

▼ TABLE 2–9

Extended ASCII characters.

SYMBOL	DEC	HEX	SYMBOL	DEC	HEX	SYMBOL	DEC	HEX	SYMBOL	DEC	HEX
Ç	128	80	á	160	A0	└	192	C0	α	224	E0
ü	129	81	í	161	A1	┴	193	C1	β	225	E1
é	130	82	ó	162	A2	┬	194	C2	Γ	226	E2
â	131	83	ú	163	A3	├	195	C3	π	227	E3
ä	132	84	ñ	164	A4	─	196	C4	Σ	228	E4
à	133	85	Ñ	165	A5	┼	197	C5	σ	229	E5
å	134	86	ª	166	A6	╞	198	C6	µ	230	E6
ç	135	87	º	167	A7	╟	199	C7	τ	231	E7
ê	136	88	¿	168	A8	╚	200	C8	Φ	232	E8
ë	137	89	⌐	169	A9	╔	201	C9	Θ	233	E9
è	138	8A	¬	170	AA	╩	202	CA	Ω	234	EA
ï	139	8B	½	171	AB	╦	203	CB	δ	235	EB
î	140	8C	¼	172	AC	╠	204	CC	∞	236	EC
ì	141	8D	¡	173	AD	═	205	CD	φ	237	ED
Ä	142	8E	«	174	AE	╬	206	CE	ε	238	EE
Å	143	8F	»	175	AF	╧	207	CF	∩	239	EF
É	144	90	░	176	B0	╨	208	D0	≡	240	F0
æ	145	91	▒	177	B1	╤	209	D1	±	241	F1
Æ	146	92	▓	178	B2	╥	210	D2	≥	242	F2
ô	147	93	│	179	B3	╙	211	D3	≤	243	F3
ö	148	94	┤	180	B4	╘	212	D4	⌠	244	F4
ò	149	95	╡	181	B5	╒	213	D5	⌡	245	F5
û	150	96	╢	182	B6	╓	214	D6	÷	246	F6
ù	151	97	╖	183	B7	╫	215	D7	≈	247	F7
ÿ	152	98	╕	184	B8	╪	216	D8	°	248	F8
Ö	153	99	╣	185	B9	┘	217	D9	•	249	F9
Ü	154	9A	║	186	BA	┌	218	DA	·	250	FA
¢	155	9B	╗	187	BB	█	219	DB	√	251	FB
£	156	9C	╝	188	BC	▄	220	DC	η	252	FC
¥	157	9D	╜	189	BD	▌	221	DD	²	253	FD
Pт	158	9E	╛	190	BE	▐	222	DE	■	254	FE
ƒ	159	9F	┐	191	BF	▀	223	DF	□	255	FF

SECTION 2–11
REVIEW

1. Convert the following binary numbers to the Gray code:
   (a) 1100   (b) 1010   (c) 11010

2. Convert the following Gray codes to binary:
   (a) 1000   (b) 1010   (c) 11101

3. What is the ASCII representation for each of the following characters? Express each as a bit pattern and in hexadecimal notation.
   (a) K   (b) r   (c) $ (d) +

2–12 ERROR DETECTION AND CORRECTION CODES

In this section, two methods for adding bits to codes to either detect a single-bit error or detect and correct a single-bit error are discussed. The parity method of error detection is introduced, and the Hamming method of single-error detection and correction is covered. When a bit in a given code word is found to be in error, it can be corrected by simply inverting it.

After completing this section, you should be able to

■ Determine if there is an error in a code based on the parity bit ■ Assign the proper parity bit to a code ■ Use the Hamming code for single-error detection and correction ■ Assign the proper parity bits for single-error correction

Parity Method for Error Detection

Many systems use a parity bit as a means for bit **error detection.** Any group of bits contain either an even or an odd number of 1s. A parity bit is attached to a group of bits to make the total number of 1s in a group always even or always odd. An even parity bit makes the total number of 1s even, and an odd parity bit makes the total odd.

A parity bit tells if the number of 1s is odd or even.

A given system operates with even or odd **parity,** but not both. For instance, if a system operates with even parity, a check is made on each group of bits received to make sure the total number of 1s in that group is even. If there is an odd number of 1s, an error has occurred.

As an illustration of how parity bits are attached to a code, Table 2–10 lists the parity bits for each BCD number for both even and odd parity. The parity bit for each BCD number is in the *P* column.

EVEN PARITY		ODD PARITY	
P	BCD	*P*	BCD
0	0000	1	0000
1	0001	0	0001
1	0010	0	0010
0	0011	1	0011
1	0100	0	0100
0	0101	1	0101
0	0110	1	0110
1	0111	0	0111
1	1000	0	1000
0	1001	1	1001

◁ **TABLE 2–10**

The BCD code with parity bits.

The parity bit can be attached to the code at either the beginning or the end, depending on system design. Notice that the total number of 1s, including the parity bit, is always even for even parity and always odd for odd parity.

Detecting an Error　A parity bit provides for the detection of a single bit error (or any odd number of errors, which is very unlikely) but cannot check for two errors in one group. For instance, let's assume that we wish to transmit the BCD code 0101. (Parity can be used with any number of bits; we are using four for illustration.) The total code transmitted, including the even parity bit, is

Even parity bit
00101
BCD code

Now let's assume that an error occurs in the third bit from the left (the 1 becomes a 0).

Even parity bit
00001
Bit errror

When this code is received, the parity check circuitry determines that there is only a single 1 (odd number), when there should be an even number of 1s. Because an even number of 1s does not appear in the code when it is received, an error is indicated.

An odd parity bit also provides in a similar manner for the detection of a single error in a given group of bits.

EXAMPLE 2–39

Assign the proper even parity bit to the following code groups:

(a) 1010   (b) 111000   (c) 101101

(d) 1000111001001   (e) 101101011111

Solution   Make the parity bit either 1 or 0 as necessary to make the total number of 1s even. The parity bit will be the left-most bit (color).

(a) 01010   (b) 1111000   (c) 0101101

(d) 0100011100101   (e) 1101101011111

Related Problem   Add an even parity bit to the 7-bit ASCII code for the letter K.

EXAMPLE 2–40

An odd parity system receives the following code groups: 10110, 11010, 110011, 110101110100, and 1100010101010. Determine which groups, if any, are in error.

Solution   Since odd parity is required, any group with an even number of 1s is incorrect. The following groups are in error: **110011** and **1100010101010.**

Related Problem   The following ASCII character is received by an odd parity system: 00110111. Is it correct?

The Hamming Error Correction Code

As you have seen, a single parity bit allows for the detection of single-bit errors in a code word. A single parity bit can indicate that there is an error in a certain group of bits. In order to correct a detected error, more information is required because the position of the bit in error must be identified before it can be corrected. More than one parity bit must be included in a group of bits to be able to correct a detected error. In a 7-bit code, there are seven possible single-bit errors. In this case, three parity bits can not only detect an error but can specify the position of the bit in error. The **Hamming code** provides for single-error correction. The following coverage illustrates the construction of a 7-bit Hamming code for single-error correction.

Number of Parity Bits　If the number of data bits is designated *d,* then the number of parity bits, *p,* is determined by the following relationship:

$$2^p \geq d + p + 1$$

Equation 2–1

For example, if we have four data bits, then *p* is found by trial and error with Equation 2–1. Let *p* = 2. Then

$$2^p = 2^2 = 4$$

and

$$d + p + 1 = 4 + 2 + 1 = 7$$

Since 2^p must be equal to or greater than $d + p + 1$, the relationship in Equation 2–1 is *not* satisfied. We have to try again. Let *p* = 3. Then

$$2^p = 2^3 = 8$$

and

$$d + p + 1 = 4 + 3 + 1 = 8$$

This value of *p* satisfies the relationship of Equation 2–1, so three parity bits are required to provide single-error correction for four data bits. It should be noted here that error detection and correction are provided for *all* bits, both parity and data, in a code group; that is, the parity bits also check themselves.

Placement of the Parity Bits in the Code　Now that we have found the number of parity bits required in our particular example, we must arrange the bits properly in the code. At this point you should realize that in this example the code is composed of the four data bits and the three parity bits. The left-most bit is designated *bit 1,* the next bit is *bit 2,* and so on as follows:

bit 1,　bit 2,　bit 3,　bit 4,　bit 5,　bit 6,　bit 7

he parity bits are located in the positions that are numbered corresponding to ascending powers of two (1, 2, 4, 8, . . .), as indicated:

P_1,　P_2,　D_1,　P_3,　D_2,　D_3,　D_4

he symbol P_n designates a particular parity bit, and D_n designates a particular data bit.

Assignment of Parity Bit Values　Finally, we must properly assign a 1 or 0 value to each parity bit. Since each parity bit provides a check on certain other bits in the total code, we must know the value of these others in order to assign the parity bit value. To find the bit values, first number each bit position in binary, that is, write the binary number for each decimal position number, as shown in the second two rows of Table 2–11. Next, indicate the parity and data bit locations, as shown in the first row of Table 2–11. Notice that the binary position number of parity bit P_1 has a 1 for its right-most digit. *This parity bit checks all bit positions, including itself, that have 1s in the same location in the binary position numbers.* Therefore, parity bit P_1 checks bit positions 1, 3, 5, and 7.

▼ **TABLE 2–11**

Bit position table for a 7-bit error correction code.

BIT DESIGNATION	P_1	P_2	D_1	P_3	D_2	D_3	D_4
BIT POSITION	1	2	3	4	5	6	7
BINARY POSITION NUMBER	001	010	011	100	101	110	111
Data bits (D_n)							
Parity bits (P_n)							

The binary position number for parity bit P_2 has a 1 for its middle bit. It checks all bit positions, including itself, that have 1s in this same position. Therefore, parity bit P_2 checks bit positions 2, 3, 6, and 7.

The binary position number for parity bit P_3 has a 1 for its left-most bit. It checks all bit positions, including itself, that have 1s in this same position. Therefore, parity bit P_3 checks bit positions 4, 5, 6, and 7.

In each case, the parity bit is assigned a value to make the quantity of 1s in the set of bits that it checks either odd or even, depending on which is specified. The following examples should make this procedure clear.

EXAMPLE 2–41

Determine the Hamming code for the BCD number 1001 (data bits), using even parity.

Solution **Step 1:** Find the number of parity bits required. Let $p = 3$. Then

$$2^p = 2^3 = 8$$
$$d + p + 1 = 4 + 3 + 1 = 8$$

Three parity bits are sufficient.

$$\text{Total code bits} = 4 + 3 = 7$$

Step 2: Construct a bit position table, as shown in Table 2–12, and enter the data bits. Parity bits are determined in the following steps.

▼ TABLE 2–12

BIT DESIGNATION BIT POSITION BINARY POSITION NUMBER	P_1 1 001	P_2 2 010	D_1 3 011	P_3 4 100	D_2 5 101	D_3 6 110	D_4 7 111
Data bits			1		0	0	1
Parity bits	0	0		1			

Step 3: Determine the parity bits as follows:

Bit P_1 checks bit positions 1, 3, 5, and 7 and must be a 0 for there to be an even number of 1s (2) in this group.

Bit P_2 checks bit positions 2, 3, 6, and 7 and must be a 0 for there to be an even number of 1s (2) in this group.

Bit P_3 checks bit positions 4, 5, 6, and 7 and must be a 1 for there to be an even number of 1s (2) in this group.

Step 4: These parity bits are entered in Table 2–12, and the resulting combined code is 0011001.

Related Problem Determine the Hamming code for the BCD number 1000 using even parity.

EXAMPLE 2–42

Determine the Hamming code for the data bits 10110 using odd parity.

Solution **Step 1:** Determine the number of parity bits required. In this case the number of data bits, *d*, is five. From the previous example we know that $p = 3$ will not work. Try $p = 4$:

$$2^p = 2^4 = 16$$
$$d + p + 1 = 5 + 4 + 1 = 10$$

Four parity bits are sufficient.

$$\text{Total code bits} = 5 + 4 = 9$$

Step 2: Construct a bit position table, Table 2–13, and enter the data bits. Parity bits are determined in the following steps. Notice that P_4 is in bit position 8.

▼ TABLE 2–13

BIT DESIGNATION	P_1	P_2	D_1	P_3	D_2	D_3	D_4	P_4	D_5
BIT POSITION	1	2	3	4	5	6	7	8	9
BINARY POSITION NUMBER	0001	0010	0011	0100	0101	0110	0111	1000	1001
Data bits			1		0	1	1		0
Parity bits	1	0		1				1	

Step 3: Determine the parity bits as follows:

Bit P_1 checks bit positions 1, 3, 5, 7, and 9 and must be a 1 for there to be an odd number of 1s (3) in this group.

Bit P_2 checks bit positions 2, 3, 6, and 7 and must be a 0 for there to be an odd number of 1s (3) in this group.

Bit P_3 checks bit positions 4, 5, 6, and 7 and must be a 1 for there to be an odd number of 1s (3) in this group.

Bit P_4 checks bit positions 8 and 9 and must be a 1 for there to be an odd number of 1s (1) in this group.

Step 4: These parity bits are entered in the Table 2–13, and the resulting combined code is 101101110.

Related Problem Determine the Hamming code for 11001 using odd parity.

Detecting and Correcting an Error with the Hamming Code

Now that the Hamming method for constructing an error-correction code has been covered, how do you use it to locate and correct an error? Each parity bit, along with its corresponding group of bits, must be checked for the proper parity. If there are three parity bits in a code word, then three parity checks are made. If there are four parity bits, four checks must be made, and so on. Each parity check will yield a good or a bad result. The total result of all the parity checks indicates the bit, if any, that is in error, as follows:

Step 1. Start with the group checked by P_1.

Step 2. Check the group for proper parity. A 0 represents a good parity check, and 1 represents a bad check.

Step 3. Repeat step 2 for each parity group.

Step 4. The binary number formed by the results of all the parity checks designates the position of the code bit that is in error. This is the *error position code*. The first parity check generates the least significant bit (LSB). If all checks are good, there is no error.

EXAMPLE 2–43

Assume that the code word in Example 2–41 (0011001) is transmitted and that 0010001 is received. The receiver does not "know" what was transmitted and must look for proper parities to determine if the code is correct. Designate any error that has occurred in transmission if even parity is used.

Solution First, make a bit position table, as indicated in Table 2–14.

▼ TABLE 2–14

BIT DESIGNATION	P_1	P_2	D_1	P_3	D_2	D_3	D_4
BIT POSITION	1	2	3	4	5	6	7
BINARY POSITION NUMBER	001	010	011	100	101	110	111
Received code	0	0	1	0	0	0	1

First parity check:
   Bit P_1 checks positions 1, 3, 5, and 7.
   There are two 1s in this group.
   Parity check is good. ⟶ 0 (LSB)

Second parity check:
   Bit P_2 checks positions 2, 3, 6, and 7.
   There are two 1s in this group.
   Parity check is good. ⟶ 0

Third parity check:
   Bit P_3 checks positions 4, 5, 6, and 7.
   There is one 1 in this group.
   Parity check is bad. ⟶ 1 (MSB)

Result:
   The error position code is 100 (binary four). This says that the bit in position 4 is in error. It is a 0 and should be a 1. The corrected code is 0011001, which agrees with the transmitted code.

Related Problem Repeat the process illustrated in the example if the received code is 0111001.

EXAMPLE 2–44

The code 101101010 is received. Correct any errors. There are four parity bits, and odd parity is used.

Solution   First, make a bit position table like Table 2–15.

▼ **TABLE 2–15**

BIT DESIGNATION	P_1	P_2	D_1	P_3	D_2	D_3	D_4	P_4	D_5
BIT POSITION	1	2	3	4	5	6	7	8	9
BINARY POSITION NUMBER	0001	0010	0011	0100	0101	0110	0111	1000	1001
Received code	1	0	1	1	0	1	0	1	0

First parity check:
Bit P_1 checks positions 1, 3, 5, 7, and 9.
There are two 1s in this group.
Parity check is bad. ⟶ 1 (LSB)

Second parity check:
Bit P_2 checks positions 2, 3, 6, and 7.
There are two 1s in this group.
Parity check is bad. ⟶ 1

Third parity check:
Bit P_3 checks positions 4, 5, 6, and 7.
There are two 1s in this group.
Parity check is bad. ⟶ 1

Fourth parity check:
Bit P_4 checks positions 8 and 9.
There is one 1 in this group.
Parity check is good. ⟶ 0 (MSB)

Result:
The error position code is 0111 (binary seven). This says that the bit in position 7 is in error. The corrected code is therefore 101101110.

Related Problem   The code 101111001 is received. Correct any error if odd parity is used.

**SECTION 2–12
REVIEW**

1. Which odd-parity code is in error?
   (a) 1011   (b) 1110   (c) 0101   (d) 1000
2. Which even-parity code is in error?
   (a) 11000110   (b) 00101000   (c) 10101010   (d) 11111011
3. Add an even parity bit to the end of each of the following codes.
   (a) 1010100   (b) 0100000   (c) 1110111   (d) 1000110
4. How many parity bits are required for data bits 11010 using the Hamming code?
5. Create the Hamming code for the data bits 0011 using even parity.

SUMMARY

- A binary number is a weighted number in which the weight of each whole number digit is a positive power of two and the weight of each fractional digit is a negative power of two. The whole number weights increase from right to left—from least significant digit to most significant.

- A binary number can be converted to a decimal number by summing the decimal values of the weights of all the 1s in the binary number.

- A decimal whole number can be converted to binary by using the sum-of-weights or the repeated division-by-2 method.

- A decimal fraction can be converted to binary by using the sum-of-weights or the repeated multiplication-by-2 method.

- The basic rules for binary addition are as follows:

  $0 + 0 = 0$
  $0 + 1 = 1$
  $1 + 0 = 1$
  $1 + 1 = 10$

- The basic rules for binary subtraction are as follows:

  $0 - 0 = 0$
  $1 - 1 = 0$
  $1 - 0 = 1$
  $10 - 1 = 1$

- The 1's complement of a binary number is derived by changing 1s to 0s and 0s to 1s.

- The 2's complement of a binary number can be derived by adding 1 to the 1's complement.

- Binary subtraction can be accomplished with addition by using the 1's or 2's complement method.

- A positive binary number is represented by a 0 sign bit.

- A negative binary number is represented by a 1 sign bit.

- For arithmetic operations, negative binary numbers are represented in 1's complement or 2's complement form.

- In an addition operation, an overflow is possible when both numbers are positive or when both numbers are negative. An incorrect sign bit in the sum indicates the occurrence of an overflow.

- The hexadecimal number system consists of 16 digits and characters, 0 through 9 followed by A through F.

- One hexadecimal digit represents a 4-bit binary number, and its primary usefulness is in simplifying bit patterns and making them easier to read.

- A decimal number can be converted to hexadecimal by the repeated division-by-16 method.

- The octal number system consists of eight digits, 0 through 7.

- A decimal number can be converted to octal by using the repeated division-by-8 method.

- Octal-to-binary conversion is accomplished by simply replacing each octal digit with its 3-bit binary equivalent. The process is reversed for binary-to-octal conversion.

- A decimal number is converted to BCD by replacing each decimal digit with the appropriate 4-bit binary code.

- The ASCII is a 7-bit alphanumeric code that is widely used in computer systems for input and output of information.

- A parity bit is used to detect an error in a code.

- The Hamming code provides for single-error detection and correction.

KEY TERMS

Key terms and other bold terms in the chapter are defined in the end-of-book glossary.

Alphanumeric Consisting of numerals, letters, and other characters.

ASCII American Standard Code for Information Interchange; the most widely used alphanumeric code.

BCD Binary coded decimal; a digital code in which each of the decimal digits, 0 through 9, is represented by a group of four bits.

Byte A group of eight bits.

Floating-point number A number representation based on scientific notation in which the number consists of an exponent and a mantissa.

Hamming code A type of error correction code.

Hexadecimal Describes a number system with a base of 16.

LSB Least significant bit; the right-most bit in a binary whole number or code.

MSB Most significant bit; the left-most bit in a binary whole number or code.

Octal Describes a number system with a base of eight.

Parity In relation to binary codes, the condition of evenness or oddness of the number of 1s in a code group.

SELF-TEST

Answers are at the end of the chapter.

1. $2 \times 10^1 + 8 \times 10^0$ is equal to
    (a) 10   (b) 280   (c) 2.8   (d) 28

2. The binary number 1101 is equal to the decimal number
    (a) 13   (b) 49   (c) 11   (d) 3

3. The binary number 11011101 is equal to the decimal number
    (a) 121   (b) 221   (c) 441   (d) 256

4. The decimal number 17 is equal to the binary number
    (a) 10010   (b) 11000   (c) 10001   (d) 01001

5. The decimal number 175 is equal to the binary number
    (a) 11001111   (b) 10101110   (c) 10101111   (d) 11101111

6. The sum of 11010 + 01111 equals
    (a) 101001   (b) 101010   (c) 110101   (d) 101000

7. The difference of 110 − 010 equals
    (a) 001   (b) 010   (c) 101   (d) 100

8. The 1's complement of 10111001 is
    (a) 01000111   (b) 01000110   (c) 11000110   (d) 10101010

9. The 2's complement of 11001000 is
    (a) 00110111   (b) 00110001   (c) 01001000   (d) 00111000

10. The decimal number +122 is expressed in the 2's complement form as
    (a) 01111010   (b) 11111010   (c) 01000101   (d) 10000101

11. The decimal number −34 is expressed in the 2's complement form as
    (a) 01011110   (b) 10100010   (c) 11011110   (d) 01011101

12. A single-precision floating-point binary number has a total of
    (a) 8 bits   (b) 16 bits   (c) 24 bits   (d) 32 bits

13. In the 2's complement form, the binary number 10010011 is equal to the decimal number
    (a) −19   (b) +109   (c) +91   (d) −109

14. The binary number 101100111001010100001 can be written in octal as
    (a) 5471230_8 (b) 5471241_8 (c) 2634521_8 (d) 23162501_8

15. The binary number 1000110101000110111 can be written in hexadecimal as
    (a) $AD467_{16}$ (b) $8C46F_{16}$ (c) $8D46F_{16}$ (d) $AE46F_{16}$

16. The binary number for $F7A9_{16}$ is
    (a) 1111011110101001   (b) 1110111110101001
    (c) 1111111010110001   (d) 1111011010101001

17. The BCD number for decimal 473 is
    (a) 111011010   (b) 110001110011   (c) 010001110011   (d) 010011110011

18. Refer to Table 2–7. The command STOP in ASCII is
    (a) 1010011101010010011111010000   (b) 1010010100110010011101010000
    (c) 1001010110110110011101010001   (d) 1010011101010010011101100100

19. The code that has an even-parity error is
    (a) 1010011   (b) 1101000   (c) 1001000   (d) 1110111

PROBLEMS

Answers to odd-numbered problems are at the end of the book.

SECTION 2–1 Decimal Numbers

1. What is the weight of the digit 6 in each of the following decimal numbers?
   (a) 1386   (b) 54,692   (c) 671,920

2. Express each of the following decimal numbers as a power of ten:
   (a) 10   (b) 100   (c) 10,000   (d) 1,000,000

3. Give the value of each digit in the following decimal numbers:
   (a) 471   (b) 9356   (c) 125,000

4. How high can you count with four decimal digits?

SECTION 2–2 Binary Numbers

5. Convert the following binary numbers to decimal:
   (a) 11   (b) 100   (c) 111   (d) 1000
   (e) 1001   (f) 1100   (g) 1011   (h) 1111

6. Convert the following binary numbers to decimal:
   (a) 1110   (b) 1010   (c) 11100   (d) 10000
   (e) 10101   (f) 11101   (g) 10111   (h) 11111

7. Convert each binary number to decimal:
   (a) 110011.11   (b) 101010.01   (c) 1000001.111
   (d) 1111000.101   (e) 1011100.10101   (f) 1110001.0001
   (g) 1011010.1010   (h) 1111111.11111

8. What is the highest decimal number that can be represented by each of the following numbers of binary digits (bits)?
   (a) two   (b) three   (c) four   (d) five   (e) six
   (f) seven   (g) eight   (h) nine   (i) ten   (j) eleven

9. How many bits are required to represent the following decimal numbers?
   (a) 17   (b) 35   (c) 49   (d) 68
   (e) 81   (f) 114   (g) 132   (h) 205

10. Generate the binary sequence for each decimal sequence:

    (a) 0 through 7   (b) 8 through 15   (c) 16 through 31

    (d) 32 through 63   (e) 64 through 75

SECTION 2–3   **Decimal-to-Binary Conversion**

11. Convert each decimal number to binary by using the sum-of-weights method:

    (a) 10   (b) 17   (c) 24   (d) 48

    (e) 61   (f) 93   (g) 125   (h) 186

12. Convert each decimal fraction to binary using the sum-of-weights method:

    (a) 0.32   (b) 0.246   (c) 0.0981

13. Convert each decimal number to binary using repeated division by 2:

    (a) 15   (b) 21   (c) 28   (d) 34

    (e) 40   (f) 59   (g) 65   (h) 73

14. Convert each decimal fraction to binary using repeated multiplication by 2:

    (a) 0.98   (b) 0.347   (c) 0.9028

SECTION 2–4   **Binary Arithmetic**

15. Add the binary numbers:

    (a) $11 + 01$ (b) $10 + 10$ (c) $101 + 11$

    (d) $111 + 110$ (e) $1001 + 101$ (f) $1101 + 1011$

16. Use direct subtraction on the following binary numbers:

    (a) $11 - 1$ (b) $101 - 100$ (c) $110 - 101$

    (d) $1110 - 11$ (e) $1100 - 1001$ (f) $11010 - 10111$

17. Perform the following binary multiplications:

    (a) 11×11 (b) 100×10 (c) 111×101

    (d) 1001×110 (e) 1101×1101 (f) 1110×1101

18. Divide the binary numbers as indicated:

    (a) $100 \div 10$ (b) $1001 \div 11$ (c) $1100 \div 100$

SECTION 2–5   **1's and 2's Complements of Binary Numbers**

19. Determine the 1's complement of each binary number:

    (a) 101   (b) 110   (c) 1010

    (d) 11010111   (e) 1110101   (f) 00001

20. Determine the 2's complement of each binary number using either method:

    (a) 10   (b) 111   (c) 1001   (d) 1101

    (e) 11100   (f) 10011   (g) 10110000   (h) 00111101

SECTION 2–6   **Signed Numbers**

21. Express each decimal number in binary as an 8-bit sign-magnitude number:

    (a) $+29$ (b) -85 (c) $+100$ (d) -123

22. Express each decimal number as an 8-bit number in the 1's complement form:

    (a) -34 (b) $+57$ (c) -99 (d) $+115$

23. Express each decimal number as an 8-bit number in the 2's complement form:

    (a) $+12$ (b) -68 (c) $+101$ (d) -125

24. Determine the decimal value of each signed binary number in the sign-magnitude form:

    (a) 10011001   (b) 01110100   (c) 10111111

25. Determine the decimal value of each signed binary number in the 1's complement form:

    (a) 10011001   **(b)** 01110100   **(c)** 10111111

26. Determine the decimal value of each signed binary number in the 2's complement form:

    (a) 10011001   **(b)** 01110100   **(c)** 10111111

27. Express each of the following sign-magnitude binary numbers in single-precision floating-point format:

    (a) 0111110000101011   **(b)** 100110000011000

28. Determine the values of the following single-precision floating-point numbers:

    (a) 1 10000001 01001001110001000000000

    (b) 0 11001100 10000111110100100000000

SECTION 2–7   **Arithmetic Operations with Signed Numbers**

29. Convert each pair of decimal numbers to binary and add using the 2's complement form:

    (a) 33 and 15   **(b)** 56 and -27 **(c)** -46 and 25   **(d)** -110 and -84

30. Perform each addition in the 2's complement form:

    (a) 00010110 + 00110011   **(b)** 01110000 + 10101111

31. Perform each addition in the 2's complement form:

    (a) 10001100 + 00111001   **(b)** 11011001 + 11100111

32. Perform each subtraction in the 2's complement form:

    (a) 00110011 − 00010000   **(b)** 01100101 − 11101000

33. Multiply 01101010 by 11110001 in the 2's complement form.

34. Divide 01000100 by 00011001 in the 2's complement form.

SECTION 2–8   **Hexadecimal Numbers**

35. Convert each hexadecimal number to binary:

    (a) 38_{16} **(b)** 59_{16} **(c)** $A14_{16}$ **(d)** $5C8_{16}$

    (e) 4100_{16} **(f)** $FB17_{16}$ **(g)** $8A9D_{16}$

36. Convert each binary number to hexadecimal:

    (a) 1110   **(b)** 10   **(c)** 10111

    (d) 10100110   **(e)** 1111110000   **(f)** 100110000010

37. Convert each hexadecimal number to decimal:

    (a) 23_{16} **(b)** 92_{16} **(c)** $1A_{16}$ **(d)** $8D_{16}$

    (e) $F3_{16}$ **(f)** EB_{16} **(g)** $5C2_{16}$ **(h)** 700_{16}

38. Convert each decimal number to hexadecimal:

    (a) 8   **(b)** 14   **(c)** 33   **(d)** 52

    (e) 284   **(f)** 2890   **(g)** 4019   **(h)** 6500

39. Perform the following additions:

    (a) $37_{16} + 29_{16}$ **(b)** $A0_{16} + 6B_{16}$ **(c)** $FF_{16} + BB_{16}$

40. Perform the following subtractions:

    (a) $51_{16} - 40_{16}$ **(b)** $C8_{16} - 3A_{16}$ **(c)** $FD_{16} - 88_{16}$

SECTION 2–9   **Octal Numbers**

41. Convert each octal number to decimal:

    (a) 12_8 **(b)** 27_8 **(c)** 56_8 **(d)** 64_8 **(e)** 103_8

    (f) 557_8 **(g)** 163_8 **(h)** 1024_8 **(i)** 7765_8

42. Convert each decimal number to octal by repeated division by 8:

    (a) 15   **(b)** 27   **(c)** 46   **(d)** 70

    (e) 100   **(f)** 142   **(g)** 219   **(h)** 435

43. Convert each octal number to binary:

  (a) 13_8 **(b)** 57_8 **(c)** 101_8 **(d)** 321_8 **(e)** 540_8

  (f) 4653_8 **(g)** 13271_8 **(h)** 45600_8 **(i)** 100213_8

44. Convert each binary number to octal:

  (a) 111   **(b)** 10   **(c)** 110111

  (d) 101010   **(e)** 1100   **(f)** 1011110

  (g) 101100011001   **(h)** 10110000011   **(i)** 111111101111000

SECTION 2–10   Binary Coded Decimal (BCD)

45. Convert each of the following decimal numbers to 8421 BCD:

  (a) 10   **(b)** 13   **(c)** 18   **(d)** 21   **(e)** 25   **(f)** 36

  (g) 44   **(h)** 57   **(i)** 69   **(j)** 98   **(k)** 125   **(l)** 156

46. Convert each of the decimal numbers in Problem 45 to straight binary, and compare the number of bits required with that required for BCD.

47. Convert the following decimal numbers to BCD:

  (a) 104   **(b)** 128   **(c)** 132   **(d)** 150   **(e)** 186

  (f) 210   **(g)** 359   **(h)** 547   **(i)** 1051

48. Convert each of the BCD numbers to decimal:

  (a) 0001   **(b)** 0110   **(c)** 1001

  (d) 00011000   **(e)** 00011001   **(f)** 00110010

  (g) 01000101   **(h)** 10011000   **(i)** 100001110000

49. Convert each of the BCD numbers to decimal:

  (a) 10000000   **(b)** 001000110111

  (c) 001101000110   **(d)** 010000100001

  (e) 011101010100   **(f)** 100000000000

  (g) 100101111000   **(h)** 0001011010000011

  (i) 1001000000011000   **(j)** 0110011001100111

50. Add the following BCD numbers:

  (a) 0010 + 0001   **(b)** 0101 + 0011

  (c) 0111 + 0010   **(d)** 1000 + 0001

  (e) 00011000 + 00010001   **(f)** 01100100 + 00110011

  (g) 01000000 + 01000111   **(h)** 10000101 + 00010011

51. Add the following BCD numbers:

  (a) 1000 + 0110   **(b)** 0111 + 0101

  (c) 1001 + 1000   **(d)** 1001 + 0111

  (e) 00100101 + 00100111   **(f)** 01010001 + 01011000

  (g) 10011000 + 10010111   **(h)** 010101100001 + 011100001000

52. Convert each pair of decimal numbers to BCD, and add as indicated:

  (a) 4 + 3   **(b)** 5 + 2   **(c)** 6 + 4   **(d)** 17 + 12

  (e) 28 + 23   **(f)** 65 + 58   **(g)** 113 + 101   **(h)** 295 + 157

SECTION 2–11   Digital Codes

53. In a certain application a 4-bit binary sequence cycles from 1111 to 0000 periodically. There are four bit changes, and because of circuit delays, these changes may not occur at the same instant. For example, if the LSB changes first, the number will appear as 1110 during the transition from 1111 to 0000 and may be misinterpreted by the system. Illustrate how the Gray code avoids this problem.

54. Convert each binary number to Gray code:

    (a) 11011   **(b)** 1001010   **(c)** 1111011101110

55. Convert each Gray code to binary:

    (a) 1010   **(b)** 00010   **(c)** 11000010001

56. Convert each of the following decimal numbers to ASCII. Refer to Table 2–7.

    (a) 1   **(b)** 3   **(c)** 6   **(d)** 10   **(e)** 18

    (f) 29   **(g)** 56   **(h)** 75   **(i)** 107

57. Determine each ASCII character. Refer to Table 2–7.

    (a) 0011000   **(b)** 1001010   **(c)** 0111101

    (d) 0100011   **(e)** 0111110   **(f)** 1000010

58. Decode the following ASCII coded message:

    1001000 1100101 1101100 1101100 1101111 0101110
    0100000 1001000 1101111 1110111 0100000 1100001
    1110010 1100101 0100000 1111001 1101111 1110101
    0111111

59. Write the message in Problem 58 in hexadecimal.

60. Convert the following computer program statement to ASCII:

    30 INPUT A, B

SECTION 2–12   Error Detection and Correction Codes

61. Determine which of the following even parity codes are in error:

    (a) 100110010   **(b)** 011101010   **(c)** 10111111010001010

62. Determine which of the following odd parity codes are in error:

    (a) 11110110   **(b)** 00110001   **(c)** 01010101010101010

63. Attach the proper even parity bit to each of the following bytes of data:

    (a) 10100100   **(b)** 00001001   **(c)** 11111110

64. Determine the even-parity Hamming code for the data bits 1100.

65. Determine the odd-parity Hamming code for the data bits 11001.

66. Correct any error in each of the following Hamming codes with even parity.

    (a) 1110100   **(b)** 1000111

67. Correct any error in each of the following Hamming codes with odd parity.

    (a) 110100011   **(b)** 100001101

ANSWERS

SECTION REVIEWS

SECTION 2–1   Decimal Numbers

1. **(a)** 1370: 10   **(b)** 6725: 100   **(c)** 7051: 1000   **(d)** 58.72: 0.1

2. **(a)** $51 = (5 \times 10) + (1 \times 1)$ **(b)** $137 = (1 \times 100) + (3 \times 10) + (7 \times 1)$ **(c)** $1492 = (1 \times 1000) + (4 \times 100) + (9 \times 10) + (2 \times 1)$ **(d)** $106.58 = (1 \times 100) + (0 \times 10) + (6 \times 1) + (5 \times 0.1) + (8 \times 0.01)$

SECTION 2–2   Binary Numbers

1. $2^8 - 1 = 255$

2. Weight is 16.

3. $10111101.011 = 189.375$

SECTION 2–3   **Decimal-to-Binary Conversion**

   1. **(a)** $23 = 10111$ **(b)** $57 = 111001$ **(c)** $45.5 = 101101.1$

   2. **(a)** $14 = 1110$ **(b)** $21 = 10101$ **(c)** $0.375 = 0.011$

SECTION 2–4   **Binary Arithmetic**

   1. **(a)** $1101 + 1010 = 10111$ **(b)** $10111 + 01101 = 100100$

   2. **(a)** $1101 - 0100 = 1001$ **(b)** $1001 - 0111 = 0010$

   3. **(a)** $110 \times 111 = 101010$ **(b)** $1100 \div 011 = 100$

SECTION 2–5   **1's and 2's Complements of Binary Numbers**

   1. **(a)** 1's comp of $00011010 = 11100101$ **(b)** 1's comp of $11110111 = 00001000$

      (c) 1's comp of $10001101 = 01110010$

   2. **(a)** 2's comp of $00010110 = 11101010$ **(b)** 2's comp of $11111100 = 00000100$

      (c) 2's comp of $10010001 = 01101111$

SECTION 2–6   **Signed Numbers**

   1. Sign-magnitude: $+9 = 00001001$

   2. 1's comp: $-33 = 11011110$

   3. 2's comp: $-46 = 11010010$

   4. Sign bit, exponent, and mantissa

SECTION 2–7   **Arithmetic Operations with Signed Numbers**

   1. Cases of addition: positive number is larger, negative number is larger, both are positive, both are negative

   2. $00100001 + 10111100 = 11011101$

   3. $01110111 - 00110010 = 01000101$

   4. Sign of product is positive.

   5. $00000101 \times 01111111 = 01001111011$

   6. Sign of quotient is negative.

   7. $00110000 \div 00001100 = 00000100$

SECTION 2–8   **Hexadecimal Numbers**

   1. **(a)** $10110011 = B3_{16}$ **(b)** $110011101000 = CE8_{16}$

   2. **(a)** $57_{16} = 01010111$ **(b)** $3A5_{16} = 001110100101$

      (c) $F8OB_{16} = 1111100000001011$

   3. $9B30_{16} = 39,728_{10}$

   4. $573_{10} = 23D_{16}$

   5. **(a)** $18_{16} + 34_{16} = 4C_{16}$ **(b)** $3F_{16} + 2A_{16} = 69_{16}$

   6. **(a)** $75_{16} - 21_{16} = 54_{16}$ **(b)** $94_{16} - 5C_{16} = 38_{16}$

SECTION 2–9   **Octal Numbers**

   1. **(a)** $73_8 = 59_{10}$ **(b)** $125_8 = 85_{10}$

   2. **(a)** $98_{10} = 142_8$ **(b)** $163_{10} = 243_8$

   3. **(a)** $46_8 = 100110$ **(b)** $723_8 = 111010011$ **(c)** $5624_8 = 101110010100$

   4. **(a)** $110101111 = 657_8$ **(b)** $1001100010 = 1142_8$ **(c)** $10111111001 = 2771_8$

SECTION 2–10   **Binary Coded Decimal (BCD)**

   1. **(a)** 0010: 2   **(b)** 1000: 8   **(c)** 0001: 1   **(d)** 0100: 4

2. (a) $6_{10} = 0110$ (b) $15_{10} = 00010101$ (c) $273_{10} = 001001110011$

(d) $849_{10} = 100001001001$

3. (a) $10001001 = 89_{10}$ (b) $001001111000 = 278_{10}$ (c) $000101010111 = 157_{10}$

4. A 4-bit sum is invalid when it is greater than 9_{10}.

SECTION 2–11 Digital Codes

1. (a) $1100_2 = 1010$ Gray (b) $1010_2 = 1111$ Gray (c) $11010_2 = 10111$ Gray

2. (a) 1000 Gray $= 1111_2$ (b) 1010 Gray $= 1100_2$ (c) 11101 Gray $= 10110_2$

3. (a) K: $1001011 \rightarrow 4B_{16}$ (b) r: $1110010 \rightarrow 72_{16}$

(c) \$: $0100100 \rightarrow 24_{16}$ (d) +: $0101011 \rightarrow 2B_{16}$

SECTION 2–12 Error Detection and Correction Codes

1. (c) 0101 has an error.

2. (d) 11111011 has an error.

3. (a) 10101001 (b) 01000001 (c) 11101110 (d) 10001101

4. Four parity bits

5. 1 0 0 0 0 1 1 (parity bits are red)

RELATED PROBLEMS FOR EXAMPLES

2–1 9 has a value of 900, 3 has a value of 30, 9 has a value of 9.

2–2 6 has a value of 60, 7 has a value of 7, 9 has a value of 9/10 (0.9), 2 has a value of 2/100 (0.02), 4 has a value of 4/1000 (0.004).

2–3 $10010001 = 128 + 16 + 1 = 145$ **2–4** $10.111 = 2 + 0.5 + 0.25 + 0.125 = 2.875$

2–5 $125 = 64 + 32 + 16 + 8 + 4 + 1 = 1111101$ **2–6** $39 = 100111$

2–7 $1111 + 1100 = 11011$ **2–8** $111 - 100 = 011$ **2–9** $110 - 101 = 001$

2–10 $1101 \times 1010 = 10000010$ **2–11** $1100 \div 100 = 11$ **2–12** 00110101

2–13 01000000 **2–14** See Table 2–16. **2–15** $01110111 = +119_{10}$

▶ TABLE 2–16

	SIGN-MAGNITUDE	1'S COMP	2'S COMP
+19	00010011	00010011	00010011
−19	10010011	11101100	11101101

2–16 $11101011 = -20_{10}$ **2–17** $11010111 = -41_{10}$

2–18 11000010001010011000000000 **2–19** 01010101 **2–20** 00010001

2–21 1001000110 **2–22** $(83)(-59) = -4897$ (10110011011111 in 2's comp)

2–23 $100 \div 25 = 4$ (0100) **2–24** $4F79C_{16}$ **2–25** 0110101111010011_2

2–26 $6BD_{16} = 011010111101 = 2^{10} + 2^9 + 2^7 + 2^5 + 2^4 + 2^3 + 2^2 + 2^0$

$= 1024 + 512 + 128 + 32 + 16 + 8 + 4 + 1 = 1725_{10}$

2–27 $60A_{16} = (6 \times 256) + (0 \times 16) + (10 \times 1) = 1546_{10}$

2–28 $2591_{10} = A1F_{16}$ **2–29** $4C_{16} + 3A_{16} = 86_{16}$

2–30 $BCD_{16} - 173_{16} = A5A_{16}$

2–31 (a) $001011_2 = 11_{10} = 13_8$ (b) $010101_2 = 21_{10} = 25_8$

(c) $001100000_2 = 96_{10} = 140_8$ (d) $111101010110_2 = 3926_{10} = 7526_8$

2–32 1250762_8 **2–33** 1001011001110011 **2–34** $82,276_{10}$

2–35 1001100101101000 **2–36** 10000010 **2–37** (a) 111011 (Gray) (b) 111010_2

2–38 The sequence of codes for 80 INPUT Y is $38_{16}30_{16}20_{16}49_{16}4E_{16}50_{16}55_{16}54_{16}20_{16}59_{16}$

2–39 01001011 **2–40** Yes **2–41** *1110000* **2–42** *001010001*

2–43 The bit in position 010 (2) is in error. Correct to 0011001.

2–44 The bit in position 0010 (2) is in error. Correct to 111111000.

SELF-TEST

1. (d)   **2.** (a)   **3.** (b)   **4.** (c)   **5.** (c)   **6.** (a)   **7.** (d)   **8.** (b)

9. (d)   **10.** (a)   **11.** (c)   **12.** (d)   **13.** (d)   **14.** (b)   **15.** (c)   **16.** (a)

17. (c)   **18.** (a)   **19.** (b)

3

LOGIC GATES

CHAPTER OBJECTIVES

■ Describe the operation of the inverter, the AND gate, and the OR gate

■ Describe the operation of the NAND gate and the NOR gate

■ Express the operation of NOT, AND, OR, NAND, and NOR gates with Boolean algebra

■ Describe the operation of the exclusive-OR and exclusive-NOR gates

■ Recognize and use both the distinctive shape logic gate symbols and the rectangular outline logic gate symbols of ANSI/IEEE Standard 91-1984

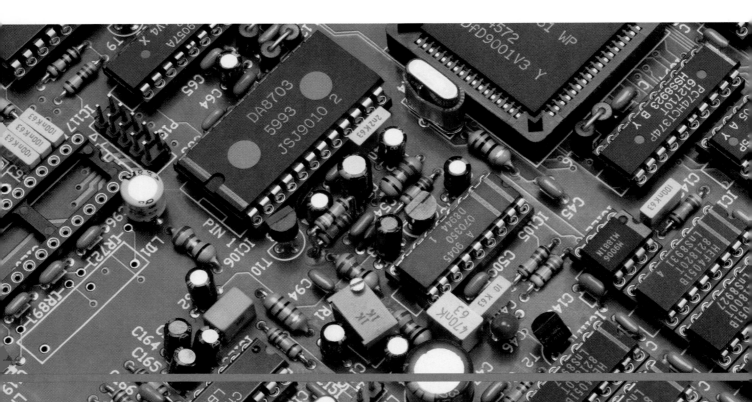

- Construct timing diagrams showing the proper time relationships of inputs and outputs for the various logic gates

- Discuss the basic concepts of programmable logic

- Make basic comparisons between the major IC technologies—CMOS and TTL

- Explain how the different series within the CMOS and TTL families differ from each other

- Define *propagation delay time, power dissipation, speed-power product,* and *fan-out* in relation to logic gates

- List specific fixed-function integrated circuit devices that contain the various logic gates

- Use each logic gate in simple applications

- Troubleshoot logic gates for opens and shorts by using the oscilloscope

KEY TERMS

Inverter	Fuse
Truth table	Antifuse
Timing diagram	EPROM
Boolean algebra	EEPROM
Complement	SRAM
AND gate	Target device
Enable	JTAG
OR gate	CMOS
NAND gate	TTL
NOR gate	Propagation delay time
Exclusive-OR gate	Fan-out
Exclusive-NOR gate	Unit load
AND array	

INTRODUCTION

The emphasis in this chapter is on the operation, application, and troubleshooting of logic gates. The relationship of input and output waveforms of a gate using timing diagrams is thoroughly covered.

Logic symbols used to represent the logic gates are in accordance with ANSI/IEEE Standard 91-1984. This standard has been adopted by private industry and the military for use in internal documentation as well as published literature.

Both programmable logic and fixed-function logic are discussed in this chapter. Because integrated circuits (ICs) are used in all applications, the logic function of a device is generally of greater importance to the technician or technologist than the details of the component-level circuit operation within the IC package. Therefore, detailed coverage of the devices at the component level can be treated as an optional topic. For those who need it and have the time, a thorough coverage of digital integrated circuit technologies is available in Chapter 14, portions of which may be referenced at appropriate points throughout the text. *Suggestion:* Review Section 1–3 before you start this chapter.

FIXED-FUNCTION LOGIC DEVICES

(CMOS AND TTL SERIES)

74XX00	74XX02	74XX04
74XX08	74XX10	74XX11
74XX20	74XX21	74XX27
74XX30	74XX32	74XX86
74XX266		

WWW. **VISIT THE COMPANION WEBSITE**
Study aids for this chapter are available at
http://www.prenhall.com/floyd

3–1   THE INVERTER

The inverter (NOT circuit) performs the operation called *inversion* or *complementation*. The inverter changes one logic level to the opposite level. In terms of bits, it changes a 1 to a 0 and a 0 to a 1.

After completing this section, you should be able to

■ Identify negation and polarity indicators ■ Identify an inverter by either its distinctive shape symbol or its rectangular outline symbol ■ Produce the truth table for an inverter ■ Describe the logical operation of an inverter

Standard logic symbols for the **inverter** are shown in Figure 3–1. Part (a) shows the *distinctive shape* symbols, and part (b) shows the *rectangular outline* symbols. In this textbook, distinctive shape symbols are generally used; however, the rectangular outline symbols are found in many industry publications, and you should become familiar with them as well. (Logic symbols are in accordance with **ANSI/IEEE** Standard 91-1984.)

▶ FIGURE 3–1

Standard logic symbols for the inverter (ANSI/IEEE Std. 91-1984).

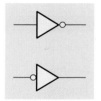
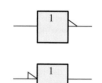
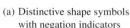

(a) Distinctive shape symbols with negation indicators

(b) Rectangular outline symbols with polarity indicators

The Negation and Polarity Indicators

The negation indicator is a "bubble" (○) that indicates **inversion** or *complementation* when it appears on the input or output of any logic element, as shown in Figure 3–1(a) for the inverter. Generally, inputs are on the left of a logic symbol and the output is on the right. When appearing on the input, the bubble means that a 0 is the active or *asserted* input state, and the input is called an active-LOW input. When appearing on the output, the bubble means that a 0 is the active or asserted output state, and the output is called an active-LOW output. The absence of a bubble on the input or output means that a 1 is the active or asserted state, and in this case, the input or output is called active-HIGH.

The polarity or level indicator is a "triangle" (◣) that indicates inversion when it appears on the input or output of a logic element, as shown in Figure 3–1(b). When appearing on the input, it means that a LOW level is the active or asserted input state. When appearing on the output, it means that a LOW level is the active or asserted output state.

Either indicator (bubble or triangle) can be used both on distinctive shape symbols and on rectangular outline symbols. Figure 3–1(a) indicates the principal inverter symbols used in this text. Note that a change in the placement of the negation or polarity indicator does not imply a change in the way an inverter operates.

▼ TABLE 3–1

Inverter truth table.

INPUT	OUTPUT
LOW (0)	HIGH (1)
HIGH (1)	LOW (0)

Inverter Truth Table

When a HIGH level is applied to an inverter input, a LOW level will appear on its output. When a LOW level is applied to its input, a HIGH will appear on its output. This operation is summarized in Table 3–1, which shows the output for each possible input in terms of levels and corresponding bits. A table such as this is called a **truth table.**

Inverter Operation

Figure 3–2 shows the output of an inverter for a pulse input, where t_1 and t_2 indicate the corresponding points on the input and output pulse waveforms.

When the input is LOW, the output is HIGH; when the input is HIGH, the output is LOW, thereby producing an inverted output pulse.

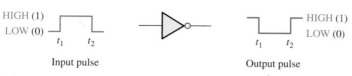

▲ **FIGURE 3–2**

Inverter operation with a pulse input. Open file F03-02 to verify inverter operation.

Timing Diagrams

Recall from Chapter 1 that a **timing diagram** is basically a graph that accurately displays the relationship of two or more waveforms with respect to each other on a time basis. For example, the time relationship of the output pulse to the input pulse in Figure 3–2 can be shown with a simple timing diagram by aligning the two pulses so that the occurrences of the pulse edges appear in the proper time relationship. The rising edge of the input pulse and the falling edge of the output pulse occur at the same time (ideally). Similarly, the falling edge of the input pulse and the rising edge of the output pulse occur at the same time (ideally). This timing relationship is shown in Figure 3–3. Timing diagrams are especially useful for illustrating the time relationship of digital waveforms with multiple pulses.

A timing diagram shows how two or more waveforms relate in time.

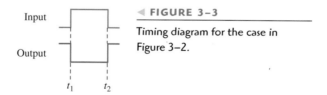

◄ **FIGURE 3–3**

Timing diagram for the case in Figure 3–2.

EXAMPLE 3–1

A waveform is applied to an inverter in Figure 3–4. Determine the output waveform corresponding to the input and show the timing diagram. According to the placement of the bubble, what is the active output state?

▶ **FIGURE 3–4**

Solution The output waveform is exactly opposite to the input (inverted), as shown in Figure 3–5, which is the basic timing diagram. The active or asserted output state is **0.**

▶ FIGURE 3–5

Input 1 / 0

Output 1 / 0

*Related Problem** If the inverter is shown with the negative indicator (bubble) on the input instead of the output, how is the timing diagram affected?

*Answers are at the end of the chapter.

Logic Expression for an Inverter

Boolean algebra uses variables and operators to describe a logic circuit.

In **Boolean algebra,** which is the mathematics of logic circuits and will be covered thoroughly in Chapter 4, a variable is designated by a letter. The **complement** of a variable is designated by a bar over the letter. A variable can take on a value of either 1 or 0. If a given variable is 1, its complement is 0 and vice versa.

The operation of an inverter (NOT circuit) can be expressed as follows: If the input variable is called A and the output variable is called X, then

$$X = \overline{A}$$

This expression states that the output is the complement of the input, so if $A = 0$, then $X = 1$, and if $A = 1$, then $X = 0$. Figure 3–6 illustrates this. The complemented variable $\overline{A}$ can be read as "A bar" or "not A."

▶ FIGURE 3–6

The inverter complements an input variable.

$A \longrightarrow\!\!\triangleright\!\!\circ\!\!\longrightarrow X = \overline{A}$

An Application

Figure 3–7 shows a circuit for producing the 1's complement of an 8-bit binary number. The bits of the binary number are applied to the inverter inputs and the 1's complement of the number appears on the outputs.

▶ FIGURE 3–7

Example of a 1's complement circuit using inverters.

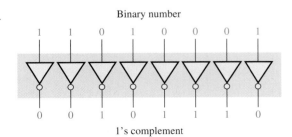

Binary number

1 1 0 1 0 0 0 1

0 0 1 0 1 1 1 0

1's complement

1. When a 1 is on the input of an inverter, what is the output?
2. An active HIGH pulse (HIGH level when asserted, LOW level when not) is required on an inverter input.
   (a) Draw the appropriate logic symbol, using the distinctive shape and the negation indicator, for the inverter in this application.
   (b) Describe the output when a positive-going pulse is applied to the input of an inverter.

3–2 THE AND GATE

The AND gate is one of the basic gates that can be combined to form any logic function. An AND gate can have two or more inputs and performs what is known as logical multiplication.

After completing this section, you should be able to

■ Identify an AND gate by its distinctive shape symbol or by its rectangular outline symbol ■ Describe the operation of an AND gate ■ Generate the truth table for an AND gate with any number of inputs ■ Produce a timing diagram for an AND gate with any specified input waveforms ■ Write the logic expression for an AND gate with any number of inputs ■ Discuss examples of AND gate applications

The term *gate* is used to describe a circuit that performs a basic logic operation. The AND gate is composed of two or more inputs and a single output, as indicated by the standard logic symbols shown in Figure 3–8. Inputs are on the left, and the output is on the right in each symbol. Gates with two inputs are shown; however, an AND gate can have any number of inputs greater than one. Although examples of both distinctive shape symbols and rectangular outline symbols are shown, the distinctive shape symbol, shown in part (a), is used predominantly in this book.

COMPUTER NOTE

Logic gates are the building blocks of computers. Most of the functions in a computer, with the exception of certain types of memory, are implemented with logic gates used on a very large scale. For example, a microprocessor, which is the main part of a computer, is made up of hundreds of thousands or even millions of logic gates.

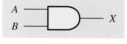
(a) Distinctive shape

(b) Rectangular outline with the AND (&) qualifying symbol

▲ **FIGURE 3–8**

Standard logic symbols for the AND gate showing two inputs (ANSI/IEEE Std. 91-1984).

Operation of an AND Gate

An **AND gate** produces a HIGH output *only* when *all* of the inputs are HIGH. When any of the inputs is LOW, the output is LOW. Therefore, the basic purpose of an AND gate is to determine when certain conditions are simultaneously true, as indicated by HIGH levels on all of its inputs, and to produce a HIGH on its output to indicate that all these conditions are true. The inputs of the 2-input AND gate in Figure 3–8 are labeled A and B, and the output is labeled X. The gate operation can be stated as follows:

An AND gate can have more than two inputs.

For a 2-input AND gate, output X is HIGH only when inputs A and B are HIGH; X is LOW when either A or B is LOW, or when both A and B are LOW.

Figure 3–9 illustrates a 2-input AND gate with all four possibilities of input combinations and the resulting output for each.

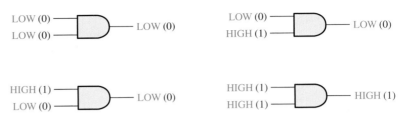

▲ FIGURE 3–9

All possible logic levels for a 2-input AND gate. Open file F03-09 to verify AND gate operation.

AND Gate Truth Table

For an AND gate, all HIGH inputs make a HIGH output.

The logical operation of a gate can be expressed with a truth table that lists all input combinations with the corresponding outputs, as illustrated in Table 3–2 for a 2-input AND gate. The truth table can be expanded to any number of inputs. Although the terms HIGH and LOW tend to give a "physical" sense to the input and output states, the truth table is shown with 1s and 0s; a HIGH is equivalent to a 1 and a LOW is equivalent to a 0 in positive logic. For any AND gate, regardless of the number of inputs, the output is HIGH *only* when *all* inputs are HIGH.

▶ TABLE 3–2

Truth table for a 2-input AND gate.

| INPUTS | | OUTPUT |
A	B	X
0	0	0
0	1	0
1	0	0
1	1	1

1 = HIGH, 0 = LOW

The total number of possible combinations of binary inputs to a gate is determined by the following formula:

Equation 3–1

$$N = 2^n$$

where N is the number of possible input combinations and n is the number of input variables. To illustrate,

For two input variables: $N = 2^2 = 4$ combinations

For three input variables: $N = 2^3 = 8$ combinations

For four input variables: $N = 2^4 = 16$ combinations

You can determine the number of input bit combinations for gates with any number of inputs by using Equation 3–1.

EXAMPLE 3–2

(a) Develop the truth table for a 3-input AND gate.

(b) Determine the total number of possible input combinations for a 4-input AND gate.

Solution **(a)** There are eight possible input combinations ($2^3 = 8$) for a 3-input AND gate. The input side of the truth table (Table 3–3) shows all eight combinations of three bits. The output side is all 0s except when all three input bits are 1s.

▶ **TABLE 3–3**

INPUTS			OUTPUT
A	B	C	X
0	0	0	0
0	0	1	0
0	1	0	0
0	1	1	0
1	0	0	0
1	0	1	0
1	1	0	0
1	1	1	1

(b) $N = 2^4 = \mathbf{16}$. There are 16 possible combinations of input bits for a 4-input AND gate.

Related Problem Develop the truth table for a 4-input AND gate.

Operation with Waveform Inputs

In most applications, the inputs to a gate are not stationary levels but are voltage waveforms that change frequently between HIGH and LOW logic levels. Now let's look at the operation of AND gates with pulse waveform inputs, keeping in mind that an AND gate obeys the truth table operation regardless of whether its inputs are constant levels or levels that change back and forth.

Let's examine the waveform operation of an AND gate by looking at the inputs with respect to each other in order to determine the output level at any given time. In Figure 3–10, inputs A and B are both HIGH (1) during the time interval, t_1, making output X HIGH (1) during this interval. During time interval t_2, input A is LOW (0) and input B is HIGH (1),

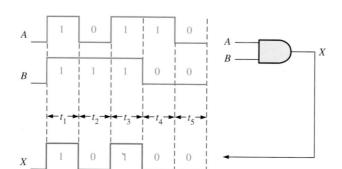

◀ **FIGURE 3–10**

Example of AND gate operation with a timing diagram showing input and output relationships.

so the output is LOW (0). During time interval t_3, both inputs are HIGH (1) again, and there-fore the output is HIGH (1). During time interval t_4, input A is HIGH (1) and input B is LOW (0), resulting in a LOW (0) output. Finally, during time interval t_5, input A is LOW (0), input B is LOW (0), and the output is therefore LOW (0). As you know, a diagram of input and output waveforms showing time relationships is called a *timing diagram.*

EXAMPLE 3–3

If two waveforms, A and B, are applied to the AND gate inputs as in Figure 3–11, what is the resulting output waveform?

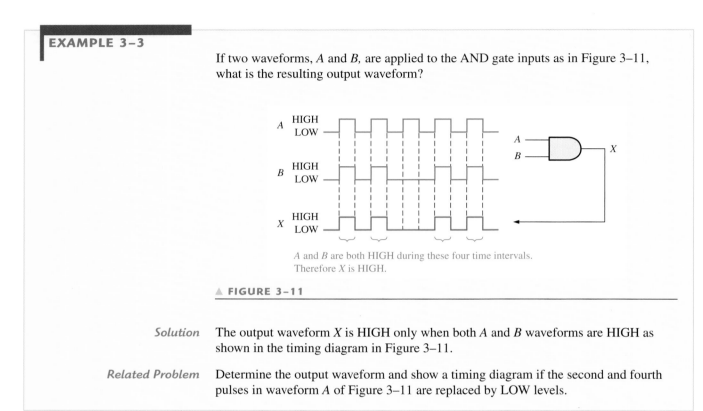

A and B are both HIGH during these four time intervals. Therefore X is HIGH.

▲ **FIGURE 3–11**

Solution   The output waveform X is HIGH only when both A and B waveforms are HIGH as shown in the timing diagram in Figure 3–11.

Related Problem   Determine the output waveform and show a timing diagram if the second and fourth pulses in waveform A of Figure 3–11 are replaced by LOW levels.

Remember, when analyzing the waveform operation of logic gates, it is important to pay careful attention to the time relationships of all the inputs with respect to each other and to the output.

EXAMPLE 3–4

For the two input waveforms, A and B, in Figure 3–12, show the output waveform with its proper relation to the inputs.

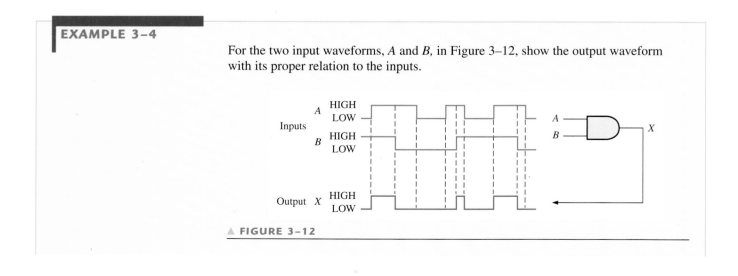

▲ **FIGURE 3–12**

Solution   The output waveform is HIGH only when both of the input waveforms are HIGH as shown in the timing diagram.

Related Problem   Show the output waveform if the B input to the AND gate in Figure 3–12 is always HIGH.

EXAMPLE 3–5

For the 3-input AND gate in Figure 3–13, determine the output waveform in relation to the inputs.

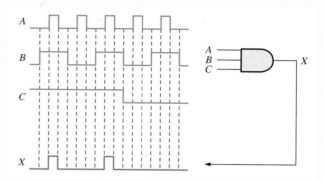

▲ FIGURE 3–13

Solution   The output waveform X of the 3-input AND gate is HIGH only when all three input waveforms A, B, and C are HIGH.

Related Problem   What is the output waveform of the AND gate in Figure 3-13 if the C input is always HIGH?

Logic Expressions for an AND Gate

The logical AND function of two variables is represented mathematically either by placing a dot between the two variables, as $A \cdot B$, or by simply writing the adjacent letters without the dot, as AB. We will normally use the latter notation because it is easier to write.

Boolean multiplication follows the same basic rules governing binary multiplication, which were discussed in Chapter 2 and are as follows:

$$0 \cdot 0 = 0$$
$$0 \cdot 1 = 0$$
$$1 \cdot 0 = 0$$
$$1 \cdot 1 = 1$$

Boolean multiplication is the same as the AND function.

The operation of a 2-input AND gate can be expressed in equation form as follows: If one input variable is A, the other input variable is B, and the output variable is X, then the Boolean expression is

$$X = AB$$

COMPUTER NOTE

Computers can utilize all of the basic logic operations when it is necessary to selectively manipulate certain bits in one or more bytes of data. Selective bit manipulations are done with a *mask*. For example, to clear (make all 0s) the right four bits in a data byte but keep the left four bits, ANDing the data byte with 11110000 will do the job. Notice that any bit ANDed with zero will be 0 and any bit ANDed with 1 will remain the same. If 10101010 is ANDed with the mask 11110000, the result is 10100000.

Figure 3–14(a) shows the AND gate logic symbol with two input variables and the output variable indicated.

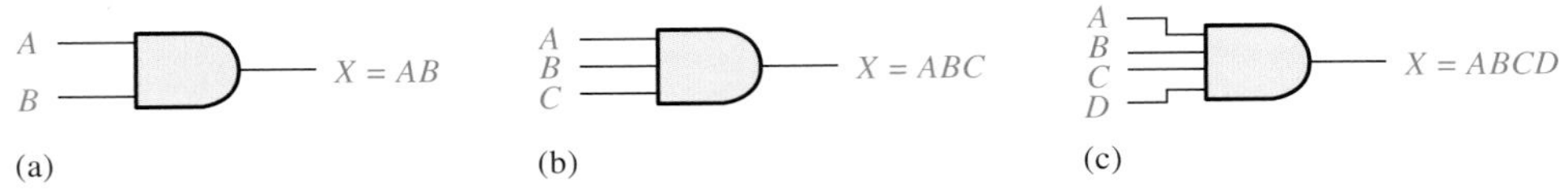

(a)　　　　　　　　　(b)　　　　　　　　　(c)

Boolean expressions for AND gates with two, three, and four inputs.

When variables are shown together like *ABC*, they are ANDed.

To extend the AND expression to more than two input variables, simply use a new letter for each input variable. The function of a 3-input AND gate, for example, can be expressed as $X = ABC$, where A, B, and C are the input variables. The expression for a 4-input AND gate can be $X = ABCD$, and so on. Parts (b) and (c) of Figure 3–14 show AND gates with three and four input variables, respectively.

You can evaluate an AND gate operation by using the Boolean expressions for the output. For example, each variable on the inputs can be either a 1 or a 0; so for the 2-input AND gate, make substitutions in the equation for the output, $X = AB$, as shown in Table 3–4. This evaluation shows that the output X of an AND gate is a 1 (HIGH) only when both inputs are 1s (HIGHs). A similar analysis can be made for any number of input variables.

▶ TABLE 3–4

A	B	AB = X
0	0	$0 \cdot 0 = 0$
0	1	$0 \cdot 1 = 0$
1	0	$1 \cdot 0 = 0$
1	1	$1 \cdot 1 = 1$

Applications

The AND Gate as an Enable/Inhibit Device   A common application of the AND gate is to **enable** (that is, to allow) the passage of a signal (pulse waveform) from one point to another at certain times and to inhibit (prevent) the passage at other times.

A simple example of this particular use of an AND gate is shown in Figure 3–15, where the AND gate controls the passage of a signal (waveform A) to a digital counter. The purpose of this circuit is to measure the frequency of waveform A. The enable pulse has a width of precisely 1 s. When the enable pulse is HIGH, waveform A passes through the gate to the counter; and when the enable pulse is LOW, the signal is prevented from passing through the gate (inhibited).

During the 1 second (1 s) interval of the enable pulse, pulses in waveform A pass through the AND gate to the counter. The number of pulses passing through during the 1 s interval is equal to the frequency of waveform A. For example, Figure 3–15 shows six pulses in one second, which is a frequency of 6 Hz. If 1000 pulses pass through the gate in the 1 s interval of the enable pulse, there are 1000 pulses/s, or a frequency of 1000 Hz.

The counter counts the number of pulses per second and produces a binary output that goes to a decoding and display circuit to produce a readout of the frequency. The enable

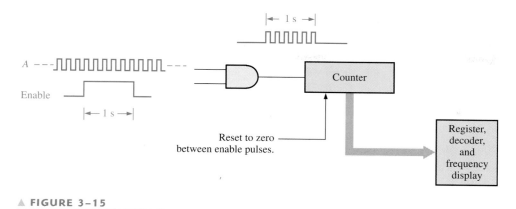

▲ FIGURE 3–15

An AND gate performing an enable/inhibit function for a frequency counter.

pulse repeats at certain intervals and a new updated count is made so that if the frequency changes, the new value will be displayed. Between enable pulses, the counter is reset so that it starts at zero each time an enable pulse occurs. The current frequency count is stored in a register so that the display is unaffected by the resetting of the counter.

A Seat Belt Alarm System   In Figure 3–16, an AND gate is used in a simple automobile seat belt alarm system to detect when the ignition switch is on *and* the seat belt is unbuckled. If the ignition switch is on, a HIGH is produced on input *A* of the AND gate. If the seat belt is not properly buckled, a HIGH is produced on input *B* of the AND gate. Also, when the ignition switch is turned on, a timer is started that produces a HIGH on input *C* for 30 s. If all three conditions exist—that is, if the ignition is on *and* the seat belt is unbuckled *and* the timer is running—the output of the AND gate is HIGH, and an audible alarm is energized to remind the driver.

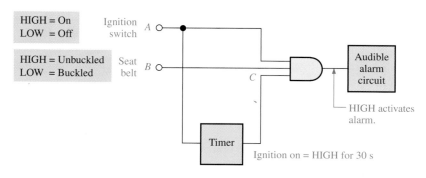

▲ FIGURE 3–16

A simple seat belt alarm circuit using an AND gate.

SECTION 3–2
REVIEW

1. When is the output of an AND gate HIGH?
2. When is the output of an AND gate LOW?
3. Describe the truth table for a 5-input AND gate.

3–3 THE OR GATE

The OR gate is another of the basic gates from which all logic functions are constructed. An OR gate can have two or more inputs and performs what is known as logical addition.

After completing this section, you should be able to

■ Identify an OR gate by its distinctive shape symbol or by its rectangular outline symbol ■ Describe the operation of an OR gate ■ Generate the truth table for an OR gate with any number of inputs ■ Produce a timing diagram for an OR gate with any specified input waveforms ■ Write the logic expression for an OR gate with any number of inputs ■ Discuss examples of OR gate applications

An OR gate can have more than two inputs.

An **OR gate** has two or more inputs and one output, as indicated by the standard logic symbols in Figure 3–17, where OR gates with two inputs are illustrated. An OR gate can have any number of inputs greater than one. Although both distinctive shape and rectangular outline symbols are shown, the distinctive shape OR gate symbol is used in this textbook.

▶ **FIGURE 3–17**

Standard logic symbols for the OR gate showing two inputs (ANSI/IEEE Std. 91-1984).

(a) Distinctive shape

(b) Rectangular outline with the OR (≥ 1) qualifying symbol

Operation of an OR Gate

An OR gate produces a HIGH on the output when *any* of the inputs is HIGH. The output is LOW only when all of the inputs are LOW. Therefore, an OR gate determines when one or more of its inputs are HIGH and produces a HIGH on its output to indicate this condition. The inputs of the 2-input OR gate in Figure 3–17 are labeled A and B, and the output is labeled X. The operation of the gate can be stated as follows:

For a 2-input OR gate, output X is HIGH when either input A or input B is HIGH, or when both A and B are HIGH; X is LOW only when both A and B are LOW.

The HIGH level is the active or asserted output level for the OR gate. Figure 3–18 illustrates the operation for a 2-input OR gate for all four possible input combinations.

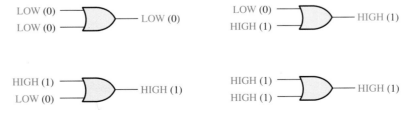

▲ **FIGURE 3–18**

All possible logic levels for a 2-input OR gate. Open file F03-18 to verify OR gate operation.

OR Gate Truth Table

The operation of a 2-input OR gate is described in Table 3–5. This truth table can be expanded for any number of inputs; but regardless of the number of inputs, the output is HIGH when one or more of the inputs are HIGH.

For an OR gate, at least one HIGH input makes a HIGH output.

| INPUTS | | OUTPUT |
A	B	X
0	0	0
0	1	1
1	0	1
1	1	1

1 = HIGH, 0 = LOW

◀ **TABLE 3–5**

Truth table for a 2-input OR gate.

Operation with Waveform Inputs

Now let's look at the operation of an OR gate with pulse waveform inputs, keeping in mind its logical operation. Again, the important thing in the analysis of gate operation with pulse waveforms is the time relationship of all the waveforms involved. For example, in Figure 3–19, inputs A and B are both HIGH (1) during time interval t_1, making output X HIGH (1). During time interval t_2, input A is LOW (0), but because input B is HIGH (1), the output is HIGH (1). Both inputs are LOW (0) during time interval t_3, so there is a LOW (0) output during this time. During time interval t_4, the output is HIGH (1) because input A is HIGH (1).

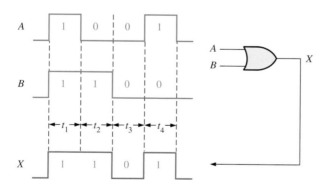

▲ **FIGURE 3–19**

Example of OR gate operation with a timing diagram showing input and output time relationships.

In this illustration, we have applied the truth table operation of the OR gate to each of the time intervals during which the levels are nonchanging. Examples 3–6 through 3–8 further illustrate OR gate operation with waveforms on the inputs.

EXAMPLE 3–6

If the two input waveforms, A and B, in Figure 3–20 are applied to the OR gate, what is the resulting output waveform?

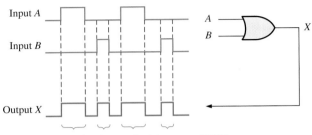

When either input or both inputs are HIGH, the output is HIGH.

▲ **FIGURE 3–20**

Solution The output waveform X of a 2-input OR gate is HIGH when either or both input waveforms are HIGH as shown in the timing diagram. In this case, both input waveforms are never HIGH at the same time.

Related Problem Determine the output waveform and show the timing diagram if input A is changed such that it is HIGH from the beginning of the existing first pulse to the end of the existing second pulse.

EXAMPLE 3–7

For the two input waveforms, A and B, in Figure 3–21, show the output waveform with its proper relation to the inputs.

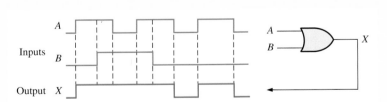

▲ **FIGURE 3–21**

Solution When either or both input waveforms are HIGH, the output is HIGH as shown by the output waveform X in the timing diagram.

Related Problem Determine the output waveform and show the timing diagram if the middle pulse of input A is replaced by a LOW level.

EXAMPLE 3–8

For the 3-input OR gate in Figure 3–22, determine the output waveform in proper time relation to the inputs.

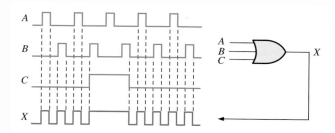

▲ **FIGURE 3–22**

Solution   The output is HIGH when one or more of the input waveforms are HIGH as indicated by the output waveform X in the timing diagram.

Related Problem   Determine the output waveform and show the timing diagram if input C is always LOW.

Logic Expressions for an OR Gate

The logical OR function of two variables is represented mathematically by a + between the two variables, for example, $A + B$.

> When variables are separated by +, they are ORed.

Addition in Boolean algebra involves variables whose values are either binary 1 or binary 0. The basic rules for **Boolean addition** are as follows:

$$0 + 0 = 0$$
$$0 + 1 = 1$$
$$1 + 0 = 1$$
$$1 + 1 = 1$$

Boolean addition is the same as the OR function.

Notice that Boolean addition differs from binary addition in the case where two 1s are added. There is no carry in Boolean addition.

The operation of a 2-input OR gate can be expressed as follows: If one input variable is A, if the other input variable is B, and if the output variable is X, then the Boolean expression is

$$X = A + B$$

Figure 3–23(a) shows the OR gate logic symbol with two input variables and the output variable labeled.

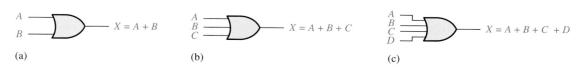

(a)   (b)   (c)

▲ **FIGURE 3–23**

Boolean expressions for OR gates with two, three, and four inputs.

To extend the OR expression to more than two input variables, a new letter is used for each additional variable. For instance, the function of a 3-input OR gate can be expressed as $X = A + B + C$. The expression for a 4-input OR gate can be written as $X = A + B + C + D$, and so on. Parts (b) and (c) of Figure 3–23 show OR gates with three and four input variables, respectively.

OR gate operation can be evaluated by using the Boolean expressions for the output X by substituting all possible combinations of 1 and 0 values for the input variables, as shown in Table 3–6 for a 2-input OR gate. This evaluation shows that the output X of an OR gate is a 1 (HIGH) when any one or more of the inputs are 1 (HIGH). A similar analysis can be extended to OR gates with any number of input variables.

▶ **TABLE 3–6**

A	B	A + B = X
0	0	0 + 0 = 0
0	1	0 + 1 = 1
1	0	1 + 0 = 1
1	1	1 + 1 = 1

An Application

A simplified portion of an intrusion detection and alarm system is shown in Figure 3–24. This system could be used for one room in a home—a room with two windows and a door. The sensors are magnetic switches that produce a HIGH output when open and a LOW output when closed. As long as the windows and the door are secured, the switches are closed and all three of the OR gate inputs are LOW. When one of the windows or the door is opened, a HIGH is produced on that input to the OR gate and the gate output goes HIGH. It then activates and latches an alarm circuit to warn of the intrusion.

▶ **FIGURE 3–24**

A simplified intrusion detection system using an OR gate.

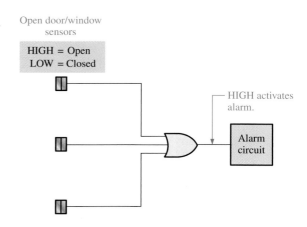

SECTION 3–3
REVIEW

1. When is the output of an OR gate HIGH?
2. When is the output of an OR gate LOW?
3. Describe the truth table for a 3-input OR gate.

3-4 THE NAND GATE

The NAND gate is a popular logic element because it can be used as a universal gate; that is, NAND gates can be used in combination to perform the AND, OR, and inverter operations. The universal property of the NAND gate will be examined thoroughly in Chapter 5.

After completing this section, you should be able to

■ Identify a NAND gate by its distinctive shape symbol or by its rectangular outline symbol ■ Describe the operation of a NAND gate ■ Develop the truth table for a NAND gate with any number of inputs ■ Produce a timing diagram for a NAND gate with any specified input waveforms ■ Write the logic expression for a NAND gate with any number of inputs ■ Describe NAND gate operation in terms of its negative-OR equivalent ■ Discuss examples of NAND gate applications

The term *NAND* is a contraction of NOT-AND and implies an AND function with a complemented (inverted) output. The standard logic symbol for a 2-input NAND gate and its equivalency to an AND gate followed by an inverter are shown in Figure 3–25(a), where the symbol ≡ means equivalent to. A rectangular outline symbol is shown in part (b).

The NAND is the same as the AND except the output is inverted.

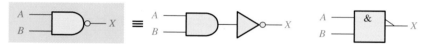

(a) Distinctive shape, 2-input NAND gate and its NOT/AND equivalent

(b) Rectangular outline, 2-input NAND gate with polarity indicator

▲ FIGURE 3–25

Standard NAND gate logic symbols (ANSI/IEEE Std. 91–1984).

Operation of a NAND Gate

A **NAND gate** produces a LOW output only when all the inputs are HIGH. When any of the inputs is LOW, the output will be HIGH. For the specific case of a 2-input NAND gate, as shown in Figure 3–25 with the inputs labeled A and B and the output labeled X, the operation can be stated as follows:

> **For a 2-input NAND gate, output X is LOW only when inputs A and B are HIGH; X is HIGH when either A or B is LOW, or when both A and B are LOW.**

Note that this operation is opposite that of the AND in terms of the output level. In a NAND gate, the LOW level (0) is the active or asserted output level, as indicated by the bubble on the output. Figure 3–26 illustrates the operation of a 2-input NAND gate for all four input combinations, and Table 3–7 is the truth table summarizing the logical operation of the 2-input NAND gate.

▲ FIGURE 3–26

Operation of a 2-input NAND gate. Open file F03-26 to verify NAND gate operation.

▶ TABLE 3–7

Truth table for a 2-input NAND gate.

| INPUTS | | OUTPUT |
A	B	X
0	0	1
0	1	1
1	0	1
1	1	0

1 = HIGH, 0 = LOW.

Operation with Waveform Inputs

Now let's look at the pulse waveform operation of a NAND gate. Remember from the truth table that the only time a LOW output occurs is when all of the inputs are HIGH.

EXAMPLE 3–9

If the two waveforms A and B shown in Figure 3–27 are applied to the NAND gate inputs, determine the resulting output waveform.

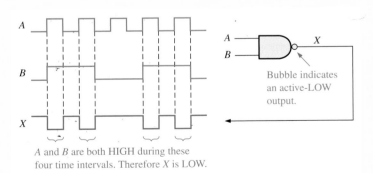

A and B are both HIGH during these four time intervals. Therefore X is LOW.

▲ FIGURE 3–27

Solution Output waveform X is LOW only during the four time intervals when both input waveforms A and B are HIGH as shown in the timing diagram.

Related Problem Determine the output waveform and show the timing diagram if input waveform B is inverted.

EXAMPLE 3–10

Show the output waveform for the 3-input NAND gate in Figure 3–28 with its proper time relationship to the inputs.

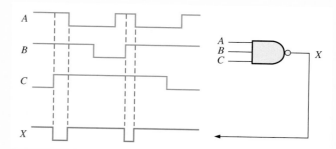

▲ FIGURE 3–28

Solution   The output waveform X is LOW only when all three input waveforms are HIGH as shown in the timing diagram.

Related Problem   Determine the output waveform and show the timing diagram if input waveform A is inverted.

Negative-OR Equivalent Operation of a NAND Gate   Inherent in a NAND gate's operation is the fact that one or more LOW inputs produce a HIGH output. Table 3–7 shows that output X is HIGH (1) when any of the inputs, A and B, is LOW (0). From this viewpoint, a NAND gate can be used for an OR operation that requires one or more LOW inputs to produce a HIGH output. This aspect of NAND operation is referred to as **negative-OR.** The term *negative* in this context means that the inputs are defined to be in the active or asserted state when LOW.

For a 2-input NAND gate performing a negative-OR operation, output X is HIGH when either input A or input B is LOW, or when both A and B are LOW.

When a NAND gate is used to detect one or more LOWs on its inputs rather than all HIGHs, it is performing the negative-OR operation and is represented by the standard logic symbol shown in Figure 3–29. Although the two symbols in Figure 3–29 represent the same physical gate, they serve to define its role or mode of operation in a particular application, as illustrated by Examples 3–11 through 3–13.

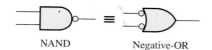

NAND   Negative-OR

◀ FIGURE 3–29

Standard symbols representing the two equivalent operations of a NAND gate.

EXAMPLE 3–11

A manufacturing plant uses two tanks to store certain liquid chemicals that are required in a manufacturing process. Each tank has a sensor that detects when the chemical level drops to 25% of full. The sensors produce a HIGH level of 5 V when the tanks are more than one-quarter full. When the volume of chemical in a tank drops to one-quarter full, the sensor puts out a LOW level of 0 V.

It is required that a single green light-emitting diode (LED) on an indicator panel show when both tanks are more than one-quarter full. Show how a NAND gate can be used to implement this function.

Solution

Figure 3–30 shows a NAND gate with its two inputs connected to the tank level sensors and its output connected to the indicator panel. The operation can be stated as follows: If tank *A and* tank *B* are above one-quarter full, the LED is on.

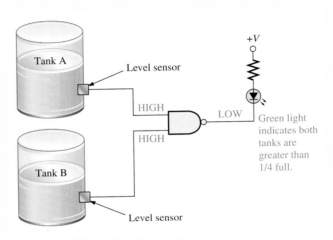

▲ **FIGURE 3–30**

As long as both sensor outputs are HIGH (5 V), indicating that both tanks are more than one-quarter full, the NAND gate output is LOW (0 V). The green LED circuit is arranged so that a LOW voltage turns it on.

Related Problem

How can the circuit of Figure 3–30 be modified to monitor the levels in three tanks rather than two?

EXAMPLE 3–12

The supervisor of the manufacturing process described in Example 3–11 has decided that he would prefer to have a red LED display come on when at least one of the tanks falls to the quarter-full level rather than have the green LED display indicate when both are above one quarter. Show how this requirement can be implemented.

Solution

Figure 3–31 shows a NAND gate operating as a negative-OR gate to detect the occurrence of at least one LOW on its inputs. A sensor puts out a LOW voltage if the volume in its tank goes to one-quarter full or less. When this happens, the gate output goes HIGH. The red LED circuit in the panel is arranged so that a HIGH voltage turns it on. The operation can be stated as follows: If tank *A or* tank *B or* both are below one-quarter full, the LED is on.

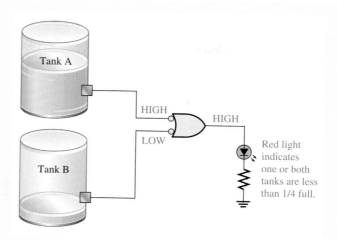

▲ FIGURE 3–31

Notice that, in this example and in Example 3–11, the same 2-input NAND gate is used, but a different gate symbol is used in the schematic, illustrating the different way in which the NAND and equivalent negative-OR operations are used.

Related Problem   How can the circuit in Figure 3–31 be modified to monitor four tanks rather than two?

EXAMPLE 3–13

For the 4-input NAND gate in Figure 3–32, operating as a negative-OR, determine the output with respect to the inputs.

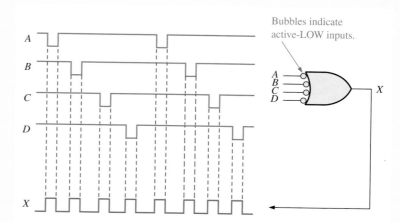

▲ FIGURE 3–32

Solution   The output waveform X is HIGH any time an input waveform is LOW as shown in the timing diagram.

Related Problem   Determine the output waveform if input waveform A is inverted before it is applied to the gate.

Logic Expressions for a NAND Gate

A bar over a variable or variables indicates an inversion.

The Boolean expression for the output of a 2-input NAND gate is

$$X = \overline{AB}$$

This expression says that the two input variables, A and B, are first ANDed and then complemented, as indicated by the bar over the AND expression. This is a description in equation form of the operation of a NAND gate with two inputs. Evaluating this expression for all possible values of the two input variables, you get the results shown in Table 3–8.

▶ TABLE 3–8

A	B	$\overline{AB} = X$
0	0	$\overline{0 \cdot 0} = \overline{0} = 1$
0	1	$\overline{0 \cdot 1} = \overline{0} = 1$
1	0	$\overline{1 \cdot 0} = \overline{0} = 1$
1	1	$\overline{1 \cdot 1} = \overline{1} = 0$

Once an expression is determined for a given logic function, that function can be evaluated for all possible values of the variables. The evaluation tells you exactly what the output of the logic circuit is for each of the input conditions, and it therefore gives you a complete description of the circuit's logic operation. The NAND expression can be extended to more than two input variables by including additional letters to represent the other variables.

SECTION 3–4 REVIEW

1. When is the output of a NAND gate LOW?
2. When is the output of a NAND gate HIGH?
3. Describe the functional differences between a NAND gate and a negative-OR gate. Do they both have the same truth table?
4. Write the output expression for a NAND gate with inputs A, B, and C.

3–5   THE NOR GATE

The NOR gate, like the NAND gate, is a useful logic element because it can also be used as a universal gate; that is, NOR gates can be used in combination to perform the AND, OR, and inverter operations. The universal property of the NOR gate will be examined thoroughly in Chapter 5.

After completing this section, you should be able to

■ Identify a NOR gate by its distinctive shape symbol or by its rectangular outline symbol ■ Describe the operation of a NOR gate ■ Develop the truth table for a NOR gate with any number of inputs ■ Produce a timing diagram for a NOR gate with any specified input waveforms ■ Write the logic expression for a NOR gate with any number of inputs ■ Describe NOR gate operation in terms of its negative-AND equivalent ■ Discuss examples of NOR gate applications

The term *NOR* is a contraction of NOT-OR and implies an OR function with an inverted (complemented) output. The standard logic symbol for a 2-input NOR gate and its equivalent OR gate followed by an inverter are shown in Figure 3–33(a). A rectangular outline symbol is shown in part (b).

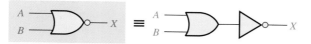

(a) Distinctive shape, 2-input NOR gate and its NOT/OR equivalent

(b) Rectangular outline, 2-input NOR gate with polarity indicator

▲ **FIGURE 3–33**

Standard NOR gate logic symbols (ANSI/IEEE Std. 91-1984).

Operation of a NOR Gate

A **NOR gate** produces a LOW output when *any* of its inputs is HIGH. Only when all of its inputs are LOW is the output HIGH. For the specific case of a 2-input NOR gate, as shown in Figure 3–33 with the inputs labeled *A* and *B* and the output labeled *X*, the operation can be stated as follows:

> **For a 2-input NOR gate, output *X* is LOW when either input *A* or input *B* is HIGH, or when both *A* and *B* are HIGH; *X* is HIGH only when both *A* and *B* are LOW.**

This operation results in an output level opposite that of the OR gate. In a NOR gate, the LOW output is the active or asserted output level as indicated by the bubble on the output. Figure 3–34 illustrates the operation of a 2-input NOR gate for all four possible input combinations, and Table 3–9 is the truth table for a 2-input NOR gate.

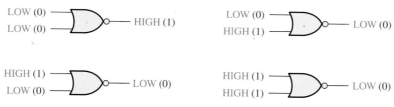

▲ **FIGURE 3–34**

Operation of a 2-input NOR gate. Open file F03-34 to verify NOR gate operation.

INPUTS		OUTPUT
A	*B*	*X*
0	0	1
0	1	0
1	0	0
1	1	0
1 = HIGH, 0 = LOW.		

◀ **TABLE 3–9**

Truth table for a 2-input NOR gate.

Operation with Waveform Inputs

The next two examples illustrate the operation of a NOR gate with pulse waveform inputs. Again, as with the other types of gates, we will simply follow the truth table operation to determine the output waveforms in the proper time relationship to the inputs.

EXAMPLE 3–14

If the two waveforms shown in Figure 3–35 are applied to a NOR gate, what is the resulting output waveform?

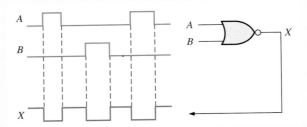

▲ FIGURE 3–35

Solution   Whenever any input of the NOR gate is HIGH, the output is LOW as shown by the output waveform *X* in the timing diagram.

Related Problem   Invert input *B* and determine the output waveform in relation to the inputs.

EXAMPLE 3–15

Show the output waveform for the 3-input NOR gate in Figure 3–36 with the proper time relation to the inputs.

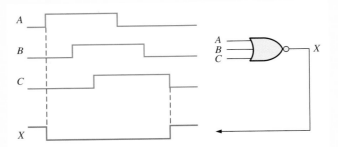

▲ FIGURE 3–36

Solution   The output *X* is LOW when any input is HIGH as shown by the output waveform *X* in the timing diagram.

Related Problem   With the *B* and *C* inputs inverted, determine the output and show the timing diagram.

Negative-AND Equivalent Operation of the NOR Gate   A NOR gate, like the NAND, has another aspect of its operation that is inherent in the way it logically functions. Table 3–9 shows that a HIGH is produced on the gate output only when all of the inputs are LOW. From this viewpoint, a NOR gate can be used for an AND operation that requires all LOW inputs to produce a HIGH output. This aspect of NOR operation is called **negative-AND.** The term *negative* in this context means that the inputs are defined to be in the active or asserted state when LOW.

> **For a 2-input NOR gate performing a negative-AND operation, output X is HIGH only when both inputs A and B are LOW.**

When a NOR gate is used to detect all LOWs on its inputs rather than one or more HIGHs, it is performing the negative-AND operation and is represented by the standard symbol in Figure 3–37. It is important to remember that the two symbols in Figure 3–37 represent the same physical gate and serve only to distinguish between the two modes of its operation. The following three examples illustrate this.

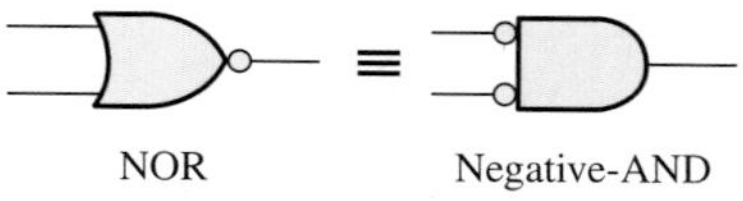

◄ **FIGURE 3–37**

Standard symbols representing the two equivalent operations of a NOR gate.

NOR   Negative-AND

EXAMPLE 3–16

A device is needed to indicate when two LOW levels occur simultaneously on its inputs and to produce a HIGH output as an indication. Specify the device.

Solution   A 2-input NOR gate operating as a negative-AND gate is required to produce a HIGH output when both inputs are LOW, as shown in Figure 3–38.

▶ **FIGURE 3–38**

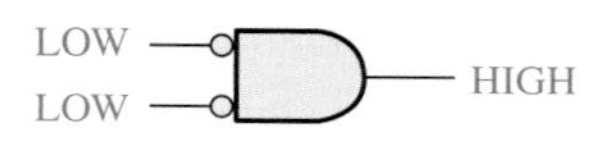

LOW
LOW — HIGH

Related Problem   A device is needed to indicate when one or two HIGH levels occur on its inputs and to produce a LOW output as an indication. Specify the device.

EXAMPLE 3–17

As part of an aircraft's functional monitoring system, a circuit is required to indicate the status of the landing gears prior to landing. A green LED display turns on if all three gears are properly extended when the "gear down" switch has been activated in preparation for landing. A red LED display turns on if any of the gears fail to extend properly prior to landing. When a landing gear is extended, its sensor produces a LOW voltage. When a landing gear is retracted, its sensor produces a HIGH voltage. Implement a circuit to meet this requirement.

Solution   Power is applied to the circuit only when the "gear down" switch is activated. Use a NOR gate for each of the two requirements as shown in Figure 3–39. One NOR gate operates as a negative-AND to detect a LOW from each of the three landing gear sensors. When all three of the gate inputs are LOW, the three landing gears are properly extended and the resulting HIGH output from the negative-AND gate turns on the green LED display. The other NOR gate operates as a NOR to detect if one or more of the landing gears remain retracted when the "gear down" switch is activated. When one or more of the landing gears remain retracted, the resulting HIGH from the

sensor is detected by the NOR gate, which produces a LOW output to turn on the red
LED warning display.

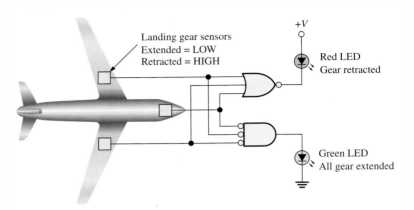

▲ FIGURE 3–39

Related Problem   What type of gate should be used to detect if all three landing gears are retracted after
takeoff, assuming a LOW output is required to activate an LED display?

HANDS
ON
TIP

When driving a load such as an LED with a logic gate, consult the manufacturer's data
sheet for maximum drive capabilities (output current). A regular IC logic gate may not be
capable of handling the current required by certain loads such as some LEDs. Logic gates
with a buffered output, such as an open–collector (OC) or open–drain (OD) output, are
available in many types of IC logic gate configurations. The output current capability of
typical IC logic gates is limited to the μA or relatively low mA range. For example, standard TTL can handle output currents up to 16 mA. Most LEDs require currents in the range
of about 10 mA to 50 mA.

EXAMPLE 3–18

For the 4-input NOR gate operating as a negative-AND in Figure 3–40, determine the
output relative to the inputs.

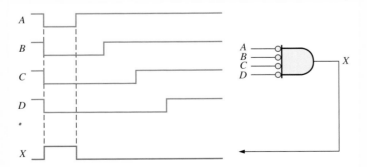

▲ FIGURE 3–40

Solution Any time all of the input waveforms are LOW, the output is HIGH as shown by output waveform X in the timing diagram.

Related Problem Determine the output with input D inverted and show the timing diagram.

Logic Expressions for a NOR Gate

The Boolean expression for the output of a 2-input NOR gate can be written as

$$X = \overline{A + B}$$

This equation says that the two input variables are first ORed and then complemented, as indicated by the bar over the OR expression. Evaluating this expression, you get the results shown in Table 3–10. The NOR expression can be extended to more than two input variables by including additional letters to represent the other variables.

A	B	$\overline{A+B}=X$
0	0	$\overline{0 + 0} = \overline{0} = 1$
0	1	$\overline{0 + 1} = \overline{1} = 0$
1	0	$\overline{1 + 0} = \overline{1} = 0$
1	1	$\overline{1 + 1} = \overline{1} = 0$

◀ **TABLE 3–10**

SECTION 3–5 REVIEW

1. When is the output of a NOR gate HIGH?
2. When is the output of a NOR gate LOW?
3. Describe the functional difference between a NOR gate and a negative-AND gate. Do they both have the same truth table?
4. Write the output expression for a 3-input NOR with input variables A, B, and C.

3–6 THE EXCLUSIVE-OR AND EXCLUSIVE-NOR GATES

Exclusive-OR and exclusive-NOR gates are formed by a combination of other gates already discussed, as you will see in Chapter 5. However, because of their fundamental importance in many applications, these gates are often treated as basic logic elements with their own unique symbols.

After completing this section, you should be able to

■ Identify the exclusive-OR and exclusive-NOR gates by their distinctive shape symbols or by their rectangular outline symbols ■ Describe the operations of exclusive-OR and exclusive-NOR gates ■ Show the truth tables for exclusive-OR and exclusive-NOR gates ■ Produce a timing diagram for an exclusive-OR or exclusive-NOR gate with any specified input waveforms ■ Discuss examples of exclusive-OR and exclusive-NOR gate applications

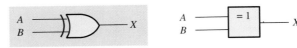

The Exclusive-OR Gate

COMPUTER NOTE

Exclusive-OR gates connected to form an adder circuit allow a computer to perform addition, subtraction, multiplication, and division in its Arithmetic Logic Unit (ALU). An exclusive-OR gate combines basic AND, OR, and NOT logic.

Standard symbols for an exclusive-OR (XOR for short) gate are shown in Figure 3–41. The XOR gate has only two inputs.

A
B
X
A
B
= 1
X

(a) Distinctive shape

(b) Rectangular outline with the XOR

▲ **FIGURE 3–41**

Standard logic symbols for the exclusive-OR gate.

For an exclusive-OR gate, opposite inputs make the output HIGH.

The output of an **exclusive-OR gate** is HIGH *only* when the two inputs are at opposite logic levels. This operation can be stated as follows with reference to inputs *A* and *B* and output *X:*

For an exclusive-OR gate, output *X* is HIGH when input *A* is LOW and input *B* is HIGH, or when input *A* is HIGH and input *B* is LOW; *X* is LOW when *A* and *B* are both HIGH or both LOW.

The four possible input combinations and the resulting outputs for an XOR gate are illustrated in Figure 3–42. The HIGH level is the active or asserted output level and occurs only when the inputs are at opposite levels. The operation of an XOR gate is summarized in the truth table shown in Table 3–11.

▶ **FIGURE 3–42**

All possible logic levels for an exclusive-OR gate. Open file F03-42 to verify XOR gate operation.

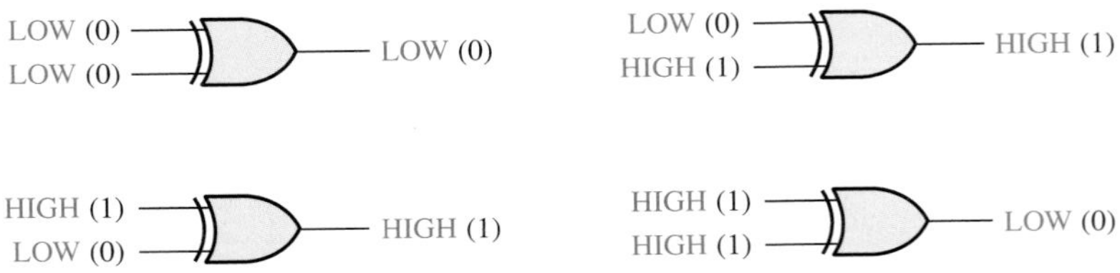

LOW (0)
LOW (0)
LOW (0)
LOW (0)
HIGH (1)
HIGH (1)
HIGH (1)
LOW (0)
HIGH (1)
HIGH (1)
HIGH (1)
LOW (0)

▶ **TABLE 3–11**

Truth table for an exclusive-OR gate.

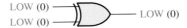

INPUTS		OUTPUT
A	*B*	*X*
0	0	0
0	1	1
1	0	1
1	1	0

EXAMPLE 3–19

A certain system contains two identical circuits operating in parallel. As long as both are operating properly, the outputs of both circuits are always the same. If one of the circuits fails, the outputs will be at opposite levels at some time. Devise a way to detect that a failure has occurred in one of the circuits.

Solution The outputs of the circuits are connected to the inputs of an XOR gate as shown in Figure 3–43. A failure in either one of the circuits produces differing outputs, which cause the XOR inputs to be at opposite levels. This condition produces a HIGH on the output of the XOR gate, indicating a failure in one of the circuits.

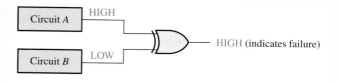

▲ FIGURE 3–43

Related Problem Will the exclusive-OR gate always detect simultaneous failures in both circuits of Figure 3–43? If not, under what condition?

The Exclusive-NOR Gate

Standard symbols for an **exclusive-NOR** (XNOR) **gate** are shown in Figure 3–44. Like the XOR gate, an XNOR has only two inputs. The bubble on the output of the XNOR symbol indicates that its output is opposite that of the XOR gate. When the two input logic levels are opposite, the output of the exclusive-NOR gate is LOW. The operation can be stated as follows (A and B are inputs, X is the output):

For an exclusive-NOR gate, output X is LOW when input A is LOW and input B is HIGH, or when A is HIGH and B is LOW; X is HIGH when A and B are both HIGH or both LOW.

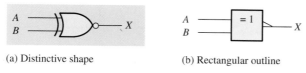

(a) Distinctive shape   (b) Rectangular outline

▲ FIGURE 3–44

Standard logic symbols for the exclusive-NOR gate.

The four possible input combinations and the resulting outputs for an XNOR gate are shown in Figure 3–45. The operation of an XNOR gate is summarized in Table 3–12. Notice that the output is HIGH when the same level is on both inputs.

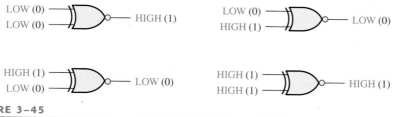

▲ FIGURE 3–45

All possible logic levels for an exclusive-NOR gate. Open file F03-45 to verify XNOR gate operation.

▶ TABLE 3–12

Truth table for an exclusive-NOR gate.

| INPUTS | | OUTPUT |
A	B	X
0	0	1
0	1	0
1	0	0
1	1	1

Operation with Waveform Inputs

As we have done with the other gates, let's examine the operation of XOR and XNOR gates with pulse waveform inputs. As before, we apply the truth table operation during each distinct time interval of the pulse waveform inputs, as illustrated in Figure 3–46 for an XOR gate. You can see that the input waveforms A and B are at opposite levels during time intervals t_2 and t_4. Therefore, the output X is HIGH during these two times. Since both inputs are at the same level, either both HIGH or both LOW, during time intervals t_1 and t_3, the output is LOW during those times as shown in the timing diagram.

▶ FIGURE 3–46

Example of exclusive-OR gate operation with pulse waveform inputs.

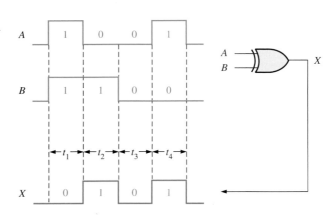

EXAMPLE 3–20

Determine the output waveforms for the XOR gate and for the XNOR gate, given the input waveforms, A and B, in Figure 3–47.

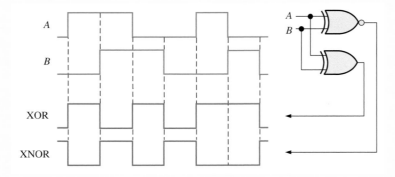

▲ FIGURE 3–47

Solution The output waveforms are shown in Figure 3–47. Notice that the XOR output is HIGH only when both inputs are at opposite levels. Notice that the XNOR output is HIGH only when both inputs are the same.

Related Problem Determine the output waveforms if the two input waveforms, *A* and *B*, are inverted.

An Application

An exclusive-OR gate can be used as a two-bit adder. Recall from Chapter 2 that the basic rules for binary addition are as follows: $0 + 0 = 0, 0 + 1 = 1, 1 + 0 = 1$, and $1 + 1 = 10$. An examination of the truth table for an XOR gate will show you that its output is the binary sum of the two input bits. In the case where the inputs are both 1s, the output is the sum 0, but you lose the carry of 1. In Chapter 6 you will see how XOR gates are combined to make complete adding circuits. Figure 3–48 illustrates an XOR gate used as a basic adder.

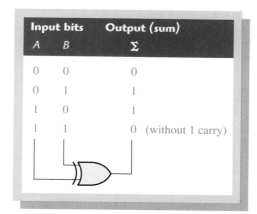

Input bits		Output (sum)
A	B	Σ
0	0	0
0	1	1
1	0	1
1	1	0 (without 1 carry)

◄ **FIGURE 3–48**

An XOR gate used to add two bits.

SECTION 3–6 REVIEW

1. When is the output of an XOR gate HIGH?
2. When is the output of an XNOR gate HIGH?
3. How can you use an XOR gate to detect when two bits are different?

3–7   PROGRAMMABLE LOGIC

Programmable logic was introduced in Chapter 1. In this section, the basic concept of the programmable AND array, which forms the basis for most programmable logic, is discussed, and the major process technologies are covered. A programmable logic device (PLD) is one that does not initially have a fixed-logic function but that can be programmed to implement just about any logic design. As you have learned, two types of PLD are the SPLD and CPLD. In addition to the PLD, the other major category of programmable logic is the FPGA. For simplicity, all of these devices will be referred to as PLDs. Also, some important concepts in programming are discussed.

After completing this section, you should be able to

■ Describe the concept of a programmable AND array ■ Discuss various process technologies ■ Discuss text entry and graphic entry as two methods for programmable logic design ■ Describe methods for downloading a design to a programmable logic device ■ Explain in-system programming

Basic Concept of the AND Array

Most types of PLDs use some form of **AND array**. Basically, this array consists of AND gates and a matrix of interconnections with programmable links at each cross point, as shown in Figure 3–49(a). The purpose of the programmable links is to either make or break a connection between a row line and a column line in the interconnection matrix. For each input to an AND gate, only one programmable link is left intact in order to connect the desired variable to the gate input. Figure 3–49(b) illustrates an array after it has been programmed.

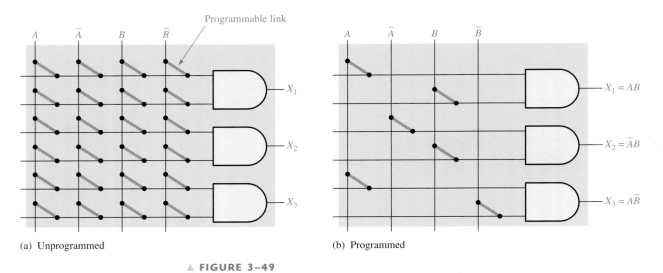

(a) Unprogrammed   (b) Programmed

▲ **FIGURE 3–49**

Basic concept of a programmable AND array.

EXAMPLE 3–21

Show the AND array in Figure 3–49(a) programmed for the following outputs:
$X_1 = A\overline{B}$, $X_2 = \overline{A}B$, and $X_3 = \overline{A}\,\overline{B}$

Solution   See Figure 3–50.

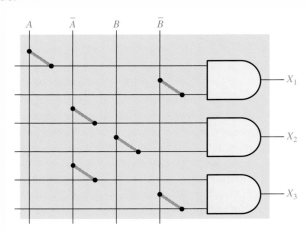

▲ **FIGURE 3–50**

Related Problem   How many rows, columns, and AND gate inputs are required for three input variables in a 3-AND gate array?

Programmable Link Process Technologies

Several different process technologies are used for programmable links in PLDs.

Fuse Technology This was the original programmable link technology. It is still used in some SPLDs. The **fuse** is a metal link that connects a row and a column in the inter-connection matrix. Before programming, there is a fused connection at each intersection. To program a device, the selected fuses are opened by passing a current through them sufficient to "blow" the fuse and break the connection. The intact fuses remain and provide a connection between the rows and columns. The fuse link is illustrated in Figure 3–51. Programmable logic devices that use fuse technology are one-time programmable (**OTP**).

(a) Fuse intact before programming

(b) Programming current

(c) Fuse open after programming

◀ **FIGURE 3–51**

The programmable fuse link.

Antifuse Technology An **antifuse** programmable link is the opposite of a fuse link. Instead of breaking the connection, a connection is made during programming. An antifuse starts out as an open circuit whereas the fuse starts out as a short circuit. Before programming, there are no connections between the rows and columns in the interconnection matrix. An antifuse is basically two conductors separated by an insulator. To program a device with antifuse technology, a programmer tool applies a sufficient voltage across selected antifuses to break down the insulation between the two conductive materials, causing the insulator to become a low-resistance link. The antifuse link is illustrated in Figure 3–52. An antifuse device is also a one-time programmable (OTP) device.

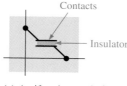

(a) Antifuse is open before programming

(b) Programming voltage breaks down insulation layer to create contact.

(c) Antifuse is effectively shorted after programming

◀ **FIGURE 3–52**

The programmable antifuse link.

EPROM Technology In certain programmable logic devices, the programmable links are similar to the memory cells in **EPROMs** (electrically programmable read-only memories). This type of PLD is programmed using a special tool known as a device programmer. The device is inserted into the programmer, which is connected to a computer running the programming software. Most EPROM-based PLDs are one-time programmable (OTP). However, those with windowed packages can be erased with UV (ultraviolet) light and reprogrammed using a standard PLD programming fixture. EPROM process technology uses a special type of MOS transistor, known as a floating-gate transistor, as the programmable link. The floating-gate device utilizes a process called Fowler-Nordheim tunneling to place electrons in the floating-gate structure.

In a programmable AND array, the floating-gate transistor acts as a switch to connect the row line to either a HIGH or a LOW, depending on the input variable. For input variables that are not used, the transistor is programmed to be permanently *off* (open). Figure 3–53 shows one AND gate in a simple array. Variable A controls the state of the transistor in the first column, and variable B controls the transistor in the third column. When a transistor is *off*, like an open switch, the input line to the AND gate is at $+V$ (HIGH). When a transistor is *on*, like a closed switch, the input line is connected to ground (LOW). When variable A or B is 0 (LOW), the transistor is *on*, keeping the input line to the AND gate LOW. When A or B is 1 (HIGH), the transistor is *off*, keeping the input line to the AND gate HIGH.

▶ **FIGURE 3–53**

A simple AND array with EPROM technology. Only one gate in the array is shown for simplicity.

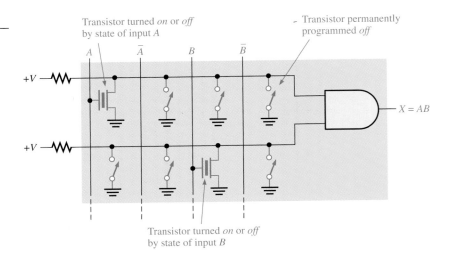

EEPROM Technology Electrically erasable programmable read-only memory technology is similar to EPROM because it also uses a type of floating-gate transistor in E²CMOS cells. The difference is that **EEPROM** can be erased and reprogrammed electrically without the need for UV light or special fixtures. An E²CMOS device can be programmed after being installed on a printed circuit board, and many can be reprogrammed while operating in a system. This is called in-system programming (**ISP**). Figure 3–53 can also be used as an example to represent an AND array with EEPROM technology.

A flash array is a type of EEPROM array that not only can be erased much faster than with standard EEPROM technology but can also result in higher density devices.

SRAM Technology Many FPGAs and some CPLDs use a process technology similar to that used in **SRAMs** (static random-access memories). The basic concept of SRAM-based programmable logic arrays is illustrated in Figure 3–54(a). A SRAM-type memory cell is used to turn a transistor *on* or *off* to connect or disconnect rows and columns. For example, when the memory cell contains a 1 (green), the transistor is *on* and connects the associated row and column lines, as shown in part (b). When the memory cell contains a 0 (blue), the transistor is *off* so there is no connection between the lines, as shown in part (c).

SRAM technology is different from the other process technologies discussed because it is a volatile technology. This means that a SRAM cell does not retain data when power is turned *off*. The programming data must be loaded into a memory; and when power is turned *on*, the data from the memory reprograms the SRAM-based PLD.

The fuse, antifuse, EPROM, and EEPROM process technologies are nonvolatile, so they retain their programming when the power is *off*. A fuse is permanently open, an antifuse is permanently closed, and floating-gate transistors used in EPROM and EEPROM-based arrays can retain their *on* or *off* state indefinitely.

COMPUTER NOTE

Most system-level designs incorporate a variety of devices such as RAMs, ROMs, controllers, and processors that are interconnected by a large quantity of general-purpose logic devices often referred to as "glue" logic. PLDs have come to replace many of the SSI and MSI "glue" devices. The use of PLDs provides a reduction in package count.

For example, in computer memory systems, PLDs can be used for memory address decoding and to generate memory write signals as well as other functions.

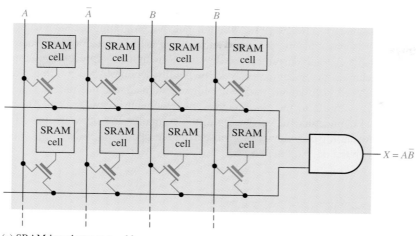

(a) SRAM-based programmable array

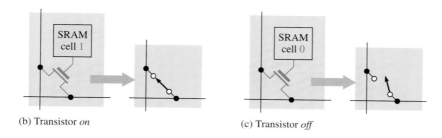

(b) Transistor *on*   (c) Transistor *off*

Device Programming

The general concept of programming was introduced in Chapter 1, and you have seen how interconnections can be made in a simple array by opening or closing the programmable links. SPLDs, CPLDs, and FPGAs are programmed in essentially the same way. The devices with OTP (one-time programmable) process technologies (fuse, antifuse, or EPROM) must be programmed with a special hardware fixture called a *programmer*. The programmer is connected to a computer by a standard interface cable, as shown in Figure 3–55. Development software is installed on the computer, and the device is inserted into the programmer socket. Most programmers have adapters, such as the one shown, that allow different types of packages to be plugged in.

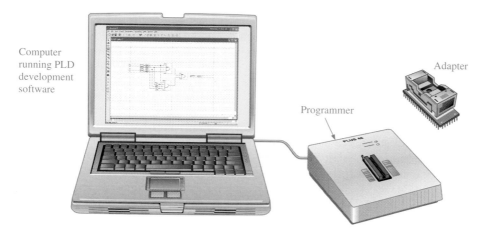

◀ **FIGURE 3–55**

Setup for programming a PLD in a programming fixture (programmer).

Computer running PLD development software

Adapter

Programmer

EEPROM and SRAM-based programmable logic devices are reprogrammable and can be reconfigured multiple times. Although a device programmer can be used for this type of device, it is generally programmed initially on a PLD development board, as shown in Figure 3–56. A logic design can be developed using this approach because any necessary changes during the design process can be readily accomplished by simply reprogramming the PLD. A PLD to which a software logic design can be downloaded is called a **target device.** In addition to the target device, development boards typically provide other circuitry and connectors for interfacing to the computer and other peripheral circuits. Also, test points and display devices for observing the operation of the programmed device are included on the development board.

▶ **FIGURE 3–56**

Programming setup for reprogrammable logic devices.

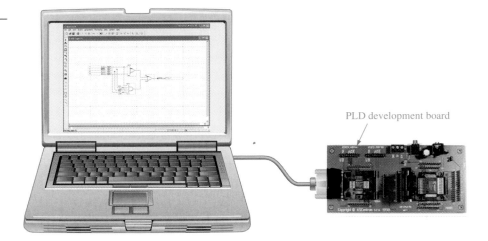

PLD development board

Design Entry As you learned in Chapter 1, design entry is where the logic design is programmed into the development software. The two main ways to enter a design are by text entry or graphic (schematic) entry, and manufacturers of programmable logic provide software packages to support their devices that allow for both methods.

Text entry in most development software, regardless of the manufacturer, supports two or more hardware development languages (**HDLs**). For example, all software packages support both IEEE standard HDLs, VHDL, and Verilog. Some software packages also support certain proprietary languages such as ABEL, CUPL, and AHDL.

In **graphic (schematic) entry,** logic symbols such as AND gates and OR gates are placed on the screen and interconnected to form the desired circuit. In this method you use the familiar logic symbols, but the software actually converts each symbol and interconnections to a text file for the computer to use; you do not see this process. A simple example of both a text entry screen and a graphic entry screen for an AND gate is shown in Figure 3–57. As a general rule, graphic entry is used for less-complex logic circuits and text entry, although it can also be used for very simple logic, is used for larger, more complex implementation.

In-System Programming (ISP)

Certain CPLDs and FPGAs can be programmed after they have been installed on a system printed circuit board (PCB). After a logic design has been developed and fully tested on a development board, it can then be programmed into a "blank" device that is already soldered onto a system board in which it will be operating. Also, if a design change is required, the device on the system board can be reconfigured to incorporate the design modifications.

In a production situation, programming a device on the system board minimizes handling and eliminates the need for keeping stocks of preprogrammed devices. It also rules

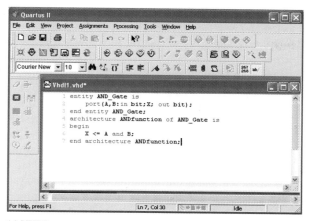

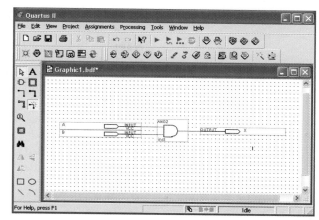

(a) VHDL text entry (b) Equivalent graphic (schematic) entry

▲ FIGURE 3–57

Examples of design entry of an AND gate.

out the possibility of wrong parts being placed in a product. Unprogrammed (blank) devices can be kept in the warehouse and programmed on-board as needed. This minimizes the capital a business needs for inventories and enhances the quality of its products.

JTAG The standard established by the Joint Test Action Group is the commonly used name for IEEE Std. 1149.1. The **JTAG** standard was developed to provide a simple method, called boundary scan, for testing programmable devices for functionality as well as testing circuit boards for bad connections—shorted pins, open pins, bad traces, and the like. More recently, JTAG has been used as a convenient way of configuring programmable devices in-system. As the demand for field-upgradable products increases, the use of JTAG as a convenient way of reprogramming CPLDs and FPGAs will continue to increase.

JTAG-compliant devices have internal dedicated hardware that interprets instructions and data provided by four dedicated signals. These signals are defined by the JTAG standard to be TDI (Test Data In), TDO (Test Data Out), TMS (Test Mode Select), and TCK (Test Clock). The dedicated JTAG hardware interprets instructions and data on the TDI and TMS signals, and drives data out on the TDO signal. The TCK signal is used to clock the process. A JTAG-compliant printed circuit board is represented in Figure 3–58.

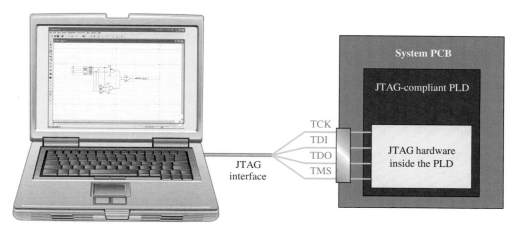

▲ FIGURE 3–58

Simplified illustration of in-system programming via a JTAG interface.

Embedded Processor Another approach to in-system programming is the use of an embedded microprocessor and memory. The processor is embedded within the system along with the CPLD or FPGA and other circuitry, and it is dedicated to the purpose of in-system configuration of the programmable device.

As you have learned, SRAM-based devices are volatile and lose their programmed data when the power is turned *off*. It is necessary to store the programming data in a PROM (programmable read-only memory), which is nonvolatile. When power is turned *on*, the embedded processor takes control of transferring the stored data from the PROM to the CPLD or FPGA.

Also, an embedded processor is sometimes used for reconfiguration of a programmable device while the system is running. In this case, design changes are done with software, and the new data are then loaded into a PROM without disturbing the operation of the system. The processor controls the transfer of the data to the device "on-the-fly" at an appropriate time. A simple block diagram of an embedded processor/programmable logic system is shown in Figure 3–59.

▶ **FIGURE 3–59**

Simplified block diagram of a PLD with an embedded processor and memory.

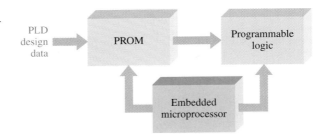

SECTION 3–7 REVIEW

1. List five process technologies used for programmable links in programmable logic.
2. What does the term *volatile* mean in relation to PLDs and which process technology is volatile?
3. What are two design entry methods for programming a PLDs and FPGAs?
4. Define JTAG.

3–8 FIXED-FUNCTION LOGIC

Two major digital integrated circuit (IC) technologies that are used to implement logic gates are CMOS and TTL. The logic operations of NOT, AND, OR, NAND, NOR, and exclusive-OR are the same regardless of the IC technology used; that is, an AND gate has the same logic function whether it is implemented with CMOS or TTL.

After completing this section, you should be able to

■ Identify the most common CMOS and TTL series ■ Compare CMOS and TTL in terms of device types and performance parameters ■ Define *propagation delay time* ■ Define *power dissipation* ■ Define *fan-out* ■ Define *speed-power product* ■ Interpret basic data sheet information

CMOS stands for Complementary Metal-Oxide Semiconductor and is implemented with a type of field-effect transistor. TTL stands for Transistor-Transistor Logic and is implemented with bipolar junction transistors. Keep in mind that CMOS and TTL differ only in the type of circuit components and values of parameters and not in the basic logic operation. A CMOS AND gate has the same logic operation as a TTL AND gate. This is true for all the other basic logic functions. The difference in CMOS and TTL is in performance characteristics such as switching speed (propagation delay), power dissipation, noise immunity, and other parameters.

CMOS

There is little disagreement about which circuit technology, CMOS or TTL, is the most widely used. It appears that CMOS has become the dominant technology and may eventually replace TTL in small- and medium-scale ICs. Although TTL dominated for many years mainly because it had faster switching speeds and a greater selection of device types, CMOS always had the advantage of much lower power dissipation although that parameter is frequency dependent. The switching speeds of CMOS have been greatly improved and are now competitive with TTL, while low power dissipation and other desirable factors have been retained as the technology has progressed.

CMOS Series   The categories of CMOS in terms of the dc supply voltage are the 5 V CMOS, the 3.3 V CMOS, the 2.5 V CMOS, and the 1.8 V CMOS. The lower-voltage CMOS families are a more recent development and are the result of an effort to reduce the power dissipation. Since power dissipation is proportional to the square of the voltage, a reduction from 5 V to 3.3 V, for example, cuts the power by 34% with other factors remaining the same.

Within each supply voltage category, several series of CMOS logic gates are available. These series within the CMOS family differ in their performance characteristics and are designated by the prefix 74 or 54 followed by a letter or letters that indicate the series and then a number that indicates the type of logic device. The prefix 74 indicates commercial grade for general use, and the prefix 54 indicates military grade for more severe environments. We will refer only to the 74-prefixed devices in this textbook. The basic CMOS series for the 5 V category and their designations include

- 74HC and 74HCT—High-speed CMOS (the "T" indicates TTL compatibility)
- 74AC and 74ACT—Advanced CMOS
- 74AHC and 74AHCT—Advanced High-speed CMOS

The basic CMOS series for the 3.3 V category and their designations include

- 74LV—Low-voltage CMOS
- 74LVC—Low-voltage CMOS
- 74ALVC—Advanced Low-voltage CMOS

In addition to the 74 series there is a 4000 series, which is an older, low-speed CMOS technology that is still available, although in limited use. In addition to the "pure" CMOS, there is a series that combines both CMOS and TTL called BiCMOS. The basic BiCMOS series and their designations are as follows:

- 74BCT—BiCMOS
- 74ABT—Advanced BiCMOS
- 74LVT—Low-voltage BiCMOS
- 74ALB—Advanced Low-voltage BiCMOS

TTL

TTL has been a popular digital IC technology for many years. One advantage of TTL is that it is not sensitive to electrostatic discharge as CMOS is and, therefore, is more practical in most laboratory experimentation and prototyping because you do not have to worry about handling precautions.

TTL Series Like CMOS, several series of TTL logic gates are available, all which operate from a 5 V dc supply. These series within the TTL family differ in their performance characteristics and are designated by the prefix 74 or 54 followed by a letter or letters that indicate the series and a number that indicates the type of logic device within the series. A TTL IC can be distinguished from CMOS by the letters that follow the 74 or 54 prefix.

The basic TTL series and their designations are as follows:

- 74—standard TTL (no letter)

- 74S—Schottky TTL

- 74AS—Advanced Schottky TTL

- 74LS—Low-power Schottky TTL

- 74ALS—Advanced Low-power Schottky TTL

- 74F—Fast TTL

Types of Fixed-Function Logic Gates

All of the basic logic operations, NOT, AND, OR, NAND, NOR, exclusive-OR (XOR), and exclusive-NOR (XNOR) are available in both CMOS and TTL. In addition to these, buffered output gates are also available for driving loads that require high currents. The types of gate configurations typically available in IC packages are identified by the last two or three digits in the series designation. For example, 74LS04 is a low-power **Schottky** hex inverter package. Some of the common logic gate configurations and their standard identifier digits are as follows:

- Quad 2-input NAND—**00**
- Quad 2-input NOR—**02**
- Hex inverter—**04**
- Quad 2-input AND—**08**
- Triple 3-input NAND—**10**
- Triple 3-input AND—**11**
- Dual 4-input NAND—**20**
- Dual 2-input AND—**21**
- Triple 3-input NOR—**27**
- Single 8-input NAND—**30**
- Quad 2-input OR—**32**
- Quad XOR—**86**
- Quad XNOR—**266**

IC Packages All of the 74 series CMOS are pin-compatible with the same types of devices in TTL. This means that a CMOS digital IC such as the 74HC00 (quad 2-input NAND), which contains four 2-input NAND gates in one IC package, has the identical package pin numbers for each input and output as does the corresponding TTL device. Typical IC gate packages, the dual in-line package (DIP) for plug-in or feedthrough mounting and the small-outline integrated circuit (SOIC) package for surface mounting, are shown in Figure 3–60. In some cases, other types of packages are also available. The SOIC package is significantly smaller than the DIP. The pin configuration diagrams for most of the fixed-function logic devices listed above are shown in Figure 3–61.

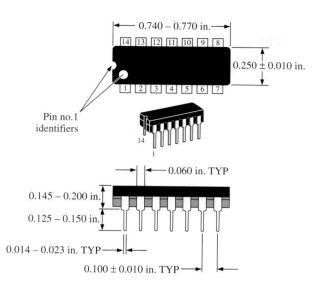
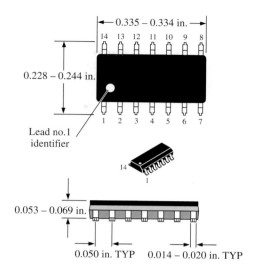

(a) 14-pin dual in-line package (DIP) for feedthrough mounting

(b) 14-pin small outline package (SOIC) for surface mounting

▲ **FIGURE 3–60**

Typical dual in-line (DIP) and small-outline (SOIC) packages showing pin numbers and basic dimensions.

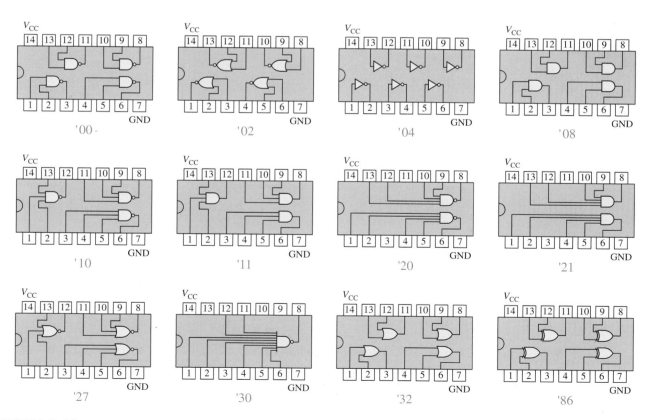

▲ **FIGURE 3–61**

Pin configuration diagrams for some common fixed-function IC gate configurations.

Single-Gate Logic   A limited selection of CMOS gates is available in single-gate packages. With one gate to a package, this series comes in tiny 5-pin packages that are intended for use in last-minute modifications for squeezing logic into tight spots where available space is limited.

Logic Symbols   The logic symbols for fixed-function integrated circuits use the standard gate symbols and show the number of gates in the IC package and the associated pin numbers for each gate as well as the pin numbers for V_{CC} and ground. An example is shown in Figure 3–62 for a hex inverter and for a quad 2-input NAND gate. Both the distinctive shape and the rectangular outline formats are shown. Regardless of the logic family, all devices with the same suffix are pin-compatible; in other words, they will have the same arrangement of pin numbers. For example, the 7400, 74S00, 74LS00, 74ALS00, 74F00, 74HC00, and 74AHC00 are all pin-compatible quad 2-input NAND gate packages.

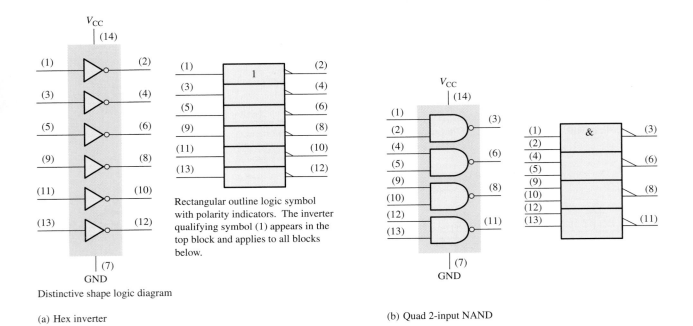

Rectangular outline logic symbol with polarity indicators. The inverter qualifying symbol (1) appears in the top block and applies to all blocks below.

Distinctive shape logic diagram

(a) Hex inverter

(b) Quad 2-input NAND

▲ **FIGURE 3–62**

Logic symbols for hex inverter (04 suffix) and quad 2-input NAND (00 suffix). The symbol applies to the same device in any CMOS or TTL series.

Performance Characteristics and Parameters

Several things define the performance of a logic circuit. These performance characteristics are the switching speed measured in terms of the propagation delay time, the power dissipation, the fan-out or drive capability, the speed-power product, the dc supply voltage, and the input/output logic levels.

High-speed logic has a short propagation delay time.

Propagation Delay Time   This parameter is a result of the limitation on switching speed or frequency at which a logic circuit can operate. The terms *low speed* and *high speed,* applied to logic circuits, refer to the propagation delay time. The shorter the propagation delay, the higher the speed of the circuit and the higher the frequency at which it can operate.

   Propagation delay time, t_P, of a logic gate is the time interval between the application of an input pulse and the occurrence of the resulting output pulse. There are two different measurements of propagation delay time associated with a logic gate that apply to all the types of basic gates:

- t_{PHL}: The time between a specified reference point on the input pulse and a corresponding reference point on the resulting output pulse, with the output changing from the HIGH level to the LOW level (HL).

- t_{PLH}: The time between a specified reference point on the input pulse and a corresponding reference point on the resulting output pulse, with the output changing from the LOW level to the HIGH level (LH).

EXAMPLE 3–22

Show the propagation delay times of the inverter in Figure 3–63(a).

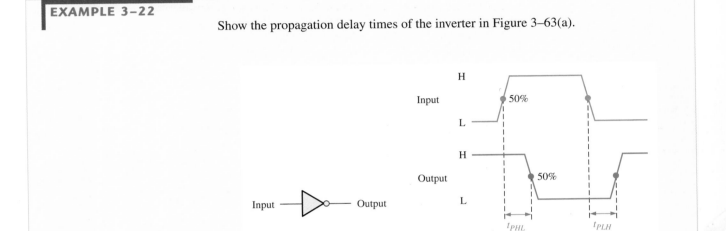

(a)   (b)

▲ **FIGURE 3–63**

Solution   The propagation delay times, t_{PHL} and t_{PLH}, are indicated in part (b) of the figure. In this case, the delays are measured between the 50% points of the corresponding edges of the input and output pulses. The values of t_{PHL} and t_{PLH} are not necessarily equal but in many cases they are the same.

Related Problem   One type of logic gate has a specified maximum t_{PLH} and t_{PHL} of 10 ns. For another type of gate the value is 4 ns. Which gate can operate at the highest frequency?

For standard-series TTL gates, the typical propagation delay is 11 ns and for F-series gates it is 3.3 ns. For HCT-series CMOS, the propagation delay is 7 ns, for the AC series it is 5 ns, and for the ALVC series it is 3 ns. All specified values are dependent on certain operating conditions as stated on a data sheet.

DC Supply Voltage (V_{CC})   The typical dc supply voltage for CMOS is either 5 V, 3.3 V, 2.5 V, or 1.8 V, depending on the category. An advantage of CMOS is that the supply voltages can vary over a wider range than for TTL. The 5 V CMOS can tolerate supply variations from 2 V to 6 V and still operate properly although propagation delay time and power dissipation are significantly affected. The 3.3 V CMOS can operate with supply voltages from 2 V to 3.6 V. The typical dc supply voltage for TTL is 5.0 V with a minimum of 4.5 V and a maximum of 5.5 V.

A lower power dissipation means less current from the dc supply.

Power Dissipation The **power dissipation,** P_D, of a logic gate is the product of the dc supply voltage and the average supply current. Normally, the supply current when the gate output is LOW is greater than when the gate output is HIGH. The manufacturer's data sheet usually designates the supply current for the LOW output state as I_{CCL} and for the HIGH state as I_{CCH}. The average supply current is determined based on a 50% duty cycle (output LOW half the time and HIGH half the time), so the average power dissipation of a logic gate is

Equation 3–2
$$P_D = V_{CC}\left(\frac{I_{CCH} + I_{CCL}}{2}\right)$$

CMOS series gates have very low power dissipations compared to the TTL series. However, the power dissipation of CMOS is dependent on the frequency of operation. At zero frequency the quiescent power is typically in the microwatt/gate range, and at the maximum operating frequency it can be in the low milliwatt range; therefore, power is sometimes specified at a given frequency. The HC series, for example, has a power of 2.75 μW/gate at 0 Hz (quiescent) and 600 μW/gate at 1 MHz.

Power dissipation for TTL is independent of frequency. For example, the ALS series uses 1.4 mW/gate regardless of the frequency and the F series uses 6 mW/gate.

Input and Output Logic Levels V_{IL} is the LOW level input voltage for a logic gate, and V_{IH} is the HIGH level input voltage. The 5 V CMOS accepts a maximum voltage of 1.5 V as V_{IL} and a minimum voltage of 3.5 V as V_{IH}. TTL accepts a maximum voltage of 0.8 V as V_{IL} and a minimum voltage of 2 V as V_{IH}.

V_{OL} is the LOW level output voltage and V_{OH} is the HIGH level output voltage. For 5 V CMOS, the maximum V_{OL} is 0.33 V and the minimum V_{OH} is 4.4 V. For TTL, the maximum V_{OL} is 0.4 V and the minimum V_{OH} is 2.4 V. All values depend on operating conditions as specified on the data sheet.

Speed-Power Product (SPP) This parameter (**speed-power product**) can be used as a measure of the performance of a logic circuit taking into account the propagation delay time and the power dissipation. It is especially useful for comparing the various logic gate series within the CMOS or TTL family or for comparing a CMOS gate to a TTL gate.

The SPP of a logic circuit is the product of the propagation delay time and the power dissipation and is expressed in joules (J), which is the unit of energy. The formula is

Equation 3–3
$$SPP = t_P P_D$$

EXAMPLE 3–23

A certain gate has a propagation delay of 5 ns and $I_{CCH} = 1$ mA and $I_{CCL} = 2.5$ mA with a dc supply voltage of 5 V. Determine the speed-power product.

Solution
$$P_D = V_{CC}\left(\frac{I_{CCH} + I_{CCL}}{2}\right) = 5\text{ V}\left(\frac{1\text{ mA} + 2.5\text{ mA}}{2}\right) = 5\text{ V}(1.75\text{ mA}) = 8.75\text{ mW}$$

$$SPP = (5\text{ ns})(8.75\text{ mW}) = \textbf{43.75 pJ}$$

Related Problem If the propagation delay of a gate is 15 ns and its *SPP* is 150 pJ, what is its average power dissipation?

Fan-Out and Loading The **fan-out** of a logic gate is the maximum number of inputs of the same series in an IC family that can be connected to a gate's output and still maintain the output voltage levels within specified limits. Fan-out is a significant parameter only for TTL because of the type of circuit technology. Since very high impedances are associated with CMOS circuits, the fan-out is very high but depends on frequency because of capacitive effects.

Fan-out is specified in terms of **unit loads.** A unit load for a logic gate equals one input to a like circuit. For example, a unit load for a 74LS00 NAND gate equals *one* input to another logic gate in the 74LS series (not necessarily a NAND gate). Because the current from a LOW input (I_{IL}) of a 74LS00 gate is 0.4 mA and the current that a LOW output (I_{OL}) can accept is 8.0 mA, the number of unit loads that a 74LS00 gate can drive in the LOW state is

A higher fan-out means that a gate output can be connected to more gate inputs.

$$\text{Unit loads} = \frac{I_{OL}}{I_{IL}} = \frac{8.0 \text{ mA}}{0.4 \text{ mA}} = 20$$

Figure 3–64 shows LS logic gates driving a number of other gates of the same circuit technology, where the number of gates depends on the particular circuit technology. For example, as you have seen, the maximum number of gate inputs (unit loads) that a 74LS series TTL gate can drive is 20.

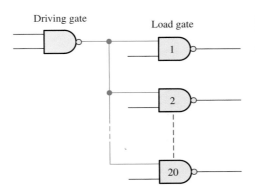

Driving gate

Load gate

◄ **FIGURE 3–64**

The LS TTL NAND gate output fans out to a maximum of 20 LS TTL gate inputs.

Data Sheets

A typical data sheet consists of an information page that shows, among other things, the logic diagram and packages, the recommended operating conditions, the electrical characteristics, and the switching characteristics. Partial data sheets for a 74LS00 and a 74HC00A are shown in Figures 3–65 and 3–66, respectively. The length of data sheets vary and some have much more information than others. Additional data sheets are provided on the Texas Instruments CD-ROM accompanying this textbook.

HANDS
ON
TIP

Unused gate inputs for TTL and CMOS should be connected to the appropriate logic level (HIGH or LOW). For AND/NAND, it is recommended that unused inputs be connected to V_{CC} (through a 1.0 kΩ resistor with TTL) and for OR/NOR, unused inputs should be connected to ground.

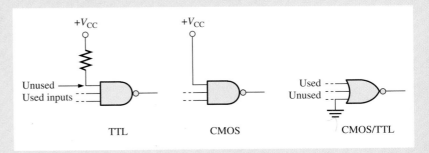

QUAD 2-INPUT NAND GATE

• ESD > 3500 Volts

SN54/74LS00

QUAD 2-INPUT NAND GATE
LOW POWER SCHOTTKY

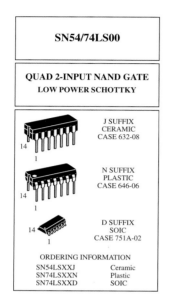

J SUFFIX
CERAMIC
CASE 632-08
14
1

N SUFFIX
PLASTIC
CASE 646-06
14
1

D SUFFIX
SOIC
CASE 751A-02
14
1

ORDERING INFORMATION

SN54LSXXJ	Ceramic
SN74LSXXN	Plastic
SN74LSXXD	SOIC

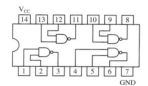

V_{CC}
14 13 12 11 10 9 8

1 2 3 4 5 6 7
GND

SN54/74LS00

DC CHARACTERISTICS OVER OPERATING TEMPERATURE RANGE (unless otherwise specified)

Symbol	Parameter		Min	Typ	Max	Unit	Test Conditions	
V_{IH}	Input HIGH Voltage		2.0			V	Guaranteed Input HIGH Voltage for All Inputs	
V_{IL}	Input LOW Voltage	54			0.7	V	Guaranteed Input LOW Voltage for All Inputs	
		74			0.8			
V_{IK}	Input Clamp Diode Voltage			−0.65	−1.5	V	V_{CC} = MIN, I_{IN} = −18 mA	
V_{OH}	Ouput HIGH Voltage	54	2.5	3.5		V	V_{CC} = MIN, I_{OH} = MAX, V_{IN} = V_{IH} or V_{IL} per Truth Table	
		74	2.7	3.5		V		
V_{OL}	Ouput LOW Voltage	54, 74		0.25	0.4	V	I_{OL} = 4.0 mA	V_{CC} = V_{CC} MIN, V_{IN} = V_{IL}
		74		0.35	0.5	V	I_{OL} = 8.0 mA	or V_{IH} per Truth Table
I_{IH}	Input HIGH Current				20	μA	V_{CC} = MAX, V_{IN} = 2.7 V	
					0.1	mA	V_{CC} = MAX, V_{IN} = 7.0 V	
I_{IL}	Input LOW Current				−0.4	mA	V_{CC} = MAX, I_{N} = 0.4 V	
I_{OS}	Short Circuit Current (Note 1)		−20		−100	mA	V_{CC} = MAX	
I_{CC}	Power Supply Current Total, Output HIGH				1.6	mA	V_{CC} = MAX	
	Total, Output LOW				4.4			

NOTE 1: Not more than one output should be shorted at a time, nor for more than 1 second.

AC CHARACTERISTICS (T_A = 25°C)

Symbol	Parameter	Min	Typ	Max	Unit	Test Conditions
t_{PLH}	Turn-Off Delay, Input to Output		9.0	15	ns	V_{CC} = 5.0 V
t_{PHL}	Turn-On Delay, Input to Output		10	15	ns	C_L = 15 pF

GUARANTEED OPERATING RANGES

Symbol	Parameter		Min	Typ	Max	Unit
V_{CC}	Supply Voltage	54	4.5	5.0	5.5	V
		74	4.75	5.0	5.25	
T_A	Operating Ambient Temperature Range	54	−55	25	125	°C
		74	0	25	70	
I_{OH}	Output Current — High	54, 74			−0.4	mA
I_{OL}	Output Current — Low	54			4.0	mA
		74			8.0	

▲ **FIGURE 3–65**

The partial data sheet for a 74LS00.

SECTION 3–8 REVIEW

1. List the two types of IC technologies that are the most widely used.

2. Identify the following IC logic designators:

    (a) LS   (b) ALS   (c) F   (d) HC   (e) AC   (f) HCT   (g) LV

3. Identify the following devices according to logic function:

    (a) 74LS04   (b) 74HC00   (c) 74LV08   (d) 74ALS10

    (e) 7432   (f) 74ACT11   (g) 74AHC02

4. Which IC technology generally has the lowest power dissipation?

5. What does the term *hex inverter* mean? What does *quad 2-input NAND* mean?

6. A positive pulse is applied to an inverter input. The time from the leading edge of the input to the leading edge of the output is 10 ns. The time from the trailing edge of the input to the trailing edge of the output is 8 ns. What are the values of t_{PLH} and t_{PHL}?

7. A certain gate has a propagation delay time of 6 ns and a power dissipation of 3 mW. Determine the speed-power product?

8. Define I_{CCL} and I_{CCH}.

9. Define V_{IL} and V_{IH}.

10. Define V_{OL} and V_{OH}.

Quad 2-Input NAND Gate High-Performance Silicon–Gate CMOS

The MC54/74HC00A is identical in pinout to the LS00. The device inputs are compatible with Standard CMOS outputs; with pullup resistors, they are compatible with LSTTL outputs.

- Output Drive Capability: 10 LSTTL Loads
- Outputs Directly Interface to CMOS, NMOS and TTL
- Operating Voltage Range: 2 to 6 V
- Low Input Current: 1 μA
- High Noise Immunity Characteristic of CMOS Devices
- In Compliance With the JEDEC Standard No. 7A Requirements
- Chip Complexity: 32 FETs or 8 Equivalent Gates

LOGIC DIAGRAM

$$Y = \overline{AB}$$

PIN 14 = V_CC
PIN 7 = GND

Pinout: 14–Load Packages (Top View)

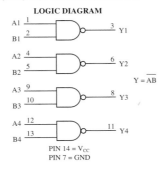

V_CC	B4	A4	Y4	B3	A3	Y3
14	13	12	11	10	9	8

1	2	3	4	5	6	7
A1	B1	Y1	A2	B2	Y2	GND

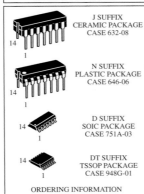

MC54/74HC00A

J SUFFIX
CERAMIC PACKAGE
CASE 632-08

N SUFFIX
PLASTIC PACKAGE
CASE 646-06

D SUFFIX
SOIC PACKAGE
CASE 751A-03

DT SUFFIX
TSSOP PACKAGE
CASE 948G-01

ORDERING INFORMATION

MC54HCXXAJ	Ceramic
MC74HCXXAN	Plastic
MC74HCXXAD	SOIC
MC74HCXXADT	TSSOP

FUNCTION TABLE

Inputs		Output
A	B	Y
L	L	H
L	H	H
H	L	H
H	H	L

MAXIMUM RATINGS*

Symbol	Parameter	Value	Unit
V_CC	DC Supply Voltage (Referenced to GND)	−0.5 to + 7.0	V
V_in	DC Input Voltage (Referenced to GND)	−0.5 to V_CC + 0.5	V
V_out	DC Output Voltage (Referenced to GND)	−0.5 to V_CC + 0.5	V
I_in	DC Input Current, per Pin	± 20	mA
I_out	DC Output Current, per Pin	± 25	mA
I_CC	DC Supply Current, V_CC and GND Pins	± 50	mA
P_D	Power Dissipation in Still Air, Plastic or Ceramic DIP†	750	mW
	SOIC Package†	500	
	TSSOP Package†	450	
T_stg	Storage Temperature	−65 to + 150	°C
T_L	Lead Temperature, 1 mm from Case for 10 Seconds		°C
	Plastic DIP, SOIC or TSSOP Package	260	
	Ceramic DIP	300	

* Maximum Ratings are those values beyond which damage to the device may occur.
Functional operation should be restricted to the Recommended Operating Conditions.
† Derating — Plastic DIP: – 10 mW/°C from 65° to 125° C
   Ceramic DIP: – 10 mW/°C from 100° to 125° C
   SOIC Package: – 7 mW/°C from 65° to 125° C
   TSSOP Package: – 6.1 mW/°C from 65° to 125° C

RECOMMENDED OPERATING CONDITIONS

Symbol	Parameter		Min	Max	Unit
V_CC	DC Supply Voltage (Referenced to GND)		2.0	6.0	V
V_in, V_out	DC Input Voltage, Output Voltage (Referenced to GND)		0	V_CC	V
T_A	Operating Temperature, All Package Types		−55	+125	°C
t_r, t_f	Input Rise and Fall Time	V_CC = 2.0 V	0	1000	ns
		V_CC = 4.5 V	0	500	
		V_CC = 6.0 V	0	400	

DC CHARACTERISTICS (Voltages Referenced to GND)

MC54/74HC00A

Symbol	Parameter	Condition	V_CC V	Guaranteed Limit −55 to 25°C	≤85°C	≤125°C	Unit
V_IH	Minimum High-Level Input Voltage	V_out = 0.1V or V_CC − 0.1V |I_out| ≤ 20μA	2.0	1.50	1.50	1.50	V
			3.0	2.10	2.10	2.10	
			4.5	3.15	3.15	3.15	
			6.0	4.20	4.20	4.20	
V_IL	Maximum Low-Level Input Voltage	V_out = 0.1V or V_CC − 0.1V |I_out| ≤ 20μA	2.0	0.50	0.50	0.50	V
			3.0	0.90	0.90	0.90	
			4.5	1.35	1.35	1.35	
			6.0	1.80	1.80	1.80	
V_OH	Minimum High-Level Output Voltage	V_in = V_IH or V_IL |I_out| ≤ 20μA	2.0	1.9	1.9	1.9	V
			4.5	4.4	4.4	4.4	
			6.0	5.9	5.9	5.9	
		V_in = V_IH or V_IL |I_out| ≤2.4mA	3.0	2.48	2.34	2.20	
		|I_out| ≤4.0mA	4.5	3.98	3.84	3.70	
		|I_out| ≤5.2mA	6.0	5.48	5.34	5.20	
V_OL	Maximum Low-Level Output Voltage	V_in = V_IH or V_IL |I_out| ≤ 20μA	2.0	0.1	0.1	0.1	V
			4.5	0.1	0.1	0.1	
			6.0	0.1	0.1	0.1	
		V_in = V_IH or V_IL |I_out| ≤2.4mA	3.0	0.26	0.33	0.40	
		|I_out| ≤4.0mA	4.5	0.26	0.33	0.40	
		|I_out| ≤5.2mA	6.0	0.26	0.33	0.40	
I_in	Maximum Input Leakage Current	V_in = V_CC or GND	6.0	±0.1	±1.0	±1.0	μA
I_CC	Maximum Quiescent Supply Current (per Package)	V_in = V_CC or GND I_out = 0μA	6.0	1.0	10	40	μA

AC CHARACTERISTICS (C_L = 50 pF, Input t_r = t_f = 6 ns)

Symbol	Parameter	V_CC V	Guaranteed Limit −55 to 25°C	≤85°C	≤125°C	Unit
t_PLH, t_PHL	Maximum Propagation Delay, Input A or B to Output Y	2.0	75	95	110	ns
		3.0	30	40	55	
		4.5	15	19	22	
		6.0	13	16	19	
t_TLH, t_THL	Maximum Output Transition Time, Any Output	2.0	75	95	110	ns
		3.0	27	32	36	
		4.5	15	19	22	
		6.0	13	16	19	
C_in	Maximum Input Capacitance		10	10	10	pF

		Typical @ 25°C, V_CC = 5.0 V, V_EE = 0 V	
C_PD	Power Dissipation Capacitance (Per Buffer)	22	pF

▲ **FIGURE 3–66**

The partial data sheet for a 74HC00A.

3-9 TROUBLESHOOTING

Troubleshooting is the process of recognizing, isolating, and correcting a fault or failure in a circuit or system. To be an effective troubleshooter, you must understand how the circuit or system is supposed to work and be able to recognize incorrect performance. For example, to determine whether or not a certain logic gate is faulty, you must know what the output should be for given inputs.

After completing this section, you should be able to

▪ Test for internally open inputs and outputs in IC gates ▪ Recognize the effects of a shorted IC input or output ▪ Test for external faults on a PC board ▪ Troubleshoot a simple frequency counter using an oscillosope

Internal Failures of IC Logic Gates

Opens and shorts are the most common types of internal gate failures. These can occur on the inputs or on the output of a gate inside the IC package. *Before attempting any troubleshooting, check for proper dc supply voltage and ground.*

Effects of an Internally Open Input   An internal open is the result of an open component on the chip or a break in the tiny wire connecting the IC chip to the package pin. An open input prevents a signal on that input from getting to the output of the gate, as illustrated in Figure 3–67(a) for the case of a 2-input NAND gate. An open TTL input acts effectively as a HIGH level, so pulses applied to the good input get through to the NAND gate output as shown in Figure 3–67(b).

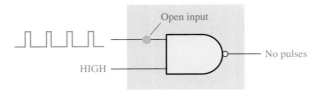

(a) Application of pulses to the open input will produce no pulses on the output.

(b) Application of pulses to the good input will produce output pulses for TTL NAND and AND gates because an open input typically acts as a HIGH. It is uncertain for CMOS.

▲ **FIGURE 3–67**

The effect of an open input on a NAND gate.

Conditions for Testing Gates   When testing a NAND gate or an AND gate, always make sure that the inputs that are not being pulsed are HIGH to enable the gate. When checking a NOR gate or an OR gate, always make sure that the inputs that are not being pulsed are LOW. When checking an XOR or XNOR gate, the level of the nonpulsed input does not matter because the pulses on the other input will force the inputs to alternate between the same level and opposite levels.

Troubleshooting an Open Input   Troubleshooting this type of failure is easily accomplished with an oscilloscope and function generator, as demonstrated in Figure 3–68 for the case of a 2-input NAND gate package. When measuring digital signals with a scope, always use dc coupling.

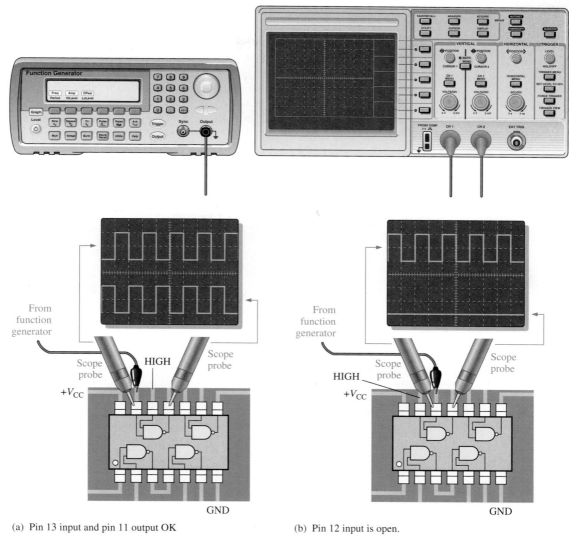

(a) Pin 13 input and pin 11 output OK   (b) Pin 12 input is open.

▲ **FIGURE 3–68**

Troubleshooting a NAND gate for an open input.

The first step in troubleshooting an IC that is suspected of being faulty is to make sure that the dc supply voltage (V_{CC}) and ground are at the appropriate pins of the IC. Next, apply continuous pulses to one of the inputs to the gate, making sure that the other input is HIGH (in the case of a NAND gate). In Figure 3–68(a), start by applying a pulse waveform to pin 13, which is one of the inputs to the suspected gate. If a pulse waveform is indicated on the output (pin 11 in this case), then the pin 13 input is not open. By the way, this also proves that the output is not open. Next, apply the pulse waveform to the other gate input (pin 12), making sure the other input is HIGH. There is no pulse waveform on the output at pin 11 and the output is LOW, indicating that the pin 12 input is open, as shown in Figure 3–68(b). The input not being pulsed must be HIGH for the case of a NAND gate or AND gate. If this were a NOR gate, the input not being pulsed would have to be LOW.

Effects of an Internally Open Output   An internally open gate output prevents a signal on any of the inputs from getting to the output. Therefore, no matter what the input conditions are, the output is unaffected. The level at the output pin of the IC will depend upon what it is externally connected to. It could be either HIGH, LOW, or floating (not fixed to any reference). In any case, there will be no signal on the output pin.

Troubleshooting an Open Output   Figure 3–69 illustrates troubleshooting an open NOR gate output. In part (a), one of the inputs of the suspected gate (pin 11 in this case) is pulsed, and the output (pin 13) has no pulse waveform. In part (b), the other input (pin 12) is pulsed and again there is no pulse waveform on the output. Under the condition that the input that is not being pulsed is at a LOW level, this test shows that the output is internally open.

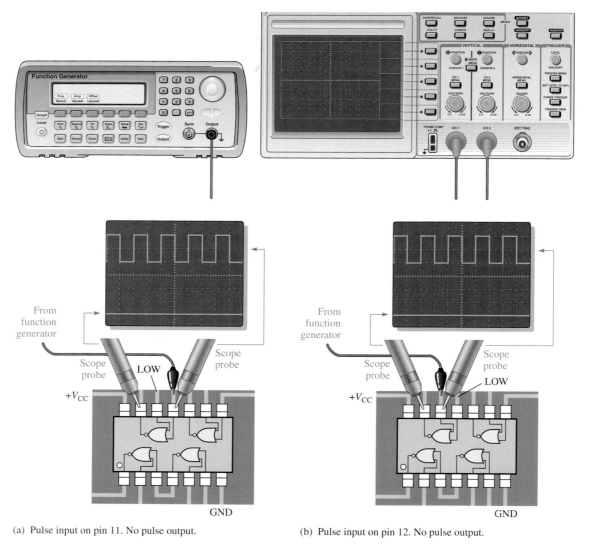

(a) Pulse input on pin 11. No pulse output.   (b) Pulse input on pin 12. No pulse output.

▲ **FIGURE 3–69**

Troubleshooting a NOR gate for an open output.

Shorted Input or Output   Although not as common as an open, an internal short to the dc supply voltage, the ground, another input, or an output can occur. When an input or output is shorted to the supply voltage, it will be stuck in the HIGH state. If an input or output is shorted to the ground, it will be stuck in the LOW state (0 V). If two inputs or an input and an output are shorted together, they will always be at the same level.

External Opens and Shorts

Many failures involving digital ICs are due to faults that are external to the IC package. These include bad solder connections, solder splashes, wire clippings, improperly etched printed circuit (PC) boards, and cracks or breaks in wires or printed circuit interconnections. These open or shorted conditions have the same effect on the logic gate as the internal faults, and troubleshooting is done in basically the same ways. A visual inspection of any circuit that is suspected of being faulty is the first thing a technician should do.

EXAMPLE 3–24

You are checking a 74LS10 triple 3-input NAND gate IC that is one of many ICs located on a PC board. You have checked pins 1 and 2 and they are both HIGH. Now you apply a pulse waveform to pin 13, and place your scope probe first on pin 12 and then on the connecting PC board trace, as indicated in Figure 3–70. Based on your observation of the scope screen, what is the most likely problem?

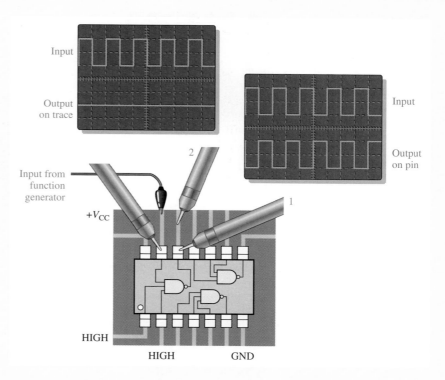

▲ FIGURE 3–70

Solution   The waveform with the probe in position 1 shows that there is pulse activity on the gate output at pin 12, but there are no pulses on the PC board trace as indicated by the probe in position 2. The gate is working properly, but the signal is not getting from pin 12 of the IC to the PC board trace.

Most likely there is a bad solder connection between pin 12 of the IC and the PC board, which is creating an open. You should resolder that point and check it again.

Related Problem   If there are no pulses at either point in Figure 3–70, what fault(s) does this indicate?

In most cases, you will be troubleshooting ICs that are mounted on PC boards or proto-type assemblies and interconnected with other ICs. As you progress through this book, you will learn how different types of digital ICs are used together to perform system functions. At this point, however, we are concentrating on individual IC gates. This limitation does not prevent us from looking at the system concept at a very basic and simplified level.

To continue the emphasis on systems, Examples 3–25 and 3–26 deal with troubleshooting the frequency counter that was introduced in Section 3–2.

After trying to operate the frequency counter shown in Figure 3–71, you find that it constantly reads out all 0s on its display, regardless of the input frequency. Determine the cause of this malfunction. The enable pulse has a width of 1 s.

Figure 3–71(a) gives an example of how the frequency counter should be working with a 12 Hz pulse waveform on the input to the AND gate. Part (b) shows that the display is improperly indicating 0 Hz.

▶ FIGURE 3–71

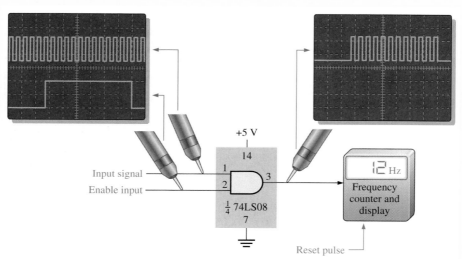

(a) The counter is working properly.

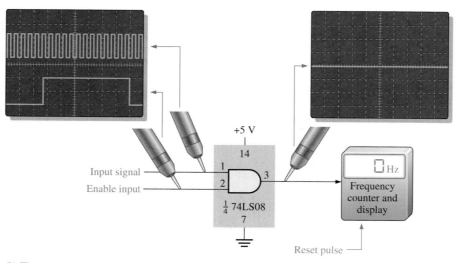

(b) The counter is not measuring a frequency.

Solution Three possible causes are

1. A constant active or asserted level on the counter reset input, which keeps the counter at zero.

2. No pulse signal on the input to the counter because of an internal open or short in the counter. This problem would keep the counter from advancing after being reset to zero.

3. No pulse signal on the input to the counter because of an open AND gate output or the absence of input signals, again keeping the counter from advancing from zero.

The first step is to make sure that V_{CC} and ground are connected to all the right places; assume that they are found to be okay. Next, check for pulses on both inputs to the AND gate. The scope indicates that there are proper pulses on both of these inputs. A check of the counter reset shows a LOW level which is known to be the unasserted level and, therefore, this is not the problem. The next check on pin 3 of the 74LS08 shows that there are no pulses on the output of the AND gate, indicating that the gate output is open. Replace the 74LS08 IC and check the operation again.

Related Problem If pin 2 of the 74LS08 AND gate is open, what indication should you see on the frequency display?

EXAMPLE 3–26

The frequency counter shown in Figure 3–72 appears to measure the frequency of input signals incorrectly. It is found that when a signal with a precisely known

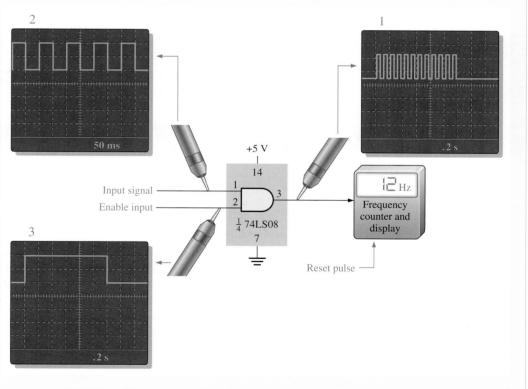

▲ **FIGURE 3–72**

frequency is applied to pin 1 of the AND gate, the oscilloscope display indicates a higher frequency. Determine what is wrong. The readings on the screen indicate sec/div.

Solution Recall from Section 3–2 that the input pulses were allowed to pass through the AND gate for exactly 1 s. The number of pulses counted in 1 s is equal to the frequency in hertz (cycles per second). Therefore, the 1 s interval, which is produced by the enable pulse on pin 2 of the AND gate, is very critical to an accurate frequency measurement. The enable pulses are produced internally by a precision oscillator circuit. The pulse must be exactly 1 s in width and in this case it occurs every 3 s to update the count. Just prior to each enable pulse, the counter is reset to zero so that it starts a new count each time.

Since the counter appears to be counting more pulses than it should to produce a frequency readout that is too high, the enable pulse is the primary suspect. Exact time-interval measurements must be made on the oscilloscope.

An input pulse waveform of exactly 10 Hz is applied to pin 1 of the AND gate and the display incorrectly shows 12 Hz. The first scope measurement, on the output of the AND gate, shows that there are 12 pulses for each enable pulse. In the second scope measurement, the input frequency is verified to be precisely 10 Hz (period = 100 ms). In the third scope measurement, the width of the enable pulse is found to be 1.2 s rather than 1 s.

The conclusion is that the enable pulse is out of calibration for some reason.

Related Problem What would you suspect if the readout were indicating a frequency less than it should be?

HANDS ON TIP

Proper grounding is very important when setting up to take measurements or work on a circuit. Properly grounding the oscilloscope protects you from shock and grounding yourself protects your circuits from damage. Grounding the oscilloscope means to connect it to earth ground by plugging the three-prong power cord into a grounded outlet. Grounding yourself means using a wrist-type grounding strap, particularly when you are working with CMOS circuits.

Also, for accurate measurements, make sure that the ground in the circuit you are testing is the same as the scope ground. This can be done by connecting the ground lead on the scope probe to a known ground point in the circuit, such as the metal chassis or a ground point on the circuit board. You can also connect the circuit ground to the GND jack on the front panel of the scope.

Troubleshooting problems that are keyed to the CD-ROM are available in the Multisim Troubleshooting Practice section of the end-of-chapter problems.

SUMMARY

- The inverter output is the complement of the input.
- The AND gate output is HIGH only when all the inputs are HIGH.
- The OR gate output is HIGH when any of the inputs is HIGH.
- The NAND gate output is LOW only when all the inputs are HIGH.
- The NAND can be viewed as a negative-OR whose output is HIGH when any input is LOW.
- The NOR gate output is LOW when any of the inputs is HIGH.
- The NOR can be viewed as a negative-AND whose output is HIGH only when all the inputs are LOW.
- The exclusive-OR gate output is HIGH when the inputs are not the same.
- The exclusive-NOR gate output is LOW when the inputs are not the same.
- Distinctive shape symbols and truth tables for various logic gates (limited to 2 inputs) are shown in Figure 3–73.

▶ FIGURE 3–73

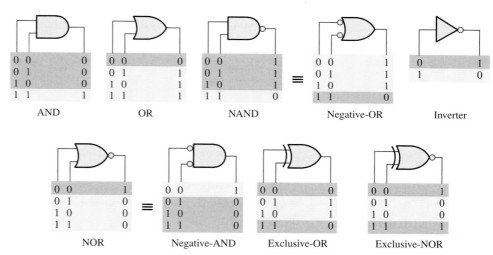

Note: Active states are shown in yellow.

- Most programmable logic devices (PLDs) are based on some form of AND array.
- Programmable link technologies are fuse, antifuse, EPROM, EEPROM, and SRAM.
- A PLD can be programmed in a hardware fixture called a programmer or mounted on a development printed circuit board.
- PLDs have an associated software development package for programming.
- Two methods of design entry using programming software are text entry (HDL) and graphic (schematic) entry.
- ISP PLDs can be programmed after they are installed in a system.
- JTAG stands for Joint Test Action Group and is an interface standard (IEEE Std. 1149.1) used for programming and testing PLDs.
- An embedded processor is used to facilitate in-system programming of PLDs.
- CMOS is made with MOS field-effect transistors.

- TTL is made with bipolar junction transistors.
- As a rule, CMOS has a lower power consumption than TTL.
- The average power dissipation of a logic gate is

$$P_D = V_{CC}\left(\frac{I_{CCH} + I_{CCL}}{2}\right)$$

- The speed-power product of a logic gate is

$$SPP = t_p P_D$$

KEY TERMS

Key terms and other bold terms in the chapter are defined in the end-of-book glossary.

AND array An array of AND gates consisting of a matrix of programmable interconnections.

AND gate A logic gate that produces a HIGH output only when all of the inputs are HIGH.

Antifuse A type of PLD nonvolatile programmable link that can be left open or can be shorted once as directed by the program.

Boolean algebra The mathematics of logic circuits.

CMOS Complementary metal-oxide semiconductor; a class of integrated logic circuits that is implemented with a type of field-effect transistor.

Complement The inverse or opposite of a number. LOW is the complement of HIGH, and 0 is the complement of 1.

EEPROM A type of nonvolatile PLD programmable link based on electrically erasable programmable read-only memory cells and can be turned on or off repeatedly by programming.

Enable To activate or put into an operational mode; an input on a logic circuit that enables its operation.

EPROM A type of PLD nonvolatile programmable link based on electrically programmable read-only memory cells and can be turned either on or off once with programming.

Exclusive-OR (XOR) gate A logic gate that produces a HIGH output only when its two inputs are at opposite levels.

Exclusive-NOR gate A logic gate that produces a LOW only when the two inputs are at opposite levels.

Fan-out The number of equivalent gate inputs of the same family series that a logic gate can drive.

Fuse A type of PLD nonvolatile programmable link that can be left shorted or can be opened once as directed by the program.

Inverter A logic circuit that inverts or complements its input.

JTAG Joint Test Action Group; an interface standard designated IEEE Std. 1149.1.

NAND gate A logic gate that produces a LOW output only when all the inputs are HIGH.

NOR gate A logic gate in which the output is LOW when one or more of the inputs are HIGH.

OR gate A logic gate that produces a HIGH output when one or more inputs are HIGH.

Propagation delay time The time interval between the occurrence of an input transition and the occurrence of the corresponding output transition in a logic circuit.

SRAM A type of PLD volatile programmable link based on static random-access memory cells and can be turned on or off repeatedly with programming.

Target device A PLD mounted on a programming fixture or development board into which a software logic design is to be downloaded.

Timing diagram A diagram of waveforms showing the proper timing relationship of all the waveforms.

Truth table A table showing the inputs and corresponding output(s) of a logic circuit.

TTL Transistor-transistor logic; a class of integrated logic circuits that uses bipolar junction transistors.

Unit load A measure of fan-out. One gate input represents one unit load to the output of a gate within the same IC family.

Answers are at the end of the chapter.

1. When the input to an inverter is HIGH (1), the output is
   (a) HIGH or 1   **(b)** LOW or 1   **(c)** HIGH or 0   **(d)** LOW or 0

2. An inverter performs an operation known as
   (a) complementation   **(b)** assertion
   (c) inversion   **(d)** both answers (a) and (c)

3. The output of an AND gate with inputs A, B, and C is a 1 (HIGH) when
   (a) $A = 1, B = 1, C = 1$ **(b)** $A = 1, B = 0, C = 1$ **(c)** $A = 0, B = 0, C = 0$

4. The output of an OR gate with inputs A, B, and C is a 1 (HIGH) when
   (a) $A = 1, B = 1, C = 1$ **(b)** $A = 0, B = 0, C = 1$ **(c)** $A = 0, B = 0, C = 0$
   (d) answers (a), (b), and (c)   **(e)** only answers (a) and (b)

5. A pulse is applied to each input of a 2-input NAND gate. One pulse goes HIGH at $t = 0$ and goes back LOW at $t = 1$ ms. The other pulse goes HIGH at $t = 0.8$ ms and goes back LOW at $t = 3$ ms. The output pulse can be described as follows:
   (a) It goes LOW at $t = 0$ and back HIGH at $t = 3$ ms.
   (b) It goes LOW at $t = 0.8$ ms and back HIGH at $t = 3$ ms.
   (c) It goes LOW at $t = 0.8$ ms and back HIGH at $t = 1$ ms.
   (d) It goes LOW at $t = 0.8$ ms and back LOW at $t = 1$ ms.

6. A pulse is applied to each input of a 2-input NOR gate. One pulse goes HIGH at $t = 0$ and goes back LOW at $t = 1$ ms. The other pulse goes HIGH at $t = 0.8$ ms and goes back LOW at $t = 3$ ms. The output pulse can be described as follows:
   (a) It goes LOW at $t = 0$ and back HIGH at $t = 3$ ms.
   (b) It goes LOW at $t = 0.8$ ms and back HIGH at $t = 3$ ms.
   (c) It goes LOW at $t = 0.8$ ms and back HIGH at $t = 1$ ms.
   (d) It goes HIGH at $t = 0.8$ ms and back LOW at $t = 1$ ms.

7. A pulse is applied to each input of an exclusive-OR gate. One pulse goes HIGH at $t = 0$ and goes back LOW at $t = 1$ ms. The other pulse goes HIGH at $t = 0.8$ ms and goes back LOW at $t = 3$ ms. The output pulse can be described as follows:
   (a) It goes HIGH at $t = 0$ and back LOW at $t = 3$ ms.
   (b) It goes HIGH at $t = 0$ and back LOW at $t = 0.8$ ms.
   (c) It goes HIGH at $t = 1$ ms and back LOW at $t = 3$ ms.
   (d) both answers (b) and (c)

8. A positive-going pulse is applied to an inverter. The time interval from the leading edge of the input to the leading edge of the output is 7 ns. This parameter is
   (a) speed-power product   **(b)** propagation delay, t_{PHL}
   (c) propagation delay, t_{PLH} **(d)** pulse width

9. The purpose of a programmable link in an AND array is to
   (a) connect an input variable to a gate input
   (b) connect a row to a column in the array matrix
   (c) disconnect a row from a column in the array matrix
   (d) do all of the above

10. The term OTP means
    (a) open test point   **(b)** one-time programmable
    (c) output test program   **(d)** output terminal positive

11. Types of PLD programmable link process technologies are
    (a) antifuse   **(b)** EEPROM
    (c) ROM   **(d)** both (a) and (b)
    (e) both (a) and (c)

12. A volatile programmable link technology is

   (a) fuse   **(b)** EPROM

   (c) SRAM   **(d)** EEPROM

13. Two ways to enter a logic design using PLD development software are

   (a) text and numeric   **(b)** text and graphic

   (c) graphic and coded   **(d)** compile and sort

14. JTAG stands for

   (a) Joint Test Action Group   **(b)** Java Top Array Group

   (c) Joint Test Array Group   **(d)** Joint Time Analysis Group

15. In-system programming of a PLD typically utilizes

   (a) an embedded clock generator   **(b)** an embedded processor

   (c) an embedded PROM   **(d)** both (a) and (b)

   (e) both (b) and (c)

16. To measure the period of a pulse waveform, you must use

   (a) a DMM   **(b)** a logic probe

   (c) an oscilloscope   **(d)** a logic pulser

17. Once you measure the period of a pulse waveform, the frequency is found by

   (a) using another setting   **(b)** measuring the duty cycle

   (c) finding the reciprocal of the period   **(d)** using another type of instrument

PROBLEMS

Answers to odd-numbered problems are at the end of the book.

SECTION 3–1 The Inverter

1. The input waveform shown in Figure 3–74 is applied to an inverter. Draw the timing diagram of the output waveform in proper relation to the input.

▶ **FIGURE 3–74**

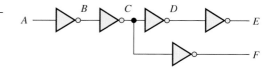

2. A network of cascaded inverters is shown in Figure 3–75. If a HIGH is applied to point A, determine the logic levels at points B through F.

▶ **FIGURE 3–75**

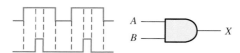

SECTION 3–2 The AND Gate

3. Determine the output, X, for a 2-input AND gate with the input waveforms shown in Figure 3–76. Show the proper relationship of output to inputs with a timing diagram.

▶ **FIGURE 3–76**

4. Repeat Problem 3 for the waveforms in Figure 3–77.

▶ FIGURE 3–77

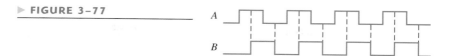

5. The input waveforms applied to a 3-input AND gate are as indicated in Figure 3–78. Show the output waveform in proper relation to the inputs with a timing diagram.

▶ FIGURE 3–78

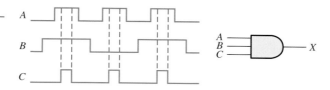

6. The input waveforms applied to a 4-input AND gate are as indicated in Figure 3–79. Show the output waveform in proper relation to the inputs with a timing diagram.

▶ FIGURE 3–79

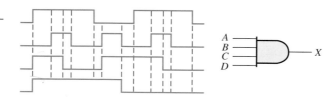

SECTION 3–3 The OR Gate

7. Determine the output for a 2-input OR gate when the input waveforms are as in Figure 3–77 and draw a timing diagram.

8. Repeat Problem 5 for a 3-input OR gate.

9. Repeat Problem 6 for a 4-input OR gate.

10. For the five input waveforms in Figure 3–80, determine the output for a 5-input AND gate and the output for a 5-input OR gate. Draw the timing diagram.

▶ FIGURE 3–80

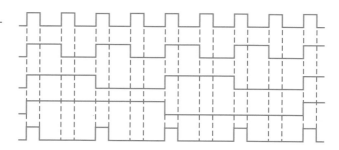

SECTION 3–4 The NAND Gate

11. For the set of input waveforms in Figure 3–81, determine the output for the gate shown and draw the timing diagram.

▶ FIGURE 3–81

12. Determine the gate output for the input waveforms in Figure 3–82 and draw the timing diagram.

▶ FIGURE 3–82

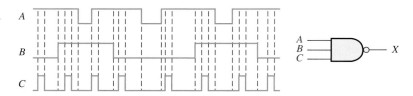

13. Determine the output waveform in Figure 3–83.

▶ FIGURE 3–83

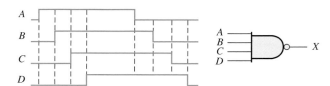

14. As you have learned, the two logic symbols shown in Figure 3–84 represent equivalent operations. The difference between the two is strictly from a functional viewpoint. For the NAND symbol, look for two HIGHs on the inputs to give a LOW output. For the negative-OR, look for at least one LOW on the inputs to give a HIGH on the output. Using these two functional viewpoints, show that each gate will produce the same output for the given inputs.

▶ FIGURE 3–84

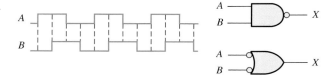

SECTION 3–5 The NOR Gate

15. Repeat Problem 11 for a 2-input NOR gate.

16. Determine the output waveform in Figure 3–85 and draw the timing diagram.

▶ FIGURE 3–85

17. Repeat Problem 13 for a 4-input NOR gate.

18. The NAND and the negative-OR symbols represent equivalent operations, but they are functionally different. For the NOR symbol, look for at least one HIGH on the inputs to give a LOW on the output. For the negative-AND, look for two LOWs on the inputs to give a HIGH output. Using these two functional points of view, show that both gates in Figure 3–86 will produce the same output for the given inputs.

▶ FIGURE 3–86

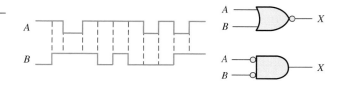

SECTION 3–6   **The Exclusive-OR and Exclusive-NOR Gates**

19. How does an exclusive-OR gate differ from an OR gate in its logical operation?

20. Repeat Problem 11 for an exclusive-OR gate.

21. Repeat Problem 11 for an exclusive-NOR gate.

22. Determine the output of an exclusive-OR gate for the inputs shown in Figure 3–77 and draw a timing diagram.

SECTION 3–7   **Programmable Logic**

23. In the simple programmed AND array with programmable links in Figure 3–87, determine the Boolean output expressions.

▶ **FIGURE 3–87**

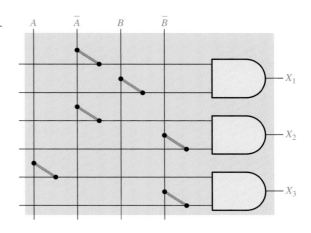

24. Determine by row and column number which fusible links must be blown in the programmable AND array of Figure 3–88 to implement each of the following product terms: $X_1 = \overline{A}BC$, $X_2 = AB\overline{C}$, $X_3 = \overline{A}B\overline{C}$.

▶ **FIGURE 3–88**

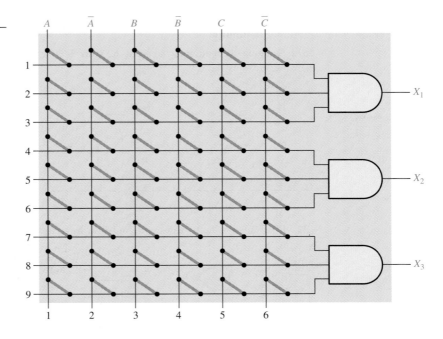

SECTION 3–8 **Fixed-Function Logic**

25. In the comparison of certain logic devices, it is noted that the power dissipation for one particular type increases as the frequency increases. Is the device TTL or CMOS?

26. Using the data sheets in Figures 3–65 and 3–66, determine the following:

   (a) 74LS00 power dissipation at maximum supply voltage and a 50% duty cycle

   (b) Minimum HIGH level output voltage for a 74LS00

   (c) Maximum propagation delay for a 74LS00

   (d) Maximum LOW level output voltage for a 74HC00A

   (e) Maximum propagation delay for a 74HC00A

27. Determine t_{PLH} and t_{PHL} from the oscilloscope display in Figure 3–89. The readings indicate V/div and sec/div for each channel.

▶ FIGURE 3–89

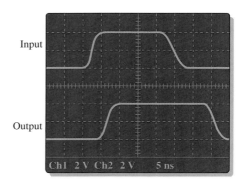

28. Gate A has $t_{PLH} = t_{PHL} = 6$ ns. Gate B has $t_{PLH} = t_{PHL} = 10$ ns. Which gate can be operated at a higher frequency?

29. If a logic gate operates on a dc supply voltage of $+5$ V and draws an average current of 4 mA, what is its power dissipation?

30. The variable I_{CCH} represents the dc supply current from V_{CC} when all outputs of an IC are HIGH. The variable I_{CCL} represents the dc supply current when all outputs are LOW. For a 74LS00 IC, determine the typical power dissipation when all four gate outputs are HIGH. (See data sheet in Figure 3–65.)

SECTION 3–9 **Troubleshooting**

31. Examine the conditions indicated in Figure 3–90, and identify the faulty gates.

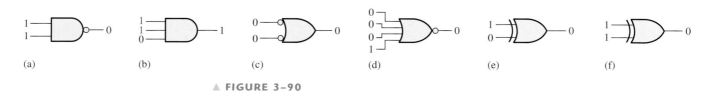

▲ FIGURE 3–90

32. Determine the faulty gates in Figure 3–91 by analyzing the timing diagrams.

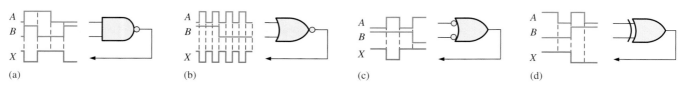

▲ FIGURE 3–91

33. Using an oscilloscope, you make the observations indicated in Figure 3–92. For each observation determine the most likely gate failure.

▶ **FIGURE 3–92**

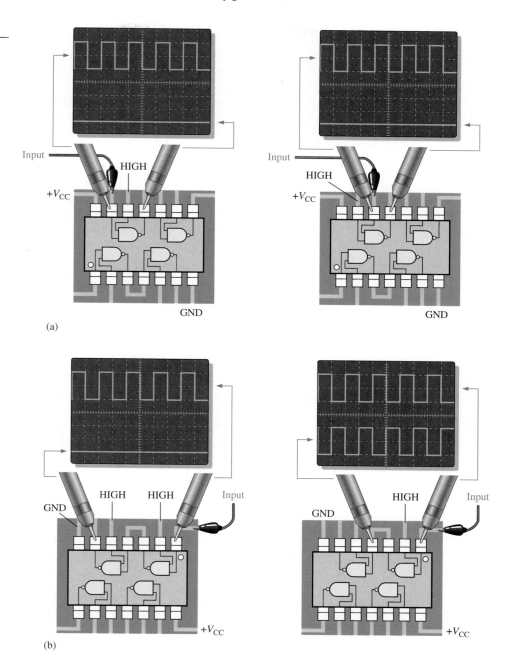

34. The seat belt alarm circuit in Figure 3–16 has malfunctioned. You find that when the ignition switch is turned on and the seat belt is unbuckled, the alarm comes on and will not go off. What is the most likely problem? How do you troubleshoot it?

35. Every time the ignition switch is turned on in the circuit of Figure 3–16, the alarm comes on for thirty seconds, even when the seat belt is buckled. What is the most probable cause of this malfunction?

36. What failure(s) would you suspect if the output of a 3-input NAND gate stays HIGH no matter what the inputs are?

Special Design Problems

37. Sensors are used to monitor the pressure and the temperature of a chemical solution stored in a vat. The circuitry for each sensor produces a HIGH voltage when a specified maximum value is exceeded. An alarm requiring a LOW voltage input must be activated when either the pressure or the temperature is excessive. Design a circuit for this application.

38. In a certain automated manufacturing process, electrical components are automatically inserted in a PC board. Before the insertion tool is activated, the PC board must be properly positioned, and the component to be inserted must be in the chamber. Each of these prerequisite conditions is indicated by a HIGH voltage. The insertion tool requires a LOW voltage to activate it. Design a circuit to implement this process.

39. Modify the frequency counter in Figure 3–15 to operate with an enable pulse that is active-LOW rather than HIGH during the 1 s interval.

40. Assume that the enable signal in Figure 3–15 has the waveform shown in Figure 3–93. Assume that waveform *B* is also available. Devise a circuit that will produce an active-HIGH reset pulse to the counter only during the time that the enable signal is LOW.

▷ **FIGURE 3–93**

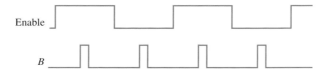

41. Design a circuit to fit in the beige block of Figure 3–94 that will cause the headlights of an automobile to be turned off automatically 15 s after the ignition switch is turned off, if the light switch is left on. Assume that a LOW is required to turn the lights off.

▷ **FIGURE 3–94**

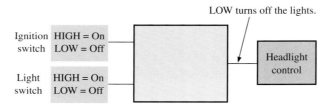

42. Modify the logic circuit for the intrusion alarm in Figure 3–24 so that two additional rooms, each with two windows and one door, can be protected.

43. Further modify the logic circuit from Problem 42 for a change in the input sensors where Open = LOW and Closed = HIGH.

Multisim Troubleshooting Practice

44. Open file P03-44, connect the Multisim logic converter to the circuit, and observe the operation of the AND gate. Based on the observed inputs and output, determine the most likely fault in the gate.

45. Open file P03-45, connect the Multisim logic converter to the circuit, and observe the operation of the NAND gate. Based on the observed inputs and output, determine the most likely fault in the gate.

46. Open file P03-46, connect the Multisim logic converter to the circuit, and observe the operation of the NOR gate. Based on the observed inputs and output, determine the most likely fault in the gate.

47. Open file P03-47, connect the Multisim logic converter to the circuit, and observe the operation of the exclusive-OR gate. Based on the observed inputs and output, determine the most likely fault in the gate.

ANSWERS

SECTION REVIEWS

SECTION 3–1 **The Inverter**

1. When the inverter input is 1, the output is 0.
2. **(a)**

(b) A negative-going pulse is on the output (HIGH to LOW and back HIGH).

SECTION 3–2 **The AND Gate**

1. An AND gate output is HIGH only when all inputs are HIGH.
2. An AND gate output is LOW when one or more inputs are LOW.
3. Five-input AND: $X = 1$ when $ABCDE = 11111$, and $X = 0$ for all other combinations of $ABCDE$.

SECTION 3–3 **The OR Gate**

1. An OR gate output is HIGH when one or more inputs are HIGH.
2. An OR gate output is LOW only when all inputs are LOW.
3. Three-input OR: $X = 0$ when $ABC = 000$, and $X = 1$ for all other combinations of ABC.

SECTION 3–4 **The NAND Gate**

1. A NAND output is LOW only when all inputs are HIGH.
2. A NAND output is HIGH when one or more inputs are LOW.
3. NAND: active-LOW output for all HIGH inputs; negative-OR: active-HIGH output for one or more LOW inputs. They have the same truth tables.
4. $X = \overline{ABC}$

SECTION 3–5 **The NOR Gate**

1. A NOR output is HIGH only when all inputs are LOW.
2. A NOR output is LOW when one or more inputs are HIGH.
3. NOR: active-LOW output for one or more HIGH inputs; negative-AND: active-HIGH output for all LOW inputs. They have the same truth tables.
4. $X = \overline{A + B + C}$

SECTION 3–6 **The Exclusive-OR and Exclusive-NOR Gates**

1. An XOR output is HIGH when the inputs are at opposite levels.
2. An XNOR output is HIGH when the inputs are at the same levels.
3. Apply the bits to the XOR inputs; when the output is HIGH, the bits are different.

SECTION 3–7 **Programmable Logic**

1. Fuse, antifuse, EPROM, EEPROM, and SRAM
2. Volatile means that all the data are lost when power is off and the PLD must be reprogrammed; SRAM-based

3. Text entry and graphic entry

4. JTAG is Joint Test Action Group; the IEEE Std. 1149.1 for programming and test interfacing.

SECTION 3–8 **Fixed-Function Logic**

1. CMOS and TTL

2. (a) LS—Low-power Schottky   (b) ALS—Advanced LS

(c) F—fast TTL   (d) HC—High-speed CMOS

(e) AC—Advanced CMOS   (f) HCT—HC CMOS TTL compatible

(g) LV—Low-voltage CMOS

3. (a) 74LS04—Hex inverter   (b) 74HC00—Quad 2-input NAND

(c) 74LV08—Quad 2-input AND   (d) 74ALS10—Triple 3-input NAND

(e) 7432—Quad 2-input OR   (f) 74ACT11—Triple 3-input AND

(g) 74AHC02—Quad 2-input NOR

4. Lowest power—CMOS

5. Six inverters in a package; four 2-input NAND gates in a package

6. $t_{PLH} = 10$ ns; $t_{PHL} = 8$ ns

7. 18 pJ

8. I_{CCL}—dc supply current for LOW output state; I_{CCH}—dc supply current for HIGH output state

9. V_{IL}—LOW input voltage; V_{IH}—HIGH input voltage

10. V_{OL}—LOW output voltage; V_{OH}—HIGH output voltage

SECTION 3–9 **Troubleshooting**

1. Opens and shorts are the most common failures.

2. An open input which effectively makes input HIGH

3. Amplitude and period

RELATED PROBLEMS FOR EXAMPLES

3–1 The timing diagram is not affected.

3–2 See Table 3–13.

▶ TABLE 3–13

INPUTS ABCD	OUTPUT X	INPUTS ABCD	OUTPUT X
0000	0	1000	0
0001	0	1001	0
0010	0	1010	0
0011	0	1011	0
0100	0	1100	0
0101	0	1101	0
0110	0	1110	0
0111	0	1111	1

3–3 See Figure 3–95.

▶ **FIGURE 3–95**

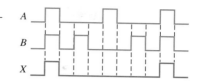

3–4 The output waveform is the same as input *A*.

3–5 See Figure 3–96.

3–6 See Figure 3–97.

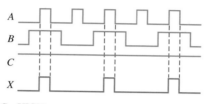

C = HIGH

▲ **FIGURE 3–96**

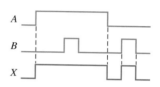

▲ **FIGURE 3–97**

3–7 See Figure 3–98.

3–8 See Figure 3–99.

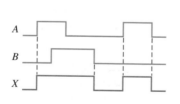

▲ **FIGURE 3–98**

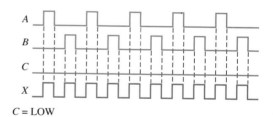

C = LOW

▲ **FIGURE 3–99**

3–9 See Figure 3–100.

3–10 See Figure 3–101.

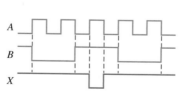

▲ **FIGURE 3–100**

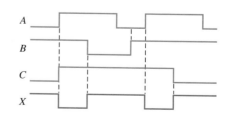

▲ **FIGURE 3–101**

3–11 Use a 3-input NAND gate.

3–12 Use a 4-input NAND gate operating as a negative-OR gate.

3–13 See Figure 3–102.

▶ **FIGURE 3–102**

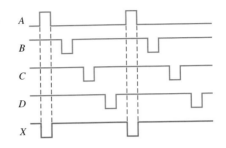

3–14 See Figure 3–103.

3–15 See Figure 3–104.

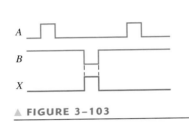

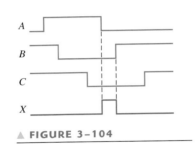

▲ **FIGURE 3–103**   ▲ **FIGURE 3–104**

3–16 Use a 2-input NOR gate.

3–17 A 3-input NAND gate.

3–18 The output is always LOW. The timing diagram is a straight line.

3–19 The exclusive-OR gate will not detect simultaneous failures if both circuits produce the same outputs.

3–20 The outputs are unaffected.

3–21 6 columns, 9 rows, and 3 AND gates with three inputs each

3–22 The gate with 4 ns t_{PLH} and t_{PHL} can operate at the highest frequency.

3–23 10 mW

3–24 The gate output or pin 13 input is internally open.

3–25 The display will show an erratic readout because the counter continues until reset.

3–26 The enable pulse is too short or the counter is reset too soon.

SELF-TEST

1. (d)   **2.** (d)   **3.** (a)   **4.** (e)   **5.** (c)   **6.** (a)   **7.** (d)   **8.** (b)   **9.** (d)

10. (b)   **11.** (d)   **12.** (c)   **13.** (b)   **14.** (a)   **15.** (d)   **16.** (c)   **17.** (c)

4

BOOLEAN ALGEBRA AND LOGIC SIMPLIFICATION

- Apply the basic laws and rules of Boolean algebra

- Apply DeMorgan's theorems to Boolean expressions

- Describe gate networks with Boolean expressions

- Evaluate Boolean expressions

- Simplify expressions by using the laws and rules of Boolean algebra

- Convert any Boolean expression into a sum-of-products (SOP) form

- Convert any Boolean expression into a product of-sums (POS) form

- Use a Karnaugh map to simplify Boolean expressions

- Use a Karnaugh map to simplify truth table functions

- Utilize "don't care" conditions to simplify logic functions

- Write a VHDL program for simple logic

- Apply Boolean algebra, the Karnaugh map method, and VHDL to a system application

KEY TERMS

Variable	Product-of-sums (POS)
Complement	Karnaugh map
Sum term	Minimization
Product term	"Don't care"
Sum-of-products (SOP)	VHDL

INTRODUCTION

In 1854, George Boole published a work titled *An Investigation of the Laws of Thought, on Which Are Founded the Mathematical Theories of Logic and Probabilities*. It was in this publication that a "logical algebra," known today as Boolean algebra, was formulated. Boolean algebra is a convenient and systematic way of expressing and analyzing the operation of logic circuits. Claude Shannon was the first to apply Boole's work to the analysis and design of logic circuits. In 1938, Shannon wrote a thesis at MIT titled *A Symbolic Analysis of Relay and Switching Circuits*.

This chapter covers the laws, rules, and theorems of Boolean algebra and their application to digital circuits. You will learn how to define a given circuit with a Boolean expression and then evaluate its operation. You will also learn how to simplify logic circuits using the methods of Boolean algebra and Karnaugh maps.

The hardware description language VHDL for programming logic devices is introduced.

■■■ DIGITAL SYSTEM APPLICATION PREVIEW

The Digital System Application illustrates concepts taught in this chapter. The 7-segment display logic in the tablet counting and control system from Chapter 1 is a good way to illustrate the application of Boolean algebra and the Karnaugh map method to obtain the simplest possible implementation in the design of logic circuits. Therefore, in this digital system application, the focus is on the BCD to 7-segment logic that drives the two system displays shown in Figure 1–58.

WWW. VISIT THE COMPANION WEBSITE

Study aids for this chapter are available at
http://www.prenhall.com/floyd

4–1 BOOLEAN OPERATIONS AND EXPRESSIONS

Boolean algebra is the mathematics of digital systems. A basic knowledge of Boolean algebra is indispensable to the study and analysis of logic circuits. In the last chapter, Boolean operations and expressions in terms of their relationship to NOT, AND, OR, NAND, and NOR gates were introduced.

After completing this section, you should be able to

■ Define *variable* ■ Define *literal* ■ Identify a sum term ■ Evaluate a sum term
■ Identify a product term ■ Evaluate a product term ■ Explain Boolean addition
■ Explain Boolean multiplication

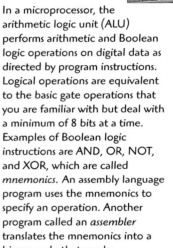

COMPUTER NOTE

In a microprocessor, the arithmetic logic unit (ALU) performs arithmetic and Boolean logic operations on digital data as directed by program instructions. Logical operations are equivalent to the basic gate operations that you are familiar with but deal with a minimum of 8 bits at a time. Examples of Boolean logic instructions are AND, OR, NOT, and XOR, which are called *mnemonics*. An assembly language program uses the mnemonics to specify an operation. Another program called an *assembler* translates the mnemonics into a binary code that can be understood by the microprocessor.

The OR gate is a Boolean adder.

Variable, complement, and *literal* are terms used in Boolean algebra. A **variable** is a symbol (usually an italic uppercase letter) used to represent a logical quantity. Any single variable can have a 1 or a 0 value. The **complement** is the inverse of a variable and is indicated by a bar over the variable (overbar). For example, the complement of the variable A is $\overline{A}$. If $A = 1$, then $\overline{A} = 0$. If $A = 0$, then $\overline{A} = 1$. The complement of the variable A is read as "not A" or "A bar." Sometimes a prime symbol rather than an overbar is used to denote the complement of a variable; for example, B' indicates the complement of B. In this book, only the overbar is used. A **literal** is a variable or the complement of a variable.

Boolean Addition

Recall from Chapter 3 that **Boolean addition** is equivalent to the OR operation and the basic rules are illustrated with their relation to the OR gate as follows:

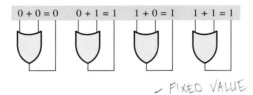

$$0 + 0 = 0 \qquad 0 + 1 = 1 \qquad 1 + 0 = 1 \qquad 1 + 1 = 1$$

FIXED VALUE

In Boolean algebra, a **sum term** is a sum of literals. In logic circuits, a sum term is produced by an OR operation with no AND operations involved. Some examples of sum terms are $A + B, A + \overline{B}, A + B + \overline{C},$ and $\overline{A} + B + C + \overline{D}$.

A sum term is equal to 1 when one or more of the literals in the term are 1. A sum term is equal to 0 only if each of the literals is 0.

EXAMPLE 4–1

Determine the values of A, B, C, and D that make the sum term $A + \overline{B} + C + \overline{D}$ equal to 0.

Solution   For the sum term to be 0, each of the literals in the term must be 0. Therefore, $A = \mathbf{0},$ $B = \mathbf{1}$ so that $\overline{B} = 0$, $C = \mathbf{0}$, and $D = \mathbf{1}$ so that $\overline{D} = 0$.

$$A + \overline{B} + C + \overline{D} = 0 + \overline{1} + 0 + \overline{1} = 0 + 0 + 0 + 0 = 0$$

Related Problem *   Determine the values of A and B that make the sum term $\overline{A} + B$ equal to 0.

---

*Answers are at the end of the chapter.

Boolean Multiplication

Also recall from Chapter 3 that **Boolean multiplication** is equivalent to the AND operation and the basic rules are illustrated with their relation to the AND gate as follows:

The AND gate is a Boolean multiplier.

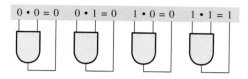

In Boolean algebra, a **product term** is the product of literals. In logic circuits, a product term is produced by an AND operation with no OR operations involved. Some examples of product terms are AB, $A\overline{B}$, ABC, and $A\overline{B}C\overline{D}$.

A product term is equal to 1 only if each of the literals in the term is 1. A product term is equal to 0 when one or more of the literals are 0.

EXAMPLE 4–2

Determine the values of A, B, C, and D that make the product term $A\overline{B}C\overline{D}$ equal to 1.

Solution   For the product term to be 1, each of the literals in the term must be 1. Therefore, $A = $ **1**, $B = $ **0** so that $\overline{B} = 1$, $C = $ **1**, and $D = $ **0** so that $\overline{D} = 1$.

$$A\overline{B}C\overline{D} = 1 \cdot \overline{0} \cdot 1 \cdot \overline{0} = 1 \cdot 1 \cdot 1 \cdot 1 = 1$$

Related Problem   Determine the values of A and B that make the product term $\overline{A}\,\overline{B}$ equal to 1.

SECTION 4–1 REVIEW

Answers are at the end of the chapter.

1. If $A = 0$, what does $\overline{A}$ equal?
2. Determine the values of A, B, and C that make the sum term $\overline{A} + B + \overline{C}$ equal to 0.
3. Determine the values of A, B, and C that make the product term $A\overline{B}C$ equal to 1.

4–2   LAWS AND RULES OF BOOLEAN ALGEBRA

As in other areas of mathematics, there are certain well-developed rules and laws that must be followed in order to properly apply Boolean algebra. The most important of these are presented in this section.

After completing this section, you should be able to

■ Apply the commutative laws of addition and multiplication ■ Apply the associative laws of addition and multiplication ■ Apply the distributive law ■ Apply twelve basic rules of Boolean algebra

Laws of Boolean Algebra

The basic laws of Boolean algebra—the **commutative laws** for addition and multiplication, the **associative laws** for addition and multiplication, and the **distributive law**—are the same as in ordinary algebra. Each of the laws is illustrated with two or three variables, but the number of variables is not limited to this.

Commutative Laws The *commutative law of addition* for two variables is written as

Equation 4–1

$$A + B = B + A$$

This law states that the order in which the variables are ORed makes no difference. Remember, in Boolean algebra as applied to logic circuits, addition and the OR operation are the same. Figure 4–1 illustrates the commutative law as applied to the OR gate and shows that it doesn't matter to which input each variable is applied. (The symbol ≡ means "equivalent to.")

▶ **FIGURE 4–1**

Application of commutative law of addition.

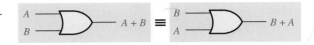

The *commutative law of multiplication* for two variables is

Equation 4–2

$$AB = BA$$

This law states that the order in which the variables are ANDed makes no difference. Figure 4–2 illustrates this law as applied to the AND gate.

▶ **FIGURE 4–2**

Application of commutative law of multiplication.

Associative Laws The *associative law of addition* is written as follows for three variables:

Equation 4–3

$$A + (B + C) = (A + B) + C$$

This law states that when ORing more than two variables, the result is the same regardless of the grouping of the variables. Figure 4–3 illustrates this law as applied to 2-input OR gates.

▶ **FIGURE 4–3**

Application of associative law of addition. Open file F04-03 to verify.

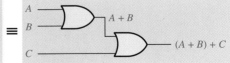

The *associative law of multiplication* is written as follows for three variables:

Equation 4–4

$$A(BC) = (AB)C$$

This law states that it makes no difference in what order the variables are grouped when ANDing more than two variables. Figure 4–4 illustrates this law as applied to 2-input AND gates.

▶ **FIGURE 4–4**

Application of associative law of multiplication. Open file F04-04 to verify.

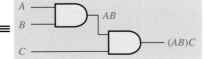

Distributive Law The distributive law is written for three variables as follows:

Equation 4–5

$$A(B + C) = AB + AC$$

This law states that ORing two or more variables and then ANDing the result with a single variable is equivalent to ANDing the single variable with each of the two or more variables and then ORing the products. The distributive law also expresses the process of *factoring* in which the common variable A is factored out of the product terms, for example, $AB + AC = A(B + C)$. Figure 4–5 illustrates the distributive law in terms of gate implementation.

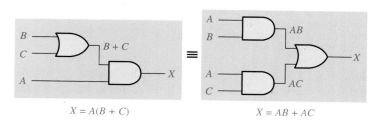

$$X = A(B + C)$$

$$X = AB + AC$$

◀ **FIGURE 4–5**

Application of distributive law. Open file F04–05 to verify.

Rules of Boolean Algebra

Table 4–1 lists 12 basic rules that are useful in manipulating and simplifying **Boolean expressions.** Rules 1 through 9 will be viewed in terms of their application to logic gates. Rules 10 through 12 will be derived in terms of the simpler rules and the laws previously discussed.

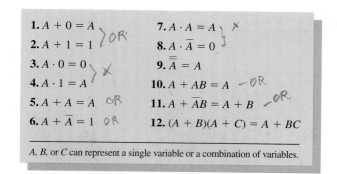

1. $A + 0 = A$	7. $A \cdot A = A$
2. $A + 1 = 1$	8. $A \cdot \overline{A} = 0$
3. $A \cdot 0 = 0$	9. $\overline{\overline{A}} = A$
4. $A \cdot 1 = A$	10. $A + AB = A$
5. $A + A = A$	11. $A + \overline{A}B = A + B$
6. $A + \overline{A} = 1$	12. $(A + B)(A + C) = A + BC$

A, B, or C can represent a single variable or a combination of variables.

◀ **TABLE 4–1**

Basic rules of Boolean algebra.

Rule 1. $A + 0 = A$ A variable ORed with 0 is always equal to the variable. If the input variable A is 1, the output variable X is 1, which is equal to A. If A is 0, the output is 0, which is also equal to A. This rule is illustrated in Figure 4–6, where the lower input is fixed at 0.

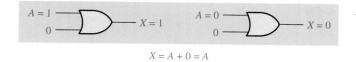

$$X = A + 0 = A$$

◀ **FIGURE 4–6**

Rule 2. $A + 1 = 1$ A variable ORed with 1 is always equal to 1. A 1 on an input to an OR gate produces a 1 on the output, regardless of the value of the variable on the other input. This rule is illustrated in Figure 4–7, where the lower input is fixed at 1.

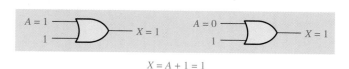

$$X = A + 1 = 1$$

◀ **FIGURE 4–7**

Rule 3. A · 0 = 0 A variable ANDed with 0 is always equal to 0. Any time one input to an AND gate is 0, the output is 0, regardless of the value of the variable on the other input. This rule is illustrated in Figure 4–8, where the lower input is fixed at 0.

▶ FIGURE 4–8

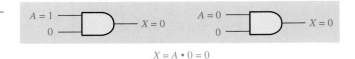

$$X = A \cdot 0 = 0$$

Rule 4. A · 1 = A A variable ANDed with 1 is always equal to the variable. If A is 0 the output of the AND gate is 0. If A is 1, the output of the AND gate is 1 because both inputs are now 1s. This rule is shown in Figure 4–9, where the lower input is fixed at 1.

▶ FIGURE 4–9

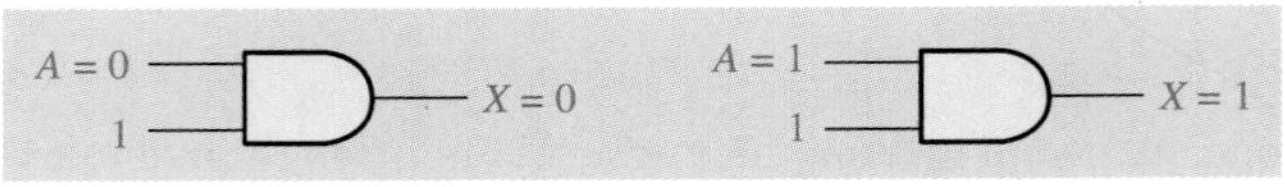

$$X = A \cdot 1 = A$$

Rule 5. A + A = A A variable ORed with itself is always equal to the variable. If A is 0, then $0 + 0 = 0$; and if A is 1, then $1 + 1 = 1$. This is shown in Figure 4–10, where both inputs are the same variable.

▶ FIGURE 4–10

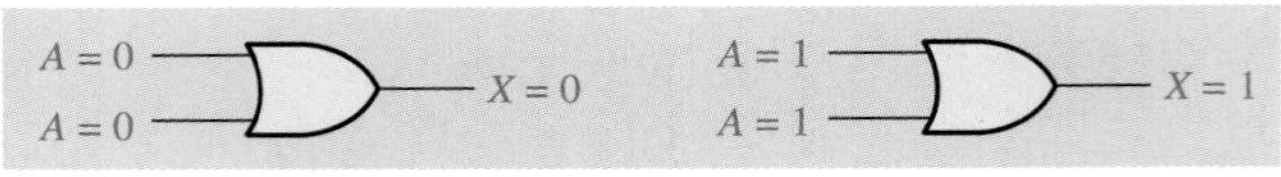

$$X = A + A = A$$

Rule 6. A + $\overline{A}$ = 1 A variable ORed with its complement is always equal to 1. If A is 0, then $0 + \overline{0} = 0 + 1 = 1$. If A is 1, then $1 + \overline{1} = 1 + 0 = 1$. See Figure 4–11, where one input is the complement of the other.

▶ FIGURE 4–11

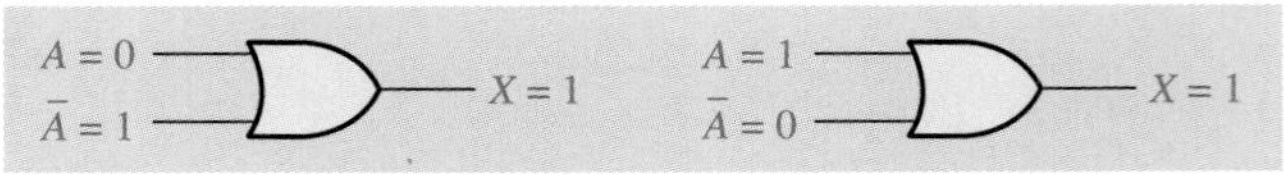

$$X = A + \overline{A} = 1$$

Rule 7. A · A = A A variable ANDed with itself is always equal to the variable. If $A = 0$, then $0 \cdot 0 = 0$; and if $A = 1$, then $1 \cdot 1 = 1$. Figure 4–12 illustrates this rule.

▶ FIGURE 4–12

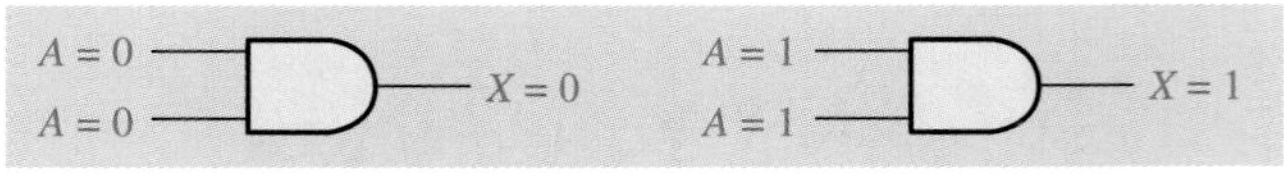

$$X = A \cdot A = A$$

Rule 8. $A \cdot \overline{A} = 0$ A variable ANDed with its complement is always equal to 0. Either A or $\overline{A}$ will always be 0; and when a 0 is applied to the input of an AND gate, the output will be 0 also. Figure 4–13 illustrates this rule.

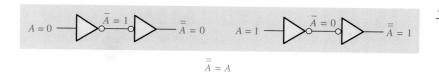

◄ FIGURE 4–13

Rule 9. $\overline{\overline{A}} = A$ The double complement of a variable is always equal to the variable. If you start with the variable A and complement (invert) it once, you get $\overline{A}$. If you then take $\overline{A}$ and complement (invert) it, you get A, which is the original variable. This rule is shown in Figure 4–14 using inverters.

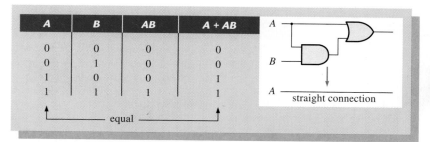

◄ FIGURE 4–14

Rule 10. $A + AB = A$ This rule can be proved by applying the distributive law, rule 2, and rule 4 as follows:

$$A + AB = A(1 + B) \quad \text{Factoring (distributive law)}$$
$$= A \cdot 1 \quad \text{Rule 2: } (1 + B) = 1$$
$$= A \quad \text{Rule 4: } A \cdot 1 = A$$

The proof is shown in Table 4–2, which shows the truth table and the resulting logic circuit simplification.

A	B	AB	A + AB
0	0	0	0
0	1	0	0
1	0	0	1
1	1	1	1

equal

◄ TABLE 4–2

Rule 10: $A + AB = A$. Open file T04-02 to verify.

Rule 11. $A + \overline{A}B = A + B$ This rule can be proved as follows:

$$A + \overline{A}B = (A + AB) + \overline{A}B \quad \text{Rule 10: } A = A + AB$$
$$= (AA + AB) + \overline{A}B \quad \text{Rule 7: } A = AA$$
$$= AA + AB + A\overline{A} + \overline{A}B \quad \text{Rule 8: adding } A\overline{A} = 0$$
$$= (A + \overline{A})(A + B) \quad \text{Factoring}$$
$$= 1 \cdot (A + B) \quad \text{Rule 6: } A + \overline{A} = 1$$
$$= A + B \quad \text{Rule 4: drop the 1}$$

The proof is shown in Table 4–3, which shows the truth table and the resulting logic circuit simplification.

▶ **TABLE 4–3**

Rule 11: $A + \overline{A}B = A + B$. Open file T04-03 to verify.

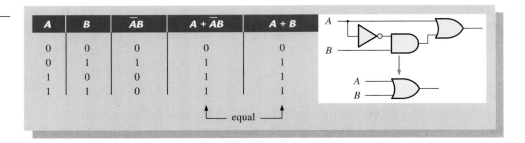

A	B	$\overline{A}B$	$A + \overline{A}B$	$A + B$
0	0	0	0	0
0	1	1	1	1
1	0	0	1	1
1	1	0	1	1

└── equal ──┘

Rule 12. $(A + B)(A + C) = A + BC$ This rule can be proved as follows:

$$
\begin{aligned}
(A + B)(A + C) &= AA + AC + AB + BC && \text{Distributive law} \\
&= A + AC + AB + BC && \text{Rule 7: } AA = A \\
&= A(1 + C) + AB + BC && \text{Factoring (distributive law)} \\
&= A \cdot 1 + AB + BC && \text{Rule 2: } 1 + C = 1 \\
&= A(1 + B) + BC && \text{Factoring (distributive law)} \\
&= A \cdot 1 + BC && \text{Rule 2: } 1 + B = 1 \\
&= A + BC && \text{Rule 4: } A \cdot 1 = A
\end{aligned}
$$

The proof is shown in Table 4–4, which shows the truth table and the resulting logic circuit simplification.

▼ **TABLE 4–4**

Rule 12: $(A + B)(A + C) = A + BC$. Open file T04-04 to verify.

A	B	C	$A + B$	$A + C$	$(A + B)(A + C)$	BC	$A + BC$
0	0	0	0	0	0	0	0
0	0	1	0	1	0	0	0
0	1	0	1	0	0	0	0
0	1	1	1	1	1	1	1
1	0	0	1	1	1	0	1
1	0	1	1	1	1	0	1
1	1	0	1	1	1	0	1
1	1	1	1	1	1	1	1

└──── equal ────┘

**SECTION 4–2
REVIEW**

1. Apply the associative law of addition to the expression $A + (B + C + D)$.

2. Apply the distributive law to the expression $A(B + C + D)$.

4–3 DEMORGAN'S THEOREMS

DeMorgan, a mathematician who knew Boole, proposed two theorems that are an important part of Boolean algebra. In practical terms, DeMorgan's theorems provide mathematical verification of the equivalency of the NAND and negative-OR gates and the equivalency of the NOR and negative-AND gates, which were discussed in Chapter 3.

After completing this section, you should be able to

■ State DeMorgan's theorems ■ Relate DeMorgan's theorems to the equivalency of the NAND and negative-OR gates and to the equivalency of the NOR and negative-AND gates ■ Apply DeMorgan's theorems to the simplification of Boolean expressions

One of DeMorgan's theorems is stated as follows:

The complement of a product of variables is equal to the sum of the complements of the variables.

Stated another way,

The complement of two or more ANDed variables is equivalent to the OR of the complements of the individual variables.

The formula for expressing this theorem for two variables is

$$\overline{XY} = \overline{X} + \overline{Y}$$

Equation 4–6

DeMorgan's second theorem is stated as follows:

The complement of a sum of variables is equal to the product of the complements of the variables.

Stated another way,

The complement of two or more ORed variables is equivalent to the AND of the complements of the individual variables.

The formula for expressing this theorem for two variables is

$$\overline{X + Y} = \overline{X}\,\overline{Y}$$

Equation 4–7

Figure 4–15 shows the gate equivalencies and truth tables for Equations 4–6 and 4–7.

◄ **FIGURE 4–15**

Gate equivalencies and the corresponding truth tables that illustrate DeMorgan's theorems. Notice the equality of the two output columns in each table. This shows that the equivalent gates perform the same logic function.

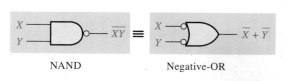

NAND   Negative-OR

Inputs		Output	
X	Y	$\overline{XY}$	$\overline{X} + \overline{Y}$
0	0	1	1
0	1	1	1
1	0	1	1
1	1	0	0

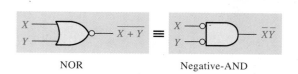

NOR   Negative-AND

Inputs		Output	
X	Y	$\overline{X + Y}$	$\overline{X}\,\overline{Y}$
0	0	1	1
0	1	0	0
1	0	0	0
1	1	0	0

As stated, DeMorgan's theorems also apply to expressions in which there are more than two variables. The following examples illustrate the application of DeMorgan's theorems to 3-variable and 4-variable expressions.

EXAMPLE 4–3

Apply DeMorgan's theorems to the expressions $\overline{XYZ}$ and $\overline{X + Y + Z}$.

Solution

$$\overline{XYZ} = \overline{X} + \overline{Y} + \overline{Z}$$

$$\overline{X + Y + Z} = \overline{X}\,\overline{Y}\,\overline{Z}$$

Related Problem Apply DeMorgan's theorem to the expression $\overline{\overline{X} + \overline{Y} + \overline{Z}}$.

EXAMPLE 4–4

Apply DeMorgan's theorems to the expressions $\overline{WXYZ}$ and $\overline{W + X + Y + Z}$.

Solution

$$\overline{WXYZ} = \overline{W} + \overline{X} + \overline{Y} + \overline{Z}$$

$$\overline{W + X + Y + Z} = \overline{W}\,\overline{X}\,\overline{Y}\,\overline{Z}$$

Related Problem Apply DeMorgan's theorem to the expression $\overline{\overline{W}\,\overline{X}\,\overline{Y}\,\overline{Z}}$.

Each variable in DeMorgan's theorems as stated in Equations 4–6 and 4–7 can also represent a combination of other variables. For example, X can be equal to the term $AB + C$, and Y can be equal to the term $A + BC$. So if you can apply DeMorgan's theorem for two variables as stated by $\overline{XY} = \overline{X} + \overline{Y}$ to the expression $\overline{(AB + C)(A + BC)}$, you get the following result:

$$\overline{(AB + C)(A + BC)} = \overline{(AB + C)} + \overline{(A + BC)}$$

Notice that in the preceding result you have two terms, $\overline{AB + C}$ and $\overline{A + BC}$, to each of which you can again apply DeMorgan's theorem $\overline{X + Y} = \overline{X}\,\overline{Y}$ individually, as follows:

$$\overline{(AB + C)} + \overline{(A + BC)} = (\overline{AB})\overline{C} + \overline{A}(\overline{BC})$$

Notice that you still have two terms in the expression to which DeMorgan's theorem can again be applied. These terms are $\overline{AB}$ and $\overline{BC}$. A final application of DeMorgan's theorem gives the following result:

$$(\overline{AB})\overline{C} + \overline{A}(\overline{BC}) = (\overline{A} + \overline{B})\overline{C} + \overline{A}(\overline{B} + \overline{C})$$

Although this result can be simplified further by the use of Boolean rules and laws, DeMorgan's theorems cannot be used any more.

Applying DeMorgan's Theorems

The following procedure illustrates the application of DeMorgan's theorems and Boolean algebra to the specific expression

$$\overline{\overline{A + B\overline{C}} + D(\overline{E + \overline{F}})}$$

Step 1. Identify the terms to which you can apply DeMorgan's theorems, and think of each term as a single variable. Let $\overline{A + B\overline{C}} = X$ and $D(\overline{E + \overline{F}}) = Y$.

Step 2. Since $\overline{X + Y} = \overline{X}\,\overline{Y}$,

$$\overline{\overline{(A + B\overline{C})} + \overline{(D(E + \overline{\overline{F}}))}} = \overline{\overline{(A + B\overline{C})}}\,\overline{\overline{(D(E + \overline{\overline{F}}))}}$$

Step 3. Use rule 9 ($\overline{\overline{A}} = A$) to cancel the double bars over the left term (this is not part of DeMorgan's theorem).

$$\overline{\overline{(A + B\overline{C})}}\,\overline{\overline{(D(\overline{E + \overline{F}}))}} = (A + B\overline{C})\overline{\overline{(D(\overline{E + \overline{F}}))}}$$

Step 4. Applying DeMorgan's theorem to the second term,

$$(A + B\overline{C})\overline{\overline{(D(\overline{E + \overline{F}}))}} = (A + B\overline{C})(\overline{D} + \overline{\overline{(E + \overline{F})}})$$

Step 5. Use rule 9 ($\overline{\overline{A}} = A$) to cancel the double bars over the $E + \overline{F}$ part of the term.

$$(A + B\overline{C})(\overline{D} + \overline{\overline{E + \overline{F}}}) = (A + B\overline{C})(\overline{D} + E + \overline{F})$$

The following three examples will further illustrate how to use DeMorgan's theorems.

EXAMPLE 4–5

Apply DeMorgan's theorems to each of the following expressions:

(a) $\overline{(A + B + C)D}$ **(b)** $\overline{ABC + DEF}$ **(c)** $\overline{A\overline{B} + \overline{C}D + EF}$

Solution   **(a)** Let $A + B + C = X$ and $D = Y$. The expression $\overline{(A + B + C)D}$ is of the form $\overline{XY} = \overline{X} + \overline{Y}$ and can be rewritten as

$$\overline{(A + B + C)D} = \overline{A + B + C} + \overline{D}$$

Next, apply DeMorgan's theorem to the term $\overline{A + B + C}$.

$$\overline{A + B + C} + \overline{D} = \overline{A}\,\overline{B}\,\overline{C} + \overline{D}$$

(b) Let $ABC = X$ and $DEF = Y$. The expression $\overline{ABC + DEF}$ is of the form $\overline{X + Y} = \overline{X}\,\overline{Y}$ and can be rewritten as

$$\overline{ABC + DEF} = (\overline{ABC})(\overline{DEF})$$

Next, apply DeMorgan's theorem to each of the terms $\overline{ABC}$ and $\overline{DEF}$.

$$(\overline{ABC})(\overline{DEF}) = (\overline{A} + \overline{B} + \overline{C})(\overline{D} + \overline{E} + \overline{F})$$

(c) Let $A\overline{B} = X$, $\overline{C}D = Y$, and $EF = Z$. The expression $\overline{A\overline{B} + \overline{C}D + EF}$ is of the form $\overline{X + Y + Z} = \overline{X}\,\overline{Y}\,\overline{Z}$ and can be rewritten as

$$\overline{A\overline{B} + \overline{C}D + EF} = (\overline{A\overline{B}})(\overline{\overline{C}D})(\overline{EF})$$

Next, apply DeMorgan's theorem to each of the terms $\overline{A\overline{B}}$, $\overline{\overline{C}D}$, and $\overline{EF}$.

$$(\overline{A\overline{B}})(\overline{\overline{C}D})(\overline{EF}) = (\overline{A} + B)(C + \overline{D})(\overline{E} + \overline{F})$$

Related Problem   Apply DeMorgan's theorems to the expression $\overline{\overline{ABC} + D + E}$.

EXAMPLE 4–6

Apply DeMorgan's theorems to each expression:

(a) $\overline{(A + B) + \overline{C}}$ (b) $\overline{(\overline{A} + B) + CD}$ (c) $\overline{(A + B)\overline{C}\overline{D} + E + \overline{F}}$

Solution

(a) $\overline{(A + B) + \overline{C}} = (\overline{A + B})\overline{\overline{C}} = (A + B)C$

(b) $\overline{(\overline{A} + B) + CD} = (\overline{\overline{A} + B})\overline{CD} = (\overline{\overline{A}}\overline{B})(\overline{C} + \overline{D}) = A\overline{B}(\overline{C} + \overline{D})$

(c) $\overline{(A + B)\overline{C}\overline{D} + E + \overline{F}} = (\overline{(A + B)\overline{C}\overline{D}})(\overline{E + \overline{F}}) = (\overline{A}\overline{B} + C + D)\overline{E}F$

Related Problem

Apply DeMorgan's theorems to the expression $\overline{\overline{A}B(C + \overline{D}) + E}$.

EXAMPLE 4–7

The Boolean expression for an exclusive-OR gate is $A\overline{B} + \overline{A}B$. With this as a starting point, use DeMorgan's theorems and any other rules or laws that are applicable to develop an expression for the exclusive-NOR gate.

Solution

Start by complementing the exclusive-OR expression and then applying DeMorgan's theorems as follows:

$$\overline{A\overline{B} + \overline{A}B} = (\overline{A\overline{B}})(\overline{\overline{A}B}) = (\overline{A} + \overline{\overline{B}})(\overline{\overline{A}} + \overline{B}) = (\overline{A} + B)(A + \overline{B})$$

Next, apply the distributive law and rule 8 ($A \cdot \overline{A} = 0$).

$$(\overline{A} + B)(A + \overline{B}) = \overline{A}A + \overline{A}\,\overline{B} + AB + B\overline{B} = \overline{A}\,\overline{B} + AB$$

The final expression for the XNOR is $\overline{A}\,\overline{B} + AB$. Note that this expression equals 1 any time both variables are 0s or both variables are 1s.

Related Problem

Starting with the expression for a 4-input NAND gate, use DeMorgan's theorems to develop an expression for a 4-input negative-OR gate.

SECTION 4–3 REVIEW

1. Apply DeMorgan's theorems to the following expressions:
   (a) $\overline{ABC + (\overline{D} + E)}$ (b) $\overline{(A + B)C}$ (c) $\overline{A + B + C + \overline{DE}}$

4–4   BOOLEAN ANALYSIS OF LOGIC CIRCUITS

Boolean algebra provides a concise way to express the operation of a logic circuit formed by a combination of logic gates so that the output can be determined for various combinations of input values.

After completing this section, you should be able to

■ Determine the Boolean expression for a combination of gates ■ Evaluate the logic operation of a circuit from the Boolean expression ■ Construct a truth table

Boolean Expression for a Logic Circuit

To derive the Boolean expression for a given logic circuit, begin at the left-most inputs and work toward the final output, writing the expression for each gate. For the example circuit in Figure 4–16, the Boolean expression is determined as follows:

A logic circuit can be described by a Boolean equation.

1. The expression for the left-most AND gate with inputs C and D is CD.

2. The output of the left-most AND gate is one of the inputs to the OR gate and B is the other input. Therefore, the expression for the OR gate is $B + CD$.

3. The output of the OR gate is one of the inputs to the right-most AND gate and A is the other input. Therefore, the expression for this AND gate is $A(B + CD)$, which is the final output expression for the entire circuit.

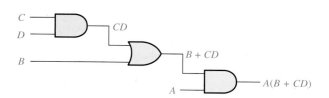

◀ FIGURE 4–16

A logic circuit showing the development of the Boolean expression for the output.

Constructing a Truth Table for a Logic Circuit

Once the Boolean expression for a given logic circuit has been determined, a truth table that shows the output for all possible values of the input variables can be developed. The procedure requires that you evaluate the Boolean expression for all possible combinations of values for the input variables. In the case of the circuit in Figure 4–16, there are four input variables (A, B, C, and D) and therefore sixteen ($2^4 = 16$) combinations of values are possible.

A logic circuit can be described by a truth table.

Evaluating the Expression   To evaluate the expression $A(B + CD)$, first find the values of the variables that make the expression equal to 1, using the rules for Boolean addition and multiplication. In this case, the expression equals 1 only if $A = 1$ and $B + CD = 1$ because

$$A(B + CD) = 1 \cdot 1 = 1$$

Now determine when the $B + CD$ term equals 1. The term $B + CD = 1$ if either $B = 1$ or $CD = 1$ or if both B and CD equal 1 because

$$B + CD = 1 + 0 = 1$$
$$B + CD = 0 + 1 = 1$$
$$B + CD = 1 + 1 = 1$$

The term $CD = 1$ only if $C = 1$ and $D = 1$.

To summarize, the expression $A(B + CD) = 1$ when $A = 1$ and $B = 1$ regardless of the values of C and D or when $A = 1$ and $C = 1$ and $D = 1$ regardless of the value of B. The expression $A(B + CD) = 0$ for all other value combinations of the variables.

Putting the Results in Truth Table Format   The first step is to list the sixteen input variable combinations of 1s and 0s in a binary sequence as shown in Table 4–5. Next, place a 1 in the output column for each combination of input variables that was determined in the evaluation. Finally, place a 0 in the output column for all other combinations of input variables. These results are shown in the truth table in Table 4–5.

▶ TABLE 4–5

Truth table for the logic circuit in Figure 4–16.

INPUTS				OUTPUT
A	B	C	D	A(B + CD)
0	0	0	0	0
0	0	0	1	0
0	0	1	0	0
0	0	1	1	0
0	1	0	0	0
0	1	0	1	0
0	1	1	0	0
0	1	1	1	0
1	0	0	0	0
1	0	0	1	0
1	0	1	0	0
1	0	1	1	1
1	1	0	0	1
1	1	0	1	1
1	1	1	0	1
1	1	1	1	1

SECTION 4–4 REVIEW

1. Replace the AND gates with OR gates and the OR gate with an AND gate in Figure 4–16 and determine the Boolean expression for the output.

2. Construct a truth table for the circuit in Question 1.

4–5 SIMPLIFICATION USING BOOLEAN ALGEBRA

Many times in the application of Boolean algebra, you have to reduce a particular expression to its simplest form or change its form to a more convenient one to implement the expression most efficiently. The approach taken in this section is to use the basic laws, rules, and theorems of Boolean algebra to manipulate and simplify an expression. This method depends on a thorough knowledge of Boolean algebra and considerable practice in its application, not to mention a little ingenuity and cleverness.

After completing this section, you should be able to

■ Apply the laws, rules, and theorems of Boolean algebra to simplify general expressions

A simplified Boolean expression uses the fewest gates possible to implement a given expression. Examples 4–8 through 4–11 illustrate Boolean simplification.

EXAMPLE 4–8

Using Boolean algebra techniques, simplify this expression:

$$AB + A(B + C) + B(B + C)$$

Solution   The following is not necessarily the only approach.

Step 1: Apply the distributive law to the second and third terms in the expression, as follows:

$$AB + AB + AC + BB + BC$$

Step 2: Apply rule 7 ($BB = B$) to the fourth term.

$$AB + AB + AC + B + BC$$

Step 3: Apply rule 5 ($AB + AB = AB$) to the first two terms.

$$AB + AC + B + BC$$

Step 4: Apply rule 10 ($B + BC = B$) to the last two terms.

$$AB + AC + B$$

Step 5: Apply rule 10 ($AB + B = B$) to the first and third terms.

$$B + AC$$

At this point the expression is simplified as much as possible. Once you gain experience in applying Boolean algebra, you can often combine many individual steps.

Related Problem   Simplify the Boolean expression $A\overline{B} + A(\overline{B + C}) + B(\overline{B + C})$.

Figure 4–17 shows that the simplification process in Example 4–8 has significantly reduced the number of logic gates required to implement the expression. Part (a) shows that five gates are required to implement the expression in its original form; however, only two gates are needed for the simplified expression, shown in part (b). It is important to realize that these two gate circuits are equivalent. That is, for any combination of levels on the *A, B,* and *C* inputs, you get the same output from either circuit.

Simplification means fewer gates for the same function.

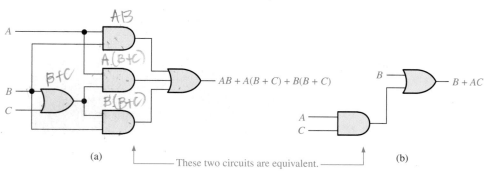

(a)   These two circuits are equivalent.   (b)

▲ **FIGURE 4–17**

Gate circuits for Example 4–8. Open file F04-17 to verify equivalency.

EXAMPLE 4–9

Simplify the following Boolean expression:

$$[A\overline{B}(C + BD) + \overline{A}\,\overline{B}]C$$

Note that brackets and parentheses mean the same thing: the term inside is multiplied (ANDed) with the term outside.

Solution

Step 1: Apply the distributive law to the terms within the brackets.

$$(A\overline{B}C + A\overline{B}BD + \overline{A}\,\overline{B})C$$

Step 2: Apply rule 8 ($\overline{B}B = 0$) to the second term within the parentheses.

$$(A\overline{B}C + A \cdot 0 \cdot D + \overline{A}\,\overline{B})C$$

Step 3: Apply rule 3 ($A \cdot 0 \cdot D = 0$) to the second term within the parentheses.

$$(A\overline{B}C + 0 + \overline{A}\,\overline{B})C$$

Step 4: Apply rule 1 (drop the 0) within the parentheses.

$$(A\overline{B}C + \overline{A}\,\overline{B})C$$

Step 5: Apply the distributive law.

$$A\overline{B}CC + \overline{A}\,\overline{B}C$$

Step 6: Apply rule 7 ($CC = C$) to the first term.

$$A\overline{B}C + \overline{A}\,\overline{B}C$$

Step 7: Factor out $\overline{B}C$.

$$\overline{B}C(A + \overline{A})$$

Step 8: Apply rule 6 ($A + \overline{A} = 1$).

$$\overline{B}C \cdot 1$$

Step 9: Apply rule 4 (drop the 1).

$$\overline{B}C$$

Related Problem

Simplify the Boolean expression $[AB(C + \overline{BD}) + \overline{AB}]CD$.

EXAMPLE 4–10

Simplify the following Boolean expression:

$$\overline{A}BC + A\overline{B}\,\overline{C} + \overline{A}\,\overline{B}\,\overline{C} + A\overline{B}C + ABC$$

Solution

Step 1: Factor BC out of the first and last terms.

$$BC(\overline{A} + A) + A\overline{B}\,\overline{C} + \overline{A}\,\overline{B}\,\overline{C} + A\overline{B}C$$

Step 2: Apply rule 6 ($\overline{A} + A = 1$) to the term in parentheses, and factor $A\overline{B}$ from the second and last terms.

$$BC \cdot 1 + A\overline{B}(\overline{C} + C) + \overline{A}\,\overline{B}\,\overline{C}$$

Step 3: Apply rule 4 (drop the 1) to the first term and rule 6 $(\overline{C} + C = 1)$ to the term in parentheses.

$$BC + A\overline{B} \cdot 1 + \overline{A}\,\overline{B}\,\overline{C}$$

Step 4: Apply rule 4 (drop the 1) to the second term.

$$BC + A\overline{B} + \overline{A}\,\overline{B}\,\overline{C}$$

Step 5: Factor $\overline{B}$ from the second and third terms.

$$BC + \overline{B}(A + \overline{A}\,\overline{C})$$

Step 6: Apply rule 11 $(A + \overline{A}\,\overline{C} = A + \overline{C})$ to the term in parentheses.

$$BC + \overline{B}(A + \overline{C})$$

Step 7: Use the distributive and commutative laws to get the following expression:

$$BC + A\overline{B} + \overline{B}\,\overline{C}$$

Related Problem   Simplify the Boolean expression $AB\overline{C} + \overline{A}\,BC + A\overline{B}C + \overline{A}\,\overline{B}\,\overline{C}$.

EXAMPLE 4–11

Simplify the following Boolean expression:

$$\overline{AB + AC} + \overline{A}\,BC$$

Solution   **Step 1:** Apply DeMorgan's theorem to the first term.

$$(\overline{AB})(\overline{AC}) + \overline{A}\,BC$$

Step 2: Apply DeMorgan's theorem to each term in parentheses.

$$(\overline{A} + \overline{B})(\overline{A} + \overline{C}) + \overline{A}\,BC$$

Step 3: Apply the distributive law to the two terms in parentheses.

$$\overline{A}\,\overline{A} + \overline{A}\,\overline{C} + \overline{A}\,\overline{B} + \overline{B}\,\overline{C} + \overline{A}\,BC$$

Step 4: Apply rule 7 $(\overline{A}\,\overline{A} = \overline{A})$ to the first term, and apply rule 10 $[\overline{A}\,\overline{B} + \overline{A}\,BC = \overline{A}\,\overline{B}(1 + C) = \overline{A}\,\overline{B}]$ to the third and last terms.

$$\overline{A} + \overline{A}\,\overline{C} + \overline{A}\,\overline{B} + \overline{B}\,\overline{C}$$

Step 5: Apply rule 10 $[\overline{A} + \overline{A}\,\overline{C} = \overline{A}(1 + \overline{C}) = \overline{A}]$ to the first and second terms.

$$\overline{A} + \overline{A}\,\overline{B} + \overline{B}\,\overline{C}$$

Step 6: Apply rule 10 $[\overline{A} + \overline{A}\,\overline{B} = \overline{A}(1 + \overline{B}) = \overline{A}]$ to the first and second terms.

$$\overline{A} + \overline{B}\,\overline{C}$$

Related Problem   Simplify the Boolean expression $\overline{AB + \overline{A}\,C} + \overline{A}\,B\overline{C}$.

SECTION 4–5
REVIEW

1. Simplify the following Boolean expressions if possible:
   (a) $A + AB + A\overline{B}C$ (b) $(\overline{A} + B)C + ABC$ (c) $A\overline{B}C(BD + CDE) + A\overline{C}$

2. Implement each expression in Question 1 as originally stated with the appropriate logic gates. Then implement the simplified expression, and compare the number of gates.

4–6   STANDARD FORMS OF BOOLEAN EXPRESSIONS

All Boolean expressions, regardless of their form, can be converted into either of two standard forms: the sum-of-products form or the product-of-sums form. Standardization makes the evaluation, simplification, and implementation of Boolean expressions much more systematic and easier.

After completing this section, you should be able to

■ Identify a sum-of-products expression ■ Determine the domain of a Boolean expression ■ Convert any sum-of-products expression to a standard form ■ Evaluate a standard sum-of-products expression in terms of binary values ■ Identify a product-of-sums expression ■ Convert any product-of-sums expression to a standard form ■ Evaluate a standard product-of-sums expression in terms of binary values ■ Convert from one standard form to the other

The Sum-of-Products (SOP) Form

An SOP expression can be implemented with one OR and two or more ANDs.

A product term was defined in Section 4–1 as a term consisting of the product (Boolean multiplication) of literals (variables or their complements). When two or more product terms are summed by Boolean addition, the resulting expression is a **sum-of-products (SOP)**. Some examples are

$$AB + ABC$$
$$ABC + CDE + \overline{B}\,\overline{C}\overline{D}$$
$$\overline{A}B + \overline{A}B\overline{C} + AC$$

Also, an SOP expression can contain a single-variable term, as in $A + \overline{A}\,BC + B\overline{C}\overline{D}$. Refer to the simplification examples in the last section, and you will see that each of the final expressions was either a single product term or in SOP form. In an SOP expression, a single overbar cannot extend over more than one variable; however, more than one variable in a term can have an overbar. For example, an SOP expression can have the term $\overline{A}\,\overline{B}\,\overline{C}$ but not $\overline{ABC}$.

Domain of a Boolean Expression   The **domain** of a general Boolean expression is the set of variables contained in the expression in either complemented or uncomplemented form. For example, the domain of the expression $\overline{A}B + A\overline{B}C$ is the set of variables A, B, C and the domain of the expression $AB\overline{C} + C\overline{D}E + \overline{B}C\overline{D}$ is the set of variables A, B, C, D, E.

AND/OR Implementation of an SOP Expression   Implementing an SOP expression simply requires ORing the outputs of two or more AND gates. A product term is produced by an AND operation, and the sum (addition) of two or more product terms is produced by an OR operation. Therefore, an SOP expression can be implemented by AND-OR logic in which the outputs of a number (equal to the number of product terms in the expression) of AND gates connect to the inputs of an OR gate, as shown in Figure 4–18 for the expression $AB + BCD + AC$. The output X of the OR gate equals the SOP expression.

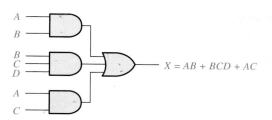

◀ FIGURE 4–18

Implementation of the SOP expression $AB + BCD + AC$.

NAND/NAND Implementation of an SOP Expression   NAND gates can be used to implement an SOP expression. Using only NAND gates, an AND/OR function can be accomplished, as illustrated in Figure 4–19. The first level of NAND gates feed into a NAND gate that acts as a negative-OR gate. The NAND and negative-OR inversions cancel and the result is effectively an AND/OR circuit.

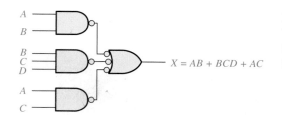

◀ FIGURE 4–19

This NAND/NAND implementation is equivalent to the AND/OR in Figure 4–18.

Conversion of a General Expression to SOP Form

Any logic expression can be changed into SOP form by applying Boolean algebra techniques. For example, the expression $A(B + CD)$ can be converted to SOP form by applying the distributive law:

$$A(B + CD) = AB + ACD$$

EXAMPLE 4–12

Convert each of the following Boolean expressions to SOP form:

(a) $AB + B(CD + EF)$ (b) $(A + B)(B + C + D)$ (c) $\overline{(\overline{A + B}) + C}$

Solution   (a) $AB + B(CD + EF) = AB + BCD + BEF$

(b) $(A + B)(B + C + D) = AB + AC + AD + BB + BC + BD$

(c) $\overline{(\overline{A + B}) + C} = (\overline{\overline{A + B}})\overline{C} = (A + B)\overline{C} = A\overline{C} + B\overline{C}$

Related Problem   Convert $\overline{A}B\overline{C} + (A + \overline{B})(B + \overline{C} + A\overline{B})$ to SOP form.

The Standard SOP Form

So far, you have seen SOP expressions in which some of the product terms do not contain all of the variables in the domain of the expression. For example, the expression $\overline{A}B\overline{C} + A\overline{B}D + \overline{A}B\overline{C}D$ has a domain made up of the variables $A, B, C,$ and D. However, notice that the complete set of variables in the domain is not represented in the first two terms of the expression; that is, D or $\overline{D}$ is missing from the first term and C or $\overline{C}$ is missing from the second term.

A *standard SOP expression* is one in which *all* the variables in the domain appear in each product term in the expression. For example, $A\overline{B}CD + \overline{A}B\overline{C}D + AB\overline{C}\overline{D}$ is a standard SOP expression. Standard SOP expressions are important in constructing truth tables, covered in Section 4–7, and in the Karnaugh map simplification method, which is covered in Section 4–8. Any nonstandard SOP expression (referred to simply as SOP) can be converted to the standard form using Boolean algebra.

Converting Product Terms to Standard SOP   Each product term in an SOP expression that does not contain all the variables in the domain can be expanded to standard form to include all variables in the domain and their complements. As stated in the following steps, a nonstandard SOP expression is converted into standard form using Boolean algebra rule 6 ($A + \overline{A} = 1$) from Table 4–1: A variable added to its complement equals 1.

Step 1.   Multiply each nonstandard product term by a term made up of the sum of a missing variable and its complement. This results in two product terms. As you know, you can multiply anything by 1 without changing its value.

Step 2.   Repeat Step 1 until all resulting product terms contain all variables in the domain in either complemented or uncomplemented form. In converting a product term to standard form, the number of product terms is doubled for each missing variable, as Example 4–13 shows.

EXAMPLE 4–13

Convert the following Boolean expression into standard SOP form:

$$A\overline{B}C + \overline{A}\,\overline{B} + AB\overline{C}D$$

Solution   The domain of this SOP expression is *A, B, C, D*. Take one term at a time. The first term, $A\overline{B}C$, is missing variable *D* or $\overline{D}$, so multiply the first term by $D + \overline{D}$ as follows:

$$A\overline{B}C = A\overline{B}C(D + \overline{D}) = A\overline{B}CD + A\overline{B}C\overline{D}$$

In this case, two standard product terms are the result.

The second term, $\overline{A}\,\overline{B}$, is missing variables *C* or $\overline{C}$ and *D* or $\overline{D}$, so first multiply the second term by $C + \overline{C}$ as follows:

$$\overline{A}\,\overline{B} = \overline{A}\,\overline{B}(C + \overline{C}) = \overline{A}\,\overline{B}C + \overline{A}\,\overline{B}\,\overline{C}$$

The two resulting terms are missing variable *D* or $\overline{D}$, so multiply both terms by $D + \overline{D}$ as follows:

$$\overline{A}\,\overline{B} = \overline{A}\,\overline{B}C + \overline{A}\,\overline{B}\,\overline{C} = \overline{A}\,\overline{B}C(D + \overline{D}) + \overline{A}\,\overline{B}\,\overline{C}(D + \overline{D})$$
$$= \overline{A}\,\overline{B}CD + \overline{A}\,\overline{B}C\overline{D} + \overline{A}\,\overline{B}\,\overline{C}D + \overline{A}\,\overline{B}\,\overline{C}\,\overline{D}$$

In this case, four standard product terms are the result.

The third term, $AB\overline{C}D$, is already in standard form. The complete standard SOP form of the original expression is as follows:

$$A\overline{B}C + \overline{A}\,\overline{B} + AB\overline{C}D = A\overline{B}CD + A\overline{B}C\overline{D} + \overline{A}\,\overline{B}CD + \overline{A}\,\overline{B}C\overline{D} + \overline{A}\,\overline{B}\,\overline{C}D + \overline{A}\,\overline{B}\,\overline{C}\,\overline{D} + AB\overline{C}D$$

Related Problem   Convert the expression $W\overline{X}Y + \overline{X}Y\overline{Z} + WX\overline{Y}$ to standard SOP form.

Binary Representation of a Standard Product Term   A standard product term is equal to 1 for only one combination of variable values. For example, the product term $A\overline{B}C\overline{D}$ is equal to 1 when $A = 1, B = 0, C = 1, D = 0$, as shown below, and is 0 for all other combinations of values for the variables.

$$A\overline{B}C\overline{D} = 1 \cdot \overline{0} \cdot 1 \cdot \overline{0} = 1 \cdot 1 \cdot 1 \cdot 1 = 1$$

In this case, the product term has a binary value of 1010 (decimal ten).

Remember, a product term is implemented with an AND gate whose output is 1 only if each of its inputs is 1. Inverters are used to produce the complements of the variables as required.

An SOP expression is equal to 1 only if one or more of the product terms in the expression is equal to 1.

EXAMPLE 4–14

Determine the binary values for which the following standard SOP expression is equal to 1:

$$ABCD + A\overline{B}\,\overline{C}D + \overline{A}\,\overline{B}\,\overline{C}\,\overline{D}$$

Solution The term $ABCD$ is equal to 1 when $A = 1$, $B = 1$, $C = 1$, and $D = 1$.

$$ABCD = 1 \cdot 1 \cdot 1 \cdot 1 = 1$$

The term $A\overline{B}\,\overline{C}D$ is equal to 1 when $A = 1$, $B = 0$, $C = 0$, and $D = 1$.

$$A\overline{B}\,\overline{C}D = 1 \cdot \overline{0} \cdot \overline{0} \cdot 1 = 1 \cdot 1 \cdot 1 \cdot 1 = 1$$

The term $\overline{A}\,\overline{B}\,\overline{C}\,\overline{D}$ is equal to 1 when $A = 0$, $B = 0$, $C = 0$, and $D = 0$.

$$\overline{A}\,\overline{B}\,\overline{C}\,\overline{D} = \overline{0} \cdot \overline{0} \cdot \overline{0} \cdot \overline{0} = 1 \cdot 1 \cdot 1 \cdot 1 = 1$$

The SOP expression equals 1 when any or all of the three product terms is 1.

Related Problem Determine the binary values for which the following SOP expression is equal to 1:

$$\overline{X}YZ + X\overline{Y}Z + XY\overline{Z} + \overline{X}Y\overline{Z} + XYZ$$

Is this a standard SOP expression?

The Product-of-Sums (POS) Form

A sum term was defined in Section 4–1 as a term consisting of the sum (Boolean addition) of literals (variables or their complements). When two or more sum terms are multiplied, the resulting expression is a **product-of-sums (POS)**. Some examples are

$$(\overline{A} + B)(A + \overline{B} + C)$$
$$(\overline{A} + \overline{B} + \overline{C})(C + \overline{D} + E)(\overline{B} + C + D)$$
$$(A + B)(A + \overline{B} + C)(\overline{A} + C)$$

A POS expression can contain a single-variable term, as in $\overline{A}(A + \overline{B} + C)(\overline{B} + \overline{C} + D)$. In a POS expression, a single overbar cannot extend over more than one variable; however, more than one variable in a term can have an overbar. For example, a POS expression can have the term $\overline{A} + \overline{B} + \overline{C}$ but not $\overline{A + B + C}$.

Implementation of a POS Expression Implementing a POS expression simply requires ANDing the outputs of two or more OR gates. A sum term is produced by an OR operation, and the product of two or more sum terms is produced by an AND operation. Therefore, a POS expression can be implemented by logic in which the outputs of a number (equal to the number of sum terms in the expression) of OR gates connect to the inputs of an AND gate, as Figure 4–20 shows for the expression $(A + B)(B + C + D)(A + C)$. The output X of the AND gate equals the POS expression.

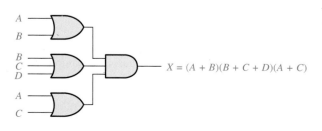

$$X = (A + B)(B + C + D)(A + C)$$

◀ **FIGURE 4–20**

Implementation of the POS expression $(A + B)(B + C + D)(A + C)$.

The Standard POS Form

So far, you have seen POS expressions in which some of the sum terms do not contain all of the variables in the domain of the expression. For example, the expression

$$(A + \overline{B} + C)(A + B + \overline{D})(A + \overline{B} + \overline{C} + D)$$

has a domain made up of the variables A, B, C, and D. Notice that the complete set of variables in the domain is not represented in the first two terms of the expression; that is, D or $\overline{D}$ is missing from the first term and C or $\overline{C}$ is missing from the second term.

A standard POS expression is one in which *all* the variables in the domain appear in each sum term in the expression. For example,

$$(\overline{A} + \overline{B} + \overline{C} + \overline{D})(A + \overline{B} + C + D)(A + B + \overline{C} + D)$$

is a standard POS expression. Any nonstandard POS expression (referred to simply as POS) can be converted to the standard form using Boolean algebra.

Converting a Sum Term to Standard POS Each sum term in a POS expression that does not contain all the variables in the domain can be expanded to standard form to include all variables in the domain and their complements. As stated in the following steps, a nonstandard POS expression is converted into standard form using Boolean algebra rule 8 ($A \cdot \overline{A} = 0$) from Table 4–1: A variable multiplied by its complement equals 0.

Step 1. Add to each nonstandard product term a term made up of the product of the missing variable and its complement. This results in two sum terms. As you know, you can add 0 to anything without changing its value.

Step 2. Apply rule 12 from Table 4–1: $A + BC = (A + B)(A + C)$

Step 3. Repeat Step 1 until all resulting sum terms contain all variables in the domain in either complemented or uncomplemented form.

EXAMPLE 4–15

Convert the following Boolean expression into standard POS form:

$$(A + \overline{B} + C)(\overline{B} + C + \overline{D})(A + \overline{B} + \overline{C} + D)$$

Solution The domain of this POS expression is A, B, C, D. Take one term at a time. The first term, $A + \overline{B} + C$, is missing variable D or $\overline{D}$, so add $D\overline{D}$ and apply rule 12 as follows:

$$A + \overline{B} + C = A + \overline{B} + C + D\overline{D} = (A + \overline{B} + C + D)(A + \overline{B} + C + \overline{D})$$

The second term, $\overline{B} + C + \overline{D}$, is missing variable A or $\overline{A}$, so add $A\overline{A}$ and apply rule 12 as follows:

$$\overline{B} + C + \overline{D} = \overline{B} + C + \overline{D} + A\overline{A} = (A + \overline{B} + C + \overline{D})(\overline{A} + \overline{B} + C + \overline{D})$$

The third term, $A + \overline{B} + \overline{C} + D$, is already in standard form. The standard POS form of the original expression is as follows:

$$(A + \overline{B} + C)(\overline{B} + C + \overline{D})(A + \overline{B} + \overline{C} + D) =$$
$$(A + \overline{B} + C + D)(A + \overline{B} + C + \overline{D})(A + \overline{B} + C + \overline{D})(\overline{A} + \overline{B} + C + \overline{D})(A + \overline{B} + \overline{C} + D)$$

Related Problem Convert the expression $(A + \overline{B})(B + C)$ to standard POS form.

Binary Representation of a Standard Sum Term A standard sum term is equal to 0 for only one combination of variable values. For example, the sum term $A + \overline{B} + C + \overline{D}$ is 0 when $A = 0$, $B = 1$, $C = 0$, and $D = 1$, as shown below, and is 1 for all other combinations of values for the variables.

$$A + \overline{B} + C + \overline{D} = 0 + \overline{1} + 0 + \overline{1} = 0 + 0 + 0 + 0 = 0$$

In this case, the sum term has a binary value of 0101 (decimal 5). Remember, a sum term is implemented with an OR gate whose output is 0 only if each of its inputs is 0. Inverters are used to produce the complements of the variables as required.

A POS expression is equal to 0 only if one or more of the sum terms in the expression is equal to 0.

EXAMPLE 4–16

Determine the binary values of the variables for which the following standard POS expression is equal to 0:

$$(A + B + C + D)(A + \overline{B} + \overline{C} + D)(\overline{A} + \overline{B} + \overline{C} + \overline{D})$$

Solution   The term $A + B + C + D$ is equal to 0 when $A = 0$, $B = 0$, $C = 0$, and $D = 0$.

$$A + B + C + D = 0 + 0 + 0 + 0 = 0$$

The term $A + \overline{B} + \overline{C} + D$ is equal to 0 when $A = 0$, $B = 1$, $C = 1$, and $D = 0$.

$$A + \overline{B} + \overline{C} + D = 0 + \overline{1} + \overline{1} + 0 = 0 + 0 + 0 + 0 = 0$$

The term $\overline{A} + \overline{B} + \overline{C} + \overline{D}$ is equal to 0 when $A = 1$, $B = 1$, $C = 1$, and $D = 1$.

$$\overline{A} + \overline{B} + \overline{C} + \overline{D} = \overline{1} + \overline{1} + \overline{1} + \overline{1} = 0 + 0 + 0 + 0 = 0$$

The POS expression equals 0 when any of the three sum terms equals 0.

Related Problem   Determine the binary values for which the following POS expression is equal to 0:

$$(X + \overline{Y} + Z)(\overline{X} + Y + Z)(X + Y + \overline{Z})(\overline{X} + \overline{Y} + \overline{Z})(X + \overline{Y} + \overline{Z})$$

Is this a standard POS expression?

Converting Standard SOP to Standard POS

The binary values of the product terms in a given standard SOP expression are not present in the equivalent standard POS expression. Also, the binary values that are not represented in the SOP expression are present in the equivalent POS expression. Therefore, to convert from standard SOP to standard POS, the following steps are taken:

Step 1. Evaluate each product term in the SOP expression. That is, determine the binary numbers that represent the product terms.

Step 2. Determine all of the binary numbers not included in the evaluation in Step 1.

Step 3. Write the equivalent sum term for each binary number from Step 2 and express in POS form.

Using a similar procedure, you can go from POS to SOP.

EXAMPLE 4–17

Convert the following SOP expression to an equivalent POS expression:

$$\overline{A}\,\overline{B}\,\overline{C} + \overline{A}B\overline{C} + \overline{A}BC + A\overline{B}C + ABC$$

Solution   The evaluation is as follows:

$$000 + 010 + 011 + 101 + 111$$

Since there are three variables in the domain of this expression, there are a total of eight (2^3) possible combinations. The SOP expression contains five of these combinations, so the POS must contain the other three which are 001, 100, and 110.

Remember, these are the binary values that make the sum term 0. The equivalent POS expression is

$$(A + B + \overline{C})(\overline{A} + B + C)(\overline{A} + \overline{B} + C)$$

Related Problem   Verify that the SOP and POS expressions in this example are equivalent by substituting binary values into each.

SECTION 4–6 REVIEW

1. Identify each of the following expressions as SOP, standard SOP, POS, or standard POS:
   (a) $AB + \overline{A}BD + \overline{A}C\overline{D}$ (b) $(A + \overline{B} + C)(A + B + \overline{C})$
   (c) $\overline{A}BC + AB\overline{C}$ (d) $A(A + \overline{C})(A + B)$
2. Convert each SOP expression in Question 1 to standard form.
3. Convert each POS expression in Question 1 to standard form.

4–7   BOOLEAN EXPRESSIONS AND TRUTH TABLES

All standard Boolean expressions can be easily converted into truth table format using binary values for each term in the expression. The truth table is a common way of presenting, in a concise format, the logical operation of a circuit. Also, standard SOP or POS expressions can be determined from a truth table. You will find truth tables in data sheets and other literature related to the operation of digital circuits.

After completing this section, you should be able to

■ Convert a standard SOP expression into truth table format ■ Convert a standard POS expression into truth table format ■ Derive a standard expression from a truth table ■ Properly interpret truth table data

Converting SOP Expressions to Truth Table Format

Recall from Section 4–6 that an SOP expression is equal to 1 only if at least one of the product terms is equal to 1. A truth table is simply a list of the possible combinations of input variable values and the corresponding output values (1 or 0). For an expression with a domain of two variables, there are four different combinations of those variables ($2^2 = 4$). For an expression with a domain of three variables, there are eight different combinations of those variables ($2^3 = 8$). For an expression with a domain of four variables, there are sixteen different combinations of those variables ($2^4 = 16$), and so on.

The first step in constructing a truth table is to list all possible combinations of binary values of the variables in the expression. Next, convert the SOP expression to standard form if it is not already. Finally, place a 1 in the output column (X) for each binary value that makes the standard SOP expression a 1 and place a 0 for all the remaining binary values. This procedure is illustrated in Example 4–18.

EXAMPLE 4–18

Develop a truth table for the standard SOP expression $\overline{A}\,\overline{B}C + A\overline{B}\,\overline{C} + ABC$.

Solution There are three variables in the domain, so there are eight possible combinations of binary values of the variables as listed in the left three columns of Table 4–6. The binary values that make the product terms in the expressions equal to 1 are $\overline{A}\,\overline{B}C$: 001; $A\overline{B}\,\overline{C}$: 100; and ABC: 111. For each of these binary values, place a 1 in the output column as shown in the table. For each of the remaining binary combinations, place a 0 in the output column.

▶ TABLE 4–6

| INPUTS | | | OUTPUT | |
A	B	C	X	PRODUCT TERM
0	0	0	0	
0	0	1	1	$\overline{A}\,\overline{B}C$
0	1	0	0	
0	1	1	0	
1	0	0	1	$A\overline{B}\,\overline{C}$
1	0	1	0	
1	1	0	0	
1	1	1	1	ABC

Related Problem Create a truth table for the standard SOP expression $\overline{A}B\overline{C} + A\overline{B}C$.

Converting POS Expressions to Truth Table Format

Recall that a POS expression is equal to 0 only if at least one of the sum terms is equal to 0. To construct a truth table from a POS expression, list all the possible combinations of binary values of the variables just as was done for the SOP expression. Next, convert the POS expression to standard form if it is not already. Finally, place a 0 in the output column (X) for each binary value that makes the expression a 0 and place a 1 for all the remaining binary values. This procedure is illustrated in Example 4–19.

EXAMPLE 4–19

Determine the truth table for the following standard POS expression:

$$(A + B + C)(A + \overline{B} + C)(A + \overline{B} + \overline{C})(\overline{A} + B + \overline{C})(\overline{A} + \overline{B} + C)$$

Solution There are three variables in the domain and the eight possible binary values are listed in the left three columns of Table 4–7. The binary values that make the sum terms in the expression equal to 0 are $A + B + C$: 000; $A + \overline{B} + C$: 010; $A + \overline{B} + \overline{C}$: 011; $\overline{A} + B + \overline{C}$: 101; and $\overline{A} + \overline{B} + C$: 110. For each of these binary values, place a 0 in the output column as shown in the table. For each of the remaining binary combinations, place a 1 in the output column.

▶ TABLE 4–7

| INPUTS | | | OUTPUT | SUM TERM |
A	B	C	X	
0	0	0	0	$(A + B + C)$
0	0	1	1	
0	1	0	0	$(A + \overline{B} + C)$
0	1	1	0	$(A + \overline{B} + \overline{C})$
1	0	0	1	
1	0	1	0	$(\overline{A} + B + \overline{C})$
1	1	0	0	$(\overline{A} + \overline{B} + C)$
1	1	1	1	

Notice that the truth table in this example is the same as the one in Example 4–18. This means that the SOP expression in the previous example and the POS expression in this example are equivalent.

Related Problem Develop a truth table for the following standard POS expression:

$$(A + \overline{B} + C)(A + B + \overline{C})(\overline{A} + \overline{B} + \overline{C})$$

Determining Standard Expressions from a Truth Table

To determine the standard SOP expression represented by a truth table, list the binary values of the input variables for which the output is 1. Convert each binary value to the corresponding product term by replacing each 1 with the corresponding variable and each 0 with the corresponding variable complement. For example, the binary value 1010 is converted to a product term as follows:

$$1010 \longrightarrow A\overline{B}C\overline{D}$$

If you substitute, you can see that the product term is 1:

$$A\overline{B}C\overline{D} = 1 \cdot \overline{0} \cdot 1 \cdot \overline{0} = 1 \cdot 1 \cdot 1 \cdot 1 = 1$$

To determine the standard POS expression represented by a truth table, list the binary values for which the output is 0. Convert each binary value to the corresponding sum term by replacing each 1 with the corresponding variable complement and each 0 with the corresponding variable. For example, the binary value 1001 is converted to a sum term as follows:

$$1001 \longrightarrow \overline{A} + B + C + \overline{D}$$

If you substitute, you can see that the sum term is 0:

$$\overline{A} + B + C + \overline{D} = \overline{1} + 0 + 0 + \overline{1} = 0 + 0 + 0 + 0 = 0$$

EXAMPLE 4–20

From the truth table in Table 4–8, determine the standard SOP expression and the equivalent standard POS expression.

► TABLE 4-8

| INPUTS | | | OUTPUT |
A	B	C	X
0	0	0	0
0	0	1	0
0	1	0	0
0	1	1	1
1	0	0	1
1	0	1	0
1	1	0	1
1	1	1	1

Solution There are four 1s in the output column and the corresponding binary values are 011, 100, 110, and 111. Convert these binary values to product terms as follows:

$$011 \longrightarrow \overline{A}BC$$
$$100 \longrightarrow A\overline{B}\,\overline{C}$$
$$110 \longrightarrow AB\overline{C}$$
$$111 \longrightarrow ABC$$

The resulting standard SOP expression for the output X is

$$X = \overline{A}BC + A\overline{B}\,\overline{C} + AB\overline{C} + ABC$$

For the POS expression, the output is 0 for binary values 000, 001, 010, and 101. Convert these binary values to sum terms as follows:

$$000 \longrightarrow A + B + C$$
$$001 \longrightarrow A + B + \overline{C}$$
$$010 \longrightarrow A + \overline{B} + C$$
$$101 \longrightarrow \overline{A} + B + \overline{C}$$

The resulting standard POS expression for the output X is

$$X = (A + B + C)(A + B + \overline{C})(A + \overline{B} + C)(\overline{A} + B + \overline{C})$$

Related Problem By substitution of binary values, show that the SOP and the POS expressions derived in this example are equivalent; that is, for any binary value they should either both be 1 or both be 0, depending on the binary value.

SECTION 4-7 REVIEW

1. If a certain Boolean expression has a domain of five variables, how many binary values will be in its truth table?

2. In a certain truth table, the output is a 1 for the binary value 0110. Convert this binary value to the corresponding product term using variables W, X, Y, and Z.

3. In a certain truth table, the output is a 0 for the binary value 1100. Convert this binary value to the corresponding sum term using variables W, X, Y, and Z.

4-8 THE KARNAUGH MAP

A Karnaugh map provides a systematic method for simplifying Boolean expressions and, if properly used, will produce the simplest SOP or POS expression possible, known as the minimum expression. As you have seen, the effectiveness of algebraic simplification depends on your familiarity with all the laws, rules, and theorems of Boolean algebra and on your ability to apply them. The Karnaugh map, on the other hand, provides a "cookbook" method for simplification.

After completing this section, you should be able to

■ Construct a Karnaugh map for three or four variables ■ Determine the binary value of each cell in a Karnaugh map ■ Determine the standard product term represented by each cell in a Karnaugh map ■ Explain cell adjacency and identify adjacent cells

The purpose of a Karnaugh map is to simplify a Boolean expression.

A **Karnaugh map** is similar to a truth table because it presents all of the possible values of input variables and the resulting output for each value. Instead of being organized into columns and rows like a truth table, the Karnaugh map is an array of **cells** in which each cell represents a binary value of the input variables. The cells are arranged in a way so that simplification of a given expression is simply a matter of properly grouping the cells. Karnaugh maps can be used for expressions with two, three, four, and five variables, but we will discuss only 3-variable and 4-variable situations to illustrate the principles. Section 4–11 deals with five variables using a 32-cell Karnaugh map. Another method, which is beyond the scope of this book, called the Quine-McClusky method can be used for higher numbers of variables.

The number of cells in a Karnaugh map is equal to the total number of possible input variable combinations as is the number of rows in a truth table. For three variables, the number of cells is $2^3 = 8$. For four variables, the number of cells is $2^4 = 16$.

The 3-Variable Karnaugh Map

The 3-variable Karnaugh map is an array of eight cells, as shown in Figure 4–21(a). In this case, *A, B,* and *C* are used for the variables although other letters could be used. Binary values of *A* and *B* are along the left side (notice the sequence) and the values of *C* are across the top. The value of a given cell is the binary values of *A* and *B* at the left in the same row combined with the value of *C* at the top in the same column. For example, the cell in the upper left corner has a binary value of 000 and the cell in the lower right corner has a binary value of 101. Figure 4–21(b) shows the standard product terms that are represented by each cell in the Karnaugh map.

▶ **FIGURE 4–21**

A 3-variable Karnaugh map showing product terms.

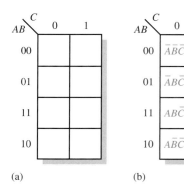

(a)   (b)

The 4-Variable Karnaugh Map

The 4-variable Karnaugh map is an array of sixteen cells, as shown in Figure 4–22(a). Binary values of *A* and *B* are along the left side and the values of *C* and *D* are across the top. The value of a given cell is the binary values of *A* and *B* at the left in the same row com-

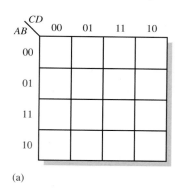

A 4-variable Karnaugh map.

bined with the binary values of C and D at the top in the same column. For example, the cell in the upper right corner has a binary value of 0010 and the cell in the lower right corner has a binary value of 1010. Figure 4–22(b) shows the standard product terms that are represented by each cell in the 4-variable Karnaugh map.

Cell Adjacency

The cells in a Karnaugh map are arranged so that there is only a single-variable change between adjacent cells. **Adjacency** is defined by a single-variable change. In the 3-variable map the 010 cell is adjacent to the 000 cell, the 011 cell, and the 110 cell. The 010 cell is not adjacent to the 001 cell, the 111 cell, the 100 cell, or the 101 cell.

Physically, each cell is adjacent to the cells that are immediately next to it on any of its four sides. A cell is not adjacent to the cells that diagonally touch any of its corners. Also, the cells in the top row are adjacent to the corresponding cells in the bottom row and the cells in the outer left column are adjacent to the corresponding cells in the outer right column. This is called "wrap-around" adjacency because you can think of the map as wrapping around from top to bottom to form a cylinder or from left to right to form a cylinder. Figure 4–23 illustrates the cell adjacencies with a 4-variable map, although the same rules for adjacency apply to Karnaugh maps with any number of cells.

Cells that differ by only one variable are adjacent.

Cells with values that differ by more than one variable are not adjacent.

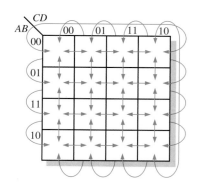

◀ FIGURE 4–23

Adjacent cells on a Karnaugh map are those that differ by only one variable. Arrows point between adjacent cells.

SECTION 4–8 REVIEW

1. In a 3-variable Karnaugh map, what is the binary value for the cell in each of the following locations:
   (a) upper left corner   (b) lower right corner
   (c) lower left corner   (d) upper right corner

2. What is the standard product term for each cell in Question 1 for variables X, Y, and Z?

3. Repeat Question 1 for a 4-variable map.

4. Repeat Question 2 for a 4-variable map using variables W, X, Y, and Z.

4–9 KARNAUGH MAP SOP MINIMIZATION

As stated in the last section, the Karnaugh map is used for simplifying Boolean expressions to their minimum form. A minimized SOP expression contains the fewest possible terms with the fewest possible variables per term. Generally, a minimum SOP expression can be implemented with fewer logic gates than a standard expression.

After completing this section, you should be able to

- Map a standard SOP expression on a Karnaugh map ■ Combine the 1s on the map into maximum groups ■ Determine the minimum product term for each group on the map ■ Combine the minimum product terms to form a minimum SOP expression ■ Convert a truth table into a Karnaugh map for simplification of the represented expression ■ Use "don't care" conditions on a Karnaugh map

Mapping a Standard SOP Expression

For an SOP expression in standard form, a 1 is placed on the Karnaugh map for each product term in the expression. Each 1 is placed in a cell corresponding to the value of a product term. For example, for the product term $A\overline{B}C$, a 1 goes in the 101 cell on a 3-variable map.

When an SOP expression is completely mapped, there will be a number of 1s on the Karnaugh map equal to the number of product terms in the standard SOP expression. The cells that do not have a 1 are the cells for which the expression is 0. Usually, when working with SOP expressions, the 0s are left off the map. The following steps and the illustration in Figure 4–24 show the mapping process.

Step 1. Determine the binary value of each product term in the standard SOP expression. After some practice, you can usually do the evaluation of terms mentally.

Step 2. As each product term is evaluated, place a 1 on the Karnaugh map in the cell having the same value as the product term.

▶ FIGURE 4–24

Example of mapping a standard SOP expression.

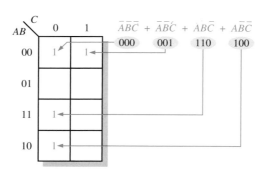

EXAMPLE 4–21

Map the following standard SOP expression on a Karnaugh map:

$$\overline{A}\,\overline{B}C + \overline{A}B\overline{C} + AB\overline{C} + ABC$$

Solution Evaluate the expression as shown below. Place a 1 on the 3-variable Karnaugh map in Figure 4–25 for each standard product term in the expression.

$$\overline{A}\,\overline{B}C + \overline{A}B\overline{C} + AB\overline{C} + ABC$$
$$001 \qquad 010 \qquad 110 \qquad 111$$

▶ FIGURE 4–25

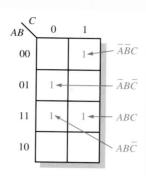

Related Problem   Map the standard SOP expression $\overline{A}BC + A\overline{B}C + A\overline{B}\,\overline{C}$ on a Karnaugh map.

EXAMPLE 4–22

Map the following standard SOP expression on a Karnaugh map:

$$\overline{A}\,\overline{B}CD + \overline{A}B\overline{C}\,\overline{D} + AB\overline{C}D + ABCD + AB\overline{C}\,\overline{D} + \overline{A}\,\overline{B}\,\overline{C}D + A\overline{B}C\overline{D}$$

Solution   Evaluate the expression as shown below. Place a 1 on the 4-variable Karnaugh map in Figure 4–26 for each standard product term in the expression.

$$\overline{A}\,\overline{B}CD + \overline{A}B\overline{C}\,\overline{D} + AB\overline{C}D + ABCD + AB\overline{C}\,\overline{D} + \overline{A}\,\overline{B}\,\overline{C}D + A\overline{B}C\overline{D}$$
$$0011 \qquad 0100 \qquad 1101 \qquad 1111 \qquad 1100 \qquad 0001 \qquad 1010$$

▶ FIGURE 4–26

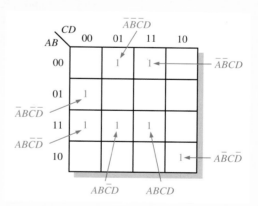

Related Problem   Map the following standard SOP expression on a Karnaugh map:

$$\overline{A}B\overline{C}\,\overline{D} + AB\overline{C}\,\overline{D} + AB\overline{C}\,\overline{D} + \overline{A}BCD$$

Mapping a Nonstandard SOP Expression

A Boolean expression must first be in standard form before you use a Karnaugh map. If an expression is not in standard form, then it must be converted to standard form by the procedure covered in Section 4–6 or by numerical expansion. Since an expression should be evaluated before mapping anyway, numerical expansion is probably the most efficient approach.

Numerical Expansion of a Nonstandard Product Term   Recall that a nonstandard product term has one or more missing variables. For example, assume that one of the product terms in a certain 3-variable SOP expression is $A\overline{B}$. This term can be expanded numerically to standard form as follows. First, write the binary value of the two variables and attach a 0 for the missing variable $\overline{C}$: 100. Next, write the binary value of the two variables and attach a 1 for the missing variable C: 101. The two resulting binary numbers are the values of the standard SOP terms $A\overline{B}\,\overline{C}$ and $A\overline{B}C$.

As another example, assume that one of the product terms in a 3-variable expression is B (remember that a single variable counts as a product term in an SOP expression). This term can be expanded numerically to standard form as follows. Write the binary value of the variable; then attach all possible values for the missing variables A and C as follows:

$$B$$
$$010$$
$$011$$
$$110$$
$$111$$

The four resulting binary numbers are the values of the standard SOP terms $\overline{A}B\overline{C}$, $\overline{A}BC$, $AB\overline{C}$, and ABC.

EXAMPLE 4–23

Map the following SOP expression on a Karnaugh map: $\overline{A} + A\overline{B} + AB\overline{C}$.

Solution   The SOP expression is obviously not in standard form because each product term does not have three variables. The first term is missing two variables, the second term is missing one variable, and the third term is standard. First expand the terms numerically as follows:

$$\overline{A} \quad + A\overline{B} \quad + AB\overline{C}$$
$$000 \qquad 100 \qquad 110$$
$$001 \qquad 101$$
$$010$$
$$011$$

Map each of the resulting binary values by placing a 1 in the appropriate cell of the 3-variable Karnaugh map in Figure 4–27.

▶ **FIGURE 4–27**

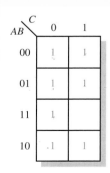

Related Problem   Map the SOP expression $BC + \overline{A}\,\overline{C}$ on a Karnaugh map.

EXAMPLE 4–24

Map the following SOP expression on a Karnaugh map:

$$\overline{B}\,\overline{C} + A\overline{B} + AB\overline{C} + A\overline{B}CD + \overline{A}\,\overline{B}\,\overline{C}D + ABCD$$

Solution   The SOP expression is obviously not in standard form because each product term does not have four variables. The first and second terms are both missing two variables, the third term is missing one variable, and the rest of the terms are standard. First expand the terms by including all combinations of the missing variables numerically as follows:

$\overline{B}\,\overline{C}$	$A\overline{B}$	$+$	$AB\overline{C}$	$+ A\overline{B}CD$	$+ \overline{A}\,\overline{B}\,\overline{C}D$	$+ ABCD$
0000	1000		1100	1010	0001	1011
0001	1001		1101			
1000	1010					
1001	1011					

Map each of the resulting binary values by placing a 1 in the appropriate cell of the 4-variable Karnaugh map in Figure 4–28. Notice that some of the values in the expanded expression are redundant.

▶ **FIGURE 4–28**

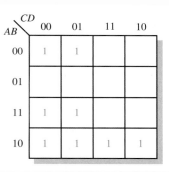

Related Problem   Map the expression $A + \overline{C}D + AC\overline{D} + \overline{A}BC\overline{D}$ on a Karnaugh map.

Karnaugh Map Simplification of SOP Expressions

The process that results in an expression containing the fewest possible terms with the fewest possible variables is called **minimization**. After an SOP expression has been mapped, a minimum SOP expression is obtained by grouping the 1s and determining the minimum SOP expression from the map.

Grouping the 1s   You can group 1s on the Karnaugh map according to the following rules by enclosing those adjacent cells containing 1s. The goal is to maximize the size of the groups and to minimize the number of groups.

1. A group must contain either 1, 2, 4, 8, or 16 cells, which are all powers of two. In the case of a 3-variable map, $2^3 = 8$ cells is the maximum group.

2. Each cell in a group must be adjacent to one or more cells in that same group, but all cells in the group do not have to be adjacent to each other.

3. Always include the largest possible number of 1s in a group in accordance with rule 1.

4. Each 1 on the map must be included in at least one group. The 1s already in a group can be included in another group as long as the overlapping groups include noncommon 1s.

EXAMPLE 4–25

Group the 1s in each of the Karnaugh maps in Figure 4–29.

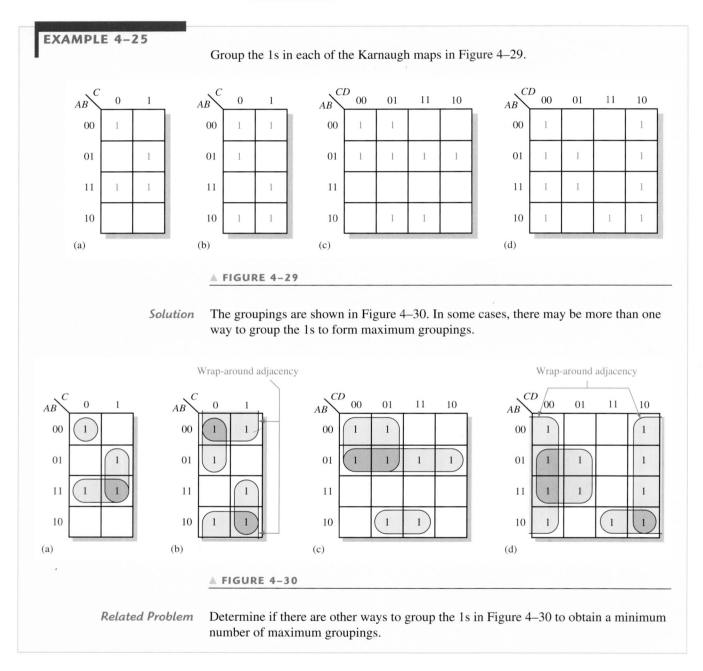

▲ FIGURE 4–29

Solution The groupings are shown in Figure 4–30. In some cases, there may be more than one way to group the 1s to form maximum groupings.

▲ FIGURE 4–30

Related Problem Determine if there are other ways to group the 1s in Figure 4–30 to obtain a minimum number of maximum groupings.

Determining the Minimum SOP Expression from the Map When all the 1s representing the standard product terms in an expression are properly mapped and grouped, the process of determining the resulting minimum SOP expression begins. The following rules are applied to find the minimum product terms and the minimum SOP expression:

1. Group the cells that have 1s. Each group of cells containing 1s creates one product term composed of all variables that occur in only one form (either uncomplemented

or complemented) within the group. Variables that occur both uncomplemented and complemented within the group are eliminated. These are called *contradictory variables.*

2. Determine the minimum product term for each group.
   a. For a 3-variable map:
      (1) A 1-cell group yields a 3-variable product term
      (2) A 2-cell group yields a 2-variable product term
      (3) A 4-cell group yields a 1-variable term
      (4) An 8-cell group yields a value of 1 for the expression
   b. For a 4-variable map:
      (1) A 1-cell group yields a 4-variable product term
      (2) A 2-cell group yields a 3-variable product term
      (3) A 4-cell group yields a 2-variable product term
      (4) An 8-cell group yields a 1-variable term
      (5) A 16-cell group yields a value of 1 for the expression

3. When all the minimum product terms are derived from the Karnaugh map, they are summed to form the minimum SOP expression.

EXAMPLE 4–26

Determine the product terms for the Karnaugh map in Figure 4–31 and write the resulting minimum SOP expression.

▶ **FIGURE 4–31**

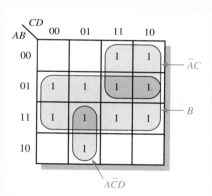

Solution   Eliminate variables that are in a grouping in both complemented and uncomplemented forms. In Figure 4–31, the product term for the 8-cell group is B because the cells within that group contain both A and $\overline{A}$, C and $\overline{C}$, and D and $\overline{D}$, which are eliminated. The 4-cell group contains B, $\overline{B}$, D, and $\overline{D}$, leaving the variables $\overline{A}$ and C, which form the product term $\overline{A}C$. The 2-cell group contains B and $\overline{B}$, leaving variables A, $\overline{C}$, and D which form the product term $A\overline{C}D$. Notice how overlapping is used to maximize the size of the groups. The resulting minimum SOP expression is the sum of these product terms:

$$B + \overline{A}C + A\overline{C}D$$

Related Problem   For the Karnaugh map in Figure 4–31, add a 1 in the lower right cell (1010) and determine the resulting SOP expression.

EXAMPLE 4–27

Determine the product terms for each of the Karnaugh maps in Figure 4–32 and write the resulting minimum SOP expression.

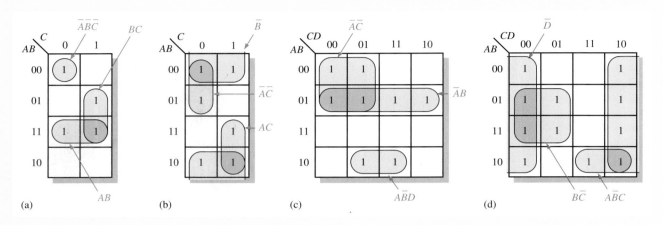

▲ **FIGURE 4–32**

Solution   The resulting minimum product term for each group is shown in Figure 4–32. The minimum SOP expressions for each of the Karnaugh maps in the figure are

(a) $AB + BC + \overline{A}\,\overline{B}\,\overline{C}$ (b) $\overline{B} + \overline{A}\,\overline{C} + AC$

(c) $\overline{A}B + \overline{A}\,\overline{C} + A\overline{B}D$ (d) $\overline{D} + A\overline{B}C + B\overline{C}$

Related Problem   For the Karnaugh map in Figure 4–32(d), add a 1 in the 0111 cell and determine the resulting SOP expression.

EXAMPLE 4–28

Use a Karnaugh map to minimize the following standard SOP expression:

$$A\overline{B}C + \overline{A}BC + \overline{A}\,\overline{B}C + \overline{A}\,\overline{B}\,\overline{C} + AB\overline{C}$$

Solution   The binary values of the expression are

$$101 + 011 + 011 + 000 + 100$$

Map the standard SOP expression and group the cells as shown in Figure 4–33.

▶ **FIGURE 4–33**

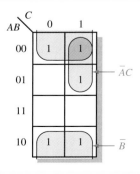

Notice the "wrap around" 4-cell group that includes the top row and the bottom row of 1s. The remaining 1 is absorbed in an overlapping group of two cells. The group of four 1s produces a single variable term, $\overline{B}$. This is determined by observing that within the group, $\overline{B}$ is the only variable that does not change from cell to cell. The group of two 1s produces a 2-variable term $\overline{A}C$. This is determined by observing that within the group, $\overline{A}$ and C do not change from one cell to the next. The product term for each group is shown. The resulting minimum SOP expression is

$$\overline{B} + \overline{A}C$$

Keep in mind that this minimum expression is equivalent to the original standard expression.

Related Problem Use a Karnaugh map to simplify the following standard SOP expression:

$$X\overline{Y}Z + XY\overline{Z} + \overline{X}YZ + \overline{X}Y\overline{Z} + X\overline{Y}\,\overline{Z} + XYZ$$

EXAMPLE 4–29

Use a Karnaugh map to minimize the following SOP expression:

$$\overline{B}\,\overline{C}D + \overline{A}B\overline{C}D + AB\overline{C}D + \overline{A}\,\overline{B}CD + A\overline{B}CD + \overline{A}\,\overline{B}\,\overline{C}D + \overline{A}BCD + ABC\overline{D} + A\overline{B}C\overline{D}$$

Solution The first term $\overline{B}\,\overline{C}\,\overline{D}$ must be expanded into $A\overline{B}\,\overline{C}\,\overline{D}$ and $\overline{A}\,\overline{B}\,\overline{C}\,\overline{D}$ to get the standard SOP expression, which is then mapped; and the cells are grouped as shown in Figure 4–34.

▶ **FIGURE 4–34**

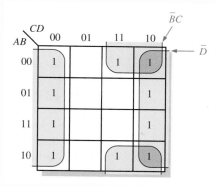

Notice that both groups exhibit "wrap around" adjacency. The group of eight is formed because the cells in the outer columns are adjacent. The group of four is formed to pick up the remaining two 1s because the top and bottom cells are adjacent. The product term for each group is shown. The resulting minimum SOP expression is

$$\overline{D} + \overline{B}C$$

Keep in mind that this minimum expression is equivalent to the original standard expression.

Related Problem Use a Karnaugh map to simplify the following SOP expression:

$$\overline{W}\,\overline{X}\,\overline{Y}\,\overline{Z} + W\overline{X}YZ + W\overline{X}\,\overline{Y}Z + \overline{W}YZ + W\overline{X}\,\overline{Y}\,\overline{Z}$$

Mapping Directly from a Truth Table

You have seen how to map a Boolean expression; now you will learn how to go directly from a truth table to a Karnaugh map. Recall that a truth table gives the output of a Boolean expression for all possible input variable combinations. An example of a Boolean expression and its truth table representation is shown in Figure 4–35. Notice in the truth table that the output X is 1 for four different input variable combinations. The 1s in the output column of the truth table are mapped directly onto a Karnaugh map into the cells corresponding to the values of the associated input variable combinations, as shown in Figure 4–35. In the figure you can see that the Boolean expression, the truth table, and the Karnaugh map are simply different ways to represent a logic function.

▶ **FIGURE 4–35**

Example of mapping directly from a truth table to a Karnaugh map.

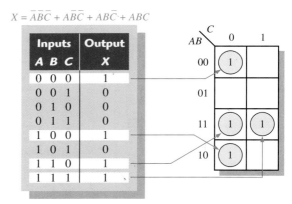

$X = \overline{A}\,\overline{B}\,\overline{C} + A\overline{B}\overline{C} + AB\overline{C} + ABC$

"Don't Care" Conditions

Sometimes a situation arises in which some input variable combinations are not allowed. For example, recall that in the BCD code covered in Chapter 2, there are six invalid combinations: 1010, 1011, 1100, 1101, 1110, and 1111. Since these unallowed states will never occur in an application involving the BCD code, they can be treated as **"don't care"** terms with respect to their effect on the output. That is, for these "don't care" terms either a 1 or a 0 may be assigned to the output; it really does not matter since they will never occur.

The "don't care" terms can be used to advantage on the Karnaugh map. Figure 4–36 shows that for each "don't care" term, an X is placed in the cell. When grouping the 1s, the Xs can be treated as 1s to make a larger grouping or as 0s if they cannot be used to advantage. The larger a group, the simpler the resulting term will be.

The truth table in Figure 4–36(a) describes a logic function that has a 1 output only when the BCD code for 7, 8, or 9 is present on the inputs. If the "don't cares" are used as 1s, the resulting expression for the function is $A + BCD$, as indicated in part (b). If the "don't cares" are not used as 1s, the resulting expression is $A\overline{B}\,\overline{C} + \overline{A}BCD$; so you can see the advantage of using "don't care" terms to get the simplest expression.

**SECTION 4–9
REVIEW**

1. Lay out Karnaugh maps for three and four variables.

2. Group the 1s and write the simplified SOP expression for the Karnaugh map in Figure 4–25.

3. Write the original standard SOP expressions for each of the Karnaugh maps in Figure 4–32.

◀ FIGURE 4–36

Example of the use of "don't care" conditions to simplify an expression.

Inputs A B C D	Output Y
0 0 0 0	0
0 0 0 1	0
0 0 1 0	0
0 0 1 1	0
0 1 0 0	0
0 1 0 1	0
0 1 1 0	0
0 1 1 1	1
1 0 0 0	1
1 0 0 1	1
1 0 1 0	X
1 0 1 1	X
1 1 0 0	X
1 1 0 1	X
1 1 1 0	X
1 1 1 1	X

Don't cares

(a) Truth table

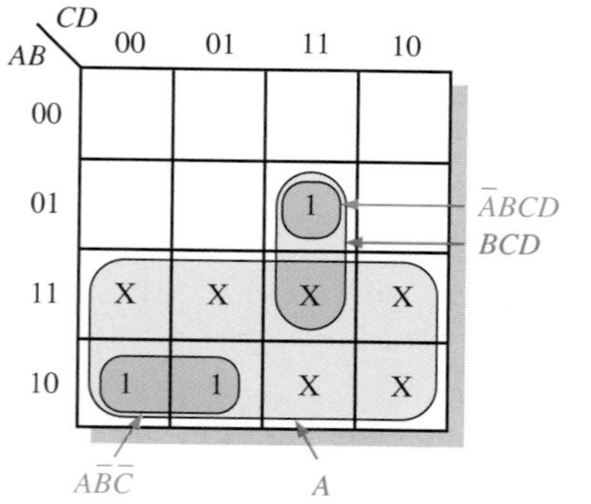

(b) Without "don't cares" $Y = A\overline{B}\,\overline{C} + \overline{A}BCD$
With "don't cares" $Y = A + BCD$

4–10 KARNAUGH MAP POS MINIMIZATION

In the last section, you studied the minimization of an SOP expression using a Karnaugh map. In this section, we will focus on POS expressions. The approaches are much the same except that with POS expressions, 0s representing the standard sum terms are placed on the Karnaugh map instead of 1s.

After completing this section, you should be able to

■ Map a standard POS expression on a Karnaugh map ■ Combine the 0s on the map into maximum groups ■ Determine the minimum sum term for each group on the map ■ Combine the minimum sum terms to form a minimum POS expression ■ Use the Karnaugh map to convert between POS and SOP

Mapping a Standard POS Expression

For a POS expression in standard form, a 0 is placed on the Karnaugh map for each sum term in the expression. Each 0 is placed in a cell corresponding to the value of a sum term. For example, for the sum term $A + \overline{B} + C$, a 0 goes in the 010 cell on a 3-variable map.

When a POS expression is completely mapped, there will be a number of 0s on the Karnaugh map equal to the number of sum terms in the standard POS expression. The cells that do not have a 0 are the cells for which the expression is 1. Usually, when working with POS expressions, the 1s are left off. The following steps and the illustration in Figure 4–37 show the mapping process.

Step 1. Determine the binary value of each sum term in the standard POS expression. This is the binary value that makes the term equal to 0.

Step 2. As each sum term is evaluated, place a 0 on the Karnaugh map in the corresponding cell.

Example of mapping a standard POS expression.

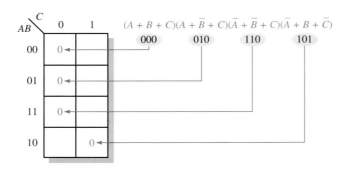

EXAMPLE 4–30

Map the following standard POS expression on a Karnaugh map:

$$(\overline{A} + \overline{B} + C + D)(\overline{A} + B + \overline{C} + \overline{D})(A + B + \overline{C} + D)(\overline{A} + \overline{B} + \overline{C} + \overline{D})(A + B + \overline{C} + D)$$

Solution   Evaluate the expression as shown below and place a 0 on the 4-variable Karnaugh map in Figure 4–38 for each standard sum term in the expression.

$$(\overline{A} + \overline{B} + C + D)(\overline{A} + B + \overline{C} + \overline{D})(A + B + \overline{C} + D)(\overline{A} + \overline{B} + \overline{C} + \overline{D})(A + B + \overline{C} + \overline{D})$$

1100	1011	0010	1111	0011

▶ FIGURE 4–38

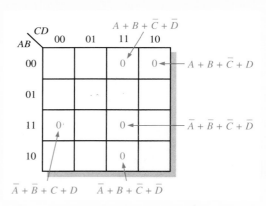

Related Problem   Map the following standard POS expression on a Karnaugh map:

$$(A + \overline{B} + \overline{C} + D)(A + B + C + \overline{D})(A + B + C + D)(\overline{A} + B + \overline{C} + D)$$

Karnaugh Map Simplification of POS Expressions

The process for minimizing a POS expression is basically the same as for an SOP expression except that you group 0s to produce minimum sum terms instead of grouping 1s to produce minimum product terms. The rules for grouping the 0s are the same as those for grouping the 1s that you learned in Section 4–9.

EXAMPLE 4–31

Use a Karnaugh map to minimize the following standard POS expression:

$$(A + B + C)(A + B + \overline{C})(A + \overline{B} + C)(A + \overline{B} + \overline{C})(\overline{A} + \overline{B} + C)$$

Also, derive the equivalent SOP expression.

Solution   The combinations of binary values of the expression are

$$(0 + 0 + 0)\,(0 + 0 + 1)\,(0 + 1 + 0)\,(0 + 1 + 1)\,(1 + 1 + 0)$$

Map the standard POS expression and group the cells as shown in Figure 4–39.

▶ **FIGURE 4–39**

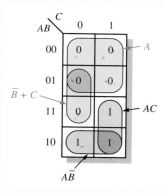

Notice how the 0 in the 110 cell is included into a 2-cell group by utilizing the 0 in the 4-cell group. The sum term for each blue group is shown in the figure and the resulting minimum POS expression is

$$A(\overline{B} + C)$$

Keep in mind that this minimum POS expression is equivalent to the original standard POS expression.

Grouping the 1s as shown by the gray areas yields an SOP expression that is equivalent to grouping the 0s.

$$AC + A\overline{B} = A(\overline{B} + C)$$

Related Problem   Use a Karnaugh map to simplify the following standard POS expression:

$$(X + \overline{Y} + Z)(X + \overline{Y} + \overline{Z})(\overline{X} + \overline{Y} + Z)(\overline{X} + Y + Z)$$

EXAMPLE 4–32

Use a Karnaugh map to minimize the following POS expression:

$$(B + C + D)(A + B + \overline{C} + D)(\overline{A} + B + C + \overline{D})(A + \overline{B} + C + D)(\overline{A} + \overline{B} + C + D)$$

Solution   The first term must be expanded into $\overline{A} + B + C + D$ and $A + B + C + D$ to get a standard POS expression, which is then mapped; and the cells are grouped as shown in

Figure 4–40. The sum term for each group is shown and the resulting minimum POS expression is

$$(C + D)\,(A + B + D)\,(\overline{A} + B + C)$$

Keep in mind that this minimum POS expression is equivalent to the original standard POS expression.

▶ FIGURE 4–40

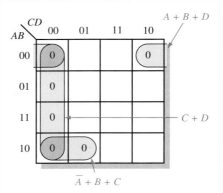

Related Problem   Use a Karnaugh map to simplify the following POS expression:

$$(W + \overline{X} + Y + \overline{Z})(W + X + Y + Z)(W + \overline{X} + \overline{Y} + Z)(\overline{W} + \overline{X} + Z)$$

Converting Between POS and SOP Using the Karnaugh Map

When a POS expression is mapped, it can easily be converted to the equivalent SOP form directly from the Karnaugh map. Also, given a mapped SOP expression, an equivalent POS expression can be derived directly from the map. This provides a good way to compare both minimum forms of an expression to determine if one of them can be implemented with fewer gates than the other.

For a POS expression, all the cells that do not contain 0s contain 1s, from which the SOP expression is derived. Likewise, for an SOP expression, all the cells that do not contain 1s contain 0s, from which the POS expression is derived. Example 4–33 illustrates this conversion.

EXAMPLE 4–33

Using a Karnaugh map, convert the following standard POS expression into a minimum POS expression, a standard SOP expression, and a minimum SOP expression.

$$(\overline{A} + \overline{B} + C + D)(A + \overline{B} + C + D)(A + B + C + \overline{D})$$

$$(A + B + \overline{C} + \overline{D})(\overline{A} + B + C + \overline{D})(A + B + \overline{C} + D)$$

Solution   The 0s for the standard POS expression are mapped and grouped to obtain the minimum POS expression in Figure 4–41(a). In Figure 4–41(b), 1s are added to the cells that do not contain 0s. From each cell containing a 1, a standard product term is obtained as indicated. These product terms form the standard SOP expression. In Figure 4–41(c), the 1s are grouped and a minimum SOP expression is obtained.

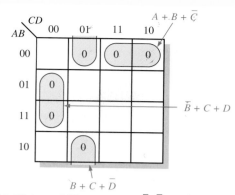

(a) Minimum POS: $(A + B + C)(\bar{B} + \bar{C} + D)(B + C + \bar{D})$

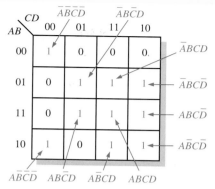

(b) Standard SOP:
$\bar{A}\bar{B}\bar{C}\bar{D} + \bar{A}\bar{B}\bar{C}D + \bar{A}BCD + \bar{A}BC\bar{D} + ABC\bar{D} + A\bar{B}C\bar{D} + A\bar{B}\bar{C}\bar{D} + AB\bar{C}D + A\bar{B}\bar{C}D + ABCD$

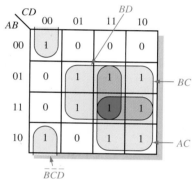

(c) Minimum SOP: $AC + BC + BD + \bar{B}\bar{C}\bar{D}$

▲ FIGURE 4–41

Related Problem Use a Karnaugh map to convert the following expression to minimum SOP form:

$$(W + \bar{X} + Y + \bar{Z})(\bar{W} + X + \bar{Y} + \bar{Z})(\bar{W} + \bar{X} + \bar{Y} + Z)(\bar{W} + \bar{X} + \bar{Z})$$

**SECTION 4–10
REVIEW**

1. What is the difference in mapping a POS expression and an SOP expression?

2. What is the standard sum term expressed with variables *A, B, C,* and *D* for a 0 in cell 1011 of the Karnaugh map?

3. What is the standard product term expressed with variables *A, B, C,* and *D* for a 1 in cell 0010 of the Karnaugh map?

4–11 FIVE-VARIABLE KARNAUGH MAPS

Boolean functions with five variables can be simplified using a 32-cell Karnaugh map. Actually, two 4-variable maps (16 cells each) are used to construct a 5-variable map. You already know the cell adjacencies within each of the 4-variable maps and how to form groups of cells containing 1s to simplify an SOP expression. All you need to learn for five variables is the cell adjacencies between the two 4-variable maps and how to group those adjacent 1s.

After completing this section, you should be able to

■ Determine cell adjacencies in a 5-variable map ■ Form maximum cell groupings in a 5-variable map ■ Minimize 5-variable Boolean expressions using the Karnaugh map

A Karnaugh map for five variables (*ABCDE*) can be constructed using two 4-variable maps with which you are already familiar. Each map contains 16 cells with all combinations of variables *B, C, D,* and *E.* One map is for *A* = 0 and the other is for *A* = 1, as shown in Figure 4–42.

▶ **FIGURE 4–42**

A 5-variable Karnaugh map.

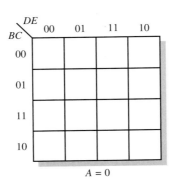 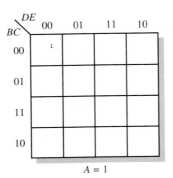

Cell Adjacencies

You already know how to determine adjacent cells within the 4-variable map. The best way to visualize cell adjacencies between the two 16-cell maps is to imagine that the *A* = 0 map is placed on top of the *A* = 1 map. Each cell in the *A* = 0 map is adjacent to the cell directly below it in the *A* = 1 map.

To illustrate, an example with four groups is shown in Figure 4–43 with the maps in a 3-dimensional arrangement. The 1s in the yellow cells form an 8-bit group (four in the *A* = 0

▶ **FIGURE 4–43**

Illustration of groupings of 1s in adjacent cells of a 5-variable map.

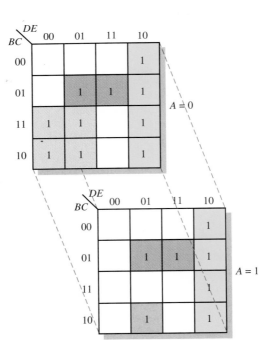

map combined with four in the $A = 1$ map). The 1s in the orange cells form a 4-bit group. The 1s in the light red cells form a 4-bit group only in the $A = 0$ map. The 1 in the gray cell in the $A = 1$ map is grouped with the 1 in the lower right light red cell in the $A = 0$ map to form a 2-bit group.

Determining the Boolean Expression The original SOP Boolean expression that is plotted on the Karnaugh map in Figure 4–43 contains seventeen 5-variable terms because there are seventeen 1s on the map. As you know, only the variables that do not change from uncomplemented to complemented or vice versa within a group remain in the expression for that group. The simplified expression taken from the map is developed as follows:

- The term for the yellow group is $D\overline{E}$.

- The term for the orange group is $\overline{B}CE$.

- The term for the light red group is $\overline{A}B\overline{D}$.

- The term for the gray cell grouped with the red cell is $B\overline{C}\,\overline{D}E$.

Combining these terms into the simplified SOP expression yields

$$X = D\overline{E} + \overline{B}CE + \overline{A}B\overline{D} + B\overline{C}\,\overline{D}E$$

EXAMPLE 4–34

Use a Karnaugh map to minimize the following standard SOP 5-variable expression:

$$X = \overline{A}\,\overline{B}\,\overline{C}D\overline{E} + \overline{A}\,BC\overline{D}\,\overline{E} + \overline{A}BCD\overline{E} + \overline{A}BC\overline{D}\,\overline{E} + \overline{A}\,\overline{B}\,\overline{C}D\overline{E} + \overline{A}BC\overline{D}E$$
$$+ \,\overline{A}BCDE + A\overline{B}\,\overline{C}\,\overline{D}\,\overline{E} + A\overline{B}\,\overline{C}DE + ABC\overline{D}E + ABCDE + A\overline{B}CDE$$

Solution Map the SOP expression. Figure 4–44 shows the groupings and their corresponding terms. Combining the terms yields the following minimized SOP expression:

$$X + \overline{A}\,D\overline{E} + \overline{B}\,\overline{C}\,\overline{D} + BCE + ACDE$$

▶ **FIGURE 4–44**

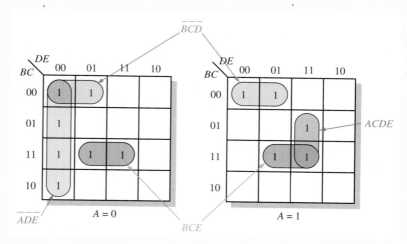

Related Problem Minimize the following expression:

$$Y = \overline{A}\,\overline{B}\,\overline{C}D\overline{E} + \overline{A}\,\overline{B}CD\overline{E} + \overline{A}BC\overline{D}\,\overline{E} + \overline{A}BC\overline{D}\,\overline{E} + A\overline{B}\,\overline{C}\,\overline{D}\,\overline{E} + A\overline{B}C\overline{D}\,\overline{E} + ABC\overline{D}\,\overline{E} + AB\overline{C}\,\overline{D}\,\overline{E}$$
$$+ \,\overline{A}\,\overline{B}\,C\overline{D}E + \overline{A}\,BCD\overline{E} + \overline{A}BCD\overline{E} + \overline{A}BC\overline{D}E + A\overline{B}\,C\overline{D}E + A\overline{B}CD\overline{E} + ABCD\overline{E} + AB\overline{C}D\overline{E}$$

4–12 VHDL (optional)

This optional section provides a brief introduction to VHDL and is not meant to teach the complete structure and syntax of the language. For more detailed information and instruction, refer to the footnote. Hardware description languages (HDLs) are tools for logic design entry, called text entry, that are used to implement logic designs in programmable logic devices. Although VHDL provides multiple ways to describe a logic circuit, only the simplest and most direct programming examples of text entry are discussed here.

After completing this section, you should be able to

■ State the essential elements of VHDL ■ Write a simple VHDL program

The V in VHDL* stands for VHSIC (*Very High Speed Integrated Circuit*) and the HDL, of course, stands for hardware description language. As mentioned, **VHDL** is a standard language adopted by the IEEE (Institute of Electrical and Electronics Engineers) and is designated IEEE Std. 1076-1993. VHDL is a complex and comprehensive language and using it to its full potential involves a lot of effort and experience.

VHDL provides three basic approaches to describing a digital circuit using software: *behavioral, data flow,* and *structural.* We will restrict this discussion to the data flow approach in which you write Boolean-type statements to describe a logic circuit. Keep in mind that VHDL, as well as the other HDLs, is a tool for implementing digital designs and is, therefore, a means to an end and not an end in itself.

It is relatively easy to write programs to describe simple logic circuits in VHDL. The logical operators are the following VHDL keywords: **and, or, not, nand, nor, xor,** and **xnor.** The two essential elements in any VHDL program are the entity and the architecture, and they must be used together. The **entity** describes a given logic function in terms of its external inputs and outputs, called ports. The **architecture** describes the internal operation of the logic function.

In its simplest form, the entity element consists of three statements: The first statement assigns a name to a logic function; the second statement, called the *port* statement which is indented, specifies the inputs and outputs; and the third statement is the *end* statement. Although you would probably not write a VHDL program for a single gate, it is instructive to start with a simple example such as an AND gate. The VHDL entity declaration for a 2-input AND gate is

Colons and semicolons must be used appropriately in all VHDL programs.

> **entity** AND_Gate2 **is**
>   **port** (A, B: **in** bit; X: **out** bit);
> **end entity** AND_Gate2;

The blue boldface terms are VHDL keywords; the other terms are identifiers that you assign; and the parentheses, colons, and semicolons are required VHDL syntax. As you can see, A and B are specified as input bits and X is specified as an output bit. The port identifiers A, B, and X as well as the entity name AND_Gate2 are user-defined and can be renamed. As in all HDLs, the placement of colons and semicolons is crucial and must be strictly adhered to.

The VHDL architecture element of the program for the 2-input AND gate described by the entity is

---

*See Floyd, Thomas. 2003. *Digital Fundamentals with VHDL.* Prentice Hall; Pellerin, David and Taylor, Douglas. 1997. *VHDL Made Easy!* Prentice Hall; Bhasker, Jayaram. 1999. *A VHDL Primer,* 3 ed. Prentice Hall.

architecture LogicFunction **of** AND_Gate2 **is**

begin

  X ⇐ A **and** B;

end architecture LogicFunction;

Again, the VHDL keywords are blue boldface, and the semicolons and the assignment operator ⇐ are required syntax. The first statement of the architecture element must reference the entity name.

The entity and the architecture are combined into a single VHDL program to describe an AND gate, as illustrated in Figure 4–45.

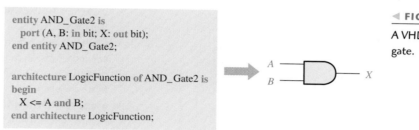

◀ **FIGURE 4–45**

A VHDL program for a 2-input AND gate.

Writing Boolean Expressions in VHDL   As you saw, the expression for a 2-input AND gate, $X = AB$, is written in VHDL as X ⇐ A **and** B;. Any Boolean expression can be written using VHDL keywords **not, and, or, nand, nor, xor,** and **xnor.** For example, the Boolean expression $X = A + B + C$ is written in VHDL as X ⇐ A **or** B **or** C;. The Boolean expression $X = A\overline{B} + \overline{C}D$ can be written as the VHDL statement X ⇐ (A **and not** B) **or** (**not** C **and** D);. As another example, the VHDL statement for a 2-input NAND gate can be written as X ⇐ **not**(A **and** B); or it can be written as X ⇐ A **nand** B;.

EXAMPLE 4–35

Write a VHDL program to describe the logic circuit in Figure 4–46.

▶ **FIGURE 4–46**

Solution   This AND/OR logic circuit is described in Boolean algebra as

$$X = AB + CD$$

The VHDL program follows. The entity name is AND_OR.

  entity AND_OR **is**

    port (A, B, C, D: **in** bit; X: **out** bit);

  end entity AND_OR;

  architecture LogicFunction **of** AND_OR **is**

  begin

    X <= (A **and** B) **or** (C **and** D);

  end architecture LogicFunction;

Related Problem   Write the VHDL statement to describe the logic circuit if a NOR gate replaces the OR gate in Figure 4–46.

SECTION 4–12 REVIEW

1. What is an HDL?
2. Name the two essential design elements in a VHDL program.
3. What does the entity do?
4. What does the architecture do?

DIGITAL SYSTEM APPLICATION

Seven-segment displays are used in many types of products. The tablet-counting and control system that was described in Chapter 1 has two 7-segment displays. These displays are used with logic circuits that decode a binary coded decimal (BCD) number and activate the appropriate digits on the display. In this digital system application, we focus on a minimum-gate design for this to illustrate an application of Boolean expressions and the Karnaugh map. As an option, VHDL is also applied.

The 7-Segment Display

Figure 4–47 shows a common display format composed of seven elements or segments. Energizing certain combinations of these segments can cause each of the ten decimal digits to be displayed. Figure 4–48 illustrates this method of digital display for each of the ten digits by using a red segment to represent one that is energized. To produce a 1, segments *b* and *c* are energized; to produce a 2, segments *a*, *b*, *g*, *e*, and *d* are used; and so on.

LED Displays One common type of 7-segment display consists of light-emitting

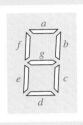

▲ FIGURE 4–47

Seven-segment display format showing arrangement of segments.

diodes (**LEDs**) arranged as shown in Figure 4–49. Each segment is an LED that emits light when there is current through it. In Figure 4–49(a) the common-anode arrangement requires the driving circuit to provide a

low-level voltage in order to activate a given segment. When a LOW is applied to a segment input, the LED is turned on, and there is current through it. In Figure 4–49(b) the common-cathode arrangement requires the driver to provide a high-level voltage to activate a segment. When a HIGH is applied to a segment input, the LED is turned on and there is current through it.

LCD Displays Another common type of 7-segment display is the liquid crystal display (**LCD**). LCDs operate by polarizing light so that a nonactivated segment reflects incident light and thus appears invisible against its background. An activated segment does not reflect incident light and thus appears dark. LCDs consume much less power than LEDs but

▲ FIGURE 4–48

Display of decimal digits with a 7-segment device.

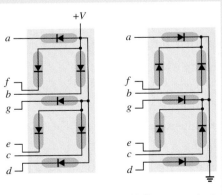

(a) Common-anode   (b) Common-cathode

◀ FIGURE 4–49

Arrangements of 7-segment LED displays.

cannot be seen in the dark, while LEDs can.

Segment Logic

Each segment is used for various decimal digits, but no one segment is used for all ten digits. Therefore, each segment must be activated by its own decoding circuit that detects the occurrence of any of the numbers in which the segment is used. From Figures 4–47 and 4–48, the segments that are required to be activated for each displayed digit are determined and listed in Table 4–9.

Truth Table for the Segment Logic

The segment decoding logic requires four binary coded decimal (BCD) inputs and seven outputs, one for each segment in the display, as indicated in the block diagram of Figure 4–50. The multiple-output truth table, shown in Table 4–10, is actually seven truth tables in one and could be separated into a separate table for each segment. A 1 in the segment output columns of the table indicates an activated segment.

Since the BCD code does not include the binary values 1010, 1011, 1100, 1101, 1110, and 1111, these combinations will never appear on the inputs and can therefore be treated as "don't care" (X) conditions, as indicated in the truth table. To conform with the practice of most IC manufacturers, A represents the least significant bit and D represents the most significant bit in this particular application.

Boolean Expressions for the Segment Logic

From the truth table, a standard SOP or POS expression can be written for each segment. For example, the standard SOP expression for segment a is

$$a = \overline{D}\,\overline{C}\,\overline{B}\,\overline{A} + \overline{D}\,\overline{C}B\overline{A} + \overline{D}\,\overline{C}BA + \overline{D}CB\overline{A} + \overline{D}CB\overline{A} + \overline{D}CBA + DC\,\overline{B}\,\overline{A} + DC\,\overline{B}A$$

and the standard SOP expression for segment e is

$$e = \overline{D}\,\overline{C}\,\overline{B}\,\overline{A} + \overline{D}C\overline{B}\,\overline{A} + \overline{D}CB\overline{A} + DC\overline{B}\,\overline{A}$$

Expressions for the other segments can be similarly developed. As you can see, the expression for segment a has eight product terms and the expression for segment e has four product terms representing each of the BCD inputs that activate that segment. This means that the standard SOP implementation of segment-a logic requires an AND-OR circuit consisting of eight 4-input AND gates and one 8-input OR gate. The implementation of segment-e logic requires four 4-input AND gates and one 4-input OR gate. In both cases, four inverters are required to produce the complement of each variable.

Karnaugh Map Minimization of the Segment Logic

Let's begin by obtaining a minimum SOP expression for segment a. A Karnaugh map for segment a is shown in Figure 4–51 and the following steps are carried out:

Step 1. The 1s are mapped directly from Table 4–10.

▶ **TABLE 4–9**

Active segments for each decimal digit.

DIGIT	SEGMENTS ACTIVATED
0	a, b, c, d, e, f
1	b, c
2	a, b, d, e, g
3	a, b, c, d, g
4	b, c, f, g
5	a, c, d, f, g
6	a, c, d, e, f, g
7	a, b, c
8	a, b, c, d, e, f, g
9	a, b, c, d, f, g

▶ **FIGURE 4–50**

Block diagram of 7-segment logic and display.

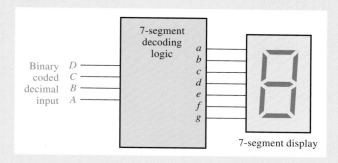

7-segment display

▶ **TABLE 4–10**

Truth table for 7-segment logic.

DECIMAL DIGIT	INPUTS				SEGMENT OUTPUTS						
	D	C	B	A	a	b	c	d	e	f	g
0	0	0	0	0	1	1	1	1	1	1	0
1	0	0	0	1	0	1	1	0	0	0	0
2	0	0	1	0	1	1	0	1	1	0	1
3	0	0	1	1	1	1	1	1	0	0	1
4	0	1	0	0	0	1	1	0	0	1	1
5	0	1	0	1	1	0	1	1	0	1	1
6	0	1	1	0	1	0	1	1	1	1	1
7	0	1	1	1	1	1	1	0	0	0	0
8	1	0	0	0	1	1	1	1	1	1	1
9	1	0	0	1	1	1	1	1	0	1	1
10	1	0	1	0	X	X	X	X	X	X	X
11	1	0	1	1	X	X	X	X	X	X	X
12	1	1	0	0	X	X	X	X	X	X	X
13	1	1	0	1	X	X	X	X	X	X	X
14	1	1	1	0	X	X	X	X	X	X	X
15	1	1	1	1	X	X	X	X	X	X	X

Output = 1 means segment is activated (on)

Output = 0 means segment is not activated (off)

Output = X means "don't care"

Step 2. All of the "don't cares" (X) are placed on the map.

Step 3. The 1s are grouped as shown. "Don't cares" and overlapping of cells are utilized to form the largest groups possible.

Step 4. Write the minimum product term for each group and sum the terms to form the minimum SOP expression.

Keep in mind that "don't cares" do not have to be included in a group, but in this case all of them are used. Also, notice that the 1s in the corner cells are grouped with a "don't care" using the "wrap around" adjacency of the corner cells.

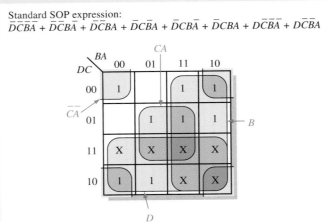

Standard SOP expression:
$\overline{D}\,\overline{C}\,\overline{B}\,\overline{A} + \overline{D}\,\overline{C}\,B\,\overline{A} + \overline{D}\,\overline{C}\,B\,A + \overline{D}\,C\,\overline{B}\,A + \overline{D}\,C\,B\,\overline{A} + \overline{D}\,C\,B\,A + D\,\overline{C}\,\overline{B}\,\overline{A} + D\,\overline{C}\,\overline{B}\,A$

Minimum SOP expression: $D + B + CA + \overline{C}\,\overline{A}$

▲ **FIGURE 4–51**

Karnaugh map minimization of the segment-*a* logic expression.

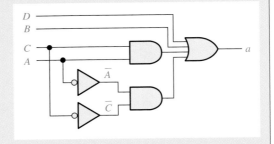

▲ **FIGURE 4–52**

The minimum logic implementation for segment *a* of the 7-segment display.

Minimum Implementation of Segment-*a* Logic   The minimum SOP expression taken from the Karnaugh map in Figure 4–52 for the segment-*a* logic is

$$D + B + CA + \overline{C}\overline{A}$$

This expression can be implemented with two 2-input AND gates, one 4-input OR gate and two inverters as shown in Figure 4–52. Compare this to the standard SOP implementation for segment-*a* logic discussed earlier; you'll see that the number of gates and inverters has been reduced from thirteen to five and, as a result, the number of interconnections has been significantly reduced.

The minimum logic for each of the remaining six segments (*b, c, d, e, f,*

and *g*) can be obtained with a similar approach.

VHDL Implementation (optional)

All of the segment logic can be described by VHDL for implementation in a programmable logic device. Segment-*a* logic can be described by the following VHDL program:

entity SEGLOGIC is

  port (A, B, C, D: in bit; SEGa: out bit);

end entity SEGLOGIC;

architecture LogicFunction of
  SEGLOGIC is

begin

  SEGa <= (A and C) or (not A
  and not C) or B or D;

end architecture LogicFunction;

System Assignment

■ *Activity 1:* Determine the minimum logic for segment *b*.

■ *Activity 2:* Determine the minimum logic for segment *c*.

■ *Activity 3:* Determine the minimum logic for segment *d*.

■ *Activity 4:* Determine the minimum logic for segment *e*.

■ *Activity 5:* Determine the minimum logic for segment *f*.

■ *Activity 6:* Determine the minimum logic for segment *g*.

■ *Optional Activity:* Complete the VHDL program for all seven segments by including each segment logic description in the architecture.

SUMMARY

■ Gate symbols and Boolean expressions for the outputs of an inverter and 2-input gates are shown in Figure 4–53.

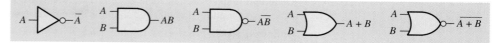

▲ **FIGURE 4–53**

■ Commutative laws: $A + B = B + A$
$$AB = BA$$

■ Associative laws: $A + (B + C) = (A + B) + C$
$$A(BC) = (AB)C$$

■ Distributive law: $A(B + C) = AB + AC$

■ Boolean rules:

1.	$A + 0 = A$	**7.**	$A \cdot A = A$
2.	$A + 1 = 1$	**8.**	$A \cdot \overline{A} = 0$
3.	$A \cdot 0 = 0$	**9.**	$\overline{\overline{A}} = A$
4.	$A \cdot 1 = A$	**10.**	$A + AB = A$
5.	$A + A = A$	**11.**	$A + \overline{A}B = A + B$
6.	$A + \overline{A} = 1$	**12.**	$(A + B)(A + C) = A + BC$

■ DeMorgan's theorems:

1. The complement of a product is equal to the sum of the complements of the terms in the product.

$$\overline{XY} = \overline{X} + \overline{Y}$$

2 The complement of a sum is equal to the product of the complements of the terms in the sum.

$$\overline{X + Y} = \overline{X}\,\overline{Y}$$

■ Karnaugh maps for 3 and 4 variables are shown in Figure 4–54. A 5-variable map is formed from two 4-variable maps.

▶ FIGURE 4–54

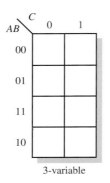

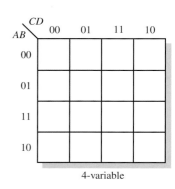

3-variable   4-variable

■ The basic design element in VHDL is an entity/architecture pair.

KEY TERMS

Key terms and other bold terms in the chapter are defined in the end-of-book glossary.

Complement The inverse or opposite of a number. In Boolean algebra, the inverse function, expressed with a bar over a variable. The complement of a 1 is 0, and vice versa.

"Don't care" A combination of input literals that cannot occur and can be used as a 1 or a 0 on a Karnaugh map for simplification.

Karnaugh map An arrangement of cells representing the combinations of literals in a Boolean expression and used for a systematic simplification of the expression.

Minimization The process that results in an SOP or POS Boolean expression that contains the fewest possible literals per term.

Product-of-sums (POS) A form of Boolean expression that is basically the ANDing of ORed terms.

Product term The Boolean product of two or more literals equivalent to an AND operation.

Sum-of-products (SOP) A form of Boolean expression that is basically the ORing of ANDed terms.

Sum term The Boolean sum of two or more literals equivalent to an OR operation.

Variable A symbol used to represent a logical quantity that can have a value of 1 or 0, usually designated by an italic letter.

VHDL A standard hardware description language. IEEE Std. 1076-1993.

SELF-TEST

Answers are at the end of the chapter.

1. The complement of a variable is always

   (a) 0   **(b)** 1   **(c)** equal to the variable   **(d)** the inverse of the variable

2. The Boolean expression $A + \overline{B} + C$ is

   (a) a sum term   **(b)** a literal term   **(c)** a product term   **(d)** a complemented term

3. The Boolean expression $\overline{A}\overline{B}C\overline{D}$ is

   (a) a sum term   **(b)** a product term   **(c)** a literal term   **(d)** always 1

4. The domain of the expression $A\bar{B}CD + A\bar{B} + \overline{C}D + B$ is

    (a) A and D (b) B only   (c) A, B, C, and D (d) none of these

5. According to the commutative law of addition,

    (a) $AB = BA$ (b) $A = A + A$

    (c) $A + (B + C) = (A + B) + C$ (d) $A + B = B + A$

6. According to the associative law of multiplication,

    (a) $B = BB$ (b) $A(BC) = (AB)C$ (c) $A + B = B + A$ (d) $B + B(B + 0)$

7. According to the distributive law,

    (a) $A(B + C) = AB + AC$ (b) $A(BC) = ABC$ (c) $A(A + 1) = A$ (d) $A + AB = A$

8. Which one of the following is *not* a valid rule of Boolean algebra?

    (a) $A + 1 = 1$ (b) $A = \bar{A}$ (c) $AA = A$ (d) $A + 0 = A$

9. Which of the following rules states that if one input of an AND gate is always 1, the output is equal to the other input?

    (a) $A + 1 = 1$ (b) $A + A = A$ (c) $A \cdot A = A$ (d) $A \cdot 1 = A$

10. According to DeMorgan's theorems, the following equality(s) is (are) correct:

    (a) $\overline{AB} = \bar{A} + \bar{B}$ (b) $\overline{XYZ} = \bar{X} + \bar{Y} + \bar{Z}$

    (c) $\overline{A + B + C} = \bar{A}\,\bar{B}\,\bar{C}$ (d) all of these

11. The Boolean expression $X = AB + CD$ represents

    (a) two ORs ANDed together   (b) a 4-input AND gate

    (c) two ANDs ORed together   (d) an exclusive-OR

12. An example of a sum-of-products expression is

    (a) $A + B(C + D)$ (b) $\bar{A}B + A\bar{C} + A\bar{B}C$

    (c) $(\bar{A} + B + C)(A + \bar{B} + C)$ (d) both answers (a) and (b)

13. An example of a product-of-sums expression is

    (a) $A(B + C) + A\bar{C}$ (b) $(A + B)(\bar{A} + B + \bar{C})$

    (c) $\bar{A} + \bar{B} + BC$ (d) both answers (a) and (b)

14. An example of a standard SOP expression is

    (a) $\bar{A}B + A\bar{B}C + AB\bar{D}$ (b) $A\bar{B}C + A\overline{CD}$

    (c) $A\bar{B} + \bar{A}B + AB$ (d) $AB\overline{CD} + A\bar{B} + \bar{A}$

15. A 3-variable Karnaugh map has

    (a) eight cells   (b) three cells   (c) sixteen cells   (d) four cells

16. In a 4-variable Karnaugh map, a 2-variable product term is produced by

    (a) a 2-cell group of 1s   (b) an 8-cell group of 1s

    (c) a 4-cell group of 1s   (d) a 4-cell group of 0s

17. On a Karnaugh map, grouping the 0s produces

    (a) a product-of-sums expression   (b) a sum-of-products expression

    (c) a "don't care" condition   (d) AND-OR logic

18. A 5-variable Karnaugh map has

    (a) sixteen cells   (b) thirty-two cells   (c) sixty-four cells

19. An SPLD that has a programmable AND array and a fixed OR array is a

    (a) PROM   (b) PLA   (c) PAL   (d) GAL

20. VHDL is a type of

    (a) programmable logic   (b) hardware description language

    (c) programmable array   (d) logical mathematics

21. In VHDL, a port is

    (a) a type of entity   (b) a type of architecture

    (c) an input or output   (d) a type of variable

PROBLEMS

Answers to odd-numbered problems are at the end of the book.

SECTION 4–1   **Boolean Operations and Expressions**

1. Using Boolean notation, write an expression that is a 1 whenever one or more of its variables (A, B, C, and D) are 1s.

2. Write an expression that is a 1 only if all of its variables (A, B, C, D, and E) are 1s.

3. Write an expression that is a 1 when one or more of its variables (A, B, and C) are 0s.

4. Evaluate the following operations:
   (a) $0 + 0 + 1$ **(b)** $1 + 1 + 1$ **(c)** $1 \cdot 0 \cdot 0$
   (d) $1 \cdot 1 \cdot 1$ **(e)** $1 \cdot 0 \cdot 1$ **(f)** $1 \cdot 1 + 0 \cdot 1 \cdot 1$

5. Find the values of the variables that make each product term 1 and each sum term 0.
   (a) AB **(b)** $A\overline{B}C$ **(c)** $A + B$ **(d)** $\overline{A} + B + \overline{C}$
   (e) $\overline{A} + \overline{B} + C$ **(f)** $\overline{A} + B$ **(g)** $A\overline{B}\,\overline{C}$

6. Find the value of X for all possible values of the variables.
   (a) $X = (A + B)C + B$ **(b)** $X = (\overline{A + B})C$ **(c)** $X = A\overline{B}C + AB$
   (d) $X = (A + B)(\overline{A} + B)$ **(e)** $X = (A + BC)(\overline{B} + \overline{C})$

SECTION 4–2   **Laws and Rules of Boolean Algebra**

7. Identify the law of Boolean algebra upon which each of the following equalities is based:
   (a) $A\overline{B} + CD + A\overline{C}D + B = B + A\overline{B} + A\overline{C}D + CD$
   (b) $AB\overline{C}D + \overline{ABC} = D\overline{C}BA + \overline{CBA}$
   (c) $AB(CD + E\overline{F} + GH) = ABCD + ABE\overline{F} + ABGH$

8. Identify the Boolean rule(s) on which each of the following equalities is based:
   (a) $\overline{AB + CD} + \overline{EF} = AB + CD + \overline{EF}$ **(b)** $A\overline{A}B + AB\overline{C} + AB\overline{B} = AB\overline{C}$
   (c) $A(BC + \overline{B}C) + AC = A(BC) + AC$ **(d)** $AB(C + \overline{C}) + AC = AB + AC$
   (e) $A\overline{B} + A\overline{B}C = A\overline{B}$ **(f)** $ABC + \overline{AB} + \overline{AB}CD = ABC + \overline{AB} + D$

SECTION 4–3   **DeMorgan's Theorems**

9. Apply DeMorgan's theorems to each expression:
   (a) $\overline{A + \overline{B}}$ **(b)** $\overline{\overline{A}B}$ **(c)** $\overline{A + B + C}$ **(d)** $\overline{ABC}$
   (e) $\overline{A(B + C)}$ **(f)** $\overline{AB + CD}$ **(g)** $\overline{A\overline{B} + C\overline{D}}$ **(h)** $\overline{(A + \overline{B})(\overline{C} + D)}$

10. Apply DeMorgan's theorems to each expression:
    (a) $\overline{A\overline{B}(C + \overline{D})}$ **(b)** $\overline{AB(CD + EF)}$
    (c) $\overline{(A + \overline{B} + C + \overline{D}) + ABC\overline{D}}$ **(d)** $\overline{(\overline{A} + B + C + D)(\overline{AB\,C}D)}$
    (e) $\overline{AB(CD + \overline{E}F)(\overline{AB} + \overline{CD})}$

11. Apply DeMorgan's theorems to the following:
    (a) $\overline{(\overline{ABC})(\overline{EFG}) + (\overline{HIJ})(\overline{KLM})}$ **(b)** $\overline{(A + \overline{BC} + CD) + \overline{BC}}$
    (c) $\overline{(\overline{A + B})(\overline{C + D})(\overline{E + F})(\overline{G + H})}$

SECTION 4–4   **Boolean Analysis of Logic Circuits**

12. Write the Boolean expression for each of the logic gates in Figure 4–55.

▶ **FIGURE 4–55**

(a)

(b)

(c)

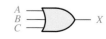

(d)

13. Write the Boolean expression for each of the logic circuits in Figure 4–56.

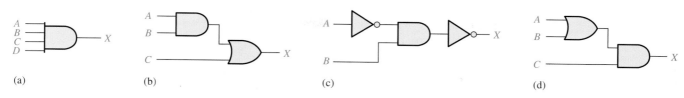

▲ **FIGURE 4–56**

14. Draw the logic circuit represented by each of the following expressions:
   (a) $A + B + C$ **(b)** ABC **(c)** $AB + C$ **(d)** $AB + CD$

15. Draw the logic circuit represented by each expression:
   (a) $A\overline{B} + \overline{A}B$ **(b)** $AB + \overline{A}\overline{B} + \overline{A}BC$
   (c) $\overline{A}B(C + \overline{D})$ **(d)** $A + B[C + D(B + \overline{C})]$

16. Construct a truth table for each of the following Boolean expressions:
   (a) $A + B$ **(b)** AB **(c)** $AB + BC$
   (d) $(A + B)C$ **(e)** $(A + B)(\overline{B} + C)$

SECTION 4–5   Simplification Using Boolean Algebra

17. Using Boolean algebra techniques, simplify the following expressions as much as possible:
   (a) $A(A + B)$ **(b)** $A(\overline{A} + AB)$ **(c)** $BC + \overline{B}C$
   (d) $A(A + \overline{A}B)$ **(e)** $A\overline{B}C + A\overline{B}C + \overline{A}\,\overline{B}C$

18. Using Boolean algebra, simplify the following expressions:
   (a) $(A + \overline{B})(A + C)$ **(b)** $\overline{A}B + \overline{A}B\overline{C} + \overline{A}BCD + \overline{A}B\overline{C}\,\overline{D}E$
   (c) $AB + \overline{A}BC + A$ **(d)** $(A + \overline{A})(AB + AB\overline{C})$
   (e) $AB + (\overline{A} + \overline{B})C + AB$

19. Using Boolean algebra, simplify each expression:
   (a) $BD + B(D + E) + \overline{D}(D + F)$ **(b)** $\overline{A}\overline{B}C + \overline{(A + B + \overline{C})} + \overline{A}\overline{B}\overline{C}D$
   (c) $(B + BC)(B + \overline{B}C)(B + D)$ **(d)** $ABCD + AB(\overline{CD}) + (\overline{AB})CD$
   (e) $ABC[AB + \overline{C}(BC + AC)]$

20. Determine which of the logic circuits in Figure 4–57 are equivalent.

▶ **FIGURE 4–57**

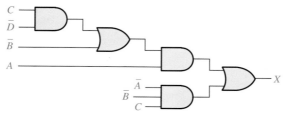

(a)

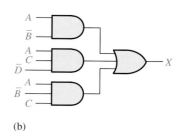

(b)

(c)

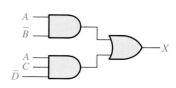

(d)

SECTION 4-6   **Standard Forms of Boolean Expressions**

21. Convert the following expressions to sum-of-product (SOP) forms:

    (a) $(A + B)(C + \overline{B})$ **(b)** $(A + \overline{B}C)C$ **(c)** $(A + C)(AB + AC)$

22. Convert the following expressions to sum-of-product (SOP) forms:

    (a) $AB + CD(A\overline{B} + CD)$ **(b)** $AB(\overline{B}\,\overline{C} + BD)$ **(c)** $A + B[AC + (B + \overline{C})D]$

23. Define the domain of each SOP expression in Problem 21 and convert the expression to standard SOP form.

24. Convert each SOP expression in Problem 22 to standard SOP form.

25. Determine the binary value of each term in the standard SOP expressions from Problem 23.

26. Determine the binary value of each term in the standard SOP expressions from Problem 24.

27. Convert each standard SOP expression in Problem 23 to standard POS form.

28. Convert each standard SOP expression in Problem 24 to standard POS form.

SECTION 4-7   **Boolean Expressions and Truth Tables**

29. Develop a truth table for each of the following standard SOP expressions:

    (a) $A\overline{B}C + \overline{A}B\overline{C} + ABC$ **(b)** $\overline{X}\,\overline{Y}Z + \overline{X}\,Y\overline{Z} + X\overline{Y}\overline{Z} + X\overline{Y}Z + \overline{X}YZ$

30. Develop a truth table for each of the following standard SOP expressions:

    (a) $\overline{A}B\overline{C}D + \overline{A}BC\overline{D} + A\overline{B}\,\overline{C}D + \overline{A}\,\overline{B}\,\overline{C}\,\overline{D}$

    (b) $WXYZ + WX\overline{Y}\overline{Z} + \overline{W}XYZ + W\overline{X}YZ + WX\overline{Y}Z$

31. Develop a truth table for each of the SOP expressions:

    (a) $\overline{A}B + AB\overline{C} + \overline{A}\,\overline{C} + A\overline{B}C$ **(b)** $\overline{X} + Y\overline{Z} + WZ + X\overline{Y}Z$

32. Develop a truth table for each of the standard POS expressions:

    (a) $(\overline{A} + \overline{B} + \overline{C})(A + B + C)(A + \overline{B} + C)$

    (b) $(\overline{A} + B + \overline{C} + D)(A + \overline{B} + C + \overline{D})(A + \overline{B} + \overline{C} + D)(\overline{A} + B + C + \overline{D})$

33. Develop a truth table for each of the standard POS expressions:

    (a) $(A + B)(A + C)(A + B + C)$

    (b) $(A + \overline{B})(A + \overline{B} + \overline{C})(B + C + D)(\overline{A} + B + \overline{C} + D)$

34. For each truth table in Figure 4–58, derive a standard SOP and a standard POS expression.

ABC	X
000	0
001	1
010	0
011	0
100	1
101	1
110	0
111	1

(a)

ABC	X
000	0
001	0
010	0
011	0
100	0
101	1
110	1
111	1

(b)

ABCD	X
0000	1
0001	1
0010	0
0011	1
0100	0
0101	1
0110	1
0111	0
1000	0
1001	1
1010	0
1011	0
1100	1
1101	0
1110	0
1111	0

(c)

ABCD	X
0000	0
0001	0
0010	1
0011	0
0100	1
0101	1
0110	0
0111	1
1000	0
1001	0
1010	0
1011	1
1100	1
1101	0
1110	0
1111	1

(d)

▲ FIGURE 4–58

SECTION 4–8 **The Karnaugh Map**

35. Draw a 3-variable Karnaugh map and label each cell according to its binary value.

36. Draw a 4-variable Karnaugh map and label each cell according to its binary value.

37. Write the standard product term for each cell in a 3-variable Karnaugh map.

SECTION 4–9 **Karnaugh MAP SOP Minimization**

38. Use a Karnaugh map to find the minimum SOP form for each expression:

   (a) $\overline{A}\,\overline{B}\,\overline{C} + \overline{A}\,\overline{B}C + A\overline{B}C$ **(b)** $AC(\overline{B} + C)$

   (c) $\overline{A}(BC + B\overline{C}) + A(BC + B\overline{C})$ **(d)** $\overline{A}\,\overline{B}\,\overline{C} + A\overline{B}\,\overline{C} + \overline{A}\,\overline{B}\,C + AB\overline{C}$

39. Use a Karnaugh map to simplify each expression to a minimum SOP form:

   (a) $\overline{A}\,\overline{B}\,\overline{C} + A\overline{B}C + \overline{A}B\overline{C} + ABC$ **(b)** $AC[\overline{B} + B(B + \overline{C})]$

   (c) $DE\overline{F} + \overline{D}E\overline{F} + \overline{D}\,\overline{E}\,\overline{F}$

40. Expand each expression to a standard SOP form:

   (a) $AB + A\overline{B}C + ABC$ **(b)** $A + BC$

   (c) $A\overline{B}\,\overline{C}D + AC\overline{D} + \overline{B}CD + \overline{A}BCD$ **(d)** $A\overline{B} + A\overline{B}\,\overline{C}D + CD + B\overline{C}D + ABCD$

41. Minimize each expression in Problem 40 with a Karnaugh map.

42. Use a Karnaugh map to reduce each expression to a minimum SOP form:

   (a) $A + B\overline{C} + CD$

   (b) $\overline{A}\,\overline{B}\,\overline{C}\,\overline{D} + \overline{A}\,\overline{B}\,C\overline{D} + ABCD + ABC\overline{D}$

   (c) $\overline{A}B(\overline{C}\,\overline{D} + \overline{C}D) + AB(\overline{C}\,\overline{D} + \overline{C}D) + A\overline{B}\,\overline{C}D$

   (d) $(\overline{A}\,\overline{B} + A\overline{B})(CD + C\overline{D})$

   (e) $\overline{A}\,\overline{B} + A\overline{B} + \overline{C}\,\overline{D} + C\overline{D}$

43. Reduce the function specified in the truth table in Figure 4–59 to its minimum SOP form by using a Karnaugh map.

44. Use the Karnaugh map method to implement the minimum SOP expression for the logic function specified in the truth table in Figure 4–60.

Inputs	Output
A B C	**X**
0 0 0	1
0 0 1	1
0 1 0	0
0 1 1	1
1 0 0	1
1 0 1	1
1 1 0	0
1 1 1	1

▲ **FIGURE 4–59**

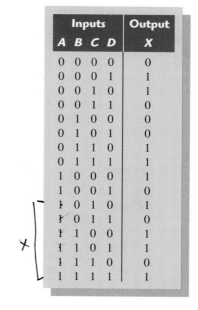

Inputs	Output
A B C D	**X**
0 0 0 0	0
0 0 0 1	1
0 0 1 0	1
0 0 1 1	0
0 1 0 0	0
0 1 0 1	0
0 1 1 0	1
0 1 1 1	1
1 0 0 0	1
1 0 0 1	0
1 0 1 0	1
1 0 1 1	0
1 1 0 0	1
1 1 0 1	1
1 1 1 0	0
1 1 1 1	1

▲ **FIGURE 4–60**

45. Solve Problem 44 for a situation in which the last six binary combinations are not allowed.

SECTION 4–10 **Karnaugh Map POS Minimization**

46. Use a Karnaugh map to find the minimum POS for each expression:

(a) $(A + B + C)(\overline{A} + \overline{B} + \overline{C})(A + \overline{B} + C)$

(b) $(X + \overline{Y})(\overline{X} + Z)(X + \overline{Y} + \overline{Z})(\overline{X} + \overline{Y} + Z)$

(c) $A(B + \overline{C})(\overline{A} + C)(A + \overline{B} + C)(\overline{A} + B + \overline{C})$

47. Use a Karnaugh map to simplify each expression to minimum POS form:

(a) $(A + \overline{B} + C + D)(\overline{A} + B + \overline{C} + D)(\overline{A} + \overline{B} + \overline{C} + \overline{D})$

(b) $(X + \overline{Y})(W + \overline{Z})(\overline{X} + \overline{Y} + \overline{Z})(W + X + Y + Z)$

48. For the function specified in the truth table of Figure 4–59, determine the minimum POS expression using a Karnaugh map.

49. Determine the minimum POS expression for the function in the truth table of Figure 4–60.

50. Convert each of the following POS expressions to minimum SOP expressions using a Karnaugh map:

(a) $(A + \overline{B})(A + \overline{C})(\overline{A} + \overline{B} + C)$

(b) $(\overline{A} + B)(\overline{A} + \overline{B} + \overline{C})(B + \overline{C} + D)(A + \overline{B} + C + \overline{D})$

SECTION 4–11 **Five-Variable Karnaugh Maps**

51. Minimize the following SOP expression using a Karnaugh map:

$$X = \overline{A}B\overline{C}D\overline{E} + \overline{A}\,\overline{B}\,\overline{C}DE + A\overline{B}\,\overline{C}DE + ABC\overline{D}E + \overline{A}BCD\overline{E} + \overline{A}BC\overline{D}E$$
$$+ \overline{A}\,\overline{B}\,\overline{C}\,\overline{D}\,\overline{E} + \overline{A}\,\overline{B}CDE + AB\overline{C}D\overline{E} + AB\overline{C}DE$$

52. Apply the Karnaugh map method to minimize the following SOP expression:

$$A = \overline{V}WXYZ + V\overline{W}XYZ + VW\overline{X}YZ + VWX\overline{Y}Z + VWXY\overline{Z} + \overline{V}\,\overline{W}\,\overline{X}\,\overline{Y}\,\overline{Z}$$
$$+ \overline{V}\,\overline{W}\,\overline{X}Y\overline{Z} + \overline{V}\,\overline{W}X\overline{Y}\overline{Z} + \overline{V}W\overline{X}\,\overline{Y}\,\overline{Z}$$

SECTION 4–12 **VHDL (optional)**

53. Write a VHDL program for the logic circuit in Figure 4–61.

▶ **FIGURE 4–61**

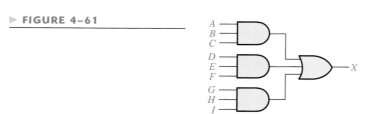

54. Write a program in VHDL for the expression

$$Y = A\overline{B}C + \overline{A}\,\overline{B}C + A\overline{B}\,\overline{C} + \overline{A}BC$$

Digital System Application

55. If you are required to choose a type of digital display for low light conditions, will you select LED or LCD 7-segment displays? Why?

56. Explain why the codes 1010, 1011, 1100, 1101, 1110, and 1111 fall into the "don't care" category in 7-segment display applications.

57. For segment b, how many fewer gates and inverters does it take to implement the minimum SOP expression than the standard SOP expression?

58. Repeat Problem 57 for the logic for segments c through g.

Special Design Problems

59. The logic for segment a in Figure 4–52 produces a HIGH output to activate the segment and so do the circuits for each of the other segments. If a type of 7-segment display is used that requires a LOW to activate each segment, modify the segment logic accordingly.

60. Redesign the logic for segment a using a minimum POS approach. Which is simpler, minimum POS or the minimum SOP?

61. Repeat Problem 60 for segments b through g.

62. Summarize the results of your redesign effort in Problems 60 and 61 and recommend the best design based on fewer ICs. Specify the types of ICs.

Multisim Troubleshooting Practice

63. Open file P04-63, apply input signals, and observe the operation of the logic circuit. Determine whether or not a fault exists.

64. Open file P04-64, apply input signals, and observe the operation of the logic circuit. Determine whether or not a fault exists.

65. Open file P04-65, apply input signals, and observe the operation of the logic circuit. Determine whether or not a fault exists.

ANSWERS

SECTION REVIEWS

SECTION 4–1   **Boolean Operations and Expressions**

  1. $\overline{A} = \overline{0} = 1$ **2.** $A = 1, B = 1, C = 0; \overline{A} + \overline{B} + C = \overline{1} + \overline{1} + 0 = 0 + 0 + 0 = 0$
  3. $A = 1, B = 0, C = 1; A\overline{B}C = 1 \cdot \overline{0} \cdot 1 = 1 \cdot 1 \cdot 1 = 1$

SECTION 4–2   **Laws and Rules of Boolean Algebra**

  1. $A + (B + C + D) = (A + B + C) + D$ **2.** $A(B + C + D) = AB + AC + AD$

SECTION 4–3   **DeMorgan's Theorems**

  1. **(a)** $\overline{ABC} + \overline{(\overline{D} + E)} = \overline{A} + \overline{B} + \overline{C} + D\overline{E}$ **(b)** $\overline{(A + B)C} = \overline{A}\overline{B} + \overline{C}$
  (c) $\overline{A + B + C + \overline{DE}} = \overline{A}\,\overline{B}\,\overline{C} + D + \overline{E}$

SECTION 4–4   **Boolean Analysis of Logic Circuits**

  1. $(C + D)B + A$

  2. Abbreviated truth table: The expression is a 1 when A is 1 or when B and C are 1s or when B and D are 1s. The expression is 0 for all other variable combinations.

SECTION 4–5   **Simplification Using Boolean Algebra**

  1. **(a)** $A + AB + A\overline{B}C = A$ **(b)** $(\overline{A} + B)C + ABC = C(\overline{A} + B)$
  (c) $A\overline{B}C(BD + CDE) + A\overline{C} = A(\overline{C} + \overline{B}DE)$

  2. **(a)** *Original:* 2 AND gates, 1 OR gate, 1 inverter; *Simplified:* No gates (straight connection)
  (b) *Original:* 2 OR gates, 2 AND gates, 1 inverter; *Simplified:* 1 OR gate, 1 AND gate, 1 inverter
  (c) *Original:* 5 AND gates, 2 OR gates, 2 inverters; *Simplified:* 2 AND gates, 1 OR gate, 2 inverters

SECTION 4–6   **Standard Forms of Boolean Expressions**

  1. **(a)** SOP   **(b)** standard POS   **(c)** standard SOP   **(d)** POS
  2. **(a)** $AB\overline{C}\,\overline{D} + AB\overline{C}D + ABC\overline{D} + ABCD + \overline{A}B\overline{C}D + \overline{A}BCD + \overline{A}\,\overline{B}CD + \overline{A}BC\overline{D}$

  (c) Already standard

  3. **(b)** Already standard
  (d) $(A + \overline{B} + \overline{C})(A + \overline{B} + C)(A + B + \overline{C})(A + B + C)$

SECTION 4–7   **Boolean Expressions and Truth Tables**

    1. $2^5 = 32$ **2.** $0110 \longrightarrow \overline{W}XY\overline{Z}$ **3.** $1100 \longrightarrow \overline{W} + \overline{X} + Y + Z$

SECTION 4–8   **The Karnaugh Map**

    1. (a) upper left cell: 000   **(b)** lower right cell: 101   **(c)** lower left cell: 100

       (d) upper right cell: 001

    2. (a) upper left cell: $\overline{X}\,\overline{Y}\,\overline{Z}$ **(b)** lower right cell: $X\overline{Y}Z$ **(c)** lower left cell: $X\overline{Y}\,\overline{Z}$

       (d) upper right cell: $\overline{X}\,\overline{Y}Z$

    3. (a) upper left cell: 0000   **(b)** lower right cell: 1010   **(c)** lower left cell: 1000

       (d) upper right cell: 0010

    4. (a) upper left cell: $\overline{W}\,\overline{X}\,\overline{Y}\,\overline{Z}$ **(b)** lower right cell: $\overline{W}XY\overline{Z}$ **(c)** lower left cell: $\overline{W}X\overline{Y}\,\overline{Z}$

       (d) upper right cell: $\overline{W}\,\overline{X}Y\overline{Z}$

SECTION 4–9   **Karnaugh Map SOP Minimization**

    1. 8-cell map for 3 variables; 16-cell map for 4 variables

    2. $AB + B\overline{C} + \overline{A}\,\overline{B}C$

    3. (a) $\overline{A}\,\overline{B}\,\overline{C} + \overline{A}BC + ABC + AB\overline{C}$

       (b) $\overline{A}\,\overline{B}\,\overline{C} + \overline{A}\,\overline{B}C + \overline{A}B\overline{C} + ABC + A\overline{B}\,\overline{C} + A\overline{B}C$

       (c) $\overline{A}\,\overline{B}\,\overline{C}\,\overline{D} + \overline{A}\,\overline{B}\,C\overline{D} + \overline{A}BC\overline{D} + \overline{A}BCD + ABCD + AB\overline{C}D + A\overline{B}\,\overline{C}D + A\overline{B}CD$

       (d) $\overline{A}\,\overline{B}\,\overline{C}\,\overline{D} + \overline{A}\,\overline{B}C\overline{D} + \overline{A}B\overline{C}\,\overline{D} + \overline{A}B\overline{C}\,\overline{D} + \overline{A}BC\overline{D} + AB\overline{C}D + A\overline{B}C\overline{D} + \overline{A}\,\overline{B}C\overline{D} +$
       $AB\overline{C}\,\overline{D} + ABC\overline{D} + A\overline{B}C\overline{D}$

SECTION 4–10   **Karnaugh Map POS Minimization**

    1. In mapping a POS expression, 0s are placed in cells whose value makes the standard sum term zero; and in mapping an SOP expression 1s are placed in cells having the same values as the product terms.

    2. 0 in the 1011 cell: $\overline{A} + B + \overline{C} + \overline{D}$ **3.** 1 in the 0010 cell: $\overline{A}\,\overline{B}C\overline{D}$

SECTION 4–11   **Five-Variable Karnaugh Maps**

    1. There are 32 combinations of 5 variables ($2^5 = 32$).

    2. $X = 1$ because the function is 1 for all possible combinations of 5 variables.

SECTION 4–12   **VHDL (optional)**

    1. An HDL is a hardware description language for programmable logic.

    2. Entity and architecture

    3. The entity specifies the inputs and outputs of a logic function.

    4. The architecture specifies operation of a logic function.

RELATED PROBLEMS FOR EXAMPLES

    4–1 $\overline{A} + B = 0$ when $A = 1$ and $B = 0$.

    4–2 $\overline{A}\,\overline{B} = 1$ when $A = 0$ and $B = 0$.   **4–3** XYZ

    4–4 $W + X + Y + Z$ **4–5** $ABC\overline{D}\,\overline{E}$ **4–6** $(A + \overline{B} + \overline{C}D)\overline{E}$

    4–7 $\overline{ABCD} = \overline{A} + \overline{B} + \overline{C} + \overline{D}$ **4–8** $A\overline{B}$ **4–9** CD

    4–10 $AB\overline{C} + \overline{A}C + \overline{A}\,\overline{B}$

    4–11 $\overline{A} + \overline{B} + \overline{C}$ **4–12** $\overline{A}BC + AB + A\overline{C} + A\overline{B} + \overline{B}\,\overline{C}$

    4–13 $W\overline{X}YZ + W\overline{X}\,\overline{Y}Z + W\overline{X}Y\overline{Z} + \overline{W}\,\overline{X}Y\overline{Z} + WX\overline{Y}Z + WX\overline{Y}\,\overline{Z}$

    4–14 011, 101, 110, 010, 111. Yes

4–15 $(A + \overline{B} + C)(A + \overline{B} + \overline{C})(A + B + C)(\overline{A} + B + C)$

4–16 010, 100, 001, 111, 011. Yes   **4–17** SOP and POS expressions are equivalent.

4–18 See Table 4–11.   **4–19** See Table 4–12.

▼ **TABLE 4–11**

A	B	C	X
0	0	0	0
0	0	1	0
0	1	0	1
0	1	1	0
1	0	0	0
1	0	1	1
1	1	0	0
1	1	1	0

▼ **TABLE 4–12**

A	B	C	X
0	0	0	1
0	0	1	0
0	1	0	0
0	1	1	1
1	0	0	1
1	0	1	1
1	1	0	1
1	1	1	0

4–20 The SOP and POS expressions are equivalent.   **4–21** See Figure 4–62.

4–22 See Figure 4–63.   **4–23** See Figure 4–64.   **4–24** See Figure 4–65.

4–25 No other ways   **4–26** $X = B + \overline{A}C + A\overline{C}D + C\overline{D}$

4–27 $X = \overline{D} + A\overline{B}C + B\overline{C} + \overline{A}B$

4–28 $Q = X + Y$

4–29 $Q = \overline{X}\,\overline{Y}\overline{Z} + W\overline{X}Z + \overline{W}YZ$

4–30 See Figure 4–66.

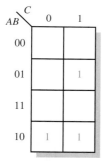

▲ **FIGURE 4–62**

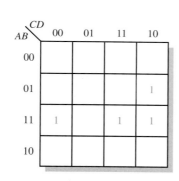

▲ **FIGURE 4–63**

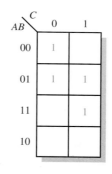

▲ **FIGURE 4–64**

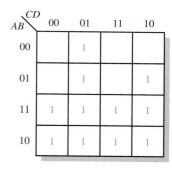

▲ **FIGURE 4–65**

▲ **FIGURE 4–66**

4–31 $Q = (X + \overline{Y})(X + \overline{Z})(\overline{X} + Y + Z)$

4–32 $Q = (\overline{X} + \overline{Y} + Z)(\overline{W} + \overline{X} + Z)(W + X + Y + Z)(W + \overline{X} + Y + \overline{Z})$

4–33 $Q = \overline{Y}Z + \overline{X}Z + \overline{W}Y + \overline{X}\,\overline{Y}Z$

4–34 $Y = \overline{D}\,\overline{E} + \overline{A}\,\overline{E} + \overline{B}\,\overline{C}\,\overline{E}$

4–35 $X \Leftarrow (A \text{ and } B) \text{ nor } (C \text{ and } D);$

SELF-TEST

1. (d)   **2.** (a)   **3.** (b)   **4.** (c)   **5.** (d)   **6.** (b)   **7.** (a)   **8.** (b)

9. (d)   **10.** (d)   **11.** (c)   **12.** (b)   **13.** (b)   **14.** (c)   **15.** (a)   **16.** (c)

17. (a)   **18.** (b)   **19.** (c)   **20.** (b)   **21.** (c)

5

COMBINATIONAL LOGIC ANALYSIS

CHAPTER OBJECTIVES

- Analyze basic combinational logic circuits, such as AND-OR, AND-OR-Invert, exclusive-OR, and exclusive-NOR

- Use AND-OR and AND-OR-Invert circuits to implement sum-of-products (SOP) and product-of-sums (POS) expressions

- Write the Boolean output expression for any combinational logic circuit

- Develop a truth table from the output expression for a combinational logic circuit

- Use the Karnaugh map to expand an output expression containing terms with missing variables into a full SOP form

- Design a combinational logic circuit for a given Boolean output expression

- Design a combinational logic circuit for a given truth table

- Simplify a combinational logic circuit to its minimum form

- Use NAND gates to implement any combinational logic function

- Use NOR gates to implement any combinational logic function

- Write VHDL programs for simple logic circuits

- Troubleshoot faulty logic circuits

- Troubleshoot logic circuits by using signal tracing and waveform analysis

- Apply combinational logic to a system application

KEY TERMS

- Universal gate
- Negative-OR
- Negative-AND
- Component

- Signal
- Node
- Signal tracing

INTRODUCTION

In Chapters 3 and 4, logic gates were discussed on an individual basis and in simple combinations. You were introduced to SOP and POS implementations, which are basic forms of combinational logic. When logic gates are connected together to produce a specified output for certain specified combinations of input variables, with no storage involved, the resulting circuit is in the category of **combinational logic.** In combinational logic, the output level is at all times dependent on the combination of input levels. This chapter expands on the material introduced in earlier chapters with a coverage of the analysis, design, and troubleshooting of various combinational logic circuits. The VHDL structural approach is introduced and applied to combinational logic.

■■■ DIGITAL SYSTEM APPLICATION PREVIEW

The Digital System Application illustrates the concepts taught in this chapter by demonstrating how combinational logic can be used for a specific purpose in a practical application. A logic circuit is used to control the level and temperature of a fluid in a storage tank. By operating inlet and outlet valves, the inflow and outflow are controlled based on level-sensor inputs. The fluid temperature is controlled by turning a heating element on or off based on temperature-sensor inputs. As an option, the use of VHDL for describing the logic is also discussed.

www. VISIT THE COMPANION WEBSITE
Study aids for this chapter are available at
http://www.prenhall.com/floyd

5–1 BASIC COMBINATIONAL LOGIC CIRCUITS

In Chapter 4, you learned that SOP expressions are implemented with an AND gate for each product term and one OR gate for summing all of the product terms. As you know, this SOP implementation is called AND-OR logic and is the basic form for realizing standard Boolean functions. In this section, the AND-OR and the AND-OR-Invert are examined; the exclusive-OR and exclusive-NOR gates, which are actually a form of AND-OR logic, are also covered.

After completing this section, you should be able to

- ■ Analyze and apply AND-OR circuits ■ Analyze and apply AND-OR-Invert circuits
- ■ Analyze and apply exclusive-OR gates ■ Analyze and apply exclusive-NOR gates

AND-OR Logic

AND–OR logic produces an SOP expression.

Figure 5–1(a) shows an AND-OR circuit consisting of two 2-input AND gates and one 2-input OR gate; Figure 5–1(b) is the ANSI standard rectangular outline symbol. The Boolean expressions for the AND gate outputs and the resulting SOP expression for the output X are shown on the diagram. In general, an AND-OR circuit can have any number of AND gates each with any number of inputs.

The truth table for a 4-input AND-OR logic circuit is shown in Table 5–1. The intermediate AND gate outputs (the AB and CD columns) are also shown in the table.

▶ **FIGURE 5–1**

An example of AND–OR logic. Open file F05-01 to verify the operation.

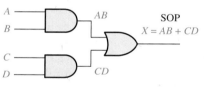 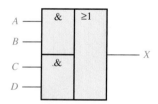

(a) Logic diagram (ANSI standard distinctive shape symbols)

(b) ANSI standard rectangular outline symbol

▶ **TABLE 5–1**

Truth table for the AND-OR logic in Figure 5–1.

INPUTS						OUTPUT
A	B	C	D	AB	CD	X
0	0	0	0	0	0	0
0	0	0	1	0	0	0
0	0	1	0	0	0	0
0	0	1	1	0	1	1
0	1	0	0	0	0	0
0	1	0	1	0	0	0
0	1	1	0	0	0	0
0	1	1	1	0	1	1
1	0	0	0	0	0	0
1	0	0	1	0	0	0
1	0	1	0	0	0	0
1	0	1	1	0	1	1
1	1	0	0	1	0	1
1	1	0	1	1	0	1
1	1	1	0	1	0	1
1	1	1	1	1	1	1

An AND-OR circuit directly implements an SOP expression, assuming the complements (if any) of the variables are available. The operation of the AND-OR circuit in Figure 5–1 is stated as follows:

For a 4-input AND-OR logic circuit, the output X is HIGH (1) if both input A and input B are HIGH (1) or both input C and input D are HIGH (1).

EXAMPLE 5–1

In a certain chemical-processing plant, a liquid chemical is used in a manufacturing process. The chemical is stored in three different tanks. A level sensor in each tank produces a HIGH voltage when the level of chemical in the tank drops below a specified point.

Design a circuit that monitors the chemical level in each tank and indicates when the level in any two of the tanks drops below the specified point.

Solution   The AND-OR circuit in Figure 5–2 has inputs from the sensors on tanks A, B, and C as shown. The AND gate G_1 checks the levels in tanks A and B, gate G_2 checks tanks A and C, and gate G_3 checks tanks B and C. When the chemical level in any two of the tanks gets too low, one of the AND gates will have HIGHs on both of its inputs, causing its output to be HIGH; and so the final output X from the OR gate is HIGH. This HIGH input is then used to activate an indicator such as a lamp or audible alarm, as shown in the figure.

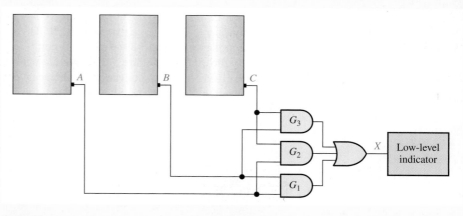

▲ **FIGURE 5–2**

*Related Problem**   Write the Boolean SOP expression for the AND-OR logic in Figure 5–2.

*Answers are at the end of the chapter.

AND-OR-Invert Logic

When the output of an AND-OR circuit is complemented (inverted), it results in an AND-OR-Invert circuit. Recall that AND-OR logic directly implements SOP expressions. POS expressions can be implemented with AND-OR-Invert logic. This is illustrated as follows, starting with a POS expression and developing the corresponding AND-OR-Invert expression.

$$X = (\overline{A} + \overline{B})(\overline{C} + \overline{D}) = (\overline{AB})(\overline{CD}) = \overline{(\overline{AB})(\overline{CD})} = \overline{\overline{AB}} + \overline{\overline{CD}} = \overline{AB + CD}$$

DEMORGAN p. 191

The logic diagram in Figure 5–3(a) shows an AND-OR-Invert circuit and the development of the POS output expression. The ANSI standard rectangular outline symbol is shown in part (b). In general, an AND-OR-Invert circuit can have any number of AND gates each with any number of inputs.

▶ **FIGURE 5–3**

An AND-OR-Invert circuit produces a POS output. Open file F05-03 to verify the operation.

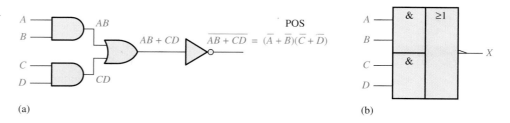

(a)　　　　　　　　　　　　　　　　　　　　　　　　(b)

The operation of the AND-OR-Invert circuit in Figure 5–3 is stated as follows:

For a 4-input AND-OR-Invert logic circuit, the output X is LOW (0) if both input A and input B are HIGH (1) or both input C and input D are HIGH (1).

A truth table can be developed from the AND-OR truth table in Table 5–1 by simply changing all 1s to 0s and all 0s to 1s in the output column.

EXAMPLE 5–2

The sensors in the chemical tanks of Example 5–1 are being replaced by a new model that produces a LOW voltage instead of a HIGH voltage when the level of the chemical in the tank drops below a critical point.

Modify the circuit in Figure 5–2 to operate with the different input levels and still produce a HIGH output to activate the indicator when the level in any two of the tanks drops below the critical point. Show the logic diagram.

Solution　　The AND-OR-Invert circuit in Figure 5–4 has inputs from the sensors on tanks A, B, and C as shown. The AND gate G_1 checks the levels in tanks A and B, gate G_2 checks tanks A and C, and gate G_3 checks tanks B and C. When the chemical level in any two of the tanks gets too low, each AND gate will have a LOW on at least one input causing its output to be LOW and, thus, the final output X from the inverter is HIGH. This HIGH output is then used to activate an indicator.

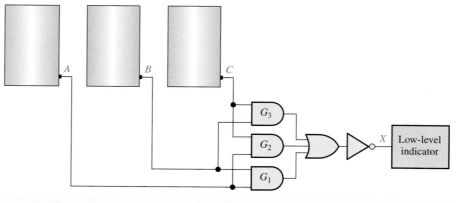

▲ **FIGURE 5–4**

Related Problem　　Write the Boolean expression for the AND-OR-Invert logic in Figure 5–4 and show that the output is HIGH (1) when any two of the inputs A, B, and C are LOW (0).

Exclusive-OR Logic

The exclusive-OR gate was introduced in Chapter 3. Although, because of its importance, this circuit is considered a type of logic gate with its own unique symbol, it is actually a combination of two AND gates, one OR gate, and two inverters, as shown in Figure 5–5(a). The two ANSI standard logic symbols are shown in parts (b) and (c).

The XOR gate is actually a combination of other gates.

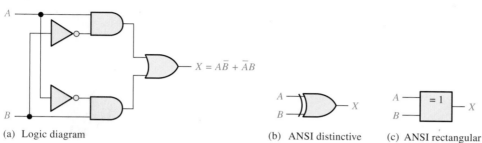

(a) Logic diagram   (b) ANSI distinctive shape symbol   (c) ANSI rectangular outline symbol

◀ **FIGURE 5–5**

Exclusive-OR logic diagram and symbols. Open file F05-05 to verify the operation.

The output expression for the circuit in Figure 5–5 is

$$X = A\overline{B} + \overline{A}B$$

Evaluation of this expression results in the truth table in Table 5–2. Notice that the output is HIGH only when the two inputs are at opposite levels. A special exclusive-OR operator ⊕ is often used, so the expression $X = A\overline{B} + \overline{A}B$ can be stated as "X is equal to A exclusive-OR B" and can be written as

$$X = A \oplus B$$

A	B	X
0	0	0
0	1	1
1	0	1
1	1	0

◀ **TABLE 5–2**

Truth table for an exclusive-OR.

Exclusive-NOR Logic

As you know, the complement of the exclusive-OR function is the exclusive-NOR, which is derived as follows:

$$X = \overline{A\overline{B} + \overline{A}B} = \overline{(A\overline{B})}\,\overline{(\overline{A}B)} = (\overline{A} + B)(A + \overline{B}) = \overline{A}\,\overline{B} + AB$$

Notice that the output X is HIGH only when the two inputs, A and B, are at the same level.

The exclusive-NOR can be implemented by simply inverting the output of an exclusive-OR, as shown in Figure 5–6(a), or by directly implementing the expression $\overline{A}\,\overline{B} + AB$, as shown in part (b).

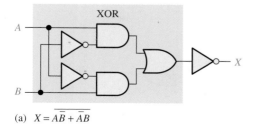

(a) $X = \overline{A\overline{B} + \overline{A}B}$

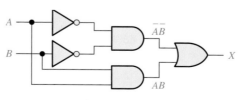

(b) $X = \overline{A}\,\overline{B} + AB$

◀ **FIGURE 5–6**

Two equivalent ways of implementing the exclusive-NOR. Open file F05-06 to verify the operation.

1. Determine the output (1 or 0) of a 4-variable AND-OR-Invert circuit for each of the following input conditions:
   (a) $A = 1, B = 0, C = 1, D = 0$ (b) $A = 1, B = 1, C = 0, D = 1$
   (c) $A = 0, B = 1, C = 1, D = 1$

2. Determine the output (1 or 0) of an exclusive-OR gate for each of the following input conditions:
   (a) $A = 1, B = 0$ (b) $A = 1, B = 1$
   (c) $A = 0, B = 1$ (d) $A = 0, B = 0$

3. Develop the truth table for a certain 3-input logic circuit with the output expression $X = \overline{A}B\overline{C} + \overline{A}BC + \overline{A}\,\overline{B}\,\overline{C} + AB\overline{C} + ABC$.

4. Draw the logic diagram for an exclusive-NOR circuit.

5–2   IMPLEMENTING COMBINATIONAL LOGIC

In this section, examples are used to illustrate how to implement a logic circuit from a Boolean expression or a truth table. Minimization of a logic circuit using the methods covered in Chapter 4 is also included.

After completing this section, you should be able to

▪ Implement a logic circuit from a Boolean expression ▪ Implement a logic circuit from a truth table ▪ Minimize a logic circuit

From a Boolean Expression to a Logic Circuit

For every Boolean expression there is a logic circuit, and for every logic circuit there is a Boolean expression.

Let's examine the following Boolean expression:

$$X = AB + CDE$$

A brief inspection shows that this expression is composed of two terms, AB and CDE, with a domain of five variables. The first term is formed by ANDing A with B, and the second term is formed by ANDing C, D, and E. The two terms are then ORed to form the output X. These operations are indicated in the structure of the expression as follows:

$$X = \overset{\downarrow\qquad\downarrow}{AB + CDE} \qquad \text{AND}$$
$$\underset{\uparrow}{\qquad\qquad} \text{OR}$$

Note that in this particular expression, the AND operations forming the two individual terms, AB and CDE, must be performed *before* the terms can be ORed.

To implement this Boolean expression, a 2-input AND gate is required to form the term AB, and a 3-input AND gate is needed to form the term CDE. A 2-input OR gate is then required to combine the two AND terms. The resulting logic circuit is shown in Figure 5–7.

▶ **FIGURE 5–7**

Logic circuit for $X = AB + CDE$.

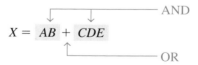

As another example, let's implement the following expression:

$$X = AB(C\overline{D} + EF)$$

A breakdown of this expression shows that the terms AB and $(C\overline{D} + EF)$ are ANDed. The term $C\overline{D} + EF$ is formed by first ANDing C and $\overline{D}$ and ANDing E and F, and then ORing these two terms. This structure is indicated in relation to the expression as follows:

$$
\begin{array}{l}
\text{AND} \\
\text{NOT} \\
\text{OR}
\end{array}
$$

$$X = AB(C\overline{D} + EF)$$

AND

Before you can implement the final expression, you must create the sum term $C\overline{D} + EF$; but before you can get this term; you must create the product terms $C\overline{D}$ and EF; but before you can get the term $C\overline{D}$, you must create $\overline{D}$. So, as you can see, the logic operations must be done in the proper order.

The logic gates required to implement $X = AB(C\overline{D} + EF)$ are as follows:

1. One inverter to form $\overline{D}$

2. Two 2-input AND gates to form $C\overline{D}$ and EF

3. One 2-input OR gate to form $C\overline{D} + EF$

4. One 3-input AND gate to form X

The logic circuit for this expression is shown in Figure 5–8(a). Notice that there is a maximum of four gates and an inverter between an input and output in this circuit (from input D to output). Often the total propagation delay time through a logic circuit is a major consideration. Propagation delays are additive, so the more gates or inverters between input and output, the greater the propagation delay time.

Unless an intermediate term, such as $C\overline{D} + EF$ in Figure 5–8(a), is required as an output for some other purpose, it is usually best to reduce a circuit to its SOP form in order to reduce the overall propagation delay time. The expression is converted to SOP as follows, and the resulting circuit is shown in Figure 5–8(b).

$$AB(C\overline{D} + EF) = ABC\overline{D} + ABEF$$

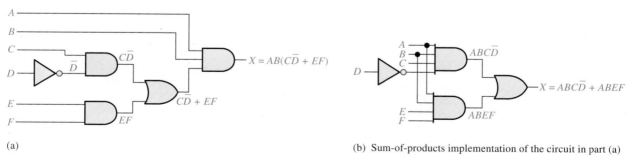

(a)

(b) Sum-of-products implementation of the circuit in part (a)

▲ FIGURE 5–8

Logic circuits for $X = AB(C\overline{D} + EF) = ABC\overline{D} + ABEF$.

From a Truth Table to a Logic Circuit

If you begin with a truth table instead of an expression, you can write the SOP expression from the truth table and then implement the logic circuit. Table 5–3 specifies a logic function.

▶ TABLE 5–3

| INPUTS | | | OUTPUT | PRODUCT TERM |
A	B	C	X	
0	0	0	0	
0	0	1	0	
0	1	0	0	
0	1	1	1	$\overline{A}BC$
1	0	0	1	$A\overline{B}\,\overline{C}$
1	0	1	0	
1	1	0	0	
1	1	1	0	

The Boolean SOP expression obtained from the truth table by ORing the product terms for which $X = 1$ is

$$X = \overline{A}BC + A\overline{B}\,\overline{C}$$

The first term in the expression is formed by ANDing the three variables $\overline{A}$, B, and C. The second term is formed by ANDing the three variables A, $\overline{B}$, and $\overline{C}$.

The logic gates required to implement this expression are as follows: three inverters to form the $\overline{A}$, $\overline{B}$, and $\overline{C}$ variables; two 3-input AND gates to form the terms $\overline{A}BC$ and $A\overline{B}\,\overline{C}$; and one 2-input OR gate to form the final output function, $\overline{A}BC + A\overline{B}\,\overline{C}$.

The implementation of this logic function is illustrated in Figure 5–9.

▶ FIGURE 5–9

Logic circuit for $X = \overline{A}BC + A\overline{B}\overline{C}$. Open file F05-09 to verify the operation.

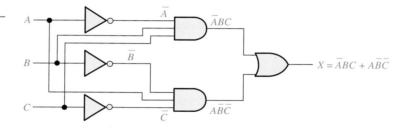

EXAMPLE 5–3

Design a logic circuit to implement the operation specified in the truth table of Table 5–4.

▼ TABLE 5–4

| INPUTS | | | OUTPUT | PRODUCT TERM |
A	B	C	X	
0	0	0	0	
0	0	1	0	
0	1	0	0	
0	1	1	1	$\overline{A}BC$
1	0	0	0	
1	0	1	1	$A\overline{B}C$
1	1	0	1	$AB\overline{C}$
1	1	1	0	

Solution Notice that $X = 1$ for only three of the input conditions. Therefore, the logic expression is

$$X = \overline{A}BC + A\overline{B}C + AB\overline{C}$$

The logic gates required are three inverters, three 3-input AND gates and one 3-input OR gate. The logic circuit is shown in Figure 5–10.

▶ **FIGURE 5–10**

Open file F05-10 to verify the operation.

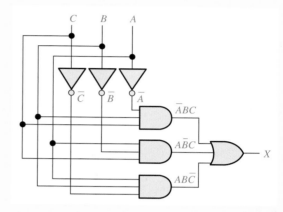

Related Problem Determine if the logic circuit of Figure 5–10 can be simplified.

EXAMPLE 5–4

Develop a logic circuit with four input variables that will only produce a 1 output when exactly three input variables are 1s.

Solution Out of sixteen possible combinations of four variables, the combinations in which there are exactly three 1s are listed in Table 5–5, along with the corresponding product term for each.

▶ **TABLE 5–5**

A	B	C	D	PRODUCT TERM
0	1	1	1	$\overline{A}BCD$
1	0	1	1	$A\overline{B}CD$
1	1	0	1	$AB\overline{C}D$
1	1	1	0	$ABC\overline{D}$

The product terms are ORed to get the following expression:

$$X = \overline{A}BCD + A\overline{B}CD + AB\overline{C}D + ABC\overline{D}$$

This expression is implemented in Figure 5–11 with AND-OR logic.

▶ **FIGURE 5–11**

Open file F05-11 to verify the operation.

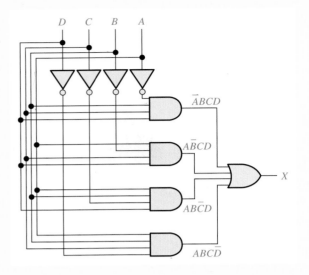

Related Problem   Determine if the logic circuit of Figure 5–11 can be simplified.

EXAMPLE 5–5

Reduce the combinational logic circuit in Figure 5–12 to a minimum form.

▶ **FIGURE 5–12**

Open file F05-12 to verify that this circuit is equivalent to the circuit in Figure 5–13.

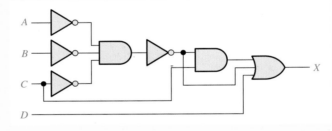

Solution   The expression for the output of the circuit is

$$X = (\overline{\overline{\overline{A}\,\overline{B}\,\overline{C}}})C + \overline{\overline{A}\,\overline{B}\,\overline{C}} + D$$

Applying DeMorgan's theorem and Boolean algebra,

$$X = (\overline{\overline{A}} + \overline{\overline{B}} + \overline{\overline{C}})C + \overline{\overline{A}} + \overline{\overline{B}} + \overline{\overline{C}} + D$$
$$= AC + BC + CC + A + B + C + D$$
$$= AC + BC + C + A + B + C + D$$
$$= C(A + B + 1) + A + B + D$$
$$X = A + B + C + D$$

The simplified circuit is a 4-input OR gate as shown in Figure 5–13.

▶ **FIGURE 5–13**

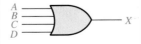

Related Problem   Verify the minimized expression $A + B + C + D$ using a Karnaugh map.

EXAMPLE 5–6

Minimize the combinational logic circuit in Figure 5–14. Inverters for the complemented variables are not shown.

▶ **FIGURE 5–14**

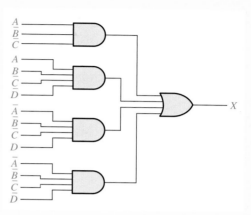

Solution The output expression is

$$X = A\overline{B}\,\overline{C} + AB\overline{C}D + \overline{A}\,\overline{B}CD + \overline{A}\,B\overline{C}\,\overline{D}$$

Expanding the first term to include the missing variables D and $\overline{D}$,

$$X = A\overline{B}\,\overline{C}(D + \overline{D}) + AB\overline{C}D + \overline{A}\,\overline{B}CD + \overline{A}\,B\overline{C}\,\overline{D}$$
$$= A\overline{B}\,\overline{C}D + A\overline{B}\,\overline{C}\,\overline{D} + AB\overline{C}D + \overline{A}\,\overline{B}CD + \overline{A}\,B\overline{C}\,\overline{D}$$

This expanded SOP expression is mapped and simplified on the Karnaugh map in Figure 5–15(a). The simplified implementation is shown in part (b). Inverters are not shown.

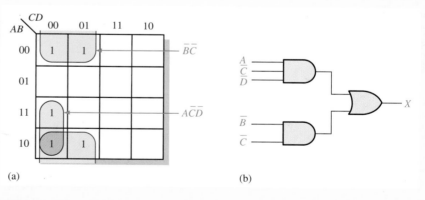

(a)   (b)

▲ **FIGURE 5–15**

Related Problem Develop the POS equivalent of the circuit in Figure 5–15(b).

SECTION 5–2 REVIEW

1. Implement the following Boolean expressions as they are stated:
   (a) $X = ABC + AB + AC$ (b) $X = AB(C + DE)$
2. Develop a logic circuit that will produce a 1 on its output only when all three inputs are 1s or when all three inputs are 0s.
3. Reduce the circuits in Question 1 to minimum SOP form.

5–3   THE UNIVERSAL PROPERTY OF NAND AND NOR GATES

Up to this point, you have studied combinational circuits implemented with AND gates, OR gates, and inverters. In this section, the universal property of the NAND gate and the NOR gate is discussed. The universality of the NAND gate means that it can be used as an inverter and that combinations of NAND gates can be used to implement the AND, OR, and NOR operations. Similarly, the NOR gate can be used to implement the inverter (NOT), AND, OR, and NAND operations.

After completing this section, you should be able to

■ Use NAND gates to implement the inverter, the AND gate, the OR gate, and the NOR gate   ■ Use NOR gates to implement the inverter, the AND gate, the OR gate, and the NAND gate

The NAND Gate as a Universal Logic Element

NAND gates can be used to produce any logic function.

The NAND gate is a **universal gate** because it can be used to produce the NOT, the AND, the OR, and the NOR functions. An inverter can be made from a NAND gate by connecting all of the inputs together and creating, in effect, a single input, as shown in Figure 5–16(a) for a 2-input gate. An AND function can be generated by the use of NAND gates alone, as shown in Figure 5–16(b). An OR function can be produced with only NAND gates, as illustrated in part (c). Finally, a NOR function is produced as shown in part (d).

▶ FIGURE 5–16

Universal application of NAND gates. Open files F05-16(a), (b), (c), and (d) to verify each of the equivalencies.

(a) One NAND gate used as an inverter

(b) Two NAND gates used as an AND gate

(c) Three NAND gates used as an OR gate

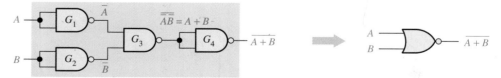

(d) Four NAND gates used as a NOR gate

In Figure 5–16(b), a NAND gate is used to invert (complement) a NAND output to form the AND function, as indicated in the following equation:

$$X = \overline{\overline{AB}} = AB$$

In Figure 5–16(c), NAND gates G_1 and G_2 are used to invert the two input variables before they are applied to NAND gate G_3. The final OR output is derived as follows by application of DeMorgan's theorem:

$$X = \overline{\overline{A}\,\overline{B}} = A + B$$

In Figure 5–16(d), NAND gate G_4 is used as an inverter connected to the circuit of part (c) to produce the NOR operation $\overline{A + B}$.

The NOR Gate as a Universal Logic Element

Like the NAND gate, the NOR gate can be used to produce the NOT, AND, OR, and NAND functions. A NOT circuit, or inverter, can be made from a NOR gate by connecting all of the inputs together to effectively create a single input, as shown in Figure 5–17(a) with a 2-input example. Also, an OR gate can be produced from NOR gates, as illustrated in Figure 5–17(b). An AND gate can be constructed by the use of NOR gates, as shown in Figure 5–17(c). In this case the NOR gates G_1 and G_2 are used as inverters, and the final output is derived by the use of DeMorgan's theorem as follows:

$$X = \overline{\overline{A} + \overline{B}} = AB$$

NOR gates can be used to produce any logic function.

Figure 5–17(d) shows how NOR gates are used to form a NAND function.

(a) One NOR gate used as an inverter

(b) Two NOR gates used as an OR gate

(c) Three NOR gates used as an AND gate

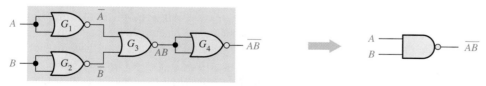

(d) Four NOR gates used as a NAND gate

◀ **FIGURE 5–17**

Universal application of NOR gates. Open files F05-17(a), (b), (c), and (d) to verify each of the equivalencies.

**SECTION 5–3
REVIEW**

1. Use NAND gates to implement each expression:
   (a) $X = \overline{A} + B$ (b) $X = A\overline{B}$
2. Use NOR gates to implement each expression:
   (a) $X = \overline{A} + B$ (b) $X = A\overline{B}$

5–4 COMBINATIONAL LOGIC USING NAND AND NOR GATES

In this section, you will see how NAND and NOR gates can be used to implement a logic function. Recall from Chapter 3 that the NAND gate also exhibits an equivalent operation called the negative-OR and that the NOR gate exhibits an equivalent operation called the negative-AND. You will see how the use of the appropriate symbols to represent the equivalent operations makes "reading" a logic diagram easier.

After completing this section, you should be able to

■ Use NAND gates to implement a logic function ■ Use NOR gates to implement a logic function ■ Use the appropriate dual symbol in a logic diagram

NAND Logic

As you have learned, a NAND gate can function as either a NAND or a negative-OR because, by DeMorgan's theorem,

$$\overline{AB} = \overline{A} + \overline{B}$$

NAND ⟶ ↑ ↑ ⟵ negative-OR

Consider the NAND logic in Figure 5–18. The output expression is developed in the following steps:

$$X = \overline{(\overline{AB})(\overline{CD})}$$
$$= \overline{(\overline{A} + \overline{B})(\overline{C} + \overline{D})}$$
$$= \overline{(\overline{A} + \overline{B})} + \overline{(\overline{C} + \overline{D})}$$
$$= \overline{\overline{A}}\,\overline{\overline{B}} + \overline{\overline{C}}\,\overline{\overline{D}}$$
$$= AB + CD$$

▶ **FIGURE 5–18**

NAND logic for $X = AB + CD$.

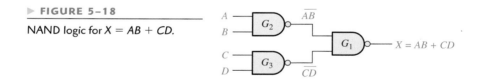

As you can see in Figure 5–18, the output expression, $AB + CD$, is in the form of two AND terms ORed together. This shows that gates G_2 and G_3 act as AND gates and that gate G_1 acts as an OR gate, as illustrated in Figure 5–19(a). This circuit is redrawn in part (b) with NAND symbols for gates G_2 and G_3 and a negative-OR symbol for gate G_1.

Notice in Figure 5–19(b) the bubble-to-bubble connections between the outputs of gates G_2 and G_3 and the inputs of gate G_1. *Since a bubble represents an inversion, two connected bubbles represent a double inversion and therefore cancel each other.* This inversion cancellation can be seen in the previous development of the output expression $AB + CD$ and is indicated by the absence of barred terms in the output expression. Thus, the circuit in Figure 5–19(b) is *effectively* an AND-OR circuit, as shown in Figure 5–19(c).

NAND Logic Diagrams Using Dual Symbols All logic diagrams using NAND gates should be drawn with each gate represented by either a NAND symbol or the equivalent negative-OR symbol to reflect the operation of the gate within the logic circuit. The NAND symbol and the **negative-OR** symbol are called *dual symbols*. When drawing a NAND logic diagram, always use the gate symbols in such a way that every connection

◀ FIGURE 5–19

Development of the AND-OR
equivalent of the circuit in
Figure 5–18.

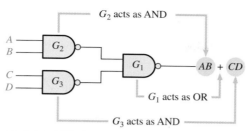

(a) Original NAND logic diagram showing effective
gate operation relative to the output expression

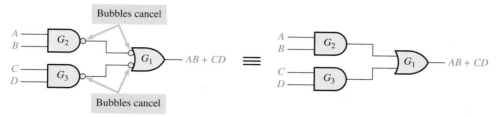

(b) Equivalent NAND/Negative-OR logic diagram   (c) AND-OR equivalent

between a gate output and a gate input is either bubble-to-bubble or nonbubble-to-nonbubble. A bubble output should not be connected to a nonbubble input or vice versa in a logic diagram.

Figure 5–20 shows an arrangement of gates to illustrate the procedure of using the appropriate dual symbols for a NAND circuit with several gate levels. Although using all NAND symbols as in Figure 5–20(a) is correct, the diagram in part (b) is much easier to "read" and is the preferred method. As shown in Figure 5–20(b), the output gate is represented with a negative-OR symbol. Then the NAND symbol is used for the level of gates right before the output gate and the symbols for successive levels of gates are alternated as you move away from the output.

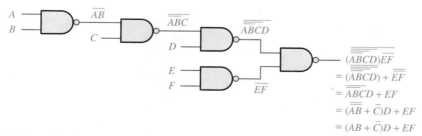

◀ FIGURE 5–20

Illustration of the use of the
appropriate dual symbols in a NAND
logic diagram.

(a) Several Boolean steps are required to arrive at final output expression.

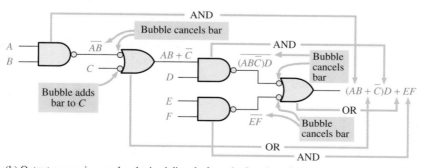

(b) Output expression can be obtained directly from the function of each gate symbol in the diagram.

The shape of the gate indicates the way its inputs will appear in the output expression and thus shows how the gate functions within the logic circuit. For a NAND symbol, the inputs appear ANDed in the output expression; and for a negative-OR symbol, the inputs appear ORed in the output expression, as Figure 5–20(b) illustrates. The dual-symbol diagram in part (b) makes it easier to determine the output expression directly from the logic diagram because each gate symbol indicates the relationship of its input variables as they appear in the output expression.

EXAMPLE 5–7

Redraw the logic diagram and develop the output expression for the circuit in Figure 5–21 using the appropriate dual symbols.

▶ **FIGURE 5–21**

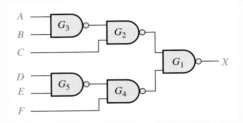

Solution Redraw the logic diagram in Figure 5–21 with the use of equivalent negative-OR symbols as shown in Figure 5–22. Writing the expression for X directly from the indicated logic operation of each gate gives $X = (\overline{A} + \overline{B})C + (\overline{D} + \overline{E})F$.

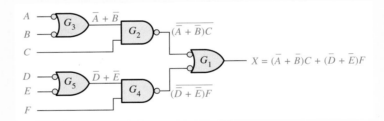

▲ **FIGURE 5–22**

Related Problem Derive the output expression from Figure 5–21 and show it is equivalent to the expression in the solution.

EXAMPLE 5–8

Implement each expression with NAND logic using appropriate dual symbols:

(a) $ABC + DE$ **(b)** $ABC + \overline{D} + \overline{E}$

Solution See Figure 5–23.

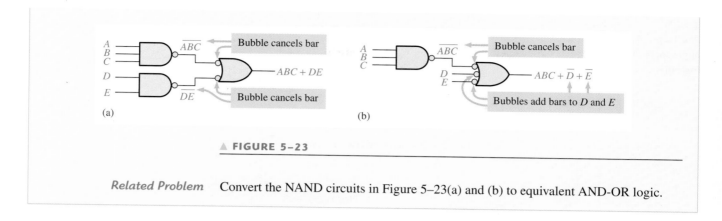

▲ FIGURE 5–23

Related Problem   Convert the NAND circuits in Figure 5–23(a) and (b) to equivalent AND-OR logic.

NOR Logic

A NOR gate can function as either a NOR or a **negative-AND,** as shown by DeMorgan's theorem.

$$\overline{A + B} = \overline{A}\,\overline{B}$$

NOR ————↑   ↑———— negative-AND

Consider the NOR logic in Figure 5–24. The output expression is developed as follows:

$$X = \overline{\overline{A + B} + \overline{C + D}} = (\overline{\overline{A + B}})(\overline{\overline{C + D}}) = (A + B)(C + D)$$

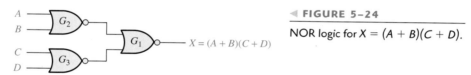

◀ FIGURE 5–24

NOR logic for $X = (A + B)(C + D)$.

As you can see in Figure 5–24, the output expression $(A + B)(C + D)$ consists of two OR terms ANDed together. This shows that gates G_2 and G_3 act as OR gates and gate G_1 acts as an AND gate, as illustrated in Figure 5–25(a). This circuit is redrawn in part (b) with a negative-AND symbol for gate G_1.

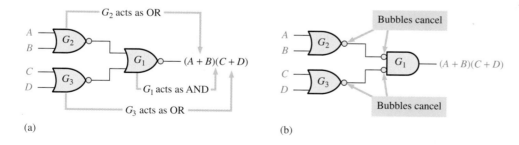

◀ FIGURE 5–25

NOR Logic Diagram Using Dual Symbols   As with NAND logic, the purpose for using the dual symbols is to make the logic diagram easier to read and analyze, as illustrated in the NOR logic circuit in Figure 5–26. When the circuit in part (a) is redrawn with dual symbols in part (b), notice that all output-to-input connections between gates are bubble-to-bubble or nonbubble-to-nonbubble. Again, you can see that the shape of each gate symbol indicates the type of term (AND or OR) that it produces in the output expression, thus making the output expression easier to determine and the logic diagram easier to analyze.

▶ **FIGURE 5–26**

Illustration of the use of the appropriate dual symbols in a NOR logic diagram.

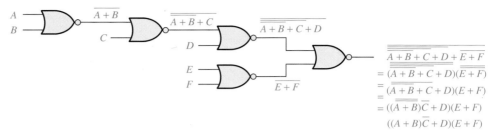

$$\overline{\overline{A + B + C + D} + \overline{E + F}}$$
$$= (\overline{A + B + C + D})(\overline{E + F})$$
$$= (\overline{\overline{A + B} + C + D})(E + F)$$
$$= ((\overline{A + B})\overline{C} + D)(E + F)$$
$$((A + B)\overline{C} + D)(E + F)$$

(a) Final output expression is obtained after several Boolean steps.

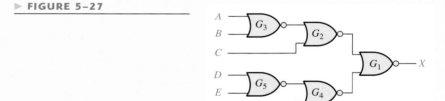

(b) Output expression can be obtained directly from the function of each gate symbol in the diagram.

EXAMPLE 5–9

Using appropriate dual symbols, redraw the logic diagram and develop the output expression for the circuit in Figure 5–27.

▶ **FIGURE 5–27**

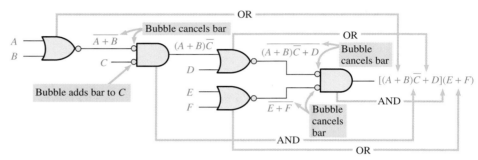

Solution   Redraw the logic diagram with the equivalent negative-AND symbols as shown in Figure 5–28. Writing the expression for X directly from the indicated operation of each gate,

$$X = (\overline{A}\,\overline{B} + C)(\overline{D}\,\overline{E} + F)$$

▶ **FIGURE 5–28**

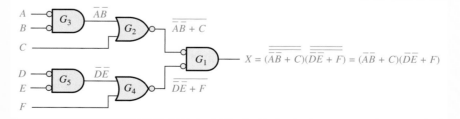

Related Problem   Prove that the output of the NOR circuit in Figure 5–27 is the same as for the circuit in Figure 5–28.

SECTION 5–4
REVIEW

1. Implement the expression $X = \overline{(\overline{A} + \overline{B} + \overline{C})DE}$ by using NAND logic.
2. Implement the expression $X = \overline{\overline{A}\,\overline{B}\,\overline{C} + (D + E)}$ with NOR logic.

5–5 LOGIC CIRCUIT OPERATION WITH PULSE WAVEFORM INPUTS

Several examples of general combinational logic circuits with pulse waveform inputs are examined in this section. Keep in mind that the operation of each gate is the same for pulse waveform inputs as for constant-level inputs. The output of a logic circuit at any given time depends on the inputs at that particular time, so the relationship of the time-varying inputs is of primary importance.

After completing this section, you should be able to

■ Analyze combinational logic circuits with pulse waveform inputs ■ Develop a timing diagram for any given combinational logic circuit with specified inputs

The operation of any gate is the same regardless of whether its inputs are pulsed or constant levels. The nature of the inputs (pulsed or constant levels) does not alter the truth table of a circuit. The examples in this section illustrate the analysis of combinational logic circuits with pulse waveform inputs.

The following is a review of the operation of individual gates for use in analyzing combinational circuits with pulse waveform inputs:

1. The output of an AND gate is HIGH only when all inputs are HIGH at the same time.

2. The output of an OR gate is HIGH only when at least one of its inputs is HIGH.

3. The output of a NAND gate is LOW only when all inputs are HIGH at the same time.

4. The output of a NOR gate is LOW only when at least one of its inputs is HIGH.

EXAMPLE 5–10

Determine the final output waveform X for the circuit in Figure 5–29, with input waveforms A, B, and C as shown.

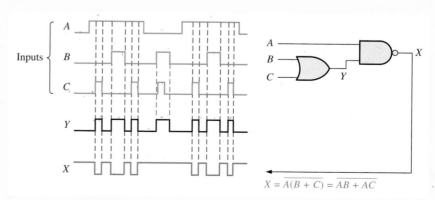

$$X = \overline{A(B + C)} = \overline{AB} + \overline{AC}$$

▲ **FIGURE 5–29**

Solution The output expression, $\overline{AB + AC}$, indicates that the output X is LOW when both A and B are HIGH or when both A and C are HIGH or when all inputs are HIGH. The output waveform X is shown in the timing diagram of Figure 5–29. The intermediate waveform Y at the output of the OR gate is also shown.

Related Problem Determine the output waveform if input A is a constant HIGH level.

EXAMPLE 5–11

Draw the timing diagram for the circuit in Figure 5–30 showing the outputs of G_1, G_2, and G_3 with the input waveforms, A, and B, as indicated.

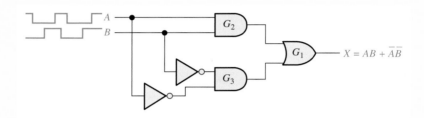

▲ FIGURE 5–30

Solution When both inputs are HIGH or when both inputs are LOW, the output X is HIGH as shown in Figure 5–31. Notice that this is an exclusive-NOR circuit. The intermediate outputs of gates G_2 and G_3 are also shown in Figure 5–31.

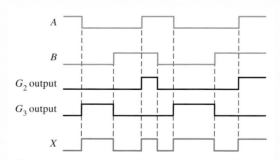

▲ FIGURE 5–31

Related Problem Determine the output X in Figure 5–30 if input B is inverted.

EXAMPLE 5–12

Determine the output waveform X for the logic circuit in Figure 5–32(a) by first finding the intermediate waveform at each of points Y_1, Y_2, Y_3, and Y_4. The input waveforms are shown in Figure 5–32(b).

▶ FIGURE 5–32

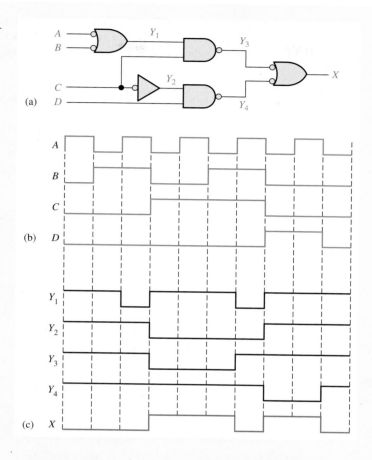

(a)

(b)

(c)

Solution All the intermediate waveforms and the final output waveform are shown in the timing diagram of Figure 5–32(c).

Related Problem Determine the waveforms Y_1, Y_2, Y_3, Y_4 and X if input waveform A is inverted.

EXAMPLE 5–13

Determine the output waveform X for the circuit in Example 5–12, Figure 5–32(a), directly from the output expression.

Solution The output expression for the circuit is developed in Figure 5–33. The SOP form indicates that the output is HIGH when A is LOW and C is HIGH or when B is LOW and C is HIGH or when C is LOW and D is HIGH.

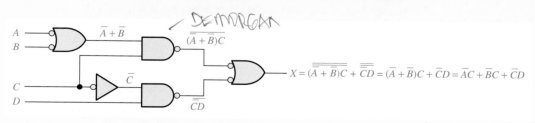

$$X = \overline{(\overline{A} + \overline{B})C} + \overline{\overline{C}D} = (\overline{A} + \overline{B})C + \overline{C}D = \overline{A}C + \overline{B}C + \overline{C}D$$

▲ FIGURE 5–33

The result is shown in Figure 5–34 and is the same as the one obtained by the intermediate-waveform method in Example 5–12. The corresponding product terms for each waveform condition that results in a HIGH output are indicated.

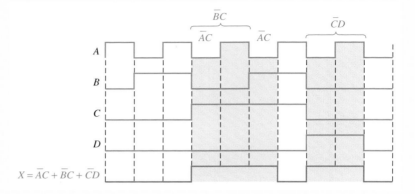

▲ FIGURE 5–34

Related Problem   Repeat this example if all the input waveforms are inverted.

SECTION 5–5 REVIEW

1. One pulse with $t_W = 50\ \mu s$ is applied to one of the inputs of an exclusive-OR circuit. A second positive pulse with $t_W = 10\ \mu s$ is applied to the other input beginning 15 μs after the leading edge of the first pulse. Show the output in relation to the inputs.

2. The pulse waveforms A and B in Figure 5–29 are applied to the exclusive-NOR circuit in Figure 5–30. Develop a complete timing diagram.

5–6   COMBINATIONAL LOGIC WITH VHDL (optional)

The purpose of describing logic using VHDL is so that it can be programmed into a PLD. The data flow approach to writing a VHDL program was described in Chapter 4. In this optional section, both the data flow approach using Boolean expressions and the structural approach are used to develop VHDL code for describing logic circuits. The VHDL component is introduced and used to illustrate structural descriptions. Some aspects of software development tools are discussed.

After completing this section, you should be able to

■ Describe a VHDL component and discuss how it is used in a program ■ Apply the structural approach and the data flow approach to writing VHDL code ■ Describe two basic software development tools

Structural Approach to VHDL Programming

The structural approach to writing a VHDL description of a logic function can be compared to installing IC devices on a circuit board and interconnecting them with wires. With the structural approach, you describe logic functions and specify how they are connected together. The VHDL **component** is a way to predefine a logic function for repeated use in a program or in other programs. The component can be used to describe anything from a simple logic gate to a complex logic function. The VHDL **signal** can be thought of as a way to specify a "wire" connection between components.

Figure 5–35 provides a simplified comparison of the structural approach to a hardware implementation on a circuit board.

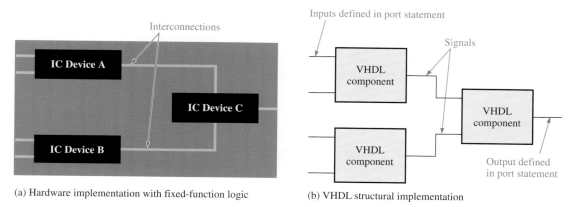

(a) Hardware implementation with fixed-function logic   (b) VHDL structural implementation

▲ **FIGURE 5–35**

Simplified comparison of the VHDL structural approach to a hardware implementation. The VHDL signals correspond to the interconnections on the circuit board, and the VHDL components correspond to the IC devices.

VHDL Components

A VHDL component describes predefined logic that can be stored as a package declaration in a VHDL library and called as many times as necessary in a program. You can use components to avoid repeating the same code over and over within a program. For example, you can create a VHDL component for an AND gate and then use it as many times as you wish without having to write a program for an AND gate every time you need one.

VHDL components are stored and are available for use when you write a program. This is similar to having, for example, a storage bin of ICs available when you are constructing a circuit. Every time you need to use one in your circuit, you reach into the storage bin and place it on the circuit board.

The VHDL program for any logic function can become a component and used whenever necessary in a larger program with the use of a component declaration of the following general form. **Component** is a VHDL keyword.

component name_of_component **is**

port (port definitions);

end component name_of_component;

For simplicity, let's assume that there are predefined VHDL data flow descriptions of a 2-input AND gate with the entity name AND_gate and a 2-input OR gate with the entity name OR_gate, as shown in Figure 5–36.

Next, assume that you are writing a program for a logic circuit that has several AND gates. Instead of rewriting the program in Figure 5–36 over and over, you can use

▶ FIGURE 5–36

Predefined programs for a 2-input AND gate and a 2-input OR gate to be used as components in the data flow approach.

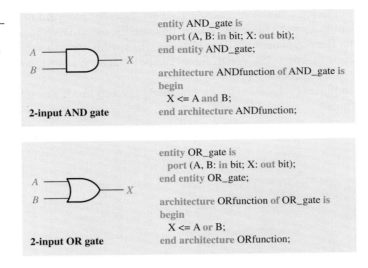

```
entity AND_gate is
    port (A, B: in bit; X: out bit);
end entity AND_gate;

architecture ANDfunction of AND_gate is
begin
    X <= A and B;
end architecture ANDfunction;
```

2-input AND gate

```
entity OR_gate is
    port (A, B: in bit; X: out bit);
end entity OR_gate;

architecture ORfunction of OR_gate is
begin
    X <= A or B;
end architecture ORfunction;
```

2-input OR gate

a component declaration to specify the AND gate. The port statement in the component declaration must correspond to the port statement in the entity declaration of the AND gate.

```
component AND_gate is

    port (A, B: in bit; X: out bit);

end component AND_gate;
```

Using Components in a Program   To use a component in a program, you must write a component instantiation statement for each instance in which the component is used. You can think of a component instantiation as a request or call for the component to be used in the main program. For example, the simple SOP logic circuit in Figure 5–37 has two AND gates and one OR gate. Therefore, the VHDL program for this circuit will have two components and three component instantiations or calls.

▶ FIGURE 5–37

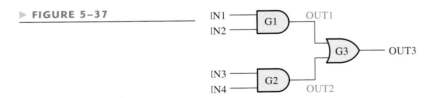

Signals   In VHDL, signals are analogous to wires that interconnect components on a circuit board. The signals in Figure 5–37 are named OUT1 and OUT2. Signals are the *internal* connections in the logic circuit and are treated differently than the inputs and outputs. Whereas the inputs and outputs are declared in the entity declaration using the port statement, the signals are declared within the architecture using the signal statement. **Signal** is a VHDL keyword.

The Program   The program for the logic in Figure 5–37 begins with an entity declaration as follows:

```
--Program for the logic circuit in Figure 5–37

entity AND_OR_Logic is

    port (IN1, IN2, IN3, IN4: in bit; OUT3: out bit);

end entity AND_OR_Logic;
```

The architecture declaration contains the component declarations for the AND gate and the OR gate, the signal definitions, and the component instantiations.

architecture LogicOperation **of** AND_OR_Logic **is**

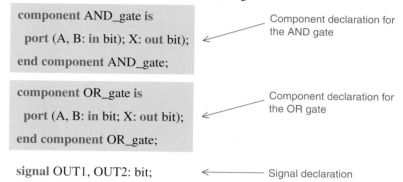

component AND_gate is

port (A, B: in bit); X: out bit); ←————— Component declaration for the AND gate

end component AND_gate;

component OR_gate is

port (A, B: in bit; X: out bit); ←————— Component declaration for the OR gate

end component OR_gate;

    signal OUT1, OUT2: bit; ←——————— Signal declaration

begin

    G1: AND_gate **port map** (A => IN1, B => IN2, X => OUT1);

    G2: AND_gate **port map** (A => IN3, B => IN4, X => OUT2); ←——— Component instantiations

    G3: OR_gate **port map** (A => OUT1, B => OUT2, X => OUT3);

end architecture LogicOperation;

Component Instantiations   Let's look at the component instantiations. First, notice that the component instantiations appear between the keyword **begin** and the **end** statement. For each instantiation an identifier is defined, such as G1, G2, and G3 in this case. Then the component name is specified. The port map essentially makes all the connections for the logic function using the operator =>. For example, the first instantiation,

    G1: AND_gate **port map** (A => IN1, B => IN2, X => OUT1);

can be explained as follows: *Input A of AND gate G1 is connected to input IN1, input B of the gate is connected to input IN2, and the output X of the gate is connected to the signal OUT1.*

    The three instantiation statements together completely describe the logic circuit in Figure 5–37, as illustrated in Figure 5–38.

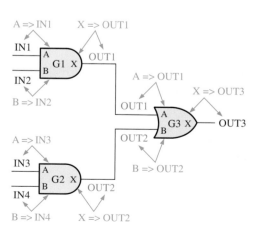

◀ **FIGURE 5–38**

Illustration of the instantiation statements and port mapping applied to the AND–OR logic. Signals are shown in red.

    Although the data flow approach using Boolean expressions would have been easier and probably the best way to describe this particular circuit, we have used this simple circuit to explain the concept of the structural approach. Example 5–14 compares the structural and data flow approaches to writing a VHDL program for an SOP logic circuit.

EXAMPLE 5-14

Write a VHDL program for the SOP logic circuit in Figure 5–39 using the structural approach. Assume that VHDL components for a 3-input NAND gate and for a 2-input NAND are available. Notice the NAND gate G4 is shown as a negative-OR.

▷ **FIGURE 5–39**

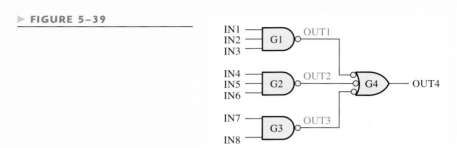

Solution   The components and component instantiations are highlighted.

```
--Program for the logic circuit in Figure 5–39
entity SOP_Logic is
    port (IN1, IN2, IN3, IN4, IN5, IN6, IN7, IN8: in bit; OUT4: out bit);
end entity SOP_Logic;

architecture LogicOperation of SOP_Logic is
-- component declaration for 3-input NAND gate

    component NAND_gate3 is
        port (A, B, C: in bit X: out bit);
    end component NAND_gate3;

-- component declaration for 2-input NAND gate

    component NAND_gate2 is
        port (A, B: in bit; X: out bit);
    end component NAND_gate;

    signal OUT1, OUT2, OUT3: bit;
begin

    G1: NAND_gate3 port map (A => IN1, B => IN2, C => IN3, X => OUT1);
    G2: NAND_gate3 port map (A => IN4, B => IN5, C => IN6, X => OUT2);
    G3: NAND_gate2 port map (A => IN7, B => IN8, X => OUT3);
    G4: NAND_gate3 port map (A => OUT1, B => OUT2, C => OUT3, X => OUT4);

end architecture LogicOperation;
```

For comparison purposes, let's write the program for the logic circuit in Figure 5–39 using the data flow approach.

```
entity SOP_Logic is
    port (IN1, IN2, IN3, IN4, IN5, IN6, IN7, IN8: in bit; OUT4: out bit);
end entity SOP_Logic;
```

> **architecture** LogicOperation **of** SOP_Logic **is**
> **begin**
>
>   OUT4 <= (IN1 **and** IN2 **and** IN3) **or** (IN4 **and** IN5 **and** IN6) **or** (IN7 **and** IN8);
> **end architecture** LogicOperation;
>
> As you can see, the data flow approach results in a much simpler code for this particular logic function. However, in situations where a logic function consists of many blocks of complex logic, the structural approach might have an advantage over the data flow approach.
>
> *Related Problem*   If another NAND gate is added to the circuit in Figure 5–39 with inputs IN9 and IN10, write a component instantiation to add to the program. Specify any other necessary changes in the program as a result.

Applying Software Development Tools

As you have learned, a software development package must be used to implement an HDL design in a target device. Once the logic has been described using an HDL and entered via a software tool called a code or text editor, it can be tested using a simulation to verify that it performs properly before actually programming the target device. Using software development tools allows for the design, development, and testing of combinational logic before it is committed to hardware. Software development tools are explored further in Chapter 11.

Typical software development tools allow you to input VHDL code on a text-based editor specific to the particular development tool that you are using. The VHDL code for a combinational logic circuit has been written using a generic text-based editor for illustration and appears on the computer screen as shown in Figure 5–40. As shown, many code editors provide enhanced features such as the highlighting of keywords.

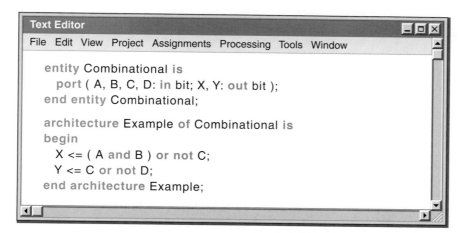

▲ **FIGURE 5–40**

A VHDL program for a combinational logic circuit after entry on a generic text editor screen that is part of a software development tool.

After the program has been written into the text editor, it is passed to the compiler. The compiler takes the high-level VHDL code and converts it into a file that can be downloaded to the target device. Once the program has been compiled, you can create a simulation for testing. Simulated input values are inserted into the logic design and allow for verification of the output(s).

You specify the input waveforms on a software tool called a waveform editor, as shown in Figure 5–41. The output waveforms are generated by a simulation of the VHDL code that you entered on the text editor in Figure 5–40. The waveform simulation provides the resulting outputs X and Y for the inputs A, B, C, and D in all sixteen combinations from $0\ 0\ 0\ 0_2$ to $1\ 1\ 1\ 1_2$.

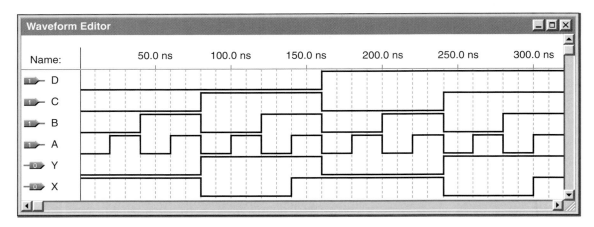

▲ FIGURE 5–41

A typical waveform editor tool showing the simulated waveforms for the logic circuit described by the VHDL code in Figure 5–40.

Recall from Chapter 3 that there are several performance characteristics of logic circuits to be considered in the creation of any digital system. Propagation delay, for example, determines the speed or frequency at which a logic circuit can operate. A timing simulation can be used to mimic the propagation delay through the logic design in the target device.

SECTION 5–6 REVIEW

1. What is a VHDL component?
2. State the purpose of a component instantiation in a program architecture.
3. How are interconnections made between components in VHDL?
4. The use of components in a VHDL program represents what approach?

5–7   TROUBLESHOOTING

The preceding sections have given you some insight into the operation of combinational logic circuits and the relationships of inputs and outputs. This type of understanding is essential when you troubleshoot digital circuits because you must know what logic levels or waveforms to look for throughout the circuit for a given set of input conditions.

In this section, an oscilloscope is used to troubleshoot a fixed-function logic circuit when a gate output is connected to several gate inputs. Also, an example of signal tracing and waveform analysis methods is presented using a scope or logic analyzer for locating a fault in a combinational logic circuit.

After completing this section, you should be able to

■ Define a circuit node ■ Use an oscilloscope to find a faulty circuit node ■ Use an oscilloscope to find an open gate output ■ Use an oscilloscope to find a shorted gate input or output ■ Use an oscilloscope or a logic analyzer for signal tracing in a combinational logic circuit

In a combinational logic circuit, the output of one gate may be connected to two or more gate inputs as shown in Figure 5–42. The interconnecting paths share a common electrical point known as a **node.**

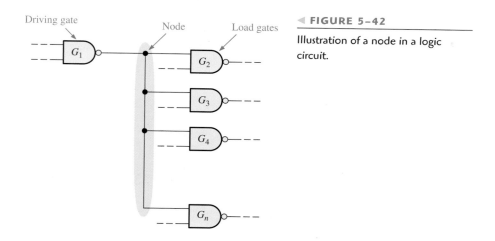

Illustration of a node in a logic circuit.

Gate G_1 in Figure 5–42 is driving the node, and the other gates represent loads connected to the node. A driving gate can drive a number of load gate inputs up to its specified fan-out. Several types of failures are possible in this situation. Some of these failure modes are difficult to isolate to a single bad gate because all the gates connected to the node are affected. Common types of failures are the following:

1. *Open output in driving gate.* This failure will cause a loss of signal to all load gates.

2. *Open input in a load gate.* This failure will not affect the operation of any of the other gates connected to the node, but it will result in loss of signal output from the faulty gate.

3. *Shorted output in driving gate.* This failure can cause the node to be stuck in the LOW state (short to ground) or in the HIGH state (short to V_{CC}).

4. *Shorted input in a load gate.* This failure can also cause the node to be stuck in the LOW state (short to ground) or in the HIGH state (short to V_{CC}).

Troubleshooting Common Faults

Open Output in Driving Gate   In this situation there is no pulse activity on the node. With circuit power on, an open node will normally result in a "floating" level, which is often indicated by noise, as illustrated in Figure 5–43.

When troubleshooting logic circuits, begin with a visual check, looking for obvious problems. In addition to components, visual inspection should include connectors. Edge connectors are frequently used to bring power, ground, and signals to a circuit board. The mating surfaces of the connector need to be clean and have a good mechanical fit. A dirty connector can cause intermittent or complete failure of the circuit. Edge connectors can be cleaned with a common pencil eraser and wiped clean with a Q-tip soaked in alcohol. Also, all connectors should be checked for loose-fitting pins.

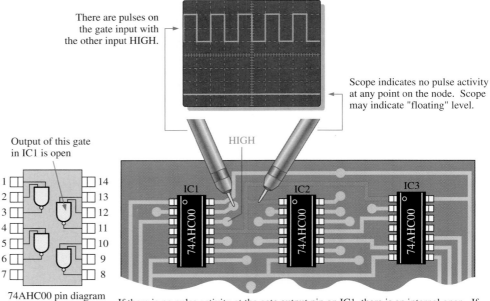

There are pulses on the gate input with the other input HIGH.

Scope indicates no pulse activity at any point on the node. Scope may indicate "floating" level.

HIGH

Output of this gate in IC1 is open

74AHC00 pin diagram from data sheet

If there is no pulse activity at the gate output pin on IC1, there is an internal open. If there is pulse activity directly on the output pin but not on the node interconnections, the connection between the pin and the board is open.

▲ **FIGURE 5–43**

Open output in driving gate. For simplicity, assume a HIGH is on one gate input.

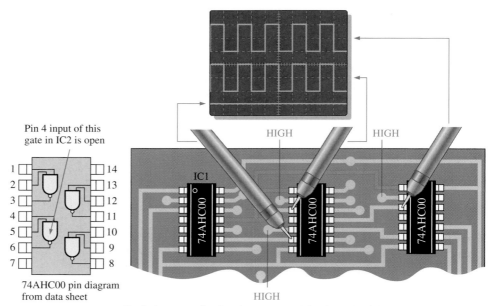

Pin 4 input of this gate in IC2 is open

HIGH   HIGH

74AHC00 pin diagram from data sheet

HIGH

Check the output pin of each gate connected to the node with other gate inputs HIGH. No pulse activity on an output indicates an open gate input or open gate output.

▲ **FIGURE 5–44**

Open input in a load gate.

Open Input in a Load Gate If the check for an open driver output is negative, then a check for an open input in a load gate should be performed. Apply the logic pulser tip to the node with all nonpulsed inputs HIGH. Then check the output of each gate for pulse activity with the logic probe, as illustrated in Figure 5–44. If one of the inputs that is normally connected to the node open, no pulses will be detected on that gate's output.

Output or Input Shorted to Ground　When the output is shorted to ground in the driving gate or the input to a load gate is shorted to ground, it will cause the node to be stuck LOW, as previously mentioned. A quick check with a scope probe will indicate this, as shown in Figure 5–45. A short to ground in the driving gate's output or in any load gate input will cause this symptom, and further checks must therefore be made to isolate the short to a particular gate.

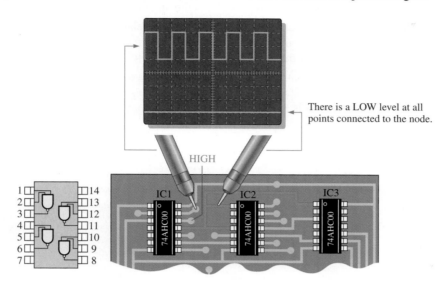

There is a LOW level at all points connected to the node.

HIGH

◀ FIGURE 5–45

Shorted output in the driving gate or shorted input in a load gate.

Signal Tracing and Waveform Analysis

Although the methods of isolating an open or a short at a node point are very useful from time to time, a more general troubleshooting technique called **signal tracing** is of value in just about every troubleshooting situation. Waveform measurement is accomplished with an oscilloscope or a logic analyzer.

Basically, the signal tracing method requires that you observe the waveforms and their time relationships at all accessible points in the logic circuit. You can begin at the inputs and, from an analysis of the waveform timing diagram for each point, determine where an incorrect waveform first occurs. With this procedure you can usually isolate the fault to a specific gate. A procedure beginning at the output and working back toward the inputs can also be used.

The general procedure for signal tracing starting at the inputs is outlined as follows:

- Within a system, define the section of logic that is suspected of being faulty.

- Start at the inputs to the section of logic under examination. We assume, for this discussion, that the input waveforms coming from other sections of the system have been found to be correct.

- For each gate, beginning at the input and working toward the output of the logic circuit, observe the output waveform of the gate and compare it with the input waveforms by using the oscilloscope or the logic analyzer.

- Determine if the output waveform is correct, using your knowledge of the logical operation of the gate.

- If the output is incorrect, the gate under test may be faulty. Pull the IC containing the gate that is suspected of being faulty, and test it out-of-circuit. If the gate is found to be faulty, replace the IC. If it works correctly, the fault is in the external circuitry or in another IC to which the tested one is connected.

- If the output is correct, go to the next gate. Continue checking each gate until an incorrect waveform is observed.

Figure 5–46 is an example that illustrates the general procedure for a specific logic circuit in the following steps:

Step 1. Observe the output of gate G_1 (test point 5) relative to the inputs. If it is correct, check the inverter next. If the output is not correct, the gate or its connections are bad; or, if the output is LOW, the input to gate G_2 may be shorted.

Step 2. Observe the output of the inverter (TP6) relative to the input. If it is correct, check gate G_2 next. If the output is not correct, the inverter or its connections are bad; or, if the output is LOW, the input to gate G_3 may be shorted.

Step 3. Observe the output of gate G_2 (TP7) relative to the inputs. If it is correct, check gate G_3 next. If the output is not correct, the gate or its connections are bad; or, if the output is LOW, the input to gate G_4 may be shorted.

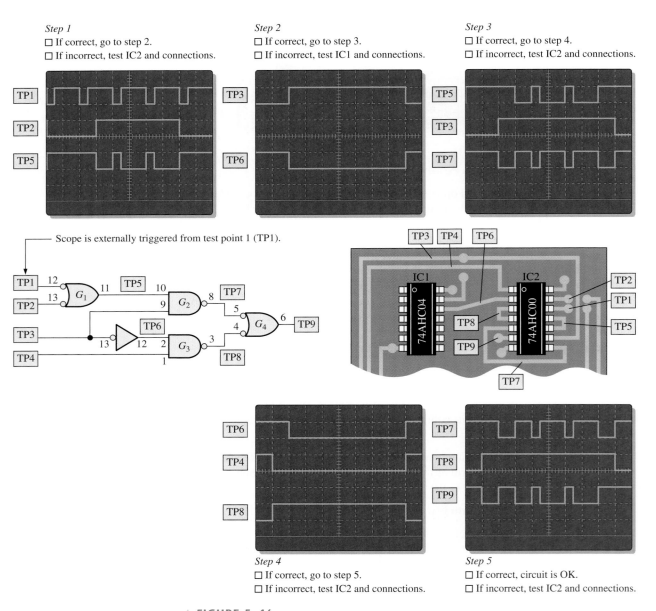

▲ **FIGURE 5–46**

Example of signal tracing and waveform analysis in a portion of a printed circuit board. TP indicates test point.

Step 4. Observe the output of gate G_3 (TP8) relative to the inputs. If it is correct, check gate G_4 next. If the output is not correct, the gate or its connections are bad; or, if the output is LOW, the input to gate G_4 (TP7) may be shorted.

Step 5. Observe the output of gate G_4 (TP9) relative to the inputs. If it is correct, the circuit is okay. If the output is not correct, the gate or its connections are bad.

EXAMPLE 5–15

Determine the fault in the logic circuit of Figure 5–47(a) by using waveform analysis. You have observed the waveforms shown in green in Figure 5–47(b). The red waveforms are correct and are provided for comparison.

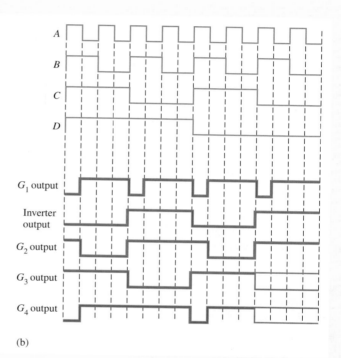

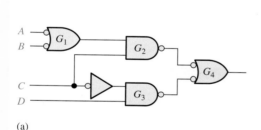

(a)

(b)

▲ FIGURE 5–47

Solution

1. Determine what the correct waveform should be for each gate. The correct waveforms are shown in red, superimposed on the actual measured waveforms, in Figure 5–47(b).

2. Compare waveforms gate by gate until you find a measured waveform that does not match the correct waveform.

In this example, everything tested is correct until gate G_3 is checked. The output of this gate is not correct as the differences in the waveforms indicate. An analysis of the waveforms indicates that if the D input to gate G_3 is open and acting as a HIGH, you will get the output waveform measured (shown in red). Notice that the output of G_4 is also incorrect due to the incorrect input from G_3.

Replace the IC containing G_3, and check the circuit's operation again.

Related Problem For the inputs in Figure 5–47(b), determine the output waveform for the logic circuit (output of G_4) if the inverter has an open output.

HANDS ON TIP

As you know, testing and troubleshooting logic circuits often require observing and comparing two digital waveforms simultaneously, such as an input and the output of a gate, on a two-channel oscilloscope. For digital waveforms, the scope should always be set to DC coupling on each channel input to avoid "shifting" the ground level. You should determine where the 0 V level is on the screen for both channels.

To compare the timing of the waveforms, the scope should be triggered from only one channel (don't use vertical mode or composite triggering). The channel selected for triggering should always be the one that has the lowest frequency waveform, if possible.

SECTION 5–7 REVIEW

1. List four common internal failures in logic gates.
2. One input of a NOR gate is externally shorted to $+V_{CC}$. How does this condition affect the gate operation?
3. Determine the output of gate G_4 in Figure 5–47(a), with inputs as shown in part (b), for the following faults:
   - (a) one input to G_1 shorted to ground
   - (b) the inverter input shorted to ground
   - (c) an open output in G_3

Troubleshooting problems that are keyed to the CD-ROM are available in the Multisim Troubleshooting Practice section of the end-of-chapter problems.

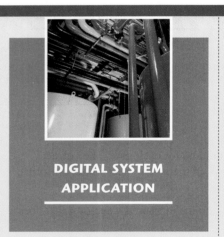

DIGITAL SYSTEM APPLICATION

In this digital system application, the digital control logic for controlling the fluid in a storage tank is developed. The purpose of the logic is to maintain an appropriate level of fluid by controlling the inlet and outlet valves. Also, the logic must control the temperature of

the fluid within a certain range and issue an alarm if any of the level or temperature sensors fail.

Basic System Operation

The outputs of the system control logic control the fluid input, fluid output, and fluid temperature. The control logic operates an inlet valve that allows fluid to flow into the tank until a high-level sensor is activated by being immersed in fluid. When the high-level sensor is immersed (activated), the control logic closes the inlet valve. The fluid in the tank must be maintained within a specified temperature range as determined by two temperature sensors. One temperature sensor indicates when the fluid is too hot, and the other indicates when the fluid is too cold. The control logic turns on a heating element if

the temperature sensors indicate the fluid is too cold. The control logic keeps the outlet valve open as long as the low-level sensor is immersed and the fluid is at a proper temperature. When the fluid level drops below the low-level sensor, the control logic closes the outlet valve.

Operational Requirements

The maximum and minimum fluid levels are determined by the positions of the level sensors in the tank. The output of each sensor is HIGH when it is immersed in the fluid and is LOW when not immersed. When the high-level sensor output is LOW, the control logic produces a HIGH and the inlet valve opens. When the high-level sensor output is HIGH, the control logic produces a LOW and the inlet valve closes.

The fluid must be within a specified temperature range before the outlet valve

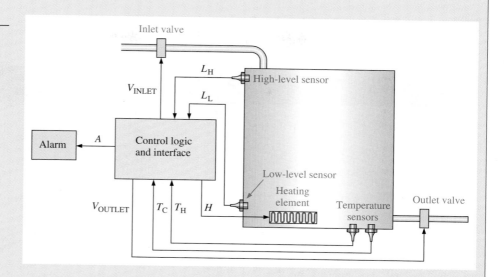

► **FIGURE 5–48**

Fluid storage tank with level and temperature sensors and controls.

► **TABLE 5–6**

Tank control logic inputs and outputs.

INPUTS TO CONTROL LOGIC			
Variable	**Description**	**Active level**	**Comments**
L_H	High-level sensor	HIGH (1)	Sensor is immersed
L_L	Low-level sensor	HIGH (1)	Sensor is immersed
T_H	High-temp sensor	HIGH (1)	Temperature too hot
T_C	Low-temp sensor	HIGH (1)	Temperature too cold

OUTPUTS FROM CONTROL LOGIC			
Variable	**Description**	**Active level**	**Comments**
V_{INLET}	Inlet valve	HIGH (1)	Valve open
V_{OUTLET}	Outlet valve	HIGH (1)	Valve open
H	Heating element	HIGH (1)	Heat on
A	Alarm	HIGH (1)	Sensor failure or too-hot condition

is opened. One sensor produces a HIGH when the temperature is too hot, and the other temperature sensor produces a HIGH when the temperature is too cold. The control logic produces a HIGH to turn on a heating element when a too-cold condition is indicated; otherwise, the heating element is turned off. When a too-hot condition is indicated, an alarm is activated.

When the low-level sensor produces a HIGH output (indicating that it is immersed) and when the output of both temperature sensors are LOW (indicating a correct temperature), the control logic opens the outlet valve. If the low-level sensor output goes LOW or if either temperature sensor outputs go LOW, the control logic closes the outlet valve.

If the control detects a failure in any of the sensors or a too-hot condition, an alarm is activated. A level-sensor failure is indicated when the high-level sensor is active and the low-level sensor is not active. A temperature-sensor failure is indicated when both are active at the same time. Figure 5–48 shows the tank control system.

The system inputs and outputs are summarized in Table 5–6, and the truth table is shown in Table 5–7.

Design of the Control Logic

There are four separate outputs: one for the inlet valve, one for the outlet valve, one for the heater, and one for the alarm. We will approach the design as four separate logic circuits.

▼ TABLE 5–7

Truth table for tank control logic.

INPUTS				OUTPUTS				COMMENTS
L_H	L_L	T_H	T_C	V_{INLET}	V_{OUTLET}	H	A	
0	0	0	0	1	0	0	0	Fill/heat off
0	0	0	1	1	0	1	0	Fill/heat on
0	0	1	0	1	0	0	1	Fill/heat off/alarm
0	0	1	1	0	0	0	1	Temp sensor fault/alarm
0	1	0	0	1	1	0	0	Fill and drain/heat off
0	1	0	1	1	0	1	0	Fill/heat on
0	1	1	0	1	0	0	1	Fill/heat off/alarm
0	1	1	1	0	0	0	1	Temp sensor fault/alarm
1	0	0	0	0	0	0	1	Level sensor fault/alarm
1	0	0	1	0	0	0	1	Level sensor fault/alarm
1	0	1	0	0	0	0	1	Level sensor fault/alarm
1	0	1	1	0	0	0	1	Multiple sensor fault/alarm
1	1	0	0	0	1	0	0	Drain/heat off
1	1	0	1	0	0	1	0	Heat on
1	1	1	0	0	0	0	1	Heat off/alarm
1	1	1	1	0	0	0	1	Temp sensor fault/alarm

▶ FIGURE 5–49

Karnaugh map simplification and implementation for the inlet valve logic.

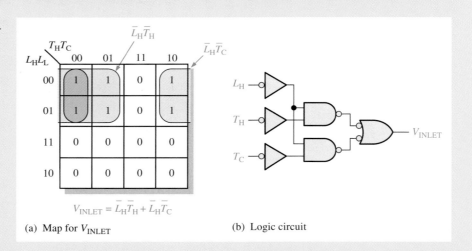

$$V_{INLET} = \bar{L}_H \bar{T}_H + \bar{L}_H \bar{T}_C$$

(a) Map for V_{INLET} (b) Logic circuit

Inlet Valve Logic Let's begin by designing the logic circuit for the inlet valve. The output of this logic circuit is the variable V_{INLET}. The first step is to transfer the data from the truth table to a Karnaugh map and develop an SOP expression.

The input variables, L_H, L_L, T_H, and T_C are map variables and the states of V_{INLET} are plotted and grouped as shown in Figure 5–49(a). The 0s on the map are for the input conditions when the inlet valve is closed, and the 1s are for the input conditions when the inlet valve is open.

The resulting SOP expression for the inlet valve logic results in the NAND implementation shown in part (b).

Outlet Valve Logic Next, let's design the logic circuit for the outlet valve. The output of this logic circuit is the variable V_{OUTLET}.

▶ FIGURE 5–50

Karnaugh map simplification and implementation for the outlet valve logic.

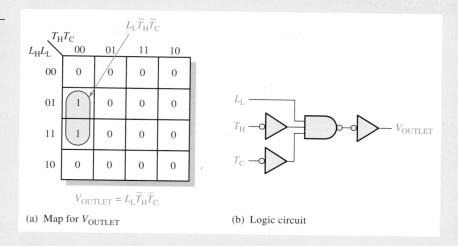

(a) Map for V_{OUTLET}

$$V_{OUTLET} = L_L \bar{T}_H \bar{T}_C$$

(b) Logic circuit

Again, the first step is to transfer the data from the truth table to a Karnaugh map and develop an SOP expression.

The input variables, L_H, L_L, T_H, and T_C are map variables and the states of V_{OUTLET} are plotted and grouped as shown in Figure 5–50(a). The 0s on the map are for the input conditions when the outlet valve is closed, and the 1s are for the input conditions when the valve is open. The resulting SOP expression for the outlet valve logic results in the NAND implementation shown in part (b).

VHDL Code for the Inlet and Outlet Valve Logic (optional)

A single entity and architecture describes the inlet valve logic and the outlet valve logic using the data flow approach as the program shows.

```
entity TankControl is
    port (LL, LH, TH, TC in bit; Vinlet,
        Voutlet: out bit);
end entity TankControl;
architecture ValveLogic of TankControl is
begin
    Vinlet <= (not LH and not TH) or
        (not LH and not TC);
    Voutlet <= LL and not TH and not TC;
end architecture ValveLogic;
```

The inlet and outlet valve logic has been designed and the VHDL code has

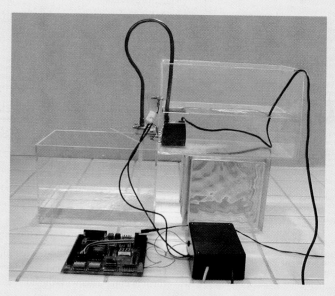

A photo of a storage tank mock-up in the electronics lab at Yuba College in California. The control logic has been programmed into a PLD on a development board and is connected to the tank fixtures to control the filling and emptying of the tank. Photo courtesy of Doug Joksch

been written. Now it's your turn to complete the remaining control logic design for the heater control and the alarm and to write the VHDL program to implement the logic in a target device.

System Assignment

■ *Activity 1* Using Table 5–7 and the Karnaugh map method, design the logic for controlling the heating element in the tank. Use NAND gates

and inverters to implement the circuit.

■ *Activity 2* Design the logic for activating the alarm.

■ *Activity 3* Combine the logic for each of the four tank control functions into a complete logic diagram.

■ *Optional Activity* Write the VHDL entity and architecture for the complete logic by modifying the code previously developed for the inlet and outlet valve logic.

SUMMARY

- AND-OR logic produces an output expression in SOP form.
- AND-OR-Invert logic produces a complemented SOP form, which is actually a POS form.
- The operational symbol for exclusive-OR is ⊕. An exclusive-OR expression can be stated in two equivalent ways:

$$A\overline{B} + \overline{A}B = A \oplus B$$

- To do an analysis of a logic circuit, start with the logic circuit, and develop the Boolean output expression or the truth table or both.
- Implementation of a logic circuit is the process in which you start with the Boolean output expressions or the truth table and develop a logic circuit that produces the output function.
- All NAND or NOR logic diagrams should be drawn using appropriate dual symbols so that bubble outputs are connected to bubble inputs and nonbubble outputs are connected to nonbubble inputs.
- When two negation indicators (bubbles) are connected, they effectively cancel each other.
- A VHDL component is a predefined logic function stored for use throughout a program or in other programs.
- A component instantiation is used to call for a component in a program.
- A VHDL signal effectively acts as an internal interconnection in a VHDL structural description.

KEY TERMS

Key terms and other bold terms in the chapter are defined in the end-of-book glossary.

Component A VHDL feature that can be used to predefine a logic function for multiple use throughout a program or programs.

Negative-AND The dual operation of a NOR gate when the inputs are active-LOW.

Negative-OR The dual operation of a NAND gate when the inputs are active-LOW.

Node A common connection point in a circuit in which a gate output is connected to one or more gate inputs.

Signal A waveform; a type of VHDL object that holds data.

Signal tracing A troubleshooting technique in which waveforms are observed in a step-by-step manner beginning at the input and working toward the output or vice versa. At each point the observed waveform is compared with the correct signal for that point.

Universal gate Either a NAND gate or a NOR gate. The term *universal* refers to the property of a gate that permits any logic function to be implemented by that gate or by a combination of gates of that kind.

SELF-TEST

Answers are at the end of the chapter.

1. The output expression for an AND-OR circuit having one AND gate with inputs A, B, C, and D and one AND gate with inputs E and F is

   (a) $ABCDEF$ **(b)** $A + B + C + D + E + F$

   (c) $(A + B + C + D)(E + F)$ **(d)** $ABCD + EF$

2. A logic circuit with an output $X = A\overline{B}C + A\overline{C}$ consists of

   (a) two AND gates and one OR gate

   (b) two AND gates, one OR gate, and two inverters

(c) two OR gates, one AND gate, and two inverters

(d) two AND gates, one OR gate, and one inverter

3. To implement the expression $\overline{A}BCD + A\overline{B}CD + AB\overline{C}\,\overline{D}$, it takes one OR gate and

    (a) one AND gate

    (b) three AND gates

    (c) three AND gates and four inverters

    (d) three AND gates and three inverters

4. The expression $\overline{A}BCD + ABC\overline{D} + A\overline{B}\,\overline{C}D$

    (a) cannot be simplified

    (b) can be simplified to $\overline{A}BC + A\overline{B}$

    (c) can be simplified to $ABC\overline{D} + \overline{A}B\overline{C}$

    (d) None of these answers is correct.

5. The output expression for an AND-OR-Invert circuit having one AND gate with inputs, *A, B, C,* and *D* and one AND gate with inputs *E* and *F* is

    (a) $ABCD + EF$

    (b) $\overline{A} + \overline{B} + \overline{C} + \overline{D} + \overline{E} + \overline{F}$

    (c) $\overline{(A + B + C + D)(E + F)}$

    (d) $(\overline{A} + \overline{B} + \overline{C} + \overline{D})(\overline{E} + \overline{F})$

6. An exclusive-OR function is expressed as

    (a) $\overline{A}\,\overline{B} + AB$ (b) $\overline{A}B + A\overline{B}$

    (c) $(\overline{A} + B)(A + \overline{B})$ (d) $(\overline{A} + \overline{B}) + (A + B)$

7. The AND operation can be produced with

    (a) two NAND gates   (b) three NAND gates

    (c) one NOR gate   (d) three NOR gates

8. The OR operation can be produced with

    (a) two NOR gates   (b) three NAND gates

    (c) four NAND gates   (d) both answers (a) and (b)

9. When using dual symbols in a logic diagram,

    (a) bubble outputs are connected to bubble inputs

    (b) the NAND symbols produce the AND operations

    (c) the negative-OR symbols produce the OR operations

    (d) All of these answers are true.

    (e) None of these answers is true.

10. All Boolean expressions can be implemented with

    (a) NAND gates only

    (b) NOR gates only

    (c) combinations of NAND and NOR gates

    (d) combinations of AND gates, OR gates, and inverters

    (e) any of these

11. A VHDL component

    (a) can be used once in each program

    (b) is a predefined description of a logic function

    (c) can be used multiple times in a program

    (d) is part of a data flow description

    (e) answers (b) and (c)

12. A component is called for use in a program by using a

(a) signal

(b) variable

(c) component instantiation

(d) architecture declaration

PROBLEMS

Answers to odd–numbered problems are at the end of the book.

SECTION 5–1 **Basic Combinational Logic Circuits**

1. Draw the ANSI distinctive shape logic diagram for a 3-wide, 4-input AND-OR-Invert circuit. Also draw the ANSI standard rectangular outline symbol.

2. Write the output expression for each circuit in Figure 5–51.

3. Write the output expression for each circuit as it appears in Figure 5–52.

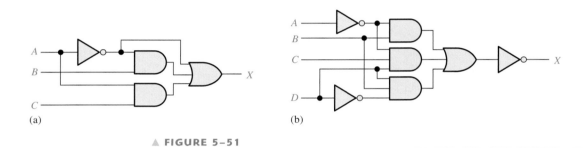

(a)　　　　　　　　　　　　　　　　(b)

▲ FIGURE 5–51

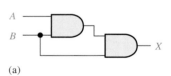 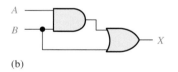 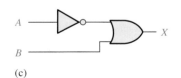

(a)　　　　　　　　(b)　　　　　　　　(c)

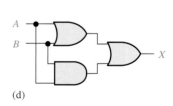 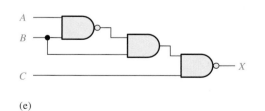 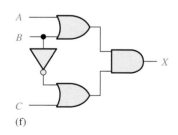

(d)　　　　　　　　(e)　　　　　　　　(f)

▲ FIGURE 5–52

4. Write the output expression for each circuit as it appears in Figure 5–53 and then change each circuit to an equivalent AND-OR configuration.

5. Develop the truth table for each circuit in Figure 5–52.

6. Develop the truth table for each circuit in Figure 5–53.

7. Show that an exclusive-NOR circuit produces a POS output.

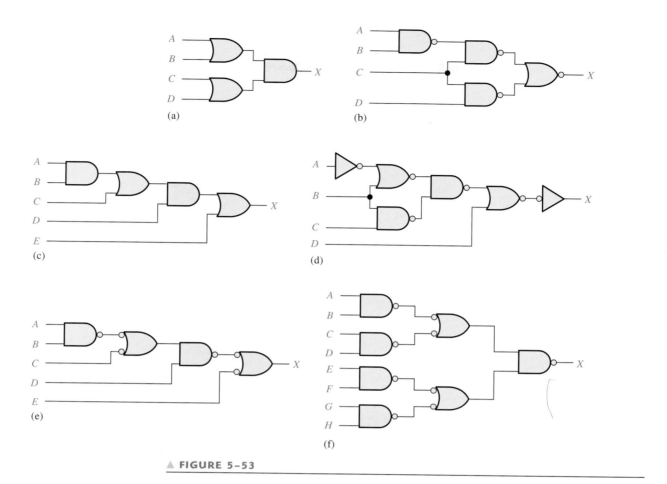

▲ FIGURE 5–53

SECTION 5–2 Implementing Combinational Logic

8. Use AND gates, OR gates, or combinations of both to implement the following logic expressions as stated:

(a) $X = AB$

(b) $X = A + B$

(c) $X = AB + C$

(d) $X = ABC + D$

(e) $X = A + B + C$

(f) $X = ABCD$

(g) $X = A(CD + B)$

(h) $X = AB(C + DEF) + CE(A + B + F)$

9. Use AND gates, OR gates, and inverters as needed to implement the following logic expressions as stated:

(a) $X = AB + \overline{B}C$ (b) $X = A(B + \overline{C})$

(c) $X = A\overline{B} + AB$ (d) $X = \overline{ABC} + B(EF + \overline{G})$

(e) $X = A[BC(A + B + C + D)]$ (f) $X = B(C\overline{D}E + \overline{E}FG)(\overline{AB} + C)$

10. Use NAND gates, NOR gates, or combinations of both to implement the following logic expressions as stated:

(a) $X = \overline{A}B + CD + (\overline{A + B})(ACD + \overline{BE})$

(b) $X = AB\overline{C}\,\overline{D} + D\overline{E}F + \overline{AF}$

(c) $X = \overline{A}[B + \overline{C}(D + E)]$

11. Implement a logic circuit for the truth table in Table 5–8.

▼ **TABLE 5–8**

| INPUTS | | | OUTPUT |
A	B	C	X
0	0	0	1
0	0	1	0
0	1	0	1
0	1	1	0
1	0	0	1
1	0	1	0
1	1	0	1
1	1	1	1

12. Implement a logic circuit for the truth table in Table 5–9.

▼ **TABLE 5–9**

| INPUTS | | | | OUTPUT |
A	B	C	D	X
0	0	0	0	0
0	0	0	1	0
0	0	1	0	1
0	0	1	1	1
0	1	0	0	1
0	1	0	1	0
0	1	1	0	0
0	1	1	1	0
1	0	0	0	1
1	0	0	1	1
1	0	1	0	1
1	0	1	1	1
1	1	0	0	0
1	1	0	1	0
1	1	1	0	0
1	1	1	1	1

13. Simplify the circuit in Figure 5–54 as much as possible, and verify that the simplified circuit is equivalent to the original by showing that the truth tables are identical.

14. Repeat Problem 13 for the circuit in Figure 5–55.

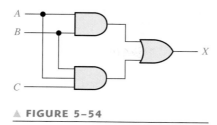

▲ FIGURE 5–54

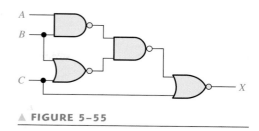

▲ FIGURE 5–55

15. Minimize the gates required to implement the functions in each part of Problem 9 in SOP form.

16. Minimize the gates required to implement the functions in each part of Problem 10 in SOP form.

17. Minimize the gates required to implement the function of the circuit in each part of Figure 5–53 in SOP form.

SECTION 5–3   The Universal Property of NAND and NOR Gates

18. Implement the logic circuits in Figure 5–51 using only NAND gates.

19. Implement the logic circuits in Figure 5–55 using only NAND gates.

20. Repeat Problem 18 using only NOR gates.

21. Repeat Problem 19 using only NOR gates.

SECTION 5–4   Combinational Logic Using NAND and NOR Gates

22. Show how the following expressions can be implemented as stated using only NOR gates:

(**a**) $X = ABC$ (**b**) $X = \overline{ABC}$ (**c**) $X = A + B$

(**d**) $X = A + B + \overline{C}$ (**e**) $X = \overline{AB} + \overline{CD}$ (**f**) $X = (A + B)(C + D)$

(**g**) $X = AB[C(\overline{DE} + \overline{AB}) + \overline{BCE}]$

23. Repeat Problem 22 using only NAND gates.

24. Implement each function in Problem 8 by using only NAND gates.

25. Implement each function in Problem 9 by using only NAND gates.

SECTION 5–5   Logic Circuit Operation with Pulse Waveform Inputs

26. Given the logic circuit and the input waveforms in Figure 5–56, draw the output waveform.

27. For the logic circuit in Figure 5–57, draw the output waveform in proper relationship to the inputs.

▶ FIGURE 5–56

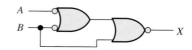

▶ FIGURE 5–57

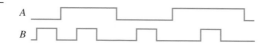

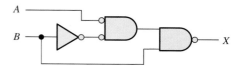

28. For the input waveforms in Figure 5–58, what logic circuit will generate the output waveform shown?

▶ FIGURE 5–58

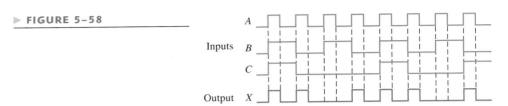

29. Repeat Problem 28 for the waveforms in Figure 5–59.

▶ FIGURE 5–59

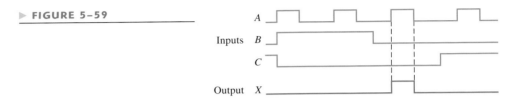

30. For the circuit in Figure 5–60, draw the waveforms at the numbered points in the proper relationship to each other.

31. Assuming a propagation delay through each gate of 10 nanoseconds (ns), determine if the *desired* output waveform X in Figure 5–61 (a pulse with a minimum $t_W = 25$ ns positioned as shown) will be generated properly with the given inputs.

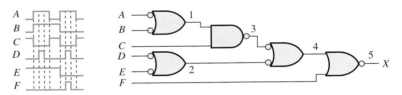

▲ FIGURE 5–60

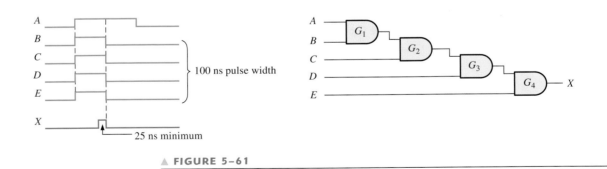

▲ FIGURE 5–61

SECTION 5–6   **Combinational Logic with VHDL (optional)**

32. Write a VHDL program using the data flow approach (Boolean expressions) to describe the logic circuit in Figure 5–51(b).

33. Write VHDL programs using the data flow approach (Boolean expressions) for the logic circuits in Figure 5–52(e) and (f).

34. Write a VHDL program using the structural approach for the logic circuit in Figure 5–53(d). Assume component declarations for each type of gate are already available.

35. Repeat Problem 34 for the logic circuit in Figure 5–53(f).

36. Describe the logic represented by the truth table in Table 5–8 using VHDL by first converting it to SOP form.

37. Develop a VHDL program for the logic in Figure 5–64 (p. 290), using both the data flow and the structural approach. Compare the resulting programs.

38. Develop a VHDL program for the logic in Figure 5–68 (p. 291), using both the data flow and the structural approach. Compare the resulting programs.

39. Given the following VHDL program, create the truth table that describes the logic circuit.

> **entity** CombLogic **is**
>
> > **port** (A, B, C, D: **in** bit; X: **out** bit);
>
> **end entity** CombLogic;
>
> **architecture** Example **of** CombLogic **is**
>
> > **begin**
> >
> > > X <= **not((not A and not B) or (not A and not C) or (not A and not D) or**
> > >
> > > > **(not B and not C) or(not B and not D) or (not D and not C));**
> >
> > **end architecture** Example;

40. Describe the logic circuit shown in Figure 5–62 with a VHDL program, using the data flow approach.

▶ FIGURE 5–62

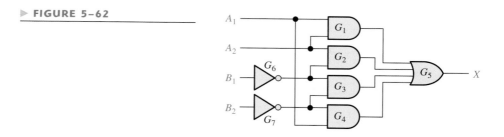

41. Repeat Problem 40 using the structural approach.

SECTION 5–7 Troubleshooting

42. For the logic circuit and the input waveforms in Figure 5–63, the indicated output waveform is observed. Determine if this is the correct output waveform.

▶ FIGURE 5–63

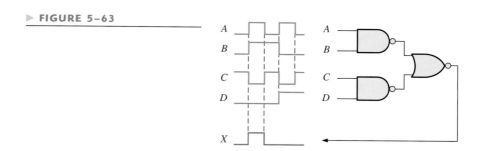

43. The output waveform in Figure 5–64 is incorrect for the inputs that are applied to the circuit. Assuming that one gate in the circuit has failed, with its output either an apparent constant HIGH or a constant LOW, determine the faulty gate and the type of failure (output open or shorted).

◄ FIGURE 5–64

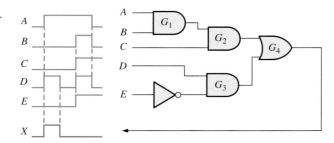

44. Repeat Problem 43 for the circuit in Figure 5–65, with input and output waveforms as shown.

45. By examining the connections in Figure 5–66, determine the driving gate and load gate(s). Specify by device and pin numbers.

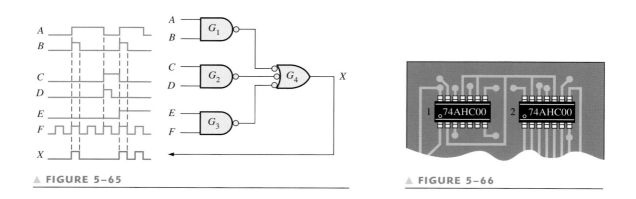

▲ FIGURE 5–65   ▲ FIGURE 5–66

46. Figure 5–67(a) is a logic circuit under test. Figure 5–67(b) shows the waveforms as observed on a logic analyzer. The output waveform is incorrect for the inputs that are applied to the circuit. Assuming that one gate in the circuit has failed, with its output either an apparent constant HIGH or a constant LOW, determine the faulty gate and the type of failure.

◄ FIGURE 5–67

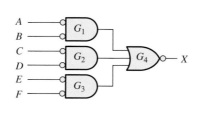   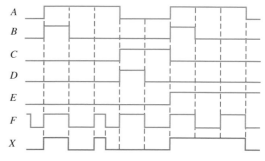

(a)   (b)

47. The logic circuit in Figure 5–68 has the input waveforms shown.

 (a) Determine the correct output waveform in relation to the inputs.

 (b) Determine the output waveform if the output of gate G_3 is open.

 (c) Determine the output waveform if the upper input to gate G_5 is shorted to ground.

▶ **FIGURE 5–68**

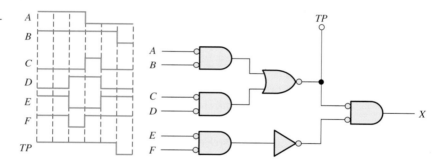

48. The logic circuit in Figure 5–69 has only one intermediate test point available besides the output, as indicated. For the inputs shown, you observe the indicated waveform at the test point. Is this waveform correct? If not, what are the possible faults that would cause it to appear as it does?

▶ **FIGURE 5–69**

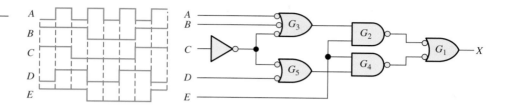

Digital System Application

49. Implement the inlet valve logic in Figure 5–49(b) using NOR gates and inverters.

50. Repeat Problem 49 for the outlet valve logic in Figure 5–50(b).

51. Implement the heater logic and the alarm logic using NOR gates and inverters.

Special Design Problems

52. Design a logic circuit to produce a HIGH output only if the input, represented by a 4-bit binary number, is greater than twelve or less than three. First develop the truth table and then draw the logic diagram.

53. Develop the logic circuit necessary to meet the following requirements:

 A battery-powered lamp in a room is to be operated from two switches, one at the back door and one at the front door. The lamp is to be on if the front switch is on and the back switch is off, or if the front switch is off and the back switch is on. The lamp is to be off if both switches are off or if both switches are on. Let a HIGH output represent the on condition and a LOW output represent the off condition.

54. Design a circuit to enable a chemical additive to be introduced into the fluid through another inlet only when the temperature is not too cold or too hot and the fluid is above the high-level sensor.

55. Develop the NAND logic for a hexadecimal keypad encoder that will convert each key closure to binary.

Multisim Troubleshooting Practice

56. Open file P05-56 and test the logic circuit to determine if there is a fault. If there is a fault, identify it if possible.

57. Open file P05-57 and test the logic circuit to determine if there is a fault. If there is a fault, identify it if possible.

58. Open file P05-58 and test the logic circuit to determine if there is a fault. If there is a fault, identify it if possible.

59. Open file P05-59 and test the logic circuit to determine if there is a fault. If there is a fault, identify it if possible.

ANSWERS

SECTION REVIEWS

SECTION 5–1   Basic Combinational Logic Circuits

1. (a) $\overline{AB + CD} = \overline{1 \cdot 0 + 1 \cdot 0} = 1$ (b) $\overline{AB + CD} = \overline{1 \cdot 1 + 0 \cdot 1} = 0$
   (c) $\overline{AB + CD} = \overline{0 \cdot 1 + 1 \cdot 1} = 0$

2. (a) $A\overline{B} + \overline{A}B = 1 \cdot \overline{0} + \overline{1} \cdot 0 = 1$ (b) $A\overline{B} + \overline{A}B = 1 \cdot \overline{1} + \overline{1} \cdot 1 = 0$
   (c) $A\overline{B} + \overline{A}B = 0 \cdot \overline{1} + \overline{0} \cdot 1 = 1$ (d) $A\overline{B} + \overline{A}B = 0 \cdot \overline{0} + \overline{0} \cdot 0 = 0$

3. $X = 1$ when $ABC = 000, 011, 101, 110,$ and $111; X = 0$ when $ABC = 001, 010,$ and 100

4. $X = AB + \overline{A}\,\overline{B}$; the circuit consists of two AND gates, one OR gate, and two inverters. See Figure 5–6(b) for diagram.

SECTION 5–2   Implementing Combinational Logic

1. (a) $X = ABC + AB + AC$: three AND gates, one OR gate
   (b) $X = AB(C + DE)$: three AND gates, one OR gate

2. $X = ABC + \overline{A}\,\overline{B}\,\overline{C}$; two AND gates, one OR gate, and three inverters

3. (a) $X = AB(C + 1) + AC = AB + AC$ (b) $X = AB(C + DE) = ABC + ABDE$

SECTION 5–3   The Universal Property of NAND and NOR Gates

1. (a) $X = \overline{A} + B$: a 2-input NAND gate with A and $\overline{B}$ on its inputs.
   (b) $X = A\overline{B}$: a 2-input NAND with A and $\overline{B}$ on its inputs, followed by one NAND used as an inverter.

2. (a) $X = \overline{A} + B$: a 2-input NOR with inputs $\overline{A}$ and B, followed by one NOR used as an inverter.
   (b) $X = A\overline{B}$: a 2-input NOR with $\overline{A}$ and B on its inputs.

SECTION 5–4   Combinational Logic Using NAND and NOR Gates

1. $X = \overline{(\overline{A} + \overline{B} + \overline{C})DE}$: a 3-input NAND with inputs, A, B, and C, with its output connected to a second 3-input NAND with two other inputs, D and E

2. $X = \overline{\overline{A}\,\overline{B}\,\overline{C} + (D + E)}$: a 3-input NOR with inputs A, B, and C, with its output connected to a second 3-input NOR with two other inputs, D and E

SECTION 5–5   Logic Circuit Operation with Pulse Waveform Inputs

1. The exclusive-OR output is a 15 μs pulse followed by a 25 μs pulse, with a separation of 10 μs between the pulses.

2. The output of the exclusive-NOR is HIGH when both inputs are HIGH or when both inputs are LOW.

SECTION 5–6   Combinational Logic with VHDL (optional)

1. A VHDL component is a predefined program describing a specified logic function.

2. A component instantiation is used to call for a specified component in a program architecture.

3. Interconnections between components are made using VHDL signals.

4. Components are used in the structural approach.

SECTION 5–7 **Troubleshooting**

1. Common gate failures are input or output open; input or output shorted to ground.

2. Input shorted to V_{CC} causes output to be stuck LOW.

3. (a) G_4 output is HIGH until rising edge of seventh pulse, then it goes LOW.

(b) G_4 output is the same as input D.

(c) G_4 output is the inverse of the G_2 output shown in Figure 5-47(b).

RELATED PROBLEMS FOR EXAMPLES

5–1 $X = AB + AC + BC$

5–2 $X = \overline{AB + AC + BC}$

If $A = 0$ and $B = 0$, $X = \overline{0 \cdot 0 + 0 \cdot 1 + 0 \cdot 1} = \overline{0} = 1$

If $A = 0$ and $C = 0$, $X = \overline{0 \cdot 1 + 0 \cdot 0 + 1 \cdot 0} = \overline{0} = 1$

If $B = 0$ and $C = 0$, $X = \overline{1 \cdot 0 + 1 \cdot 0 + 0 \cdot 0} = \overline{0} = 1$

5–3 Cannot be simplified   **5–4** Cannot be simplified

5–5 $X = A + B + C + D$ is valid.

5–6 See Figure 5–70.

▶ **FIGURE 5–70**

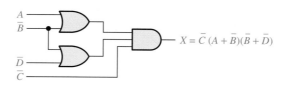

$$X = \overline{C}\,(A + \overline{B})(\overline{B} + \overline{D})$$

5–7 $X = \overline{(\overline{ABC})(\overline{DEF})} = (\overline{AB})C + (\overline{DE})F = (\overline{A} + \overline{B})C + (\overline{D} + \overline{E})F$

5–8 See Figure 5–71.

▶ **FIGURE 5–71**

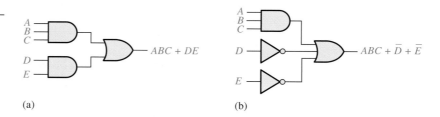

(a)   (b)

5–9 $X = \overline{\overline{(\overline{A + B + C})} + \overline{(\overline{D + E + F})}} = (\overline{A + B} + C)(\overline{D + E} + F) = (\overline{A}\,\overline{B} + C)(\overline{D}\,\overline{E} + F)$

5–10 See Figure 5–72.   **5–11** See Figure 5–73.

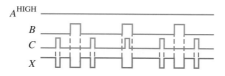

▲ **FIGURE 5–72**   ▲ **FIGURE 5–73**

5–12 See Figure 5–74. **5–13** See Figure 5–75.

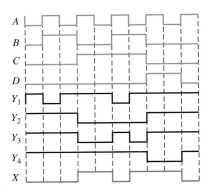

▲ **FIGURE 5–74**

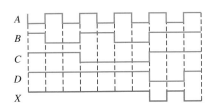

▲ **FIGURE 5–75**

5–14 G5: NAND_gate2 **port map** (A => IN9, B => IN10, X => OUT4);

5–15 See Figure 5–76.

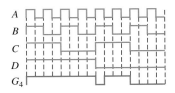

▲ **FIGURE 5–76**

SELF-TEST

1. (d) **2.** (b) **3.** (c) **4.** (a) **5.** (d) **6.** (b) **7.** (a) **8.** (d)

9. (d) **10.** (e) **11.** (e) **12.** (c)

6

FUNCTIONS OF COMBINATIONAL LOGIC

CHAPTER OBJECTIVES

■ Distinguish between half-adders and full-adders

■ Use full-adders to implement multibit parallel binary adders

■ Explain the differences between ripple carry and look-ahead carry parallel adders

■ Use the magnitude comparator to determine the relationship between two binary numbers and use cascaded comparators to handle the comparison of larger numbers

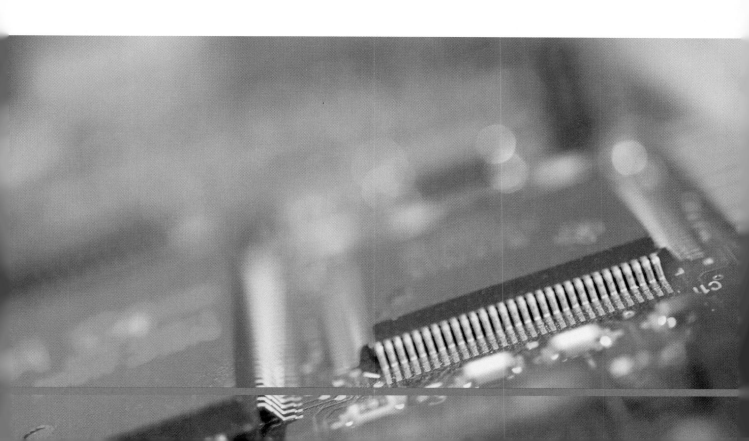

- Implement a basic binary decoder
- Use BCD-to-7-segment decoders in display systems
- Apply a decimal-to-BCD priority encoder in a simple keyboard application
- Convert from binary to Gray code, and Gray code to binary by using logic devices
- Apply multiplexers in data selection, multiplexed displays, logic function generation, and simple communications systems
- Use decoders as demultiplexers
- Explain the meaning of parity
- Use parity generators and checkers to detect bit errors in digital systems
- Implement a simple data communications system
- Identify glitches, common bugs in digital systems

KEY TERMS

- Half-adder
- Full-adder
- Cascading
- Ripple carry
- Look-ahead carry
- Decoder

- Encoder
- Priority encoder
- Multiplexer (MUX)
- Demultiplexer (DEMUX)
- Parity bit
- Glitch

INTRODUCTION

In this chapter, several types of combinational logic circuits are introduced including adders, comparators, decoders, encoders, code converters, multiplexers (data selectors), demultiplexers, and parity generators/checkers. Examples of fixed-function IC devices are included.

FIXED-FUNCTION LOGIC DEVICES

74XX42	74XX47	74XX85
74XX138	74XX139	74XX147
74XX148	74XX151	74XX154
74XX157	74XX280	74XX283

■■■ DIGITAL SYSTEM APPLICATION PREVIEW

The Digital System Application illustrates concepts from this chapter and deals with one portion of a traffic light control system. The system applications in Chapters 6, 7, and 8 focus on various parts of the traffic light control system. Basically, this system controls the traffic light at the intersection of a busy street and a lightly traveled side street. The system includes a combinational logic section to which the topics in this chapter apply, a timing circuit section to which Chapter 7 applies, and a sequential logic section to which Chapter 8 applies.

WWW. VISIT THE COMPANION WEBSITE
Study aids for this chapter are available at
http://www.prenhall.com/floyd

6–1 BASIC ADDERS

Adders are important in computers and also in other types of digital systems in which numerical data are processed. An understanding of the basic adder operation is fundamental to the study of digital systems. In this section, the half-adder and the full-adder are introduced.

After completing this section, you should be able to

- Describe the function of a half-adder ■ Draw a half-adder logic diagram ■ Describe the function of the full-adder ■ Draw a full-adder logic diagram using half-adders ■ Implement a full-adder using AND-OR logic

The Half-Adder

A half-adder adds two bits and produces a sum and a carry output.

Recall the basic rules for binary addition as stated in Chapter 2.

$$0 + 0 = 0$$
$$0 + 1 = 1$$
$$1 + 0 = 1$$
$$1 + 1 = 10$$

The operations are performed by a logic circuit called a **half-adder.**

The half-adder accepts two binary digits on its inputs and produces two binary digits on its outputs, a sum bit and a carry bit.

A half-adder is represented by the logic symbol in Figure 6–1.

▶ **FIGURE 6–1**

Logic symbol for a half-adder. Open file F06-01 to verify operation.

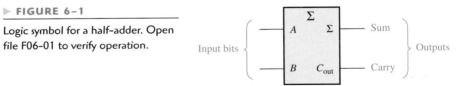

Half-Adder Logic From the operation of the half-adder as stated in Table 6–1, expressions can be derived for the sum and the output carry as functions of the inputs. Notice that the output carry (C_{out}) is a 1 only when both A and B are 1s; therefore, C_{out} can be expressed as the AND of the input variables.

Equation 6–1

$$C_{out} = AB$$

▶ **TABLE 6–1**

Half-adder truth table.

A	B	C_{out}	Σ
0	0	0	0
0	1	0	1
1	0	0	1
1	1	1	0

$Σ$ = sum
C_{out} = output carry
A and B = input variables (operands)

Now observe that the sum output (Σ) is a 1 only if the input variables, *A* and *B*, are not equal. The sum can therefore be expressed as the exclusive-OR of the input variables.

$$\Sigma = A \oplus B$$

Equation 6–2

From Equations 6–1 and 6–2, the logic implementation required for the half-adder function can be developed. The output carry is produced with an AND gate with *A* and *B* on the inputs, and the sum output is generated with an exclusive-OR gate, as shown in Figure 6–2. Remember that the exclusive-OR is implemented with AND gates, an OR gate, and inverters.

$\Sigma = A \oplus B = A\bar{B} + \bar{A}B$

$C_{out} = AB$

A
B

◀ **FIGURE 6–2**

Half-adder logic diagram.

The Full-Adder

The second category of adder is the **full-adder.**

> **The full-adder accepts two input bits and an input carry and generates a sum output and an output carry.**

A full-adder has an input carry while the half-adder does not.

The basic difference between a full-adder and a half-adder is that the full-adder accepts an input carry. A logic symbol for a full-adder is shown in Figure 6–3, and the truth table in Table 6–2 shows the operation of a full-adder.

Input bits {
A
B

Input carry C_{in}

Σ

Σ — Sum

C_{out} — Output carry

◀ **FIGURE 6–3**

Logic symbol for a full-adder. Open file F06-03 to verify operation.

◀ **TABLE 6–2**

Full-adder truth table.

A	B	C_{in}	C_{out}	Σ
0	0	0	0	0
0	0	1	0	1
0	1	0	0	1
0	1	1	1	0
1	0	0	0	1
1	0	1	1	0
1	1	0	1	0
1	1	1	1	1

C_{in} = input carry, sometimes designated as *CI*

C_{out} = output carry, sometimes designated as *CO*

Σ = sum

A and *B* = input variables (operands)

Full-Adder Logic The full-adder must add the two input bits and the input carry. From the half-adder you know that the sum of the input bits *A* and *B* is the exclusive-OR of those two variables, $A \oplus B$. For the input carry (C_{in}) to be added to the input bits, it must be exclusive-ORed with $A \oplus B$, yielding the equation for the sum output of the full-adder.

Equation 6–3

$$\Sigma = (A \oplus B) \oplus C_{in}$$

This means that to implement the full-adder sum function, two 2-input exclusive-OR gates can be used. The first must generate the term $A \oplus B$, and the second has as its inputs the output of the first XOR gate and the input carry, as illustrated in Figure 6–4(a).

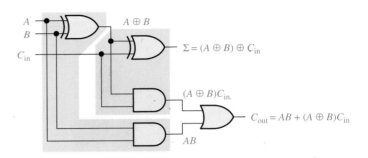

(a) Logic required to form the sum of three bits

(b) Complete logic circuit for a full-adder (each half-adder is enclosed by a shaded area)

▲ **FIGURE 6–4**

Full-adder logic. Open file F06-04 to verify operation.

The output carry is a 1 when both inputs to the first XOR gate are 1s or when both inputs to the second XOR gate are 1s. You can verify this fact by studying Table 6–2. The output carry of the full-adder is therefore produced by the inputs A ANDed with B and $A \oplus B$ ANDed with C_{in}. These two terms are ORed, as expressed in Equation 6–4. This function is implemented and combined with the sum logic to form a complete full-adder circuit, as shown in Figure 6–4(b).

Equation 6–4

$$C_{out} = AB + (A \oplus B)C_{in}$$

Notice in Figure 6–4(b) there are two half-adders, connected as shown in the block diagram of Figure 6–5(a), with their output carries ORed. The logic symbol shown in Figure 6–5(b) will normally be used to represent the full-adder.

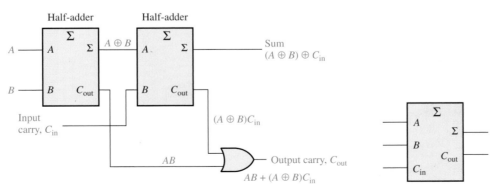

(a) Arrangement of two half-adders to form a full-adder

(b) Full-adder logic symbol

▲ **FIGURE 6–5**

Full-adder implemented with half-adders.

EXAMPLE 6–1

For each of the three full-adders in Figure 6–6, determine the outputs for the inputs shown.

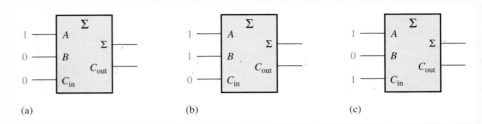

▲ FIGURE 6–6

Solution (a) The input bits are $A = 1$, $B = 0$, and $C_{in} = 0$.

$$1 + 0 + 0 = 1 \text{ with no carry}$$

Therefore, $\Sigma = \mathbf{1}$ and $C_{out} = \mathbf{0}$.

(b) The input bits are $A = 1$, $B = 1$, and $C_{in} = 0$.

$$1 + 1 + 0 = 0 \text{ with a carry of } 1$$

Therefore, $\Sigma = \mathbf{0}$ and $C_{out} = \mathbf{1}$.

(c) The input bits are $A = 1$, $B = 0$, and $C_{in} = 1$.

$$1 + 0 + 1 = 0 \text{ with a carry of } 1$$

Therefore, $\Sigma = \mathbf{0}$ and $C_{out} = \mathbf{1}$.

*Related Problem** What are the full-adder outputs for $A = 1$, $B = 1$, and $C_{in} = 1$?

*Answers are at the end of the chapter.

**SECTION 6–1
REVIEW**

Answers are at the end of the
chapter.

1. Determine the sum (Σ) and the output carry (C_{out}) of a half-adder for each set of input bits:
   (a) 01   (b) 00   (c) 10   (d) 11
2. A full-adder has $C_{in} = 1$. What are the sum (Σ) and the output carry (C_{out}) when $A = 1$ and $B = 1$?

6–2   PARALLEL BINARY ADDERS

Two or more full-adders are connected to form parallel binary adders. In this section, you will learn the basic operation of this type of adder and its associated input and output functions.

After completing this section, you should be able to

■ Use full-adders to implement a parallel binary adder ■ Explain the addition process in a parallel binary adder ■ Use the truth table for a 4-bit parallel adder ■ Apply two 74LS283s for the addition of two 4-bit numbers ■ Expand the 4-bit adder to accommodate 8-bit or 16-bit addition

As you saw in Section 6–1, a single full-adder is capable of adding two 1-bit numbers and an input carry. To add binary numbers with more than one bit, you must use additional full-adders. When one binary number is added to another, each column generates a sum bit and a 1 or 0 carry bit to the next column to the left, as illustrated here with 2-bit numbers.

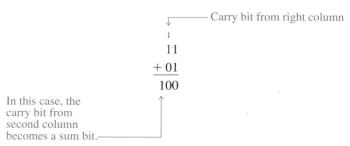

To add two binary numbers, a full-adder is required for each bit in the numbers. So for 2-bit numbers, two adders are needed; for 4-bit numbers, four adders are used; and so on. The carry output of each adder is connected to the carry input of the next higher-order adder, as shown in Figure 6–7 for a 2-bit adder. Notice that either a half-adder can be used for the least significant position or the carry input of a full-adder can be made 0 (grounded) because there is no carry input to the least significant bit position.

▶ **FIGURE 6–7**

Block diagram of a basic 2-bit parallel adder using two full-adders. Open file F06–07 to verify operation.

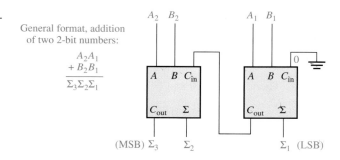

In Figure 6–7 the least significant bits (LSB) of the two numbers are represented by A_1 and B_1. The next higher-order bits are represented by A_2 and B_2. The three sum bits are Σ_1, Σ_2, and Σ_3. Notice that the output carry from the left-most full-adder becomes the most significant bit (MSB) in the sum, Σ_3.

EXAMPLE 6–2

Determine the sum generated by the 3-bit parallel adder in Figure 6–8 and show the intermediate carries when the binary numbers 101 and 011 are being added.

▶ **FIGURE 6–8**

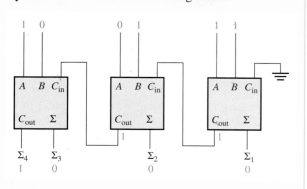

Solution The LSBs of the two numbers are added in the right-most full-adder. The sum bits and the intermediate carries are indicated in blue in Figure 6–8.

Related Problem What are the sum outputs when 111 and 101 are added by the 3-bit parallel adder?

Four-Bit Parallel Adders

A group of four bits is called a **nibble.** A basic 4-bit parallel adder is implemented with four full-adder stages as shown in Figure 6–9. Again, the LSBs (A_1 and B_1) in each number being added go into the right-most full-adder; the higher-order bits are applied as shown to the successively higher-order adders, with the MSBs (A_4 and B_4) in each number being applied to the left-most full-adder. The carry output of each adder is connected to the carry input of the next higher-order adder as indicated. These are called *internal carries.*

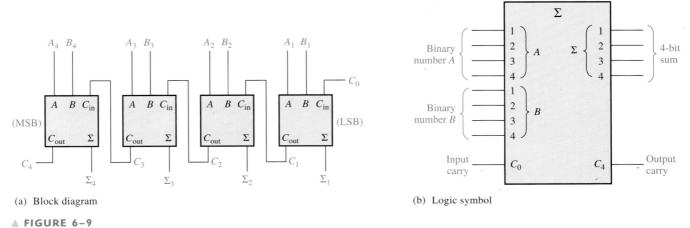

(a) Block diagram   (b) Logic symbol

▲ **FIGURE 6–9**

A 4–bit parallel adder.

In keeping with most manufacturers' data sheets, the input labeled C_0 is the input carry to the least significant bit adder; C_4, in the case of four bits, is the output carry of the most significant bit adder; and Σ_1 (LSB) through Σ_4 (MSB) are the sum outputs. The logic symbol is shown in Figure 6–9(b).

In terms of the method used to handle carries in a parallel adder, there are two types: the *ripple carry* adder and the *carry look-ahead* adder. These are discussed in Section 6–3.

Truth Table for a 4-Bit Parallel Adder

Table 6–3 is the truth table for a 4-bit adder. On some data sheets, truth tables may be called *function tables* or *functional truth tables.* The subscript *n* represents the adder bits and can

◀ **TABLE 6–3**

Truth table for each stage of a 4-bit parallel adder.

C_{n-1}	A_n	B_n	Σ_n	C_n
0	0	0	0	0
0	0	1	1	0
0	1	0	1	0
0	1	1	0	1
1	0	0	1	0
1	0	1	0	1
1	1	0	0	1
1	1	1	1	1

be 1, 2, 3, or 4 for the 4-bit adder. C_{n-1} is the carry from the previous adder. Carries C_1, C_2, and C_3 are generated internally. C_0 is an external carry input and C_4 is an output. Example 6–3 illustrates how to use Table 6–3.

EXAMPLE 6–3

Use the 4-bit parallel adder truth table (Table 6–3) to find the sum and output carry for the addition of the following two 4-bit numbers if the input carry (C_{n-1}) is 0:

$$A_4A_3A_2A_1 = 1100 \quad \text{and} \quad B_4B_3B_2B_1 = 1100$$

Solution   For $n = 1$: $A_1 = 0$, $B_1 = 0$, and $C_{n-1} = 0$. From the 1st row of the table,

$$\Sigma_1 = \mathbf{0} \quad \text{and} \quad C_1 = 0$$

For $n = 2$: $A_2 = 0$, $B_2 = 0$, and $C_{n-1} = 0$. From the 1st row of the table,

$$\Sigma_2 = \mathbf{0} \quad \text{and} \quad C_2 = 0$$

For $n = 3$: $A_3 = 1$, $B_3 = 1$, and $C_{n-1} = 0$. From the 4th row of the table,

$$\Sigma_3 = \mathbf{0} \quad \text{and} \quad C_3 = 1$$

For $n = 4$: $A_4 = 1$, $B_4 = 1$, and $C_{n-1} = 1$. From the last row of the table,

$$\Sigma_4 = \mathbf{1} \quad \text{and} \quad C_4 = \mathbf{1}$$

C_4 becomes the output carry; the sum of 1100 and 1100 is 11000.

Related Problem   Use the truth table (Table 6–3) to find the result of adding the binary numbers 1011 and 1010.

THE 74LS283 4-BIT PARALLEL ADDER

An example of a 4-bit parallel adder that is available in IC form is the 74LS283. For the 74LS283, V_{CC} is pin 16 and ground is pin 8, which is a standard configuration. The pin diagram and logic symbol for this device are shown, with pin numbers in parentheses on the logic symbol, in Figure 6–10. This device may be available in other TTL or CMOS families. Check the Texas Instruments website at www.ti.com or the TI CD-ROM accompanying this book.

▶ **FIGURE 6–10**

Four-bit parallel adder.

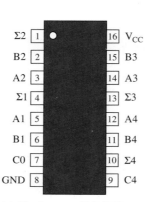

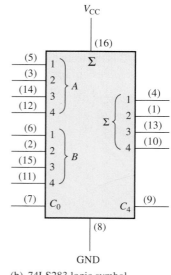

(a) Pin diagram of 74LS283   (b) 74LS283 logic symbol

IC Data Sheet Characteristics Recall that logic gates have one specified propagation delay time, t_P, from an input to the output. For IC logic, there may be several different specifications for t_P. The 4-bit parallel adder has the four t_P specifications shown in Figure 6–11, which is part of a 74LS283 data sheet.

Symbol	Parameter	Min	Typ	Max	Unit
			Limits		
t_{PLH} t_{PHL}	Propagation delay, C_0 input to any Σ output		16 15	24 24	ns
t_{PLH} t_{PHL}	Propagation delay, any A or B input to Σ outputs		15 15	24 24	ns
t_{PLH} t_{PHL}	Propagation delay, C_0 input to C_4 output		11 11	17 22	ns
t_{PLH} t_{PHL}	Propagation delay, any A or B input to C_4 output		11 12	17 17	ns

◀ **FIGURE 6–11**

Propagation delay characteristics for the 74LS283.

Adder Expansion

The 4-bit parallel adder can be expanded to handle the addition of two 8-bit numbers by using two 4-bit adders. The carry input of the low-order adder (C_0) is connected to ground because there is no carry into the least significant bit position, and the carry output of the low-order adder is connected to the carry input of the high-order adder, as shown in Figure 6–12(a). This

Adders can be expanded to handle more bits by cascading.

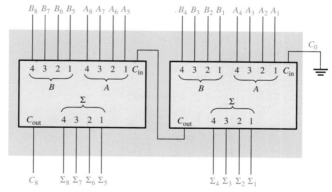

(a) Cascading of two 4-bit adders to form an 8-bit adder

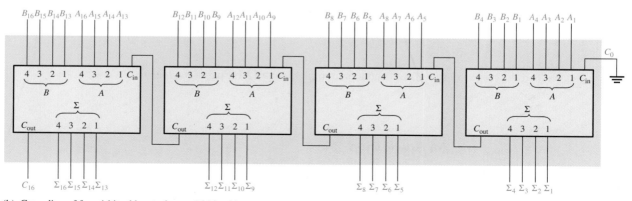

(b) Cascading of four 4-bit adders to form a 16-bit adder

▲ **FIGURE 6–12**

Examples of adder expansion.

process is known as **cascading.** Notice that, in this case, the output carry is designated C_8 because it is generated from the eighth bit position. The low-order adder is the one that adds the lower or less significant four bits in the numbers, and the high-order adder is the one that adds the higher or more significant four bits in the 8-bit numbers.

Similarly, four 4-bit adders can be cascaded to handle two 16-bit numbers as shown in Figure 6–12(b). Notice that the output carry is designated C_{16} because it is generated from the sixteenth bit position.

EXAMPLE 6–4

Show how two 74LS283 adders can be connected to form an 8-bit parallel adder. Show output bits for the following 8-bit input numbers:

$$A_8A_7A_6A_5A_4A_3A_2A_1 = 1011\,1001 \quad \text{and} \quad B_8B_7B_6B_5B_4B_3B_2B_1 = 1001\,1110$$

Solution   Two 74LS283 4-bit parallel adders are used to implement the 8-bit adder. The only connection between the two 74LS283s is the carry output (pin 9) of the low-order adder to the carry input (pin 7) of the high-order adder, as shown in Figure 6–13. Pin 7 of the low-order adder is grounded (no carry input).

The sum of the two 8-bit numbers is

$$\Sigma_9\Sigma_8\Sigma_7\Sigma_6\Sigma_5\Sigma_4\Sigma_3\Sigma_2\Sigma_1 = 101010111$$

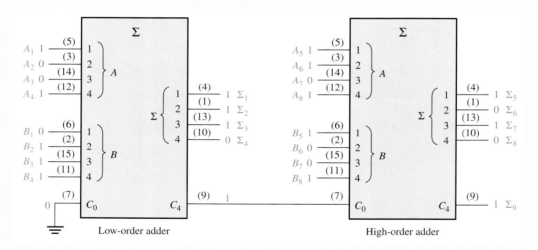

▲ **FIGURE 6–13**

Two 74LS283 adders connected as an 8-bit parallel adder (pin numbers are in parentheses).

Related Problem   Use 74LS283 adders to implement a 12-bit parallel adder.

An Application

An example of full-adder and parallel adder application is a simple voting system that can be used to simultaneously provide the number of "yes" votes and the number of "no" votes. This type of system can be used where a group of people are assembled and there is a need for immediately determining opinions (for or against), making decisions, or voting on certain issues or other matters.

In its simplest form, the system includes a switch for "yes" or "no" selection at each position in the assembly and a digital display for the number of yes votes and one for the number of no votes. The basic system is shown in Figure 6–14 for a 6-position setup, but it can be expanded to any number of positions with additional 6-position modules and additional parallel adder and display circuits.

In Figure 6–14 each full-adder can produce the sum of up to three votes. The sum and output carry of each full-adder then goes to the two lower-order inputs of a parallel binary adder. The two higher-order inputs of the parallel adder are connected to ground (0) because there is never a case where the binary input exceeds 0011 (decimal 3). For this

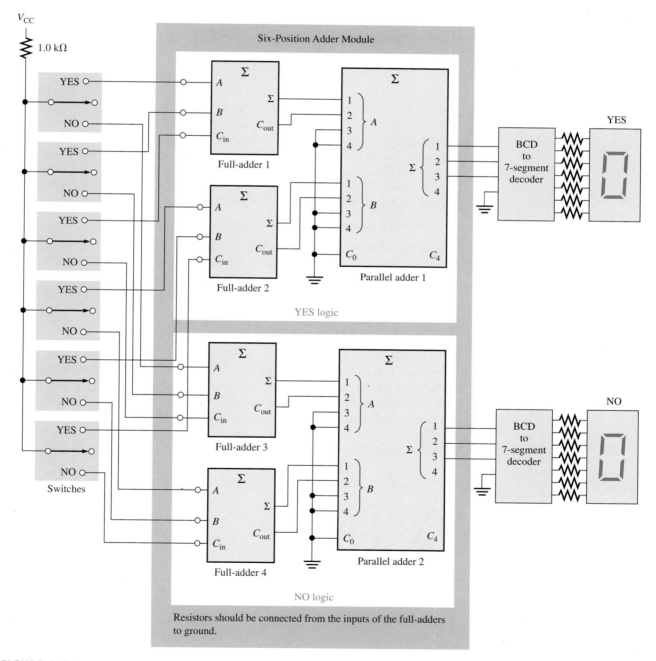

▲ FIGURE 6–14

A voting system using full-adders and parallel binary adders.

basic 6-position system, the outputs of the parallel adder go to a BCD-to-7-segment decoder that drives the 7-segment display. As mentioned, additional circuits must be included when the system is expanded.

The resistors from the inputs of each full-adder to ground assure that each input is LOW when the switch is in the neutral position (CMOS logic is used). When a switch is moved to the "yes" or to the "no" position, a HIGH level (V_{CC}) is applied to the associated full-adder input.

SECTION 6–2 REVIEW

1. Two 4-bit numbers (1101 and 1011) are applied to a 4-bit parallel adder. The input carry is 1. Determine the sum (Σ) and the output carry.

2. How many 74LS283 adders would be required to add two binary numbers each representing decimal numbers up through 1000_{10}?

6–3 RIPPLE CARRY VERSUS LOOK-AHEAD CARRY ADDERS

As mentioned in the last section, parallel adders can be placed into two categories based on the way in which internal carries from stage to stage are handled. Those categories are ripple carry and look-ahead carry. Externally, both types of adders are the same in terms of inputs and outputs. The difference is the speed at which they can add numbers. The look-ahead carry adder is much faster than the ripple carry adder.

After completing this section, you should be able to

■ Discuss the difference between a ripple carry adder and a look-ahead carry adder
■ State the advantage of look-ahead carry addition ■ Define *carry generation* and *carry propagation* and explain the difference ■ Develop look-ahead carry logic ■ Explain why cascaded 74LS283s exhibit both ripple carry and look-ahead carry properties

The Ripple Carry Adder

A **ripple carry** adder is one in which the carry output of each full-adder is connected to the carry input of the next higher-order stage (a stage is one full-adder). The sum and the output carry of any stage cannot be produced until the input carry occurs; this causes a time delay in the addition process, as illustrated in Figure 6–15. The carry propagation delay for each full-adder is the time from the application of the input carry until the output carry occurs, assuming that the *A* and *B* inputs are already present.

▶ **FIGURE 6–15**

A 4-bit parallel ripple carry adder showing "worst-case" carry propagation delays.

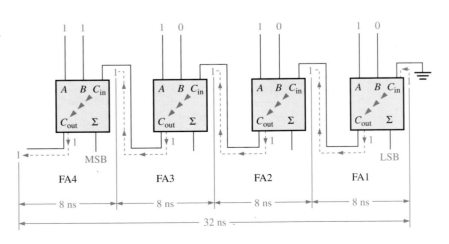

Full-adder 1 (FA1) cannot produce a potential output carry until an input carry is applied. Full-adder 2 (FA2) cannot produce a potential output carry until full-adder 1 produces an output carry. Full-adder 3 (FA3) cannot produce a potential output carry until an output carry is produced by FA1 followed by an output carry from FA2, and so on. As you can see in Figure 6–15, the input carry to the least significant stage has to ripple through all the adders before a final sum is produced. The cumulative delay through all the adder stages is a "worst-case" addition time. The total delay can vary, depending on the carry bit produced by each full-adder. If two numbers are added such that no carries (0) occur between stages, the addition time is simply the propagation time through a single full-adder from the application of the data bits on the inputs to the occurrence of a sum output.

The Look-Ahead Carry Adder

The speed with which an addition can be performed is limited by the time required for the carries to propagate, or ripple, through all the stages of a parallel adder. One method of speeding up the addition process by eliminating this ripple carry delay is called **look-ahead carry** addition. The look-ahead carry adder anticipates the output carry of each stage, and based on the inputs, produces the output carry by either carry generation or carry propagation.

Carry generation occurs when an output carry is produced (generated) internally by the full-adder. A carry is generated only when both input bits are 1s. The generated carry, C_g, is expressed as the AND function of the two input bits, A and B.

$$C_g = AB$$

Equation 6–5

Carry propagation occurs when the input carry is rippled to become the output carry. An input carry may be propagated by the full-adder when either or both of the input bits are 1s. The propagated carry, C_p, is expressed as the OR function of the input bits.

$$C_p = A + B$$

Equation 6–6

The conditions for carry generation and carry propagation are illustrated in Figure 6–16. The three arrowheads symbolize ripple (propagation).

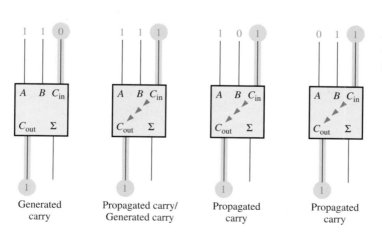

◄ FIGURE 6–16

Illustration of conditions for carry generation and carry propagation.

Generated carry Propagated carry/Generated carry Propagated carry Propagated carry

The output carry of a full-adder can be expressed in terms of both the generated carry (C_g) and the propagated carry (C_p). The output carry (C_{out}) is a 1 if the generated carry is a 1 OR if the propagated carry is a 1 AND the input carry (C_{in}) is a 1. In other words, we get an output carry of 1 if it is generated by the full-adder ($A = 1$ AND $B = 1$) or if the adder propagates the input carry ($A = 1$ OR $B = 1$) AND $C_{in} = 1$. This relationship is expressed as

$$C_{out} = C_g + C_p C_{in}$$

Equation 6–7

Now let's see how this concept can be applied to a parallel adder, whose individual stages are shown in Figure 6–17 for a 4-bit example. For each full-adder, the output carry is dependent on the generated carry (C_g), the propagated carry (C_p), and its input carry (C_{in}). The C_g and C_p functions for each stage are *immediately* available as soon as the input bits A and B and the input carry to the LSB adder are applied because they are dependent only on these bits. The input carry to each stage is the output carry of the previous stage.

▶ FIGURE 6–17

Carry generation and carry propagation in terms of the input bits to a 4-bit adder.

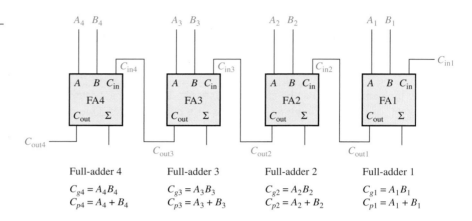

Full-adder 4
$$C_{g4} = A_4 B_4$$
$$C_{p4} = A_4 + B_4$$

Full-adder 3
$$C_{g3} = A_3 B_3$$
$$C_{p3} = A_3 + B_3$$

Full-adder 2
$$C_{g2} = A_2 B_2$$
$$C_{p2} = A_2 + B_2$$

Full-adder 1
$$C_{g1} = A_1 B_1$$
$$C_{p1} = A_1 + B_1$$

Based on this analysis, we can now develop expressions for the output carry, C_{out}, of each full-adder stage for the 4-bit example.

Full-adder 1:

$$C_{out1} = C_{g1} + C_{p1} C_{in1}$$

Full-adder 2:

$$C_{in2} = C_{out1}$$
$$C_{out2} = C_{g2} + C_{p2} C_{in2} = C_{g2} + C_{p2} C_{out1} = C_{g2} + C_{p2}(C_{g1} + C_{p1} C_{in1})$$
$$= C_{g2} + C_{p2} C_{g1} + C_{p2} C_{p1} C_{in1}$$

Full-adder 3:

$$C_{in3} = C_{out2}$$
$$C_{out3} = C_{g3} + C_{p3} C_{in3} = C_{g3} + C_{p3} C_{out2} = C_{g3} + C_{p3}(C_{g2} + C_{p2} C_{g1} + C_{p2} C_{p1} C_{in1})$$
$$= C_{g3} + C_{p3} C_{g2} + C_{p3} C_{p2} C_{g1} + C_{p3} C_{p2} C_{p1} C_{in1}$$

Full-adder 4:

$$C_{in4} = C_{out3}$$
$$C_{out4} = C_{g4} + C_{p4} C_{in4} = C_{g4} + C_{p4} C_{out3}$$
$$= C_{g4} + C_{p4}(C_{g3} + C_{p3} C_{g2} + C_{p3} C_{p2} C_{g1} + C_{p3} C_{p2} C_{p1} C_{in1})$$
$$= C_{g4} + C_{p4} C_{g3} + C_{p4} C_{p3} C_{g2} + C_{p4} C_{p3} C_{p2} C_{g1} + C_{p4} C_{p3} C_{p2} C_{p1} C_{in1}$$

Notice that in each of these expressions, the output carry for each full-adder stage is dependent only on the initial input carry (C_{in1}), the C_g and C_p functions of that stage, and the C_g and C_p functions of the preceding stages. Since each of the C_g and C_p functions can be expressed in terms of the A and B inputs to the full-adders, all the output carries are immediately available (except for gate delays), and you do not have to wait for a carry to ripple through all the stages before a final result is achieved. Thus, the look-ahead carry technique speeds up the addition process.

The C_{out} equations are implemented with logic gates and connected to the full-adders to create a 4-bit look-ahead carry adder, as shown in Figure 6–18.

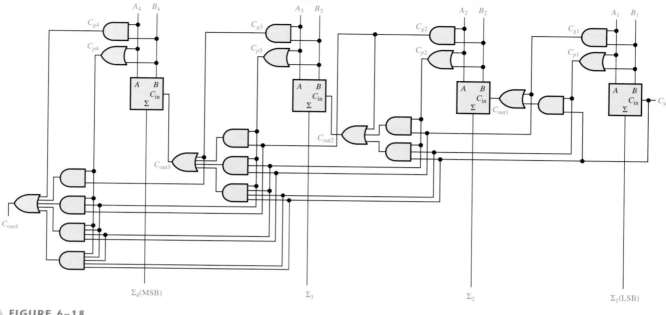

▲ FIGURE 6–18

Logic diagram for a 4-stage look-ahead carry adder.

Combination Look-Ahead and Ripple Carry Adders

The 74LS283 4-bit adder that was introduced in Section 6–2 is a look-ahead carry adder. When these adders are cascaded to expand their capability to handle binary numbers with more than four bits, the output carry of one adder is connected to the input carry of the next. This creates a ripple carry condition between the 4-bit adders so that when two or more 74LS283s are cascaded, the resulting adder is actually a combination look-ahead and ripple carry adder. The look-ahead carry operation is internal to each MSI adder and the ripple carry feature comes into play when there is a carry out of one of the adders to the next one.

**SECTION 6–3
REVIEW**

1. The input bits to a full-adder are $A = 1$ and $B = 0$. Determine C_g and C_p.
2. Determine the output carry of a full-adder when $C_{in} = 1$, $C_g = 0$, and $C_p = 1$.

6–4 COMPARATORS

The basic function of a comparator is to compare the magnitudes of two binary quantities to determine the relationship of those quantities. In its simplest form, a comparator circuit determines whether two numbers are equal.

After completing this section, you should be able to

■ Use the exclusive-OR gate as a basic comparator ■ Analyze the internal logic of a magnitude comparator that has both equality and inequality outputs ■ Apply the 74HC85 comparator to compare the magnitudes of two 4-bit numbers ■ Cascade 74HC85s to expand a comparator to eight or more bits

Equality

As you learned in Chapter 3, the exclusive-OR gate can be used as a basic comparator because its output is a 1 if the two input bits are not equal and a 0 if the input bits are equal. Figure 6–19 shows the exclusive-OR gate as a 2-bit comparator.

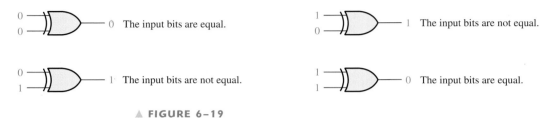

▲ **FIGURE 6–19**

Basic comparator operation.

In order to compare binary numbers containing two bits each, an additional exclusive-OR gate is necessary. The two least significant bits (LSBs) of the two numbers are compared by gate G_1, and the two most significant bits (MSBs) are compared by gate G_2, as shown in Figure 6–20. If the two numbers are equal, their corresponding bits are the same, and the output of each exclusive-OR gate is a 0. If the corresponding sets of bits are not equal, a 1 occurs on that exclusive-OR gate output.

▶ **FIGURE 6–20**

Logic diagram for equality comparison of two 2-bit numbers. Open file F06-20 to verify operation.

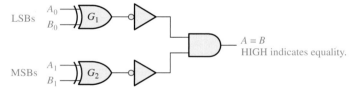

General format: Binary number $A \rightarrow A_1 A_0$
Binary number $B \rightarrow B_1 B_0$

A comparator determines if two binary numbers are equal or unequal.

In order to produce a single output indicating an equality or inequality of two numbers, two inverters and an AND gate can be used, as shown in Figure 6–20. The output of each exclusive-OR gate is inverted and applied to the AND gate input. When the two input bits for each exclusive-OR are equal, the corresponding bits of the numbers are equal, producing a 1 on both inputs to the AND gate and thus a 1 on the output. When the two numbers are not equal, one or both sets of corresponding bits are unequal, and a 0 appears on at least one input to the AND gate to produce a 0 on its output. Thus, the output of the AND gate indicates equality (1) or inequality (0) of the two numbers.

Example 6–5 illustrates this operation for two specific cases. The exclusive-OR gate and inverter are replaced by an exclusive-NOR symbol.

EXAMPLE 6–5

Apply each of the following sets of binary numbers to the comparator inputs in Figure 6–21, and determine the output by following the logic levels through the circuit.

(a) 10 and 10   **(b)** 11 and 10

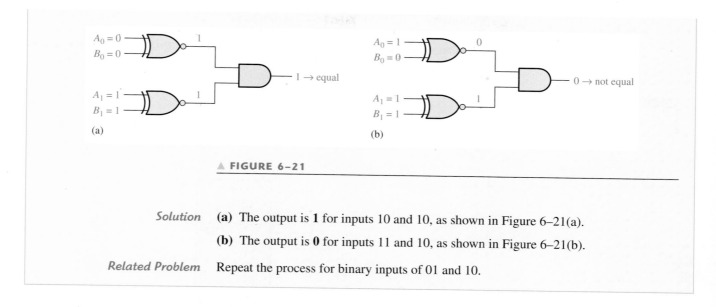

(a)

(b)

▲ FIGURE 6–21

Solution   **(a)** The output is **1** for inputs 10 and 10, as shown in Figure 6–21(a).

(b) The output is **0** for inputs 11 and 10, as shown in Figure 6–21(b).

Related Problem   Repeat the process for binary inputs of 01 and 10.

As you know from Chapter 3, the basic comparator can be expanded to any number of bits. The AND gate sets the condition that all corresponding bits of the two numbers must be equal if the two numbers themselves are equal.

Inequality

In addition to the equality output, many IC comparators provide additional outputs that indicate which of the two binary numbers being compared is the larger. That is, there is an output that indicates when number A is greater than number B ($A > B$) and an output that indicates when number A is less than number B ($A < B$), as shown in the logic symbol for a 4-bit comparator in Figure 6–22.

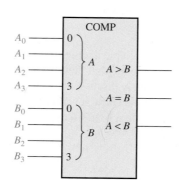

◀ FIGURE 6–22

Logic symbol for a 4-bit comparator with inequality indication.

In a computer, the *cache* is a very fast intermediate memory between the central processing unit (CPU) and the slower main memory. The CPU requests data by sending out its *address* (unique location) in memory. Part of this address is called a *tag*. The *tag address comparator* compares the tag from the CPU with the tag from the cache directory. If the two agree, the addressed data is already in the cache and is retrieved very quickly. If the tags disagree, the data must be retrieved from the main memory at a much slower rate.

To determine an inequality of binary numbers A and B, you first examine the highest-order bit in each number. The following conditions are possible:

1. If $A_3 = 1$ and $B_3 = 0$, number A is greater than number B.

2. If $A_3 = 0$ and $B_3 = 1$, number A is less than number B.

3. If $A_3 = B_3$, then you must examine the next lower bit position for an inequality.

These three operations are valid for each bit position in the numbers. The general procedure used in a comparator is to check for an inequality in a bit position, starting with the highest-order bits (MSBs). When such an inequality is found, the relationship of the two numbers is established, and any other inequalities in lower-order bit positions must be ignored because it is possible for an opposite indication to occur; *the highest-order indication must take precedence.*

EXAMPLE 6–6

Determine the $A = B$, $A > B$, and $A < B$ outputs for the input numbers shown on the comparator in Figure 6–23.

▶ FIGURE 6–23

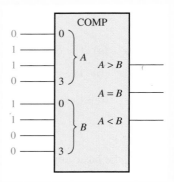

Solution   The number on the A inputs is 0110 and the number on the B inputs is 0011. The **$A > B$ output is HIGH and the other outputs are LOW.**

Related Problem   What are the comparator outputs when $A_3A_2A_1A_0 = 1001$ and $B_3B_2B_1B_0 = 1010$?

THE 74HC85 4-BIT MAGNITUDE COMPARATOR

The 74HC85 is a comparator that is also available in other IC families. The pin diagram and logic symbol are shown in Figure 6–24. Notice that this device has all the inputs and outputs of the generalized comparator previously discussed and, in addition, has three cascading inputs: $A < B$, $A = B$, $A > B$. These inputs allow several comparators to be cascaded for comparison of any number of bits greater than four. To expand the comparator, the $A < B$,

▶ FIGURE 6–24

Pin diagram and logic symbol for the 74HC85 4-bit magnitude comparator (pin numbers are in parentheses).

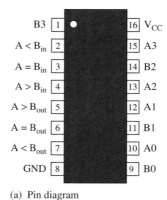

(a) Pin diagram   (b) Logic symbol

$A = B$, and $A > B$ outputs of the lower-order comparator are connected to the corresponding cascading inputs of the next higher-order comparator. The lowest-order comparator must have a HIGH on the $A = B$ input and LOWs on the $A < B$ and $A > B$ inputs. This device may be available in other CMOS or TTL families. Check the Texas Instruments website at www.ti.com or the TI CD-ROM accompanying this book.

EXAMPLE 6–7

Use 74HC85 comparators to compare the magnitudes of two 8-bit numbers. Show the comparators with proper interconnections.

Solution   Two 74HC85s are required to compare two 8-bit numbers. They are connected as shown in Figure 6–25 in a cascaded arrangement.

▶ **FIGURE 6–25**

An 8-bit magnitude comparator using two 74HC85s.

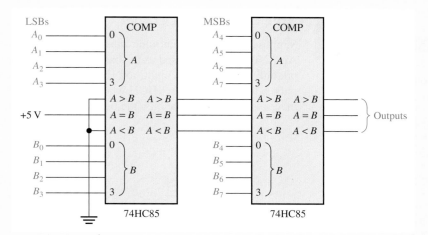

Related Problem   Expand the circuit in Figure 6–25 to a 16-bit comparator.

**SECTION 6–4
REVIEW**

1. The binary numbers $A = 1011$ and $B = 1010$ are applied to the inputs of a 74HC85. Determine the outputs.

2. The binary numbers $A = 11001011$ and $B = 11010100$ are applied to the 8-bit comparator in Figure 6–25. Determine the states of output pins 5, 6, and 7 on each 74HC85.

Most CMOS devices contain protection circuitry to guard against damage from high static voltages or electric fields. However, precautions must be taken to avoid applications of any voltages higher than maximum rated voltages. For proper operation, input and output voltages should be between ground and V_{CC}. Also, remember that unused inputs must always be connected to an appropriate logic level (ground or V_{CC}). Unused outputs may be left open.

6–5   DECODERS

A **decoder** is a digital circuit that detects the presence of a specified combination of bits (code) on its inputs and indicates the presence of that code by a specified output level. In its general form, a decoder has n input lines to handle n bits and from one to 2^n output lines to indicate the presence of one or more n-bit combinations. In this section, several decoders are introduced. The basic principles can be extended to other types of decoders.

After completing this section, you should be able to

- Define *decoder* ■ Design a logic circuit to decode any combination of bits ■ Describe the 74HC154 binary-to-decimal decoder ■ Expand decoders to accommodate larger numbers of bits in a code ■ Describe the 74LS47 BCD-to-7-segment decoder ■ Discuss zero suppression in 7-segment displays ■ Apply decoders to specific applications

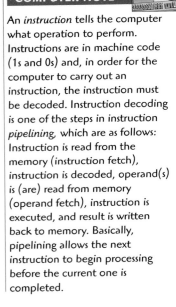

COMPUTER NOTE

An *instruction* tells the computer what operation to perform. Instructions are in machine code (1s and 0s) and, in order for the computer to carry out an instruction, the instruction must be decoded. Instruction decoding is one of the steps in instruction *pipelining,* which are as follows: Instruction is read from the memory (instruction fetch), instruction is decoded, operand(s) is (are) read from memory (operand fetch), instruction is executed, and result is written back to memory. Basically, pipelining allows the next instruction to begin processing before the current one is completed.

The Basic Binary Decoder

Suppose you need to determine when a binary 1001 occurs on the inputs of a digital circuit. An AND gate can be used as the basic decoding element because it produces a HIGH output only when all of its inputs are HIGH. Therefore, you must make sure that all of the inputs to the AND gate are HIGH when the binary number 1001 occurs; this can be done by inverting the two middle bits (the 0s), as shown in Figure 6–26.

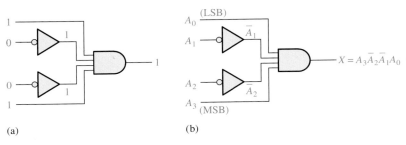

(a)   (b)

▲ **FIGURE 6–26**

Decoding logic for the binary code 1001 with an active-HIGH output.

The logic equation for the decoder of Figure 6–26(a) is developed as illustrated in Figure 6–26(b). You should verify that the output is 0 except when $A_0 = 1$, $A_1 = 0$, $A_2 = 0$, and $A_3 = 1$ are applied to the inputs. A_0 is the LSB and A_3 is the MSB. *In the representation of a binary number or other weighted code in this book, the LSB is the right-most bit in a horizontal arrangement and the topmost bit in a vertical arrangement, unless specified otherwise.*

If a NAND gate is used in place of the AND gate in Figure 6–26, a LOW output will indicate the presence of the proper binary code, which is 1001 in this case.

EXAMPLE 6–8

Determine the logic required to decode the binary number 1011 by producing a HIGH level on the output.

Solution   The decoding function can be formed by complementing only the variables that appear as 0 in the desired binary number, as follows:

$$X = A_3 \overline{A_2} A_1 A_0 \qquad (1011)$$

This function can be implemented by connecting the true (uncomplemented) variables A_0, A_1, and A_3 directly to the inputs of an AND gate, and inverting the variable A_2 before applying it to the AND gate input. The decoding logic is shown in Figure 6–27.

▶ **FIGURE 6–27**

Decoding logic for producing a HIGH output when 1011 is on the inputs.

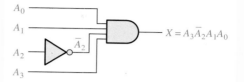

$X = A_3\overline{A_2}A_1A_0$

Related Problem   Develop the logic required to detect the binary code 10010 and produce an active-LOW output.

The 4-Bit Decoder

In order to decode all possible combinations of four bits, sixteen decoding gates are required ($2^4 = 16$). This type of decoder is commonly called either a *4-line-to-16-line decoder* because there are four inputs and sixteen outputs or a *1-of-16 decoder* because for any given code on the inputs, one of the sixteen outputs is activated. A list of the sixteen binary codes and their corresponding decoding functions is given in Table 6–4.

If an active-LOW output is required for each decoded number, the entire decoder can be implemented with NAND gates and inverters. In order to decode each of the sixteen binary codes, sixteen NAND gates are required (AND gates can be used to produce active-HIGH outputs).

▼ **TABLE 6–4**

Decoding functions and truth table for a 4-line-to-16-line (1-of-16) decoder with active-LOW outputs.

DECIMAL DIGIT	BINARY INPUTS A_3	A_2	A_1	A_0	DECODING FUNCTION	OUTPUTS 0	1	2	3	4	5	6	7	8	9	10	11	12	13	14	15
0	0	0	0	0	$\overline{A_3}\,\overline{A_2}\,\overline{A_1}\,\overline{A_0}$	0	1	1	1	1	1	1	1	1	1	1	1	1	1	1	1
1	0	0	0	1	$\overline{A_3}\,\overline{A_2}\,\overline{A_1}A_0$	1	0	1	1	1	1	1	1	1	1	1	1	1	1	1	1
2	0	0	1	0	$\overline{A_3}\,\overline{A_2}A_1\overline{A_0}$	1	1	0	1	1	1	1	1	1	1	1	1	1	1	1	1
3	0	0	1	1	$\overline{A_3}\,\overline{A_2}A_1A_0$	1	1	1	0	1	1	1	1	1	1	1	1	1	1	1	1
4	0	1	0	0	$\overline{A_3}A_2\overline{A_1}\,\overline{A_0}$	1	1	1	1	0	1	1	1	1	1	1	1	1	1	1	1
5	0	1	0	1	$\overline{A_3}A_2\overline{A_1}A_0$	1	1	1	1	1	0	1	1	1	1	1	1	1	1	1	1
6	0	1	1	0	$\overline{A_3}A_2A_1\overline{A_0}$	1	1	1	1	1	1	0	1	1	1	1	1	1	1	1	1
7	0	1	1	1	$\overline{A_3}A_2A_1A_0$	1	1	1	1	1	1	1	0	1	1	1	1	1	1	1	1
8	1	0	0	0	$A_3\overline{A_2}\,\overline{A_1}\,\overline{A_0}$	1	1	1	1	1	1	1	1	0	1	1	1	1	1	1	1
9	1	0	0	1	$A_3\overline{A_2}\,\overline{A_1}A_0$	1	1	1	1	1	1	1	1	1	0	1	1	1	1	1	1
10	1	0	1	0	$A_3\overline{A_2}A_1\overline{A_0}$	1	1	1	1	1	1	1	1	1	1	0	1	1	1	1	1
11	1	0	1	1	$A_3\overline{A_2}A_1A_0$	1	1	1	1	1	1	1	1	1	1	1	0	1	1	1	1
12	1	1	0	0	$A_3A_2\overline{A_1}\,\overline{A_0}$	1	1	1	1	1	1	1	1	1	1	1	1	0	1	1	1
13	1	1	0	1	$A_3A_2\overline{A_1}A_0$	1	1	1	1	1	1	1	1	1	1	1	1	1	0	1	1
14	1	1	1	0	$A_3A_2A_1\overline{A_0}$	1	1	1	1	1	1	1	1	1	1	1	1	1	1	0	1
15	1	1	1	1	$A_3A_2A_1A_0$	1	1	1	1	1	1	1	1	1	1	1	1	1	1	1	0

A logic symbol for a 4-line-to-16-line (1-of-16) decoder with active-LOW outputs is shown in Figure 6–28. The BIN/DEC label indicates that a binary input makes the corresponding decimal output active. The input labels 8, 4, 2, and 1 represent the binary weights of the input bits ($2^3 2^2 2^1 2^0$).

▶ **FIGURE 6–28**

Logic symbol for a 4-line-to-16-line (1-of-16) decoder. Open file F06-28 to verify operation.

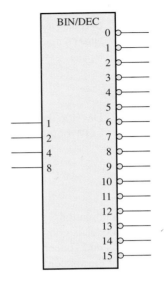

THE 74HC154 1-OF-16 DECODER

The 74HC154 is a good example of an IC decoder. The logic symbol is shown in Figure 6–29. There is an enable function (*EN*) provided on this device, which is implemented with a NOR gate used as a negative-AND. A LOW level on each chip select input, $\overline{CS}_1$ and $\overline{CS}_2$, is required in order to make the enable gate output (*EN*) HIGH. The enable gate output is

▶ **FIGURE 6–29**

Pin diagram and logic symbol for the 74HC154 1-of-16 decoder.

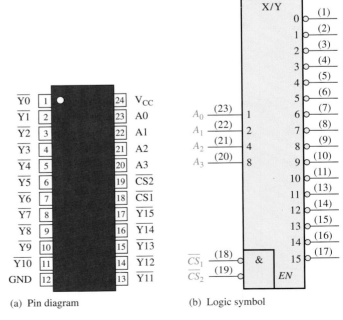

(a) Pin diagram (b) Logic symbol

connected to an input of *each* NAND gate in the decoder, so it must be HIGH for the NAND gates to be enabled. If the enable gate is not activated by a LOW on both inputs, then all sixteen decoder outputs (Y) will be HIGH regardless of the states of the four input variables, $A_0, A_1, A_2,$ and A_3. This device may be available in other CMOS or TTL families. Check the Texas Instruments website at www.ti.com or the TI CD-ROM accompanying this book.

EXAMPLE 6–9

A certain application requires that a 5-bit number be decoded. Use 74HC154 decoders to implement the logic. The binary number is represented by the format $A_4A_3A_2A_1A_0$.

Solution Since the 74HC154 can handle only four bits, two decoders must be used to decode five bits. The fifth bit, A_4, is connected to the chip select inputs, $\overline{CS}_1$ and $\overline{CS}_2$, of one decoder, and $\overline{A}_4$ is connected to the $\overline{CS}_1$ and $\overline{CS}_2$ inputs of the other decoder, as shown in Figure 6–30. When the decimal number is 15 or less, $A_4 = 0$, the low-order decoder is enabled, and the high-order decoder is disabled. When the decimal number is greater than 15, $A_4 = 1$ so $\overline{A}_4 = 0$, the high-order decoder is enabled, and the low-order decoder is disabled.

▶ **FIGURE 6–30**

A 5-bit decoder using 74HC154s.

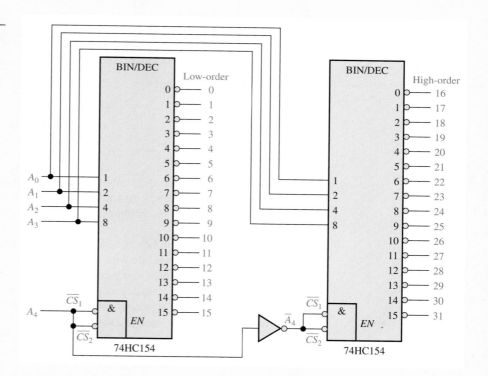

Related Problem Determine the output in Figure 6–30 that is activated for the binary input 10110.

An Application

Decoders are used in many types of applications. One example is in computers for input/output selection as depicted in the general diagram of Figure 6–31.

Computers must communicate with a variety of external devices called *peripherals* by sending and/or receiving data through what is known as input/output (I/O) ports. These

▶ FIGURE 6–31

A simplified computer I/O port system with a port address decoder with only four address lines shown.

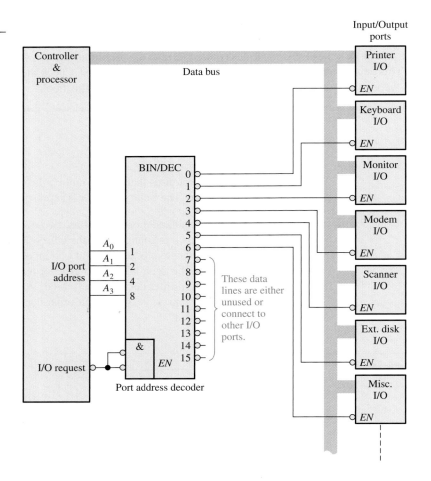

external devices include printers, modems, scanners, external disk drives, keyboard, video monitors, and other computers. As indicated in Figure 6–31, a decoder is used to select the I/O port as determined by the computer so that data can be sent or received from a specific external device.

Each I/O port has a number, called an address, which uniquely identifies it. When the computer wants to communicate with a particular device, it issues the appropriate address code for the I/O port to which that particular device is connected. This binary port address is decoded and the appropriate decoder output is activated to enable the I/O port.

As shown in Figure 6–31, binary data are transferred within the computer on a data bus, which is a set of parallel lines. For example, an 8-bit bus consists of eight parallel lines that can carry one byte of data at a time. The data bus goes to all of the I/O ports, but any data coming in or going out will only pass through the port that is enabled by the port address decoder.

The BCD-to-Decimal Decoder

The BCD-to-decimal decoder converts each BCD code (8421 code) into one of ten possible decimal digit indications. It is frequently referred as a *4-line-to-10-line decoder* or a *1-of-10 decoder*.

The method of implementation is the same as for the 1-of-16 decoder previously discussed, except that only ten decoding gates are required because the BCD code represents only the ten decimal digits 0 through 9. A list of the ten BCD codes and their corresponding decoding functions is given in Table 6–5. Each of these decoding functions is implemented with NAND gates to provide active-LOW outputs. If an active-HIGH output is

▶ TABLE 6–5

BCD decoding functions.

DECIMAL DIGIT	BCD CODE				DECODING FUNCTION
	A_3	A_2	A_1	A_0	
0	0	0	0	0	$\overline{A_3}\,\overline{A_2}\,\overline{A_1}\,\overline{A_0}$
1	0	0	0	1	$\overline{A_3}\,\overline{A_2}\,\overline{A_1}\,A_0$
2	0	0	1	0	$\overline{A_3}\,\overline{A_2}\,A_1\,\overline{A_0}$
3	0	0	1	1	$\overline{A_3}\,\overline{A_2}\,A_1\,A_0$
4	0	1	0	0	$\overline{A_3}\,A_2\,\overline{A_1}\,\overline{A_0}$
5	0	1	0	1	$\overline{A_3}\,A_2\,\overline{A_1}\,A_0$
6	0	1	1	0	$\overline{A_3}\,A_2\,A_1\,\overline{A_0}$
7	0	1	1	1	$\overline{A_3}\,A_2\,A_1\,A_0$
8	1	0	0	0	$A_3\,\overline{A_2}\,\overline{A_1}\,\overline{A_0}$
9	1	0	0	1	$A_3\,\overline{A_2}\,\overline{A_1}\,A_0$

required, AND gates are used for decoding. The logic is identical to that of the first ten decoding gates in the 1-of-16 decoder (see Table 6–4).

EXAMPLE 6–10

The 74HC42 is an integrated circuit BCD-to-decimal decoder. The logic symbol is shown in Figure 6–32. If the input waveforms in Figure 6–33(a) are applied to the inputs of the 74HC42, show the output waveforms.

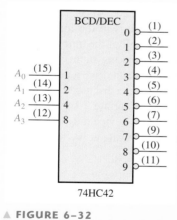

▲ FIGURE 6–32

The 74HC42 BCD-to-decimal decoder.

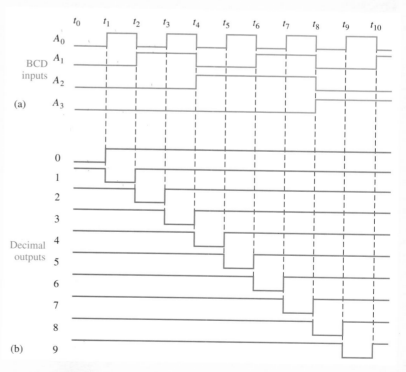

▲ FIGURE 6–33

Solution   The output waveforms are shown in Figure 6–33(b). As you can see, the inputs are sequenced through the BCD for digits 0 through 9. The output waveforms in the timing diagram indicate that sequence on the decimal-value outputs.

Related Problem   Construct a timing diagram showing input and output waveforms for the case where the BCD inputs sequence through the decimal numbers as follows: 0, 2, 4, 6, 8, 1, 3, 5, and 9.

The BCD-to-7-Segment Decoder

The BCD-to-7-segment decoder accepts the BCD code on its inputs and provides outputs to drive 7-segment display devices to produce a decimal readout. The logic diagram for a basic 7-segment decoder is shown in Figure 6–34.

▶ **FIGURE 6–34**

Logic symbol for a BCD-to-7-segment decoder/driver with active-LOW outputs. Open file F06-34 to verify operation.

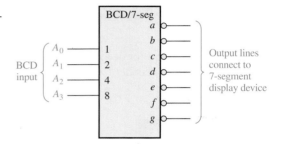

THE 74LS47 BCD-TO-7-SEGMENT DECODER/DRIVER

The 74LS47 is an example of an IC device that decodes a BCD input and drives a 7-segment display. In addition to its decoding and segment drive capability, the 74LS47 has several additional features as indicated by the $\overline{LT}$, $\overline{RBI}$, $\overline{BI}/\overline{RBO}$ functions in the logic symbol of Figure 6–35. As indicated by the bubbles on the logic symbol, all of the outputs (*a* through *g*) are active-LOW as are the $\overline{LT}$ (lamp test), *RBI* (ripple blanking input), and $\overline{BI}/\overline{RBO}$ (blanking input/ripple blanking output) functions. The outputs can drive a common-anode 7-segment display directly. Recall that 7-segment displays were discussed in Chapter 4. In addition to decoding a BCD input and producing the appropriate 7-segment outputs, the 74LS47 has lamp test and zero suppression capability. This device may be available in other TTL or CMOS families. Check the Texas Instruments website at www.ti.com or the TI CD-ROM accompanying this book.

▶ **FIGURE 6–35**

Pin diagram and logic symbol for the 74LS47 BCD-to-7-segment decoder/driver.

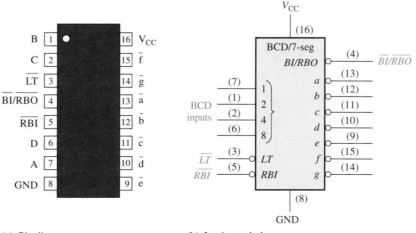

(a) Pin diagram   (b) Logic symbol

Lamp Test When a LOW is applied to the $\overline{LT}$ input and the $\overline{BI}/\overline{RBO}$ is HIGH, all of the 7 segments in the display are turned on. Lamp test is used to verify that no segments are burned out.

Zero Suppression **Zero suppression** is a feature used for multidigit displays to blank out unnecessary zeros. For example, in a 6-digit display the number 6.4 may be displayed as 006.400 if the zeros are not blanked out. Blanking the zeros at the front of a number is called *leading zero suppression* and blanking the zeros at the back of the number is called *trailing zero suppression*. Keep in mind that only nonessential zeros are blanked. With zero suppression, the number 030.080 will be displayed as 30.08 (the essential zeros remain).

Zero suppression in the 74LS47 is accomplished using the $\overline{RBI}$ and $\overline{BI}/\overline{RBO}$ functions. $\overline{RBI}$ is the ripple blanking input and $\overline{RBO}$ is the ripple blanking output on the 74LS47; these are used for zero suppression. $\overline{BI}$ is the blanking input that shares the same pin with $\overline{RBO}$; in other words, the $\overline{BI}/\overline{RBO}$ pin can be used as an input or an output. When used as a $\overline{BI}$ (blanking input), all segment outputs are HIGH (nonactive) when $\overline{BI}$ is LOW, which overrides all other inputs. The $\overline{BI}$ function is not part of the zero suppression capability of the device.

All of the segment outputs of the decoder are nonactive (HIGH) if a zero code (0000) is on its BCD inputs and if its $\overline{RBI}$ is LOW. This causes the display to be blank and produces a LOW $\overline{RBO}$.

The logic diagram in Figure 6–36(a) illustrates leading zero suppression for a whole number. The highest-order digit position (left-most) is always blanked if a zero code is on

> Zero suppression results in leading or trailing zeros in a number not showing on a display.

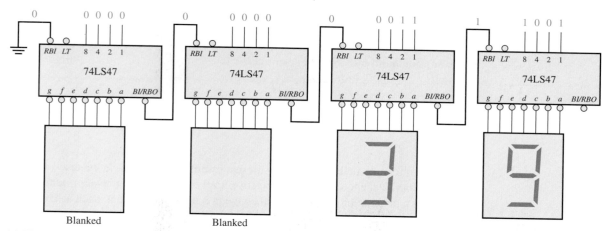

(a) Illustration of leading zero suppression

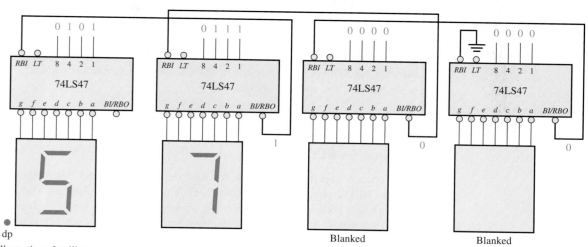

(b) Illustration of trailing zero suppression

▲ **FIGURE 6–36**

Examples of zero suppression using the 74LS47 BCD to 7-segment decoder/driver.

its BCD inputs because the $\overline{RBI}$ of the most-significant decoder is made LOW by connecting it to ground. The $\overline{RBO}$ of each decoder is connected to the $\overline{RBI}$ of the next lowest-order decoder so that all zeros to the left of the first nonzero digit are blanked. For example, in part (a) of the figure the two highest-order digits are zeros and therefore are blanked. The remaining two digits, 3 and 9 are displayed.

The logic diagram in Figure 6–36(b) illustrates trailing zero suppression for a fractional number. The lowest-order digit (right-most) is always blanked if a zero code is on its BCD inputs because the $\overline{RBI}$ is connected to ground. The $\overline{RBO}$ of each decoder is connected to the $\overline{RBI}$ of the next highest-order decoder so that all zeros to the right of the first nonzero digit are blanked. In part (b) of the figure, the two lowest-order digits are zeros and therefore are blanked. The remaining two digits, 5 and 7 are displayed. To combine both leading and trailing zero suppression in one display and to have decimal point capability, additional logic is required.

SECTION 6–5 REVIEW

1. A 3-line-to-8-line decoder can be used for octal-to-decimal decoding. When a binary 101 is on the inputs, which output line is activated?

2. How many 74HC154 1-of-16 decoders are necessary to decode a 6-bit binary number?

3. Would you select a decoder/driver with active-HIGH or active-LOW outputs to drive a common-cathode 7-segment LED display?

6–6 ENCODERS

An **encoder** is a combinational logic circuit that essentially performs a "reverse" decoder function. An encoder accepts an active level on one of its inputs representing a digit, such as a decimal or octal digit, and converts it to a coded output, such as BCD or binary. Encoders can also be devised to encode various symbols and alphabetic characters. The process of converting from familiar symbols or numbers to a coded format is called *encoding*.

After completing this section, you should be able to

■ Determine the logic for a decimal encoder ■ Explain the purpose of the priority feature in encoders ■ Describe the 74HC147 decimal-to-BCD priority encoder ■ Describe the 74LS148 octal-to-binary priority encoder ■ Expand an encoder ■ Apply the encoder to a specific application

The Decimal-to-BCD Encoder

This type of encoder has ten inputs—one for each decimal digit—and four outputs corresponding to the BCD code, as shown in Figure 6–37. This is a basic 10-line-to-4-line encoder.

The BCD (8421) code is listed in Table 6–6. From this table you can determine the relationship between each BCD bit and the decimal digits in order to analyze the logic. For instance, the most significant bit of the BCD code, A_3, is always a 1 for decimal digit 8 or 9. An OR expression for bit A_3 in terms of the decimal digits can therefore be written as

$$A_3 = 8 + 9$$

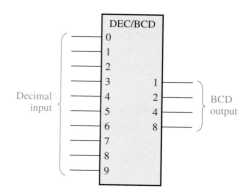

Logic symbol for a decimal-to-BCD encoder.

◀ TABLE 6–6

DECIMAL DIGIT	BCD CODE			
	A_3	A_2	A_1	A_0
0	0	0	0	0
1	0	0	0	1
2	0	0	1	0
3	0	0	1	1
4	0	1	0	0
5	0	1	0	1
6	0	1	1	0
7	0	1	1	1
8	1	0	0	0
9	1	0	0	1

Bit A_2 is always a 1 for decimal digit 4, 5, 6 or 7 and can be expressed as an OR function as follows:

$$A_2 = 4 + 5 + 6 + 7$$

Bit A_1 is always a 1 for decimal digit 2, 3, 6, or 7 and can be expressed as

$$A_1 = 2 + 3 + 6 + 7$$

Finally, A_0 is always a 1 for decimal digit 1, 3, 5, 7, or 9. The expression for A_0 is

$$A_0 = 1 + 3 + 5 + 7 + 9$$

Now let's implement the logic circuitry required for encoding each decimal digit to a BCD code by using the logic expressions just developed. It is simply a matter of ORing the appropriate decimal digit input lines to form each BCD output. The basic encoder logic resulting from these expressions is shown in Figure 6–38.

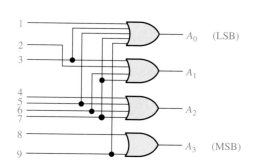

◀ FIGURE 6–38

Basic logic diagram of a decimal-to-BCD encoder. A 0-digit input is not needed because the BCD outputs are all LOW when there are no HIGH inputs.

COMPUTER NOTE

An *assembler* can be thought of as a software encoder because it interprets the mnemonic instructions with which a program is written and carries out the applicable encoding to convert each mnemonic to a machine code instruction (series of 1s and 0s) which the computer can understand. Examples of mnemonic instructions for a microprocessor are ADD, MOV (move data), MUL (multiply), XOR, JMP (jump), and OUT (output to a port).

The basic operation of the circuit in Figure 6–38 is as follows: When a HIGH appears on *one* of the decimal digit input lines, the appropriate levels occur on the four BCD output lines. For instance, if input line 9 is HIGH (assuming all other input lines are LOW), this condition will produce a HIGH on outputs A_0 and A_3 and LOWs on outputs A_1 and A_2, which is the BCD code (1001) for decimal 9.

The Decimal-to-BCD Priority Encoder   This type of encoder performs the same basic encoding function as previously discussed. A **priority encoder** also offers additional flexibility in that it can be used in applications that require priority detection. The priority function means that the encoder will produce a BCD output corresponding to the *highest-order decimal digit* input that is active and will ignore any other lower-order active inputs. For instance, if the 6 and the 3 inputs are both active, the BCD output is 0110 (which represents decimal 6).

THE 74HC147 DECIMAL-TO-BCD ENCODER

The 74HC147 is a priority encoder with active-LOW inputs (0) for decimal digits 1 through 9 and active-LOW BCD outputs as indicated in the logic symbol in Figure 6–39. A BCD zero output is represented when none of the inputs is active. The device pin numbers are in parentheses. This device may be available in other CMOS or TTL families. Check the Texas Instruments website at www.ti.com or the TI CD-ROM accompanying this book.

▶ **FIGURE 6–39**

Pin diagram and logic symbol for the 74HC147 decimal-to-BCD priority encoder (HPRI means highest value input has priority).

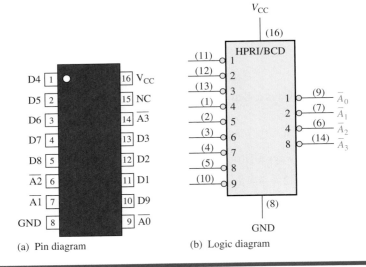

(a) Pin diagram   (b) Logic diagram

THE 74LS148 8-LINE-TO-3-LINE ENCODER

The 74LS148 is a priority encoder that has eight active-LOW inputs and three active-LOW binary outputs, as shown in Figure 6–40. This device can be used for converting octal inputs (recall that the octal digits are 0 through 7) to a 3-bit binary code. To enable the device, the *EI* (enable input) must be LOW. It also has the *EO* (enable output) and *GS* output for expansion purposes. The *EO* is LOW when the *EI* is LOW and none of the inputs (0 through 7) is active. *GS* is LOW when *EI* is LOW and any of the inputs is active. This device may be available in other TTL or CMOS families. Check the Texas Instruments website at www.ti.com or the TI CD-ROM accompanying this book.

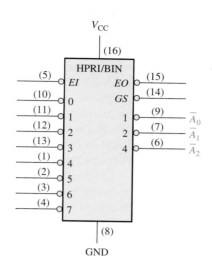

► **FIGURE 6–40**

Logic symbol for the 74LS148 8-line-to-3-line encoder.

The 74LS148 can be expanded to a 16-line-to-4-line encoder by connecting the *EO* of the higher-order encoder to the *EI* of the lower-order encoder and negative-ORing the corresponding binary outputs as shown in Figure 6–41. The *EO* is used as the fourth and most-significant bit. This particular configuration produces active-HIGH outputs for the 4-bit binary number.

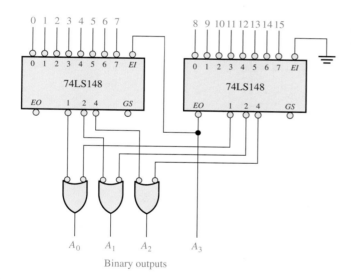

► **FIGURE 6–41**

A 16-line-to-4 line encoder using 74LS148s and external logic.

EXAMPLE 6–11

If LOW levels appear on pins, 1, 4, and 13 of the 74HC147 shown in Figure 6–39, indicate the state of the four outputs. All other inputs are HIGH.

Solution Pin 4 is the highest-order decimal digit input having a LOW level and represents decimal 7. Therefore, the output levels indicate the BCD code for decimal 7 where $\overline{A}_0$ is the LSB and $\overline{A}_3$ is the MSB. Output $\overline{A}_0$ is LOW, $\overline{A}_1$ is LOW, $\overline{A}_2$ is LOW, and $\overline{A}_3$ is HIGH.

Related Problem What are the outputs of the 74HC147 if all its inputs are LOW? If all its inputs are HIGH?

An Application

A classic application example is a keyboard encoder. The ten decimal digits on the keyboard of a computer, for example, must be encoded for processing by the logic circuitry. When one of the keys is pressed, the decimal digit is encoded to the corresponding BCD code. Figure 6–42 shows a simple keyboard encoder arrangement using a 74HC147 priority encoder. They keys are represented by ten push-button switches, each with a **pull-up resistor** to $+V$. The pull-up resistor ensures that the line is HIGH when a key is not depressed. When a key is depressed, the line is connected to ground, and a LOW is applied to the corresponding encoder input. The zero key is not connected because the BCD output represents zero when none of the other keys is depressed.

The BCD complement output of the encoder goes into a storage device, and each successive BCD code is stored until the entire number has been entered. Methods of storing BCD numbers and binary data are covered in later chapters.

▶ FIGURE 6–42

A simplified keyboard encoder.

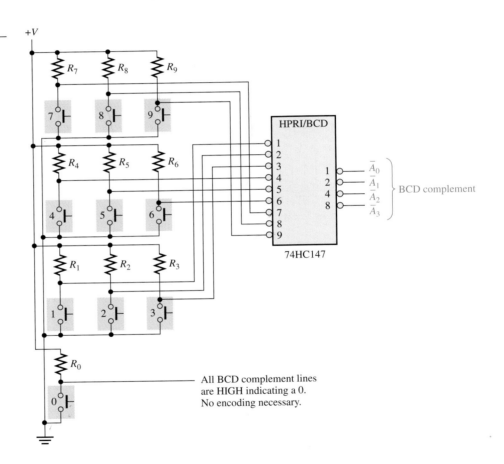

All BCD complement lines are HIGH indicating a 0. No encoding necessary.

SECTION 6–6
REVIEW

1. Suppose the HIGH levels are applied to the 2 input and the 9 input of the circuit in Figure 6–38.

   (a) What are the states of the output lines?

   (b) Does this represent a valid BCD code?

   (c) What is the restriction on the encoder logic in Figure 6–38?

2. (a) What is the $\overline{A_3}\overline{A_2}\overline{A_1}\overline{A_0}$ output when LOWs are applied to pins 1 and 5 of the 74HC147 in Figure 6–39?

   (b) What does this output represent?

6–7 CODE CONVERTERS

In this section, we will examine some methods of using combinational logic circuits to convert from one code to another.

After completing this section, you should be able to

- Explain the process for converting BCD to binary ■ Use exclusive-OR gates for conversions between binary and Gray codes

BCD-to-Binary Conversion

One method of BCD-to-binary code conversion uses adder circuits. The basic conversion process is as follows:

1. The value, or weight, of each bit in the BCD number is represented by a binary number.

2. All of the binary representations of the weights of bits that are 1s in the BCD number are added.

3. The result of this addition is the binary equivalent of the BCD number.

A more concise statement of this operation is

The binary numbers representing the weights of the BCD bits are summed to produce the total binary number.

Let's examine an 8-bit BCD code (one that represents a 2-digit decimal number) to understand the relationship between BCD and binary. For instance, you already know that the decimal number 87 can be expressed in BCD as

$$\underbrace{1000}_{8} \quad \underbrace{0111}_{7}$$

The left-most 4-bit group represents 80, and the right-most 4-bit group represents 7. That is, the left-most group has a weight of 10, and the right-most group has a weight of 1. Within each group, the binary weight of each bit is as follows:

	Tens Digit				Units Digit			
Weight:	80	40	20	10	8	4	2	1
Bit designation:	B_3	B_2	B_1	B_0	A_3	A_2	A_1	A_0

The binary equivalent of each BCD bit is a binary number representing the weight of that bit within the total BCD number. This representation is given in Table 6–7.

BCD BIT	BCD WEIGHT	(MSB) 64	32	16	8	4	2	(LSB) 1
A_0	1	0	0	0	0	0	0	1
A_1	2	0	0	0	0	0	1	0
A_2	4	0	0	0	0	1	0	0
A_3	8	0	0	0	1	0	0	0
B_0	10	0	0	0	1	0	1	0
B_1	20	0	0	1	0	1	0	0
B_2	40	0	1	0	1	0	0	0
B_3	80	1	0	1	0	0	0	0

◀ TABLE 6–7

Binary representations of BCD bit weights.

If the binary representations for the weights of all the 1s in the BCD number are added, the result is the binary number that corresponds to the BCD number. Example 6–12 illustrates this.

EXAMPLE 6–12

Convert the BCD numbers 00100111 (decimal 27) and 10011000 (decimal 98) to binary.

Solution   Write the binary representations of the weights of all 1s appearing in the numbers, and then add them together.

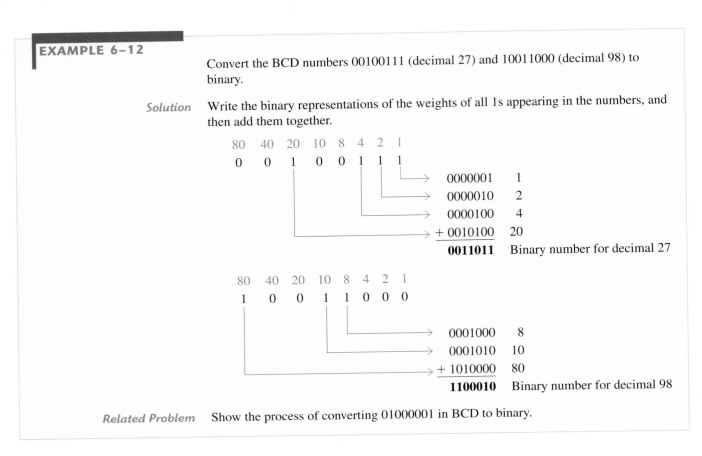

$$
\begin{array}{cccccccc}
80 & 40 & 20 & 10 & 8 & 4 & 2 & 1 \\
0 & 0 & 1 & 0 & 0 & 1 & 1 & 1
\end{array}
$$

```
                    0000001    1
                    0000010    2
                    0000100    4
                 + 0010100     20
                   0011011     Binary number for decimal 27
```

$$
\begin{array}{cccccccc}
80 & 40 & 20 & 10 & 8 & 4 & 2 & 1 \\
1 & 0 & 0 & 1 & 1 & 0 & 0 & 0
\end{array}
$$

```
                    0001000    8
                    0001010    10
                 + 1010000     80
                   1100010     Binary number for decimal 98
```

Related Problem   Show the process of converting 01000001 in BCD to binary.

With this basic procedure in mind, let's see how the process can be implemented with logic circuits. Once the binary representation for each 1 in the BCD number is determined, adder circuits can be used to add the 1s in each column of the binary representation. The 1s occur in a given column only when the corresponding BCD bit is a 1. The occurrence of a BCD 1 can therefore be used to generate the proper binary 1 in the appropriate column of the adder structure. To handle a two-decimal-digit (two-decade) BCD code, eight BCD input lines and seven binary outputs are required. (It takes seven bits to represent binary numbers through ninety-nine.)

Binary-to-Gray and Gray-to-Binary Conversion

The basic process for Gray-binary conversions was covered in Chapter 2. Exclusive-OR gates can be used for these conversions. Programmable logic devices (PLDs) can also be programmed for these code conversions. Figure 6–43 shows a 4-bit binary-to-Gray code converter, and Figure 6–44 illustrates a 4-bit Gray-to-binary converter.

▶ **FIGURE 6–43**

Four-bit binary-to-Gray conversion logic. Open file F06-43 to verify operation.

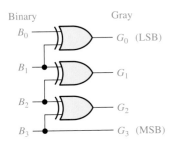

Binary   Gray

B_0 ———▷ G_0 (LSB)

B_1 ———▷ G_1

B_2 ———▷ G_2

B_3 ———▷ G_3 (MSB)

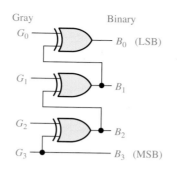

◄ FIGURE 6–44

Four-bit Gray-to-binary conversion logic. Open file F06-44 to verify operation.

EXAMPLE 6–13

    (a) Convert the binary number 0101 to Gray code with exclusive-OR gates.

    (b) Convert the Gray code 1011 to binary with exclusive-OR gates.

Solution   **(a)** 0101_2 is 0111 Gray. See Figure 6–45(a).

    (b) 1011 Gray is 1101_2. See Figure 6–45(b).

▶ FIGURE 6–45

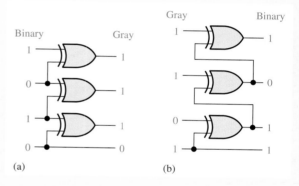

(a)   (b)

Related Problem   How many exclusive-OR gates are required to convert 8-bit binary to Gray?

SECTION 6–7 REVIEW

1. Convert the BCD number 10000101 to binary.
2. Draw the logic diagram for converting an 8-bit binary number to Gray code.

6–8 MULTIPLEXERS (DATA SELECTORS)

A **multiplexer (MUX)** is a device that allows digital information from several sources to be routed onto a single line for transmission over that line to a common destination. The basic multiplexer has several data-input lines and a single output line. It also has data-select inputs, which permit digital data on any one of the inputs to be switched to the output line. Multiplexers are also known as data selectors.

After completing this section, you should be able to

■ Explain the basic operation of a multiplexer ■ Describe the 74LS151 and the 74HC157 multiplexers ■ Expand a multiplexer to handle more data inputs ■ Use the multiplexer as a logic function generator

In a multiplexer, data goes from several lines to one line.

A logic symbol for a 4-input multiplexer (MUX) is shown in Figure 6–46. Notice that there are two data-select lines because with two select bits, any one of the four data-input lines can be selected.

▶ **FIGURE 6–46**

Logic symbol for a 1-of-4 data selector/multiplexer.

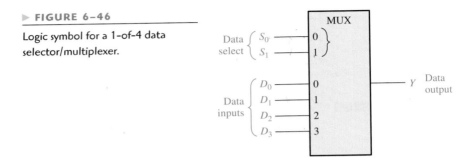

In Figure 6–46, a 2-bit code on the data-select (S) inputs will allow the data on the selected data input to pass through to the data output. If a binary 0 ($S_1 = 0$ and $S_0 = 0$) is applied to the data-select lines, the data on input D_0 appear on the data-output line. If a binary 1 ($S_1 = 0$ and $S_0 = 1$) is applied to the data-select lines, the data on input D_1 appear on the data output. If a binary 2 ($S_1 = 1$ and $S_0 = 0$) is applied, the data on D_2 appear on the output. If a binary 3 ($S_1 = 1$ and $S_0 = 1$) is applied, the data on D_3 are switched to the output line. A summary of this operation is given in Table 6–8.

▶ **TABLE 6–8**

Data selection for a 1-of-4-multiplexer.

DATA-SELECT INPUTS		INPUT SELECTED
S_1	S_0	
0	0	D_0
0	1	D_1
1	0	D_2
1	1	D_3

COMPUTER NOTE

A *bus* is an internal pathway along which electrical signals are sent from one part of a computer to another. In computer networks, a *shared bus* is one that is connected to all the microprocessors in the system in order to exchange data. A shared bus may contain memory and input/output devices that can be accessed by all the microprocessors in the system. Access to the shared bus is controlled by a *bus arbiter* (a multiplexer of sorts) that allows only one microprocessor at a time to use the system's shared bus.

Now let's look at the logic circuitry required to perform this multiplexing operation. The data output is equal to the state of the *selected* data input. You can therefore, derive a logic expression for the output in terms of the data input and the select inputs.

The data output is equal to D_0 only if $S_1 = 0$ and $S_0 = 0$: $Y = D_0\overline{S_1}\,\overline{S_0}$.

The data output is equal to D_1 only if $S_1 = 0$ and $S_0 = 1$: $Y = D_1\overline{S_1}S_0$.

The data output is equal to D_2 only if $S_1 = 1$ and $S_0 = 0$: $Y = D_2 S_1\overline{S_0}$.

The data output is equal to D_3 only if $S_1 = 1$ and $S_0 = 1$: $Y = D_3 S_1 S_0$.

When these terms are ORed, the total expression for the data output is

$$Y = D_0\overline{S_1}\,\overline{S_0} + D_1\overline{S_1}S_0 + D_2 S_1\overline{S_0} + D_3 S_1 S_0$$

The implementation of this equation requires four 3-input AND gates, a 4-input OR gate, and two inverters to generate the complements of S_1 and S_0, as shown in Figure 6–47. Because data can be selected from any one of the input lines, this circuit is also referred to as a **data selector.**

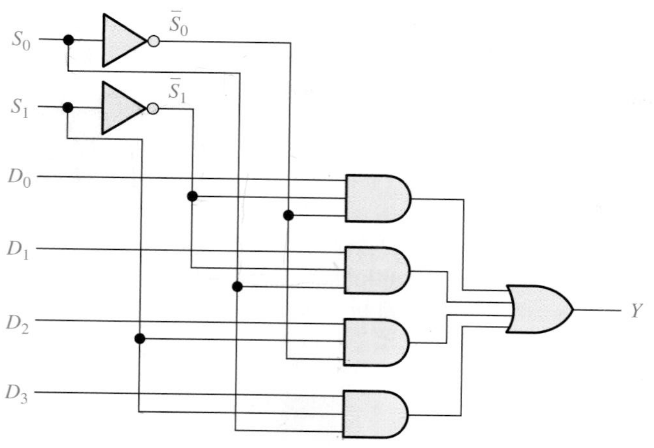

FIGURE 6–47

Logic diagram for a 4-input multiplexer. Open file F06-47 to verify operation.

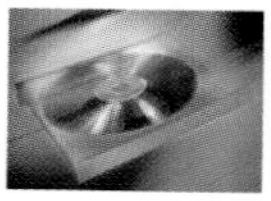

EXAMPLE 6–14

The data-input and data-select waveforms in Figure 6–48(a) are applied to the multiplexer in Figure 6–47. Determine the output waveform in relation to the inputs.

FIGURE 6–48

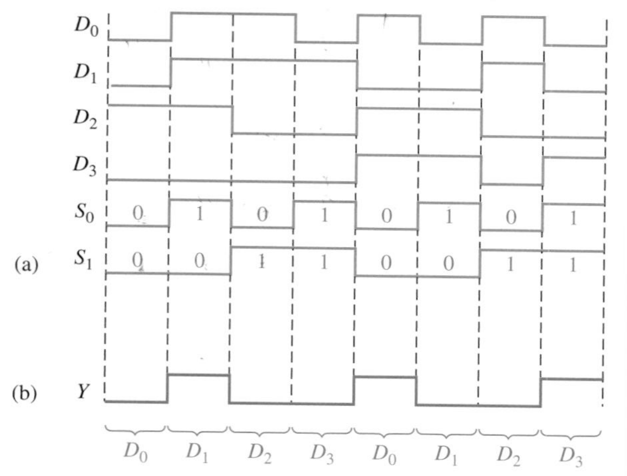

Solution The binary state of the data-select inputs during each interval determines which data input is selected. Notice that the data-select inputs go through a repetitive binary sequence 00, 01, 10, 11, 00, 01, 10, 11, and so on. The resulting output waveform is shown in Figure 6–48(b).

Related Problem Construct a timing diagram showing all inputs and the output if the S_0 and S_1 waveforms in Figure 6–48 are interchanged.

THE 74HC157 QUAD 2-INPUT DATA SELECTOR/MULTIPLEXER

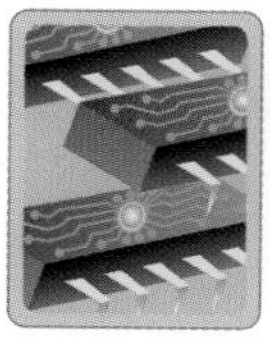

The 74HC157, as well as its LS version, consists of four separate 2-input multiplexers. Each of the four multiplexers shares a common data-select line and a common *Enable*. Because there are only two inputs to be selected in each multiplexer, a single data-select input is sufficient.

A LOW on the $\overline{Enable}$ input allows the selected input data to pass through to the output. A HIGH on the $\overline{Enable}$ input prevents data from going through to the output; that is, it disables the multiplexers. This device may be available in other CMOS or TTL families. Check the Texas Instruments website at www.ti.com or the TI CD-ROM accompanying this book.

The ANSI/IEEE Logic Symbol The pin diagram for the 74HC157 is shown in Figure 6–49(a). The ANSI/IEEE logic symbol for the 74HC157 is shown in Figure 6–49(b). Notice that the four multiplexers are indicated by the partitioned outline and that the inputs common to all four multiplexers are indicated as inputs to the notched block at the top, which is called the *common control block*. All labels within the upper MUX block apply to the other blocks below it.

▶ **FIGURE 6–49**

Pin diagram and logic symbol for the 74HC157 quadruple 2-input data selector/multiplexer.

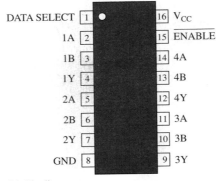

(a) Pin diagram

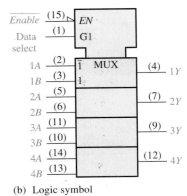

(b) Logic symbol

Notice the 1 and $\overline{1}$ labels in the MUX blocks and the G1 label in the common control block. These labels are an example of the **dependency notation** system specified in the ANSI/IEEE Standard 91-1984. In this case G1 indicates an AND relationship between the data-select input and the data inputs with 1 or $\overline{1}$ labels. (The $\overline{1}$ means that the AND relationship applies to the complement of the G1 input.) In other words, when the data-select input is HIGH, the *B* inputs of the multiplexers are selected; and when the data-select input is LOW, the *A* inputs are selected. A "G" is always used to denote AND dependency. Other aspects of dependency notation are introduced as appropriate throughout the book.

THE 74LS151 8-INPUT DATA SELECTOR/MULTIPLEXER

The 74LS151 has eight data inputs (D_0–D_7) and, therefore, three data-select or address input lines (S_0–S_2). Three bits are required to select any one of the eight data inputs ($2^3 = 8$). A LOW on the $\overline{Enable}$ input allows the selected input data to pass through to the output. Notice that the data output and its complement are both available. The pin diagram is shown in Figure 6–50(a), and the ANSI/IEEE logic symbol is shown in part (b). In this case there is no need for a common control block on the logic symbol because there is only one multiplexer to be controlled, not four as in the 74HC157. The $G\frac{0}{7}$ label within the logic symbol indicates the AND relationship between the data-select inputs and each of the data inputs 0 through 7. This device may be available in other TTL or CMOS families. Check the Texas Instruments website at www.ti.com or the TI CD-ROM accompanying this book.

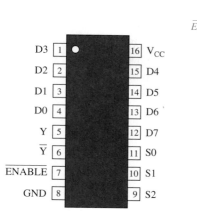 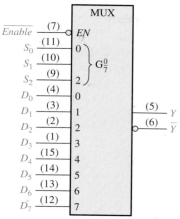

(a) Pin diagram

(b) Logic symbol

◄ FIGURE 6–50

Pin diagram and logic symbol for the 74LS151 8-input data selector/multiplexer.

EXAMPLE 6–15

Use 74LS151s and any other logic necessary to multiplex 16 data lines onto a single data-output line.

Solution   An implementation of this system is shown in Figure 6–51. Four bits are required to select one of 16 data inputs ($2^4 = 16$). In this application the $\overline{Enable}$ input is used as the most significant data-select bit. When the MSB in the data-select code is LOW, the left 74LS151 is enabled, and one of the data inputs (D_0 through D_7) is selected by the other three data-select bits. When the data-select MSB is HIGH, the right 74LS151 is enabled, and one of the data inputs (D_8 through D_{15}) is selected. The selected input data are then passed through to the negative-OR gate and onto the single output line.

► **FIGURE 6–51**

A 16-input multiplexer.

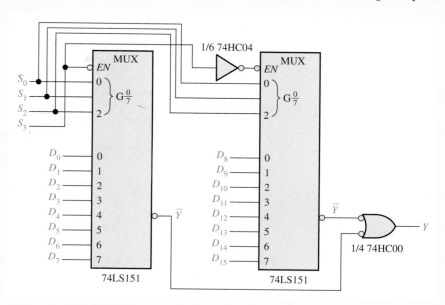

Related Problem   Determine the codes on the select inputs required to select each of the following data inputs: D_0, D_4, D_8, and D_{13}.

Applications

A 7–Segment Display Multiplexer Figure 6–52 shows a simplified method of multiplexing BCD numbers to a 7-segment display. In this example, 2-digit numbers are displayed on the 7-segment readout by the use of a single BCD-to-7-segment decoder. This basic method of display multiplexing can be extended to displays with any number of digits.

► FIGURE 6–52

Simplified 7-segment display multiplexing logic.

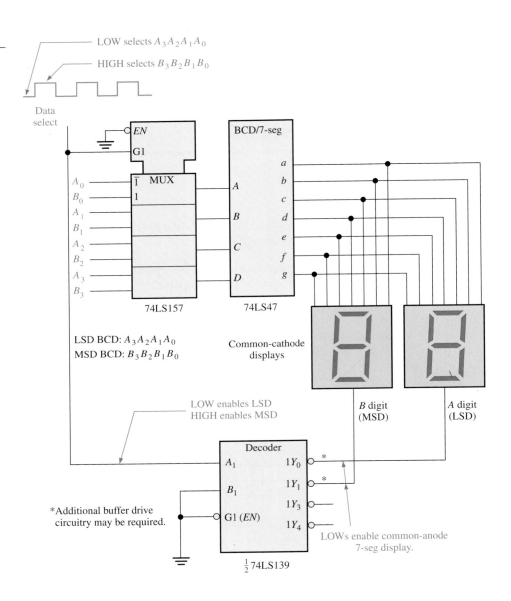

The basic operation is as follows. Two BCD digits ($A_3A_2A_1A_0$ and $B_3B_2B_1B_0$) are applied to the multiplexer inputs. A square wave is applied to the data-select line, and when it is LOW, the *A* bits ($A_3A_2A_1A_0$) are passed through to the inputs of the 74LS47 BCD-to-7-segment decoder. The LOW on the data-select also puts a LOW on the A_1 input of the 74LS139 2-line-to-4-line decoder, thus activating its 0 output and enabling the *A*-digit display by effectively connecting its common terminal to ground. The *A* digit is now *on* and the *B* digit is *off*.

When the data-select line goes HIGH, the B bits $(B_3B_2B_1B_0)$ are passed through to the inputs of the BCD-to-7-segment decoder. Also, the 74LS139 decoder's 1 output is activated, thus enabling the B-digit display. The B digit is now *on* and the A digit is *off.* The cycle repeats at the frequency of the data-select square wave. This frequency must be high enough (about 30 Hz) to prevent visual flicker as the digit displays are multiplexed.

A Logic Function Generator   A useful application of the data selector/multiplexer is in the generation of combinational logic functions in sum-of-products form. When used in this way, the device can replace discrete gates, can often greatly reduce the number of ICs, and can make design changes much easier.

To illustrate, a 74LS151 8-input data selector/multiplexer can be used to implement any specified 3-variable logic function if the variables are connected to the data-select inputs and each data input is set to the logic level required in the truth table for that function. For example, if the function is a 1 when the variable combination is $\overline{A}_2A_1\overline{A}_0$, the 2 input (selected by 010) is connected to a HIGH. This HIGH is passed through to the output when this particular combination of variables occurs on the data-select lines. An example will help clarify this application.

EXAMPLE 6–16

Implement the logic function specified in Table 6–9 by using a 74LS151 8-input data selector/multiplexer. Compare this method with a discrete logic gate implementation.

▼ TABLE 6–9

INPUTS			OUTPUT
A_2	A_1	A_0	Y
0	0	0	0
0	0	1	1
0	1	0	0
0	1	1	1
1	0	0	0
1	0	1	1
1	1	0	1
1	1	1	0

Solution   Notice from the truth table that Y is a 1 for the following input variable combinations: 001, 011, 101, and 110. For all other combinations, Y is 0. For this function to be implemented with the data selector, the data input selected by each of the above-mentioned combinations must be connected to a HIGH (5 V). All the other data inputs must be connected to a LOW (ground), as shown in Figure 6–53.

The implementation of this function with logic gates would require four 3-input AND gates, one 4-input OR gate, and three inverters unless the expression can be simplified.

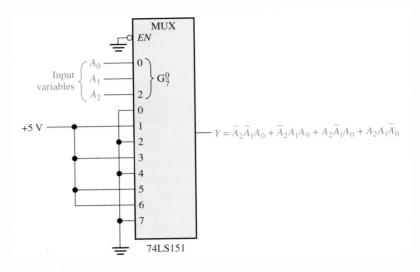

► FIGURE 6–53

Data selector/multiplexer connected as a 3-variable logic function generator.

$$Y = \overline{A}_2\overline{A}_1A_0 + \overline{A}_2A_1A_0 + A_2\overline{A}_1A_0 + A_2A_1\overline{A}_0$$

Related Problem   Use the 74LS151 to implement the following expression:

$$Y = \overline{A}_2\overline{A}_1\overline{A}_0 + A_2\overline{A}_1\overline{A}_0 + \overline{A}_2A_1\overline{A}_0$$

Example 6–16 illustrated how the 8-input data selector can be used as a logic function generator for three variables. Actually, this device can be also used as a 4-variable logic function generator by the utilization of one of the bits (A_0) in conjunction with the data inputs.

A 4-variable truth table has sixteen combinations of input variables. When an 8-bit data selector is used, each input is selected twice: the first time when A_0 is 0 and the second time when A_0 is 1. With this in mind, the following rules can be applied (Y is the output, and A_0 is the least significant bit):

1. If $Y = 0$ both times a given data input is selected by a certain combination of the input variables, $A_3A_2A_1$, connect that data input to ground (0).

2. If $Y = 1$ both times a given data input is selected by a certain combination of the input variables, $A_3A_2A_1$, connect the data input to $+V(1)$.

3. If Y is different the two times a given data input is selected by a certain combination of the input variables, $A_3A_2A_1$, and if $Y = A_0$, connect that data input to A_0.

4. If Y is different the two times a given data input is selected by a certain combination of the input variables, $A_3A_2A_1$, and if $Y=\overline{A}_0$, connect that data input to $\overline{A}_0$.

EXAMPLE 6–17

Implement the logic function in Table 6–10 by using a 74LS151 8-input data selector/multiplexer. Compare this method with a discrete logic gate implementation.

Solution   The data-select inputs are $A_3A_2A_1$. In the first row of the table, $A_3A_2A_1 = 000$ and $Y = A_0$. In the second row, where $A_3A_2A_1$ again is 000, $Y = A_0$. Thus, A_0 is connected to the 0 input. In the third row of the table, $A_3A_2A_1 = 001$ and $Y=\overline{A}_0$. Also, in the fourth row, when $A_3A_2A_1$ again is 001, $Y=\overline{A}_0$. Thus, A_0 is inverted and

▶ TABLE 6–10

DECIMAL DIGIT	INPUTS				OUTPUT Y
	A_3	A_2	A_1	A_0	
0	0	0	0	0	0
1	0	0	0	1	1
2	0	0	1	0	1
3	0	0	1	1	0
4	0	1	0	0	0
5	0	1	0	1	1
6	0	1	1	0	1
7	0	1	1	1	1
8	1	0	0	0	1
9	1	0	0	1	0
10	1	0	1	0	1
11	1	0	1	1	0
12	1	1	0	0	1
13	1	1	0	1	1
14	1	1	1	0	0
15	1	1	1	1	1

connected to the 1 input. This analysis is continued until each input is properly connected according to the specified rules. The implementation is shown in Figure 6–54.

▶ FIGURE 6–54

Data selector/multiplexer connected as a 4-variable logic function generator.

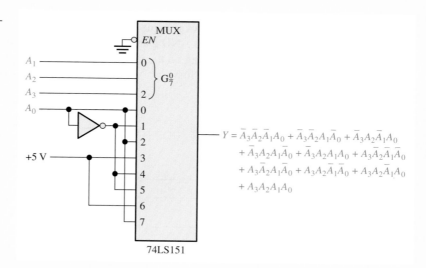

$$Y = \bar{A}_3\bar{A}_2\bar{A}_1A_0 + \bar{A}_3\bar{A}_2A_1\bar{A}_0 + \bar{A}_3A_2\bar{A}_1A_0$$
$$+ \bar{A}_3A_2A_1\bar{A}_0 + \bar{A}_3A_2A_1A_0 + A_3\bar{A}_2\bar{A}_1\bar{A}_0$$
$$+ A_3\bar{A}_2A_1\bar{A}_0 + A_3A_2\bar{A}_1\bar{A}_0 + A_3A_2\bar{A}_1A_0$$
$$+ A_3A_2A_1A_0$$

If implemented with logic gates, the function would require as many as ten 4-input AND gates, one 10-input OR gate, and four inverters, although possible simplification would reduce this requirement.

Related Problem In Table 6–10, if $Y = 0$ when the inputs are all zeros and is alternately a 1 and a 0 for the remaining rows in the table, use a 74LS151 to implement the resulting logic function.

1. In Figure 6–47, $D_0 = 1$, $D_1 = 0$, $D_2 = 1$, $D_3 = 0$, $S_0 = 1$, and $S_1 = 0$. What is the output?

2. Identify each device.
   (a) 74LS157   (b) 74LS151

3. A 74LS151 has alternating LOW and HIGH levels on its data inputs beginning with $D_0 = 0$. The data-select lines are sequenced through a binary count (000, 001, 010, and so on) at a frequency of 1 kHz. The enable input is LOW. Describe the data output waveform.

4. Briefly describe the purpose of each of the following devices in Figure 6–52:
   (a) 74LS157   (b) 74LS47   (c) 74LS139

6–9   DEMULTIPLEXERS

A **demultiplexer (DEMUX)** basically reverses the multiplexing function. It takes digital information from one line and distributes it to a given number of output lines. For this reason, the demultiplexer is also known as a data distributor. As you will learn, decoders can also be used as demultiplexers.

After completing this section, you should be able to

■ Explain the basic operation of a demultiplexer ■ Describe how the 74HC154 4-line-to-16-line decoder can be used as a demultiplexer ■ Develop the timing diagram for a demultiplexer with specified data and data selection inputs

In a demultiplexer, data goes from one line to several lines.

Figure 6–55 shows a 1-line-to-4-line demultiplexer (DEMUX) circuit. The data-input line goes to all of the AND gates. The two data-select lines enable only one gate at a time, and the data appearing on the data-input line will pass through the selected gate to the associated data-output line.

▶ **FIGURE 6–55**

A 1-line-to-4-line demultiplexer.

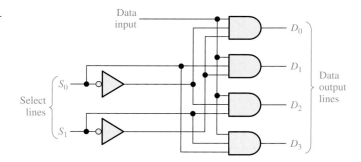

EXAMPLE 6–18

The serial data-input waveform (Data in) and data-select inputs (S_0 and S_1) are shown in Figure 6–56. Determine the data-output waveforms on D_0 through D_3 for the demultiplexer in Figure 6–55.

▶ **FIGURE 6–56**

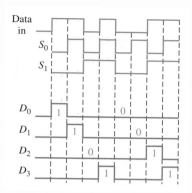

Solution   Notice that the select lines go through a binary sequence so that each successive input bit is routed to D_0, D_1, D_2, and D_3 in sequence, as shown by the output waveforms in Figure 6–56.

Related Problem   Develop the timing diagram for the demultiplexer if the S_0 and S_1 waveforms are both inverted.

THE 74HC154 DEMULTIPLEXER

We have already discussed the 74HC154 decoder in its application as a 4-line-to-16-line decoder (Section 6–5). This device and other decoders can also be used in demultiplexing applications. The logic symbol for this device when used as a demultiplexer is shown in Figure 6–57. In demultiplexer applications, the input lines are used as the data-select lines. One of the chip select inputs is used as the data-input line, with the other chip select input held LOW to enable the internal negative-AND gate at the bottom of the diagram. This device may be available in other CMOS or TTL families. Check the Texas Instruments website at www.ti.com or the TI CD-ROM accompanying this book.

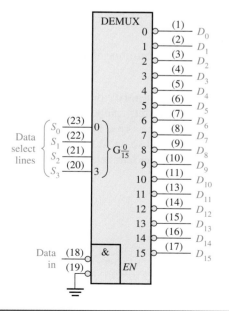

◀ **FIGURE 6–57**

The 74HC154 decoder used as a demultiplexer.

1. Generally, how can an decoder be used as a demultiplexer?
2. The 74HC154 demultiplexer in Figure 6–57 has a binary code of 1010 on the data-select lines, and the data-input line is LOW. What are the states of the output lines?

6–10   PARITY GENERATORS/CHECKERS

Errors can occur as digital codes are being transferred from one point to another within a digital system or while codes are being transmitted from one system to another. The errors take the form of undesired changes in the bits that make up the coded information; that is, a 1 can change to a 0, or a 0 to a 1, because of component malfunctions or electrical noise. In most digital systems, the probability that even a single bit error will occur is very small, and the likelihood that more than one will occur is even smaller. Nevertheless, when an error occurs undetected, it can cause serious problems in a digital system.

After completing this section, you should be able to

■ Explain the concept of parity   ■ Implement a basic parity circuit with exclusive-OR gates   ■ Describe the operation of basic parity generating and checking logic   ■ Discuss the 74LS280 9-bit parity generator/checker   ■ Discuss how error detection can be implemented in a data transmission

The parity method of error detection in which a **parity bit** is attached to a group of information bits in order to make the total number of 1s either even or odd (depending on the system) was covered in Chapter 2. In addition to parity bits, several specific codes also provide inherent error detection.

Basic Parity Logic

A parity bit indicates if the number of 1s in a code is even or odd for the purpose of error detection.

In order to check for or to generate the proper parity in a given code, a basic principle can be used:

The sum (disregarding carries) of an even number of 1s is always 0, and the sum of an odd number of 1s is always 1.

Therefore, to determine if a given code has **even parity** or **odd parity,** all the bits in that code are summed. As you know, the sum of two bits can be generated by an exclusive-OR gate, as shown in Figure 6–58(a); the sum of four bits can be formed by three exclusive-OR gates connected as shown in Figure 6–58(b); and so on. When the number of 1s on the inputs is even, the output X is 0 (LOW). When the number of 1s is odd, the output X is 1 (HIGH).

▶ FIGURE 6–58

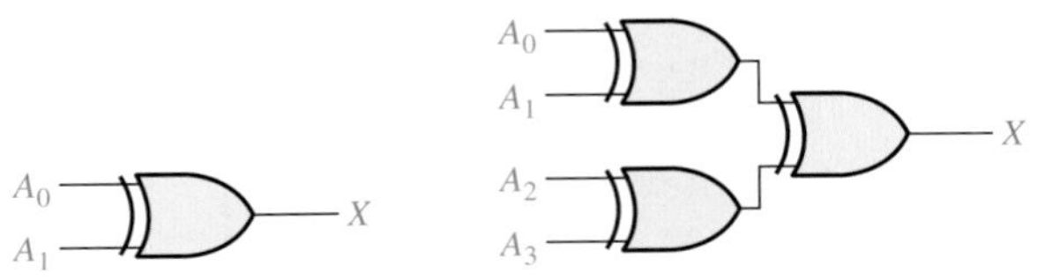

(a) Summing of two bits   (b) Summing of four bits

THE 74LS280 9-BIT PARITY GENERATOR/CHECKER

The logic symbol and function table for a 74LS280 are shown in Figure 6–59. This particular device can be used to check for odd or even parity on a 9-bit code (eight data bits and one parity bit) or it can be used to generate a parity bit for a binary code with up to nine bits. The inputs are A through I; when there is an even number of 1s on the inputs, the Σ Even output is HIGH and the Σ Odd output is LOW. This device may be available in other TTL or CMOS families. Check the Texas Instruments website at www.ti.com or the TI CD-ROM accompanying this book.

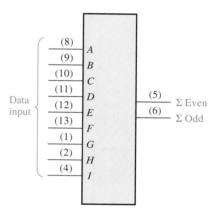

(a) Traditional logic symbol

Number of Inputs	Outputs	
A–I That Are HIGH	Σ Even	Σ Odd
0, 2, 4, 6, 8	H	L
1, 3, 5, 7, 9	L	H

(b) Function table

▲ **FIGURE 6–59**

The 74LS280 9-bit parity generator/checker.

Parity Checker When this device is used as an even parity checker, the number of input bits should always be even; and when a parity error occurs, the Σ Even output goes LOW and the Σ Odd output goes HIGH. When it is used as an odd parity checker, the number of input bits should always bé odd; and when a parity error occurs, the Σ Odd output goes LOW and the Σ Even output goes HIGH.

Parity Generator If this device is used as an even parity generator, the parity bit is taken at the Σ Odd output because this output is a 0 if there is an even number of input bits and it is a 1 if there is an odd number. When used as an odd parity generator, the parity bit is taken at the Σ Even output because it is a 0 when the number of inputs bits is odd.

A Data Transmission System with Error Detection

A simplified data transmission system is shown in Figure 6–60 to illustrate an application of parity generators/checkers, as well as multiplexers and demultiplexers, and to illustrate the need for data storage in some applications.

In this application, digital data from seven sources are multiplexed onto a single line for transmission to a distant point. The seven data bits (D_0 through D_6) are applied to the multiplexer data inputs and, at the same time, to the even parity generator inputs. The Σ Odd output of the parity generator is used as the even parity bit. This bit is 0 if the number of 1s on the inputs A through I is even and is a 1 if the number of 1s on A through I is odd. This bit is D_7 of the transmitted code.

The data-select inputs are repeatedly cycled through a binary sequence, and each data bit, beginning with D_0, is serially passed through and onto the transmission line ($\overline{Y}$). In

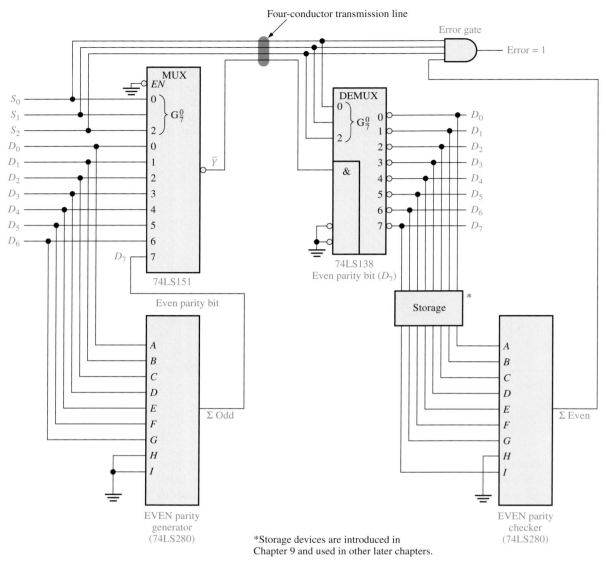

▲ **FIGURE 6–60**

Simplified data transmission system with error detection.

The Pentium microprocessor performs internal parity checks as well as parity checks of the external data and address buses. In a read operation, the external system can transfer the parity information together with the data bytes. The Pentium checks whether the resulting parity is even and sends out the corresponding signal. When it sends out an address code, the Pentium does not perform an address parity check, but it does generate an even parity bit for the address.

this example, the transmission line consists of four conductors: one carries the serial data and three carry the timing signals (data selects). There are more sophisticated ways of sending the timing information, but we are using this direct method to illustrate a basic principle.

At the demultiplexer end of the system, the data-select signals and the serial data stream are applied to the demultiplexer. The data bits are distributed by the demultiplexer onto the output lines in the order in which they occurred on the multiplexer inputs. That is, D_0 comes out on the D_0 output, D_1 comes out on the D_1 output, and so on. The parity bit comes out on the D_7 output. These eight bits are temporarily stored and applied to the even parity checker. Not all of the bits are present on the parity checker inputs until the parity bit D_7 comes out and is stored. At this time, the error gate is enabled by the data-select code 111. If the parity is correct, a 0 appears on the Σ Even output, keeping the Error output at 0. If the parity is incorrect, all 1s appear on the error gate inputs, and a 1 on the Error output results.

This particular application has demonstrated the need for data storage so that you will be better able to appreciate the usefulness of the storage devices that will be introduced in Chapter 7 and used in other later chapters.

The timing diagram in Figure 6–61 illustrates a specific case in which two 8-bit words are transmitted, one with correct parity and one with an error.

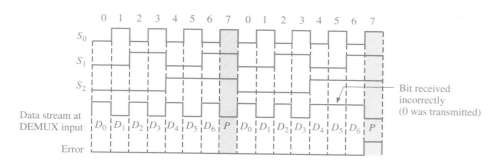

◄ **FIGURE 6–61**

Example of data transmission with and without error for the system in Figure 6–60.

SECTION 6–10 REVIEW

1. Add an even parity bit to each of the following codes:
   (a) 110100   (b) 01100011
2. Add an odd parity bit to each of the following codes:
   (a) 1010101   (b) 1000001
3. Check each of the even parity codes for an error.
   (a) 100010101   (b) 1110111001

6–11 TROUBLESHOOTING

In this section, the problem of decoder glitches is introduced and examined from a troubleshooting standpoint. A **glitch** is any undesired voltage or current spike (pulse) of very short duration. A glitch can be interpreted as a valid signal by a logic circuit and may cause improper operation.

After completing this section, you should be able to

■ Explain what a glitch is ■ Determine the cause of glitches in a decoder application
■ Use the method of output strobing to eliminate glitches

The 74LS138 was used as a DEMUX in the data transmission system in Figure 6–60. Now the 74HC138 is used as a 3-line-to-8-line decoder (binary-to-octal) in Figure 6–62 to illustrate how glitches occur and how to identify their cause. The $A_2A_1A_0$ inputs of the decoder are sequenced through a binary count, and the resulting waveforms of the inputs and outputs can be displayed on the screen of a logic analyzer, as shown in Figure 6–62. A_2 transitions are delayed from A_1 transitions and A_1 transitions are delayed from A_0 transitions. This commonly occurs when waveforms are generated by a binary counter, as you will learn in Chapter 8.

The output waveforms are correct except for the glitches that occur on some of the output signals. A logic analyzer or an oscilloscope can be used to display glitches, which are normally very difficult to see. Generally, the logic analyzer is preferred, especially for low repetition rates (less than 10 kHz) and/or irregular occurrence because most logic analyzers have a *glitch capture* capability. Oscilloscopes can be used to observe glitches with

▶ **FIGURE 6–62**

Decoder waveforms with output glitches.

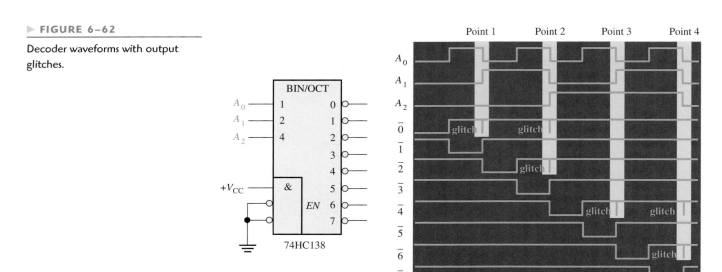

reasonable success, particularly if the glitches occur at a regular high repetition rate (greater than 10 kHz).

The points of interest indicated by the highlighted areas on the input waveforms in Figure 6–62 are displayed as shown in Figure 6–63. At point 1 there is a transitional state

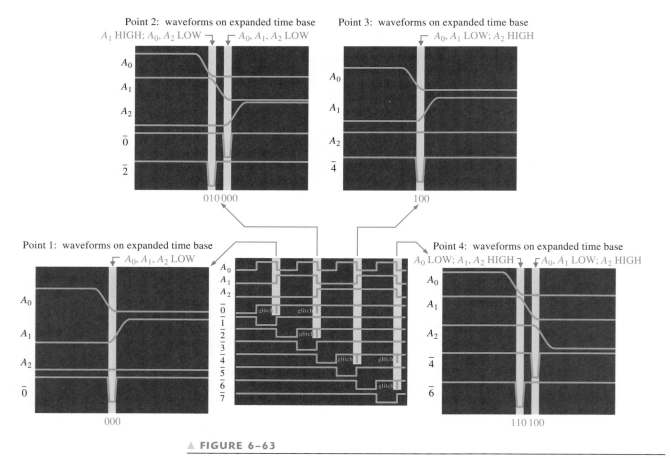

▲ **FIGURE 6–63**

Decoder waveform displays showing how transitional input states produce glitches in the output waveforms.

of 000 due to delay differences in the waveforms. This causes the first glitch on the $\bar{0}$ output of the decoder. At point 2 there are two transitional states, 010 and 000. These cause the glitch on the $\bar{2}$ output of the decoder and the second glitch on the $\bar{0}$ output, respectively. At point 3 the transitional state is 100, which causes the first glitch on the $\bar{4}$ output of the decoder. At point 4 the two transitional states, 110 and 100, result in the glitch on the $\bar{6}$ output and the second glitch on the $\bar{4}$ output, respectively.

One way to eliminate the glitch problem is a method called **strobing,** in which the decoder is enabled by a strobe pulse only during the times when the waveforms are not in transition. This method is illustrated in Figure 6–64.

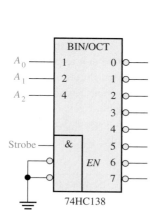

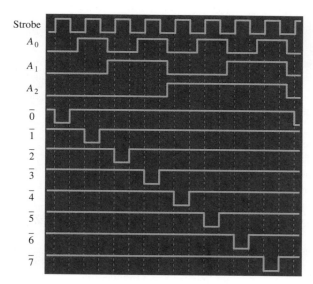

◀ FIGURE 6–64

Application of a strobe waveform to eliminate glitches on decoder outputs.

SECTION 6–11
REVIEW

1. Define the term *glitch*.
2. Explain the basic cause of glitches in decoder logic.
3. Define the term *strobe*.

Troubleshooting problems that are keyed to the CD-ROM are available in the Multisim Troubleshooting Practice section of the end-of-chapter problems.

HANDS ON TIP

In addition to glitches that are the result of propagation delays, as you have seen in the case of a decoder, other types of unwanted noise spikes can also be a problem. Current and voltage spikes on the V_{CC} and ground lines are caused by the fast switching waveforms in digital circuits. This problem can be minimized by proper printed circuit board layout. Switching spikes can be absorbed by decoupling the circuit board with a 1 μF capacitor from V_{CC} to ground. Also, smaller decoupling capacitors (0.022 μF to 0.1 μF) should be distributed at various points between V_{CC} and ground over the circuit board. Decoupling should be done especially near devices that are switching at higher rates or driving more loads such as oscillators, counters, buffers, and bus drivers.

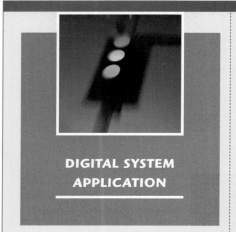

DIGITAL SYSTEM APPLICATION

In this digital system application, you begin working with a traffic light control system. In this section, the system requirements are established, a general block diagram is developed, and a state diagram is created to define the sequence of operation. A portion of the system involving combinational logic is designed and methods of testing are considered. The timing and sequential portions of the system will be dealt with in Chapters 7 and 8.

General System Requirements

A digital controller is required to control a traffic light at the intersection of a busy main street and an occasionally used side street. The main street is to have a green light for a minimum of 25 s or as long as there is no vehicle on the side street. The side street is to have a green light until there is no vehicle on the side street or for a maximum of 25 s. There is to be a 4 s caution light (yellow) between changes from green to red on both the main street and on the side street. These requirements are illustrated in the pictorial diagram in Figure 6–65.

Developing a Block Diagram of the System

From the requirements, you can develop a block diagram of the system. First, you know that the system must control six different pairs of lights. These are the red, yellow, and green lights for both directions on the main street and the red, yellow, and green lights for both directions on the side street. Also, you know that there is one external input (other than power) from a side street vehicle sensor. Figure 6–66 is a minimal block diagram showing these requirements.

Using the minimal system block diagram, you can begin to fill in the details. The system has four states, as indicated in Figure 6–65, so a logic circuit is needed to control the sequence of states (sequential logic). Also, circuits are needed to generate the proper time intervals of 25 s and 4 s that are required in the system and to generate a clock signal for cycling the system (timing circuits). The time intervals (long and short) and the vehicle sensor are inputs to the sequential logic because the sequencing of states is a function of these variables. Logic circuits are also needed to determine which of the four states the system is in at any given time, to generate the proper outputs to the lights (state decoder and light output logic), and to initiate the long and short time intervals. Interface circuits are included in the traffic light and interface unit to convert the output levels of the light output logic to the voltages and currents required to turn on each of the lights. Figure 6–67 is a more detailed block diagram showing these essential elements.

The State Diagram

A state diagram graphically shows the sequence of states in a system and the conditions for each state and for transitions from one state to the next. Actually, Figure 6–65 is a form of state diagram because it shows the sequence of states and the conditions.

Definition of Variables   Before a traditional state diagram can be developed,

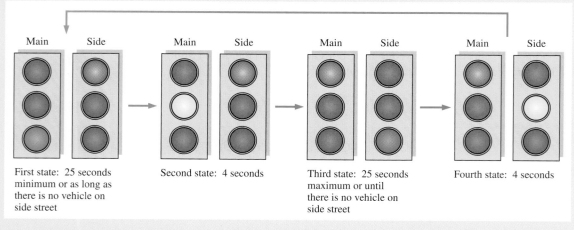

First state: 25 seconds minimum or as long as there is no vehicle on side street

Second state: 4 seconds

Third state: 25 seconds maximum or until there is no vehicle on side street

Fourth state: 4 seconds

▲ **FIGURE 6–65**

Requirements for the traffic light sequence.

► FIGURE 6-66

A minimal system block diagram.

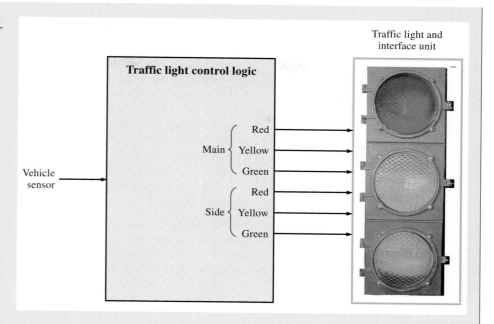

▲ FIGURE 6-67

System block diagram showing the essential elements.

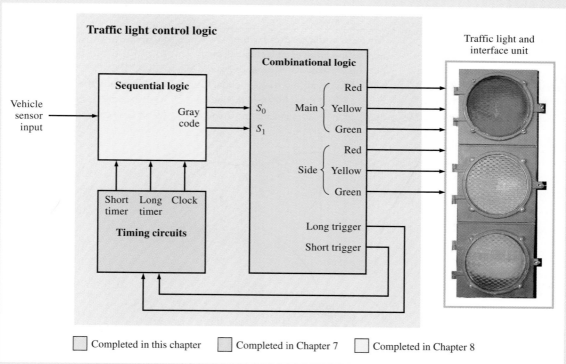

Completed in this chapter   Completed in Chapter 7   Completed in Chapter 8

the variables that determine how the system sequences through its states must be defined. These variables and their symbols are listed as follows:

■ Vehicle present on side street = V_s

■ 25 s timer (long timer) is on = T_L

■ 4 s timer (short timer) is on = T_s

The use of complemented variables indicates the opposite conditions. For example, $\overline{V_s}$ indicates that there is no vehicle on the side street, $\overline{T_L}$ indicates the long timer is off, $\overline{T}_S$ indicates the short timer is off.

Description of the State Diagram A state diagram is shown in Figure 6–68. Each of the four states is labeled according to the 2-bit Gray code sequence, as indicated by the circles. The looping arrow at each state indicates that the system remains in that state under the condition defined by the associated variable or expression. Each of the arrows going from one state to the next indicates a state transition under the condition defined by the associated variable or expression.

First state The Gray code for this state is 00. The main street light is green and the side street light is red. The system remains in this state for at least 25 s when the long timer is *on* or as long as there is no vehicle on the side street ($T_L + \overline{V}_s$). The system goes to the next state when the 25 s timer is *off* and there is a vehicle on the side street ($\overline{T}_L V_s$).

Second state The Gray code for this state is 01. The main street light is yellow (caution) and the side street light is red. The system remains in this state for 4 s when the short timer is *on* (T_S) and goes to the next state when the short timer goes *off* ($\overline{T}_S$).

Third state The Gray code for this state is 11. The main street light is red and the side street light is green. The system remains in this state when the long timer is *on* and there is a vehicle on the side street ($T_L V_s$). The system goes to the next state when the 25 s have elapsed or when there is no vehicle on the side street, whichever comes first ($\overline{T}_L + \overline{V}_s$).

Fourth state The Gray code for this state is 10. The main street light is red and the side street light is yellow. The system remains in this state for 4 s when the short timer is *on* (T_S) and goes back to the first state when the short timer goes *off* ($\overline{T}_S$).

The Combinational Logic

The focus in this chapter's system application is the combinational logic portion of the block diagram of Figure 6–67. The timing and the sequential logic circuits will be the subjects of the system application sections in Chapters 7 and 8.

A block diagram for the combinational logic portion of the system is developed as the first step in designing the logic. The three functions that this logic must perform are defined as follows, and the resulting diagram with a block for each of the three functions is shown in Figure 6–69:

■ **State Decoder** Decodes the 2-bit Gray code from the sequential logic to determine which of the four states the system is in.

■ **Light Output Logic** Uses the decoded state to activate the appropriate traffic lights for the main and side street light units.

■ **Trigger Logic** Uses the decoded states to produce signals for properly initiating (triggering) the long timer and the short timer.

Implementation of the Combinational Logic

Implementing the Decoder Logic The state decoder portion has two inputs (2-bit Gray code) and an output for each

▶ **FIGURE 6–68**

State diagram for the traffic light control system showing the Gray code sequence.

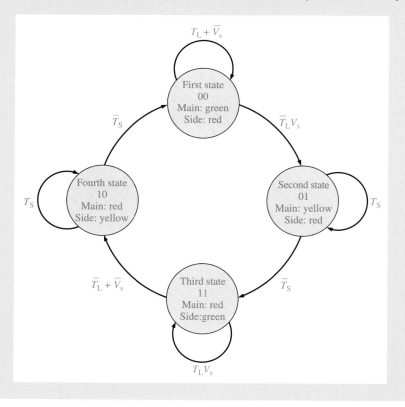

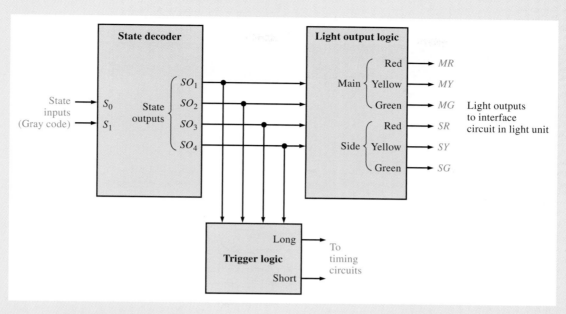

▲ **FIGURE 6–69**

Block diagram of the combinational logic.

▶ **FIGURE 6–70**

The state decoder logic.

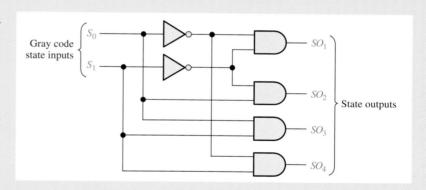

of the four states, as shown in Figure 6–70. The two Gray code inputs are designated S_0 and S_1 and the four state outputs are labeled SO_1, SO_2, SO_3, and SO_4. The Boolean expressions for the state outputs are as follows:

$$SO_1 = \bar{S}_1\bar{S}_0$$
$$SO_2 = \bar{S}_1 S_0$$
$$SO_3 = S_1 S_0$$
$$SO_4 = S_1 \bar{S}_0$$

The truth table for this state decoder logic is shown in Table 6–11.

Implementing the Light Output Logic The light output logic takes the four state outputs and produces six outputs for activating the traffic lights. These outputs are designated MR, MY, MG (for main red, main yellow, and main green) and SR, SY, SG (for side red, side yellow, and side green). Referring to the truth table in Table 6–11, you can see that the traffic light outputs can be expressed as

$$MR = SO_3 + SO_4$$
$$MY = SO_2$$
$$MG = SO_1$$

$$SR = SO_1 + SO_2$$
$$SY = SO_4$$
$$SG = SO_3$$

The output logic is implemented as shown in Figure 6–71.

Implementing the Trigger Logic The trigger logic produces two outputs. The *long* output is a LOW-to-HIGH transition that triggers the 25 s timing circuit when the system goes into the first (00) or third states (11). The *short* output is a LOW-to-HIGH transition that triggers the 4 s timing

▼ TABLE 6–11

Truth table for the combinational logic.

STATE INPUTS		STATE OUTPUTS				LIGHT OUTPUTS						TRIGGER OUTPUTS	
S_1	S_0	SO_1	SO_2	SO_3	SO_4	MR	MY	MG	SR	SY	SG	LONG	SHORT
0	0	1	0	0	0	0	0	1	1	0	0	1	0
0	1	0	1	0	0	0	1	0	1	0	0	0	1
1	1	0	0	1	0	1	0	0	0	0	1	1	0
1	0	0	0	0	1	1	0	0	0	1	0	0	1

State outputs are active-HIGH and light outputs are active-HIGH. *MR* stands for main street red, *SG* for side street green, etc.

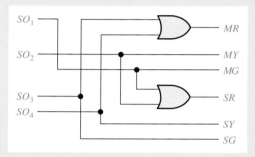

▲ FIGURE 6–71

The light output logic.

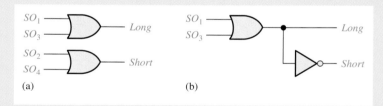

(a)   (b)

▲ FIGURE 6–72

The trigger logic.

circuit when the system goes into the second (01) or fourth (10) states. The trigger outputs are shown in Table 6–11 and in equation form as follows:

$$Long\ trigger = SO_1 + SO_3$$
$$Short\ trigger = SO_2 + SO_4$$

The trigger logic is shown in Figure 6–72(a). Table 6–11 also shows that the *Long* output and the *Short* output are complements, so the logic can also be implemented with one OR gate and one inverter, as shown in part (b).

Figure 6–73 shows the complete combinational logic that combines the state decoder, light output logic, and trigger logic.

System Assignment

■ *Activity 1* Apply waveforms for the 2-bit Gray code on the S_0 and S_1 inputs of the combinational logic and develop all of the output waveforms.

■ *Activity 2* Show how you would implement the combinational logic with 74XX functions.

■ *Optional Activity* Write a VHDL program describing the combinational logic.

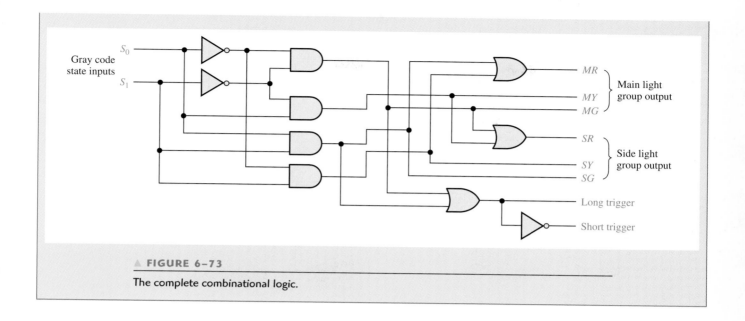

▲ **FIGURE 6–73**

The complete combinational logic.

SUMMARY

- Half-adder and full-adder operations are summarized in Figure 6–74.
- Logic symbols with pin numbers for the ICs used in this chapter are shown in Figure 6–75. Pin designations may differ from some manufacturers' data sheets.
- Standard logic functions from the 74XX series are available for use in a programmable logic design.

▶ **FIGURE 6–74**

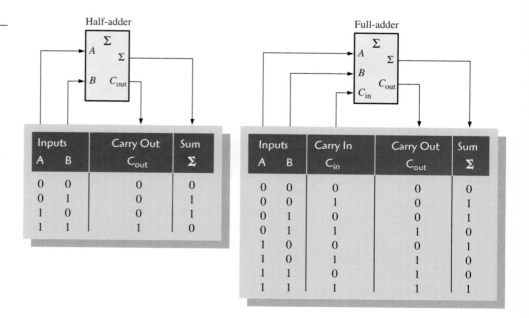

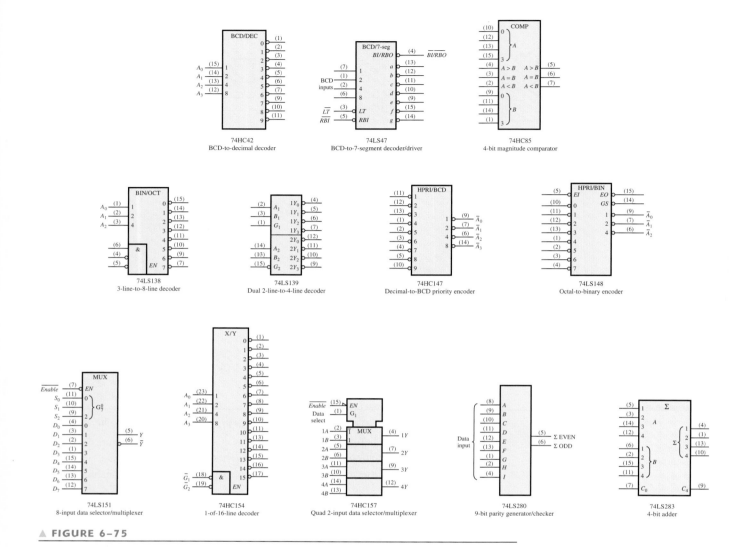

▲ FIGURE 6–75

Key terms and other bold terms in the chapter are defined in the end-of-book glossary.

Cascading Connecting the output of one device to the input of a similar device, allowing one device to drive another in order to expand the operational capability.

Decoder A digital circuit that converts coded information into a familiar or noncoded form.

Demultiplexer (DEMUX) A circuit that switches digital data from one input line to several output lines in a specified time sequence.

Encoder A digital circuit that converts information to a coded form.

Full-adder A digital circuit that adds two bits and an input carry to produce a sum and an output carry.

Glitch A voltage or current spike of short duration, usually unintentionally produced and unwanted.

Half-adder A digital circuit that adds two bits and produces a sum and an output carry. It cannot handle input carries.

Look-ahead carry A method of binary addition whereby carries from preceding adder stages are anticipated, thus eliminating carry propagation delays.

Multiplexer (MUX) A circuit that switches digital data from several input lines onto a single output line in a specified time sequence.

Parity bit A bit attached to each group of information bits to make the total number of 1s odd or even for every group of bits.

Priority encoder An encoder in which only the highest value input digit is encoded and any other active input is ignored.

Ripple carry A method of binary addition in which the output carry from each adder becomes the input carry of the next higher-order adder.

SELF-TEST

Answers are at the end of the chapter.

1. A half-adder is characterized by
   (a) two inputs and two outputs
   (b) three inputs and two outputs
   (c) two inputs and three outputs
   (d) two inputs and one output

2. A full-adder is characterized by
   (a) two inputs and two outputs
   (b) three inputs and two outputs
   (c) two inputs and three outputs
   (d) two inputs and one output

3. The inputs to a full-adder are $A = 1$, $B = 1$, $C_{in} = 0$. The outputs are
   (a) $\Sigma = 1$, $C_{out} = 1$
   (b) $\Sigma = 1$, $C_{out} = 0$
   (c) $\Sigma = 0$, $C_{out} = 1$
   (d) $\Sigma = 0$, $C_{out} = 0$

4. A 4-bit parallel adder can add
   (a) two 4-bit binary numbers
   (b) two 2-bit binary numbers
   (c) four bits at a time
   (d) four bits in sequence

5. To expand a 4-bit parallel adder to an 8-bit parallel adder, you must
   (a) use four 4-bit adders with no interconnections
   (b) use two 4-bit adders and connect the sum outputs of one to the bit inputs of the other
   (c) use eight 4-bit adders with no interconnections
   (d) use two 4-bit adders with the carry output of one connected to the carry input of the other

6. If a 74HC85 magnitude comparator has $A = 1011$ and $B = 1001$ on its inputs, the outputs are
   (a) $A > B = 0, A < B = 1, A = B = 0$
   (b) $A > B = 1, A < B = 0, A = B = 0$
   (c) $A > B = 1, A < B = 1, A = B = 0$
   (d) $A > B = 0, A < B = 0, A = B = 1$

7. If a 1-of-16 decoder with active-LOW outputs exhibits a LOW on the decimal 12 output, what are the inputs?
   (a) $A_3A_2A_1A_0 = 1010$
   (b) $A_3A_2A_1A_0 = 1110$
   (c) $A_3A_2A_1A_0 = 1100$
   (d) $A_3A_2A_1A_0 = 0100$

8. A BCD-to-7 segment decoder has 0100 on its inputs. The active outputs are
   (a) a, c, f, g
   (b) b, c, f, g
   (c) b, c, e, f
   (d) b, d, e, g

9. If an octal-to-binary priority encoder has its 0, 2, 5, and 6 inputs at the active level, the active-HIGH binary output is
   (a) 110
   (b) 010
   (c) 101
   (d) 000

10. In general, a multiplexer has
    (a) one data input, several data outputs, and selection inputs
    (b) one data input, one data output, and one selection input
    (c) several data inputs, several data outputs, and selection inputs
    (d) several data inputs, one data output, and selection inputs

11. Data selectors are basically the same as
    (a) decoders
    (b) demultiplexers
    (c) multiplexers
    (d) encoders

12. Which of the following codes exhibit even parity?
    (a) 10011000
    (b) 01111000
    (c) 11111111
    (d) 11010101
    (e) all
    (f) both answers (b) and (c)

PROBLEMS

Answers to odd-numbered problems are at the end of the book.

SECTION 6–1 **Basic Adders**

1. For the full-adder of Figure 6–4, determine the logic state (1 or 0) at each gate output for the following inputs:

   (a) $A = 1, B = 1, C_{in} = 1$ (b) $A = 0, B = 1, C_{in} = 1$ (c) $A = 0, B = 1, C_{in} = 0$

2. What are the full-adder inputs that will produce each of the following outputs:

   (a) $\Sigma = 0, C_{out} = 0$ (b) $\Sigma = 1, C_{out} = 0$
   (c) $\Sigma = 1, C_{out} = 1$ (d) $\Sigma = 0, C_{out} = 1$

3. Determine the outputs of a full-adder for each of the following inputs:

   (a) $A = 1, B = 0, C_{in} = 0$ (b) $A = 0, B = 0, C_{in} = 1$
   (c) $A = 0, B = 1, C_{in} = 1$ (d) $A = 1, B = 1, C_{in} = 1$

SECTION 6–2 **Parallel Binary Adders**

4. For the parallel adder in Figure 6–76, determine the complete sum by analysis of the logical operation of the circuit. Verify your result by longhand addition of the two input numbers.

▶ **FIGURE 6–76**

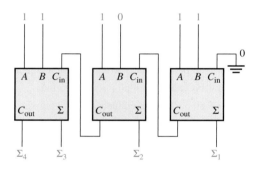

5. Repeat Problem 4 for the circuit and input conditions in Figure 6–77.

▶ **FIGURE 6–77**

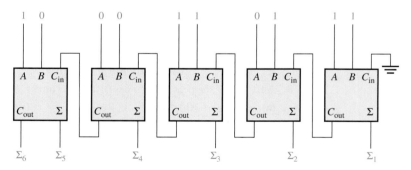

6. The input waveforms in Figure 6–78 are applied to a 2-bit adder. Determine the waveforms for the sum and the output carry in relation to the inputs by constructing a timing diagram.

▶ **FIGURE 6–78**

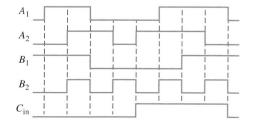

7. The following sequences of bits (right-most bit first) appear on the inputs to a 4-bit parallel adder. Determine the resulting sequence of bits on each sum output.

A_1 1001
A_2 1110
A_3 0000
A_4 1011
B_1 1111
B_2 1100
B_3 1010
B_4 0010

8. In the process of checking a 74LS283 4-bit parallel adder, the following voltage levels are observed on its pins: 1-HIGH, 2-HIGH, 3-HIGH, 4-HIGH, 5-LOW, 6-LOW, 7-LOW, 9-HIGH, 10-LOW, 11-HIGH, 12-LOW, 13-HIGH, 14-HIGH, and 15-HIGH. Determine if the IC is functioning properly.

SECTION 6–3   Ripple Carry Versus Look-Ahead Carry Adders

9. Each of the eight full-adders in an 8-bit parallel ripple carry adder exhibits the following propagation delays:

A to Σ and C_{out}:   40 ns
B to Σ and C_{out}:   40 ns
C_{in} to Σ: 35 ns
C_{in} to C_{out}:   25 ns

Determine the maximum total time for the addition of two 8-bit numbers.

10. Show the additional logic circuitry necessary to make the 4-bit look-ahead carry adder in Figure 6–18 into a 5-bit adder.

SECTION 6–4   Comparators

11. The waveforms in Figure 6–79 are applied to the comparator as shown. Determine the output $(A = B)$ waveform.

▶ **FIGURE 6–79**

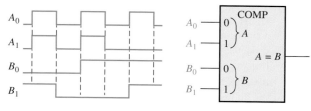

12. For the 4-bit comparator in Figure 6–80, plot each output waveform for the inputs shown. The outputs are active-HIGH.

▶ **FIGURE 6–80**

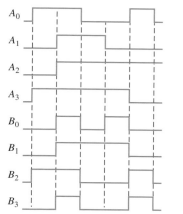
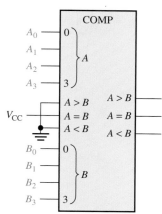

74HC85

13. For each set of binary numbers, determine the output states for the comparator of Figure 6–22.

(a) $A_3A_2A_1A_0 = 1100$ **(b)** $A_3A_2A_1A_0 = 1000$ **(c)** $A_3A_2A_1A_0 = 0100$

$B_3B_2B_1B_0 = 1001$ $B_3B_2B_1B_0 = 1011$ $B_3B_2B_1B_0 = 0100$

SECTION 6–5   Decoders

14. When a HIGH is on the output of each of the decoding gates in Figure 6–81, what is the binary code appearing on the inputs? The MSB is A_3.

▶ **FIGURE 6–81**

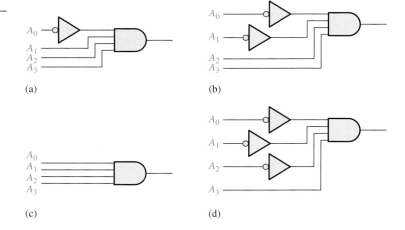

(a)   (b)

(c)   (d)

15. Show the decoding logic for each of the following codes if an active-HIGH (1) output is required:

(a) 1101   **(b)** 1000   **(c)** 11011   **(d)** 11100

(e) 101010   **(f)** 111110   **(g)** 000101   **(h)** 1110110

16. Solve Problem 15, given that an active-LOW (0) output is required.

17. You wish to detect only the presence of the codes 1010, 1100, 0001, and 1011. An active-HIGH output is required to indicate their presence. Develop the minimum decoding logic with a single output that will indicate when any one of these codes is on the inputs. For any other code, the output must be LOW.

18. If the input waveforms are applied to the decoding logic as indicated in Figure 6–82, sketch the output waveform in proper relation to the inputs.

▶ **FIGURE 6–82**

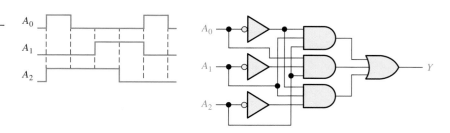

19. BCD numbers are applied sequentially to the BCD-to-decimal decoder in Figure 6–83. Draw a timing diagram, showing each output in the proper relationship with the others and with the inputs.

▶ **FIGURE 6–83**

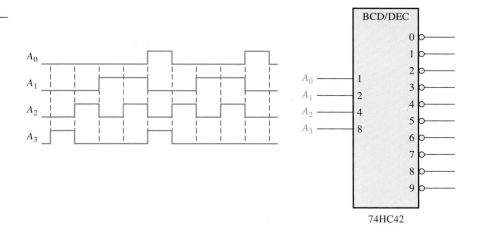

20. A 7-segment decoder/driver drives the display in Figure 6–84. If the waveforms are applied as indicated, determine the sequence of digits that appears on the display.

▶ **FIGURE 6–84**

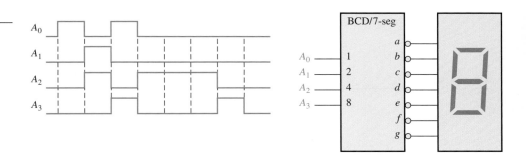

SECTION 6–6 **Encoders**

21. For the decimal-to-BCD encoder logic of Figure 6–38, assume that the 9 input and the 3 input are both HIGH. What is the output code? Is it a valid BCD (8421) code?

22. A 74HC147 encoder has LOW levels on pins 2, 5, and 12. What BCD code appears on the outputs if all the other inputs are HIGH?

SECTION 6–7 **Code Converters**

23. Convert the following decimal numbers to BCD and then to binary.

(a) 2   (b) 8   (c) 13   (d) 26   (e) 33

24. Show the logic required to convert a 10-bit binary number to Gray code, and use that logic to convert the following binary numbers to Gray code:

(a) 1010101010   (b) 1111100000   (c) 0000001110   (d) 1111111111

25. Show the logic required to convert a 10-bit Gray code to binary, and use that logic to convert the following Gray code words to binary:

(a) 1010000000   (b) 0011001100   (c) 1111000111   (d) 0000000001

SECTION 6–8 Multiplexers (Data Selectors)

26. For the multiplexer in Figure 6–85, determine the output for the following input states:
$D_0 = 0, D_1 = 1, D_2 = 1, D_3 = 0, S_0 = 1, S_1 = 0$.

▶ FIGURE 6–85

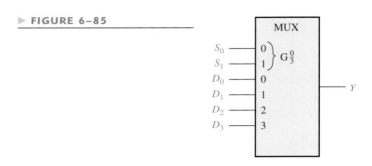

27. If the data-select inputs to the multiplexer in Figure 6–85 are sequenced as shown by the waveforms in Figure 6–86, determine the output waveform with the data inputs specified in Problem 26.

▶ FIGURE 6–86

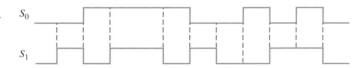

28. The waveforms in Figure 6–87 are observed on the inputs of a 74LS151 8-input multiplexer. Sketch the Y output waveform.

▶ FIGURE 6–87

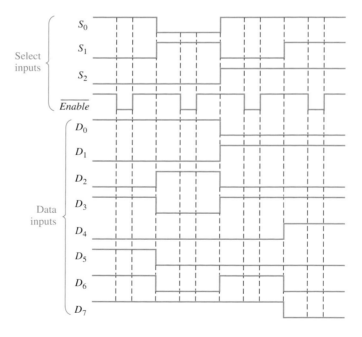

SECTION 6–9 Demultiplexers

29. Develop the total timing diagram (inputs and outputs) for a 74HC154 used in a demultiplexing application in which the inputs are as follows: The data-select inputs are repetitively sequenced through a straight binary count beginning with 0000, and the data input is a serial data stream carrying BCD data representing the decimal number 2468. The least significant digit (8) is first in the sequence, with its LSB first, and it should appear in the first 4-bit positions of the output.

SECTION 6-10 **Parity Generators/Checkers**

30. The waveforms in Figure 6–88 are applied to the 4-bit parity logic. Determine the output waveform in proper relation to the inputs. For how many bit times does even parity occur, and how is it indicated? The timing diagram includes eight bit times.

▶ FIGURE 6–88

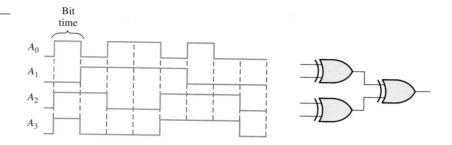

31. Determine the Σ Even and the Σ Odd outputs of a 74LS280 9-bit parity generator/checker for the inputs in Figure 6–89. Refer to the function table in Figure 6–59.

▶ FIGURE 6–89

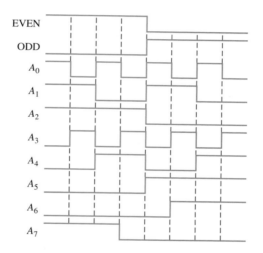

SECTION 6-11 **Troubleshooting**

32. The full-adder in Figure 6–90 is tested under all input conditions with the input waveforms shown. From your observation of the Σ and C_{out} waveforms, is it operating properly, and if not, what is the most likely fault?

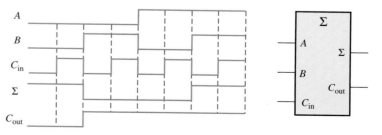

▲ FIGURE 6–90

33. List the possible faults for each decoder/display in Figure 6–91.

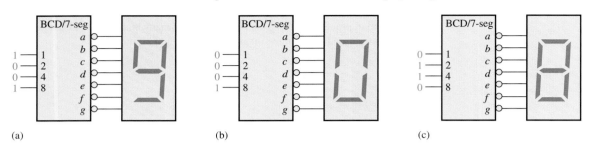

(a)   (b)   (c)

▲ **FIGURE 6–91**

34. Develop a systematic test procedure to check out the complete operation of the keyboard encoder in Figure 6–42.

35. You are testing a BCD-to-binary converter consisting of 4-bit adders as shown in Figure 6–92. First verify that the circuit converts BCD to binary. The test procedure calls for applying BCD numbers in sequential order beginning with 0_{10} and checking for the correct binary output. What symptom or symptoms will appear on the binary outputs in the event of each of the following faults? For what BCD number is each fault *first* detected?

(a) The A_1 input is open (top adder).

(b) The C_{out} is open (top adder).

(c) The Σ_4 output is shorted to ground (top adder).

(d) The 32 output is shorted to ground (bottom adder).

▶ **FIGURE 6–92**

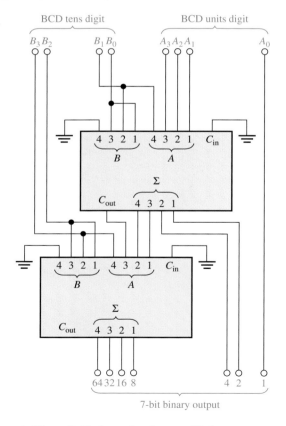

36. For the display multiplexing system in Figure 6–52, determine the most likely cause or causes for each of the following symptoms:

(a) The *B*-digit (MSD) display does not turn on at all.

(b) Neither 7-segment display turns on.

(c) The *f*-segment of both displays appears to be on all the time.

(d) There is a visible flicker on the displays.

37. Develop a systematic procedure to fully test the 74LS151 data selector IC.

38. During the testing of the data transmission system in Figure 6–60, a code is applied to the D_0 through D_6 inputs that contains an odd number of 1s. A single bit error is deliberately introduced on the serial data transmission line between the MUX and the DEMUX, but the system does not indicate an error (error output = 0). After some investigation, you check the inputs to the even parity checker and find that D_0 through D_6 contain an even number of 1s, as you would expect. Also, you find that the D_7 parity bit is a 1. What are the possible reasons for the system not indicating the error?

39. In general, describe how you would fully test the data transmission system in Figure 6–60, and specify a method for the introduction of parity errors.

Digital System Application

40. The light output logic can be implemented in the system application with fixed-function logic using a 74LS08 with the AND gates operating as negative-NOR gates. Use a 74LS00 (quad NAND gates) and any other devices that may be required to produce active-LOW outputs for the given inputs.

41. Implement the light output logic with the 74LS00 if active-LOW outputs are required.

Special Design Problems

42. Modify the design of the 7-segment display multiplexing system in Figure 6–52 to accommodate two additional digits.

43. Using Table 6–2, write the SOP expressions for the Σ and C_{out} of a full-adder. Use a Karnaugh map to minimize the expressions and then implement them with inverters and AND-OR logic. Show how you can replace the AND-OR logic with 74LS151 data selectors.

44. Implement the logic function specified in Table 6–12 by using a 74LS151 data selector.

▶ **TABLE 6–12**

INPUTS				OUTPUT
A_3	A_2	A_1	A_0	Y
0	0	0	0	0
0	0	0	1	0
0	0	1	0	1
0	0	1	1	1
0	1	0	0	0
0	1	0	1	0
0	1	1	0	1
0	1	1	1	1
1	0	0	0	1
1	0	0	1	0
1	0	1	0	1
1	0	1	1	1
1	1	0	0	0
1	1	0	1	1
1	1	1	0	0
1	1	1	1	1

45. Using two of the 6-position adder modules from Figure 6–14, design a 12-position voting system.

46. The adder block in the tablet-counting and control system in Figure 6–93 performs the addition of the 8-bit binary number from the counter and the 16-bit binary number from Register B. The result from the adder goes back into Register B. Use 74LS283s to implement this function and draw a complete logic diagram including pin numbers. Refer to Chapter 1 system application to review the operation.

47. Use 74HC85s to implement the comparator block in the tablet counting and control system in Figure 6–93 and draw a complete logic diagram including pin numbers. The comparator compares the 8-bit binary number (actually only seven bits are required) from the BCD-to-binary converter with the 8-bit binary number from the counter.

48. Two BCD-to-7-segment decoders are used in the tablet-counting and control system in Figure 6–93. One is required to drive the 2-digit *tablets/bottle* display and the other to drive the 5-digit *total tablets bottled* display. Use 74LS47s to implement each decoder and draw a complete logic diagram including pin numbers.

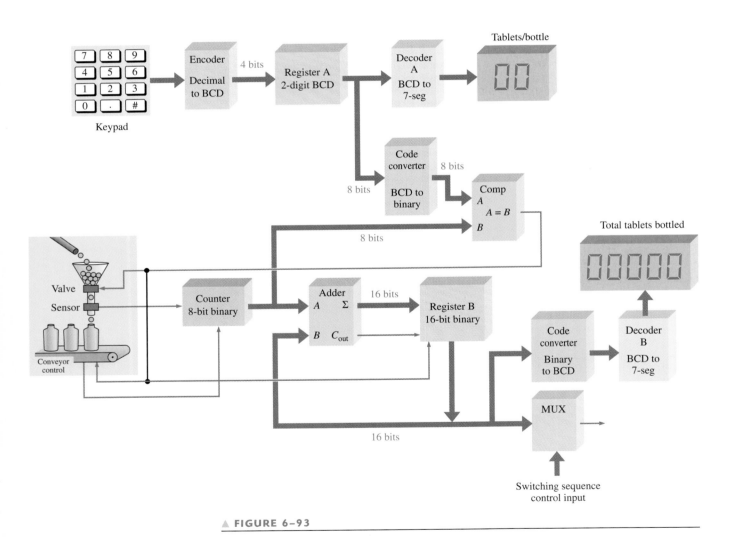

▲ FIGURE 6–93

49. The encoder shown in the system block diagram of Figure 6–93 encodes each decimal key closure and converts it to BCD. Use a 74HC147 to implement this function and draw a complete logic diagram including pin numbers.

50. The system in Figure 6–93 requires two code converters. The BCD-to-binary converter changes the 2-digit BCD number in Register A to an 8-bit binary code (actually only 7 bits are required because the MSB is always 0). Use appropriate IC code converters to implement the BCD-to-binary converter function and draw a complete logic diagram including pin numbers.

MULTISIM TROUBLESHOOTING PRACTICE

51. Open file P06-51 and test the logic circuit to determine if there is a fault. If there is a fault, identify it if possible.

52. Open file P06-52 and test the logic circuit to determine if there is a fault. If there is a fault, identify it if possible.

53. Open file P06-53 and test the logic circuit to determine if there is a fault. If there is a fault, identify it if possible.

54. Open file P06-54 and test the logic circuit to determine if there is a fault. If there is a fault, identify it if possible.

ANSWERS

SECTION REVIEWS

SECTION 6–1 **Basic Adders**

1. (a) $\Sigma = 1, C_{out} = 0$ (b) $\Sigma = 0, C_{out} = 0$
   (c) $\Sigma = 1, C_{out} = 0$ (d) $\Sigma = 0, C_{out} = 1$
2. $\Sigma = 1, C_{out} = 1$

SECTION 6–2 **Parallel Binary Adders**

1. $C_{out}\Sigma_4\Sigma_3\Sigma_2\Sigma_1 = 11001$
2. Three 74LS283s are required to add two 10-bit numbers.

SECTION 6–3 **Ripple Carry vs. Look-Ahead Carry Adders**

1. $C_g = 0, C_p = 1$
2. $C_{out} = 1$

SECTION 6–4 **Comparators**

1. $A > B = 1, A < B = 0, A = B = 0$ when $A = 1011$ and $B = 1010$
2. Right comparator: pin 7: $A < B = 1$; pin 6: $A = B = 0$; pin 5: $A > B = 0$
   Left comparator: pin 7: $A < B = 0$; pin 6: $A = B = 0$; pin 5: $A > B = 1$

SECTION 6–5 **Decoders**

1. Output 5 is active when 101 is on the inputs.
2. Four 74HC154s are used to decode a 6-bit binary number.
3. Active-LOW output drives a common-cathode LED display.

SECTION 6–6 **Encoders**

1. (a) $A_0 = 1, A_1 = 1, A_2 = 0, A_3 = 1$
   (b) No, this is not a valid BCD code.
   (c) Only one input can be active for a valid output.
2. (a) $\overline{A}_3 = 0, \overline{A}_2 = 1, \overline{A}_1 = 1, \overline{A}_0 = 1$
   (b) The output is 0111, which is the complement of 1000 (8).

SECTION 6–7   **Code Converters**

1. 10000101 (BCD) $= 1010101_2$

2. An 8-bit binary-to-Gray converter consists of seven exclusive-OR gates in an arrangement like that in Figure 6–43.

SECTION 6–8   **Multiplexers (Data Selectors)**

1. The output is 0.

2. **(a)** 74LS157: Quad 2-input data selector

   (b) 74LS151: 8-input data selector

3. The data output alternates between LOW and HIGH as the data-select inputs sequence through the binary states.

4. **(a)** The 74HC157 multiplexes the two BCD codes to the 7-segment decoder.

   (b) The 74LS47 decodes the BCD to energize the display.

   (c) The 74LS139 enables the 7-segment displays alternately.

SECTION 6–9   **Demultiplexers**

1. A decoder can be used as a multiplexer by using the input lines for data selection and an Enable line for data input.

2. The outputs are all HIGH except D_{10}, which is LOW.

SECTION 6–10   **Parity Generators/Checkers**

1. **(a)** Even parity: $\underline{1}110100$ **(b)** Even parity: $\underline{0}01100011$

2. **(a)** Odd parity: $\underline{1}1010101$ **(b)** Odd parity: $\underline{1}1000001$

3. **(a)** Code is correct, four 1s.   **(b)** Code is in error, seven 1s

SECTION 6–11   **Troubleshooting**

1. A glitch is a very short-duration voltage spike (usually unwanted).

2. Glitches are caused by transition states.

3. Strobe is the enabling of a device for a specified period of time when the device is not in transition.

RELATED PROBLEMS FOR EXAMPLES

6–1 $\Sigma = 1$, $C_{out} = 1$ **6–2** $\Sigma_1 = 0$, $\Sigma_2 = 0$, $\Sigma_3 = 1$, $\Sigma_4 = 1$

6–3 $1011 + 1010 = 10101$ **6–4**   See Figure 6–94.

▶ FIGURE 6–94

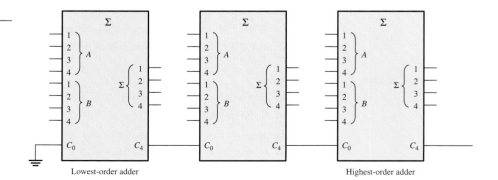

Lowest-order adder   Highest-order adder

6–5 See Figure 6–95.

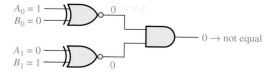

6–6 $A > B = 0, A = B = 0, A < B = 1$

6–7 See Figure 6–96.

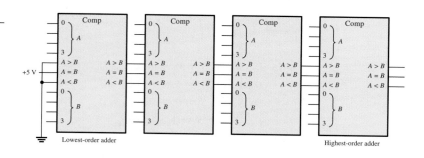

6–8 See Figure 6–97.

6–9 Output 22

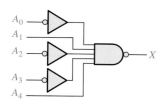

6–10 See Figure 6–98.

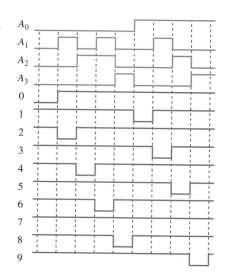

6–11 All inputs LOW: $\overline{A}_0 = 0$, $\overline{A}_1 = 1$, $\overline{A}_2 = 1$, $\overline{A}_3 = 0$

All inputs HIGH: All outputs HIGH.

6–12 BCD 01000001

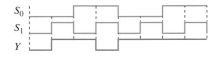

00000001	1
00101000	40
Binary 00101001	41

6–13 Seven exclusive-OR gates

6–14 See Figure 6–99.

▶ **FIGURE 6–99**

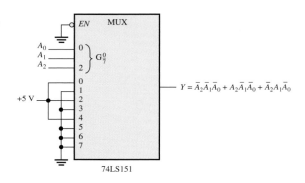

6–15 D_0: $S_3 = 0$, $S_2 = 0$, $S_1 = 0$, $S_0 = 0$

D_4: $S_3 = 0$, $S_2 = 1$, $S_1 = 0$, $S_0 = 0$

D_8: $S_3 = 1$, $S_2 = 0$, $S_1 = 0$, $S_0 = 0$

D_{13}: $S_3 = 1$, $S_2 = 1$, $S_1 = 0$, $S_0 = 1$

6–16 See Figure 6–100.

▶ **FIGURE 6–100**

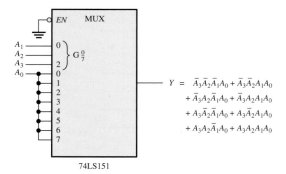

$Y = \overline{A}_2\overline{A}_1\overline{A}_0 + A_2\overline{A}_1\overline{A}_0 + \overline{A}_2A_1\overline{A}_0$

74LS151

6–17 See Figure 6–101.

▶ **FIGURE 6–101**

$Y = \overline{A}_3\overline{A}_2\overline{A}_1A_0 + \overline{A}_3\overline{A}_2A_1A_0$
$+ \overline{A}_3A_2\overline{A}_1A_0 + \overline{A}_3A_2A_1A_0$
$+ A_3\overline{A}_2\overline{A}_1A_0 + A_3\overline{A}_2A_1A_0$
$+ A_3A_2\overline{A}_1A_0 + A_3A_2A_1A_0$

74LS151

6–18 See Figure 6–102.

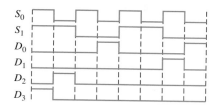

SELF-TEST

1. (a) **2.** (b) **3.** (c) **4.** (a) **5.** (d) **6.** (b) **7.** (c) **8.** (b)

9. (a) **10.** (d) **11.** (c) **12.** (f)

7

LATCHES, FLIP-FLOPS, AND TIMERS

CHAPTER OBJECTIVES

- Use logic gates to construct basic latches

- Explain the difference between an S-R latch and a D latch

- Recognize the difference between a latch and a flip-flop

- Explain how S-R, D, and J-K flip-flops differ

- Understand the significance of propagation delays, set-up time, hold time, maximum operating frequency, minimum clock pulse widths, and power dissipation in the application of flip-flops

- Apply flip-flops in basic applications

- Explain how retriggerable and nonretriggerable one-shots differ
- Connect a 555 timer to operate as either an astable multivibrator or a one-shot
- Troubleshoot basic flip-flop circuits

KEY TERMS

- Latch
- Bistable
- SET
- RESET
- Clock
- Edge-triggered flip-flop
- Synchronous
- D flip-flop
- J-K flip-flop
- Toggle

- Preset
- Clear
- Propagation delay time
- Set-up time
- Hold time
- Power dissipation
- One-shot
- Monostable
- Timer
- Astable

INTRODUCTION

This chapter begins a study of the fundamentals of sequential logic. Bistable, monostable, and astable logic devices called *multivibrators* are covered. Two categories of bistable devices are the latch and the flip-flop. Bistable devices have two stable states, called SET and RESET; they can retain either of these states indefinitely, making them useful as storage devices. The basic difference between latches and flip-flops is the way in which they are changed from one state to the other. The flip-flop is a basic building block for counters, registers, and other sequential control logic and is used in certain types of memories. The monostable multivibrator, commonly known as the one-shot, has only one stable state. A one-shot produces a single controlled-width pulse when activated or triggered. The astable multivibrator has no stable state and is used primarily as an oscillator, which is a self-sustained waveform generator. Pulse oscillators are used as the sources for timing waveforms in digital systems.

FIXED-FUNCTION LOGIC DEVICES

74XX74	74XX279	74XX122
555	74121	74XX75
74XX112		

■■■ DIGITAL SYSTEM APPLICATION PREVIEW

The Digital System Application continues with the traffic light control system from Chapter 6. The focus in this chapter is the timing circuit portion of the system that produces the clock, the long time interval for the red and green lights, and the short time interval for the caution light. The clock is used as the basic system timing signal for advancing the sequential logic through its states. The sequential logic will be developed in Chapter 8.

WWW. VISIT THE COMPANION WEBSITE
Study aids for this chapter are available at
http://www.prenhall.com/floyd

7–1   LATCHES

The **latch** is a type of temporary storage device that has two stable states (bistable) and is normally placed in a category separate from that of flip-flops. Latches are similar to flip-flops because they are bistable devices that can reside in either of two states using a feedback arrangement, in which the outputs are connected back to the opposite inputs. The main difference between latches and flip-flops is in the method used for changing their state.

After completing this section, you should be able to

- ■ Explain the operation of a basic S-R latch   ■ Explain the operation of a gated S-R latch   ■ Explain the operation of a gated D latch   ■ Implement an S-R or D latch with logic gates   ■ Describe the 74LS279 and 74LS75 quad latches

COMPUTER NOTE

Latches are sometimes used in computer systems for multiplexing data onto a bus. For example, data being input to a computer from an external source have to share the data bus with data from other sources. When the data bus becomes unavailable to the external source, the existing data must be temporarily stored, and latches placed between the external source and the data bus may be used to do this. When the data bus is unavailable to the external source, the latches must be disconnected from the bus using a method known as tristating. When the data bus becomes available, the external data pass through the latches, thus the term *transparent latch*. The gated D latch performs this function because when it is enabled, the data on its input appear on the output just as though there were a direct connection. Data on the input are stored as soon as the latch is disabled.

The S-R (SET-RESET) Latch

A latch is a type of **bistable** logic device or **multivibrator.** An active-HIGH input S-R (SET-RESET) latch is formed with two cross-coupled NOR gates, as shown in Figure 7–1(a); an active-LOW input $\overline{S}$-$\overline{R}$ latch is formed with two cross-coupled NAND gates, as shown in Figure 7–1(b). Notice that the output of each gate is connected to an input of the opposite gate. This produces the regenerative **feedback** that is characteristic of all latches and flip-flops.

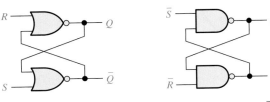

(a) Active-HIGH input S-R latch   (b) Active-LOW input $\overline{S}$-$\overline{R}$ latch

▲ **FIGURE 7–1**

Two versions of SET-RESET (S-R) latches. Open file F07-01 and verify the operation of both latches.

To explain the operation of the latch, we will use the NAND gate $\overline{S}$-$\overline{R}$ latch in Figure 7–1(b). This latch is redrawn in Figure 7–2 with the negative-OR equivalent symbols used for the NAND gates. This is done because LOWs on the $\overline{S}$ and $\overline{R}$ lines are the activating inputs.

The latch in Figure 7–2 has two inputs, $\overline{S}$ and $\overline{R}$, and two outputs, Q and $\overline{Q}$. Let's start by assuming that both inputs and the Q output are HIGH. Since the Q output is connected back to an input of gate G_2, and the $\overline{R}$ input is HIGH, the output of G_2 must be LOW. This LOW output is coupled back to an input of gate G_1, ensuring that its output is HIGH.

▶ **FIGURE 7–2**

Negative-OR equivalent of the NAND gate $\overline{S}$-$\overline{R}$ latch in Figure 9–1(b).

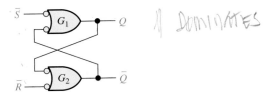

When the Q output is HIGH, the latch is in the **SET** state. It will remain in this state indefinitely until a LOW is temporarily applied to the $\overline{R}$ input. With a LOW on the $\overline{R}$ input and a HIGH on $\overline{S}$, the output of gate G_2 is forced HIGH. This HIGH on the $\overline{Q}$ output is coupled back to an input of G_1, and since the $\overline{S}$ input is HIGH, the output of G_1 goes LOW. This

LOW on the Q output is then coupled back to an input of G_2, ensuring that the $\overline{Q}$ output remains HIGH even when the LOW on the $\overline{R}$ input is removed. When the Q output is LOW, the latch is in the **RESET** state. Now the latch remains indefinitely in the RESET state until a LOW is applied to the $\overline{S}$ input.

In normal operation, the outputs of a latch are always complements of each other.

When Q is HIGH, $\overline{Q}$ is LOW, and when Q is LOW, $\overline{Q}$ is HIGH.

An invalid condition in the operation of an active-LOW input $\overline{S}$-$\overline{R}$ latch occurs when LOWs are applied to both $\overline{S}$ and $\overline{R}$ at the same time. As long as the LOW levels are simultaneously held on the inputs, both the Q and $\overline{Q}$ outputs are forced HIGH, thus violating the basic complementary operation of the outputs. Also, if the LOWs are released simultaneously, both outputs will attempt to go LOW. Since there is always some small difference in the propagation delay time of the gates, one of the gates will dominate in its transition to the LOW output state. This, in turn, forces the output of the slower gate to remain HIGH. In this situation, you cannot reliably predict the next state of the latch.

Figure 7–3 illustrates the active-LOW input $\overline{S}$-$\overline{R}$ latch operation for each of the four possible combinations of levels on the inputs. (The first three combinations are valid, but the last is not.) Table 7–1 summarizes the logic operation in truth table form. Operation of the active-HIGH input NOR gate latch in Figure 7–1(a) is similar but requires the use of opposite logic levels.

A latch can reside in either of its two states, SET or RESET.

SET means that the Q output is HIGH.

RESET means that the Q output is LOW.

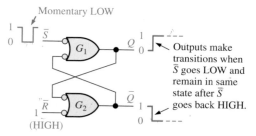

Latch starts out RESET ($Q = 0$).

Outputs make transitions when $\overline{S}$ goes LOW and remain in same state after $\overline{S}$ goes back HIGH.

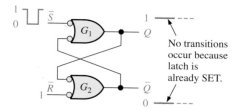

Latch starts out SET ($Q = 1$).

No transitions occur because latch is already SET.

(a) Two possibilities for the SET operation

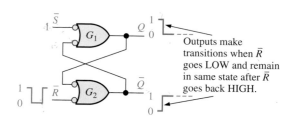

Latch starts out SET ($Q = 1$).

Outputs make transitions when $\overline{R}$ goes LOW and remain in same state after $\overline{R}$ goes back HIGH.

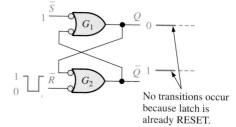

Latch starts out RESET ($Q = 0$).

No transitions occur because latch is already RESET.

(b) Two possibilities for the RESET operation

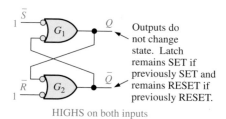

HIGHS on both inputs

Outputs do not change state. Latch remains SET if previously SET and remains RESET if previously RESET.

(c) No-change condition

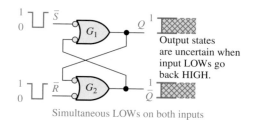

Simultaneous LOWs on both inputs

Output states are uncertain when input LOWs go back HIGH.

(d) Invalid condition

▲ **FIGURE 7–3**

The three modes of basic $\overline{S}$-$\overline{R}$ latch operation (SET, RESET, no-change) and the invalid condition.

INPUTS		OUTPUTS		COMMENTS
S̄	R̄	Q	Q̄	
1	1	NC	NC	No change. Latch remains in present state.
0	1	1	0	Latch SET.
1	0	0	1	Latch RESET.
0	0	1	1	Invalid condition

Logic symbols for both the active-HIGH input and the active-LOW input latches are shown in Figure 7–4.

► FIGURE 7–4

Logic symbols for the S-R and S̄-R̄ latch.

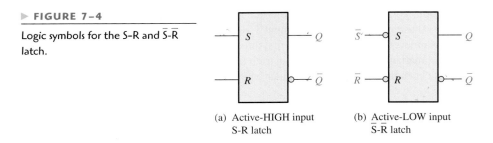

(a) Active-HIGH input
    S-R latch

(b) Active-LOW input
    S̄-R̄ latch

Example 7–1 illustrates how an active-LOW input S̄-R̄ latch responds to conditions on its inputs. LOW levels are pulsed on each input in a certain sequence and the resulting Q output waveform is observed. The S̄ = 0, R̄ = 0 condition is avoided because it results in an invalid mode of operation and is a major drawback of any SET-RESET type of latch.

EXAMPLE 7–1

If the S̄ and R̄ waveforms in Figure 7–5(a) are applied to the inputs of the latch in Figure 7–4(b), determine the waveform that will be observed on the Q output. Assume that Q is initially LOW.

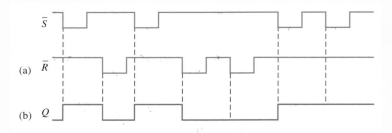

(a) S̄

     R̄

(b) Q

▲ FIGURE 7–5

Solution   See Figure 7–5(b).

Related Problem*   Determine the Q output of an active-HIGH input S-R latch if the waveforms in Figure 7–5(a) are inverted and applied to the inputs.

*Answers are at the end of the chapter.

An Application

The Latch as a Contact-Bounce Eliminator   A good example of an application of an $\overline{S}\text{-}\overline{R}$ latch is in the elimination of mechanical switch contact "bounce." When the pole of a switch strikes the contact upon switch closure, it physically vibrates or bounces several times before finally making a solid contact. Although these bounces are very short in duration, they produce voltage spikes that are often not acceptable in a digital system. This situation is illustrated in Figure 7–6(a).

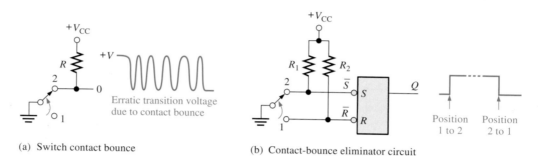

(a) Switch contact bounce   (b) Contact-bounce eliminator circuit

▲ **FIGURE 7–6**

The $\overline{S}\text{-}\overline{R}$ latch used to eliminate switch contact bounce.

An $\overline{S}\text{-}\overline{R}$ latch can be used to eliminate the effects of switch bounce as shown in Figure 7–6(b). The switch is normally in position 1, keeping the $\overline{R}$ input LOW and the latch RESET. When the switch is thrown to position 2, $\overline{R}$ goes HIGH because of the pull-up resistor to V_{CC}, and $\overline{S}$ goes LOW on the first contact. Although $\overline{S}$ remains LOW for only a very short time before the switch bounces, this is sufficient to set the latch. Any further voltage spikes on the $\overline{S}$ input due to switch bounce do not affect the latch, and it remains SET. Notice that the Q output of the latch provides a clean transition from LOW to HIGH, thus eliminating the voltage spikes caused by contact bounce. Similarly, a clean transition from HIGH to LOW is made when the switch is thrown back to position 1.

THE 74LS279 SET-RESET LATCH

The 74LS279 is a quad $\overline{S}\text{-}\overline{R}$ latch represented by the logic diagram of Figure 7–7(a) and the pin diagram in part (b). Notice that two of the latches each have two $\overline{S}$ inputs.

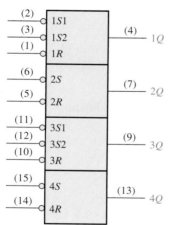

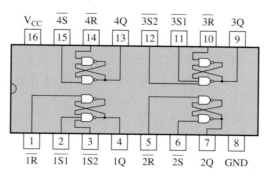

(a) Logic diagram

(b) Pin diagram

▲ **FIGURE 7–7**

The 74LS279 quad $\overline{S}\text{-}\overline{R}$ latch.

The Gated S-R Latch

A gated latch requires an enable input, *EN* (*G* is also used to designate an enable input). The logic diagram and logic symbol for a gated S-R latch are shown in Figure 7–8. The *S* and *R* inputs control the state to which the latch will go when a HIGH level is applied to the *EN* input. The latch will not change until *EN* is HIGH; but as long as it remains HIGH, the output is controlled by the state of the *S* and *R* inputs. In this circuit, the invalid state occurs when both *S* and *R* are simultaneously HIGH.

▶ FIGURE 7–8

A gated S-R latch.

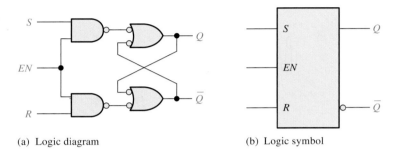

(a) Logic diagram   (b) Logic symbol

EXAMPLE 7–2

Determine the *Q* output waveform if the inputs shown in Figure 7–9(a) are applied to a gated S-R latch that is initially RESET.

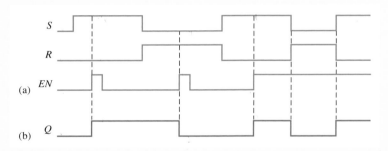

▲ FIGURE 7–9

Solution   The *Q* waveform is shown in Figure 7–9(b). When *S* is HIGH and *R* is LOW, a HIGH on the *EN* input sets the latch. When *S* is LOW and *R* is HIGH, a HIGH on the *EN* input resets the latch.

Related Problem   Determine the *Q* output of a gated S-R latch if the *S* and *R* inputs in Figure 7–9(a) are inverted.

The Gated D Latch

Another type of gated latch is called the D latch. It differs from the S-R latch because it has only one input in addition to *EN*. This input is called the *D* (data) input. Figure 7–10 contains a logic diagram and logic symbol of a D latch. When the *D* input is HIGH and the *EN* input is HIGH, the latch will set. When the *D* input is LOW and *EN* is HIGH, the latch will reset. Stated another way, the output *Q* follows the input *D* when *EN* is HIGH.

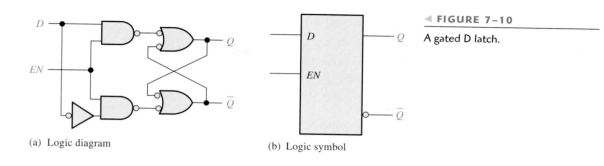

(a) Logic diagram

(b) Logic symbol

◀ **FIGURE 7–10**

A gated D latch.

EXAMPLE 7–3

Determine the Q output waveform if the inputs shown in Figure 7–11(a) are applied to a gated D latch, which is initially RESET.

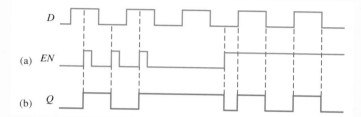

(a) D
     EN

(b) Q

▲ **FIGURE 7–11**

·Solution   The Q waveform is shown in Figure 7–11(b). When D is HIGH and EN is HIGH, Q goes HIGH. When D is LOW and EN is HIGH, Q goes LOW. When EN is LOW, the state of the latch is not affected by the D input.

Related Problem   Determine the Q output of the gated D latch if the D input in Figure 7–11(a) is inverted.

THE 74LS75 D LATCH

An example of a gated D latch is the 74LS75 represented by the logic symbol in Figure 7–12(a). This device has four latches. Notice that each active-HIGH EN input is shared by two latches and is designated as a control input (C). The truth table for each latch is shown in Figure 7–12(b). The X in the truth table represents a "don't care" condition. In this case, when the EN input is LOW, it does not matter what the D input is because the outputs are unaffected and remain in their prior states.

◀ **FIGURE 7–12**

The 74LS75 quad gated D latches.

(a) Logic symbol

Inputs		Outputs		
D	EN	Q	$\overline{Q}$	Comments
0	1	0	1	RESET
1	1	1	0	SET
X	0	Q_0	$\overline{Q}_0$	No change

Note: Q_0 is the prior output level before the indicated input conditions were established.

(b) Truth table (each latch)

1. List three types of latches.
2. Develop the truth table for the active-HIGH input S-R latch in Figure 7–1(a).
3. What is the Q output of a D latch when *EN* = 1 and *D* = 1?

7–2   EDGE-TRIGGERED FLIP-FLOPS

Flip-flops are synchronous bistable devices, also known as *bistable multivibrators.* In this case, the term *synchronous* means that the output changes state only at a specified point on the triggering input called the **clock** (CLK), which is designated as a control input, *C;* that is, changes in the output occur in synchronization with the clock.

After completing this section, you should be able to

■ Define *clock* ■ Define *edge-triggered flip-flop* ■ Explain the difference between a flip-flop and a latch ■ Identify an edge-triggered flip-flop by its logic symbol ■ Discuss the difference between a positive and a negative edge-triggered flip-flop ■ Discuss and compare the operation of S-R, D, and J-K edge-triggered flip-flops and explain the differences in their truth tables. ■ Discuss the asynchronous inputs of a flip-flop ■ Describe the 74AHC74 and the 74HC112 flip-flops

The dynamic input indicator ▷ means the flip-flop changes state only on the edge of a clock pulse.

An **edge-triggered flip-flop** changes state either at the positive edge (rising edge) or at the negative edge (falling edge) of the clock pulse and is sensitive to its inputs only at this transition of the clock. Three types of edge-triggered flip-flops are covered in this section: S-R, D, and J-K. Although the S-R flip-flop is not available in IC form, it is the basis for the D and J-K flip-flops. The logic symbols for all of these flip-flops are shown in Figure 7–13. Notice that each type can be either positive edge-triggered (no bubble at *C* input) or negative edge-triggered (bubble at *C* input). The key to identifying an edge-triggered flip-flop by its logic symbol is the small triangle inside the block at the clock (*C*) input. This triangle is called the *dynamic input indicator.*

▷ **FIGURE 7–13**

Edge-triggered flip-flop logic symbols (top: positive edge-triggered; bottom: negative edge-triggered).

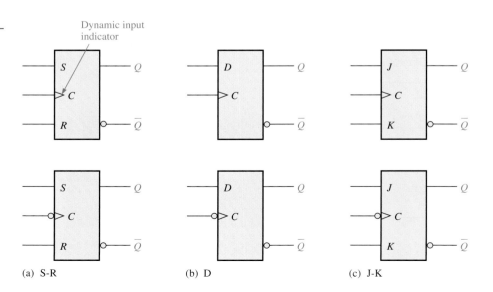

(a) S-R   (b) D   (c) J-K

The Edge-Triggered S-R Flip-Flop

The S and R inputs of the **S-R flip-flop** are called synchronous inputs because data on these inputs are transferred to the flip-flop's output only on the triggering edge of the clock pulse. When S is HIGH and R is LOW, the Q output goes HIGH on the triggering edge of the clock pulse, and the flip-flop is SET. When S is LOW and R is HIGH, the Q output goes LOW on the triggering edge of the clock pulse, and the flip-flop is RESET. When both S and R are LOW, the output does not change from its prior state. An invalid condition exists when both S and R are HIGH.

This basic operation of a positive edge-triggered flip-flop is illustrated in Figure 7–14, and Table 7–2 is the truth table for this type of flip-flop. Remember, *the flip-flop cannot change state except on the triggering edge of a clock pulse.* The S and R inputs can be changed at any time when the clock input is LOW or HIGH (except for a very short interval around the triggering transition of the clock) without affecting the output.

An S-R flip-flop cannot have both S and R inputs HIGH at the same time.

COMPUTER NOTE

Semiconductor memories in computers consist of large numbers of individual cells. Each storage cell holds a 1 or a 0. One type of memory is the Static Random Access Memory or SRAM, which uses flip-flops for the storage cells because a flip-flop will retain either of its two states indefinitely as long as dc power is applied, thus the term *static*. This type of memory is classified as a *volatile* memory because all the stored data are lost when power is turned off. Another type of memory, the Dynamic Random Access Memory or DRAM, uses capacitance rather than flip-flops as the basic storage element and must be periodically refreshed in order to maintain the stored data.

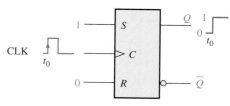

(a) $S = 1$, $R = 0$ flip-flop SETS on positive clock edge. (If already SET, it remains SET.)

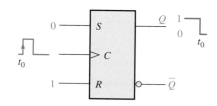

(b) $S = 0$, $R = 1$ flip-flop RESETS on positive clock edge. (If already RESET, it remains RESET.)

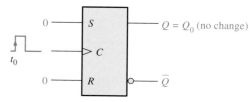

(c) $S = 0$, $R = 0$ flip-flop does not change. (If SET, it remains SET; if RESET, it remains RESET.)

▲ **FIGURE 7–14**

Operation of a positive edge-triggered S-R flip-flop.

INPUTS			OUTPUTS		
S	R	CLK	Q	$\overline{Q}$	COMMENTS
0	0	X	Q_0	$\overline{Q}_0$	No change
0	1	↑	0	1	RESET
1	0	↑	1	0	SET
1	1	↑	?	?	Invalid

↑ = clock transition LOW to HIGH

X = irrelevant ("don't care")

Q_0 = output level prior to clock transition

◀ **TABLE 7–2**

Truth table for a positive edge-triggered S-R flip-flop.

The operation and truth table for a negative edge-triggered S-R flip-flop are the same as those for a positive edge-triggered device except that the falling edge of the clock pulse is the triggering edge.

EXAMPLE 7-4

Determine the Q and $\overline{Q}$ output waveforms of the flip-flop in Figure 7–15 for the S, R, and CLK inputs in Figure 7–16(a). Assume that the positive edge-triggered flip-flop is initially RESET.

▶ **FIGURE 7–15**

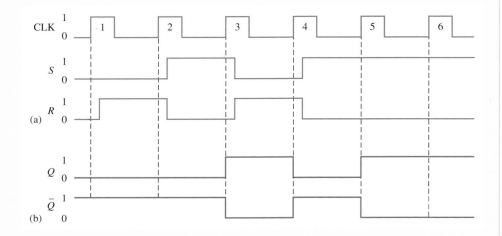

▲ **FIGURE 7–16**

Solution

1. At clock pulse 1, S is LOW and R is LOW, so Q does not change.

2. At clock pulse 2, S is LOW and R is HIGH, so Q remains LOW (RESET).

3. At clock pulse 3, S is HIGH and R is LOW, so Q goes HIGH (SET).

4. At clock pulse 4, S is LOW and R is HIGH, so Q goes LOW (RESET).

5. At clock pulse 5, S is HIGH and R is LOW, so Q goes HIGH (SET).

6. At clock pulse 6, S is HIGH and R is LOW, so Q stays HIGH.

Once Q is determined, $\overline{Q}$ is easily found since it is simply the complement of Q. The resulting waveforms for Q and $\overline{Q}$ are shown in Figure 7–16(b) for the input waveforms in part (a).

Related Problem

Determine Q and $\overline{Q}$ for the S and R inputs in Figure 7–16(a) if the flip-flop is a negative edge-triggered device.

A Method of Edge-Triggering

A simplified implementation of an edge-triggered S-R flip-flop is illustrated in Figure 7–17(a) and is used to demonstrate the concept of edge-triggering. This coverage of the S-R flip-flop does not imply that it is the most important type. Actually, the D flip-flop and the J-K flip-flop are available in IC form and more widely used than the S-R type. However, understanding the S-R is important because both the D and the J-K flip-flops are derived from the S-R flip-flop. Notice that the S-R flip-flop differs from the gated S-R latch only in that it has a pulse transition detector.

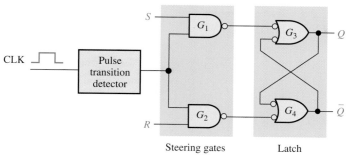

◀ **FIGURE 7–17**

Edge triggering.

(a) A simplified logic diagram for a positive edge-triggered S-R flip-flop

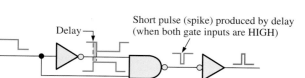

(b) A type of pulse transition detector

One basic type of pulse transition detector is shown in Figure 7–17(b). As you can see, there is a small delay on one input to the NAND gate so that the inverted clock pulse arrives at the gate input a few nanoseconds after the true clock pulse. This circuit produces a very short-duration spike on the positive-going transition of the clock pulse. In a negative edge-triggered flip-flop the clock pulse is inverted first, thus producing a narrow spike on the negative-going edge.

The circuit in Figure 7–17 is partitioned into two sections, one labeled Steering gates and the other labeled Latch. The steering gates direct, or steer, the clock spike either to the input to gate G_3 or to the input to gate G_4, depending on the state of the S and R inputs. To understand the operation of this flip-flop, begin with the assumptions that it is in the RESET state ($Q = 0$) and that the S, R, and CLK inputs are all LOW. For this condition, the outputs of gate G_1 and gate G_2 are both HIGH. The LOW on the Q output is coupled back into one input of gate G_4, making the $\overline{Q}$ output HIGH. Because $\overline{Q}$ is HIGH, both inputs to gate G_3 are HIGH (remember, the output of gate G_1 is HIGH), holding the Q output LOW. If a pulse is applied to the CLK input, the outputs of gates G_1 and G_2 remain HIGH because they are disabled by the LOWs on the S input and the R input; therefore, there is no change in the state of the flip-flop—it remains in the RESET state.

Let's now make S HIGH, leave R LOW, and apply a clock pulse. Because the S input to gate G_1 is now HIGH, the output of gate G_1 goes LOW for a very short time (spike) when CLK goes HIGH, causing the Q output to go HIGH. Both inputs to gate G_4 are now HIGH (remember, gate G_2 output is HIGH because R is LOW), forcing the $\overline{Q}$ output LOW. This LOW on $\overline{Q}$ is coupled back into one input of gate G_3, ensuring that the Q output will remain HIGH. The flip-flop is now in the SET state. Figure 7–18 illustrates the logic level transitions that take place within the flip-flop for this condition.

Next, let's make S LOW and R HIGH and apply a clock pulse. Because the R input is now HIGH, the positive-going edge of the clock produces a negative-going spike on the output of gate G_2, causing the $\overline{Q}$ output to go HIGH. Because of this HIGH on $\overline{Q}$, both inputs to gate G_3 are now HIGH (remember, the output of gate G_1 is HIGH because of the LOW on S), forcing the Q output to go LOW. This LOW on Q is coupled back into one input of gate G_4, ensuring that $\overline{Q}$ will remain HIGH. The flip-flop is now in the RESET state.

COMPUTER NOTE

All logic operations that are performed with hardware can also be implemented in software. For example, the operation of a J-K flip-flop can be performed with specific computer instructions. If two bits were used to represent the J and K inputs, the computer would do nothing for 00, a data bit representing the Q output would be set (1) for 10, the Q data bit would be cleared (0) for 01, and the Q data bit would be complemented for 11. Although it may be unusual to use a computer to simulate a flip-flop, the point is that all hardware operations can be simulated using software.

▶ FIGURE 7–18

Flip-flop making a transition from the RESET state to the SET state on the positive-going edge of the clock pulse.

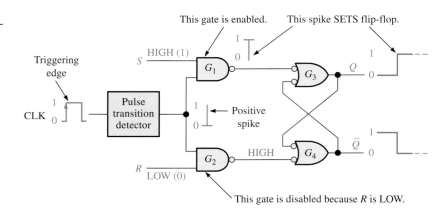

Figure 7–19 illustrates the logic level transitions that occur within the flip-flop for this condition. As with the gated latch, an invalid condition exists if a clock pulse occurs when both S and R are HIGH at the same time. This is the major drawback of the S-R flip-flop.

▶ FIGURE 7–19

Flip-flop making a transition from the SET state to the RESET state on the positive-going edge of the clock pulse.

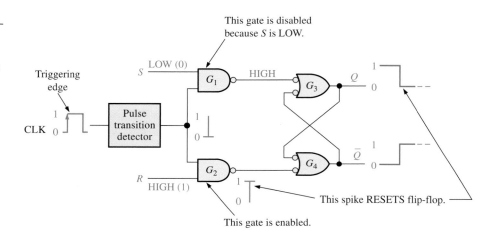

The Edge-Triggered D Flip-Flop

The Q output of a D flip-flop assumes the state of the D input on the triggering edge of the clock.

The **D flip-flop** is useful when a single data bit (1 or 0) is to be stored. The addition of an inverter to an S-R flip-flop creates a basic D flip-flop, as in Figure 7–20, where a positive edge-triggered type is shown.

▶ FIGURE 7–20

A positive edge-triggered D flip-flop formed with an S-R flip-flop and an inverter.

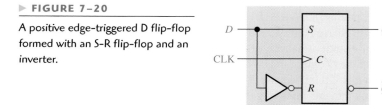

Notice that the flip-flop in Figure 7–20 has only one input, the D input, in addition to the clock. If there is a HIGH on the D input when a clock pulse is applied, the flip-flop will set, and the HIGH on the D input is stored by the flip-flop on the positive-going edge of the clock pulse. If there is a LOW on the D input when the clock pulse is applied, the flip-flop will reset, and the LOW on the D input is stored by the flip-flop on the leading edge of the clock pulse. In the SET state the flip-flop is storing a 1, and in the RESET state it is storing a 0.

The logical operation of the positive edge-triggered D flip-flop is summarized in Table 7–3. The operation of a negative edge-triggered device is, of course, the same, except that triggering occurs on the falling edge of the clock pulse. Remember, Q follows D at the active or triggering clock edge.

◀ TABLE 7–3

Truth table for a positive edge-triggered D flip-flop.

| INPUTS | | OUTPUTS | | |
D	CLK	Q	$\bar{Q}$	COMMENTS
1	↑	1	0	SET (stores a 1)
0	↑	0	1	RESET (stores a 0)

↑ = clock transition LOW to HIGH

EXAMPLE 7–5

Given the waveforms in Figure 7–21(a) for the D input and the clock, determine the Q output waveform if the flip-flop starts out RESET.

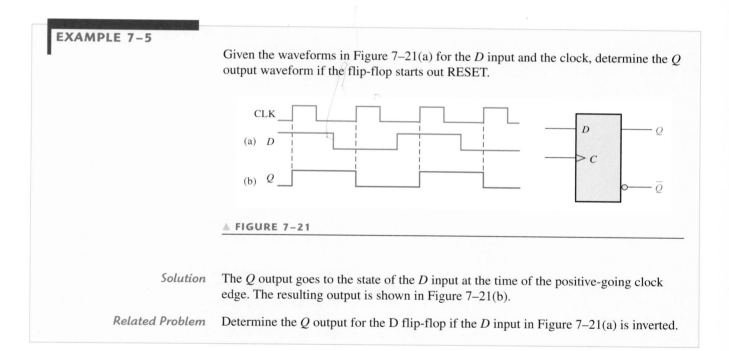

▲ FIGURE 7–21

Solution   The Q output goes to the state of the D input at the time of the positive-going clock edge. The resulting output is shown in Figure 7–21(b).

Related Problem   Determine the Q output for the D flip-flop if the D input in Figure 7–21(a) is inverted.

The Edge-Triggered J-K Flip-Flop

The **J-K flip-flop** is versatile and is a widely used type of flip-flop. The functioning of the J-K flip-flop is identical to that of the S-R flip-flop in the SET, RESET, and no-change conditions of operation. The difference is that the J-K flip-flop has no invalid state as does the S-R flip-flop.

Figure 7–22 shows the basic internal logic for a positive edge-triggered J-K flip-flop. It differs from the S-R edge-triggered flip-flop in that the Q output is connected back to the input of gate G_2, and the $\bar{Q}$ output is connected back to the input of gate G_1. The two control inputs are labeled J and K in honor of Jack Kilby, who invented the integrated circuit. A J-K flip-flop can also be of the negative edge-triggered type, in which case the clock input is inverted.

▶ **FIGURE 7–22**

A simplified logic diagram for a positive edge-triggered J-K flip-flop.

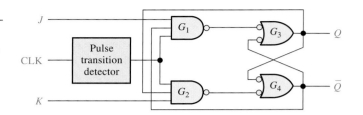

Let's assume that the flip-flop in Figure 7–23 is RESET and that the J input is HIGH and the K input is LOW rather than as shown. When a clock pulse occurs, a leading-edge spike indicated by ① is passed through gate G_1 because $\overline{Q}$ is HIGH and J is HIGH. This will cause the latch portion of the flip-flop to change to the SET state. The flip-flop is now SET.

▶ **FIGURE 7–23**

Transitions illustrating the toggle operation when $J = 1$ and $K = 1$.

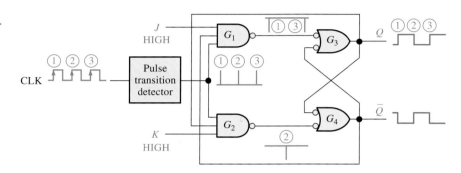

In the toggle mode, a J-K flip-flop changes state on every clock pulse.

If you make J LOW and K HIGH, the next clock spike indicated by ② will pass through gate G_2 because Q is HIGH and K is HIGH. This will cause the latch portion of the flip-flop to change to the RESET state.

If you apply a LOW to both the J and K inputs, the flip-flop will stay in its present state when a clock pulse occurs. A LOW on both J and K results in a *no-change* condition.

So far, the logical operation of the J-K flip-flop is the same as that of the S-R type in the SET, RESET, and no-change modes. The difference in operation occurs when both the J and K inputs are HIGH. To see this, assume that the flip-flop is RESET. The HIGH on the $\overline{Q}$ enables gate G_1, so the clock spike indicated by ③ passes through to set the flip-flop. Now there is a HIGH on Q, which allows the next clock spike to pass through gate G_2 and reset the flip-flop.

As you can see, on each successive clock spike, the flip-flop changes to the opposite state. This mode is called **toggle** operation. Figure 7–23 illustrates the transitions when the flip-flop is in the toggle mode. A J-K flip-flop connected for toggle operation is sometimes called a *T flip-flop*.

Table 7–4 summarizes the logical operation of the edge-triggered J-K flip-flop in truth table form. Notice that there is no invalid state as there is with an S-R flip-flop. The truth table for a negative edge-triggered device is identical except that it is triggered on the falling edge of the clock pulse.

▶ **TABLE 7–4**

Truth table for a positive edge-triggered J-K flip-flop.

INPUTS			OUTPUTS		
J	K	CLK	Q	$\overline{Q}$	COMMENTS
0	0	↑	Q_0	$\overline{Q}_0$	No change
0	1	↑	0	1	RESET
1	0	↑	1	0	SET
1	1	↑	$\overline{Q}_0$	Q_0	Toggle

↑ = clock transition LOW to HIGH
Q_0 = output level prior to clock transition

EXAMPLE 7–6

The waveforms in Figure 7–24(a) are applied to the J, K, and clock inputs as indicated. Determine the Q output, assuming that the flip-flop is initially RESET.

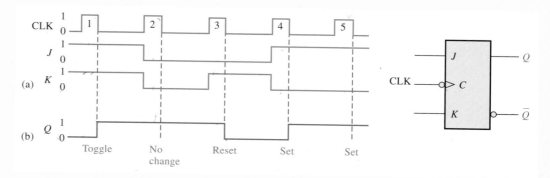

(a)

(b)

Toggle   No change   Reset   Set   Set

▲ **FIGURE 7–24**

Solution

1. First, since this is a negative edge-triggered flip-flop, as indicated by the "bubble" at the clock input, the Q output will change only on the negative-going edge of the clock pulse.

2. At the first clock pulse, both J and K are HIGH; and because this is a toggle condition, Q goes HIGH.

3. At clock pulse 2, a no-change condition exists on the inputs, keeping Q at a HIGH level.

4. When clock pulse 3 occurs, J is LOW and K is HIGH, resulting in a RESET condition; Q goes LOW.

5. At clock pulse 4, J is HIGH and K is LOW, resulting in a SET condition; Q goes HIGH.

6. A SET condition still exists on J and K when clock pulse 5 occurs, so Q will remain HIGH.

The resulting Q waveform is indicated in Figure 7–24(b).

Related Problem

Determine the Q output of the J-K flip-flop if the J and K inputs in Figure 7–24(a) are inverted.

EXAMPLE 7–7

The waveforms in Figure 7–25(a) are applied to the flip-flop as shown. Determine the Q output, starting in the RESET state.

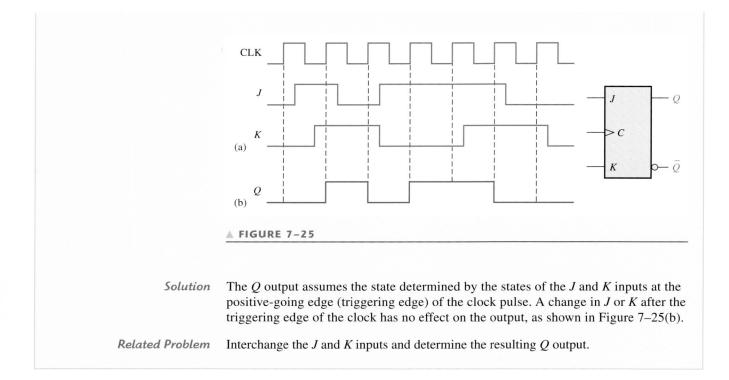

▲ **FIGURE 7–25**

Solution   The Q output assumes the state determined by the states of the J and K inputs at the positive-going edge (triggering edge) of the clock pulse. A change in J or K after the triggering edge of the clock has no effect on the output, as shown in Figure 7–25(b).

Related Problem   Interchange the J and K inputs and determine the resulting Q output.

Asynchronous Preset and Clear Inputs

An active preset input makes the Q output HIGH (SET).

For the flip-flops just discussed, the *S-R, D,* and *J-K* inputs are called *synchronous inputs* because data on these inputs are transferred to the flip-flop's output only on the triggering edge of the clock pulse; that is, the data are transferred synchronously with the clock.

Most integrated circuit flip-flops also have **asynchronous** inputs. These are inputs that affect the state of the flip-flop *independent of the clock.* They are normally labeled **preset** (*PRE*) and **clear** (*CLR*), or *direct set* (S_D) and *direct reset* (R_D) by some manufacturers. An active level on the preset input will set the flip-flop, and an active level on the clear input will reset it. A logic symbol for a J-K flip-flop with preset and clear inputs is shown in Figure 7–26. These inputs are active-LOW, as indicated by the bubbles. These preset and clear inputs must both be kept HIGH for synchronous operation.

An active clear input makes the Q output LOW (RESET).

▶ **FIGURE 7–26**

Logic symbol for a J-K flip-flop with active-LOW preset and clear inputs.

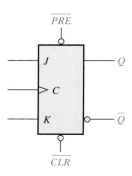

Figure 7–27 shows the logic diagram for an edge-triggered J-K flip-flop with active-LOW preset $(\overline{PRE})$ and clear $(\overline{CLR})$ inputs. This figure illustrates basically how these inputs work. As you can see, they are connected so that they override the effect of the synchronous inputs, *J, K,* and the clock.

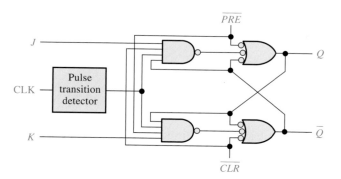

◄ **FIGURE 7–27**

Logic diagram for a basic J-K flip-flop with active-LOW preset and clear inputs.

EXAMPLE 7–8

For the positive edge-triggered J-K flip-flop with preset and clear inputs in Figure 7–28, determine the Q output for the inputs shown in the timing diagram in part (a) if Q is initially LOW.

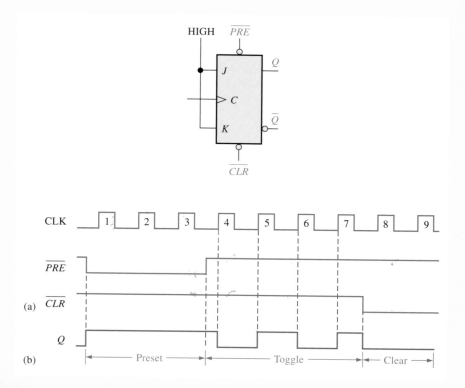

▲ **FIGURE 7–28**

Open file F07-28 to verify the operation.

Solution   **1.** During clock pulses 1, 2, and 3, the preset $(\overline{PRE})$ is LOW, keeping the flip-flop SET regardless of the synchronous *J* and *K* inputs.

2. For clock pulses 4, 5, 6, and 7, toggle operation occurs because *J* is HIGH, *K* is HIGH, and both $\overline{PRE}$ and $\overline{CLR}$ are HIGH.

3. For clock pulses 8 and 9, the clear ($\overline{CLR}$) input is LOW, keeping the flip-flop RESET regardless of the synchronous inputs.

The resulting *Q* output is shown in Figure 7–28(b).

Related Problem If you interchange the $\overline{PRE}$ and $\overline{CLR}$ waveforms in Figure 7–28(a), what will the *Q* output look like?

Let's look at two specific edge-triggered flip-flops. They are representative of the various types of flip-flops available in IC form and, like most other devices, are available in CMOS and in TTL logic families.

THE 74AHC74 DUAL D FLIP-FLOP

This CMOS device contains two identical D flip-flops that are independent of each other except for sharing V_{CC} and ground. The flip-flops are positive edge-triggered and have active-LOW asynchronous preset and clear inputs. The logic symbols for the individual flip-flops within the package are shown in Figure 7–29(a), and an ANSI/IEEE standard single block symbol that represents the entire device is shown in part (b). The pin numbers are shown in parentheses.

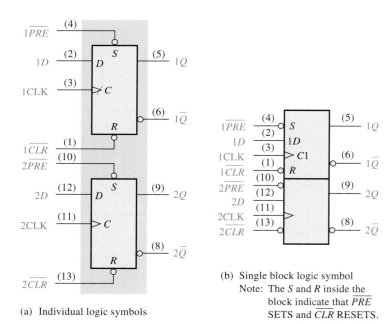

(a) Individual logic symbols

(b) Single block logic symbol
Note: The *S* and *R* inside the block indicate that $\overline{PRE}$ SETS and $\overline{CLR}$ RESETS.

▲ FIGURE 7–29

Logic symbols for the 74AHC74 dual positive edge-triggered D flip-flop.

THE 74HC112 DUAL J-K FLIP-FLOP

This device has two identical flip-flops that are negative edge-triggered and have active-LOW asynchronous preset and clear inputs. The logic symbols are shown in Figure 7–30.

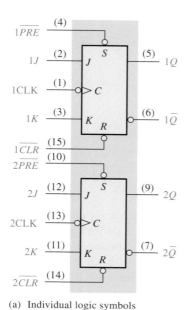

(a) Individual logic symbols

◀ **FIGURE 7–30**

Logic symbols for the 74HC112 dual negative edge-triggered J-K flip-flop.

(b) Single block logic symbol

EXAMPLE 7–9

The 1*J*, 1*K*, 1CLK, 1$\overline{PRE}$, and 1$\overline{CLR}$ waveforms in Figure 7–31(a) are applied to one of the negative edge-triggered flip-flops in a 74HC112 package. Determine the 1*Q* output waveform.

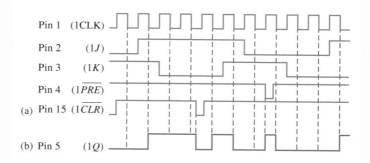

Pin 1 (1CLK)
Pin 2 (1*J*)
Pin 3 (1*K*)
Pin 4 (1$\overline{PRE}$)
(a) Pin 15 (1$\overline{CLR}$)

(b) Pin 5 (1*Q*)

▲ **FIGURE 7–31**

Solution The resulting 1*Q* waveform is shown in Figure 7–31(b). Notice that each time a LOW is applied to the 1$\overline{PRE}$ or 1$\overline{CLR}$, the flip-flop is set or reset regardless of the states of the other inputs.

Related Problem Determine the 1*Q* output waveform if the waveforms for 1$\overline{PRE}$ and 1$\overline{CLR}$ are interchanged.

SECTION 7-2 REVIEW

1. Describe the main difference between a gated S-R latch and an edge-triggered S-R flip-flop.

2. How does a J-K flip-flop differ from an S-R flip-flop in its basic operation?

3. Assume that the flip-flop in Figure 7-21 is negative edge-triggered. Describe the output waveform for the same CLK and D waveforms.

7-3 FLIP-FLOP OPERATING CHARACTERISTICS

The performance, operating requirements, and limitations of flip-flops are specified by several operating characteristics or parameters found on the data sheet for the device. Generally, the specifications are applicable to all CMOS and TTL flip-flops.

After completing this section, you should be able to

■ Define *propagation delay time* ■ Explain the various propagation delay time specifications ■ Define *set-up time* and discuss how it limits flip-flop operation ■ Define *hold time* and discuss how it limits flip-flop operation ■ Discuss the significance of maximum clock frequency ■ Discuss the various pulse width specifications ■ Define *power dissipation* and calculate its value for a specific device ■ Compare various series of flip-flops in terms of their operating parameters

Propagation Delay Times

A **propagation delay time** is the interval of time required after an input signal has been applied for the resulting output change to occur. Four categories of propagation delay times are important in the operation of a flip-flop:

1. Propagation delay t_{PLH} as measured from the triggering edge of the clock pulse to the LOW-to-HIGH transition of the output. This delay is illustrated in Figure 7-32(a).

2. Propagation delay t_{PHL} as measured from the triggering edge of the clock pulse to the HIGH-to-LOW transition of the output. This delay is illustrated in Figure 7-32(b).

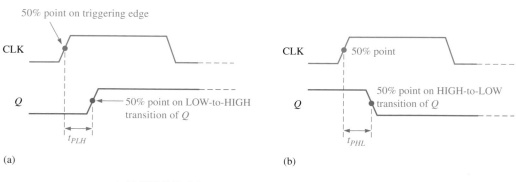

(a) (b)

▲ **FIGURE 7-32**

Propagation delays, clock to output.

3. Propagation delay t_{PLH} as measured from the leading edge of the preset input to the LOW-to-HIGH transition of the output. This delay is illustrated in Figure 7–33(a) for an active-LOW preset input.

4. Propagation delay t_{PHL} as measured from the leading edge of the clear input to the HIGH-to-LOW transition of the output. This delay is illustrated in Figure 7–33(b) for an active-LOW clear input.

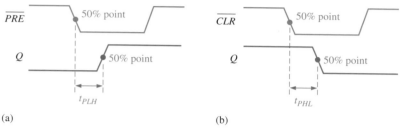

(a)   (b)

▲ FIGURE 7–33

Propagation delays, preset input to output and clear input to output.

Set-up Time

The **set-up time** (t_s) is the minimum interval required for the logic levels to be maintained constantly on the inputs (J and K, or S and R, or D) prior to the triggering edge of the clock pulse in order for the levels to be reliably clocked into the flip-flop. This interval is illustrated in Figure 7–34 for a D flip-flop.

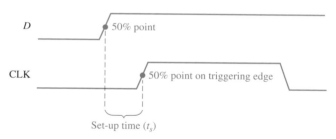

▲ FIGURE 7–34

Set-up time (t_s). The logic level must be present on the D input for a time equal to or greater than t_s before the triggering edge of the clock pulse for reliable data entry.

An advantage of CMOS is that it can operate over a wider range of dc supply voltages (typically 2 V to 6 V) than TTL and, therefore, less expensive power supplies that do not have precise regulation can be used. Also, batteries can be used as secondary or primary sources for CMOS circuits. In addition, lower voltages mean that the IC dissipates less power. The drawback is that the performance of CMOS is degraded with lower supply voltages. For example, the guaranteed maximum clock frequency of a CMOS flip-flop is much less at $V_{CC} = 2$ V than at $V_{CC} = 6$ V.

Hold Time

The **hold time** (t_h) is the minimum interval required for the logic levels to remain on the inputs after the triggering edge of the clock pulse in order for the levels to be reliably clocked into the flip-flop. This is illustrated in Figure 7–35 for a D flip-flop.

▶ FIGURE 7-35

Hold time (t_h). The logic level must remain on the D input for a time equal to or greater than t_h after the triggering edge of the clock pulse for reliable data entry.

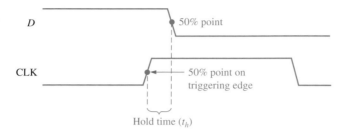

Maximum Clock Frequency

The maximum clock frequency (f_{max}) is the highest rate at which a flip-flop can be reliably triggered. At clock frequencies above the maximum, the flip-flop would be unable to respond quickly enough, and its operation would be impaired.

Pulse Widths

Minimum pulse widths (t_W) for reliable operation are usually specified by the manufacturer for the clock, preset, and clear inputs. Typically, the clock is specified by its minimum HIGH time and its minimum LOW time.

Power Dissipation

The **power dissipation** of any digital circuit is the total power consumption of the device. For example, if the flip-flop operates on a $+5$ V dc source and draws 5 mA of current, the power dissipation is

$$P = V_{CC} \times I_{CC} = 5 \text{ V} \times 5 \text{ mA} = 25 \text{ mW}$$

The power dissipation is very important in most applications in which the capacity of the dc supply is a concern. As an example, let's assume that you have a digital system that requires a total of ten flip-flops, and each flip-flop dissipates 25 mW of power. The total power requirement is

$$P_T = 10 \times 25 \text{ mW} = 250 \text{ mW} = 0.25 \text{ W}$$

This tells you the output capacity required of the dc supply. If the flip-flops operate on $+5$ V dc, then the amount of current that the supply must provide is

$$I = \frac{250 \text{ mW}}{5 \text{ V}} = 50 \text{ mA}$$

You must use a $+5$ V dc supply that is capable of providing at least 50 mA of current.

Comparison of Specific Flip-Flops

Table 7–5 provides a comparison, in terms of the operating parameters discussed in this section, of four CMOS and TTL flip-flops of the same type.

▼ TABLE 7–5

Comparison of operating parameters for four IC families of flip-flops of the same type at 25°C.

PARAMETER	CMOS		TTL	
	74HC74A	74AHC74	74LS74A	74F74
t_{PHL} (CLK to Q)	17 ns	4.6 ns	40 ns	6.8 ns
t_{PLH} (CLK to Q)	17 ns	4.6 ns	25 ns	8.0 ns
t_{PHL} ($\overline{CLR}$ to Q)	18 ns	4.8 ns	40 ns	9.0 ns
t_{PLH} ($\overline{PRE}$ to Q)	18 ns	4.8 ns	25 ns	6.1 ns
t_s (set-up time)	14 ns	5.0 ns	20 ns	2.0 ns
t_h (hold time)	3.0 ns	0.5 ns	5 ns	1.0 ns
t_W (CLK HIGH)	10 ns	5.0 ns	25 ns	4.0 ns
t_W (CLK LOW)	10 ns	5.0 ns	25 ns	5.0 ns
$t_W(\overline{CLR/PRE})$	10 ns	5.0 ns	25 ns	4.0 ns
f_{max}	35 MHz	170 MHz	25 MHz	100 MHz
Power, quiescent	0.012 mW	1.1 mW		
Power, 50% duty cycle			44 mW	88 mW

SECTION 7–3
REVIEW

1. Define the following:

   (a) set-up time   (b) hold time

2. Which specific flip-flop in Table 7–5 can be operated at the highest frequency?

7–4   FLIP-FLOP APPLICATIONS

In this section, three general applications of flip-flops are discussed to give you an idea of how they can be used. In Chapters 8 and 9, flip-flop applications in counters and registers are covered in detail.

After completing this section, you should be able to

■ Discuss the application of flip-flops in data storage ■ Describe how flip-flops are used for frequency division ■ Explain how flip-flops are used in basic counter applications

Parallel Data Storage

A common requirement in digital systems is to store several bits of data from parallel lines simultaneously in a group of flip-flops. This operation is illustrated in Figure 7–36(a) using four flip-flops. Each of the four parallel data lines is connected to the D input of a flip-flop. The clock inputs of the flip-flops are connected together, so that each flip-flop is triggered by the same clock pulse. In this example, positive edge-triggered flip-flops are used, so the data on the D inputs are stored simultaneously by the flip-flops on the positive edge of the clock, as indicated in the timing diagram in Figure 7–36(b). Also, the asynchronous reset (R) inputs are connected to a common $\overline{CLR}$ line, which initially resets all the flip-flops.

▶ FIGURE 7–36

Example of flip-flops used in a basic register for parallel data storage.

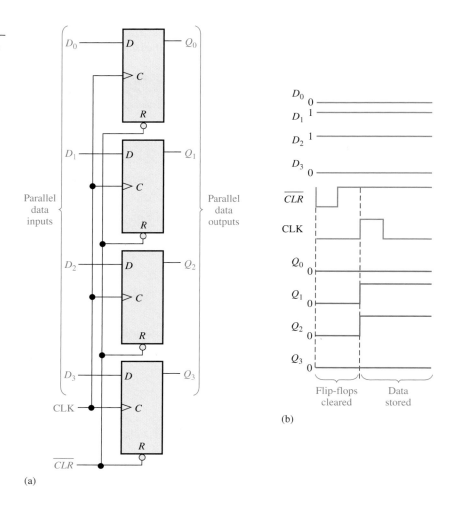

(a)

(b)

This group of four flip-flops is an example of a basic register used for data storage. In digital systems, data are normally stored in groups of bits (usually eight or multiples thereof) that represent numbers, codes, or other information. Registers are covered in detail in Chapter 9.

Frequency Division

Another application of a flip-flop is dividing (reducing) the frequency of a periodic waveform. When a pulse waveform is applied to the clock input of a J-K flip-flop that is connected to toggle ($J = K = 1$), the Q output is a square wave with one-half the frequency of the clock input. Thus, a single flip-flop can be applied as a divide-by-2 device, as is illustrated in Figure 7–37. As you can see, the flip-flop changes state on each triggering clock

▶ FIGURE 7–37

The J-K flip-flop as a divide-by-2 device. Q is one-half the frequency of CLK.

edge (positive edge-triggered in this case). This results in an output that changes at half the frequency of the clock waveform.

Further division of a clock frequency can be achieved by using the output of one flip-flop as the clock input to a second flip-flop, as shown in Figure 7–38. The frequency of the Q_A output is divided by 2 by flip-flop B. The Q_B output is, therefore, one-fourth the frequency of the original clock input. Propagation delay times are not shown on the timing diagrams.

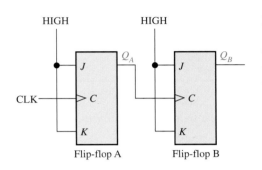

◄ FIGURE 7–38

Example of two J-K flip-flops used to divide the clock frequency by 4. Q_A is one-half and Q_B is one-fourth the frequency of CLK.

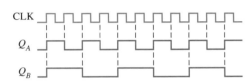

By connecting flip-flops in this way, a frequency division of 2^n is achieved, where n is the number of flip-flops. For example, three flip-flops divide the clock frequency by $2^3 = 8$; four flip-flops divide the clock frequency by $2^4 = 16$; and so on.

EXAMPLE 7–10

Develop the f_{out} waveform for the circuit in Figure 7–39 when an 8 kHz square wave input is applied to the clock input of flip-flop A.

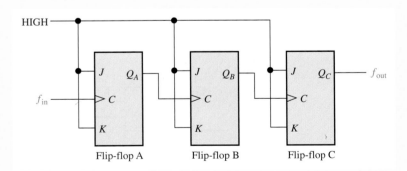

▲ FIGURE 7–39

Solution The three flip-flops are connected to divide the input frequency by eight ($2^3 = 8$) and the f_{out} waveform is shown in Figure 7–40. Since these are positive edge-triggered flip-flops, the outputs change on the positive-going clock edge. There is one output pulse for every eight input pulses, so the output frequency is 1 kHz. Waveforms of Q_A and Q_B are also shown.

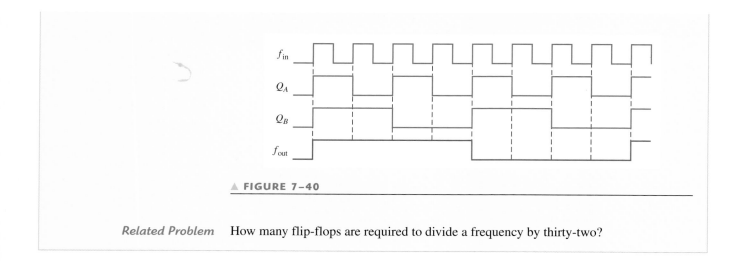

▲ **FIGURE 7–40**

Related Problem   How many flip-flops are required to divide a frequency by thirty-two?

Counting

Another important application of flip-flops is in digital counters, which are covered in detail in Chapter 8. The concept is illustrated in Figure 7–41. The flip-flops are negative edge-triggered J-Ks. Both flip-flops are initially RESET. Flip-flop A toggles on the negative-going transition of each clock pulse. The Q output of flip-flop A clocks flip-flop B, so each time Q_A makes a HIGH-to-LOW transition, flip-flop B toggles. The resulting Q_A and Q_B waveforms are shown in the figure.

▶ **FIGURE 7–41**

Flip-flops used to generate a binary count sequence. Two repetitions (00, 01, 10, 11) are shown.

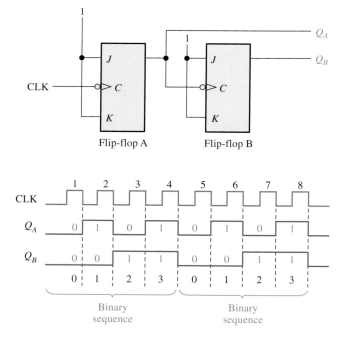

Observe the sequence of Q_A and Q_B in Figure 7–41. Prior to clock pulse 1, $Q_A = 0$ and $Q_B = 0$; after clock pulse 1, $Q_A = 1$ and $Q_B = 0$; after clock pulse 2, $Q_A = 0$ and $Q_B = 1$; and after clock pulse 3, $Q_A = 1$ and $Q_B = 1$. If we take Q_A as the least significant bit, a 2-bit sequence is produced as the flip-flops are clocked. This binary sequence repeats every four clock pulses, as shown in the timing diagram of Figure 7–41. Thus, the flip-flops are counting in sequence from 0 to 3 (00, 01, 10, 11) and then recycling back to 0 to begin the sequence again.

EXAMPLE 7–11

Determine the output waveforms in relation to the clock for Q_A, Q_B, and Q_C in the circuit of Figure 7–42 and show the binary sequence represented by these waveforms.

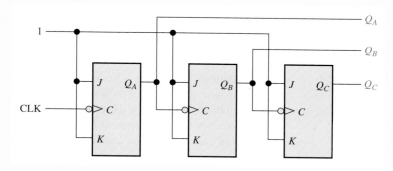

▲ FIGURE 7–42

Solution   The output timing diagram is shown in Figure 7–43. Notice that the outputs change on the negative-going edge of the clock pulses. The outputs go through the binary sequence 000, 001, 010, 011, 100, 101, 110, and 111 as indicated.

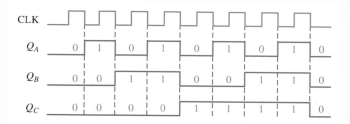

▲ FIGURE 7–43

Related Problem   How many flip-flops are required to produce a binary sequence representing decimal numbers 0 through 15?

SECTION 7–4 REVIEW

1. What is a group of flip-flops used for data storage called?

2. How must a J-K flip-flop be connected to function as a divide-by-2 device?

3. How many flip-flops are required to produce a divide-by-64 device?

7–5   ONE-SHOTS

The **one-shot** is a **monostable** multivibrator, a device with only one stable state. A one-shot is normally in its stable state and will change to its unstable state only when triggered. Once it is triggered, the one-shot remains in its unstable state for a predetermined length of time and then automatically returns to its stable state. The time that the device stays in its unstable state determines the pulse width of its output.

After completing this section, you should be able to

▪ Describe the basic operation of a one-shot ▪ Explain how a nonretriggerable one-shot works ▪ Explain how a retriggerable one-shot works ▪ Set up the 74121 and the 74LS122 one-shots to obtain a specified output pulse width ▪ Recognize a Schmitt trigger symbol and explain basically what it means

A one-shot produces a single pulse each time it is triggered.

Figure 7–44 shows a basic one-shot (monostable multivibrator) that is composed of a logic gate and an inverter. When a pulse is applied to the **trigger** input, the output of gate G_1 goes LOW. This HIGH-to-LOW transition is coupled through the capacitor to the input of inverter G_2. The apparent LOW on G_2 makes its output go HIGH. This HIGH is connected back into G_1, keeping its output LOW. Up to this point the trigger pulse has caused the output of the one-shot, Q, to go HIGH.

▶ **FIGURE 7–44**

A simple one-shot circuit.

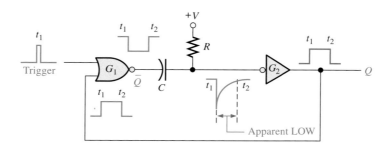

The capacitor immediately begins to charge through R toward the high voltage level. The rate at which it charges is determined by the RC time constant. When the capacitor charges to a certain level, which appears as a HIGH to G_2, the output goes back LOW.

To summarize, the output of inverter G_2 goes HIGH in response to the trigger input. It remains HIGH for a time set by the RC time constant. At the end of this time, it goes LOW. A single narrow trigger pulse produces a single output pulse whose time duration is controlled by the RC time constant. This operation is illustrated in Figure 7–44.

A typical one-shot logic symbol is shown in Figure 7–45(a), and the same symbol with an external R and C is shown in Figure 7–45(b). The two basic types of IC one-shots are nonretriggerable and retriggerable.

▶ **FIGURE 7–45**

Basic one-shot logic symbols. *CX* and *RX* stand for external components.

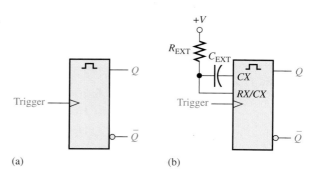

(a)   (b)

A nonretriggerable one-shot will not respond to any additional trigger pulses from the time it is triggered into its unstable state until it returns to its stable state. In other words, it will ignore any trigger pulses occurring before it times out. The time that the one-shot remains in its unstable state is the pulse width of the output.

Figure 7–46 shows the nonretriggerable one-shot being triggered at intervals greater than its pulse width and at intervals less than the pulse width. Notice that in the second case, the additional pulses are ignored.

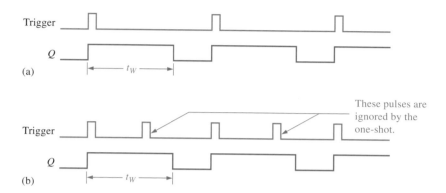

◀ **FIGURE 7–46**

Nonretriggerable one-shot action.

A retriggerable one-shot can be triggered before it times out. The result of retriggering is an extension of the pulse width as illustrated in Figure 7–47.

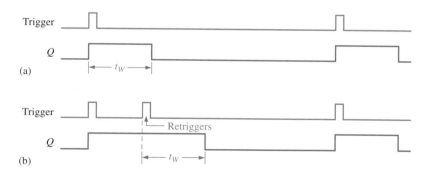

◀ **FIGURE 7–47**

Retriggerable one-shot action.

THE 74121 NONRETRIGGERABLE ONE-SHOT

The 74121 is an example of a nonretriggerable IC one-shot. It has provisions for external R and C, as shown in Figure 7–48. The inputs labeled A_1, A_2, and B are gated trigger inputs. The R_{INT} input connects to a 2 kΩ internal timing resistor.

Setting the Pulse Width   A typical pulse width of about 30 ns is produced when no external timing components are used and the internal timing resistor (R_{INT}) is connected to V_{CC}, as shown in Figure 7–49(a). The pulse width can be set anywhere between about 30 ns and 28 s by the use of external components. Figure 7–49(b) shows the configuration using the internal resistor (2 kΩ) and an external capacitor. Part (c) shows the configuration using an external resistor and an external capacitor. The output pulse width is set by the values of the resistor (R_{INT} = 2 kΩ, and R_{EXT} is selected) and the capacitor according to the following formula:

$$t_W = 0.7RC_{EXT}$$

Equation 7–1

where R is either R_{INT} or R_{EXT}. When R is in kilohms (kΩ) and C_{EXT} is in picofarads (pF), the output pulse width t_W is in nanoseconds (ns).

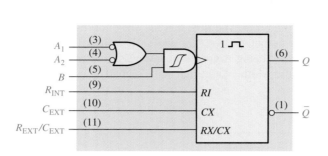
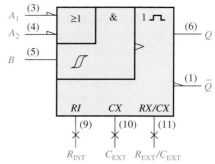

(a) Traditional logic symbol

(b) ANSI/IEEE std. 91–1984 logic symbol ($\times$ = nonlogic connection). "1 ⊓" is the qualifying symbol for a nonretriggerable one-shot.

▲ **FIGURE 7–48**

Logic symbols for the 74121 nonretriggerable one-shot.

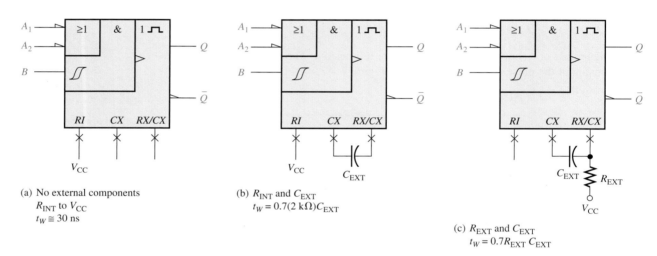

(a) No external components
R_{INT} to V_{CC}
$t_W \cong 30$ ns

(b) R_{INT} and C_{EXT}
$t_W = 0.7(2 \text{ k}\Omega)C_{EXT}$

(c) R_{EXT} and C_{EXT}
$t_W = 0.7R_{EXT}C_{EXT}$

▲ **FIGURE 7–49**

Three ways to set the pulse width of a 74121.

The Schmitt-Trigger Symbol The symbol $\int$ indicates a Schmitt-trigger input. This type of input uses a special threshold circuit that produces **hysteresis,** a characteristic that prevents erratic switching between states when a slow-changing trigger voltage hovers around the critical input level. This allows reliable triggering to occur even when the input is changing as slowly as 1 volt/second.

THE 74LS122 RETRIGGERABLE ONE-SHOT

The 74LS122 is an example of a retriggerable IC one-shot with a clear input. It also has provisions for external R and C, as shown in Figure 7–50. The inputs labeled A_1, A_2, B_1, and B_2 are the gated trigger inputs.

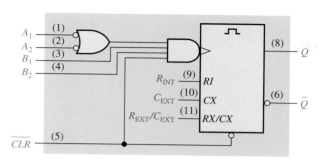

(a) Traditional logic symbol

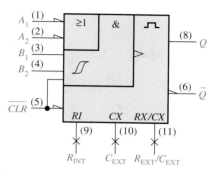

(b) ANSI/IEEE std. 91–1984 logic symbol
(× = nonlogic connection). ⊓ is the
qualifying symbol for a retriggerable
one-shot.

▲ **FIGURE 7–50**

Logic symbol for the 74LS122 retriggerable one-shot.

A minimum pulse width of approximately 45 ns is obtained with no external components. Wider pulse widths are achieved by using external components. A general formula for calculating the values of these components for a specified pulse width (t_W) is

$$t_W = 0.32 R C_{EXT} \left(1 + \frac{0.7}{R} \right)$$

Equation 7–2

where 0.32 is a constant determined by the particular type of one-shot, R is in kΩ and is either the internal or the external resistor, C_{EXT} is in pF, and t_W is in ns. The internal resistance is 10 kΩ and can be used instead of an external resistor. (Notice the difference between this formula and that for the 74121, shown in Equation 7–1.)

EXAMPLE 7–12

A certain application requires a one-shot with a pulse width of approximately 100 ms. Using a 74121, show the connections and the component values.

Solution Arbitrarily select $R_{EXT} = $ **39 kΩ** and calculate the necessary capacitance.

$$t_W = 0.7 R_{EXT} C_{EXT}$$

$$C_{EXT} = \frac{t_W}{0.7 R_{EXT}}$$

where C_{EXT} is in pF, R_{EXT} is in kΩ, and t_W is in ns. Since 100 ms = 1×10^8 ns,

$$C_{EXT} = \frac{1 \times 10^8 \text{ ns}}{0.7(39 \text{ kΩ})} = 3.66 \times 10^{-6} \text{ pF} = \textbf{3.66 } \boldsymbol{\mu}\textbf{F}$$

A standard 3.3 μF capacitor will give an output pulse width of 91 ms. The proper connections are shown in Figure 7–51. To achieve a pulse width closer to 100 ms, other combinations of values for R_{EXT} and C_{EXT} can be tried. For example, $R_{EXT} = $ 68 kΩ and $C_{EXT} = $ 2.2 μF gives a pulse width of 105 ms.

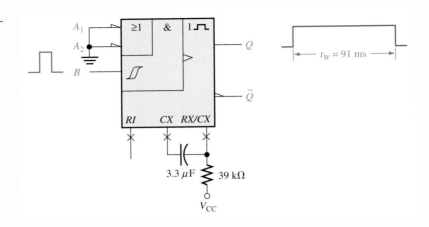

Related Problem Use an external capacitor in conjunction with R_{INT} to produce an output pulse width of 10 μs from the 74121.

EXAMPLE 7–13

Determine the values of R_{EXT} and C_{EXT} that will produce a pulse width of 1 μs when connected to a 74LS122.

Solution

Assume a value of $C_{\text{EXT}} = $ **560 pF** and then solve for R_{EXT}. The pulse width must be expressed in ns and C_{EXT} in pF. R_{EXT} will be in kΩ.

$$t_W = 0.32 R_{\text{EXT}} C_{\text{EXT}}\left(1 + \frac{0.7}{R_{\text{EXT}}}\right) = 0.32 R_{\text{EXT}} C_{\text{EXT}} + 0.7\left(\frac{0.32 R_{\text{EXT}} C_{\text{EXT}}}{R_{\text{EXT}}}\right)$$

$$= 0.32 R_{\text{EXT}} C_{\text{EXT}} + (0.7)(0.32)C_{\text{EXT}}$$

$$R_{\text{EXT}} = \frac{t_W - (0.7)(0.32)C_{\text{EXT}}}{0.32 C_{\text{EXT}}} = \frac{t_W}{0.32 C_{\text{EXT}}} - 0.7$$

$$= \frac{1000 \text{ ns}}{(0.32)560 \text{ pF}} - 0.7 = \textbf{4.88 k}\boldsymbol{\Omega}$$

Use a standard value of **4.7 kΩ**.

Related Problem Show the connections and component values for a 74LS122 one-shot with an output pulse width of 5 μs. Assume $C_{\text{EXT}} = $ 560 pF.

An Application

One practical one-shot application is a sequential timer that can be used to illuminate a series of lights. This type of circuit can be used, for example, in a lane change directional indicator for highway construction projects or in sequential turn signals on automobiles.

Figure 7–52 shows three 74LS122 one-shots connected as a sequential timer. This particular circuit produces a sequence of three 1 s pulses. The first one-shot is triggered by a switch closure or a low-frequency pulse input, producing a 1 s output pulse. When the first one-shot (OS 1) times out and the 1 s pulse goes LOW, the second one-shot (OS 2) is triggered, also producing a 1 s output pulse. When this second pulse goes LOW, the third one-shot (OS 3) is triggered and the third 1 s pulse is produced. The output timing is illustrated in the figure. Variations of this basic arrangement can be used to produce a variety of timed outputs.

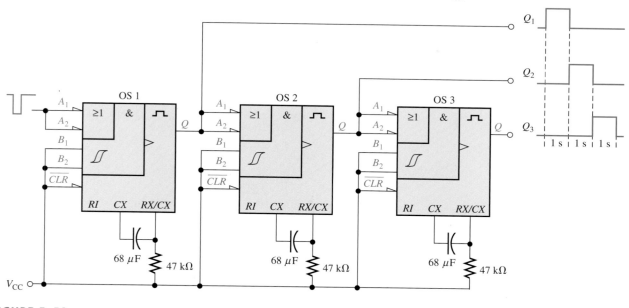

▲ **FIGURE 7–52**

A sequential timing circuit using three 74LS122 one-shots.

7–6   THE 555 TIMER

The 555 **timer** is a versatile and widely used IC device because it can be configured in two different modes as either a monostable multivibrator (one-shot) or as an astable multivibrator (oscillator). An astable multivibrator has no stable states and therefore changes back and forth (oscillates) between two unstable states without any external triggering.

After completing this section, you should be able to

■ Describe the basic elements in a 555 timer   ■ Set up a 555 timer as a one-shot
■ Set up a 555 timer as an oscillator

Basic Operation

A functional diagram showing the internal components of a 555 timer is shown in Figure 7–53. The comparators are devices whose outputs are HIGH when the voltage on the positive (+) input is greater than the voltage on the negative (−) input and LOW when the − input voltage is greater than the + input voltage. The voltage divider consisting of three 5 kΩ resistors provides a trigger level of $\frac{1}{3}V_{CC}$ and a threshold level of $\frac{2}{3}V_{CC}$. The control voltage input (pin 5) can be used to externally adjust the trigger and threshold levels to other values if necessary. When the normally HIGH trigger input momentarily goes below $\frac{1}{3}V_{CC}$, the

A 555 timer can operate as either a one-shot (monostable) or as an oscillator (astable).

output of comparator B switches from LOW to HIGH and sets the S-R latch, causing the output (pin 3) to go HIGH and turning the discharge transistor Q_1 off. The output will stay HIGH until the normally LOW threshold input goes above $\frac{2}{3}V_{CC}$ and causes the output of comparator A to switch from LOW to HIGH. This resets the latch, causing the output to go back LOW and turning the discharge transistor on. The external reset input can be used to reset the latch independent of the threshold circuit. The trigger and threshold inputs (pins 2 and 6) are controlled by external components connected to produce either monostable or astable action.

▶ **FIGURE 7–53**

Internal functional diagram of a 555 timer (pin numbers are in parenthesis).

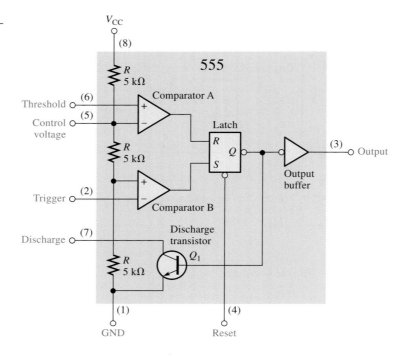

Monostable (One-Shot) Operation

An external resistor and capacitor connected as shown in Figure 7–54 are used to set up the 555 timer as a nonretriggerable one-shot. The pulse width of the output is determined by the time constant of R_1 and C_1 according to the following formula:

Equation 7–3

$$t_W = 1.1R_1C_1$$

▶ **FIGURE 7–54**

The 555 timer connected as a one-shot.

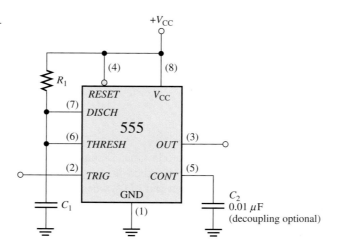

The control voltage input is not used and is connected to a decoupling capacitor C_2 to prevent noise from affecting the trigger and threshold levels.

Before a trigger pulse is applied, the output is LOW and the discharge transistor Q_1 is *on*, keeping C_1 discharged as shown in Figure 7–55(a). When a negative-going trigger pulse is applied at t_0, the output goes HIGH and the discharge transistor turns *off*, allowing capacitor C_1 to begin charging through R_1 as shown in part (b). When C_1 charges to $\frac{1}{3}V_{CC}$, the output goes back LOW at t_1 and Q_1 turns *on* immediately, discharging C_1 as shown in part (c). As you can see, the charging rate of C_1 determines how long the output is HIGH.

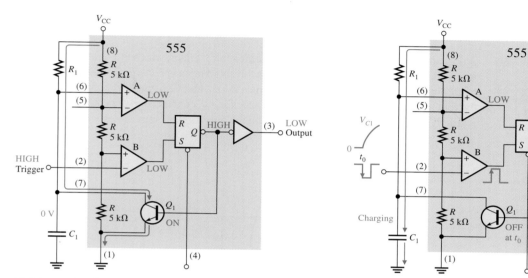

(a) Prior to triggering. (The current path is indicated by the red arrow.) (b) When triggered

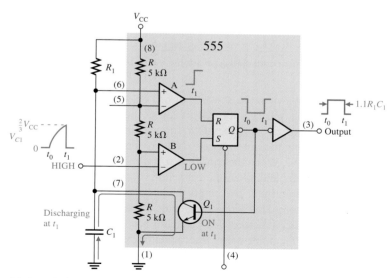

(c) At end of charging interval

▲ **FIGURE 7–55**

One-shot operation of the 555 timer.

EXAMPLE 7–14

What is the output pulse width for a 555 monostable circuit with $R_1 = 2.2\ k\Omega$ and $C_1 = 0.01\ \mu F$?

Solution   From Equation 7–3 the pulse width is

$$t_W = 1.1R_1C_1 = 1.1(2.2\ k\Omega)(0.01\ \mu F) = \textbf{24.2 } \boldsymbol{\mu s}$$

Related Problem   For $C_1 = 0.01\ \mu F$, determine the value of R_1 for a pulse width of 1 ms.

Astable Operation

COMPUTER NOTE

All computers require a timing source to provide accurate clock waveforms. The timing section controls all system timing and is responsible for the proper operation of the system hardware. The timing section usually consists of a crystal-controlled oscillator and counters for frequency division. Using a high-frequency oscillator divided down to a lower frequency provides for greater accuracy and frequency stability.

A 555 timer connected to operate as an **astable** multivibrator, which is a nonsinusoidal **oscillator,** is shown in Figure 7–56. Notice that the threshold input (*THRESH*) is now connected to the trigger input (*TRIG*). The external components R_1, R_2, and C_1 form the timing network that sets the frequency of oscillation. The 0.01 μF capacitor, C_2, connected to the control (*CONT*) input is strictly for decoupling and has no effect on the operation; in some cases it can be left off.

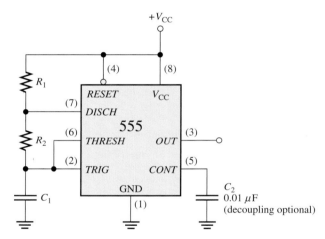

▲ **FIGURE 7–56**

The 555 timer connected as an astable multivibrator (oscillator).

Initially, when the power is turned on, the capacitor (C_1) is uncharged and thus the trigger voltage (pin 2) is at 0 V. This causes the output of comparator B to be HIGH and the output of comparator A to be LOW, forcing the output of the latch, and thus the base of Q_1, LOW and keeping the transistor off. Now, C_1 begins charging through R_1 and R_2, as indicated in Figure 7–57. When the capacitor voltage reaches $\frac{1}{3}V_{CC}$, comparator B switches to its LOW output state; and when the capacitor voltage reaches $\frac{2}{3}V_{CC}$, comparator A switches to its HIGH output state. This resets the latch, causing the base of Q_1 to go HIGH and turning on the transistor. This sequence creates a discharge path for the capacitor through R_2 and the transistor, as indicated. The capacitor now begins to discharge, causing comparator A to go LOW. At the point where the capacitor discharges down to $\frac{1}{3}V_{CC}$, comparator B switches HIGH; this sets the latch, making the base of Q_1 LOW and turning off the transistor. Another charging cycle begins, and the entire process repeats. The result is a rectangular wave output whose duty cycle depends on the values of R_1 and R_2. The frequency of oscillation is given by the following formula, or it can be found using the graph in Figure 7–58.

Equation 7–4

$$f = \frac{1.44}{(R_t + 2R_2)C_1}$$

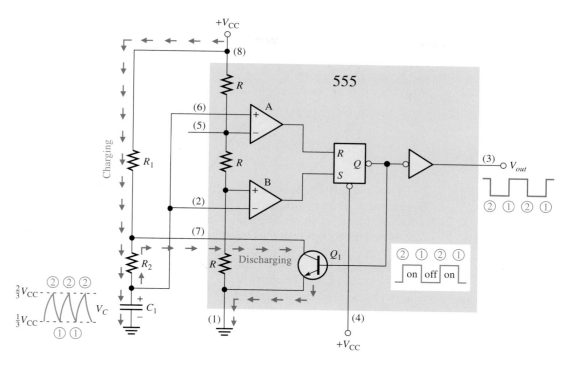

▲ **FIGURE 7–57**

Operation of the 555 timer in the astable mode.

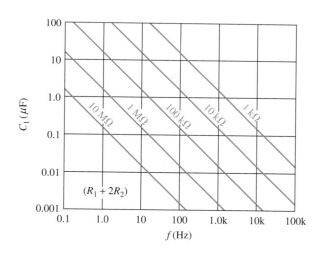

◀ **FIGURE 7–58**

Frequency of oscillation as a function of C_1 and $R_1 + 2R_2$. The sloped lines are values of $R_1 + 2R_2$.

By selecting R_1 and R_2, the duty cycle of the output can be adjusted. Since C_1 charges through $R_1 + R_2$ and discharges only through R_2, duty cycles approaching a minimum of 50 percent can be achieved if $R_2 >> R_1$ so that the charging and discharging times are approximately equal.

An expression for the duty cycle is developed as follows. The time that the output is HIGH (t_H) is how long it takes C_1 to charge from $\frac{1}{3}V_{CC}$ to $\frac{2}{3}V_{CC}$. It is expressed as

$$t_H = 0.7(R_1 + R_2)C_1$$

Equation 7–5 ✓

The time that the output is LOW (t_L) is how long it takes C_1 to discharge from $\frac{1}{3}V_{CC}$ to $\frac{2}{3}V_{CC}$. It is expressed as

$$t_L = 0.7R_2C_1$$

Equation 7–6 ✓

The period, T, of the output waveform is the sum of t_H and t_L. This is the reciprocal of f in Equation 7–4.

$$T = t_H + t_L = 0.7(R_1 + 2R_2)C_1$$ ✓

Finally, the duty cycle is

$$\text{Duty cycle} = \frac{t_H}{T} = \frac{t_H}{t_H + t_L}$$

Equation 7–7

$$\text{Duty cycle} = \left(\frac{R_1 + R_2}{R_1 + 2R_2}\right)100\%$$ ✓

To achieve duty cycles of less than 50 percent, the circuit in Figure 7–56 can be modified so that C_1 charges through only R_1 and discharges through R_2. This is achieved with a diode, D_1, placed as shown in Figure 7–59. The duty cycle can be made less than 50 percent by making R_1 less than R_2. Under this condition, the expression for the duty cycle is

Equation 7–8

$$\text{Duty cycle} = \left(\frac{R_1}{R_1 + R_2}\right)100\%$$ ✓

▶ **FIGURE 7–59**

The addition of diode D_1 allows the duty cycle of the output to be adjusted to less than 50 percent by making $R_1 < R_2$.

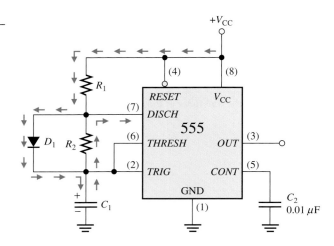

EXAMPLE 7–15

A 555 timer configured to run in the astable mode (oscillator) is shown in Figure 7–60. Determine the frequency of the output and the duty cycle.

▶ **FIGURE 7–60**

Open file F07-60 to verify operation.

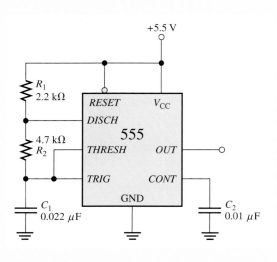

Solution   Use Equations 7–4 and 7–7.

$$f = \frac{1.44}{(R_1 + 2R_2)C_1} = \frac{1.44}{(2.2 \text{ k}\Omega + 9.4 \text{ k}\Omega)0.022 \text{ }\mu\text{F}} = \textbf{5.64 kHz}$$

$$\text{Duty cycle} = \left(\frac{R_1 + R_2}{R_1 + 2R_2}\right)100\% = \left(\frac{2.2 \text{ k}\Omega + 4.7 \text{ k}\Omega}{2.2 \text{ k}\Omega + 9.4 \text{ k}\Omega}\right)100\% = \textbf{59.5\%}$$

Related Problem   Determine the duty cycle in Figure 7–60 if a diode is connected across R_2 as indicated in Figure 7–59.

SECTION 7–6
REVIEW

1. Explain the difference in operation between an astable multivibrator and a monostable multivibrator.
2. For a certain astable multivibrator, $t_H = 15$ ms and $T = 20$ ms. What is the duty cycle of the output?

7–7   TROUBLESHOOTING

It is standard practice to test a new circuit design to be sure that it is operating as specified. New fixed-function designs are "breadboarded" and tested before the design is finalized. The term *breadboard* refers to a method of temporarily hooking up a circuit so that its operation can be verified and any faults (bugs) worked out before a prototype unit is built.

After completing this section, you should be able to

■ Describe how the timing of a circuit can produce erroneous glitches ■ Approach the troubleshooting of a new design with greater insight and awareness of potential problems

The circuit shown in Figure 7–61(a) generates two clock waveforms (CLK A and CLK B) that have an alternating occurrence of pulses. Each waveform is to be one-half the frequency of the original clock (CLK), as shown in the ideal timing diagram in part (b).

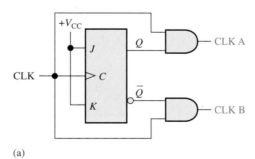

(a)

◄ **FIGURE 7–61**

Two-phase clock generator with ideal waveforms. Open file F07–61 and verify the operation.

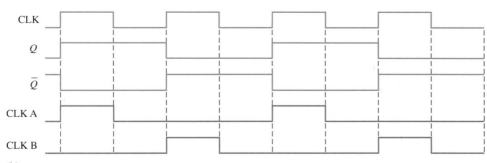

(b)

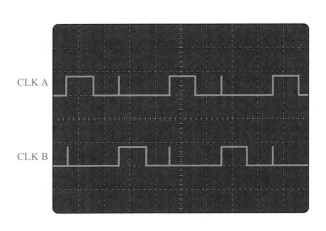

(a) Oscilloscope display of CLK A and CLK B waveforms with glitches indicated by the "spikes".

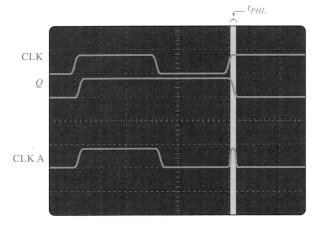

(b) Oscilloscope display showing propagation delay that creates glitch on CLK A waveform

▲ FIGURE 7–62

Oscilloscope displays for the circuit in Figure 7–61.

When the circuit is tested with an oscilloscope or logic analyzer, the CLK A and CLK B waveforms appear on the display screen as shown in Figure 7–62(a). Since glitches occur on both waveforms, something is wrong with the circuit either in its basic design or in the way it is connected. Further investigation reveals that the glitches are caused by a **race** condition between the CLK signal and the Q and $\overline{Q}$ signals at the inputs of the AND gates. As displayed in Figure 7–62(b), the propagation delays between CLK and Q and $\overline{Q}$ create a short-duration coincidence of HIGH levels at the leading edges of alternate clock pulses. Thus, there is a basic design flaw.

The problem can be corrected by using a negative edge-triggered flip-flop in place of the positive edge-triggered device, as shown in Figure 7–63(a). Although the propagation delays between CLK and Q and $\overline{Q}$ still exist, they are initiated on the trailing edges of the clock (CLK), thus eliminating the glitches, as shown in the timing diagram of Figure 7–63(b).

▶ FIGURE 7–63

Two-phase clock generator using negative edge-triggered flip-flop to eliminate glitches. Open file F07-63 and verify the operation.

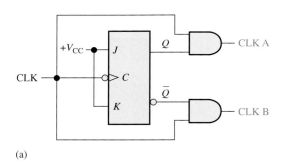

(a)

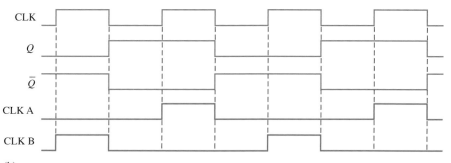

(b)

SECTION 7–7 REVIEW

1. Can a negative edge-triggered D flip-flop be used in the circuit of Figure 7–63?
2. What device can be used to provide the clock for the circuit in Figure 7–63?

Troubleshooting problems that are keyed to the CD-ROM are available in the Multisim Troubleshooting Practice section of the end-of-chapter problems.

DIGITAL SYSTEM APPLICATION

The traffic light control system that was started in Chapter 6 is continued in this chapter. In the last chapter, the combinational logic was developed.

In this chapter, the timing circuits are developed. These circuits produce a 4 s time interval for the caution light and a 25 s time interval for the red and green lights. Also, a 4 Hz clock signal is produced by the timing circuits. The overall traffic light control system block diagram that

was introduced in Chapter 6 is shown again in Figure 7–64 for reference.

Timing Circuits Requirements

The timing circuits consist of three parts— the 4 s timer, the 25 s timer, and the 10 kHz oscillator—as shown in the block diagram in Figure 7–65. The 4 s timer and the 25 s timer are implemented with 74121 one-shots as shown in Figure 7–66 (a) and (b). The 10 kHz oscillator is

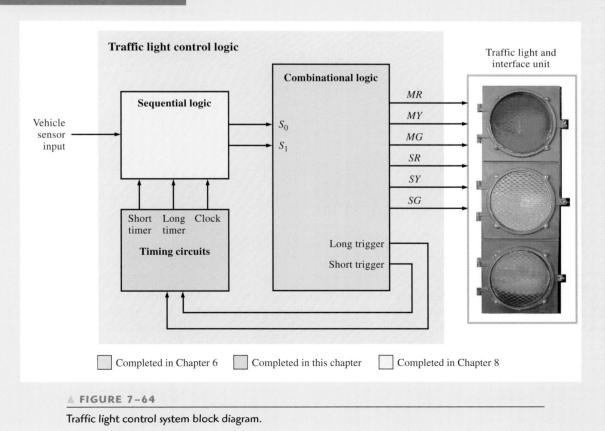

◢ **FIGURE 7–64**

Traffic light control system block diagram.

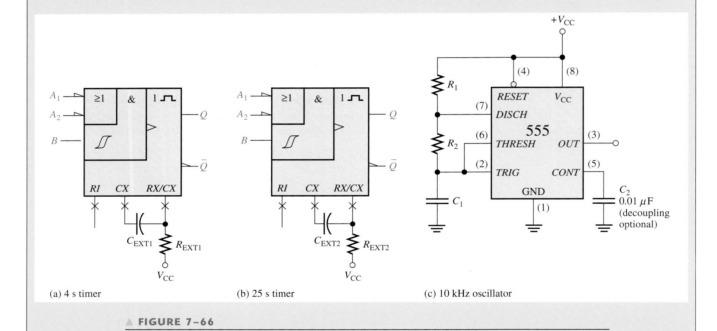

▲ **FIGURE 7–65**

Block diagram of the timing circuits.

implemented with a 555 timer as shown in Figure 7–66(c).

System Assignment

■ *Activity 1* Determine the external R and C values for the 4 s timer in Figure 7–66(a).

■ *Activity 2* Determine the external R and C values for the 25 s timer in Figure 7–66(b).

■ *Activity 3* Determine the R and C values for the 10 kHz 555 oscillator in Figure 7–66(c).

(a) 4 s timer   (b) 25 s timer   (c) 10 kHz oscillator

▲ **FIGURE 7–66**

The timing circuits.

HANDS ON TIP

Glitches that occur in digital systems are very fast (extremely short in duration) and can be difficult to see on an oscilloscope, particularly at lower sweep rates. A logic analyzer, however, can show a glitch easily. To look for glitches using a logic analyzer, select "latch" mode or (if available) transitional sampling. In the latch mode, the analyzer looks for a voltage level change. When a change occurs, even if it is of extremely short duration (a few nanoseconds), the information is "latched" into the analyzer's memory as another sampled data point. When the data are displayed, the glitch will show as an obvious change in the sampled data, making it easy to identify.

SUMMARY

■ Symbols for latches and flip-flops are shown in Figure 7–67.

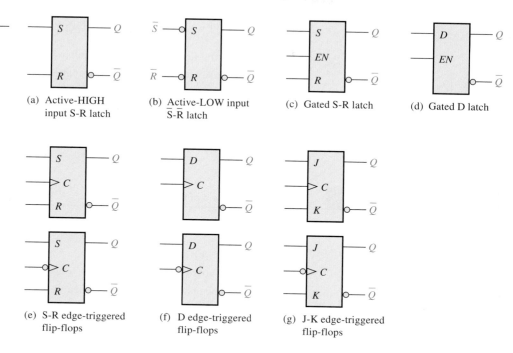

(a) Active-HIGH input S-R latch

(b) Active-LOW input $\overline{S}$-$\overline{R}$ latch

(c) Gated S-R latch

(d) Gated D latch

(e) S-R edge-triggered flip-flops

(f) D edge-triggered flip-flops

(g) J-K edge-triggered flip-flops

■ Latches are bistable devices whose state normally depends on asynchronous inputs.

■ Edge-triggered flip-flops are bistable devices with synchronous inputs whose state depends on the inputs only at the triggering transition of a clock pulse. Changes in the outputs occur at the triggering transition of the clock.

■ Monostable multivibrators (one-shots) have one stable state. When the one-shot is triggered, the output goes to its unstable state for a time determined by an *RC* circuit.

■ Astable multivibrators have no stable states and are used as oscillators to generate timing waveforms in digital systems.

KEY TERMS

Key terms and other bold terms in the chapter are defined in the end-of-book glossary.

Astable Having no stable state. An astable multivibrator oscillates between two quasi-stable states.

Bistable Having two stable states. Flip-flops and latches are bistable multivibrators.

Clear An asynchronous input used to reset a flip-flop (make the *Q* output 0).

Clock The triggering input of a flip-flop.

D flip-flop A type of bistable multivibrator in which the output assumes the state of the *D* input on the triggering edge of a clock pulse.

Edge-triggered flip-flop A type of flip-flop in which the data are entered and appear on the output on the same clock edge.

Hold time The time interval required for the control levels to remain on the inputs to a flip-flop after the triggering edge of the clock in order to reliably activate the device.

J-K flip-flop A type of flip-flop that can operate in the SET, RESET, no-change, and toggle modes.

Latch A bistable digital circuit used for storing a bit.

Monostable Having only one stable state. A monostable multivibrator, commonly called a *one-shot*, produces a single pulse in response to a triggering input.

One-shot A monostable multivibrator.

Power dissipation The amount of power required by a circuit.

Preset An asynchronous input used to set a flip-flop (make the Q output 1).

Propagation delay time The interval of time required after an input signal has been applied for the resulting output change to occur.

RESET The state of a flip-flop or latch when the output is 0; the action of producing a RESET state.

SET The state of a flip-flop or latch when the output is 1; the action of producing a SET state.

Set-up time The time interval required for the control levels to be on the inputs to a digital circuit, such as a flip-flop, prior to the triggering edge of a clock pulse.

Synchronous Having a fixed time relationship.

Timer A circuit that can be used as a one-shot or as an oscillator.

Toggle The action of a flip-flop when it changes state on each clock pulse.

SELF-TEST

Answers are at the end of the chapter.

1. If an S-R latch has a 1 on the S input and a 0 on the R input and then the S input goes to 0, the latch will be
   - (a) set
   - (b) reset
   - (c) invalid
   - (d) clear

2. The invalid state of an S-R latch occurs when
   - (a) $S = 1, R = 0$
   - (b) $S = 0, R = 1$
   - (c) $S = 1, R = 1$
   - (d) $S = 0, R = 0$

3. For a gated D latch, the Q output always equals the D input
   - (a) before the enable pulse
   - (b) during the enable pulse
   - (c) immediately after the enable pulse
   - (d) answers (b) and (c)

4. Like the latch, the flip-flop belongs to a category of logic circuits known as
   - (a) monostable multivibrators
   - (b) bistable multivibrators
   - (c) astable multivibrators
   - (d) one-shots

5. The purpose of the clock input to a flip-flop is to
   - (a) clear the device
   - (b) set the device
   - (c) always cause the output to change states
   - (d) cause the output to assume a state dependent on the controlling (S-R, J-K, or D) inputs.

6. For an edge-triggered D flip-flop,
   - (a) a change in the state of the flip-flop can occur only at a clock pulse edge
   - (b) the state that the flip-flop goes to depends on the D input
   - (c) the output follows the input at each clock pulse
   - (d) all of these answers

7. A feature that distinguishes the J-K flip-flop from the S-R flip-flop is the
   - (a) toggle condition
   - (b) preset input
   - (c) type of clock
   - (d) clear input

8. A flip-flop is in the toggle condition when
   - (a) $J = 1, K = 0$
   - (b) $J = 1, K = 1$
   - (c) $J = 0, K = 0$
   - (d) $J = 0, K = 1$

9. A J-K flip-flop with $J = 1$ and $K = 1$ has a 10 kHz clock input. The Q output is
   - (a) constantly HIGH
   - (b) constantly LOW
   - (c) a 10 kHz square wave
   - (d) a 5 kHz square wave

10. A one-shot is a type of

    (a) monostable multivibrator   **(b)** astable multivibrator   **(c)** timer

    (d) answers (a) and (c)   **(e)** answers (b) and (c)

11. The output pulse width of a nonretriggerable one-shot depends on

    (a) the trigger intervals   **(b)** the supply voltage

    (c) a resistor and capacitor   **(d)** the threshold voltage

12. An astable multivibrator

    (a) requires a periodic trigger input   **(b)** has no stable state

    (c) is an oscillator   **(d)** produces a periodic pulse output

    (e) answers (a), (b), (c), and (d)   **(f)** answers (b), (c), and (d) only

---

PROBLEMS

Answers to odd-numbered problems are at the end of the book.

SECTION 7–1   Latches

1. If the waveforms in Figure 7–68 are applied to an active-LOW input S-R latch, draw the resulting Q output waveform in relation to the inputs. Assume that Q starts LOW.

▶ **FIGURE 7–68**

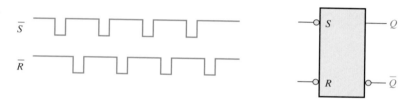

2. Solve Problem 1 for the input waveforms in Figure 7–69 applied to an active-HIGH S-R latch.

▶ **FIGURE 7–69**

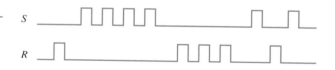

3. Solve Problem 1 for the input waveforms in Figure 7–70.

▶ **FIGURE 7–70**

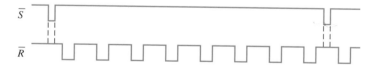

4. For a gated S-R latch, determine the Q and $\overline{Q}$ outputs for the inputs in Figure 7–71. Show them in proper relation to the enable input. Assume that Q starts LOW.

▶ **FIGURE 7–71**

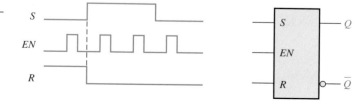

5. Solve Problem 4 for the inputs in Figure 7–72.

6. Solve Problem 4 for the inputs in Figure 7–73.

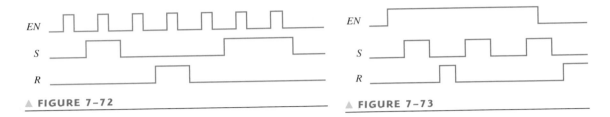

▲ FIGURE 7–72   ▲ FIGURE 7–73

7. For a gated D latch, the waveforms shown in Figure 7–74 are observed on its inputs. Draw the timing diagram showing the output waveform you would expect to see at Q if the latch is initially RESET.

▶ FIGURE 7–74

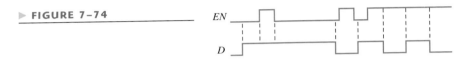

SECTION 7–2 Edge-Triggered Flip-Flops

8. Two edge-triggered S-R flip-flops are shown in Figure 7–75. If the inputs are as shown, draw the Q output of each flip-flop relative to the clock, and explain the difference between the two. The flip-flops are initially RESET.

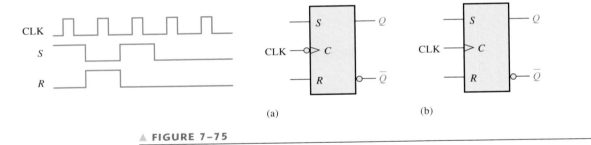

(a)   (b)

▲ FIGURE 7–75

9. The Q output of an edge-triggered S-R flip-flop is shown in relation to the clock signal in Figure 7–76. Determine the input waveforms on the S and R inputs that are required to produce this output if the flip-flop is a positive edge-triggered type.

▶ FIGURE 7–76

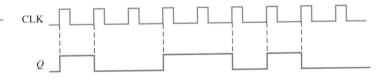

10. Draw the Q output relative to the clock for a D flip-flop with the inputs as shown in Figure 7–77. Assume positive edge-triggering and Q initially LOW.

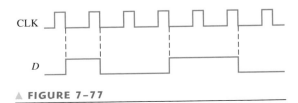

▲ FIGURE 7–77

11. Solve Problem 10 for the inputs in Figure 7–78.

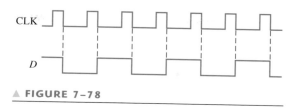

▲ FIGURE 7–78

12. For a positive edge-triggered J-K flip-flop with inputs as shown in Figure 7–79, determine the Q output relative to the clock. Assume that Q starts LOW.

▶ FIGURE 7–79

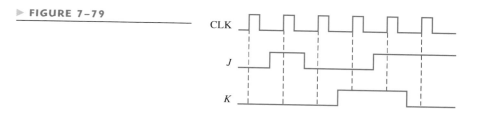

13. Solve Problem 12 for the inputs in Figure 7–80.

14. Determine the Q waveform relative to the clock if the signals shown in Figure 7–81 are applied to the inputs of the J-K flip-flop. Assume that Q is initially LOW.

15. For a negative edge-triggered J-K flip-flop with the inputs in Figure 7–82, develop the Q output waveform relative to the clock. Assume that Q is initially LOW.

▶ FIGURE 7–80

▶ FIGURE 7–81

▶ FIGURE 7–82

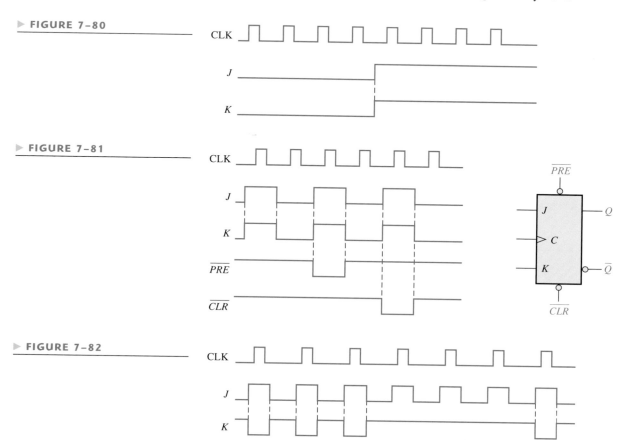

16. The following serial data are applied to the flip-flop through the AND gates as indicated in Figure 7–83. Determine the resulting serial data that appear on the Q output. There is one clock pulse for each bit time. Assume that Q is initially 0 and that $\overline{PRE}$ and $\overline{CLR}$ are HIGH. Rightmost bits are applied first.

J_1: 1 0 1 0 0 1 1

J_2: 0 1 1 1 0 1 0

J_3: 1 1 1 1 0 0 0

K_1: 0 0 0 1 1 1 0

K_2: 1 1 0 1 1 0 0

K_3: 1 0 1 0 1 0 1

▶ FIGURE 7–83

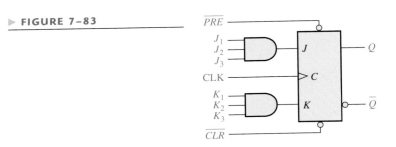

17. For the circuit in Figure 7–83, complete the timing diagram in Figure 7–84 by showing the Q output (which is initially LOW). Assume $\overline{PRE}$ and $\overline{CLR}$ remain HIGH.

▶ FIGURE 7–84

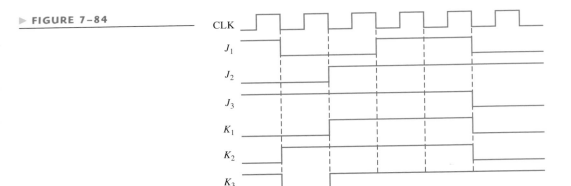

18. Solve Problem 17 with the same J and K inputs but with the $\overline{PRE}$ and $\overline{CLR}$ inputs as shown in Figure 7–85 in relation to the clock.

▶ FIGURE 7–85

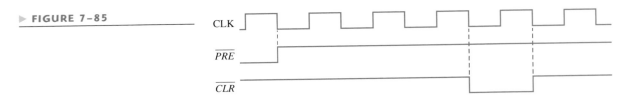

SECTION 7–3 Flip-Flop Operating Characteristics

19. What determines the power dissipation of a flip-flop?

20. Typically, a manufacturer's data sheet specifies four different propagation delay times associated with a flip-flop. Name and describe each one.

21. The data sheet of a certain flip-flop specifies that the minimum HIGH time for the clock pulse is 30 ns and the minimum LOW time is 37 ns. What is the maximum operating frequency?

22. The flip-flop in Figure 7–86 is initially RESET. Show the relation between the Q output and the clock pulse if propagation delay t_{PLH} (clock to Q) is 8 ns.

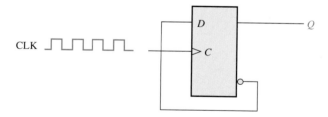

23. The direct current required by a particular flip-flop that operates on a +5 V dc source is found to be 10 mA. A certain digital device uses 15 of these flip-flops. Determine the current capacity required for the +5 V dc supply and the total power dissipation of the system.

24. For the circuit in Figure 7–87, determine the maximum frequency of the clock signal for reliable operation if the set-up time for each flip-flop is 2 ns and the propagation delays (t_{PLH} and t_{PHL}) from clock to output are 5 ns for each flip-flop.

▶ **FIGURE 7–87**

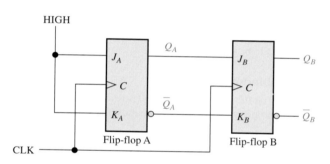

SECTION 7–4 Flip-Flop Applications

25. A D flip-flop is connected as shown in Figure 7–88. Determine the Q output in relation to the clock. What specific function does this device perform?

▶ **FIGURE 7–88**

26. For the circuit in Figure 7–87, develop a timing diagram for eight clock pulses, showing the Q_A and Q_B outputs in relation to the clock.

SECTION 7–5 One-Shots

27. Determine the pulse width of a 74121 one-shot if the external resistor is 3.3 kΩ and the external capacitor is 2000 pF.

28. An output pulse of 5 μs duration is to be generated by a 74LS122 one-shot. Using a capacitor of 10,000 pF, determine the value of external resistance required.

SECTION 7–6 The 555 Timer

29. Create a one-shot, using a 555 timer that will produce a 0.25 s output pulse.

30. A 555 timer is configured to run as an astable multivibrator as shown in Figure 7–89. Determine its frequency.

▶ **FIGURE 7–89**

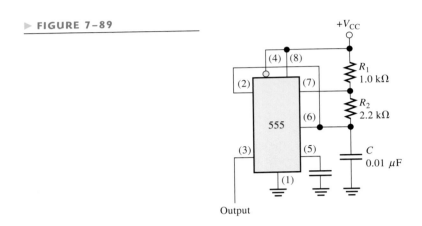

31. Determine the values of the external resistors for a 555 timer used as an astable multivibrator with an output frequency of 20 kHz, if the external capacitor C is 0.002 μF and the duty cycle is to be approximately 75%.

SECTION 7–7 Troubleshooting

32. The flip-flop in Figure 7–90 is tested under all input conditions as shown. Is it operating properly? If not, what is the most likely fault?

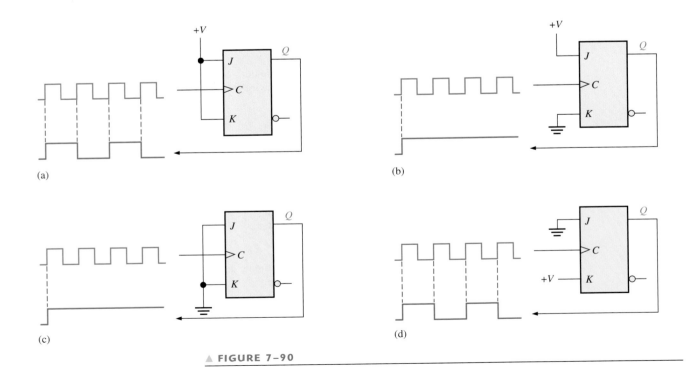

▲ **FIGURE 7–90**

33. A 74HC00 quad NAND gate IC is used to construct a gated S-R latch on a protoboard in the lab as shown in Figure 7–91. The schematic in part (a) is used to connect the circuit in part (b). When you try to operate the latch, you find that the Q output stays HIGH no matter what the inputs are. Determine the problem.

34. Determine if the flip-flop in Figure 7–92 is operating properly, and if not, identify the most probable fault.

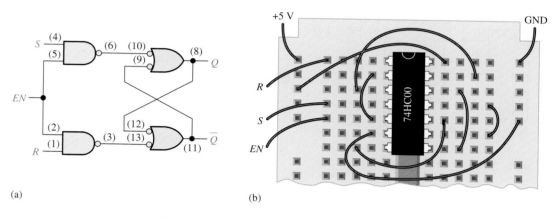

(a)   (b)

▲ **FIGURE 7–91**

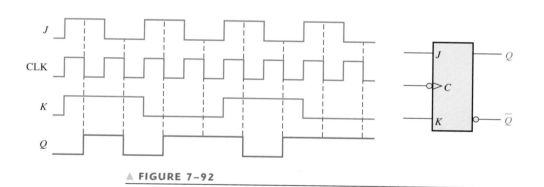

▲ **FIGURE 7–92**

35. The parallel data storage circuit in Figure 7–36 does not operate properly. To check it out, you first make sure that V_{CC} and ground are connected, and then you apply LOW levels to all the D inputs and pulse the clock line. You check the Q outputs and find them all to be LOW; so far, so good. Next you apply HIGHs to all the D inputs and again pulse the clock line. When you check the Q outputs, they are still all LOW. What is the problem, and what procedure will you use to isolate the fault to a single device?

36. The flip-flop circuit in Figure 7–93(a) is used to generate a binary count sequence. The gates form a decoder that is supposed to produce a HIGH when a binary zero or a binary three state occurs (00 or 11). When you check the Q_A and Q_B outputs, you get the display shown in part (b), which reveals glitches on the decoder output (X) in addition to the correct pulses. What is causing these glitches, and how can you eliminate them?

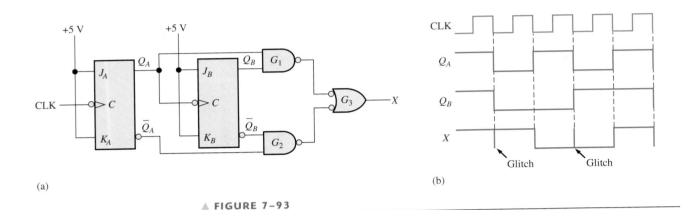

(a)

(b)

▲ **FIGURE 7–93**

37. Determine the Q_A, Q_B and X outputs over six clock pulses in Figure 7–93(a) for each of the following faults in the TTL circuits. Start with both Q_A and Q_B LOW.

   (a) J_A input open   **(b)** K_B input open

   (c) Q_B output open   **(d)** clock input to flip-flop B shorted

   (e) gate G_2 output open

38. Two 74121 one-shots are connected on a circuit board as shown in Figure 7–94. After observing the oscilloscope display, do you conclude that the circuit is operating properly? If not, what is the most likely problem?

▶ **FIGURE 7–94**

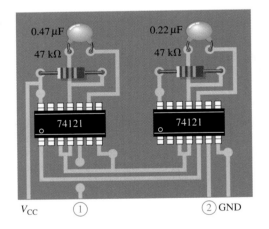

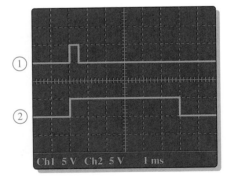

Digital System Application

39. Use 555 timers to implement the 4 s and the 25 s one-shots for the timing circuits portion of the traffic light control system. The trigger input to the 555 cannot stay LOW after its negative-going transition, so you will have to develop a circuit to produce very short negative-going pulses to trigger the long and short timers when the system goes into each state.

Special Design Problems

40. Devise a basic counting circuit that produces a binary sequence from zero through seven by using negative edge-triggered J-K flip-flops.

41. In the shipping department of a softball factory, the balls roll down a conveyor and through a chute single file into boxes for shipment. Each ball passing through the chute activates a switch circuit that produces an electrical pulse. The capacity of each box is 32 balls. Design a logic circuit to indicate when a box is full so that an empty box can be moved into position.

42. List the design changes that would be necessary in the traffic light control system to add a 15 s left turn arrow for the main street. The turn arrow will occur after the red light and prior to the green light. Modify the state diagram from Chapter 6 to show these changes.

Multisim Troubleshooting Practice

43. Open file P07-43 and test the latches to determine which one is faulty.

44. Open file P07-44 and test the J-K flip-flops to determine which one is faulty.

45. Open file P07-45 and test the D flip-flops to determine which one is faulty.

46. Open file P07-46 and test the one-shots to determine which one is faulty.

47. Open file P07-47 and test the divide-by-four circuit to determine if there is a fault. If there is a fault, identify it if possible.

ANSWERS

SECTION REVIEWS

SECTION 7–1 Latches

1. Three types of latches are S-R, gated S-R, and gated D.
2. $SR = 00$, NC; $SR = 01$, $Q = 0$; $SR = 10$, $Q = 1$; $SR = 11$, invalid
3. $Q = 1$

SECTION 7–2 Edge-Triggered Flip-Flops

1. The output of a gated S-R latch can change any time the gate enable (EN) input is active. The output of an edge-triggered S-R flip-flop can change only on the triggering edge of a clock pulse.
2. The J-K flip-flop does not have an invalid state as does the S-R flip-flop.
3. Output Q goes HIGH on the trailing edge of the first clock pulse, LOW on the trailing edge of the second pulse, HIGH on the trailing edge of the third pulse, and LOW on the trailing edge of the fourth pulse.

SECTION 7–3 Flip-Flop Operating Characteristics

1. **(a)** Set-up time is the time required for input data to be present before the triggering edge of the clock pulse.
   (b) Hold time is the time required for data to remain on the inputs after the triggering edge of the clock pulse.
2. The 74AHC74 can be operated at the highest frequency, according to Table 7–5.

SECTION 7–4 Flip-Flop Applications

1. A group of data storage flip-flops is a register.
2. For divide-by-2 operation, the flip-flop must toggle ($J = 1$, $K = 1$).
3. Six flip-flops are used in a divide-by-64 device.

SECTION 7–5 **One-Shots**

1. A nonretriggerable one-shot times out before it can respond to another trigger input. A retriggerable one-shot responds to each trigger input.

2. Pulse width is set with external R and C components.

SECTION 7–6 **The 555 Timer**

1. An astable multivibrator has no stable state. A monostable multivibrator has one stable state.

2. Duty cycle = (15 ms/20 ms)100% = 75%

SECTION 7–7 **Troubleshooting**

1. Yes, a negative edge-triggered D flip-flop can be used.

2. An astable multivibrator using a 555 timer can be used to provide the clock.

RELATED PROBLEMS FOR EXAMPLES

7–1 The Q output is the same as shown in Figure 7–5(b).

7–2 See Figure 7–95.   **7–3** See Figure 7–96.

7–4 See Figure 7–97.   **7–5** See Figure 7–98.

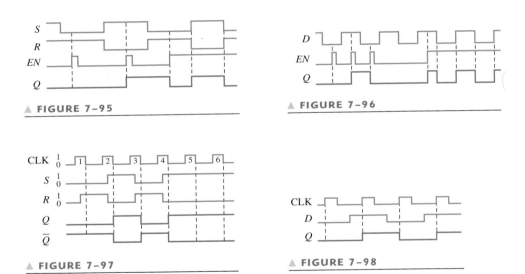

▲ FIGURE 7–95

▲ FIGURE 7–96

▲ FIGURE 7–97

▲ FIGURE 7–98

7–6 See Figure 7–99. **7–7** See Figure 7–100.
7–8 See Figure 7–101. **7–9** See Figure 7–102.

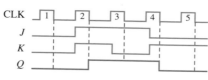

▲ **FIGURE 7–99**

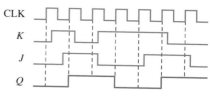

▲ **FIGURE 7–100**

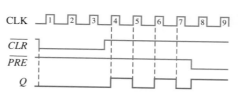

▲ **FIGURE 7–101**

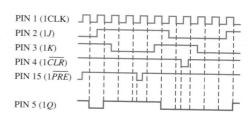

▲ **FIGURE 7–102**

7–10 $2^5 = 32$. Five flip-flops are required.

7–11 Sixteen states require four flip-flops ($2^4 = 16$).

7–12 $C_{EXT} = 7143$ pF connected from CX to RX/CX of the 74121.

7–13 $C_{EXT} = 560$ pF, $R_{EXT} = 27$ kΩ. See Figure 7–103.

7–14 $R_1 = 91$ kΩ

7–15 Duty cycle $\cong 32\%$

▶ **FIGURE 7–103**

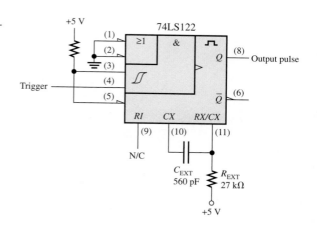

SELF-TEST

1. (a) **2.** (c) **3.** (d) **4.** (b) **5.** (d) **6.** (d) **7.** (a) **8.** (b)
9. (d) **10.** (d) **11.** (c) **12.** (f)

8

COUNTERS

CHAPTER OUTLINE

CHAPTER OBJECTIVES

- Describe the difference between an asynchronous and a synchronous counter

- Analyze counter timing diagrams

- Analyze counter circuits

- Explain how propagation delays affect the operation of a counter

- Determine the modulus of a counter

- Modify the modulus of a counter

- Recognize the difference between a 4-bit binary counter and a decade counter

- Use an up/down counter to generate forward and reverse binary sequences
- Determine the sequence of a counter
- Use IC counters in various applications
- Design a counter that will have any specified sequence of states
- Use cascaded counters to achieve a higher modulus
- Use logic gates to decode any given state of a counter
- Eliminate glitches in counter decoding
- Explain how a digital clock operates
- Interpret counter logic symbols that use dependency notation
- Troubleshoot counters for various types of faults

KEY TERMS

- Asynchronous
- Recycle
- Modulus
- Decade
- Synchronous
- Terminal count
- State machine
- State diagram
- Cascade

INTRODUCTION

As you learned in Chapter 7, flip-flops can be connected together to perform counting operations. Such a group of flip-flops is a counter. The number of flip-flops used and the way in which they are connected determine the number of states (called the modulus) and also the specific sequence of states that the counter goes through during each complete cycle.

Counters are classified into two broad categories according to the way they are clocked: asynchronous and synchronous. In asynchronous counters, commonly called *ripple counters,* the first flip-flop is clocked by the external clock pulse and then each successive flip-flop is clocked by the output of the preceding flip-flop. In synchronous counters, the clock input is connected to all of the flip-flops so that they are clocked simultaneously. Within each of these two categories, counters are classified primarily by the type of sequence, the number of states, or the number of flip-flops in the counter.

FIXED-FUNCTION DEVICES

74XX93	74XX161	74XX162
74XX163	74XX190	74XX47

■■■ DIGITAL SYSTEM APPLICATION PREVIEW

The Digital System Application illustrates the concepts from this chapter. It continues the traffic light control system from the last two chapters. The focus in this chapter is the sequential logic portion of the system that produces the traffic light sequence based on inputs from the timing circuits and the vehicle sensor. The portions of the system developed in Chapters 6 and 7 are combined with the sequential logic to complete the system.

WWW. VISIT THE COMPANION WEBSITE
Study aids for this chapter are available at
http://www.prenhall.com/floyd

8-1 ASYNCHRONOUS COUNTER OPERATION

The term **asynchronous** refers to events that do not have a fixed time relationship with each other and, generally, do not occur at the same time. An **asynchronous counter** is one in which the flip-flops (FF) within the counter do not change states at exactly the same time because they do not have a common clock pulse.

After completing this section, you should be able to

■ Describe the operation of a 2-bit asynchronous binary counter ■ Describe the operation of a 3-bit asynchronous binary counter ■ Define *ripple* in relation to counters ■ Describe the operation of an asynchronous decade counter ■ Develop counter timing diagrams ■ Discuss the 74LS93 4-bit asynchronous binary counter

A 2-Bit Asynchronous Binary Counter

The clock input of an asynchronous counter is always connected only to the LSB flip-flop.

Figure 8–1 shows a 2-bit counter connected for asynchronous operation. Notice that the clock (CLK) is applied to the clock input (C) of *only* the first flop-flop, FF0, which is always the least significant bit (LSB). The second flip-flop, FF1, is triggered by the $\overline{Q}_0$ output of FF0. FF0 changes state at the positive-going edge of each clock pulse, but FF1 changes only when triggered by a positive-going transition of the $\overline{Q}_0$ output of FF0. Because of the inherent propagation delay time through a flip-flop, a transition of the input clock pulse (CLK) and a transition of the $\overline{Q}_0$ output of FF0 can never occur at exactly the same time. Therefore, the two flip-flops are never simultaneously triggered, so the counter operation is asynchronous.

▶ **FIGURE 8–1**

A 2-bit asynchronous binary counter. Open file F08-01 to verify operation.

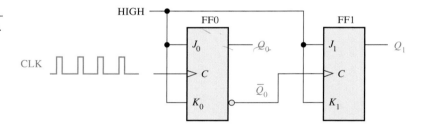

The Timing Diagram   Let's examine the basic operation of the asynchronous counter of Figure 8–1 by applying four clock pulses to FF0 and observing the Q output of each flip-flop. Figure 8–2 illustrates the changes in the state of the flip-flop outputs in response to the clock pulses. Both flip-flops are connected for toggle operation ($J = 1$, $K = 1$) and are assumed to be initially RESET (Q LOW).

▶ **FIGURE 8–2**

Timing diagram for the counter of Figure 8–1. As in previous chapters, output waveforms are shown in green.

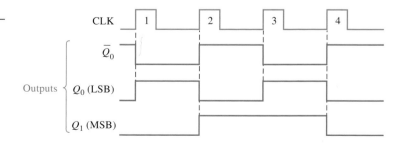

Asynchronous counters are also known as ripple counters.

The positive-going edge of CLK1 (clock pulse 1) causes the Q_0 output of FF0 to go HIGH, as shown in Figure 8–2. At the same time the $\overline{Q}_0$ output goes LOW, but it has no effect on FF1 because a positive-going transition must occur to trigger the flip-flop. After the leading edge of CLK1, $Q_0 = 1$ and $Q_1 = 0$. The positive-going edge of CLK2 causes Q_0 to go LOW. Output $\overline{Q}_0$ goes HIGH and triggers FF1, causing Q_1 to go HIGH. After the leading edge of CLK2, $Q_0 = 0$ and $Q_1 = 1$. The positive-going edge of CLK3 causes Q_0 to go HIGH again. Output $\overline{Q}_0$ goes LOW and has no effect on FF1. Thus, after the leading edge of CLK3, $Q_0 = 1$ and $Q_1 = 1$. The positive-going edge of CLK4 causes Q_0 to go LOW, while $\overline{Q}_0$ goes HIGH and triggers FF1, causing Q_1 to go LOW. After the leading edge of CLK4, $Q_0 = 0$ and $Q_1 = 0$. The counter has now recycled to its original state (both flip-flops are RESET).

In the timing diagram, the waveforms of the Q_0 and Q_1 outputs are shown relative to the clock pulses as illustrated in Figure 8–2. For simplicity, the transitions of Q_0, Q_1, and the clock pulses are shown as simultaneous even though this is an asynchronous counter. There is, of course, some small delay between the CLK and the Q_0 transition and between the $\overline{Q}_0$ transition and the Q_1 transition.

Note in Figure 8–2 that the 2-bit counter exhibits four different states, as you would expect with two flip-flops ($2^2 = 4$). Also, notice that if Q_0 represents the least significant bit (LSB) and Q_1 represents the most significant bit (MSB), the sequence of counter states represents a sequence of binary numbers as listed in Table 8–1.

In digital logic, Q_0 is always the LSB unless otherwise specified.

CLOCK PULSE	Q_1	Q_0
Initially	0	0
1	0	1
2	1	0
3	1	1
4 (recycles)	0	0

◀ TABLE 8–1

Binary state sequence for the counter in Figure 8–1.

Since it goes through a binary sequence, the counter in Figure 8–1 is a binary counter. It actually counts the number of clock pulses up to three, and on the fourth pulse it recycles to its original state ($Q_0 = 0$, $Q_1 = 0$). The term **recycle** is commonly applied to counter operation; it refers to the transition of the counter from its final state back to its original state.

A 3-Bit Asynchronous Binary Counter The state sequence for a 3-bit binary counter is listed in Table 8–2, and a 3-bit asynchronous binary counter is shown in Figure 8–3(a). The basic operation is the same as that of the 2-bit counter except that the 3-bit counter has eight

CLOCK PULSE	Q_2	Q_1	Q_0
Initially	0	0	0
1	0	0	1
2	0	1	0
3	0	1	1
4	1	0	0
5	1	0	1
6	1	1	0
7	1	1	1
8 (recycles)	0	0	0

◀ TABLE 8–2

State sequence for a 3-bit binary counter.

Three-bit asynchronous binary counter and its timing diagram for one cycle. Open file F08-03 to verify operation.

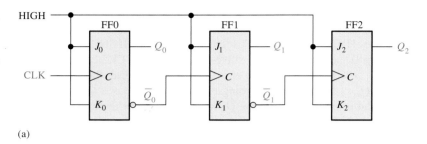

(a)

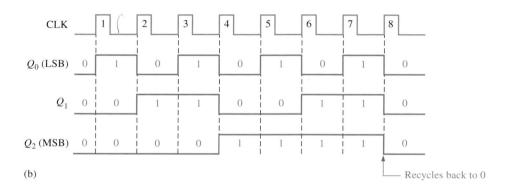

(b)

Recycles back to 0

states, due to its three flip-flops. A timing diagram is shown in Figure 8–3(b) for eight clock pulses. Notice that the counter progresses through a binary count of zero through seven and then recycles to the zero state. This counter can be easily expanded for higher count, by connecting additional toggle flip-flops.

Propagation Delay   Asynchronous counters are commonly referred to as **ripple counters** for the following reason: The effect of the input clock pulse is first "felt" by FF0. This effect cannot get to FF1 immediately because of the propagation delay through FF0. Then there is the propagation delay through FF1 before FF2 can be triggered. Thus, the effect of an input clock pulse "ripples" through the counter, taking some time, due to propagation delays, to reach the last flip-flop.

   To illustrate, notice that all three flip-flops in the counter of Figure 8–3 change state on the leading edge of CLK4. This ripple clocking effect is shown in Figure 8–4 for the first four clock pulses, with the propagation delays indicated. The LOW-to-HIGH transition of Q_0 occurs one delay time (t_{PLH}) after the positive-going transition of the clock pulse. The

Propagation delays in a 3-bit asynchronous (ripple-clocked) binary counter.

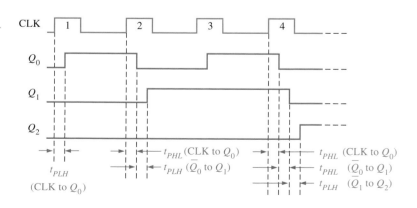

LOW-to-HIGH transition of Q_1 occurs one delay time (t_{PLH}) after the positive-going transition of $\overline{Q}_0$. The LOW-to-HIGH transition of Q_2 occurs one delay time (t_{PLH}) after the positive-going transition of $\overline{Q}_1$. As you can see, FF2 is not triggered until two delay times after the positive-going edge of the clock pulse, CLK4. Thus, it takes three propagation delay times for the effect of the clock pulse, CLK4, to ripple through the counter and change Q_2 from LOW to HIGH.

This cumulative delay of an asynchronous counter is a major disadvantage in many applications because it limits the rate at which the counter can be clocked and creates decoding problems. The maximum cumulative delay in a counter must be less than the period of the clock waveform.

EXAMPLE 8–1

A 4-bit asynchronous binary counter is shown in Figure 8–5(a). Each flip-flop is negative edge-triggered and has a propagation delay for 10 nanoseconds (ns). Develop a timing diagram showing the Q output of each flip-flop, and determine the total propagation delay time from the triggering edge of a clock pulse until a corresponding change can occur in the state of Q_3. Also determine the maximum clock frequency at which the counter can be operated.

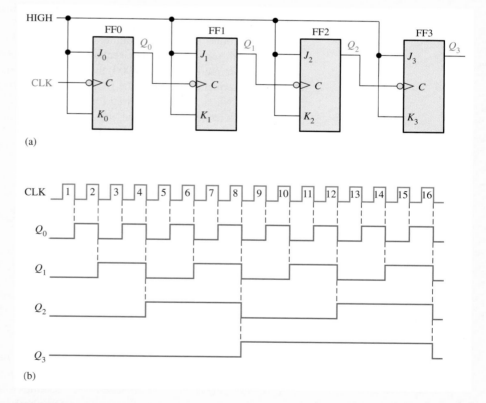

(a)

(b)

▲ **FIGURE 8–5**

Four-bit asynchronous binary counter and its timing diagram. Open file F08-05 and verify the operation.

Solution   The timing diagram with delays omitted is as shown in Figure 8–5(b). For the total delay time, the effect of CLK8 or CLK16 must propagate through four flip-flops before Q_3 changes, so

$$t_{p(tot)} = 4 \times 10 \text{ ns} = \textbf{40 ns}$$

The maximum clock frequency is

$$f_{\text{max}} = \frac{1}{t_{p(tot)}} = \frac{1}{40 \text{ ns}} = \textbf{25 MHz}$$

Related Problem *   Show the timing diagram if all of the flip-flops in Figure 8–5(a) are positive edge-triggered.

_____

*Answers are at the end of the chapter.

Asynchronous Decade Counters

A counter can have 2^n states, where *n* is the number of flip-flops.

The **modulus** of a counter is the number of unique states through which the counter will sequence. The maximum possible number of states (maximum modulus) of a counter is 2^n, where *n* is the number of flip-flops in the counter. Counters can be designed to have a number of states in their sequence that is less than the maximum of 2^n. This type of sequence is called a *truncated sequence.*

One common modulus for counters with truncated sequences is ten (called MOD10). Counters with ten states in their sequence are called **decade** counters. A decade counter with a count sequence of zero (0000) through nine (1001) is a BCD decade counter because its ten-state sequence produces the BCD code. This type of counter is useful in display applications in which BCD is required for conversion to a decimal readout.

To obtain a truncated sequence, it is necessary to force the counter to recycle before going through all of its possible states. For example, the BCD decade counter must recycle back to the 0000 state after the 1001 state. A decade counter requires four flip-flops (three flip-flops are insufficient because $2^3 = 8$).

Let's use a 4-bit asynchronous counter such as the one in Example 8–1 and modify its sequence to illustrate the principle of truncated counters. One way to make the counter recycle after the count of nine (1001) is to decode count ten (1010) with a NAND gate and connect the output of the NAND gate to the clear $(\overline{CLR})$ inputs of the flip-flops, as shown in Figure 8–6(a).

Partial Decoding   Notice in Figure 8–6(a) that only Q_1 and Q_3 are connected to the NAND gate inputs. This arrangement is an example of *partial decoding,* in which the two unique states ($Q_1 = 1$ and $Q_3 = 1$) are sufficient to decode the count of ten because none of the other states (zero through nine) have both Q_1 and Q_3 HIGH at the same time. When the counter goes into count ten (1010), the decoding gate output goes LOW and asynchronously resets all the flip-flops.

The resulting timing diagram is shown in Figure 8–6(b). Notice that there is a glitch on the Q_1 waveform. The reason for this glitch is that Q_1 must first go HIGH before the count of ten can be decoded. Not until several nanoseconds after the counter goes to the count of ten does the output of the decoding gate go LOW (both inputs are HIGH). Thus, the counter is in the 1010 state for a short time before it is reset to 0000, thus producing the glitch on Q_1 and the resulting glitch on the $\overline{CLR}$ line that resets the counter.

Other truncated sequences can be implemented in a similar way, as Example 8–2 shows.

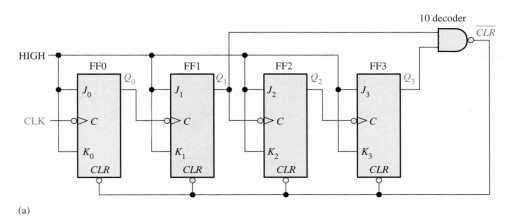

(a)

◀ FIGURE 8-6

An asynchronously clocked decade counter with asynchronous recycling.

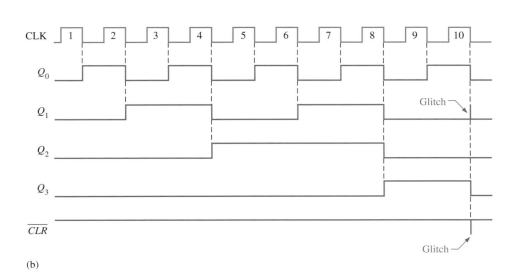

(b)

EXAMPLE 8-2

Show how an asynchronous counter can be implemented having a modulus of twelve with a straight binary sequence from 0000 through 1011.

Solution   Since three flip-flops can produce a maximum of eight states, four flip-flops are required to produce any modulus greater than eight but less than or equal to sixteen.

When the counter gets to its last state, 1011, it must recycle back to 0000 rather than going to its normal next state of 1100, as illustrated in the following sequence chart:

Q_3	Q_2	Q_1	Q_0	
0	0	0	0	←
.	.	.	.	
.	.	.	.	Recycles
.	.	.	.	
1	0	1	1	←
1	1	0	0	← Normal next state

Observe that Q_0 and Q_1 both go to 0 anyway, but Q_2 and Q_3 must be forced to 0 on the twelfth clock pulse. Figure 8–7(a) shows the modulus-12 counter. The NAND gate partially decodes count twelve (1100) and resets flip-flop 2 and flip-flop 3. Thus, on the twelfth clock pulse, the counter is forced to recycle from count eleven to count zero, as shown in the timing diagram of Figure 8–7(b). (It is in count twelve for only a few nanoseconds before it is reset by the glitch on $\overline{CLR}$.)

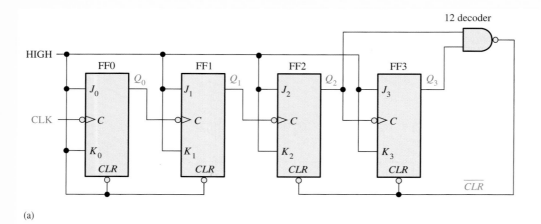

(a)

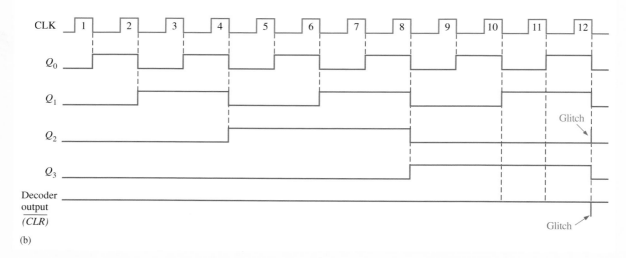

(b)

▲ **FIGURE 8–7**

Asynchronously clocked modulus–12 counter with asynchronous recycling.

Related Problem   How can the counter in Figure 8–7(a) be modified to make it a modulus-13 counter?

THE 74LS93 4-BIT ASYNCHRONOUS BINARY COUNTER

The 74LS93 is an example of a specific integrated circuit asynchronous counter. As the logic diagram in Figure 8–8 shows, this device actually consists of a single flip-flop and a 3-bit asynchronous counter. This arrangement is for flexibility. It can be used as a divide-by-2 device if only the single flip-flop is used, or it can be used as a modulus-8 counter if only the 3-bit counter portion is used. This device also provides gated reset in-

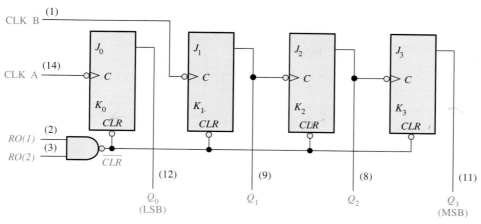

▲ FIGURE 8–8

The 74LS93 4–bit asynchronous binary counter logic diagram. (Pin numbers are in parentheses, and all J and K inputs are internally connected HIGH.)

puts, $RO(1)$ and $RO(2)$. When both of these inputs are HIGH, the counter is reset to the 0000 state $\overline{CLR}$.

Additionally, the 74LS93 can be used as a 4-bit modulus-16 counter (counts 0 through 15) by connecting the Q_0 output to the CLK B input as shown in Figure 8–9(a). It can also be configured as a decade counter (counts 0 through 9) with asynchronous recycling by using the gated reset inputs for partial decoding of count ten, as shown in Figure 8–9(b).

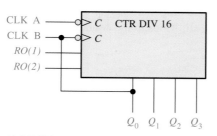

(a) 74LS93 connected as a modulus-16 counter

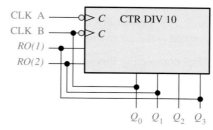

(b) 74LS93 connected as a decade counter

▲ FIGURE 8–9

Two configurations of the 74LS93 asynchronous counter. (The qualifying label, CTR DIV *n*, indicates a counter with *n* states.)

EXAMPLE 8–3

Show how the 74LS93 can be used as a modulus-12 counter.

Solution   Use the gated reset inputs, $RO(1)$ and $RO(2)$, to partially decode count 12 (remember, there is an internal NAND gate associated with these inputs). The count-12 decoding is accomplished by connecting Q_3 to $RO(1)$ and Q_2 to $RO(2)$, as shown in Figure 8–10. Output Q_0 is connected to CLK B to create a 4-bit counter.

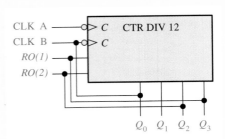

▶ FIGURE 8–10

74LS93 connected as a modulus-12 counter.

Immediately after the counter goes to count 12 (1100), it is reset to 0000. The recycling, however, results in a glitch on Q_2 because the counter must go into the 1100 state for several nanoseconds before recycling.

Related Problem   Show how the 74LS93 can be connected as a modulus-13 counter.

**SECTION 8–1
REVIEW**
Answers are at the end of the chapter.

1. What does the term *asynchronous* mean in relation to counters?
2. How many states does a modulus-14 counter have? What is the minimum number of flip-flops required?

8–2 SYNCHRONOUS COUNTER OPERATION

The term **synchronous** refers to events that have a fixed time relationship with each other. A **synchronous counter** is one in which all the flip-flops in the counter are clocked at the same time by a common clock pulse.

After completing this section, you should be able to

■ Describe the operation of a 2-bit synchronous binary counter ■ Describe the operation of a 3-bit synchronous binary counter ■ Describe the operation of a 4-bit synchronous binary counter ■ Describe the operation of a synchronous decade counter ■ Develop counter timing diagrams ■ Discuss the 74HC163 4-bit binary counter and the 74F162 BCD decade counter

A 2-Bit Synchronous Binary Counter

Figure 8–11 shows a 2-bit synchronous binary counter. Notice that an arrangement different from that for the asynchronous counter must be used for the J_1 and K_1 inputs of FF1 in order to achieve a binary sequence.

▶ FIGURE 8–11

A 2-bit synchronous binary counter.

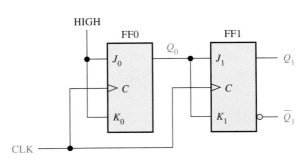

The operation of this synchronous counter is as follows: First, assume that the counter is initially in the binary 0 state; that is, both flip-flops are RESET. When the positive edge of the first clock pulse is applied, FF0 will toggle and Q_0 will therefore go HIGH. What happens to FF1 at the positive-going edge of CLK1? To find out, let's look at the input conditions of FF1. Inputs J_1 and K_1 are both LOW because Q_0, to which they are connected, has not yet gone HIGH. Remember, there is a propagation delay from the triggering edge of the clock pulse until the Q output actually makes a transition. So, $J = 0$ and $K = 0$ when the leading edge of the first clock pulse is applied. This is a no-change condition, and therefore FF1 does not change state. A timing detail of this portion of the counter operation is shown in Figure 8–12(a).

The clock input goes to each flip-flop in a synchronous counter.

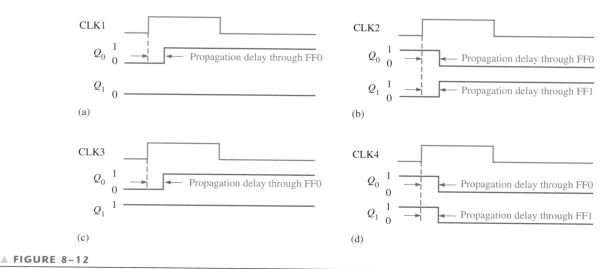

▲ FIGURE 8–12

Timing details for the 2-bit synchronous counter operation (the propagation delays of both flip-flops are assumed to be equal).

After CLK1, $Q_0 = 1$ and $Q_1 = 0$ (which is the binary 1 state). When the leading edge of CLK2 occurs, FF0 will toggle and Q_0 will go LOW. Since FF1 has a HIGH ($Q_0 = 1$) on its J_1 and K_1 inputs at the triggering edge of this clock pulse, the flip-flop toggles and Q_1 goes HIGH. Thus, after CLK2, $Q_0 = 0$ and $Q_1 = 1$ (which is a binary 2 state). The timing detail for this condition is shown in Figure 8–12(b).

When the leading edge of CLK3 occurs, FF0 again toggles to the SET state ($Q_0 = 1$), and FF1 remains SET ($Q_1 = 1$) because its J_1 and K_1 inputs are both LOW ($Q_0 = 0$). After this triggering edge, $Q_0 = 1$ and $Q_1 = 1$ (which is a binary 3 state). The timing detail is shown in Figure 8–12(c).

Finally, at the leading edge of CLK4, Q_0 and Q_1 go LOW because they both have a toggle condition on their J and K inputs. The timing detail is shown in Figure 8–12(d). The counter has now recycled to its original state, binary 0.

The complete timing diagram for the counter in Figure 8–11 is shown in Figure 8–13. Notice that all the waveform transitions appear coincident; that is, the propagation delays are not indicated. Although the delays are an important factor in the synchronous counter operation, in an overall timing diagram they are normally omitted for simplicity. Major waveform

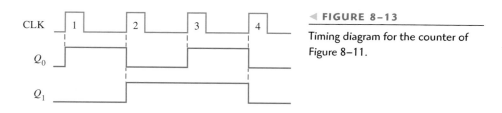

◀ FIGURE 8–13

Timing diagram for the counter of Figure 8–11.

relationships resulting from the normal operation of a circuit can be conveyed completely without showing small delay and timing differences. However, in high-speed digital circuits, these small delays are an important consideration in design and troubleshooting.

A 3-Bit Synchronous Binary Counter

A 3-bit synchronous binary counter is shown in Figure 8–14, and its timing diagram is shown in Figure 8–15. You can understand this counter operation by examining its sequence of states as shown in Table 8–3.

▶ **FIGURE 8–14**

A 3-bit synchronous binary counter. Open file F08-14 to verify the operation.

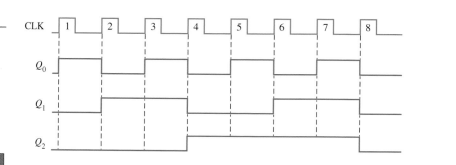

▶ **FIGURE 8–15**

Timing diagram for the counter of Figure 8–14.

▶ **TABLE 8–3**

Binary state sequence for a 3-bit binary counter.

CLOCK PULSE	Q_2	Q_1	Q_0
Initially	0	0	0
1	0	0	1
2	0	1	0
3	0	1	1
4	1	0	0
5	1	0	1
6	1	1	0
7	1	1	1
8 (recycles)	0	0	0

First, let's look at Q_0. Notice that Q_0 changes on each clock pulse as the counter progresses from its original state to its final state and then back to its original state. To produce this operation, FF0 must be held in the toggle mode by constant HIGHs on its J_0 and K_0 inputs. Notice that Q_1 goes to the opposite state following each time Q_0 is a 1. This change occurs at CLK2, CLK4, CLK6, and CLK8. The CLK8 pulse causes the counter to recycle. To produce this operation, Q_0 is connected to the J_1 and K_1 inputs of FF1.

When Q_0 is a 1 and a clock pulse occurs, FF1 is in the toggle mode and therefore changes state. The other times, when Q_0 is a 0, FF1 is in the no-change mode and remains in its present state.

Next, let's see how FF2 is made to change at the proper times according to the binary sequence. Notice that both times Q_2 changes state, it is preceded by the unique condition in which both Q_0 and Q_1 are HIGH. This condition is detected by the AND gate and applied to the J_2 and K_2 inputs of FF2. Whenever both Q_0 and Q_1 are HIGH, the output of the AND gate makes the J_2 and K_2 inputs of FF2 HIGH, and FF2 toggles on the following clock pulse. At all other times, the J_2 and K_2 inputs of FF2 are held LOW by the AND gate output, and FF2 does not change state.

A 4-Bit Synchronous Binary Counter

Figure 8–16(a) shows a 4-bit synchronous binary counter, and Figure 8–16(b) shows its timing diagram. This particular counter is implemented with negative edge-triggered flip-flops. The reasoning behind the J and K input control for the first three flip-flops is the same as previously discussed for the 3-bit counter. The fourth stage, FF3, changes only twice in the sequence. Notice that both of these transitions occur following the times that Q_0, Q_1, and Q_2 are all HIGH. This condition is decoded by AND gate G_2 so that when a clock pulse occurs, FF3 will change state. For all other times the J_3 and K_3 inputs of FF3 are LOW, and it is in a no-change condition.

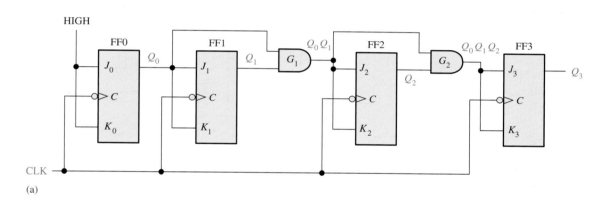

(a)

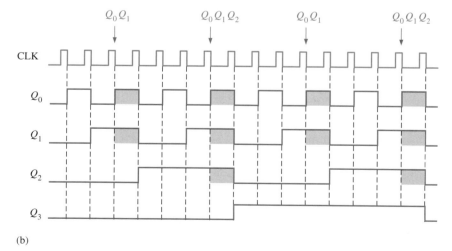

(b)

▲ **FIGURE 8–16**

A 4-bit synchronous binary counter and timing diagram. Points where the AND gate outputs are HIGH are indicated by the shaded areas.

A 4-Bit Synchronous Decade Counter

A decade counter has ten states.

As you know, a BCD decade counter exhibits a truncated binary sequence and goes from 0000 through the 1001 state. Rather than going from the 1001 state to the 1010 state, it recycles to the 0000 state. A synchronous BCD decade counter is shown in Figure 8–17. The timing diagram for the decade counter is shown in Figure 8–18.

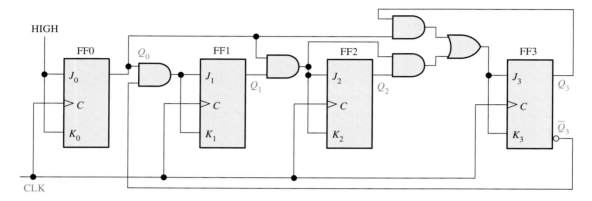

▲ FIGURE 8–17

A synchronous BCD decade counter. Open file F08-17 to verify operation.

▶ FIGURE 8–18

Timing diagram for the BCD decade counter (Q_0 is the LSB).

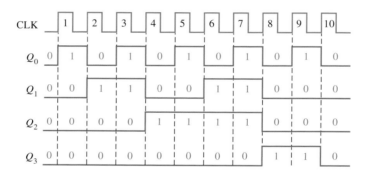

You can understand the counter operation by examining the sequence of states in Table 8–4 and by following the implementation in Figure 8–17. First, notice that FF0 (Q_0) toggles on each clock pulse, so the logic equation for its J_0 and K_0 inputs is

$$J_0 = K_0 = 1$$

This equation is implemented by connecting J_0 and K_0 to a constant HIGH level.

Next, notice in Table 8–4 that FF1 (Q_1) changes on the next clock pulse each time $Q_0 = 1$ and $Q_3 = 0$, so the logic equation for the J_1 and K_1 inputs is

$$J_1 = K_1 = Q_0\overline{Q}_3$$

This equation is implemented by ANDing Q_0 and $\overline{Q}_3$ and connecting the gate output to the J_1 and K_1 inputs of FF1.

Flip-flop 2 (Q_2) changes on the next clock pulse each time both $Q_0 = 1$ and $Q_1 = 1$. This requires an input logic equation as follows:

$$J_2 = K_2 = Q_0Q_1$$

This equation is implemented by ANDing Q_0 and Q_1 and connecting the gate output to the J_2 and K_2 inputs of FF2.

CLOCK PULSE	Q_3	Q_2	Q_1	Q_0
Initially	0	0	0	0
1	0	0	0	1
2	0	0	1	0
3	0	0	1	1
4	0	1	0	0
5	0	1	0	1
6	0	1	1	0
7	0	1	1	1
8	1	0	0	0
9	1	0	0	1
10 (recycles)	0	0	0	0

Finally, FF3 (Q_3) changes to the opposite state on the next clock pulse each time $Q_0 = 1$, $Q_1 = 1$, and $Q_2 = 1$ (state 7), or when $Q_0 = 1$ and $Q_3 = 1$ (state 9). The equation for this is as follows:

$$J_3 = K_3 = Q_0Q_1Q_2 + Q_0Q_3$$

This function is implemented with the AND/OR logic connected to the J_3 and K_3 inputs of FF3 as shown in the logic diagram in Figure 8–17. Notice that the differences between this decade counter and the modulus-16 binary counter in Figure 8–16 are the $Q_0\overline{Q_3}$ AND gate, the Q_0Q_3 AND gate, and the OR gate; this arrangement detects the occurrence of the 1001 state and causes the counter to recycle properly on the next clock pulse.

THE 74HC163 4-BIT SYNCHRONOUS BINARY COUNTER

The 74HC163 is an example of an integrated circuit 4-bit synchronous binary counter. A logic symbol is shown in Figure 8–19 with pin numbers in parentheses. This counter has several features in addition to the basic functions previously discussed for the general synchronous binary counter.

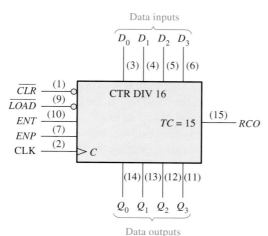

◀ FIGURE 8–19

The 74HC163 4-bit synchronous binary counter. (The qualifying label CTR DIV 16 indicates a counter with sixteen states.)

First, the counter can be synchronously preset to any 4-bit binary number by applying the proper levels to the parallel data inputs. When a LOW is applied to the $\overline{LOAD}$ input, the counter will assume the state of the data inputs on the next clock pulse. Thus, the counter sequence can be started with any 4-bit binary number.

Also, there is an active-LOW clear input ($\overline{CLR}$), which synchronously resets all four flip-flops in the counter. There are two enable inputs, *ENP* and *ENT*. These inputs must both be HIGH for the counter to sequence through its binary states. When at least one input is LOW, the counter is disabled. The ripple clock output (*RCO*) goes HIGH when the counter reaches the last state in its sequence of fifteen, called the **terminal count** ($TC = 15$). This output, in conjunction with the enable inputs, allows these counters to be cascaded for higher count sequences.

Figure 8–20 shows a timing diagram of this counter being preset to twelve (1100) and then counting up to its terminal count, fifteen (1111). Input D_0 is the least significant input bit, and Q_0 is the least significant output bit.

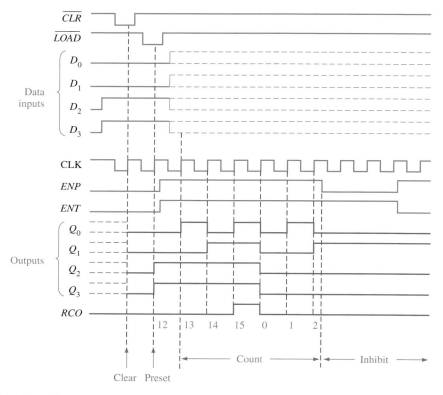

▲ FIGURE 8-20

Timing example for a 74HC163.

Let's examine this timing diagram in detail. This will aid you in interpreting timing diagrams in this chapter or on manufacturers' data sheets. To begin, the LOW level pulse on the $\overline{CLR}$ input causes all the outputs (Q_0, Q_1, Q_2, and Q_3) to go LOW.

Next, the LOW level pulse on the $\overline{LOAD}$ input synchronously enters the data on the data inputs (D_0, D_1, D_2, and D_3) into the counter. These data appear on the Q outputs at the time of the first positive-going clock edge after $\overline{LOAD}$ goes LOW. This is the preset operation. In this particular example, Q_0 is LOW, Q_1 is LOW, Q_2 is HIGH, and Q_3 is HIGH. This, of course, is a binary 12 (Q_0 is the LSB).

The counter now advances through states 13, 14, and 15 on the next three positive-going clock edges. It then recycles to 0, 1, 2 on the following clock pulses. Notice that both *ENP* and *ENT* inputs are HIGH during the state sequence. When *ENP* goes LOW, the counter is inhibited and remains in the binary 2 state.

THE 74F162 SYNCHRONOUS BCD DECADE COUNTER

The 74F162 is an example of a decade counter. It can be preset to any BCD count by the use of the data inputs and a LOW on the $\overline{PE}$ input. A LOW on the asynchronous $\overline{SR}$ will reset the counter. The enable inputs CEP and CET must both be HIGH for the counter to advance through its sequence of states in response to a positive transition on the CLK input. The enable inputs in conjunction with the terminal count, TC (1001), provide for cascading several decade counters. Figure 8–21 shows the logic symbol for the 74F162 counter, and Figure 8–22 is a timing diagram showing the counter being preset to count 7 (0111). Cascaded counters will be discussed in Section 8–5.

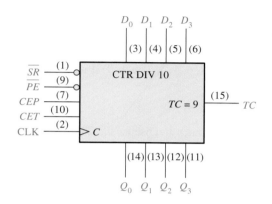

◄ **FIGURE 8–21**

The 74F162 synchronous BCD decade counter. (The qualifying label CTR DIV 10 indicates a counter with ten states.)

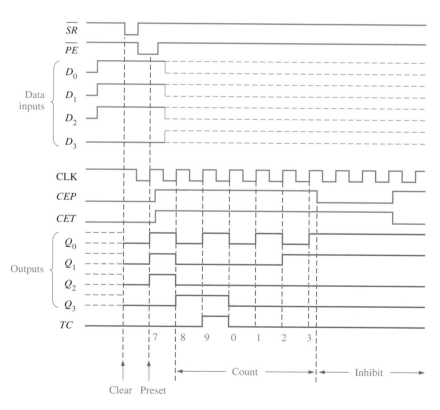

▲ **FIGURE 8–22**

Timing example for a 74F162.

8–3   UP/DOWN SYNCHRONOUS COUNTERS

An **up/down counter** is one that is capable of progressing in either direction through a certain sequence. An up/down counter, sometimes called a bidirectional counter, can have any specified sequence of states. A 3-bit binary counter that advances upward through its sequence (0, 1, 2, 3, 4, 5, 6, 7) and then can be reversed so that it goes through the sequence in the opposite direction (7, 6, 5, 4, 3, 2, 1, 0) is an illustration of up/down sequential operation.

After completing this section, you should be able to

■ Explain the basic operation of an up/down counter ■ Discuss the 74HC190 up/down decade counter

In general, most up/down counters can be reversed at any point in their sequence. For instance, the 3-bit binary counter can be made to go through the following sequence:

$$\overbrace{\text{UP}}\qquad\qquad\overbrace{\text{UP}}$$
$$0,\ 1,\ 2,\ 3,\ 4,\ 5,\ \underbrace{4,\ 3,\ 2,}_{\text{DOWN}}\ \underbrace{3,\ 4,\ 5,\ 6,\ 7,}\ \underbrace{6,\ 5,}_{\text{DOWN}}\ \text{etc.}$$

Table 8–5 shows the complete up/down sequence for a 3-bit binary counter. The arrows indicate the state-to-state movement of the counter for both its UP and its DOWN modes of operation. An examination of Q_0 for both the up and down sequences shows that FF0 toggles on each clock pulse. Thus, the J_0 and K_0 inputs of FF0 are

$$J_0 = K_0 = 1$$

For the up sequence, Q_1 changes state on the next clock pulse when $Q_0 = 1$. For the down sequence, Q_1 changes on the next clock pulse when $Q_0 = 0$. Thus, the J_1 and K_1 inputs of FF1 must equal 1 under the conditions expressed by the following equation:

$$J_1 = K_1 = (Q_0 \cdot \text{UP}) + (\overline{Q_0} \cdot \text{DOWN})$$

▶ TABLE 8–5

Up/Down sequence for a 3-bit binary counter.

CLOCK PULSE	UP	Q_2	Q_1	Q_0	DOWN
0		0	0	0	
1		0	0	1	
2		0	1	0	
3		0	1	1	
4		1	0	0	
5		1	0	1	
6		1	1	0	
7		1	1	1	

For the up sequence, Q_2 changes state on the next clock pulse when $Q_0 = Q_1 = 1$. For the down sequence, Q_2 changes on the next clock pulse when $Q_0 = Q_1 = 0$. Thus, the J_2 and K_2 inputs of FF2 must equal 1 under the conditions expressed by the following equation:

$$J_2 = K_2 = (Q_0 \cdot Q_1 \cdot UP) + (\overline{Q}_0 \cdot \overline{Q}_1 \cdot DOWN)$$

Each of the conditions for the J and K inputs of each flip-flop produces a toggle at the appropriate point in the counter sequence.

Figure 8–23 shows a basic implementation of a 3-bit up/down binary counter using the logic equations just developed for the J and K inputs of each flip-flop. Notice that the $UP/\overline{DOWN}$ control input is HIGH for UP and LOW for DOWN.

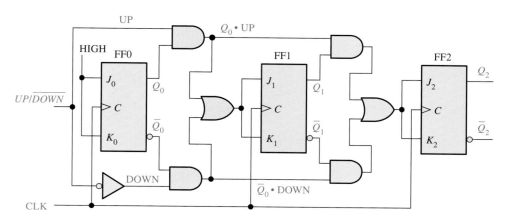

◀ **FIGURE 8–23**

A basic 3-bit up/down synchronous counter. Open file F08-23 to verify operation.

EXAMPLE 8–4

Show the timing diagram and determine the sequence of a 4-bit synchronous binary up/down counter if the clock and $UP/\overline{DOWN}$ control inputs have waveforms as shown in Figure 8–24(a). The counter starts in the all 0s state and is positive edge-triggered.

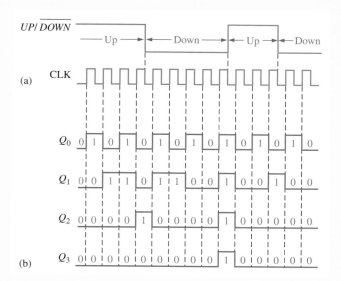

▲ **FIGURE 8–24**

Solution The timing diagram showing the Q outputs is shown in Figure 8–24(b). From these waveforms, the counter sequence is as shown in Table 8–6.

▶ TABLE 8–6

Q_3	Q_2	Q_1	Q_0	
0	0	0	0	
0	0	0	1	
0	0	1	0	UP
0	0	1	1	
0	1	0	0	
0	0	1	1	
0	0	1	0	
0	0	0	1	DOWN
0	0	0	0	
1	1	1	1	
0	0	0	0	
0	0	0	1	UP
0	0	1	0	
0	0	0	1	DOWN
0	0	0	0	

Related Problem   Show the timing diagram if the $UP/\overline{DOWN}$ control waveform in Figure 8–24(a) is inverted.

THE 74HC190 UP/DOWN DECADE COUNTER

Figure 8–25 shows a logic diagram for the 74HC190, an example of an integrated circuit up/down synchronous counter. The direction of the count is determined by the level of the up/down input $(D/\overline{U})$. When this input is HIGH, the counter counts down; when it is LOW, the counter counts up. Also, this device can be preset to any desired BCD digit as determined by the states of the data inputs when the $\overline{LOAD}$ input is LOW.

▶ FIGURE 8–25

The 74HC190 up/down synchronous decade counter.

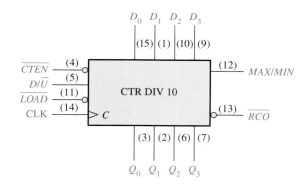

The *MAX/MIN* output produces a HIGH pulse when the terminal count nine (1001) is reached in the UP mode or when the terminal count zero (0000) is reached in the DOWN mode. This *MAX/MIN* output, along with the ripple clock output $(\overline{RCO})$ and the count enable input $(\overline{CTEN})$, is used when cascading counters. (Cascaded counters are discussed in Section 8–5.)

Figure 8–26 is a timing diagram that shows the 74HC190 counter preset to seven (0111) and then going through a count-up sequence followed by a count-down sequence. The

MAX/MIN output is HIGH when the counter is in either the all-0s state (*MIN*) or the 1001 state (*MAX*).

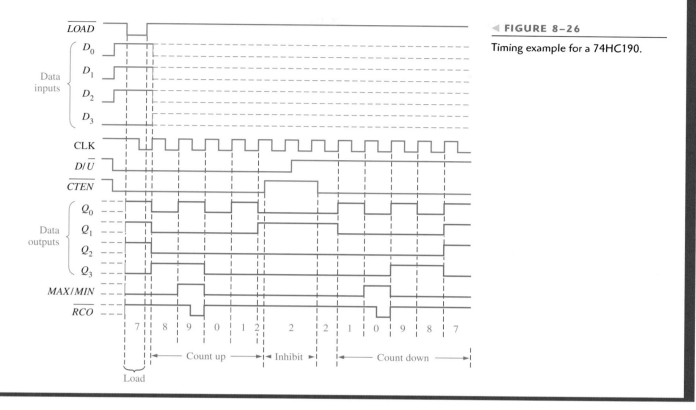

◀ **FIGURE 8–26**

Timing example for a 74HC190.

SECTION 8–3 REVIEW

1. A 4-bit up/down binary counter is in the DOWN mode and in the 1010 state. On the next clock pulse, to what state does the counter go?

2. What is the terminal count of a 4-bit binary counter in the UP mode? In the DOWN mode? What is the next state after the terminal count in the DOWN mode?

8–4   DESIGN OF SYNCHRONOUS COUNTERS

In this section, you will see how sequential circuit design techniques can be applied specifically to counter design. In general, sequential circuits can be classified into two types: (1) those in which the output or outputs depend only on the present internal state (called *Moore circuits*) and (2) those in which the output or outputs depend on both the present state and the input or inputs (called *Mealy circuits*). This section is recommended for those who want an introduction to counter design or to state machine design in general. It is not a prerequisite for any other material.

After completing this section, you should be able to

■ Describe a general sequential circuit in terms of its basic parts and its inputs and outputs ■ Develop a state diagram for a given sequence ■ Develop a next-state table for a specified counter sequence ■ Create a flip-flop transition table ■ Use the Karnaugh map method to derive the logic requirements for a synchronous counter ■ Implement a counter to produce a specified sequence of states

General Model of a Sequential Circuit

Before proceeding with a specific counter design technique, let's begin with a general definition of a **sequential circuit** or **state machine:** A general sequential circuit consists of a combinational logic section and a memory section (flip-flops), as shown in Figure 8–27. In a clocked sequential circuit, there is a clock input to the memory section as indicated.

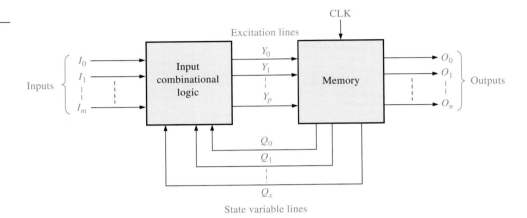

The information stored in the memory section, as well as the inputs to the combinational logic $(I_0, I_1, \ldots, I_m)$, is required for proper operation of the circuit. At any given time, the memory is in a state called the *present state* and will advance to a *next state* on a clock pulse as determined by conditions on the excitation lines $(Y_0, Y_1, \ldots, Y_p)$. The present state of the memory is represented by the state variables $(Q_0, Q_1, \ldots, Q_x)$. These state variables, along with the inputs $(I_0, I_1, \ldots, I_m)$, determine the system outputs $(O_0, O_1, \ldots, O_n)$.

Not all sequential circuits have input and output variables as in the general model just discussed. However, all have excitation variables and state variables. Counters are a special case of clocked sequential circuits. In this section, a general design procedure for sequential circuits is applied to synchronous counters in a series of steps.

Step 1: State Diagram

The first step in the design of a counter is to create a state diagram. A **state diagram** shows the progression of states through which the counter advances when it is clocked. As an example, Figure 8–28 is a state diagram for a basic 3-bit Gray code counter. This particular circuit has no inputs other than the clock and no outputs other than the outputs taken off each flip-flop in the counter. You may wish to review the coverage of the Gray code in Chapter 2 at this time.

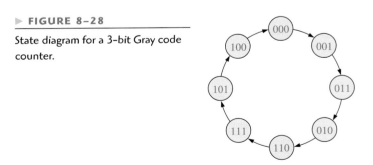

Step 2: Next-State Table

Once the sequential circuit is defined by a state diagram, the second step is to derive a next-state table, which lists each state of the counter (present state) along with the corresponding next state. *The next state is the state that the counter goes to from its present state upon ap-*

plication of a clock pulse. The next-state table is derived from the state diagram and is shown in Table 8–7 for the 3-bit Gray code counter. Q_0 is the least significant bit.

PRESENT STATE			NEXT STATE		
Q_2	Q_1	Q_0	Q_2	Q_1	Q_0
0	0	0	0	0	1
0	0	1	0	1	1
0	1	1	0	1	0
0	1	0	1	1	0
1	1	0	1	1	1
1	1	1	1	0	1
1	0	1	1	0	0
1	0	0	0	0	0

▸ **TABLE 8–7**

Next-state table for 3-bit Gray code counter.

Step 3: Flip-Flop Transition Table

Table 8–8 is a transition table for the J-K flip-flop. All possible output transitions are listed by showing the Q output of the flip-flop going from present states to next states. Q_N is the present state of the flip-flop (before a clock pulse) and Q_{N+1} is the next state (after a clock pulse). For each output transition, the J and K inputs that will cause the transition to occur are listed. An X indicates a "don't care" (the input can be either a 1 or a 0).

OUTPUT TRANSITIONS		FLIP-FLOP INPUTS	
Q_N	Q_{N+1}	J	K
0 →	0	0	X
0 →	1	1	X
1 →	0	X	1
1 →	1	X	0

Q_N: present state

Q_{N+1}: next state

X: "don't care"

▸ **TABLE 8–8**

Transition table for a J-K flip-flop.

To design the counter, the transition table is applied to each of the flip-flops in the counter, based on the next-state table (Table 8–7). For example, for the present state 000, Q_0 goes from a present state of 0 to a next state of 1. To make this happen, J_0 must be a 1 and you don't care what K_0 is ($J_0 = 1$, $K_0 = X$), as you can see in the transition table (Table 8–8). Next, Q_1 is 0 in the present state and remains a 0 in the next state. For this transition, $J_1 = 0$ and $K_1 = X$. Finally, Q_2 is 0 in the present state and remains a 0 in the next state. Therefore, $J_2 = 0$ and $K_2 = X$. This analysis is repeated for each present state in Table 8–7.

Step 4: Karnaugh Maps

Karnaugh maps can be used to determine the logic required for the J and K inputs of each flip-flop in the counter. There is a Karnaugh map for the J input and a Karnaugh map for the K input of each flip-flop. In this design procedure, each cell in a Karnaugh map represents one of the present states in the counter sequence listed in Table 8–7.

From the J and K states in the transition table (Table 8–8) a 1, 0, or X is entered into each present state cell on the maps depending on the transition of the Q output for a particular flip-flop. To illustrate this procedure, two sample entries are shown for the J_0 and the K_0 inputs to the least significant flip-flop (Q_0) in Figure 8–29.

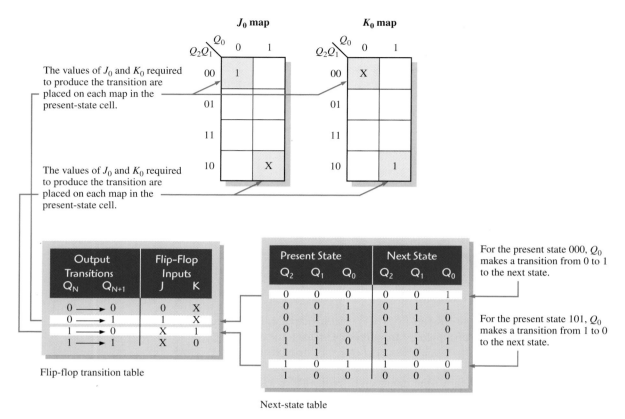

The values of J_0 and K_0 required to produce the transition are placed on each map in the present-state cell.

The values of J_0 and K_0 required to produce the transition are placed on each map in the present-state cell.

For the present state 000, Q_0 makes a transition from 0 to 1 to the next state.

For the present state 101, Q_0 makes a transition from 1 to 0 to the next state.

Flip-flop transition table

Next-state table

▲ FIGURE 8–29

Examples of the mapping procedure for the counter sequence represented in Table 8–7 and Table 8–8.

The completed Karnaugh maps for all three flip-flops in the counter are shown in Figure 8–30. The cells are grouped as indicated and the corresponding Boolean expressions for each group are derived.

▶ FIGURE 8–30

Karnaugh maps for present-state J and K inputs.

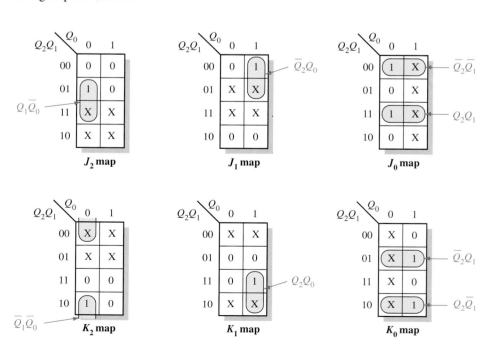

Step 5: Logic Expressions for Flip-Flop Inputs

From the Karnaugh maps of Figure 8–30 you obtain the following expressions for the J and K inputs of each flip-flop:

$$J_0 = Q_2Q_1 + \overline{Q_2}\,\overline{Q_1} = \overline{Q_2 \oplus Q_1}$$
$$K_0 = Q_2\overline{Q_1} + \overline{Q_2}Q_1 = Q_2 \oplus Q_1$$
$$J_1 = \overline{Q_2}Q_0$$
$$K_1 = Q_2Q_0$$
$$J_2 = Q_1\overline{Q_0}$$
$$K_2 = \overline{Q_1}\,\overline{Q_0}$$

Step 6: Counter Implementation

The final step is to implement the combinational logic from the expressions for the J and K inputs and connect the flip-flops to form the complete 3-bit Gray code counter as shown in Figure 8–31.

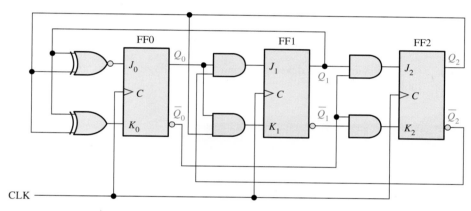

▲ **FIGURE 8–31**

Three-bit Gray code counter. Open file F08-31 to verify operation.

A summary of steps used in the design of this counter follows. In general, these steps can be applied to any sequential circuit.

1. Specify the counter sequence and draw a state diagram.

2. Derive a next-state table from the state diagram.

3. Develop a transition table showing the flip-flop inputs required for each transition. The transition table is always the same for a given type of flip-flop.

4. Transfer the J and K states from the transition table to Karnaugh maps. There is a Karnaugh map for each input of each flip-flop.

5. Group the Karnaugh map cells to generate and derive the logic expression for each flip-flop input.

6. Implement the expressions with combinational logic, and combine with the flip-flops to create the counter.

This procedure is now applied to the design of other synchronous counters in Examples 8–5 and 8–6.

EXAMPLE 8–5

Design a counter with the irregular binary count sequence shown in the state diagram of Figure 8–32. Use J-K flip-flops.

▶ **FIGURE 8–32**

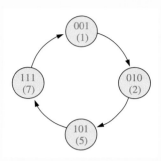

Solution **Step 1:** The state diagram is as shown. Although there are only four states, a 3-bit counter is required to implement this sequence because the maximum binary count is seven. Since the required sequence does not include all the possible binary states, the invalid states (0, 3, 4, and 6) can be treated as "don't cares" in the design. However, if the counter should erroneously get into an invalid state, you must make sure that it goes back to a valid state.

Step 2: The next-state table is developed from the state diagram and is given in Table 8–9.

▶ **TABLE 8–9**

Next-state table.

PRESENT STATE			NEXT STATE		
Q_2	Q_1	Q_0	Q_2	Q_1	Q_0
0	0	1	0	1	0
0	1	0	1	0	1
1	0	1	1	1	1
1	1	1	0	0	1

Step 3: The transition table for the J-K flip-flop is repeated in Table 8–10.

▶ **TABLE 8–10**

Transition table for a J-K flip-flop.

OUTPUT TRANSITIONS			FLIP-FLOP INPUTS	
Q_N		Q_{N+1}	J	K
0	⟶	0	0	X
0	⟶	1	1	X
1	⟶	0	X	1
1	⟶	1	X	0

Step 4: The *J* and *K* inputs are plotted on the present-state Karnaugh maps in Figure 8–33. Also "don't cares" can be placed in the cells corresponding to the invalid states of 000, 011, 100, and 110, as indicated by the red Xs.

▶ FIGURE 8–33

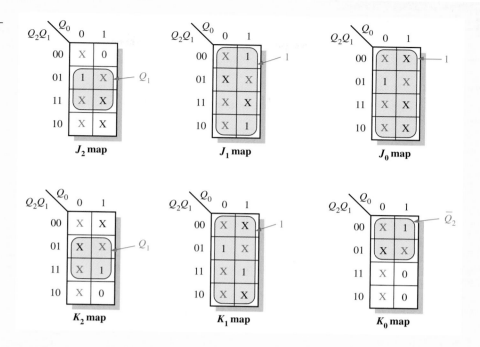

Step 5: Group the 1s, taking advantage of as many of the "don't care" states as possible for maximum simplification, as shown in Figure 8–33. Notice that when *all* cells in a map are grouped, the expression is simply equal to 1. The expression for each J and K input taken from the maps is as follows:

$$J_0 = 1, K_0 = \overline{Q}_2$$
$$J_1 = K_1 = 1$$
$$J_2 = K_2 = Q_1$$

Step 6: The implementation of the counter is shown in Figure 8–34.

▶ FIGURE 8–34

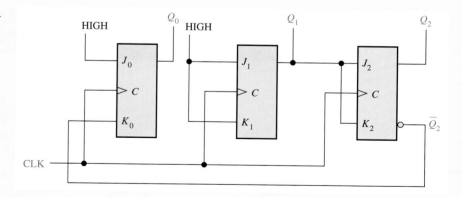

An analysis shows that if the counter, by accident, gets into one of the invalid states (0, 3, 4, 6), it will always return to a valid state according to the following sequences: $0 \rightarrow 3 \rightarrow 4 \rightarrow 7$, and $6 \rightarrow 1$.

Related Problem   Verify the analysis that proves the counter will always return (eventually) to a valid state from an invalid state.

EXAMPLE 8-6

Develop a synchronous 3-bit up/down counter with a Gray code sequence. The counter should count up when an UP/$\overline{\text{DOWN}}$ control input is 1 and count down when the control input is 0.

Solution

Step 1: The state diagram is shown in Figure 8–35. The 1 or 0 beside each arrow indicates the state of the UP/$\overline{\text{DOWN}}$ control input, Y.

▶ **FIGURE 8-35**

State diagram for a 3-bit up/down Gray code counter.

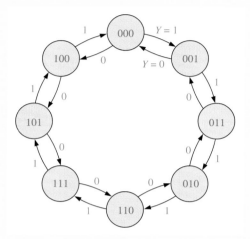

Step 2: The next-state table is derived from the state diagram and is shown in Table 8–11. Notice that for each present state there are two possible next states, depending on the UP/$\overline{\text{DOWN}}$ control variable, Y.

▼ **TABLE 8-11**

Next-state table for 3-bit up/down Gray code counter.

PRESENT STATE			NEXT STATE					
			$Y = 0$ (DOWN)			$Y = 1$ (UP)		
Q_2	Q_1	Q_0	Q_2	Q_1	Q_0	Q_2	Q_1	Q_0
0	0	0	1	0	0	0	0	1
0	0	1	0	0	0	0	1	1
0	1	1	0	0	1	0	1	0
0	1	0	0	1	1	1	1	0
1	1	0	0	1	0	1	1	1
1	1	1	1	1	0	1	0	1
1	0	1	1	1	1	1	0	0
1	0	0	1	0	1	0	0	0

$Y = $ UP/$\overline{\text{DOWN}}$ control input.

Step 3: The transition table for the J-K flip-flops is repeated in Table 8–12.

▶ **TABLE 8–12**

Transition table for a J-K flip-flop.

OUTPUT TRANSITIONS		FLIP-FLOP INPUTS	
Q_N	Q_{N+1}	J	K
0	⟶ 0	0	X
0	⟶ 1	1	X
1	⟶ 0	X	1
1	⟶ 1	X	0

Step 4: The Karnaugh maps for the J and K inputs of the flip-flops are shown in Figure 8–36. The UP/$\overline{\text{DOWN}}$ control input, Y, is considered one of the state variables along with Q_0, Q_1, and Q_2. Using the next-state table, the information in the "Flip-Flop Inputs" column of Table 8–12 is transferred onto the maps as indicated for each present state of the counter.

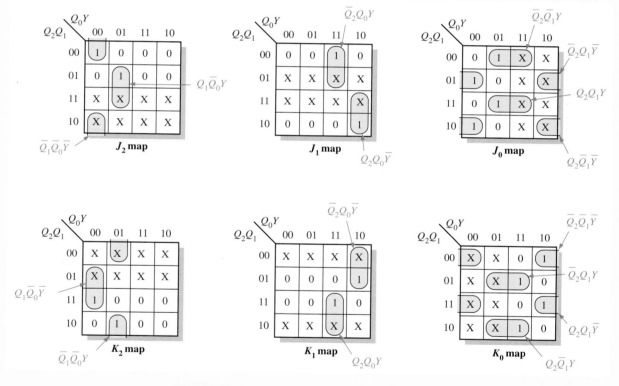

▲ **FIGURE 8–36**

J and K maps for Table 8–11. The UP/$\overline{\text{DOWN}}$ control input, Y, is treated as a fourth variable.

Step 5: The 1s are combined in the largest possible groupings, with "don't cares" (Xs) used where possible. The groups are factored, and the expressions for the J and K inputs are as follows:

$$J_0 = Q_2Q_1Y + Q_2\overline{Q_1}\,\overline{Y} + \overline{Q_2}\,\overline{Q_1}Y + \overline{Q_2}Q_1\overline{Y} \qquad K_0 = \overline{Q_2}\,\overline{Q_1}\,\overline{Y} + \overline{Q_2}Q_1Y + Q_2\overline{Q_1}Y + Q_2Q_1\overline{Y}$$

$$J_1 = \overline{Q_2}Q_0Y + Q_2Q_0\overline{Y} \qquad\qquad\qquad K_1 = \overline{Q_2}Q_0\overline{Y} + Q_2Q_0Y$$

$$J_2 = Q_1\overline{Q_0}Y + \overline{Q_1}\,\overline{Q_0}\,\overline{Y} \qquad\qquad\qquad K_2 = Q_1\overline{Q_0}\,\overline{Y} + \overline{Q_1}\,\overline{Q_0}Y$$

Step 6: The *J* and *K* equations are implemented with combinational logic, and the complete counter is shown in Figure 8–37.

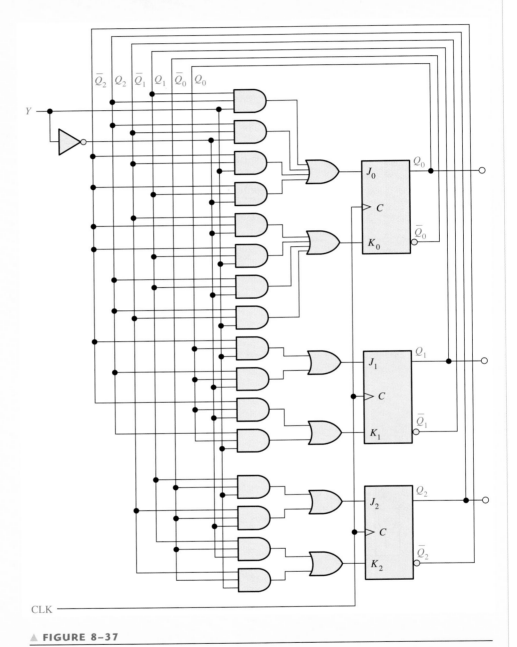

▲ **FIGURE 8–37**

Three-bit up/down Gray code counter.

Related Problem   Verify that the logic in Figure 8–37 agrees with the expressions in Step 5.

SECTION 8–4
REVIEW

1. A flip-flop is presently in the RESET state and must go to the SET state on the next clock pulse. What must J and K be?

2. A flip-flop is presently in the SET state and must remain SET on the next clock pulse. What must J and K be?

3. A binary counter is in the $Q_3\overline{Q_2}Q_1\overline{Q_0} = 1010$ state.

   (a) What is its next state?

   (b) What condition must exist on each flip-flop input to ensure that it goes to the proper next state on the clock pulse?

8–5 CASCADED COUNTERS

Counters can be connected in cascade to achieve higher-modulus operation. In essence, **cascading** means that the last-stage output of one counter drives the input of the next counter.

After completing this section, you should be able to

■ Determine the overall modulus of cascaded counters ■ Analyze the timing diagram of a cascaded counter configuration ■ Use cascaded counters as a frequency divider ■ Use cascaded counters to achieve specified truncated sequences

An example of two counters connected in cascade is shown in Figure 8–38 for a 2-bit and a 3-bit ripple counter. The timing diagram is shown in Figure 8–39. Notice that the

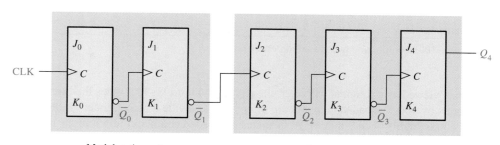

Modulus-4 counter   Modulus-8 counter

◀ FIGURE 8–38

Two cascaded counters (all J and K inputs are HIGH).

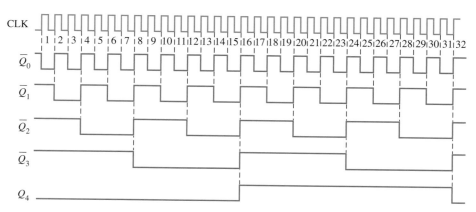

◀ FIGURE 8–39

Timing diagram for the cascaded counter configuration of Figure 8–38.

The overall modulus of cascaded counters is equal to the product of the individual moduli.

final output of the modulus-8 counter, Q_4, occurs once for every 32 input clock pulses. The overall modulus of the cascaded counters is 32; that is, they act as a divide-by-32 counter.

When operating synchronous counters in a cascaded configuration, it is necessary to use the count enable and the terminal count functions to achieve higher-modulus operation. On some devices the count enable is labeled simply *CTEN* (or some other designation such as *G*), and terminal count (*TC*) is analogous to ripple clock output (*RCO*) on some IC counters.

Figure 8–40 shows two decade counters connected in cascade. The terminal count (*TC*) output of counter 1 is connected to the count enable (*CTEN*) input of counter 2. Counter 2 is inhibited by the LOW on its *CTEN* input until counter 1 reaches its last, or terminal, state and its terminal count output goes HIGH. This HIGH now enables counter 2, so that when the first clock pulse after counter 1 reaches its terminal count (CLK10), counter 2 goes from its initial state to its second state. Upon completion of the entire second cycle of counter 1 (when counter 1 reaches terminal count the second time), counter 2 is again enabled and advances to its next state. This sequence continues. Since these are decade counters, counter 1 must go through ten complete cycles before counter 2 completes its first cycle. In other words, for every ten cycles of counter 1, counter 2 goes through one cycle. Thus, counter 2 will complete one cycle after one hundred clock pulses. The overall modulus of these two cascaded counters is $10 \times 10 = 100$.

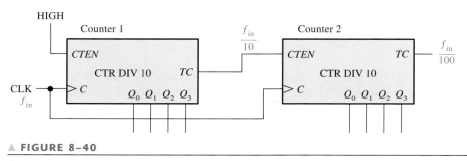

▲ **FIGURE 8–40**

A modulus-100 counter using two cascaded decade counters.

When viewed as a frequency divider, the circuit of Figure 8–40 divides the input clock frequency by 100. Cascaded counters are often used to divide a high-frequency clock signal to obtain highly accurate pulse frequencies. Cascaded counter configurations used for such purposes are sometimes called *countdown chains*. For example, suppose that you have a basic clock frequency of 1 MHz and you wish to obtain 100 kHz, 10 kHz, and 1 kHz; a series of cascaded decade counters can be used. If the 1 MHz signal is divided by 10, the output is 100 kHz. Then if the 100 kHz signal is divided by 10, the output is 10 kHz. Another division by 10 produces the 1 kHz frequency. The general implementation of this countdown chain is shown in Figure 8–41.

▶ **FIGURE 8–41**

Three cascaded decade counters forming a divide-by-1000 frequency divider with intermediate divide-by-10 and divide-by-100 outputs.

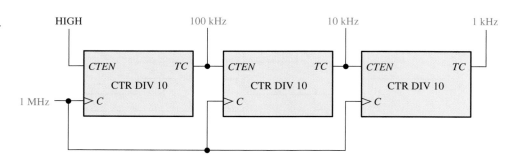

EXAMPLE 8–7

Determine the overall modulus of the two cascaded counter configurations in Figure 8–42.

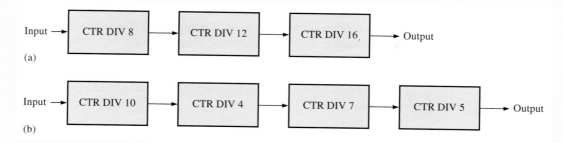

(a)

(b)

▲ **FIGURE 8–42**

Solution   In Figure 8–42(a), the overall modulus for the 3-counter configuration is

$$8 \times 12 \times 16 = \mathbf{1536}$$

In Figure 8–42(b), the overall modulus for the 4-counter configuration is

$$10 \times 4 \times 7 \times 5 = \mathbf{1400}$$

Related Problem   How many cascaded decade counters are required to divide a clock frequency by 100,000?

EXAMPLE 8–8

Use 74F162 decade counters to obtain a 10 kHz waveform from a 1 MHz clock. Show the logic diagram.

Solution   To obtain 10 kHz from a 1 MHz clock requires a division factor of 100. Two 74F162 counters must be cascaded as shown in Figure 8–43. The left counter produces a *TC* pulse for every 10 clock pulses. The right counter produces a *TC* pulse for every 100 clock pulses.

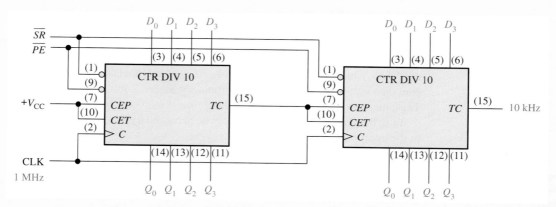

▲ **FIGURE 8–43**

A divide-by-100 counter using two 74F162 decade counters.

Related Problem   Determine the frequency of the waveform at the Q_0 output of the second counter (the one on the right) in Figure 8–43.

Cascaded Counters with Truncated Sequences

The preceding discussion has shown how to achieve an overall modulus (divide-by-factor) that is the product of the individual moduli of all the cascaded counters. This can be considered *full-modulus cascading*.

Often an application requires an overall modulus that is less than that achieved by full-modulus cascading. That is, a truncated sequence must be implemented with cascaded counters. To illustrate this method, we will use the cascaded counter configuration in Figure 8–44. This particular circuit uses four 74HC161 4-bit synchronous binary counters. If these four counters (sixteen bits total) were cascaded in a full-modulus arrangement, the modulus would be

$$2^{16} = 65,536$$

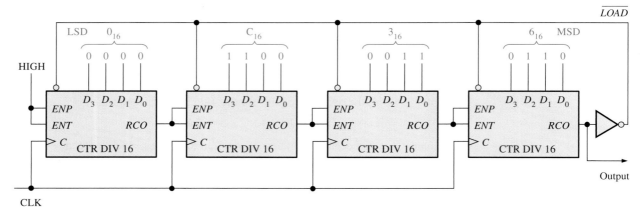

▲ FIGURE 8–44

A divide-by-40,000 counter using 74HC161 4-bit binary counters. Note that each of the parallel data inputs is shown in binary order (the right-most bit D_0 is the LSB in each counter).

Let's assume that a certain application requires a divide-by-40,000 counter (modulus 40,000). The difference between 65,536 and 40,000 is 25,536, which is the number of states that must be *deleted* from the full-modulus sequence. The technique used in the circuit of Figure 8–44 is to preset the cascaded counter to 25,536 (63C0 in hexadecimal) each time it recycles, so that it will count from 25,536 up to 65,535 on each full cycle. Therefore, each full cycle of the counter consists of 40,000 states.

Notice in Figure 8–44 that the *RCO* output of the right-most counter is inverted and applied to the $\overline{LOAD}$ input of each 4-bit counter. Each time the count reaches its terminal value of 65,535, which is 1111111111111111_2, *RCO* goes HIGH and causes the number on the parallel data inputs ($63C0_{16}$) to be synchronously loaded into the counter with the clock pulse. Thus, there is one *RCO* pulse from the right-most 4-bit counter for every 40,000 clock pulses.

With this technique any modulus can be achieved by synchronous loading of the counter to the appropriate initial state on each cycle.

**SECTION 8–5
REVIEW**

1. How many decade counters are necessary to implement a divide-by-1000 (modulus-1000) counter? A divide-by-10,000?

2. Show with general block diagrams how to achieve each of the following, using a flip-flop, a decade counter, and a 4-bit binary counter, or any combination of these:

   (a) Divide-by-20 counter   (b) Divide-by-32 counter

   (c) Divide-by-160 counter   (d) Divide-by-320 counter

8–6 COUNTER DECODING

In many applications, it is necessary that some or all of the counter states be decoded. The decoding of a counter involves using decoders or logic gates to determine when the counter is in a certain binary state in its sequence. For instance, the terminal count function previously discussed is a single decoded state (the last state) in the counter sequence.

After completing this section, you should be able to

■ Implement the decoding logic for any given state in a counter sequence ■ Explain why glitches occur in counter decoding logic ■ Use the method of strobing to eliminate decoding glitches

Suppose that you wish to decode binary state 6 (110) of a 3-bit binary counter. When $Q_2 = 1$, $Q_1 = 1$, and $Q_0 = 0$, a HIGH appears on the output of the decoding gate, indicating that the counter is at state 6. This can be done as shown in Figure 8–45. This is called *active-HIGH decoding*. Replacing the AND gate with a NAND gate provides active-LOW decoding.

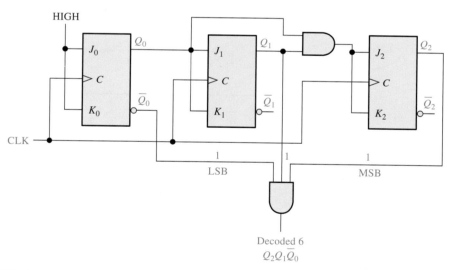

▲ FIGURE 8–45

Decoding of state 6 (110). Open file F08-45 to verify operation.

EXAMPLE 8–9

Implement the decoding of binary state 2 and binary state 7 of a 3-bit synchronous counter. Show the entire counter timing diagram and the output waveforms of the decoding gates. Binary 2 = $\overline{Q_2}Q_1\overline{Q_0}$ and binary 7 = $Q_2Q_1Q_0$.

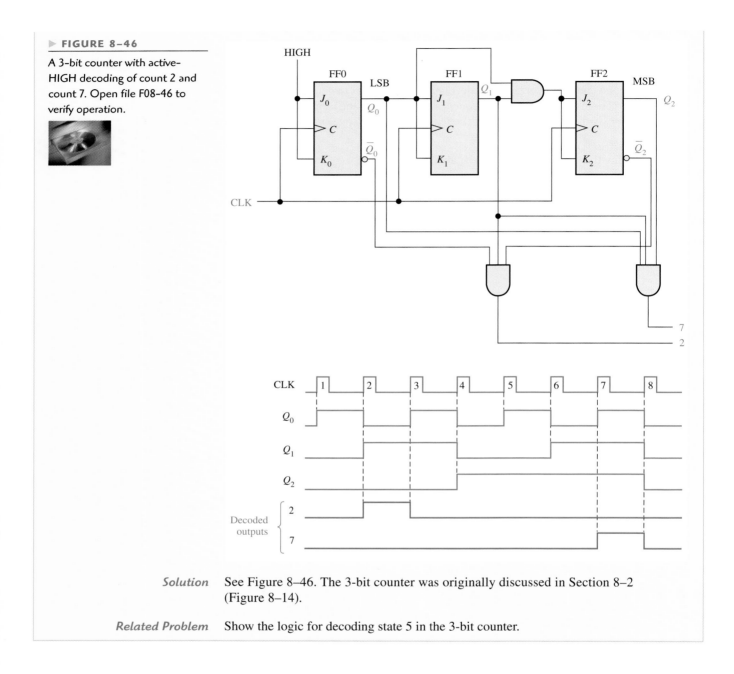

▶ FIGURE 8–46

A 3-bit counter with active-HIGH decoding of count 2 and count 7. Open file F08–46 to verify operation.

Solution	See Figure 8–46. The 3-bit counter was originally discussed in Section 8–2 (Figure 8–14).	
Related Problem	Show the logic for decoding state 5 in the 3-bit counter.	

Decoding Glitches

A glitch is an unwanted spike of voltage.

The problem of glitches produced by the decoding process was discussed in Chapter 6. As you have learned, the propagation delays due to the ripple effect in asynchronous counters create transitional states in which the counter outputs are changing at slightly different times. These transitional states produce undesired voltage spikes of short duration (glitches) on the outputs of a decoder connected to the counter. The glitch problem can also occur to some degree with synchronous counters because the propagation delays from the clock to the Q outputs of each flip-flop in a counter can vary slightly.

Figure 8–47 shows a basic asynchronous BCD decade counter connected to a BCD-to-decimal decoder. To see what happens in this case, let's look at a timing diagram in which the propagation delays are taken into account, as shown in Figure 8–48. Notice that these delays cause false states of short duration. The value of the false binary state at each crit-

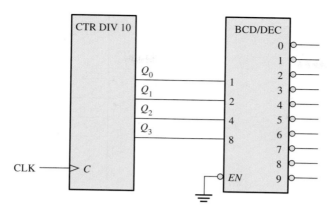

A basic decade (BCD) counter and decoder.

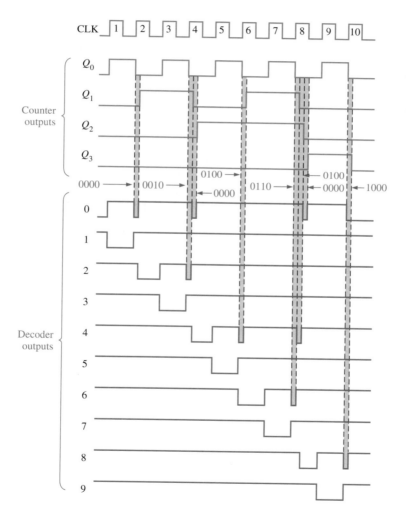

Outputs with glitches from the decoder in Figure 8–47. Glitch widths are exaggerated for illustration and are usually only a few nanoseconds wide.

ical transition is indicated on the diagram. The resulting glitches can be seen on the decoder outputs.

One way to eliminate the glitches is to enable the decoded outputs at a time after the glitches have had time to disappear. This method is known as *strobing* and can be accomplished in the case of an active-HIGH clock by using the LOW level of the clock to enable the decoder, as shown in Figure 8–49. The resulting improved timing diagram is shown in Figure 8–50.

▶ FIGURE 8–49

The basic decade counter and decoder with strobing to eliminate glitches.

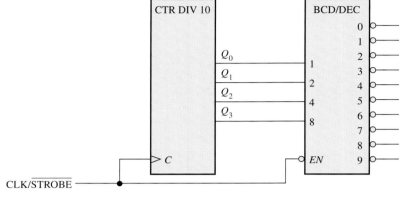

▶ FIGURE 8–50

Strobed decoder outputs for the circuit of Figure 8–49.

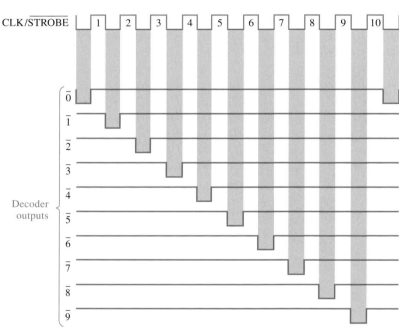

SECTION 8–6 REVIEW

1. What transitional states are possible when a 4-bit asynchronous binary counter changes from

(a) count 2 to count 3 (b) count 3 to count 4

(c) count 10_{10} to count 11_{10} (d) count 15 to count 0

8–7 COUNTER APPLICATIONS

The digital counter is a useful and versatile device that is found in many applications. In this section, some representative counter applications are presented.

After completing this section, you should be able to

■ Describe how counters are used in a basic digital clock system ■ Explain how a divide-by-60 counter is implemented and how it is used in a digital clock ■ Explain how the hours counter is implemented ■ Discuss the application of a counter in an automobile parking control system ■ Describe how a counter is used in the process of parallel-to-serial data conversion

A Digital Clock

A common example of a counter application is in timekeeping systems. Figure 8–51 is a simplified logic diagram of a digital clock that displays seconds, minutes, and hours. First, a 60 Hz sinusoidal ac voltage is converted to a 60 Hz pulse waveform and divided down to a 1 Hz pulse waveform by a divide-by-60 counter formed by a divide-by-10 counter followed by a divide-by-6 counter. Both the *seconds* and *minutes* counts are also produced by divide-by-60 counters, the details of which are shown in Figure 8–52. These counters count from 0 to 59 and then recycle to 0; synchronous decade counters are used in this particular implementation. Notice that the divide-by-6 portion is formed with a decade counter with a truncated sequence achieved by using the decoder count 6 to asynchronously clear the counter. The terminal count, 59, is also decoded to enable the next counter in the chain.

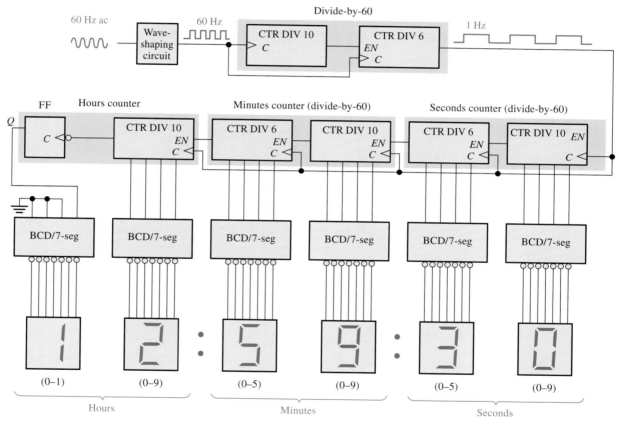

▲ FIGURE 8–51

Simplified logic diagram for a 12-hour digital clock. Logic details using specific devices are shown in Figures 8–52 and 8–53.

The *hours* counter is implemented with a decade counter and a flip-flop as shown in Figure 8–53. Consider that initially both the decade counter and the flip-flop are RESET, and the decode-12 gate and decode-9 gate outputs are HIGH. The decade counter advances through all of its states from zero to nine, and on the clock pulse that recycles it from nine back to zero, the flip-flop goes to the SET state ($J = 1$, $K = 0$). This illuminates a 1 on the tens-of-hours display. The total count is now ten (the decade counter is in the zero state and the flip-flop is SET).

Next, the total count advances to eleven and then to twelve. In state 12 the Q_2 output of the decade counter is HIGH, the flip-flop is still SET, and thus the decode-12 gate output is

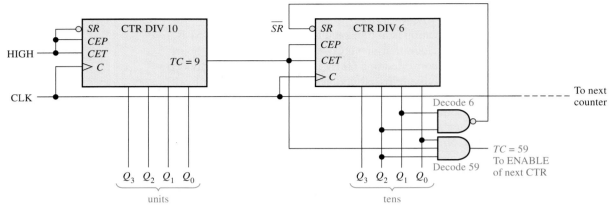

Logic diagram of typical divide-by-60 counter using 74F162 synchronous decade counters. Note that the outputs are in binary order (the right-most bit is the LSB).

▶ **FIGURE 8–53**

Logic diagram for hours counter and decoders. Note that on the counter inputs and outputs, the right-most bit is the LSB.

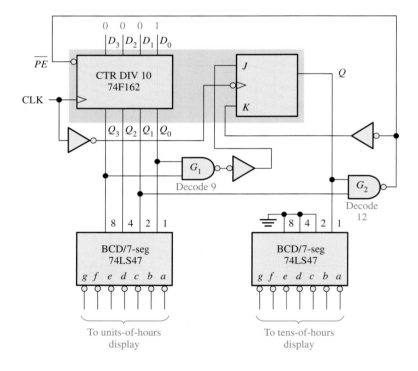

LOW. This activates the $\overline{PE}$ input of the decade counter. On the next clock pulse, the decade counter is preset to state 1 by the data inputs, and the flip-flop is RESET ($J = 0$, $K = 1$). As you can see, this logic always causes the counter to recycle from twelve back to one rather than back to zero.

Automobile Parking Control

This counter example illustrates the use of an up/down counter to solve an everyday problem. The problem is to devise a means of monitoring available spaces in a one-hundred-space parking garage and provide for an indication of a full condition by illuminating a display sign and lowering a gate bar at the entrance.

A system that solves this problem consists of (1) optoelectronic sensors at the entrance and exit of the garage, (2) an up/down counter and associated circuitry, and (3) an interface circuit that uses the counter output to turn the FULL sign on or off as required and lower or raise the gate bar at the entrance. A general block diagram of this system is shown in Figure 8–54.

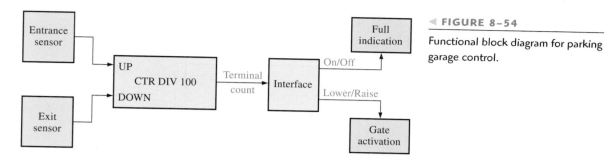

◀ FIGURE 8–54

Functional block diagram for parking garage control.

A logic diagram of the up/down counter is shown in Figure 8–55. It consists of two cascaded 74HC190 up/down decade counters. The operation is described in the following paragraphs.

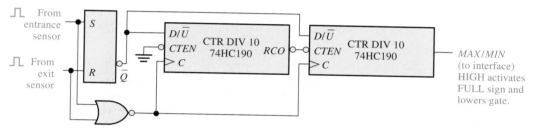

▲ FIGURE 8–55

Logic diagram for modulus-100 up/down counter for automobile parking control.

The counter is initially preset to 0 using the parallel data inputs, which are not shown. Each automobile entering the garage breaks a light beam, activating a sensor that produces an electrical pulse. This positive pulse sets the S-R latch on its leading edge. The LOW on the $\overline{Q}$ output of the latch puts the counter in the UP mode. Also, the sensor pulse goes through the NOR gate and clocks the counter on the LOW-to-HIGH transition of its trailing edge. Each time an automobile enters the garage, the counter is advanced by one **(incremented).** When the one-hundredth automobile enters, the counter goes to its last state (100_{10}). The *MAX/MIN* output goes HIGH and activates the interface circuit (no detail), which lights the FULL sign and lowers the gate bar to prevent further entry.

Incrementing a counter increases its count by one.

When an automobile exits, an optoelectronic sensor produces a positive pulse, which resets the S-R latch and puts the counter in the DOWN mode. The trailing edge of the clock decreases the count by one **(decremented).** If the garage is full and an automobile leaves, the *MAX/MIN* output of the counter goes LOW, turning off the FULL sign and raising the gate.

Decrementing a counter decreases its count by one.

Parallel-to-Serial Data Conversion (Multiplexing)

A simplified example of data transmission using multiplexing and demultiplexing techniques was introduced in Chapter 6. Essentially, the parallel data bits on the multiplexer inputs are converted to serial data bits on the single transmission line. A group of bits appearing simultaneously on parallel lines is called *parallel data*. A group of bits appearing on a single line in a time sequence is called *serial data*.

Parallel-to-serial conversion is normally accomplished by the use of a counter to provide a binary sequence for the data-select inputs of a data selector/multiplexer, as illustrated in Figure 8–56. The Q outputs of the modulus-8 counter are connected to the data-select inputs of an 8-bit multiplexer.

▶ FIGURE 8–56

Parallel-to-serial data conversion logic.

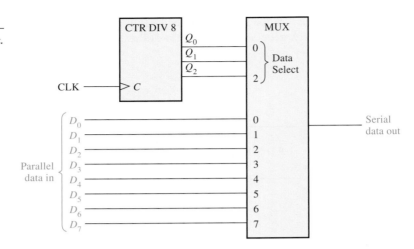

Figure 8–57 is a timing diagram illustrating the operation of this circuit. The first byte (eight-bit group) of parallel data is applied to the multiplexer inputs. As the counter goes through a binary sequence from zero to seven, each bit, beginning with D_0, is sequentially selected and passed through the multiplexer to the output line. After eight clock pulses the

▶ FIGURE 8–57

Example of parallel-to-serial conversion timing for the circuit in Figure 8–56.

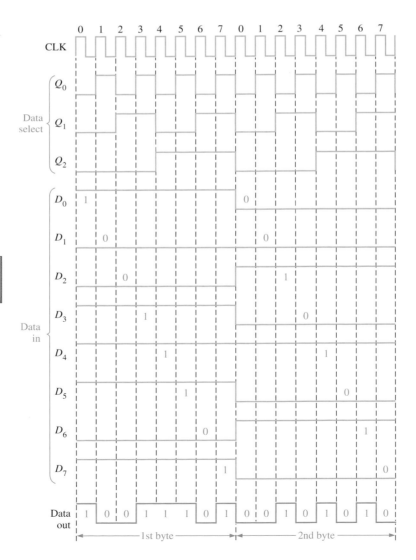

COMPUTER NOTE

Computers contain an internal counter that can be programmed for various frequencies and tone durations, thus producing "music." To select a particular tone, the programmed instruction selects a divisor that is sent to the counter. The divisor sets the counter up to divide the basic peripheral clock frequency to produce an audio tone. The duration of a tone can also be set by a programmed instruction; thus, a basic counter is used to produce melodies by controlling the frequency and duration of tones.

data byte has been converted to a serial format and sent out on the transmission line. When the counter recycles back to 0, the next byte is applied to the data inputs and is sequentially converted to serial form as the counter cycles through its eight states. This process continues repeatedly as each parallel byte is converted to a serial byte.

**SECTION 8–7
REVIEW**

1. Explain the purpose of each NAND gate in Figure 8–53.
2. Identify the two recycle conditions for the hours counter in Figure 8–51, and explain the reason for each.

8–8 LOGIC SYMBOLS WITH DEPENDENCY NOTATION

Up to this point, the logic symbols with dependency notation specified in ANSI/IEEE Standard 91-1984 have been introduced on a limited basis. In many cases, the new symbols do not deviate greatly from the traditional symbols. A significant departure from what we are accustomed to does occur, however, for some devices, including counters and other more complex devices. Although we will continue to use primarily the more traditional and familiar symbols throughout this book, a brief coverage of logic symbols with dependency notation is provided. A specific IC counter is used as an example.

After completing this section, you should be able to

■ Interpret logic symbols that include dependency notation ■ Identify the common block and the individual elements of a counter symbol ■ Interpret the qualifying symbol ■ Discuss control dependency ■ Discuss mode dependency ■ Discuss AND dependency

Dependency notation is fundamental to the ANSI/IEEE standard. Dependency notation is used in conjunction with the logic symbols to specify the relationships of inputs and outputs so that the logical operation of a given device can be determined entirely from its logic symbol without a prior knowledge of the details of its internal structure and without a detailed logic diagram for reference. This coverage of a specific logic symbol with dependency notation is intended to aid in the interpretation of other such symbols that you may encounter in the future.

The 74HC163 4-bit synchronous binary counter is used for illustration. For comparison, Figure 8–58 shows a traditional block symbol and the ANSI/IEEE symbol with dependency notation. Basic descriptions of the symbol and the dependency notation follow.

Common Control Block   The upper block with notched corners in Figure 8–58(b) has inputs and an output that are considered common to all elements in the device and not unique to any one of the elements.

Individual Elements   The lower block in Figure 8–58(b), which is partitioned into four abutted sections, represents the four storage elements (D flip-flops) in the counter, with inputs D_0, D_1, D_2, and D_3 and outputs Q_0, Q_1, Q_2, and Q_3.

Qualifying Symbol   The label "CTR DIV 16" in Figure 8–58(b) identifies the device as a counter (CTR) with sixteen states (DIV 16).

Control Dependency (C)   As shown in Figure 8–58(b), the letter C denotes control dependency. Control inputs usually enable or disable the data inputs (D, J, K, S, and R) of a storage element. The C input is usually the clock input. In this case the digit 5 following C ($C5/2,3,4+$) indicates that the inputs labeled with a 5 prefix are dependent on the clock

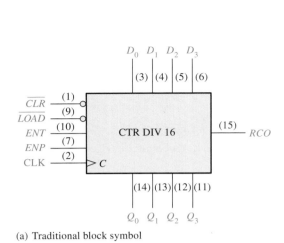

(a) Traditional block symbol

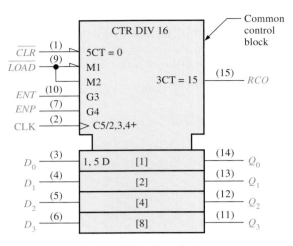

(b) ANSI/IEEE Std. 91-1984 logic symbol

▲ FIGURE 8–58

The 74HC163 4-bit synchronous counter.

(synchronous with the clock). For example, 5CT = 0 on the $\overline{CLR}$ input indicates that the clear function is dependent on the clock; that is, it is a synchronous clear. When the $\overline{CLR}$ input is LOW (0), the counter is reset to zero ($CT = 0$) on the triggering edge of the clock pulse. Also, the 5 D label at the input of storage element [1] indicates that the data storage is dependent on (synchronous with) the clock. All labels in the [1] storage element apply to the [2], [4], and [8] elements below it since they are not labeled differently.

Mode Dependency (M)   As shown in Figure 8–58(b), the letter *M* denotes mode dependency. This label is used to indicate how the functions of various inputs or outputs depend on the mode in which the device is operating. In this case the device has two modes of operation. When the $\overline{LOAD}$ input is LOW (0), as indicated by the triangle input, the counter is in a preset mode (M1) in which the input data ($D_0, D_1, D_2,$ and D_3) are synchronously loaded into the four flip-flops. The digit 1 following *M* in M1 and the 1 in the label 1, 5 D show a dependency relationship and indicate that input data are stored only when the device is in the preset mode (M1), in which $\overline{LOAD} = 0$. When the $\overline{LOAD}$ input is HIGH (1), the counter advances through its normal binary sequence, as indicated by M2 and the 2 in C5/2,3,4+.

AND Dependency (G)   As shown in Figure 8–58(b), the letter *G* denotes AND dependency, indicating that an input designated with *G* followed by a digit is ANDed with any other input or output having the same digit as a prefix in its label. In this particular example, the G3 at the *ENT* input and the 3CT = 15 at the *RCO* output are related, as indicated by the 3, and that relationship is an AND dependency, indicated by the *G*. This tells us that *ENT* must be HIGH (no triangle on the input) *and* the count must be fifteen ($CT = 15$) for the *RCO* output to be HIGH.

Also, the digits 2, 3, and 4 in the label C5/2,3,4+ indicate that the counter advances through its states when $\overline{LOAD} = 1$, as indicated by the mode dependency label M2, and when $ENT = 1$ and $ENP = 1$, as indicated by the AND dependency labels G3 and G4. The + indicates that the counter advances by one count when these conditions exist.

SECTION 8–8 REVIEW

1. In dependency notation, what do the letters *C, M,* and *G* stand for?

2. By what letter is data storage denoted?

8–9 TROUBLESHOOTING

The troubleshooting of counters can be simple or quite involved, depending on the type of counter and the type of fault. This section will give you some insight into how to approach the troubleshooting of sequential circuits.

After completing this section, you should be able to

■ Detect a faulty counter ■ Isolate faults in maximum-modulus cascaded counters
■ Isolate faults in cascaded counters with truncated sequences ■ Determine faults in counters implemented with individual flip-flops

Counters

For a counter with a straightforward sequence that is not controlled by external logic, about the only thing to check (other than V_{CC} and ground) is the possibility of open or shorted inputs or outputs. An IC counter almost never alters its sequence of states because of an internal fault, so you need only check for pulse activity on the Q outputs to detect the existence of an open or a short. The absence of pulse activity on one of the Q outputs indicates an internal open or a short on the line, which may be internal or external to the IC. Absence of pulse activity on all the Q outputs indicates that the clock input is faulty or the clear input is stuck in its active state.

To check the clear input, apply a constant active level while the counter is clocked. You will observe a LOW on each of the Q outputs if it is functioning properly.

A synchronous parallel load feature on a counter can be checked by activating the parallel load input and exercising each state as follows: Apply LOWs to the parallel data inputs, pulse the clock input once, and check for LOWs on all the Q outputs. Next, apply HIGHs to all the parallel data inputs, pulse the clock input once, and check for HIGHs on all the Q outputs.

Cascaded Counters with Maximum Modulus

A failure in one of the counters in a chain of cascaded counters can affect all the counters that follow it. For example, if a count enable input opens, it effectively acts as a HIGH (for TTL), and the counter is always enabled. This type of failure in one of the counters will cause that counter to run at the full clock rate and will also cause all the succeeding counters to run at higher than normal rates. This is illustrated in Figure 8–59 for a divide-by-1000 cascaded counter arrangement where an open enable (*CTEN*) input acts as a TTL HIGH and continuously enables the second counter. Other faults that can affect "downstream" counter stages are open or shorted clock inputs or terminal count outputs. In some of these situations, pulse activity can be observed, but it may be at the wrong frequency. Exact frequency or frequency ratio measurements must be made.

Cascaded Counters with Truncated Sequences

The count sequence of a cascaded counter with a truncated sequence, such as that in Figure 8–60, can be affected by other types of faults in addition to those mentioned for maximum-modulus cascaded counters. For example, a failure in one of the parallel data inputs, the $\overline{LOAD}$ input, or the inverter can alter the preset count and thus change the modulus of the counter.

For example, suppose the D_3 input of the most significant counter in Figure 8–60 is open and acts as a HIGH. Instead of 6_{16} (0110) being preset into the counter, E_{16} (1110) is preset in. So, instead of beginning with $63C0_{16}$ ($25,536_{10}$) each time the counter recycles,

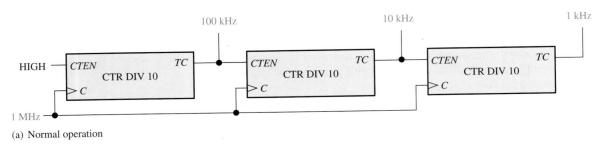

(a) Normal operation

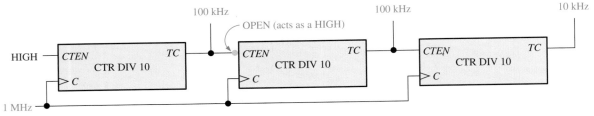

(b) Count Enable (*CTEN*) input of second counter open

▲ FIGURE 8–59

Example of a failure that affects following counters in a cascaded arrangement.

the sequence will begin with $E3C0_{16}$ ($58,304_{10}$). This changes the modulus of the counter from 40,000 to $65,536 - 58,304 = 7232$.

To check this counter, apply a known clock frequency, for example 1 MHz, and measure the output frequency at the final terminal count output. If the counter is operating properly, the output frequency is

$$f_{out} = \frac{f_{in}}{modulus} = \frac{1 \text{ MHz}}{40,000} = 25 \text{ Hz}$$

In this case, the specific failure described in the preceding paragraph will cause the output frequency to be

$$f_{out} = \frac{f_{in}}{modulus} = \frac{1 \text{ MHz}}{7232} = 138.3 \text{ Hz}$$

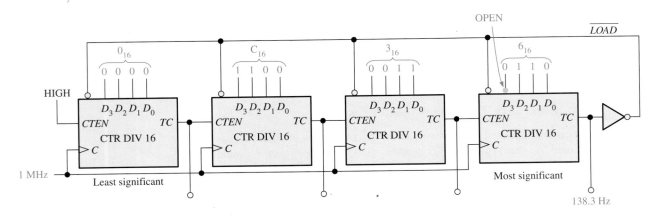

▲ FIGURE 8–60

Example of a failure in a cascaded counter with a truncated sequence.

EXAMPLE 8–10

Frequency measurements are made on the truncated counter in Figure 8–61 as indicated. Determine if the counter is working properly, and if not, isolate the fault.

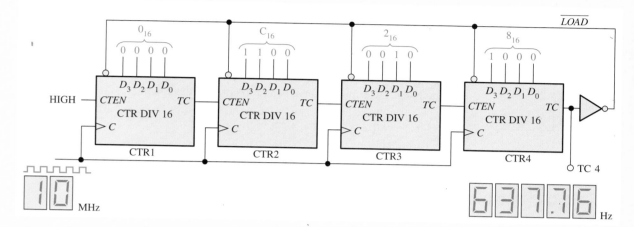

▲ FIGURE 8–61

Solution   Check to see if the frequency measured at TC 4 is correct. If it is, the counter is working properly.

$$\text{truncated modulus} = \text{full modulus} - \text{preset count}$$
$$= 16^4 - 82C0_{16}$$
$$= 65,536 - 33,472 = 32,064$$

The correct frequency at TC 4 is

$$f_4 = \frac{10 \text{ MHz}}{32,064} = 311.88 \text{ Hz}$$

Uh oh! There is a problem. The measured frequency of 637.76 Hz does not agree with the correct calculated frequency of 311.88 Hz.

To find the faulty counter, determine the actual truncated modulus as follows:

$$\text{modulus} = \frac{f_{\text{in}}}{f_{\text{out}}} = \frac{10 \text{ MHz}}{637.76 \text{ Hz}} = 15,680$$

Because the truncated modulus should be 32,064, most likely the counter is being preset to the wrong count when it recycles. The actual preset count is determined as follows:

$$\text{truncated modulus} = \text{full modulus} - \text{preset count}$$
$$\text{preset count} = \text{full modulus} - \text{truncated modulus}$$
$$= 65,536 - 15,680$$
$$= 49,856$$
$$= C2C0_{16}$$

This shows that the counter is being preset to $C2C0_{16}$ instead of $82C0_{16}$ each time it recycles.

Counters 1, 2, and 3 are being preset properly but counter 4 is not. Since $C_{16} = 1100_2$, the D_2 input to counter 4 is HIGH when it should be LOW. This is most likely caused by an **open input.** Check for an external open caused by a bad solder connection, a broken conductor, or a bent pin on the IC. If none can be found, replace the IC and the counter should work properly.

Related Problem   Determine what the output frequency at TC 4 would be if the D_3 input of counter 3 were open.

Counters Implemented with Individual Flip-Flops

Counters implemented with individual flip-flop and gate ICs are sometimes more difficult to troubleshoot because there are many more inputs and outputs with external connections than there are in an IC counter. The sequence of a counter can be altered by a single open or short on an input or output, as Example 8–11 illustrates.

EXAMPLE 8–11

Suppose that you observe the output waveforms that are indicated for the counter in Figure 8–62. Determine if there is a problem with the counter.

▶ **FIGURE 8–62**

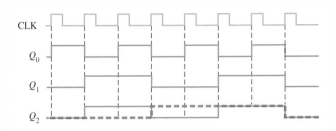

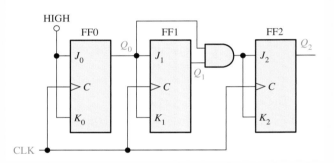

Solution   The Q_2 waveform is incorrect. The correct waveform is shown as a red dashed line. You can see that the Q_2 waveform looks exactly like the Q_1 waveform, so whatever is causing FF1 to toggle appears to also be controlling FF2.

Checking the J and K inputs to FF2, you find a waveform that looks like Q_0. This result indicates that Q_0 is somehow getting through the AND gate. The only way this can happen is if the Q_1 input to the AND gate is always HIGH. However, you have seen that Q_1 has a correct waveform. This observation leads to the conclusion that the lower input to the AND gate must be internally open and acting as a HIGH. Replace the AND gate and retest the circuit.

Related Problem   Describe the Q_2 output of the counter in Figure 8–62 if the Q_1 output of FF1 is open.

**SECTION 8–9
REVIEW**

1. What failures can cause the counter in Figure 8–59 to have no pulse activity on any of the *TC* outputs?

2. What happens if the inverter in Figure 8–61 develops an open output?

Troubleshooting problems that are keyed to the CD-ROM are available in the Multisim Troubleshooting Practice section of the end-of-chapter problems.

DIGITAL SYSTEM APPLICATION

The traffic light control system that was started in Chapter 6 and continued in Chapter 7 is completed in this chapter. In Chapter 6, the combinational logic was developed.

In Chapter 7, the timing circuits were developed.

In this chapter, the sequential logic is developed and all the blocks are connected to produce the complete traffic control system. The overall system block diagram is shown again in Figure 8–63.

Sequential Logic Requirements

The sequential logic controls the sequencing of the traffic lights based on inputs from the timing circuits and the vehicle sensor. The sequential logic will produce a 2-bit Gray code sequence for the four states of the system that are indicated in Figure 8–64.

Block Diagram The sequential logic consists of a 2-bit Gray code counter and associated input logic, as shown in Figure 8–65.

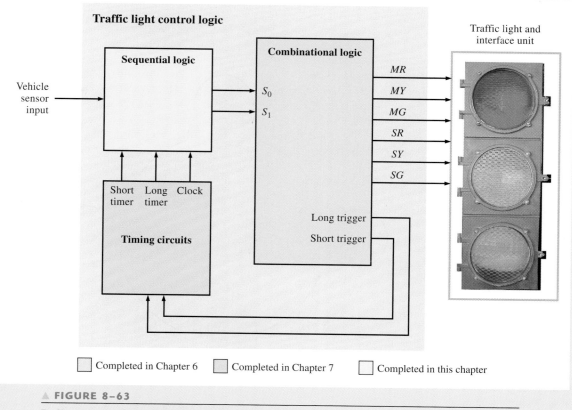

Traffic light control logic

Sequential logic

Combinational logic

Vehicle sensor input

S_0
S_1

Traffic light and interface unit

MR
MY
MG
SR
SY
SG

Short timer   Long timer   Clock

Timing circuits

Long trigger
Short trigger

☐ Completed in Chapter 6   ☐ Completed in Chapter 7   ☐ Completed in this chapter

▲ **FIGURE 8–63**

Traffic light control system block diagram.

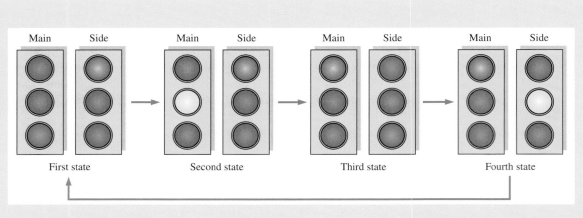

Sequence of traffic light states.

▶ **FIGURE 8–65**

Block diagram of the sequential logic.

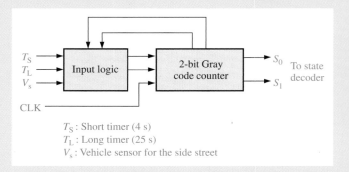

T_S : Short timer (4 s)
T_L : Long timer (25 s)
V_s : Vehicle sensor for the side street

▶ **FIGURE 8–66**

State diagram for the traffic light control system.

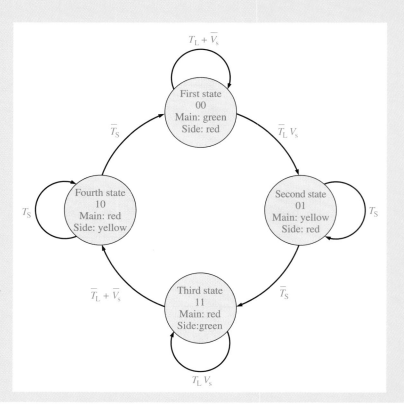

The counter produces a sequence of four states. Transitions from one state to the next are determined by the 4 s timer, the 25 s timer, and the vehicle sensor input. The clock for the counter is the 10 kHz signal produced by the oscillator in the timing circuits.

State Diagram   The state diagram for the traffic light control system was introduced in Chapter 6 and is shown again in Figure 8–66. Based on this state diagram the sequential logic operation is described as follows.

First state: The Gray code for this state is 00. The main street light is green and the side street light is red. The system remains in this state for at least 25 s when the long timer is *on* or as long as there is no vehicle on the side street. This is expressed as $T_L + \overline{V}_s$. The system goes to the next state when the long timer is *off* and there is a vehicle on the side street. This is expressed as $(\overline{T}_L V_s)$.

Second state: The Gray code for this state is 01. The main street light is yellow and the side street light is red. The system remains in this state for 4 s when the short timer is *on* (T_S) and goes to the next state when the short timer goes *off* $(\overline{T}_S)$.

Third state: The Gray code for this state is 11. The main street light is red and the side street light is green. The system remains in this state when the long timer is *on* and there is a vehicle on the side street. This is expressed as $T_L V_s$. The system goes to the next state when the long timer goes *off* or when there is no vehicle on the side street. This is expressed as $\overline{T}_L + \overline{V}_s$.

Fourth state: The Gray code for this state is 10. The main street light is red and the side street light is yellow. The system remains in this state for 4 s when the short timer is *on* (T_S) and goes back to the first state when the short timer goes *off* $(\overline{T}_S)$.

Sequential Logic Implementation   The diagram in Figure 8–67, shows that two D flip-flops are used to implement the Gray counter. Outputs from the input logic provide the D inputs to the flip-flops and the counter is clocked by the 10 kHz clock from the oscillator. The input logic has five input variables: Q_0, Q_1, T_L, T_S, and V_s.

The D flip-flop transition table is shown in Table 8–13. From the state diagram, a next-state table can be developed as shown in Table 8–14. The input conditions for T_L, T_S, and V_s for each present-state/next-state combination are listed in the table.

From Table 8–13 and Table 8–14, the logic conditions required for each flip-flop to go to the 1 state can be determined. For example, Q_0 goes from 0 to 1 when the present state is 00 and the input condition is $\overline{T}_L V_s$, as indicated on the second row of Table 8–13. D_0 must be a 1 to make Q_0 go to a 1 or to remain a 1 on the next clock pulse. For D_0 to be a 1, a logic expression can be written from Table 8–14:

$$\begin{aligned} D_0 &= \overline{Q}_1 \overline{Q}_0 \overline{T}_L V_s + \overline{Q}_1 Q_0 T_S \\ &\quad + \overline{Q}_1 Q_0 \overline{T}_S + Q_1 Q_0 T_L V_s \\ &= \overline{Q}_1 \overline{Q}_0 \overline{T}_L V_s + \overline{Q}_1 Q_0 + Q_1 Q_0 T_L V_s \end{aligned}$$

▶ **FIGURE 8–67**

Sequential logic diagram.

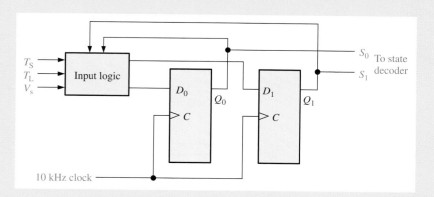

▼ **TABLE 8–13**

D flip-flop transition table.

OUTPUT TRANSITIONS		FLIP-FLOP INPUT
Q_N	Q_{N+1}	D
0 ⟶	0	0
0 ⟶	1	1
1 ⟶	0	0
1 ⟶	1	1

You can use a Karnaugh map to reduce the D_0 expression further to

$$D_0 = \overline{Q}_1\overline{T}_LV_s + \overline{Q}_1Q_0 + Q_0T_LV_s$$

Also, from Table 8–14, the expression for D_1 can be developed.

$$D_1 = \overline{Q}_1Q_0\overline{T}_S + Q_1Q_0T_LV_s$$
$$\quad + Q_1Q_0\overline{T}_L + Q_1Q_0\overline{V}_s + Q_1\overline{Q}_0T_S$$

$$= \overline{Q}_1Q_0\overline{T}_S + Q_1Q_0(T_LV_s + \overline{T}_L)$$
$$\quad + Q_1Q_0\overline{V}_s + Q_1\overline{Q}_0T_S$$

$$= \overline{Q}_1Q_0\overline{T}_S + Q_1Q_0(V_s + \overline{T}_L)$$
$$\quad + Q_1Q_0\overline{V}_s + Q_1\overline{Q}_0T_S$$

$$= \overline{Q}_1Q_0\overline{T}_S + Q_1Q_0(V_s + \overline{T}_L + \overline{V}_s)$$
$$\quad + Q_1\overline{Q}_0T_S$$

$$= \overline{Q}_1Q_0\overline{T}_S + Q_1Q_0 + Q_1\overline{Q}_0T_S$$

You can use a Karnaugh map to reduce the D_1 expression further to

$$D_1 = Q_0\overline{T}_S + Q_1T_S$$

D_0 and D_1 are implemented as shown in Figure 8–68.

Combining the input logic with the 2-bit counter, the complete sequential logic diagram is shown in Figure 8–69.

Next-state table for the sequential logic transitions.

PRESENT STATE		NEXT STATE		INPUT CONDITIONS	FF INPUTS	
Q_1	Q_0	Q_1	Q_0		D_1	D_0
0	0	0	0	$T_L + \overline{V}_s$	0	0
0	0	0	1	$\overline{T}_LV_s$	0	1
0	1	0	1	T_S	0	1
0	1	1	1	$\overline{T}_S$	1	1
1	1	1	1	$T_L V_s$	1	1
1	1	1	0	$\overline{T}_L + \overline{V}_s$	1	0
1	0	1	0	T_S	1	0
1	0	0	0	$\overline{T}_S$	0	0

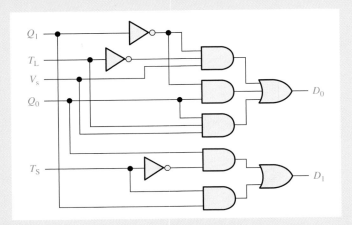

▲ FIGURE 8–68

Input logic for the 2-bit Gray code counter.

▶ FIGURE 8–69

The sequential logic.

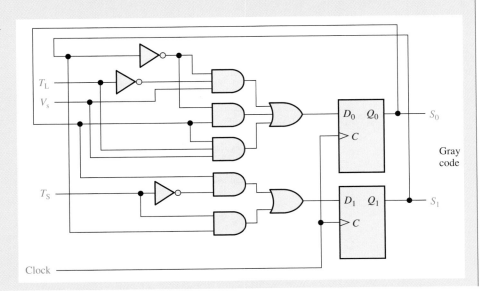

The Complete Traffic Light Control System

Now that we have all three blocks (combinational logic, timing circuits, and sequential logic), we combine them to form a complete system, as shown in the block diagram of Figure 8–70.

The Interface Circuits
Interface circuits are necessary because the logic cannot drive the lights directly due to the current and voltage requirements. There are several possible ways to provide an interface but two possible designs are provided in Appendix B.

System Assignment

- *Activity 1* Use a Karnaugh map to confirm that the simplified expression for D_0 is correct.
- *Activity 2* Use a Karnaugh map to confirm that the simplified expression for D_1 is correct.

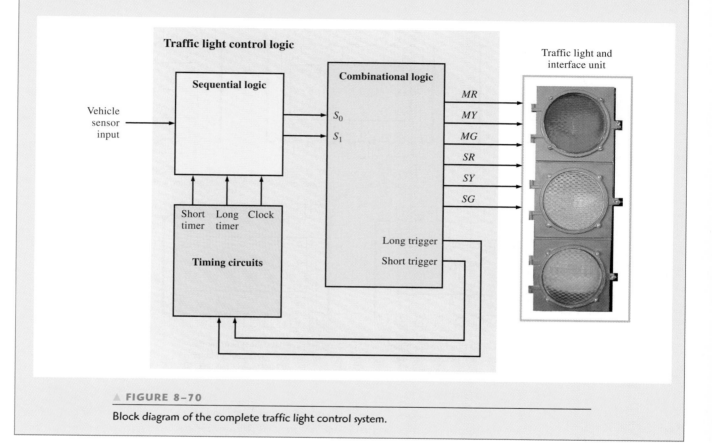

▲ FIGURE 8–70

Block diagram of the complete traffic light control system.

SUMMARY

- Asynchronous and synchronous counters differ only in the way in which they are clocked, as shown in Figure 8–71. Synchronous counters can run at faster clock rates than asynchronous counters.

▶ FIGURE 8–71

Comparison of asynchronous and synchronous counters.

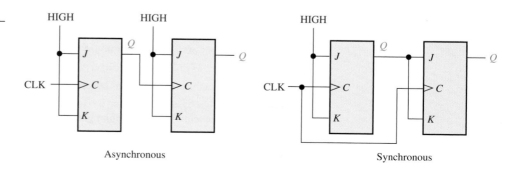

- Connection diagrams for the IC counters introduced in this chapter are shown in Figure 8–72.
- The maximum modulus of a counter is the maximum number of possible states and is a function of the number of stages (flip-flops). Thus,

$$\text{Maximum modulus} = 2^n$$

where n is the number of stages in the counter. The modulus of a counter is the *actual* number of states in its sequence and can be equal to or less than the maximum modulus.

- The overall modulus of cascaded counters is equal to the product of the moduli of the individual counters.

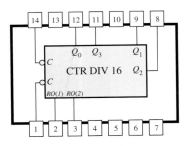

74LS93 4-bit asynchronous binary counter

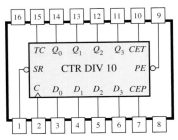

74F162 synchronous BCD decade counter with asynchronous clear

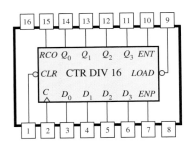

74HC161 4-bit synchronous binary counter with asynchronous clear

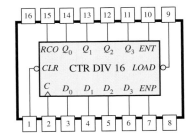

74HC163 4-bit synchronous binary counter with synchronous clear

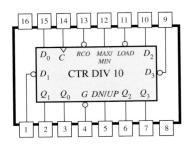

74HC190 synchronous up/down decade counter (*G* is Count Enable.)

▲ FIGURE 8–72

Note that the labels (names of inputs and outputs) are consistent with text but may differ from the particular manufacturer's data book you are using. The devices shown are functionally the same and pin compatible with the same device types in other available CMOS and TTL IC families.

KEY TERMS

Key terms and other bold terms in the chapter are defined in the end-of-book glossary.

Asynchronous Not occurring at the same time.

Cascade To connect "end-to-end" as when several counters are connected from the terminal count output of one counter to the enable input of the next counter.

Decade Characterized by ten states or values.

Modulus The number of unique states through which a counter will sequence.

Recycle To undergo transition (as in a counter) from the final or terminal state back to the initial state.

State diagram A graphic depiction of a sequence of states or values.

State machine A logic system exhibiting a sequence of states conditioned by internal logic and external inputs; any sequential circuit exhibiting a specified sequence of states.

Synchronous Occurring at the same time.

Terminal count The final state in a counter's sequence.

SELF-TEST

Answers are at the end of the chapter.

1. Asynchronous counters are known as
   (a) ripple counters   (b) multiple clock counters
   (c) decade counters   (d) modulus counters

2. An asynchronous counter differs from a synchronous counter in
   (a) the number of states in its sequence
   (b) the method of clocking
   (c) the type of flip-flops used
   (d) the value of the modulus

3. The modulus of a counter is
   (a) the number of flip-flops
   (b) the actual number of states in its sequence
   (c) the number of times it recycles in a second
   (d) the maximum possible number of states

4. A 3-bit binary counter has a maximum modulus of
   (a) 3   (b) 6   (c) 8   (d) 16

5. A 4-bit binary counter has a maximum modulus of
   (a) 16   (b) 32   (c) 8   (d) 4

6. A modulus-12 counter must have
   (a) 12 flip-flops   (b) 3 flip-flops
   (c) 4 flip-flops   (d) synchronous clocking

7. Which one of the following is an example of a counter with a truncated modulus?
   (a) Modulus 8   (b) Modulus 14
   (c) Modulus 16   (d) Modulus 32

8. A 4-bit ripple counter consists of flip-flops that each have a propagation delay from clock to Q output of 12 ns. For the counter to recycle from 1111 to 0000, it takes a total of
   (a) 12 ns   (b) 24 ns   (c) 48 ns   (d) 36 ns

9. A BCD counter is an example of
   (a) a full-modulus counter   (b) a decade counter
   (c) a truncated-modulus counter   (d) answers (b) and (c)

10. Which of the following is an invalid state in an 8421 BCD counter?
    (a) 1100   (b) 0010   (c) 0101   (d) 1000

11. Three cascaded modulus-10 counters have an overall modulus of
    (a) 30   (b) 100   (c) 1000   (d) 10,000

12. A 10 MHz clock frequency is applied to a cascaded counter consisting of a modulus-5 counter, a modulus-8 counter, and two modulus-10 counters. The lowest output frequency possible is
    (a) 10 kHz   (b) 2.5 kHz   (c) 5 kHz   (d) 25 kHz

13. A 4-bit binary up/down counter is in the binary state of zero. The next state in the DOWN mode is
    (a) 0001   (b) 1111   (c) 1000   (d) 1110

14. The terminal count of a modulus-13 binary counter is
    (a) 0000   (b) 1111   (c) 1101   (d) 1100

PROBLEMS

Answers to odd-numbered problems are at the end of the book.

SECTION 8–1   Asynchronous Counter Operation

1. For the ripple counter shown in Figure 8–73, show the complete timing diagram for eight clock pulses, showing the clock, Q_0, and Q_1 waveforms.

▶ FIGURE 8–73

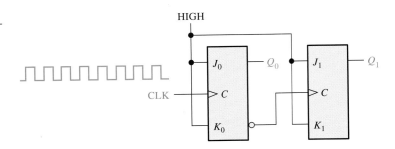

2. For the ripple counter in Figure 8–74, show the complete timing diagram for sixteen clock pulses. Show the clock, Q_0, Q_1, and Q_2 waveforms.

▶ FIGURE 8–74

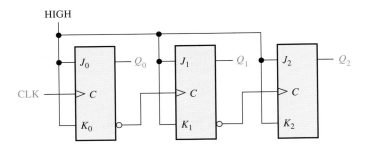

3. In the counter of Problem 2, assume that each flip-flop has a propagation delay from the triggering edge of the clock to a change in the Q output of 8 ns. Determine the worst-case (longest) delay time from a clock pulse to the arrival of the counter in a given state. Specify the state or states for which this worst-case delay occurs.

4. Show how to connect a 74LS93 4-bit asynchronous counter for each of the following moduli:

(a) 9   (b) 11   (c) 13   (d) 14   (e) 15

SECTION 8–2   **Synchronous Counter Operation**

5. If the counter of Problem 3 were synchronous rather than asynchronous, what would be the longest delay time?

6. Show the complete timing diagram for the 5-stage synchronous binary counter in Figure 8–75. Verify that the waveforms of the Q outputs represent the proper binary number after each clock pulse.

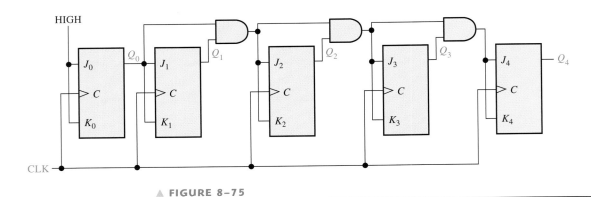

▲ FIGURE 8–75

7. By analyzing the J and K inputs to each flip-flop prior to each clock pulse, prove that the decade counter in Figure 8–76 progresses through a BCD sequence. Explain how these conditions in each case cause the counter to go to the next proper state.

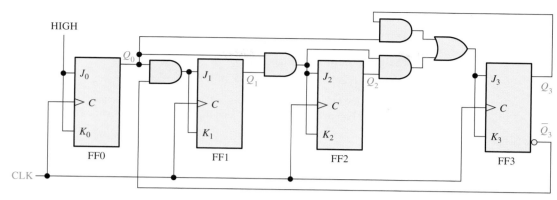

▲ FIGURE 8–76

8. The waveforms in Figure 8–77 are applied to the count enable, clear, and clock inputs as indicated. Show the counter output waveforms in proper relation to these inputs. The clear input is asynchronous.

▶ FIGURE 8–77

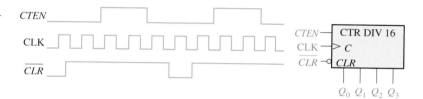

9. A BCD decade counter is shown in Figure 8–78. The waveforms are applied to the clock and clear inputs as indicated. Determine the waveforms for each of the counter outputs (Q_0, Q_1, Q_2, and Q_3). The clear is synchronous, and the counter is initially in the binary 1000 state.

▶ FIGURE 8–78

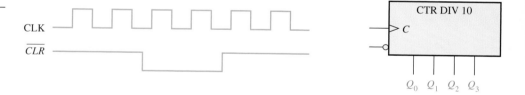

10. The waveforms in Figure 8–79 are applied to a 74HC163 counter. Determine the Q outputs and the RCO. The inputs are $D_0 = 1$, $D_1 = 1$, $D_2 = 0$, and $D_3 = 1$.

▶ FIGURE 8–79

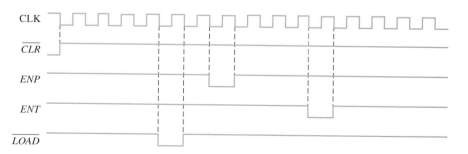

11. The waveforms in Figure 8–79 are applied to a 74F162 counter. Determine the Q outputs and the TC. The inputs are $D_0 = 1$, $D_1 = 0$, $D_2 = 0$, and $D_3 = 1$.

SECTION 8–3   **Up/Down Synchronous Counters**

12. Show a complete timing diagram for a 3-bit up/down counter that goes through the following sequence. Indicate when the counter is in the UP mode and when it is in the DOWN mode. Assume positive edge-triggering.

0, 1, 2, 3, 2, 1, 2, 3, 4, 5, 6, 5, 4, 3, 2, 1, 0

13. Develop the Q output waveforms for a 74HC190 up/down counter with the input waveforms shown in Figure 8–80. A binary 0 is on the data inputs. Start with a count of 0000.

▶ **FIGURE 8–80**

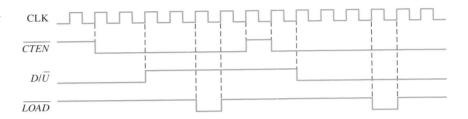

SECTION 8–4   **Design of Synchronous Counters**

14. Determine the sequence of the counter in Figure 8–81.

▶ **FIGURE 8–81**

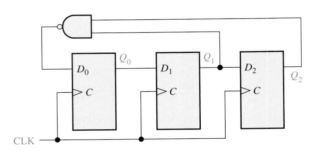

15. Determine the sequence of the counter in Figure 8–82. Begin with the counter cleared.

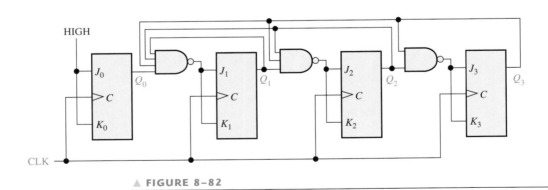

▲ **FIGURE 8–82**

16. Design a counter to produce the following sequence. Use J-K flip-flops.

00, 10, 01, 11, 00, . . .

17. Design a counter to produce the following binary sequence. Use J-K flip-flops.

1, 4, 3, 5, 7, 6, 2, 1, . . .

18. Design a counter to produce the following binary sequence. Use J-K flip-flops.

0, 9, 1, 8, 2, 7, 3, 6, 4, 5, 0, . . .

19. Design a binary counter with the sequence shown in the state diagram of Figure 8–83.

▶ **FIGURE 8–83**

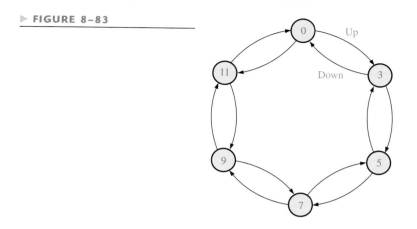

SECTION 8–5 **Cascaded Counters**

20. For each of the cascaded counter configurations in Figure 8–84, determine the frequency of the waveform at each point indicated by a circled number, and determine the overall modulus.

▶ **FIGURE 8–84**

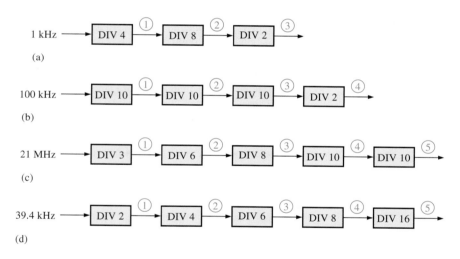

21. Expand the counter in Figure 8–41 to create a divide-by-10,000 counter and a divide-by-100,000 counter.

22. With general block diagrams, show how to obtain the following frequencies from a 10 MHz clock by using single flip-flops, modulus-5 counters, and decade counters:

(a) 5 MHz (b) 2.5 MHz (c) 2 MHz (d) 1 MHz (e) 500 kHz

(f) 250 kHz (g) 62.5 kHz (h) 40 kHz (i) 10 kHz (j) 1 kHz

SECTION 8–6 **Counter Decoding**

23. Given a BCD decade counter with only the Q outputs available, show what decoding logic is required to decode each of the following states and how it should be connected to the counter. A HIGH output indication is required for each decoded state. The MSB is to the left.

(a) 0001 (b) 0011 (c) 0101 (d) 0111 (e) 1000

24. For the 4-bit binary counter connected to the decoder in Figure 8–85, determine each of the decoder output waveforms in relation to the clock pulses.

25. If the counter in Figure 8–85 is asynchronous, determine where the decoding glitches occur on the decoder output waveforms.

▶ FIGURE 8–85

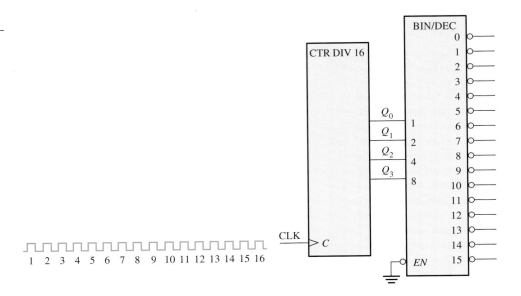

26. Modify the circuit in Figure 8–85 to eliminate decoding glitches.

27. Analyze the counter in Figure 8–45 for the occurrence of glitches on the decode gate output. If glitches occur, suggest a way to eliminate them.

28. Analyze the counter in Figure 8–46 for the occurrence of glitches on the outputs of the decoding gates. If glitches occur, make a design change that will eliminate them.

SECTION 8–7 Counter Applications

29. Assume that the digital clock of Figure 8–51 is initially reset to 12 o'clock. Determine the binary state of each counter after sixty-two 60 Hz pulses have occurred.

30. What is the output frequency of each counter in the digital clock circuit of Figure 8–51?

31. For the automobile parking control system in Figure 8–54, a pattern of entrance and exit sensor pulses during a given 24-hour period are shown in Figure 8–86. If there were 53 cars already in the garage at the beginning of the period, what is the state of the counter at the end of the 24 hours?

▶ FIGURE 8–86

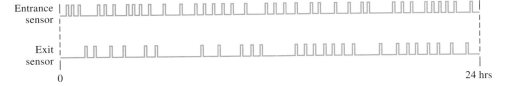

32. The binary number for decimal 57 appears on the parallel data inputs of the parallel-to-serial converter in Figure 8–56 (D_0 is the LSB). The counter initially contains all zeros and a 10 kHz clock is applied. Develop the timing diagram showing the clock, the counter outputs, and the serial data output.

SECTION 8–9 Troubleshooting

33. For the counter in Figure 8–1, show the timing diagram for the Q_0 and Q_1 waveforms for each of the following faults (assume Q_0 and Q_1 are initially LOW):

   (a) clock input to FF0 shorted to ground

   (b) Q_0 output open

   (c) clock input to FF1 open

   (d) J input to FF0 open

   (e) K input to FF1 shorted to ground

34. Solve Problem 33 for the counter in Figure 8–11.

35. Isolate the fault in the counter in Figure 8–3 by analyzing the waveforms in Figure 8–87.

36. From the waveform diagram in Figure 8–88, determine the most likely fault in the counter of Figure 8–14.

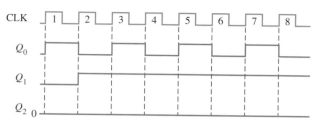

▲ FIGURE 8–87

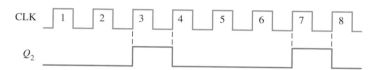

▲ FIGURE 8–88

37. Solve Problem 36 if the Q_2 output has the waveform observed in Figure 8–89. Outputs Q_0 and Q_1 are the same as in Figure 8–88.

▷ FIGURE 8–89

38. You apply a 5 MHz clock to the cascaded counter in Figure 8–44 and measure a frequency of 76.2939 Hz at the last *RCO* output. Is this correct, and if not, what is the most likely problem?

39. Develop a table for use in testing the counter in Figure 8–44 that will show the frequency at the final *RCO* output for all possible open failures of the parallel data inputs (D_0, D_1, D_2, and D_3) taken one at a time. Use 10 MHz as the test frequency for the clock.

40. The tens-of-hours 7-segment display in the digital clock system of Figure 8–51 continuously displays a 1. All the other digits work properly. What could be the problem?

41. What would be the visual indication of an open Q_1 output in the tens portion of the minutes counter in Figure 8–51? Also see Figure 8–52.

42. One day (perhaps a Monday) complaints begin flooding in from patrons of a parking garage that uses the control system depicted in Figures 8–54 and 8–55. The patrons say that they enter the garage because the gate is up and the FULL sign is off but that, once in, they can find no empty space. As the technician in charge of this facility, what do you think the problem is, and how will you troubleshoot and repair the system as quickly as possible?

Digital System Application

43. Implement the input logic in the sequential circuit portion of the traffic light control system using only NAND gates.

44. Replace the D flip-flops in the 2-bit Gray code state counter in Figure 8–67 with J-K flip-flops.

45. Specify how you would change the time interval for the green light from 25 s to 60 s.

Special Design Problems

46. Design a modulus-1000 counter by using 74F162 decade counters.

47. Modify the design of the counter in Figure 8–44 to achieve a modulus of 30,000.

48. Repeat Problem 47 for a modulus of 50,000.

49. Modify the digital clock in Figures 8–51, 8–52, and 8–53 so that it can be preset to any desired time.

50. Design an alarm circuit for the digital clock that can detect a predetermined time (hours and minutes only) and produce a signal to activate an audio alarm.

51. Modify the design of the circuit in Figure 8–55 for a 1000-space parking garage and a 3000-space parking garage.

52. Implement the parallel-to-serial data conversion logic in Figure 8–56 with specific fixed-function devices.

53. In Problem 15 you found that the counter locks up and alternates between two states. It turns out that this operation is the result of a design flaw. Redesign the counter so that when it goes into the second of the lock-up states, it will recycle to the all-0s state on the next clock pulse.

54. Modify the block diagram of the traffic light control system in Figure 8–63 to reflect the addition of a 15 s left turn signal on the main street immediately preceding the green light.

Multisim Troubleshooting Practice

55. Open file P08-55 and test the 4-bit asynchronous counter to determine if there is a fault. If there is a fault, identify it if possible.

56. Open file P08-56 and test the 3-bit synchronous counter to determine if there is a fault. If there is a fault, identify it if possible.

57. Open file P08-57 and test the BCD counter to determine if there is a fault. If there is a fault, identify it if possible.

58. Open file P08-58 and test the 74163 4-bit binary counter to determine if there is a fault. If there is a fault, identify it if possible.

59. Open file P08-59 and test the 74190 Up/Down decade counter to determine if there is a fault. If there is a fault, identify it if possible.

ANSWERS

SECTION REVIEWS

SECTION 8–1 **Asynchronous Counter Operation**

    1. Asynchronous means that each flip-flop after the first one is enabled by the output of the preceding flip-flop.

    2. A modulus-14 counter has fourteen states requiring four flip-flops.

SECTION 8–2 **Synchronous Counter Operation**

    1. All flip-flops in a synchronous counter are clocked simultaneously.

    2. The counter can be preset (initialized) to any given state.

    3. Counter is enabled when *ENP* and *ENT* are both HIGH; *RCO* goes HIGH when final state in sequence is reached.

SECTION 8–3 **Up/Down Synchronous Counters**

    1. The counter goes to 1001.

    2. UP: 1111: DOWN: 0000; the next state is 1111.

SECTION 8–4 **Design of Synchronous Counters**

    1. $J = 1, K = X$ ("don't care")

    2. $J = X$ ("don't care"), $K = 0$

    3. **(a)** The next state is 1011.

        (b) Q_3 (MSB): no-change or SET; Q_2: no-change or RESET; Q_1: no change or SET; Q_0 (LSB): SET or toggle

SECTION 8–5 **Cascaded Counters**

    1. Three decade counters produce ÷ 1000; 4 decade counters produce ÷ 10,000.

    2. **(a)** ÷ 20: flip-flop and DIV 10   **(b)** ÷ 32: flip-flop and DIV 16

        (c) ÷ 160: DIV 16 and DIV 10   **(d)** ÷ 320: DIV 16 and DIV 10 and flip-flop

SECTION 8–6 **Counter Decoding**

1. (a) No transitional states because there is a single bit change

   (b) 0000, 0001, 0010, 0101, 0110, 0111

   (c) No transitional states because there is a single bit change

   (d) 0001, 0010, 0011, 0100, 0101, 0110, 0111, 1000, 1001, 1010, 1011, 1100, 1101, 1110

SECTION 8–7 **Counter Applications**

1. Gate G_1 resets flip-flop on first clock pulse after count 12. Gate G_2 decodes count 12 to preset counter to 0001.

2. The hours decade counter advances through each state from zero to nine, and as it recycles from nine back to zero, the flip-flop is toggled to the SET state. This produces a ten (10) on the display. When the hours decade counter is in state 12, the decode NAND gate causes the counter to recycle to state 1 on the next clock pulse. The flip-flop resets. This results in a one (01) on the display.

SECTION 8–8 **Logic Symbols with Dependency Rotation**

1. C: control, usually clock; M: mode; G: AND

2. D indicates data storage.

SECTION 8–9 **Troubleshooting**

1. No pulses on TC outputs: $CTEN$ of first counter shorted to ground or to a LOW; clock input of first counter open; clock line shorted to ground or to a LOW; TC output of first counter shorted to ground or to a LOW.

2. With inverter output open, the counter does not recycle at the preset count but acts as a full-modulus counter.

RELATED PROBLEMS FOR EXAMPLES

8–1 See Figure 8–90.

▶ FIGURE 8–90

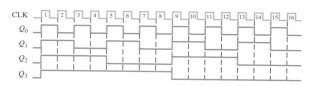

8–2 Connect Q_0 to the NAND gate as a third input (Q_2 and Q_3 are two of the inputs). Connect the $\overline{CLR}$ line to the $\overline{CLR}$ input of FF0 as well as FF2 and FF3.

8–3 See Figure 8–91.

8–4 See Figure 8–92.

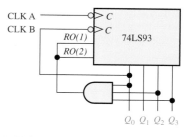

▲ FIGURE 8–91

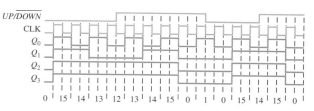

▲ FIGURE 8–92

▼ TABLE 8–15

PRESENT INVALID STATE			J-K INPUTS						NEXT STATE			
Q_2	Q_1	Q_0	J_2	K_2	J_1	K_1	J_0	K_0	Q_2	Q_1	Q_0	
0	0	0	0	0	1	1	1	1	0	1	1	
0	1	1	1	1	1	1	1	1	1	0	0	
1	0	0	0	0	1	1	1	0	1	1	1	valid state
1	1	0	1	1	1	1	1	0	0	0	1	valid state

8–5 See Table 8–15.

8–6 Application of Boolean algebra to the logic in Figure 8–37 shows that the output of each OR gate agrees with the expression in Step 5.

8–7 Five decade counters are required. $10^5 = 100,000$

8–8 $f_{Q0} = 1 \text{ MHz}/[(10)(2)] = 50 \text{ kHz}$

8–9 See Figure 8–93.

8–10 $8AC0_{16}$ would be loaded. $16^4 - 8AC0_{16} = 65,536 - 32,520 = 30,016$
$f_{TC4} = 10 \text{ MHz}/30,016 = 333.2 \text{ Hz}$

8–11 See Figure 8–94.

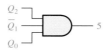

▲ FIGURE 8–93

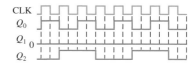

▲ FIGURE 8–94

SELF-TEST

1. (a)   **2.** (b)   **3.** (b)   **4.** (c)   **5.** (a)   **6.** (c)   **7.** (b)   **8.** (c)

9. (d)   **10.** (a)   **11.** (c)   **12.** (b)   **13.** (b)   **14.** (d)

9

SHIFT REGISTERS

CHAPTER OBJECTIVES

▪ Identify the basic forms of data movement in shift registers

▪ Explain how serial in/serial out, serial in/parallel out, parallel in/serial out, and parallel in/parallel out shift registers operate

▪ Describe how a bidirectional shift register operates

▪ Determine the sequence of a Johnson counter

▪ Set up a ring counter to produce a specified sequence

▪ Construct a ring counter from a shift register

▪ Use a shift register as a time-delay device

▪ Use a shift register to implement a serial-to-parallel data converter

- Implement a basic shift-register-controlled keyboard encoder
- Interpret ANSI/IEEE Standard 91-1984 shift register symbols with dependency notation
- Use shift registers in a system application

KEY TERMS

- Register
- Stage
- Shift
- Load
- Bidirectional

INTRODUCTION

Shift registers are a type of sequential logic circuit closely related to digital counters. Registers are used primarily for the storage of digital data and typically do not possess a characteristic internal sequence of states as do counters. There are exceptions, however, and these are covered in Section 9–7.

In this chapter, the basic types of shift registers are studied and several applications are presented. Also, a troubleshooting method is introduced.

FIXED-FUNCTION LOGIC DEVICES

74XX164   74XX165   74XX174
74XX194   74XX195

■ ■ ■   DIGITAL SYSTEM APPLICATION PREVIEW

The Digital System Application illustrates the concepts from this chapter. A security entry system for controlling the alarms in a building is introduced. The system uses two types of shift registers as well as other types of devices covered in previous chapters. The system also includes a memory that will be the focus of the digital system application in Chapter 10.

WWW.   VISIT THE COMPANION WEBSITE
Study aids for this chapter are available at
http://www.prenhall.com/floyd

9–1 BASIC SHIFT REGISTER FUNCTIONS

Shift registers consist of arrangements of flip-flops and are important in applications involving the storage and transfer of data in a digital system. A register, unlike a counter, has no specified sequence of states, except in certain very specialized applications. A register, in general, is used solely for storing and shifting data (1s and 0s) entered into it from an external source and typically possesses no characteristic internal sequence of states.

After completing this section, you should be able to

■ Explain how a flip-flop stores a data bit ■ Define the storage capacity of a shift register ■ Describe the shift capability of a register

A register can consist of one or more flip-flops used to store and shift data.

A **register** is a digital circuit with two basic functions: data storage and data movement. The storage capability of a register makes it an important type of memory device. Figure 9–1 illustrates the concept of storing a 1 or a 0 in a D flip-flop. A 1 is applied to

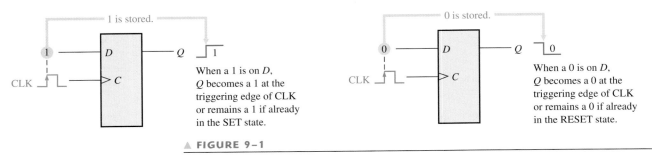

When a 1 is on D, Q becomes a 1 at the triggering edge of CLK or remains a 1 if already in the SET state.

When a 0 is on D, Q becomes a 0 at the triggering edge of CLK or remains a 0 if already in the RESET state.

▲ FIGURE 9–1

The flip-flop as a storage element.

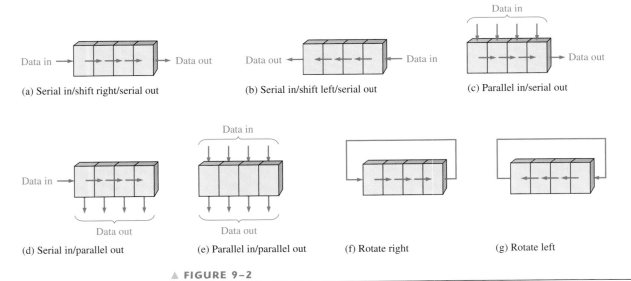

(a) Serial in/shift right/serial out

(b) Serial in/shift left/serial out

(c) Parallel in/serial out

(d) Serial in/parallel out

(e) Parallel in/parallel out

(f) Rotate right

(g) Rotate left

▲ FIGURE 9–2

Basic data movement in shift registers. (Four bits are used for illustration. The bits move in the direction of the arrows.)

the data input as shown, and a clock pulse is applied that stores the 1 by *setting* the flip-flop. When the 1 on the input is removed, the flip-flop remains in the SET state, thereby storing the 1. A similar procedure applies to the storage of a 0 by *resetting* the flip-flop, as also illustrated in Figure 9–1.

The *storage capacity* of a register is the total number of bits (1s and 0s) of digital data it can retain. Each **stage** (flip-flop) in a shift register represents one bit of storage capacity; therefore, the number of stages in a register determines its storage capacity.

The **shift** capability of a register permits the movement of data from stage to stage within the register or into or out of the register upon application of clock pulses. Figure 9–2 illustrates the types of data movement in shift registers. The block represents any arbitrary 4-bit register, and the arrows indicate the direction of data movement.

SECTION 9–1 REVIEW

Answers are at the end of the chapter.

1. Generally, what is the difference between a counter and a shift register?
2. What two principal functions are performed by a shift register?

9–2   SERIAL IN/SERIAL OUT SHIFT REGISTERS

The serial in/serial out shift register accepts data serially—that is, one bit at a time on a single line. It produces the stored information on its output also in serial form.

After completing this section, you should be able to

■ Explain how data bits are serially entered into a shift register ■ Describe how data bits are shifted through the register ■ Explain how data bits are serially taken out of a shift register ■ Develop and analyze timing diagrams for serial in/serial out registers

Let's first look at the serial entry of data into a typical shift register. Figure 9–3 shows a 4-bit device implemented with D flip-flops. With four stages, this register can store up to four bits of data.

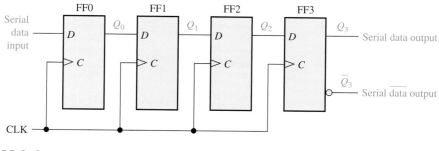

▲ **FIGURE 9–3**

Serial in/serial out shift register.

COMPUTER NOTE

Frequently, it is necessary to *clear* an internal register in a computer. For example, a register may be cleared prior to an arithmetic or other operation. One way that registers in a computer are cleared is using software to subtract the contents of the register from itself. The result of course, will always be zero. For example, a computer instruction that performs this operation is SUB AL,AL. With this instruction, the register named AL is cleared.

Figure 9–4 illustrates entry of the four bits 1010 into the register, beginning with the right-most bit. The register is initially clear. The 0 is put onto the data input line, making $D = 0$ for FF0. When the first clock pulse is applied, FF0 is reset, thus storing the 0.

Next the second bit, which is a 1, is applied to the data input, making $D = 1$ for FF0 and $D = 0$ for FF1 because the D input of FF1 is connected to the Q_0 output. When the second

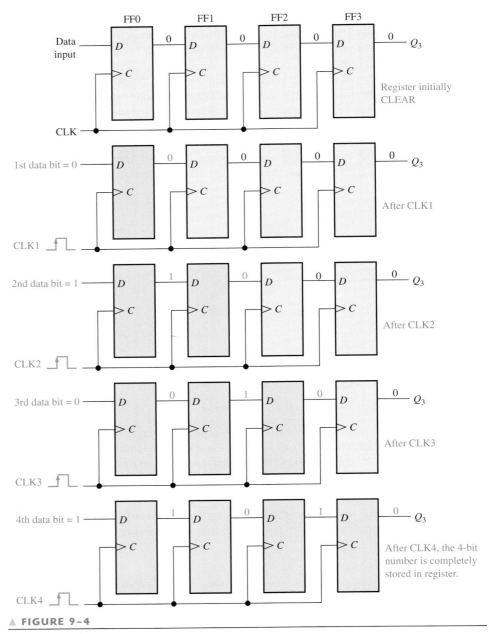

▲ FIGURE 9–4

Four bits (1010) being entered serially into the register.

clock pulse occurs, the 1 on the data input is shifted into FF0, causing FF0 to set; and the 0 that was in FF0 is shifted into FF1.

The third bit, a 0, is now put onto the data-input line, and a clock pulse is applied. The 0 is entered into FF0, the 1 stored in FF0 is shifted into FF1, and the 0 stored in FF1 is shifted into FF2.

The last bit, a 1, is now applied to the data input, and a clock pulse is applied. This time the 1 is entered into FF0, the 0 stored in FF0 is shifted into FF1, the 1 stored in FF1 is shifted into FF2, and the 0 stored in FF2 is shifted into FF3. This completes the serial entry of the four bits into the shift register, where they can be stored for any length of time as long as the flip-flops have dc power.

For serial data, one bit at a time is transferred.

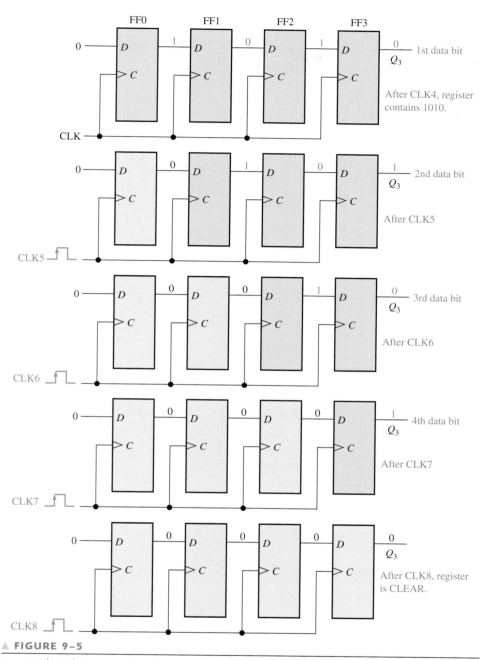

▲ FIGURE 9–5

Four bits (1010) being serially shifted out of the register and replaced by all zeros.

If you want to get the data out of the register, the bits must be shifted out serially and taken off the Q_3 output, as Figure 9–5 illustrates. After CLK4 in the data-entry operation just described, the right-most bit, 0, appears on the Q_3 output. When clock pulse CLK5 is applied, the second bit appears on the Q_3 output. Clock pulse CLK6 shifts the third bit to the output, and CLK7 shifts the fourth bit to the output. While the original four bits are being shifted out, more bits can be shifted in. All zeros are shown being shifted in.

EXAMPLE 9–1

Show the states of the 5-bit register in Figure 9–6(a) for the specified data input and clock waveforms. Assume that the register is initially cleared (all 0s).

▶ FIGURE 9–6

Open file F09-06 to verify operation.

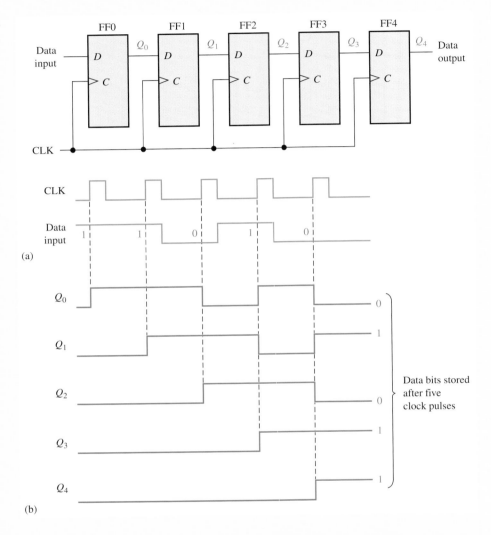

Solution   The first data bit (1) is entered into the register on the first clock pulse and then shifted from left to right as the remaining bits are entered and shifted. The register contains $Q_4Q_3Q_2Q_1Q_0 = 11010$ after five clock pulses. See Figure 9–6(b).

Related Problem*   Show the states of the register if the data input is inverted. The register is initially cleared.

*Answers are at the end of the chapter.

A traditional logic block symbol for an 8-bit serial in/serial out shift register is shown in Figure 9–7. The "SRG 8" designation indicates a shift register (SRG) with an 8-bit capacity.

▶ FIGURE 9–7

Logic symbol for an 8-bit serial in/serial out shift register.

1. Develop the logic diagram for the shift register in Figure 9–3, using J-K flip-flops to replace the D flip-flops.

2. How many clock pulses are required to enter a byte of data serially into an 8-bit shift register?

9–3   SERIAL IN/PARALLEL OUT SHIFT REGISTERS

Data bits are entered serially (right-most bit first) into this type of register in the same manner as discussed in Section 9–2. The difference is the way in which the data bits are taken out of the register; in the parallel output register, the output of each stage is available. Once the data are stored, each bit appears on its respective output line, and all bits are available simultaneously, rather than on a bit-by-bit basis as with the serial output.

After completing this section, you should be able to

■ Explain how data bits are taken out of a shift register in parallel ■ Compare serial output to parallel output ■ Discuss the 74HC164 8-bit shift register ■ Develop and analyze timing diagrams for serial in/parallel out registers

Figure 9–8 shows a 4-bit serial in/parallel out shift register and its logic block symbol.

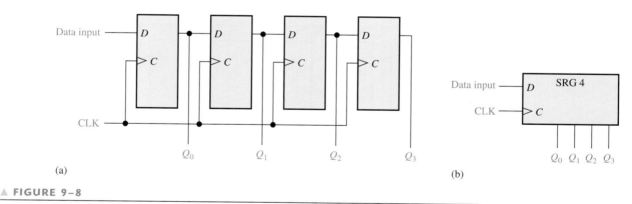

▲ FIGURE 9–8

A serial in/parallel out shift register.

EXAMPLE 9–2

Show the states of the 4-bit register (SRG 4) for the data input and clock waveforms in Figure 9–9(a). The register initially contains all 1s.

Solution   The register contains 0110 after four clock pulses. See Figure 9–9(b).

Related Problem   If the data input remains 0 after the fourth clock pulse, what is the state of the register after three additional clock pulses?

▶ FIGURE 9–9

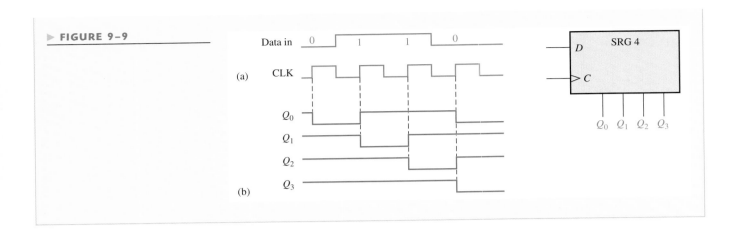

THE 74HC164 8-BIT SERIAL IN/PARALLEL OUT SHIFT REGISTER

The 74HC164 is an example of an IC shift register having serial in/parallel out operation. The logic diagram is shown in Figure 9–10(a), and a typical logic block symbol is shown in part (b). Notice that this device has two gated serial inputs, A and B, and a clear $(\overline{CLR})$ input that is active-LOW. The parallel outputs are Q_0 through Q_7.

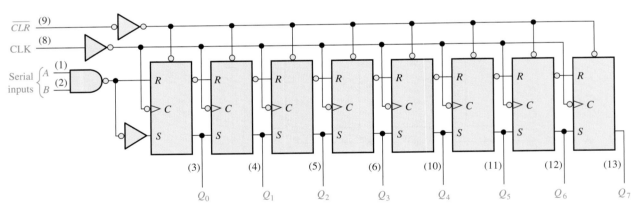

(a) Logic diagram

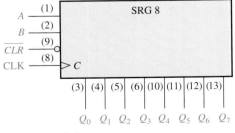

(b) Logic symbol

▲ FIGURE 9–10

The 74HC164 8-bit serial in/parallel out shift register.

A sample timing diagram for the 74HC164 is shown in Figure 9–11. Notice that the serial input data on input A are shifted into and through the register after input B goes HIGH.

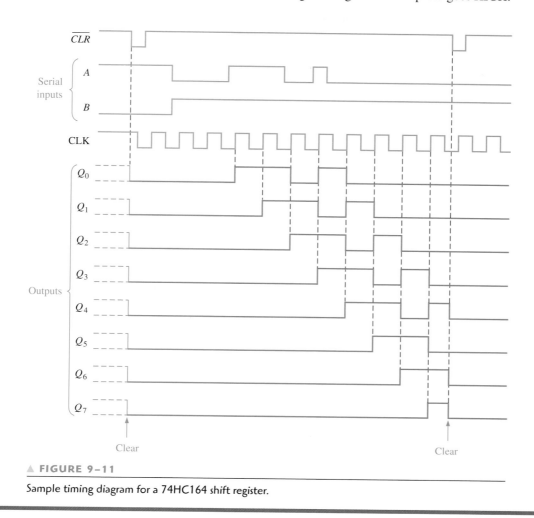

▲ FIGURE 9–11

Sample timing diagram for a 74HC164 shift register.

SECTION 9–3
REVIEW

1. The bit sequence 1101 is serially entered (right–most bit first) into a 4–bit parallel out shift register that is initially clear. What are the Q outputs after two clock pulses?

2. How can a serial in/parallel out register be used as a serial in/serial out register?

9–4   PARALLEL IN/SERIAL OUT SHIFT REGISTERS

For a register with parallel data inputs, the bits are entered simultaneously into their respective stages on parallel lines rather than on a bit-by-bit basis on one line as with serial data inputs. The serial output is the same as described in Section 9–2, once the data are completely stored in the register.

After completing this section, you should be able to

■ Explain how data bits are entered into a shift register in parallel ■ Compare serial input to parallel input ■ Discuss the 74HC165 8-bit parallel-load shift register ■ Develop and analyze timing diagrams for parallel in/serial out registers

For parallel data, multiple bits are transferred at one time.

Figure 9–12 illustrates a 4-bit parallel in/serial out shift register and a typical logic symbol. Notice that there are four data-input lines, D_0, D_1, D_2, and D_3, and a $SHIFT/\overline{LOAD}$ input, which allows four bits of data to **load** in parallel into the register. When $SHIFT/\overline{LOAD}$ is LOW, gates G_1 through G_4 are enabled, allowing each data bit to be applied to the D input of its respective flip-flop. When a clock pulse is applied, the flip-flops with $D = 1$ will set and those with $D = 0$ will reset, thereby storing all four bits simultaneously.

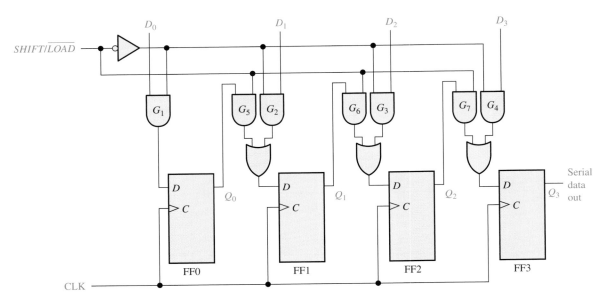

(a) Logic diagram

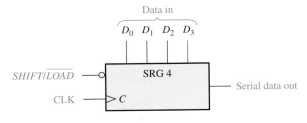

(b) Logic symbol

▲ **FIGURE 9–12**

A 4-bit parallel in/serial out shift register. Open file F09-12 to verify operation.

When $SHIFT/\overline{LOAD}$ is HIGH, gates G_1 through G_4 are disabled and gates G_5 through G_7 are enabled, allowing the data bits to shift right from one stage to the next. The OR gates allow either the normal shifting operation or the parallel data-entry operation, depending on which AND gates are enabled by the level on the $SHIFT/\overline{LOAD}$ input. Notice that FF0 has a single AND to disable the parallel input, D_0. It does not require an AND/OR arrangement because there is no serial data in.

EXAMPLE 9–3

Show the data-output waveform for a 4-bit register with the parallel input data and the clock and $SHIFT/\overline{LOAD}$ waveforms given in Figure 9–13(a). Refer to Figure 9–12(a) for the logic diagram.

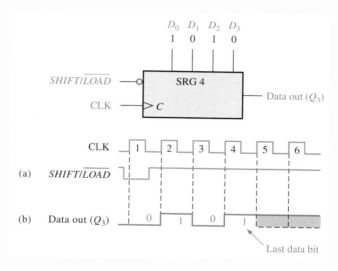

▲ FIGURE 9–13

Solution   On clock pulse 1, the parallel data ($D_0D_1D_2D_3 = 1010$) are loaded into the register, making Q_3 a 0. On clock pulse 2 the 1 from Q_2 is shifted onto Q_3; on clock pulse 3 the 0 is shifted onto Q_3; on clock pulse 4 the last data bit (1) is shifted onto Q_3; and on clock pulse 5, all data bits have been shifted out, and only 1s remain in the register (assuming the D input remains a 1). See Figure 9–13(b).

Related Problem   Show the data-output waveform for the clock and $SHIFT/\overline{LOAD}$ inputs shown in Figure 9–13(a) if the parallel data are $D_0D_1D_2D_3 = 0101$.

THE 74HC165 8-BIT PARALLEL LOAD SHIFT REGISTER

The 74HC165 is an example of an IC shift register that has a parallel in/serial out operation (it can also be operated as serial in/serial out). Figure 9–14(a) shows the internal logic diagram for this device, and part (b) shows a typical logic block symbol. A LOW on the $SHIFT/\overline{LOAD}$ input ($SH/\overline{LD}$) enables all the NAND gates for parallel loading. When an input data bit is a 1, the flip-flop is asynchronously set by a LOW out of the upper gate.

When an input data bit is a 0, the flip-flop is asynchronously reset by a LOW out of the lower gate. Additionally, data can be entered serially on the *SER* input. Also, the clock can be inhibited anytime with a HIGH on the *CLK INH* input. The serial data outputs of the register are Q_7 and its complement $\overline{Q_7}$. This implementation is different from the synchronous method of parallel loading previously discussed, demonstrating that there are usually several ways to accomplish the same function.

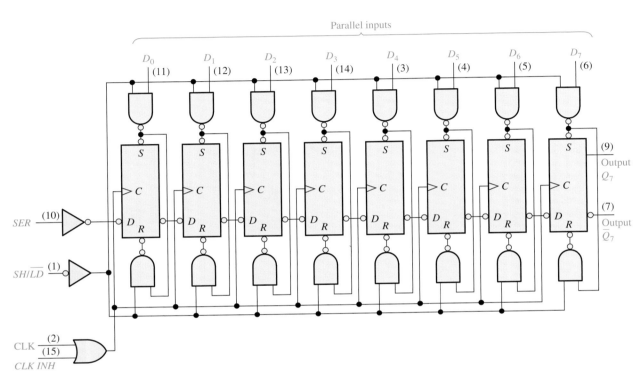

(a) Logic diagram

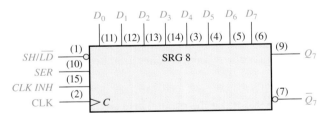

(b) Logic symbol

▲ **FIGURE 9–14**

The 74HC165 8-bit parallel load shift register.

Figure 9–15 is a timing diagram showing an example of the operation of a 74HC165 shift register.

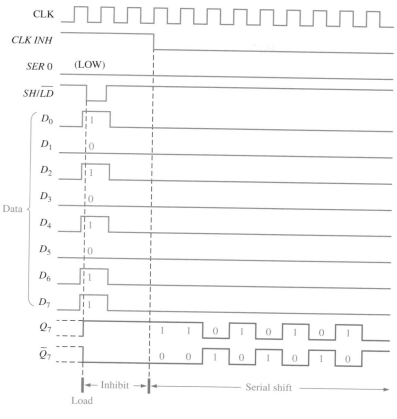

▲ FIGURE 9–15

Sample timing diagram for a 74HC165 shift register.

1. Explain the function of the *SHIFT/$\overline{LOAD}$* input.
2. Is the parallel load operation in a 74HC165 shift register synchronous or asynchronous? What does this mean?

9–5   PARALLEL IN/PARALLEL OUT SHIFT REGISTERS

Parallel entry of data was described in Section 9–4, and parallel output of data has also been discussed previously. The parallel in/parallel out register employs both methods. Immediately following the simultaneous entry of all data bits, the bits appear on the parallel outputs.

After completing this section, you should be able to

■ Discuss the 74HC195 4-bit parallel-access shift register ■ Develop and analyze timing diagrams for parallel in/parallel out registers

Figure 9–16 shows a parallel in/parallel out register.

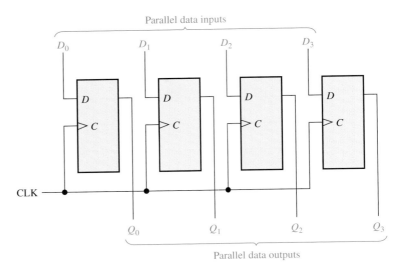

▲ FIGURE 9–16

A parallel in/parallel out register.

THE 74HC195 4-BIT PARALLEL-ACCESS SHIFT REGISTER

The 74HC195 can be used for parallel in/parallel out operation. Because it also has a serial input, it can be used for serial in/serial out and serial in/parallel out operations. It can be used for parallel in/serial out operation by using Q_3 as the output. A typical logic block symbol is shown in Figure 9–17.

▶ FIGURE 9–17

The 74HC195 4-bit parallel access shift register.

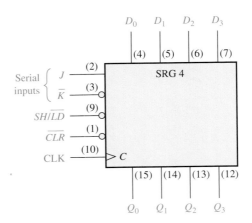

When the $SHIFT/\overline{LOAD}$ input ($SH/\overline{LD}$) is LOW, the data on the parallel inputs are entered synchronously on the positive transition of the clock. When $SH/\overline{LD}$ is HIGH, stored data will shift right (Q_0 to Q_3) synchronously with the clock. Inputs J and $\overline{K}$ are the serial data inputs to the first stage of the register (Q_0); Q_3 can be used for serial output data. The active-LOW clear input is asynchronous.

The timing diagram in Figure 9–18 illustrates the operation of this register.

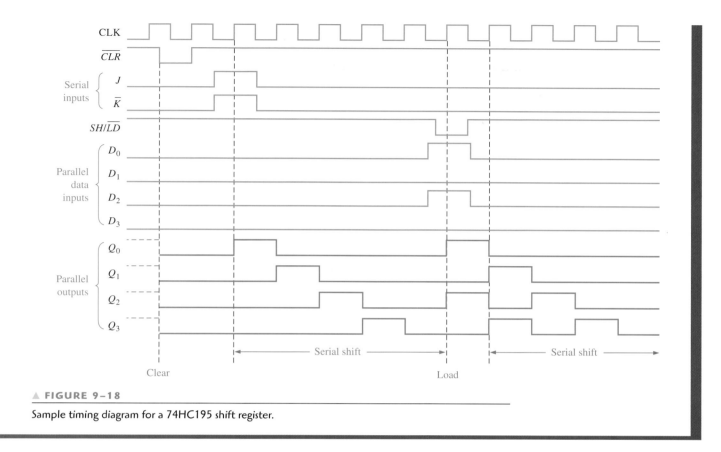

▲ FIGURE 9–18

Sample timing diagram for a 74HC195 shift register.

**SECTION 9–5
REVIEW**

1. In Figure 9–16, $D_0 = 1$, $D_1 = 0$, $D_2 = 0$, and $D_3 = 1$. After three clock pulses, what are the data outputs?
2. For a 74HC195, $SH/\overline{LD} = 1$, $J = 1$, and $\overline{K} = 1$. What is Q_0 after one clock pulse?

9–6   BIDIRECTIONAL SHIFT REGISTERS

A bidirectional shift register is one in which the data can be shifted either left or right. It can be implemented by using gating logic that enables the transfer of a data bit from one stage to the next stage to the right or to the left, depending on the level of a control line.

After completing this section, you should be able to

■ Explain the operation of a bidirectional shift register ■ Discuss the 74HC194 4-bit bidirectional universal shift register ■ Develop and analyze timing diagrams for bidirectional shift registers

A 4-bit **bidirectional** shift register is shown in Figure 9–19. A HIGH on the $RIGHT/\overline{LEFT}$ control input allows data bits inside the register to be shifted to the right, and a LOW enables data bits inside the register to be shifted to the left. An examination of the gating logic will make the operation apparent. When the $RIGHT/\overline{LEFT}$ control input is HIGH, gates G_1 through G_4 are enabled, and the state of the Q output of each flip-flop is passed through to the *D* input of the *following* flip-flop. When a clock pulse occurs, the data bits are shifted one place to the *right*. When the $RIGHT/\overline{LEFT}$ control input is LOW, gates G_5 through G_8 are enabled, and the *Q* output of each flip-flop is passed through to the *D* input of the *preceding* flip-flop. When a clock pulse occurs, the data bits are then shifted one place to the *left*.

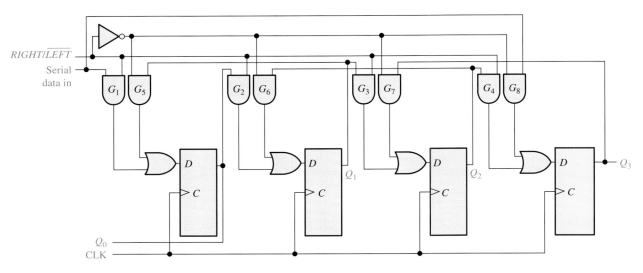

▲ FIGURE 9–19

Four-bit bidirectional shift register. Open file F09-19 to verify the operation.

EXAMPLE 9–4

Determine the state of the shift register of Figure 9–19 after each clock pulse for the given $RIGHT/\overline{LEFT}$ control input waveform in Figure 9–20(a). Assume that $Q_0 = 1$, $Q_1 = 1$, $Q_2 = 0$, and $Q_3 = 1$ and that the serial data-input line is LOW.

► FIGURE 9–20

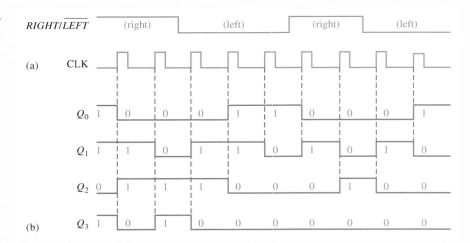

Solution   See Figure 9–20(b).

Related Problem   Invert the $RIGHT/\overline{LEFT}$ waveform, and determine the state of the shift register in Figure 9–19 after each clock pulse.

THE 74HC194 4-BIT BIDIRECTIONAL UNIVERSAL SHIFT REGISTER

The 74HC194 is an example of a universal bidirectional shift register in integrated circuit form. A **universal shift register** has both serial and parallel input and output capability. A logic block symbol is shown in Figure 9–21, and a sample timing diagram is shown in Figure 9–22.

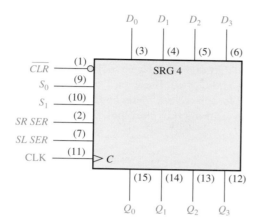

◀ **FIGURE 9–21**

The 74HC194 4-bit bidirectional universal shift register.

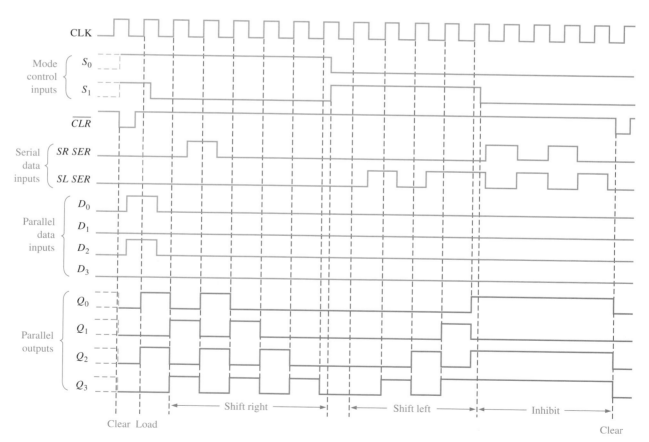

▲ **FIGURE 9–22**

Sample timing diagram for a 74HC194 shift register.

Parallel loading, which is synchronous with a positive transition of the clock, is accomplished by applying the four bits of data to the parallel inputs and a HIGH to the S_0 and S_1 inputs. Shift right is accomplished synchronously with the positive edge of the clock when S_0 is HIGH and S_1 is LOW. Serial data in this mode are entered at the shift-right serial input (*SR SER*). When S_0 is LOW and S_1 is HIGH, data bits shift left synchronously with the clock, and new data are entered at the shift-left serial input (*SL SER*). Input *SR SER* goes into the Q_0 stage, and *SL SER* goes into the Q_3 stage.

SECTION 9–6 REVIEW

1. Assume that the 4-bit bidirectional shift register in Figure 9–19 has the following contents: $Q_0 = 1$, $Q_1 = 1$, $Q_2 = 0$, and $Q_3 = 0$. There is a 1 on the serial data-input line. If *RIGHT/LEFT* is HIGH for three clock pulses and LOW for two more clock pulses, what are the contents after the fifth clock pulse?

9–7 SHIFT REGISTER COUNTERS

A shift register counter is basically a shift register with the serial output connected back to the serial input to produce special sequences. These devices are often classified as counters because they exhibit a specified sequence of states. Two of the most common types of shift register counters, the Johnson counter and the ring counter, are introduced in this section.

After completing this section, you should be able to

■ Discuss how a shift register counter differs from a basic shift register ■ Explain the operation of a Johnson counter ■ Specify a Johnson sequence for any number of bits ■ Explain the operation of a ring counter and determine the sequence of any specific ring counter

The Johnson Counter

In a **Johnson counter** the complement of the output of the last flip-flop is connected back to the D input of the first flip-flop (it can be implemented with other types of flip-flops as well). This feedback arrangement produces a characteristic sequence of states, as shown in Table 9–1 for a 4-bit device and in Table 9–2 for a 5-bit device. Notice that the 4-bit sequence has a total of eight states, or bit patterns, and that the 5-bit sequence has a total of ten states. In general, a Johnson counter will produce a modulus of $2n$, where n is the number of stages in the counter.

The implementations of the 4-stage and 5-stage Johnson counters are shown in Figure 9–23. The implementation of a Johnson counter is very straightforward and is the same regardless of the number of stages. The Q output of each stage is connected to the D input of the next stage (assuming that D flip-flops are used). The single exception is that the $\overline{Q}$ output of the last stage is connected back to the D input of the first stage. As the sequences in Table 9–1 and 9–2 show, the counter will "fill up" with 1s from left to right, and then it will "fill up" with 0s again.

◀ **TABLE 9–1**

Four-bit Johnson sequence.

CLOCK PULSE	Q_0	Q_1	Q_2	Q_3
0	0	0	0	0
1	1	0	0	0
2	1	1	0	0
3	1	1	1	0
4	1	1	1	1
5	0	1	1	1
6	0	0	1	1
7	0	0	0	1

◀ **TABLE 9–2**

Five-bit Johnson sequence.

CLOCK PULSE	Q_0	Q_1	Q_2	Q_3	Q_4
0	0	0	0	0	0
1	1	0	0	0	0
2	1	1	0	0	0
3	1	1	1	0	0
4	1	1	1	1	0
5	1	1	1	1	1
6	0	1	1	1	1
7	0	0	1	1	1
8	0	0	0	1	1
9	0	0	0	0	1

◀ **FIGURE 9–23**

Four-bit and 5-bit Johnson counters.

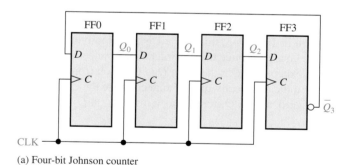

(a) Four-bit Johnson counter

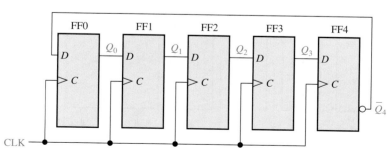

(b) Five-bit Johnson counter

Diagrams of the timing operations of the 4-bit and 5-bit counters are shown in Figures 9–24 and 9–25, respectively.

▶ **FIGURE 9–24**

Timing sequence for a 4-bit Johnson counter.

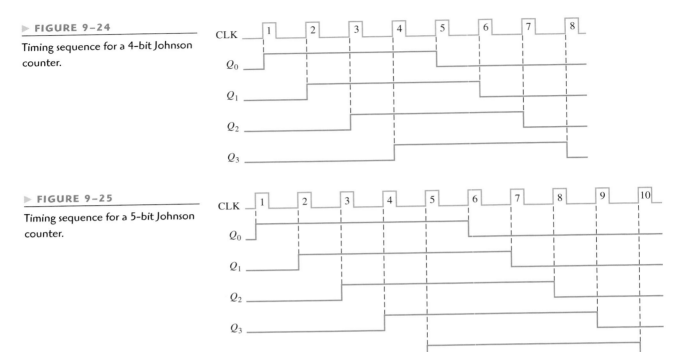

▶ **FIGURE 9–25**

Timing sequence for a 5-bit Johnson counter.

The Ring Counter

The **ring counter** utilizes one flip-flop for each state in its sequence. It has the advantage that decoding gates are not required. In the case of a 10-bit ring counter, there is a unique output for each decimal digit.

A logic diagram for a 10-bit ring counter is shown in Figure 9–26. The sequence for this ring counter is given in Table 9–3. Initially, a 1 is preset into the first flip-flop, and the rest of the flip-flops are cleared. Notice that the interstage connections are the same as those for a Johnson counter, except that Q rather than $\overline{Q}$ is fed back from the last stage. The ten outputs of the counter indicate directly the decimal count of the clock pulse. For instance, a 1 on Q_0 represents a zero, a 1 on Q_1 represents a one, a 1 on Q_2 represents a two, a 1 on Q_3

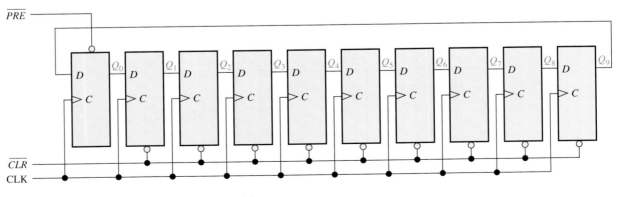

▲ **FIGURE 9–26**

A 10-bit ring counter. Open file F09-26 to verify operation.

CLOCK PULSE	Q_0	Q_1	Q_2	Q_3	Q_4	Q_5	Q_6	Q_7	Q_8	Q_9
0	1	0	0	0	0	0	0	0	0	0
1	0	1	0	0	0	0	0	0	0	0
2	0	0	1	0	0	0	0	0	0	0
3	0	0	0	1	0	0	0	0	0	0
4	0	0	0	0	1	0	0	0	0	0
5	0	0	0	0	0	1	0	0	0	0
6	0	0	0	0	0	0	1	0	0	0
7	0	0	0	0	0	0	0	1	0	0
8	0	0	0	0	0	0	0	0	1	0
9	0	0	0	0	0	0	0	0	0	1

◄ **TABLE 9–3**

Ten-bit ring counter sequence.

represents a three, and so on. You should verify for yourself that the 1 is always retained in the counter and simply shifted "around the ring," advancing one stage for each clock pulse.

Modified sequences can be achieved by having more than a single 1 in the counter, as illustrated in Example 9–5.

EXAMPLE 9–5

If a 10-bit ring counter similar to Figure 9–26 has the initial state 1010000000, determine the waveform for each of the Q outputs.

Solution   See Figure 9–27.

▶ **FIGURE 9–27**

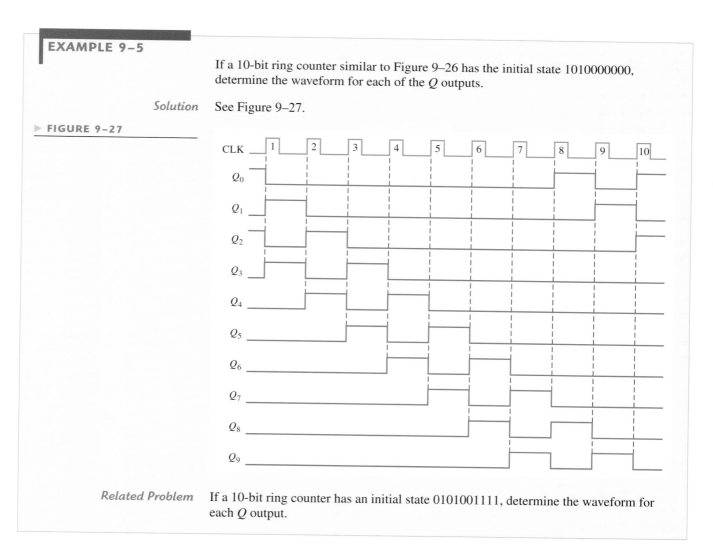

Related Problem   If a 10-bit ring counter has an initial state 0101001111, determine the waveform for each Q output.

SECTION 9–7 REVIEW

1. How many states are there in an 8-bit Johnson counter sequence?
2. Write the sequence of states for a 3-bit Johnson counter starting with 000.

9–8 SHIFT REGISTER APPLICATIONS

Shift registers are found in many types of applications, a few of which are presented in this section.

After completing this section, you should be able to

■ Use a shift register to generate a time delay ■ Implement a specified ring counter sequence using a 74HC195 shift register ■ Discuss how shift registers are used for serial-to-parallel conversion of data ■ Define *UART* ■ Explain the operation of a keyboard encoder and how registers are used in this application

COMPUTER NOTE

The general-purpose registers in the Pentium are all 32-bit registers that can be used for temporary data storage as well as specific uses. Four of these registers are as follows. The *accumulator* (EAX) is used mainly for temporary storage of data and instruction operands. The *base register* (EBX) is used to store a value temporarily. The *count register* (ECX) is mainly used to determine the number of repetitions in certain loop, string, shift, or rotate operations. The *data register* (EDX), is normally used for the temporary storage of data.

Time Delay

The serial in/serial out shift register can be used to provide a time delay from input to output that is a function of both the number of stages (n) in the register and the clock frequency.

When a data pulse is applied to the serial input as shown in Figure 9–28 (A and B connected together), it enters the first stage on the triggering edge of the clock pulse. It is then shifted from stage to stage on each successive clock pulse until it appears on the serial output n clock periods later. This time-delay operation is illustrated in Figure 9–28, in which an 8-bit serial in/serial out shift register is used with a clock frequency of 1 MHz to achieve a time delay (t_d) of 8 μs (8 × 1 μs). This time can be adjusted up or down by changing the clock frequency. The time delay can also be increased by cascading shift registers and decreased by taking the outputs from successively lower stages in the register if the outputs are available, as illustrated in Example 9–6.

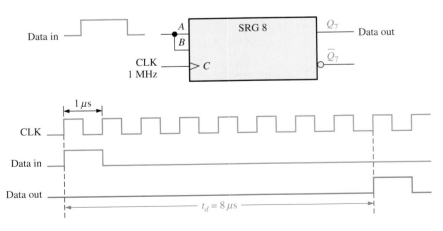

▲ **FIGURE 9–28**

The shift register as a time-delay device.

EXAMPLE 9-6

Determine the amount of time delay between the serial input and each output in Figure 9–29. Show a timing diagram to illustrate.

▶ **FIGURE 9–29**

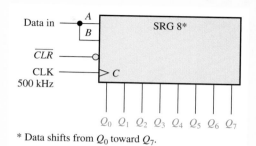

* Data shifts from Q_0 toward Q_7.

Solution The clock period is 2 μs. Thus, the time delay can be increased or decreased in 2 μs increments from a minimum of 2 μs to a maximum of 16 μs, as illustrated in Figure 9–30.

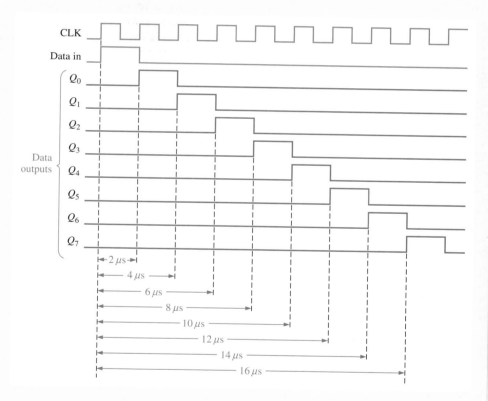

▲ **FIGURE 9–30**

Timing diagram showing time delays for the register in Figure 9–29.

Related Problem Determine the clock frequency required to obtain a time delay of 24 μs to the Q_7 output in Figure 9–29.

A RING COUNTER USING THE 74HC195 SHIFT REGISTER

If the output is connected back to the serial input, a shift register can be used as a ring counter. Figure 9–31 illustrates this application with a 74HC195 4-bit shift register.

▶ **FIGURE 9–31**

74HC195 connected as a ring counter.

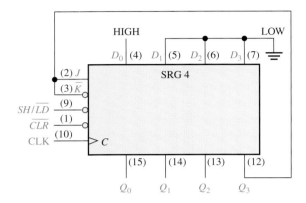

Initially, a bit pattern of 1000 (or any other pattern) can be synchronously preset into the counter by applying the bit pattern to the parallel data inputs, taking the $SH/\overline{LD}$ input LOW, and applying a clock pulse. After this initialization, the 1 continues to circulate through the ring counter, as the timing diagram in Figure 9–32 shows.

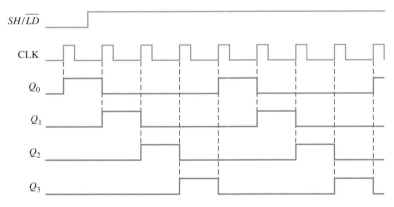

▲ **FIGURE 9–32**

Timing diagram showing two complete cycles of the ring counter in Figure 9–31 when it is initially preset to 1000.

Serial-to-Parallel Data Converter

Serial data transmission from one digital system to another is commonly used to reduce the number of wires in the transmission line. For example, eight bits can be sent serially over one wire, but it takes eight wires to send the same data in parallel.

A computer or microprocessor-based system commonly requires incoming data to be in parallel format, thus the requirement for serial-to-parallel conversion. A simplified serial-to-parallel data converter, in which two types of shift registers are used, is shown in Figure 9–33.

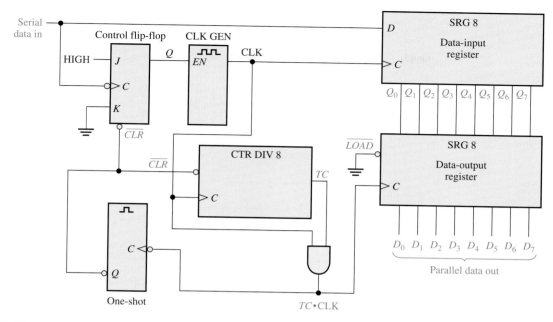

▲ **FIGURE 9–33**

Simplified logic diagram of a serial-to-parallel converter.

To illustrate the operation of this serial-to-parallel converter, the serial data format shown in Figure 9–34 is used. It consists of eleven bits. The first bit (start bit) is always 0 and always begins with a HIGH-to-LOW transition. The next eight bits (D_7 through D_0) are the data bits (one of the bits can be parity), and the last two bits (stop bits) are always 1s. When no data are being sent, there is a continuous 1 on the serial data line.

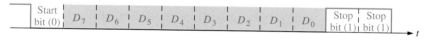

▲ **FIGURE 9–34**

Serial data format.

The HIGH-to-LOW transition of the start bit sets the control flip-flop, which enables the clock generator. After a fixed delay time, the clock generator begins producing a pulse waveform, which is applied to the data-input register and to the divide-by-8 counter. The clock has a frequency precisely equal to that of the incoming serial data, and the first clock pulse after the start bit occurs during the first data bit.

The timing diagram in Figure 9–35 illustrates the following basic operation: The eight data bits (D_7 through D_0) are serially shifted into the data-input register. After the eighth clock pulse, a HIGH-to-LOW transition of the terminal count (TC) output of the counter ANDed with the clock ($TC \cdot$CLK) loads the eight bits that are in the data-input register into the data-output register. This same transition also triggers the one-shot, which produces a short-duration pulse to clear the counter and reset the control flip-flop and thus disable the clock generator. The system is now ready for the next group of eleven bits, and it waits for the next HIGH-to-LOW transition at the beginning of the start bit.

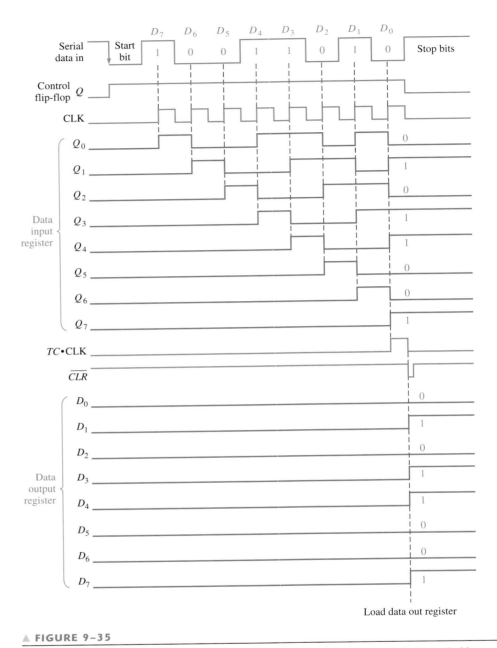

▲ **FIGURE 9–35**

Timing diagram illustrating the operation of the serial-to-parallel data converter in Figure 9–33.

By reversing the process just stated, parallel-to-serial data conversion can be accomplished. However, since the serial data format must be produced, additional requirements must be taken into consideration.

Universal Asynchronous Receiver Transmitter (UART)

As mentioned, computers and microprocessor-based systems often send and receive data in a parallel format. Frequently, these systems must communicate with external devices that send and/or receive serial data. An interfacing device used to accomplish these conversions is the UART (Universal Asynchronous Receiver Transmitter). Figure 9–36 illustrates the UART in a general microprocessor-based system application.

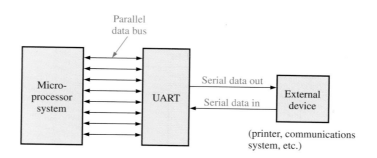

UART interface.

A UART includes a serial-to-parallel data converter such as we have discussed and a parallel-to-serial converter, as shown in Figure 9–37. The data bus is basically a set of parallel conductors along which data move between the UART and the microprocessor system. Buffers interface the data registers with the data bus.

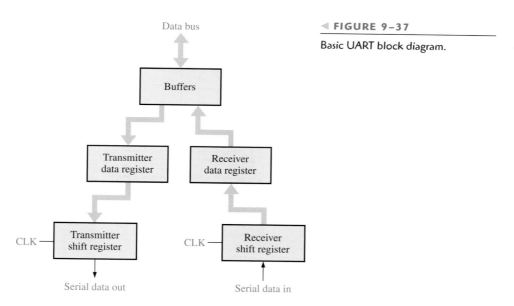

◀ FIGURE 9–37

Basic UART block diagram.

The UART receives data in serial format, converts the data to parallel format, and places them on the data bus. The UART also accepts parallel data from the data bus, converts the data to serial format, and transmits them to an external device.

Keyboard Encoder

The keyboard encoder is a good example of the application of a shift register used as a ring counter in conjunction with other devices. Recall that a simplified computer keyboard encoder without data storage was presented in Chapter 6.

Figure 9–38 shows a simplified keyboard encoder for encoding a key closure in a 64-key matrix organized in eight rows and eight columns. Two 74HC195 4-bit shift registers are connected as an 8-bit ring counter with a fixed bit pattern of seven 1s and one 0 preset into it when the power is turned on. Two 74HC147 priority encoders (introduced in Chapter 6) are used as eight-line-to-three-line encoders (9 input HIGH, 8 output unused) to encode the ROW and COLUMN lines of the keyboard matrix. The 74HC174 (hex flip-flops) is used as a parallel in/parallel out register in which the ROW/COLUMN code from the priority encoders is stored.

The basic operation of the keyboard encoder in Figure 9–38 is as follows: The ring counter "scans" the rows for a key closure as the clock signal shifts the 0 around the counter

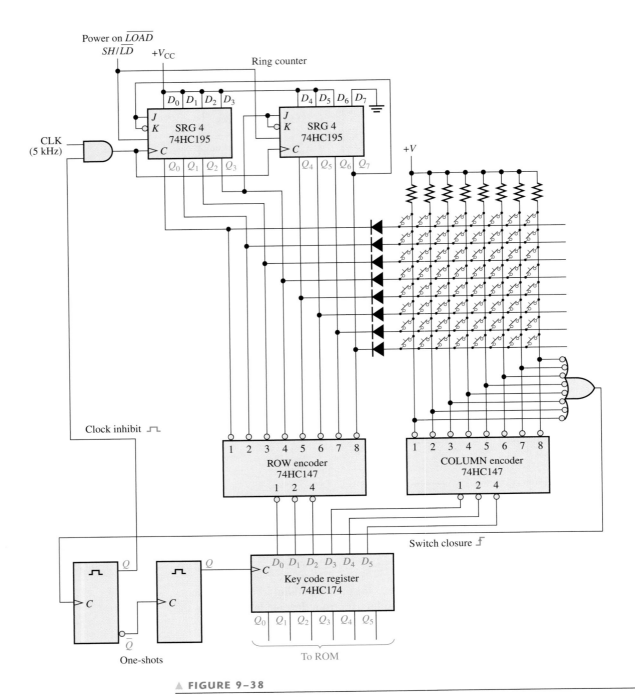

Simplified keyboard encoding circuit.

at a 5 kHz rate. The 0 (LOW) is sequentially applied to each ROW line, while all other ROW lines are HIGH. All the ROW lines are connected to the ROW encoder inputs, so the 3-bit output of the ROW encoder at any time is the binary representation of the ROW line that is LOW. When there is a key closure, one COLUMN line is connected to one ROW line. When the ROW line is taken LOW by the ring counter, that particular COLUMN line is also pulled LOW. The COLUMN encoder produces a binary output corresponding to the COLUMN in which the key is closed. The 3-bit ROW code plus the 3-bit COLUMN code uniquely identifies the key that is closed. This 6-bit code is applied to the inputs of the key code register. When a key is closed, the two one-shots produce a delayed clock pulse to parallel-load the

6-bit code into the key code register. This delay allows the contact bounce to die out. Also, the first one-shot output inhibits the ring counter to prevent it from scanning while the data are being loaded into the key code register.

The 6-bit code in the key code register is now applied to a ROM (read-only memory) to be converted to an appropriate alphanumeric code that identifies the keyboard character. ROMs are studied in Chapter 10.

SECTION 9–8 REVIEW

1. In the keyboard encoder, how many times per second does the ring counter scan the keyboard?

2. What is the 6-bit ROW/COLUMN code (key code) for the top row and the left-most column in the keyboard encoder?

3. What is the purpose of the diodes in the keyboard encoder? What is the purpose of the resistors?

9–9 LOGIC SYMBOLS WITH DEPENDENCY NOTATION

Two examples of ANSI/IEEE Standard 91-1984 symbols with dependency notation for shift registers are presented. Two specific IC shift registers are used as examples.

After completing this section, you should be able to

■ Understand and interpret the logic symbols with dependency notation for the 74HC164 and the 74HC194 shift registers

The logic symbol for a 74HC164 8-bit parallel output serial shift register is shown in Figure 9–39. The common control inputs are shown on the notched block. The clear ($\overline{CLR}$) input is indicated by an R (for RESET) inside the block. Since there is no dependency prefix to link R with the clock (C1), the clear function is asynchronous. The right arrow symbol after C1 indicates data flow from Q_0 to Q_7. The A and B inputs are ANDed, as indicated by the embedded AND symbol, to provide the synchronous data input, 1D, to the first stage (Q_0). Note the dependency of D on C, as indicated by the 1 suffix on C and the 1 prefix on D.

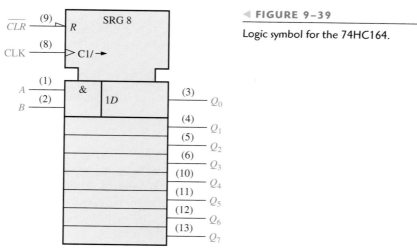

◀ **FIGURE 9–39**

Logic symbol for the 74HC164.

Figure 9–40 is the logic symbol for the 74HC194 4-bit bidirectional universal shift register. Starting at the top left side of the control block, note that the $\overline{CLR}$ input is active-LOW and is asynchronous (no prefix link with C). Inputs S_0 and S_1 are mode inputs that determine the *shift-right, shift-left,* and *parallel load* modes of operation, as indicated by the $\frac{0}{3}$ dependency designation following the *M*. The $\frac{0}{3}$ represents the binary states of 0, 1, 2, and 3 on the S_0 and S_1 inputs. When one of these digits is used as a prefix for another input, a dependency is established. The $1{\rightarrow}/2{\leftarrow}$ symbol on the clock input indicates the following: $1{\rightarrow}$ indicates that a right shift (Q_0 toward Q_3) occurs when the mode inputs (S_0, S_1) are in the binary 1 state ($S_0 = 1$, $S_1 = 0$), $2{\leftarrow}$ indicates that a left shift (Q_3 toward Q_0) occurs when the mode inputs are in the binary 2 state ($S_0 = 0$, $S_1 = 1$). The shift-right serial input (*SR SER*) is both mode-dependent and clock-dependent, as indicated by 1, 4D. The parallel inputs (D_0, D_1, D_2, and D_3) are all mode-dependent (prefix 3 indicates parallel load mode) and clock-dependent, as indicated by 3, 4D. The shift-left serial input (*SL SER*) is both mode-dependent and clock-dependent, as indicated by 2, 4D.

The four modes for the 74HC194 are summarized as follows:

Do nothing:	$S_0 = 0$, $S_1 = 0$	(mode 0)
Shift right:	$S_0 = 1$, $S_1 = 0$	(mode 1, as in 1, 4D)
Shift left:	$S_0 = 0$, $S_1 = 1$	(mode 2, as in 2, 4D)
Parallel load:	$S_0 = 1$, $S_1 = 1$	(mode 3, as in 3, 4D)

▶ **FIGURE 9–40**

Logic symbol for the 74HC194.

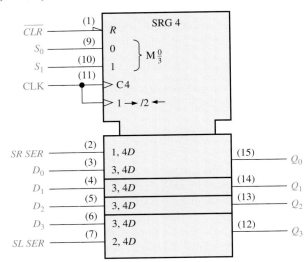

SECTION 9–9 REVIEW

1. In Figure 9–40, are there any inputs that are dependent on the mode inputs being in the 0 state?

2. Is the parallel load synchronous with the clock?

9–10 TROUBLESHOOTING

A traditional method of troubleshooting sequential logic and other more complex systems uses a procedure of "exercising" the circuit under test with a known input waveform (stimulus) and then observing the output for the correct bit pattern.

After completing this section, you should be able to

■ Explain the procedure of "exercising" as a troubleshooting technique ■ Discuss exercising of a serial-to-parallel converter

The serial-to-parallel data converter in Figure 9–33 is used to illustrate the "exercising" procedure. The main objective in exercising the circuit is to force all elements (flip-flops and gates) into all of their states to be certain that nothing is stuck in a given state as a result of a fault. The input test pattern, in this case, must be designed to force each flip-flop in the registers into both states, to clock the counter through all of its eight states, and to take the control flip-flop, the clock generator, the one-shot, and the AND gate through their paces.

The input test pattern that accomplishes this objective for the serial-to-parallel data converter is based on the serial data format in Figure 9–34. It consists of the pattern 10101010 in one serial group of data bits followed by 01010101 in the next group, as shown in Figure 9–41. These patterns are generated on a repetitive basis by a special test-pattern generator. The basic test setup is shown in Figure 9–42.

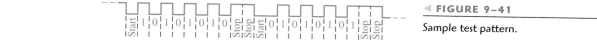

◀ **FIGURE 9–41**

Sample test pattern.

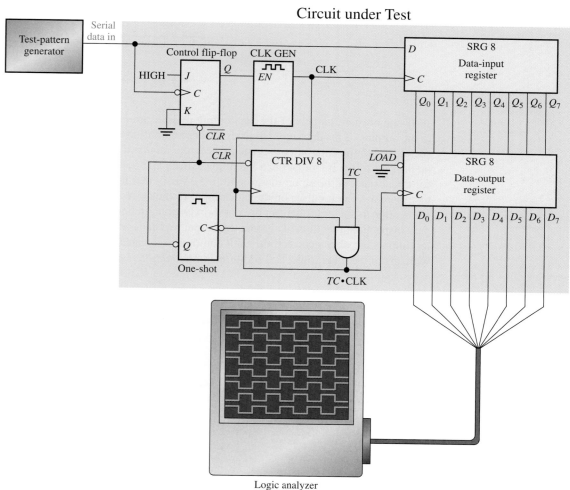

▲ **FIGURE 9–42**

Basic test setup for the serial-to-parallel data converter of Figure 9-33.

After both patterns have been run through the circuit under test, all the flip-flops in the data-input register and in the data-output register have resided in both SET and RESET states, the counter has gone through its sequence (once for each bit pattern), and all the other devices have been exercised.

To check for proper operation, each of the parallel data outputs is observed for an alternating pattern of 1s and 0s as the input test patterns are repetitively shifted into the data-input register and then loaded into the data-output register. The proper timing diagram is shown in Figure 9–43. The outputs can be observed in pairs with a dual-trace oscilloscope, or all eight outputs can be observed simultaneously with a logic analyzer configured for timing analysis.

▶ FIGURE 9–43

Proper outputs for the circuit under test in Figure 9–42. The input test pattern is shown.

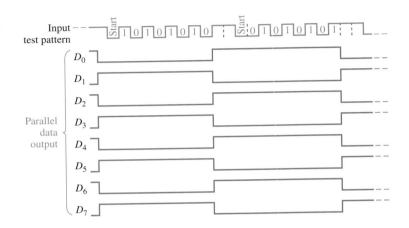

If one or more outputs of the data-output register are incorrect, then you must back up to the outputs of the data-input register. If these outputs are correct, then the problem is associated with the data-output register. Check the inputs to the data-output register directly on the pins of the IC for an open input line. Check that power and ground are correct (look for the absence of noise on the ground line). Verify that the load line is a solid LOW and that there are clock pulses on the clock input of the correct amplitude. Make sure that the connection to the logic analyzer did not short two output lines together. If all of these checks pass inspection, then it is likely that the output register is defective. If the data-input register outputs are also incorrect, the fault could be associated with the input register itself or with any of the other logic, and additional investigation is necessary to isolate the problem.

HANDS ON TIP

When measuring digital signals with an oscilloscope, you should always use dc coupling, rather than ac coupling. The reason that ac coupling is not best for viewing digital signals is that the 0 V level of the signal will appear at the *average* level of the signal, not at true ground or 0 V level. It is much easier to find a "floating" ground or incorrect logic level with dc coupling. If you suspect an open ground in a digital circuit, increase the sensitivity of the scope to the maximum possible. A good ground will never appear to have noise under this condition, but an open will likely show some noise, which appears as a random fluctuation in the 0 V level.

SECTION 9–10 REVIEW

1. What is the purpose of providing a test input to a sequential logic circuit?
2. Generally, when an output waveform is found to be incorrect, what is the next step to be taken?

Troubleshooting problems that are keyed to the CD-ROM are available in the Multisim Troubleshooting Practice section of the end-of-chapter problems.

DIGITAL SYSTEM APPLICATION

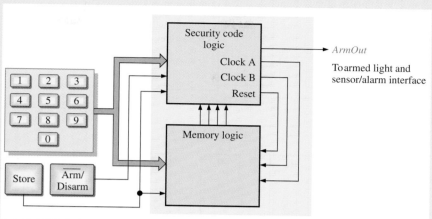

▲ FIGURE 9–44

Basic block diagram of the security system.

In this system application, a relatively simple system is developed to control the security of a room or building. The system can be programmed with a 4-digit security code by entering the four digits, one at a time, from a keypad, in the *disarm* mode. Once the security code has been entered and stored, the system is switched to the *arm* mode. To disarm the system, you must enter the correct 4-digit code on the keypad.

Basic Operation

A basic block diagram is shown in Figure 9–44. The system logic consists of the security code logic and the memory logic. In this chapter, the focus is on the code entry logic. The memory logic will be developed in Chapter 10, and the two logic sections will be combined to form the complete system logic.

The secure $\overline{\text{Arm}}$/Disarm switch places the security system in either *arm* mode or *disarm* mode. Programming is accomplished by first putting the system into the *disarm* mode and then pressing the secure *store* switch followed by the digit key for each of the four digits to be entered. After this process, the memory contains the BCD codes for each of the four security code digits. When the system is switched to the *arm* mode, the *ArmOut* enables the alarm system sensors and lights an LED to indicate the system is armed. To enter the room or building, the system must be switched to the *disarm* mode, and the correct 4-digit security code must be entered on the keypad.

Security Code Logic

The security code logic controls the arming, disarming, programming, and entry. The basic logic diagram is shown in Figure 9–45. When the system is first armed by placing the $\overline{\text{Arm}}$/Disarm switch in the *Arm* position, shift register C contains 00010000 so that there is a LOW on *ArmOut* which activates the system sensors, the alarm circuits, and the ARMED indicator. Also, a reset pulse is generated by OSE for the memory address counter.

Entry To deactivate the system so that you can enter the secured area, you must enter the correct 4-digit code that matches the code stored in the memory. The first digit of the security code is entered from the keypad. The decimal-to-BCD encoder produces the BCD code representing the digit that was pressed on the keypad. One-shot A (OSA) is triggered through gate G1 producing a pulse that clocks the 4-bit BCD code from the encoder into shift register A and the stored code from the first memory address into shift register B. Once the codes are in registers A and B, they are applied to the inputs of the comparator. When a correct digit is entered on the keypad, the 4 bits on the *A* inputs of the comparator and the 4 bits on the *B* inputs are the same, resulting in a HIGH (1) on the *A* = *B* output of the comparator and putting shift register C into the *Shift* (SH) mode.

The trailing edge of the OSA output pulse triggers OSB which, in turn, triggers OSC on the trailing edge of its output pulse. The output of OSC provides *Clock B* to the memory address counter and clocks shift register C to shift the 00010000 to the right so that the register now contains 00001000. Since there is still a 0 (LOW) on the serial output *ArmOut*, the system remains armed.

When the second correct code digit is entered on the keypad, the contents of shift register C are shifted to 00000100, and the system remains armed. When the third code digit is entered on the keypad, the contents of shift register C are shifted to 00000010. When the fourth and last code digit is entered, the contents of shift register C are shifted to 00000001. Now, the 1 (HIGH) on the serial output *ArmOut* disarms the system and permits entry.

If an incorrect code digit is entered at any time, the comparator output goes LOW, producing a LOW on the $SH/\overline{LD}$ and triggered OSF to send a reset pulse to the memory address counter. Shift register C is now in the *Parallel Load* mode. OSC then clocks the register and loads the prewired code 00010000 into the register. At this point, you must start over and reenter the entire four code digits.

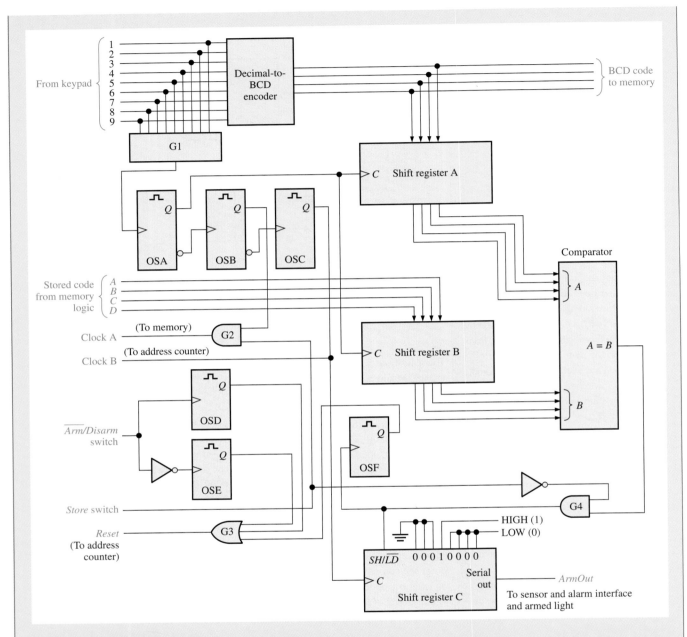

Basic logic diagram of the security code logic.

Programming   To program a 4-digit code into the system, the $\overline{Arm}$/Disarm switch is placed in the *disarm* position. This triggers one-shot OSD which sends a reset pulse through G3 to the memory address counter and resets it to 00, the first address in the memory. The Store switch is placed in the *Store* position which disables

the $A = B$ output of the comparator via gate G4 and enables the output of OSB via G2 to provide a clock to the memory during the programming of a code into the memory.

Next, the first digit of the desired security code is entered from the keypad. OSA is triggered through gate G1 as a result

of the key closure and, in turn, triggers OSB which produces clock A to store the code in the memory. OSB triggers OSC producing clock B for the memory address counter and advancing it to the second address (01). The second digit of the code is entered from the keypad, and the sequence described for the first digit is repeated. After the fourth and last code digit is entered, the memory contains the 4-digit security code. If an incorrect digit is accidentally entered, you must finish entering four digits or reactivate the STORE switch to ensure that the memory counter contains the first address again. Once the programming is done, the system is switched to *armed* mode.

System Assignment

- *Activity 1* Describe the purpose of Shift register A.

- *Activity 2* Describe the purpose of Shift register B.

- *Activity 3* Describe the purpose of Shift register C.

- *Activity 4* Describe the purpose of the comparator.

- *Optional Activity* Using 74XX logic ICS and the other components, implement the security code logic in Figure 9–45. Make any changes necessary to accommodate the devices being used. Debug and test the logic and describe any design flaws (if any) that you find.

SUMMARY

- The basic types of data movement in shift registers are illustrated in Figure 9–46.

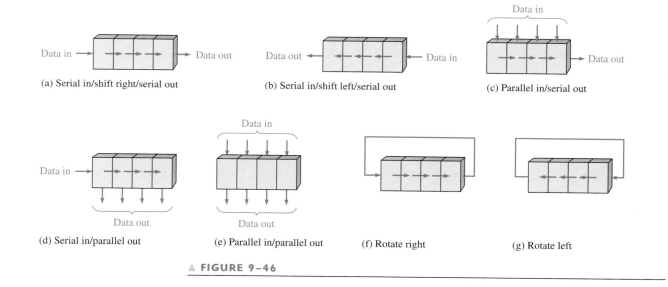

(a) Serial in/shift right/serial out
(b) Serial in/shift left/serial out
(c) Parallel in/serial out
(d) Serial in/parallel out
(e) Parallel in/parallel out
(f) Rotate right
(g) Rotate left

▲ **FIGURE 9–46**

- Shift register counters are shift registers with feedback that exhibit special sequences. Examples are the Johnson counter and the ring counter.

- The Johnson counter has $2n$ states in its sequence, where n is the number of stages.

- The ring counter has n states in its sequence.

KEY TERMS

Key terms and other bold terms in the chapter are defined in the end-of-book glossary.

Bidirectional Having two directions. In a bidirectional shift register, the stored data can be shifted right or left.

Load To enter data into a shift register.

Register One or more flip-flops used to store and shift data.

Shift To move binary data from stage to stage within a shift register or other storage device or to move binary data into or out of the device.

Stage One storage element in a register.

Answers are at the end of the chapter.

1. A stage in a shift register consists of
   (a) a latch   (b) a flip-flop   (c) a byte of storage   (d) four bits of storage

2. To serially shift a byte of data into a shift register, there must be
   (a) one clock pulse   (b) one load pulse
   (c) eight clock pulses   (d) one clock pulse for each 1 in the data

3. To parallel load a byte of data into a shift register with a synchronous load, there must be
   (a) one clock pulse   (b) one clock pulse for each 1 in the data
   (c) eight clock pulses   (d) one clock pulse for each 0 in the data

4. The group of bits 10110101 is serially shifted (right-most bit first) into an 8-bit parallel output shift register with an initial state of 11100100. After two clock pulses, the register contains
   (a) 01011110   (b) 10110101   (c) 01111001   (d) 00101101

5. With a 100 kHz clock frequency, eight bits can be serially entered into a shift register in
   (a) 80 μs   (b) 8 μs   (c) 80 ms   (d) 10 μs

6. With a 1 MHz clock frequency, eight bits can be parallel entered into a shift register
   (a) in 8 μs   (b) in the propagation delay time of eight flip-flops
   (c) in 1 μs   (d) in the propagation delay time of one flip-flop

7. A modulus-10 Johnson counter requires
   (a) ten flip-flops   (b) four flip-flops
   (c) five flip-flops   (d) twelve flip-flops

8. A modulus-10 ring counter requires a minimum of
   (a) ten flip-flops   (b) five flip-flops
   (c) four flip-flops   (d) twelve flip-flops

9. When an 8-bit serial in/serial out shift register is used for a 24 μs time delay, the clock frequency must be
   (a) 41.67 kHz   (b) 333 kHz   (c) 125 kHz   (d) 8 MHz

10. The purpose of the ring counter in the keyboard encoding circuit of Figure 9–38 is
   (a) to sequentially apply a HIGH to each row for detection of key closure
   (b) to provide trigger pulses for the key code register
   (c) to sequentially apply a LOW to each row for detection of key closure
   (d) to sequentially reverse bias the diodes in each row

Answers to odd-numbered problems are at the end of the book.

SECTION 9–1   **Basic Shift Register Functions**

1. Why are shift registers considered basic memory devices?

2. What is the storage capacity of a register that can retain two bytes of data?

SECTION 9–2   **Serial In/Serial Out Shift Registers**

3. For the data input and clock in Figure 9–47, determine the states of each flip-flop in the shift register of Figure 9–3 and show the Q waveforms. Assume that the register contains all 1s initially.

▶ FIGURE 9–47

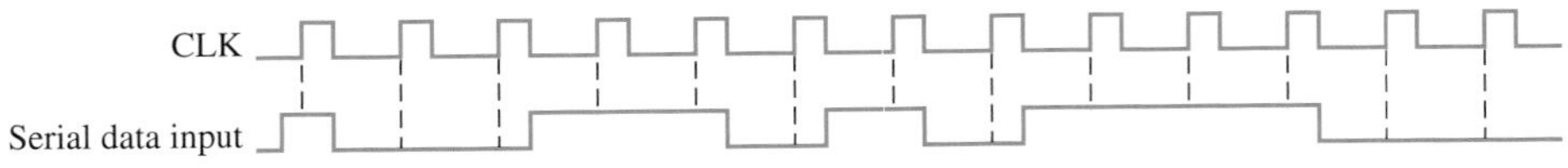

4. Solve Problem 3 for the waveforms in Figure 9–48.

▶ FIGURE 9–48

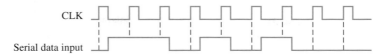

5. What is the state of the register in Figure 9–49 after each clock pulse if it starts in the 101001111000 state?

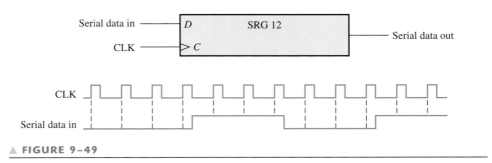

▲ FIGURE 9–49

6. For the serial in/serial out shift register, determine the data-output waveform for the data-input and clock waveforms in Figure 9–50. Assume that the register is initially cleared.

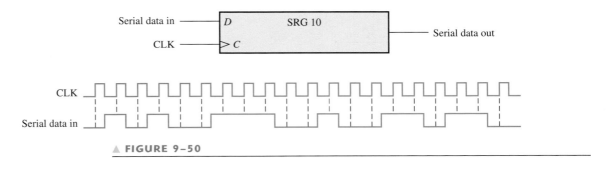

▲ FIGURE 9–50

7. Solve Problem 6 for the waveforms in Figure 9–51.

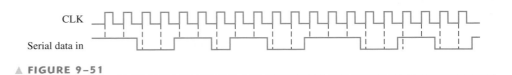

▲ FIGURE 9–51

8. A leading-edge clocked serial in/serial out shift register has a data-output waveform as shown in Figure 9–52. What binary number is stored in the 8-bit register if the first data bit out (left-most) is the LSB?

▶ FIGURE 9–52

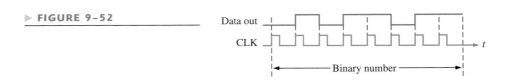

SECTION 9–3 **Serial In/Parallel Out Shift Registers**

9. Show a complete timing diagram showing the parallel outputs for the shift register in Figure 9–8. Use the waveforms in Figure 9–50 with the register initially clear.

10. Solve Problem 9 for the input waveforms in Figure 9–51.

11. Develop the Q_0 through Q_7 outputs for a 74HC164 shift register with the input waveforms shown in Figure 9–53.

▶ FIGURE 9–53

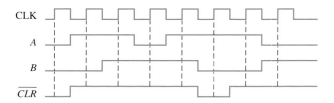

SECTION 9–4 **Parallel In/Serial Out Shift Registers**

12. The shift register in Figure 9–54(a) has $SHIFT/\overline{LOAD}$ and CLK inputs as shown in part (b). The serial data input (SER) is a 0. The parallel data inputs are $D_0 = 1$, $D_1 = 0$, $D_2 = 1$, and $D_3 = 0$ as shown. Develop the data-output waveform in relation to the inputs.

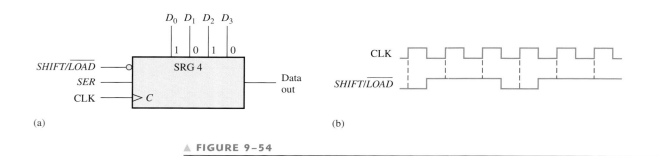

(a)   (b)

▲ FIGURE 9–54

13. The waveforms in Figure 9–55 are applied to a 74HC165 shift register. The parallel inputs are all 0. Determine the Q_7 waveform.

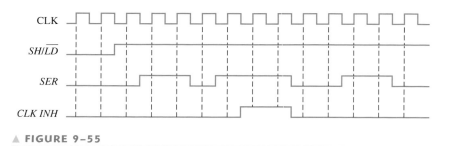

▲ FIGURE 9–55

14. Solve Problem 13 if the parallel inputs are all 1.

15. Solve Problem 13 if the SER input is inverted.

► FIGURE 9–56

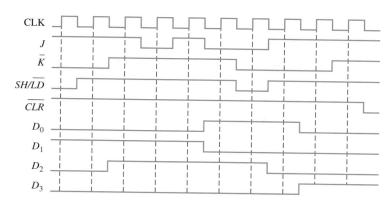

SECTION 9–5 Parallel In/Parallel Out Shift Registers

16. Determine all the Q output waveforms for a 74HC195 4-bit shift register when the inputs are as shown in Figure 9–56.

17. Solve Problem 16 if the $SH/\overline{LD}$ input is inverted and the register is initially clear.

18. Use two 74HC195 shift registers to form an 8-bit shift register. Show the required connections.

SECTION 9–6 Bidirectional Shift Registers

19. For the 8-bit bidirectional register in Figure 9–57, determine the state of the register after each clock pulse for the $RIGHT/\overline{LEFT}$ control waveform given. A HIGH on this input enables a shift to the right, and a LOW enables a shift to the left. Assume that the register is initially storing the decimal number seventy-six in binary, with the right-most position being the LSB. There is a LOW on the data-input line.

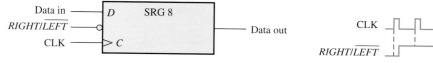

▲ FIGURE 9–57

20. Solve Problem 19 for the waveforms in Figure 9–58.

► FIGURE 9–58

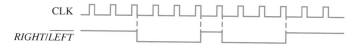

21. Use two 74HC194 4-bit bidirectional shift registers to create an 8-bit bidirectional shift register. Show the connections.

22. Determine the Q outputs of a 74HC194 with the inputs shown in Figure 9–59. Inputs D_0, D_1, D_2, and D_3 are all HIGH.

► FIGURE 9–59

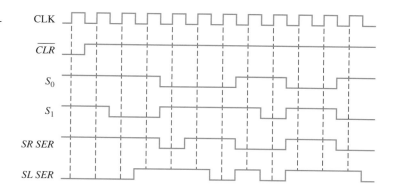

SECTION 9–7 Shift Register Counters

23. How many flip-flops are required to implement each of the following in a Johnson counter configuration:

(a) modulus-6   (b) modulus-10

(c) modulus-14   (d) modulus-16

24. Draw the logic diagram for a modulus-18 Johnson counter. Show the timing diagram and write the sequence in tabular form.

25. For the ring counter in Figure 9–60, show the waveforms for each flip-flop output with respect to the clock. Assume that FF0 is initially SET and that the rest are RESET. Show at least ten clock pulses.

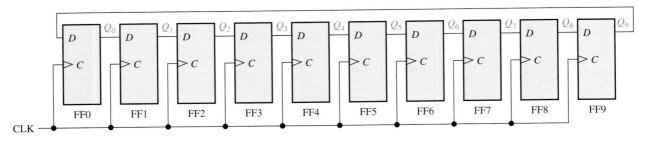

▲ **FIGURE 9–60**

26. The waveform pattern in Figure 9–61 is required. Devise a ring counter, and indicate how it can be preset to produce this waveform on its Q_9 output. At CLK16 the pattern begins to repeat.

▶ **FIGURE 9–61**

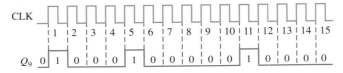

SECTION 9–8 Shift Register Applications

27. Use 74HC195 4-bit shift registers to implement a 16-bit ring counter. Show the connections.

28. What is the purpose of the power-on $\overline{LOAD}$ input in Figure 9–38?

29. What happens when two keys are pressed simultaneously in Figure 9–38?

SECTION 9–10 Troubleshooting

30. Based on the waveforms in Figure 9–62(a), determine the most likely problem with the register in part (b) of the figure.

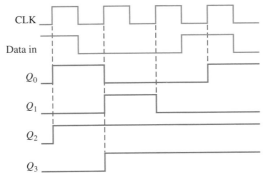

(a)

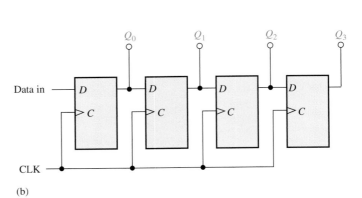

(b)

▲ **FIGURE 9–62**

31. Refer to the parallel in/serial out shift register in Figure 9–12. The register is in the state where $Q_0Q_1Q_2Q_3 = 1001$, and $D_0D_1D_2D_3 = 1010$ is loaded in. When the *SHIFT/LOAD* input is taken HIGH, the data shown in Figure 9–63 are shifted out. Is this operation correct? If not, what is the most likely problem?

▶ **FIGURE 9–63**

32. You have found that the bidirectional register in Figure 9–19 will shift data right but not left. What is the most likely fault?

33. For the keyboard encoder in Figure 9–38, list the possible faults for each of the following symptoms:

   (a) The state of the key code register does not change for any key closure.

   (b) The state of the key code register does not change when any key in the third row is closed. A proper code occurs for all other key closures.

   (c) The state of the key code register does not change when any key in the first column is closed. A proper code occurs for all other key closures.

   (d) When any key in the second column is closed, the left three bits of the key code ($Q_0Q_1Q_2$) are correct, but the right three bits are all 1s.

34. Develop a test procedure for exercising the keyboard encoder in Figure 9–38. Specify the procedure on a step-by-step basis, indicating the output code from the key code register that should be observed at each step in the test.

35. What symptoms are observed for the following failures in the serial-to-parallel converter in Figure 9–33:

   (a) AND gate output stuck in HIGH state

   (b) clock generator output stuck in LOW state

   (c) third stage of data-input register stuck in SET state

   (d) terminal count output of counter stuck in HIGH state

Digital System Application

36. What is the major purpose of the security code logic?

37. Assume the entry code is 1939. Determine the states of shift register A and shift register C after the second correct digit has been entered.

38. Assume the entry code is 7646 and the digits 7645 are entered. Determine the states of shift register A and shift register C after each of the digits is entered.

Special Design Problems

39. Specify the devices that can be used to implement the serial-to-parallel data converter in Figure 9–33. Develop the complete logic diagram, showing any modifications necessary to accommodate the specific devices used.

40. Modify the serial-to-parallel converter in Figure 9–33 to provide 16-bit conversion.

41. Design an 8-bit parallel-to-serial data converter that produces the data format in Figure 9–34. Show a logic diagram and specify the devices.

42. Design a power-on $\overline{LOAD}$ circuit for the keyboard encoder in Figure 9–38. This circuit must generate a short-duration LOW pulse when the power switch is turned on.

43. Implement the test-pattern generator used in Figure 9–42 to troubleshoot the serial-to-parallel converter.

44. Review the tablet-counting and control system that was introduced in Chapter 1. (a) Utilizing the knowledge gained in this chapter, implement registers A and B in that system using specific fixed-function IC devices. (b) Implement the system using your development software.

Multisim Troubleshooting Practice

45. Open file P09-45 and test the 4-bit shift register to determine if there is a fault. Identify the fault if possible.

46. Open file P09-46 and test the 74164 8-bit serial in/parallel out shift register to determine if there is a fault. If there is a fault, identify it if possible.

47. Open file P09-47 and test the 74165 8-bit parallel load shift register to determine if there is a fault. If there is a fault, identify it if possible.

48. Open file P09-48 and test the 74195 4-bit parallel access shift register to determine if there is a fault. If there is a fault, identify it if possible.

49. Open file P09-49 and test the 10-bit ring counter to determine if there is a fault. If there is a fault, identify it if possible.

ANSWERS

SECTION REVIEWS

SECTION 9–1 **Basic Shift Register Functions**

1. A counter has a specified sequence of states, but a shift register does not.

2. Storage and data movement are two functions of a shift register.

SECTION 9–2 **Serial In/Serial Out Shift Registers**

1. FF0: data input to J_0, $\overline{\text{data input}}$ to K_0; FF1: Q_0 to J_1, $\overline{Q_0}$ to K_1; FF2: Q_1 to J_2, $\overline{Q_1}$ to K_2; FF3: Q_2 to J_3, $\overline{Q_2}$ to K_3

2. Eight clock pulses

SECTION 9–3 **Serial In/Parallel Out Shift Registers**

1. 0100 after 2 clock pulses

2. Take the serial output from the right-most flip-flop for serial out operation.

SECTION 9–4 **Parallel In/Serial Out Shift Registers**

1. When $SHIFT/\overline{LOAD}$ is HIGH, the data are shifted right one bit per clock pulse. When $SHIFT/\overline{LOAD}$ is LOW, the data on the parallel inputs are loaded into the register.

2. The parallel load operation is asynchronous, so it is not dependent on the clock.

SECTION 9–5 **Parallel In/Parallel Out Shift Registers**

1. The data outputs are 1001.

2. $Q_0 = 1$ after one clock pulse

SECTION 9–6 **Bidirectional Shift Registers**

1. 1111 after the fifth clock pulse

SECTION 9–7 **Shift Register Counters**

1. Sixteen states are in an 8-bit Johnson counter sequence.

2. For a 3-bit Johnson counter: 000, 100, 110, 111, 011, 001, 000

SECTION 9–8 **Shift Register Applications**

1. 625 scans/second

2. $Q_5Q_4Q_3Q_2Q_1Q_0 = 011011$

3. The diodes provide unidirectional paths for pulling the ROWs LOW and preventing HIGHs on the ROW lines from being connected to the switch matrix. The resistors pull the COLUMN lines HIGH.

SECTION 9–9 **Logic Symbols with Dependency Notation**

1. No inputs are dependent on the mode inputs being in the 0 state.

2. Yes, the parallel load is synchronous with the clock as indicated by the $4D$ label.

SECTION 9–10 Troubleshooting

1. A test input is used to sequence the circuit through all of its states.

2. Check the input to that portion of the circuit. If the signal on that input is correct, the fault is isolated to the circuitry between the good input and the bad output.

RELATED PROBLEMS FOR EXAMPLES

9–1 See Figure 9–64.

▶ **FIGURE 9–64**

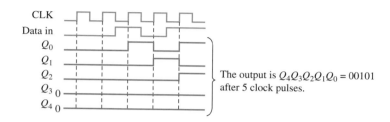

The output is $Q_4Q_3Q_2Q_1Q_0 = 00101$ after 5 clock pulses.

9–2 The state of the register after three additional clock pulses is 0000.

9–3 See Figure 9–65.

▶ **FIGURE 9–65**

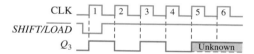

9–4 See Figure 9–66.

▶ **FIGURE 9–66**

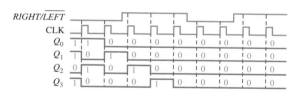

9–5 See Figure 9–67.

▶ **FIGURE 9–67**

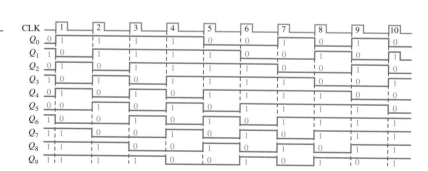

9–6 $f = 1/3\ \mu s = 333$ kHz

SELF-TEST

1. (b) **2.** (c) **3.** (a) **4.** (c) **5.** (a) **6.** (d) **7.** (c) **8.** (a)
9. (b) **10.** (c)

10

MEMORY AND STORAGE

CHAPTER OBJECTIVES

- Define the basic memory characteristics

- Explain what a RAM is and how it works

- Explain the difference between static RAMs (SRAMs) and dynamic RAMs (DRAMs)

- Explain what a ROM is and how it works

- Describe the various types of PROMs

- Discuss the characteristics of a flash memory

- Describe the expansion of ROMs and RAMs to increase word length and word capacity

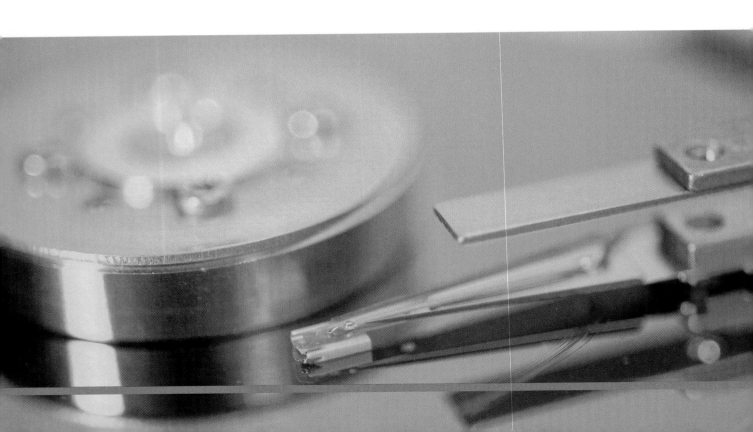

- Discuss special types of memories such as FIFO and LIFO
- Describe the basic organization of magnetic disks and magnetic tapes
- Describe the basic operation of magneto-optical disks and optical disks
- Describe basic methods for memory testing
- Develop flowcharts for memory testing
- Apply a memory device in a system application

KEY TERMS

- Byte
- Word
- Cell
- Address
- Capacity
- Write
- Read
- RAM
- ROM

- SRAM
- Bus
- DRAM
- PROM
- EPROM
- Flash memory
- FIFO
- LIFO
- Hard disk

INTRODUCTION

Chapter 9 covered shift registers, which are a type of storage device; in fact, a shift register is essentially a small-scale memory. The memory devices covered in this chapter are generally used for longer-term storage of larger amounts of data than registers can provide.

Computers and other types of systems require the permanent or semipermanent storage of large amounts of binary data. Microprocessor-based systems rely on storage devices and memories for their operation because of the necessity for storing programs and for retaining data during processing.

In computer terminology, *memory* usually refers to RAM and ROM and *storage* refers to hard disk, floppy disk, and CD-ROM. In this chapter semiconductor memories and magnetic and optical storage media are covered.

■ ■ ■ DIGITAL SYSTEM APPLICATION PREVIEW

The digital system application at the end of the chapter completes the security system from Chapter 9. The focus in this chapter is the memory logic portion of the system, which stores the entry code. Once the memory logic is developed, it is interfaced with the security code logic from Chapter 9 to form the complete system.

WWW. **VISIT THE COMPANION WEBSITE**
Study aids for this chapter are available at
http://www.prenhall.com/floyd

10–1 BASICS OF SEMICONDUCTOR MEMORY

Memory is the portion of a system for storing binary data in large quantities. Semiconductor memories consist of arrays of elements that are generally either latches or capacitors.

After completing this chapter, you should be able to

■ Explain how a memory stores binary data ■ Discuss the basic organization of a memory ■ Describe the write operation ■ Describe the read operation ■ Describe the addressing operation ■ Explain what RAMs and ROMs are

COMPUTER NOTE

The general definition of *word* is a complete unit of information consisting of a unit of binary data. When applied to computer instructions, a word is more specifically defined as two bytes (16 bits). As a very important part of assembly language used in computers, the DW (Define Word) directive means to define data in 16-bit units. This definition is independent of the particular microprocessor or the size of its data bus. Assembly language also allows definitions of bytes (8 bits) with the DB directive, double words (32 bits) with the DD directive, and quad–words (64 bits) with the QD directive.

Units of Binary Data: Bits, Bytes, Nibbles, and Words

As a rule, memories store data in units that have from one to eight bits. The smallest unit of binary data, as you know, is the **bit.** In many applications, data are handled in an 8-bit unit called a **byte** or in multiples of 8-bit units. The byte can be split into two 4-bit units that are called **nibbles.** A complete unit of information is called a **word** and generally consists of one or more bytes. Some memories store data in 9-bit groups; a 9-bit group consists of a byte plus a parity bit.

The Basic Semiconductor Memory Array

Each storage element in a memory can retain either a 1 or a 0 and is called a **cell.** Memories are made up of arrays of cells, as illustrated in Figure 10–1 using 64 cells as an example. Each block in the **memory array** represents one storage cell, and its location can be identified by specifying a row and a column.

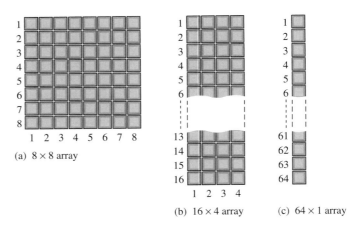

(a) 8 × 8 array

(b) 16 × 4 array

(c) 64 × 1 array

▲ **FIGURE 10–1**

A 64-cell memory array organized in three different ways.

The 64-cell array can be organized in several ways based on units of data. Figure 10–1(a) shows an 8 × 8 array, which can be viewed as either a 64-bit memory or an 8-byte memory. Part (b) shows a 16 × 4 array, which is a 16-nibble memory, and part (c) shows a 64 × 1 array, which is a 64-bit memory. A memory is identified by the number of words it can store times the word size. For example, a 16k × 8 memory can store 16,384 words of eight bits each. The inconsistency here is common in memory terminology. The actual number of words is always a power of 2, which, in this case, is $2^{14} = 16,384$. However, it is common practice to state the number to the nearest thousand, in this case, 16k.

Memory Address and Capacity

The location of a unit of data in a memory array is called its **address.** For example, in Figure 10–2(a), the address of a bit in the 2-dimensional array is specified by the row and column as shown. In Figure 10–2(b), the address of a byte is specified only by the row. So, as you can see, the address depends on how the memory is organized into units of data. Personal computers have random-access memories organized in bytes. This means that the smallest group of bits that can be addressed is eight.

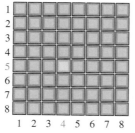

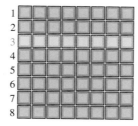

▪ **FIGURE 10–2**

Examples of memory address in a 2-dimensional array.

(a) The address of the blue bit is row 5, column 4.

(b) The address of the blue byte is row 3.

In Figure 10–3, the address of a byte in the three-dimensional array is specified by the row and column, as shown. In this case, the smallest group of bits that can be accessed is eight.

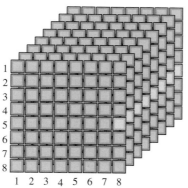

▪ **FIGURE 10–3**

Example of memory address in a 3-dimensional array.

The address of the blue byte is row 5, column 8.

The **capacity** of a memory is the total number of data units that can be stored. For example, in the bit-organized memory array in Figure 10–2(a), the capacity is 64 bits. In the byte-organized memory array in Figure 10–2(b), the capacity is 8 bytes, which is also 64 bits. In Figure 10–3, the capacity is 64 bytes. Computer memories typically have 256 MB (MB is megabyte) or more of internal memory.

Basic Memory Operations

Since a memory stores binary data, data must be put into the memory and data must be copied from the memory when needed. The **write** operation puts data into a specified address in the memory, and the **read** operation copies data out of a specified address in the memory. The addressing operation, which is part of both the write and the read operations, selects the specified memory address.

Data units go into the memory during a write operation and come out of the memory during a read operation on a set of lines called the *data bus.* As indicated in Figure 10–4, the data bus is bidirectional, which means that data can go in either direction (into the memory or out of the memory). In this case of byte-organized memories, the data bus has at least eight lines so that all eight bits in a selected address are transferred in parallel. For a write or a read operation, an address is selected by placing a binary code representing the desired address on a set of lines called the *address bus*. The address code is decoded internally, and the appropriate address is selected. In the case of the 3-dimensional memory array in Figure 10–4(b) there are two decoders, one for the rows and one for the columns. The number of lines in the address bus depends on the capacity of the memory. For example, a 15-bit address code can select 32,768 locations (2^{15}) in the memory, a 16-bit address code can select 65,536 locations (2^{16}) in the memory, and so on. In personal computers a 32-bit address bus can select 4,294,967,296 locations (2^{32}), expressed as 4G.

▶ FIGURE 10–4

Block diagram of a 2-dimensional memory and a 3-dimensional memory showing address bus, address decoder(s), bidirectional data bus, and read/write inputs.

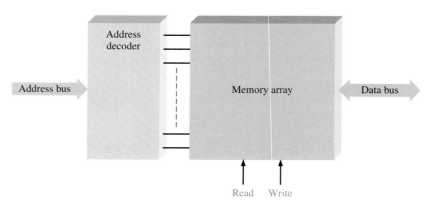

(a) 2-dimensional memory array

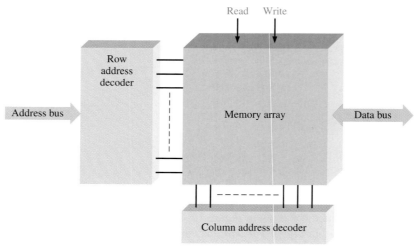

(b) 3-dimensional memory array

The Write Operation   A simplified write operation is illustrated in Figure 10–5. To store a byte of data in the memory, a code held in the address register is placed on the address bus. Once the address code is on the bus, the address decoder decodes the address and selects the specified location in the memory. The memory then gets a write command, and the data byte held in the data register is placed on the data bus and stored in the selected memory address, thus completing the write operation. When a new data byte is written into a memory address, the current data byte stored at that address is overwritten (replaced with a new data byte).

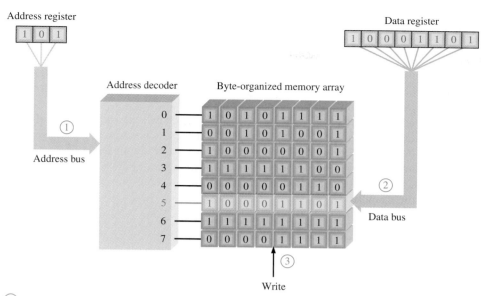

Illustration of the write operation.

① Address code 101 is placed on the address bus and address 5 is selected.

② Data byte is placed on the data bus.

③ Write command causes the data byte to be stored in address 5, replacing previous data.

The Read Operation   A simplified read operation is illustrated in Figure 10–6. Again, a code held in the address register is placed on the address bus. Once the address code is on the bus, the address decoder decodes the address and selects the specified location in the memory. The memory then gets a read command, and a "copy" of the data byte that is stored in the selected memory address is placed on the data bus and loaded into the data register, thus completing the read operation. When a data byte is read from a memory address, it also remains stored at that address. This is called *nondestructive read*.

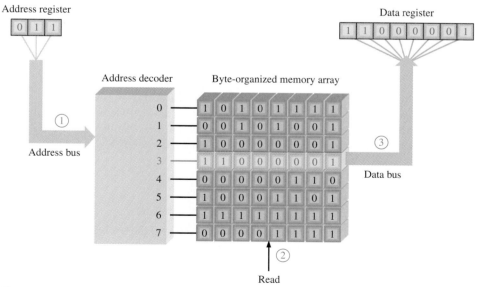

Illustration of the read operation.

① Address code 011 is placed on the address bus and address 3 is selected.

② Read command is applied.

③ The contents of address 3 is placed on the data bus and shifted into data register. The contents of address 3 is not erased by the read operation.

RAMs and ROMs

The two major categories of semiconductor memories are the RAM and the ROM. **RAM** (random-access memory) is a type of memory in which all addresses are accessible in an equal amount of time and can be selected in any order for a read or write operation. All RAMs have both *read* and *write* capability. Because RAMs lose stored data when the power is turned off, they are **volatile** memories.

ROM (read-only memory) is a type of memory in which data are stored permanently or semipermanently. Data can be read from a ROM, but there is no write operation as in the RAM. The ROM, like the RAM, is a random-access memory but the term *RAM* tradition-ally means a random-access *read/write* memory. Several types of RAMs and ROMs will be covered in this chapter. Because ROMs retain stored data even if power is turned off, they are **nonvolatile** memories.

SECTION 10–1 **REVIEW** Answers are at the end of the chapter.	1. What is the smallest unit of data that can be stored in a memory? 2. What is the bit capacity of a memory that can store 256 bytes of data? 3. What is a write operation? 4. What is a read operation? 5. How is a given unit of data located in a memory? 6. Describe the difference between a RAM and a ROM.

10–2 RANDOM-ACCESS MEMORIES (RAMs)

RAMs are read/write memories in which data can be written into or read from any selected address in any sequence. When a data unit is written into a given address in the RAM, the data unit previously stored at that address is replaced by the new data unit. When a data unit is read from a given address in the RAM, the data unit remains stored and is not erased by the read operation. This nondestructive read operation can be viewed as copying the content of an address while leaving the content intact. A RAM is typically used for short-term data storage because it cannot retain stored data when power is turned off.

After completing this section, you should be able to

■ Name the two categories of RAM ■ Explain what a SRAM is ■ Describe the SRAM storage cell ■ Explain the difference between an asynchronous SRAM and a synchronous burst SRAM ■ Explain what a DRAM is ■ Describe the DRAM storage cells ■ Discuss the types of DRAM ■ Compare the SRAM with the DRAM

The RAM Family

The two categories of RAM are the *static RAM* (SRAM) and the *dynamic RAM* (DRAM). Static RAMs generally use latches as storage elements and can therefore store data indefi-nitely *as long as dc power is applied*. Dynamic RAMs use capacitors as storage elements and cannot retain data very long without the capacitors being recharged by a process called **refreshing.** Both SRAMs and DRAMs will lose stored data when dc power is removed and, therefore, are classified as volatile memories.

Data can be read much faster from SRAMs than from DRAMs. However, DRAMs can store much more data than SRAMs for a given physical size and cost because the

DRAM cell is much simpler, and more cells can be crammed into a given chip area than in the SRAM.

The basic types of SRAM are the *asynchronous SRAM* and the *synchronous SRAM* with a burst feature. The basic types of DRAM are the *Fast Page Mode DRAM* (FPM DRAM), the *Extended Data Out DRAM* (EDO DRAM), the *Burst EDO DRAM* (BEDO DRAM), and the *synchronous DRAM* (SDRAM). These are shown in Figure 10–7.

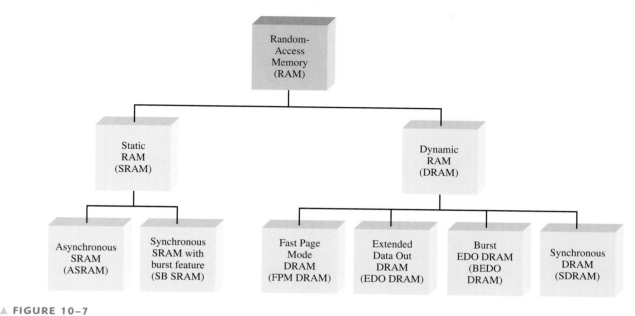

▲ **FIGURE 10-7**

The RAM family.

Static RAMs (SRAMs)

Memory Cell All static RAMs are characterized by latch memory cells. As long as dc power is applied to a **static memory** cell, it can retain a 1 or 0 state indefinitely. If power is removed, the stored data bit is lost.

Figure 10–8 shows a basic **SRAM** latch memory cell. The cell is selected by an active level on the Select line and a data bit (1 or 0) is written into the cell by placing it on the Data in line. A data bit is read by taking it off the Data out line.

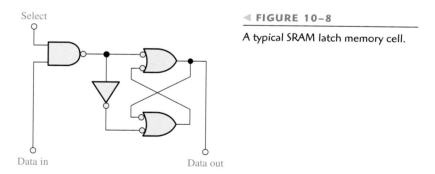

◄ **FIGURE 10-8**

A typical SRAM latch memory cell.

Basic Static Memory Cell Array The memory cells in a SRAM are organized in rows and columns, as illustrated in Figure 10–9 for the case of an $n \times 4$ array. All the cells in a row share the same Row Select line. Each set of Data in and Data out lines go to each cell in a given column and are connected to a single data line that serves as both an input and output (Data I/O) through the data input and data output buffers.

▶ **FIGURE 10–9**

Basic SRAM array.

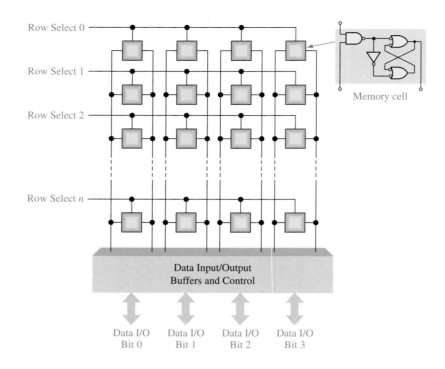

To write a data unit, in this case a nibble, into a given row of cells in the memory array, the Row Select line is taken to its active state and four data bits are placed on the Data I/O lines. The Write line is then taken to its active state, which causes each data bit to be stored in a selected cell in the associated column. To read a data unit, the Read line is taken to its active state, which causes the four data bits stored in the selected row to appear on the Data I/O lines.

Basic Asynchronous SRAM Organization

An asynchronous SRAM is one in which the operation is not synchronized with a system clock. To illustrate the general organization of a SRAM, a 32k × 8 bit memory is used. A logic symbol for this memory is shown in Figure 10–10.

▶ **FIGURE 10–10**

Logic diagram for an asynchronous 32k × 8 SRAM.

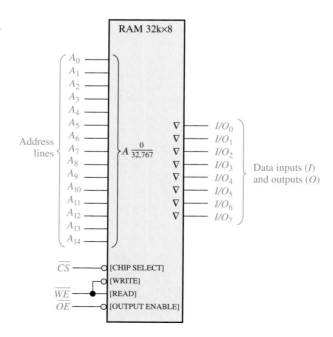

In the READ mode, the eight data bits that are stored in a selected address appear on the data output lines. In the WRITE mode, the eight data bits that are applied to the data input lines are stored at a selected address. The data input and data output lines (I/O_0 through I/O_7) share the same lines. During READ, they act as output lines (O_0 through O_7) and during WRITE they act as input lines (I_0 through I_7).

Tristate Outputs and Buses   Tristate buffers in a memory allow the data lines to act as either input or output lines and connect the memory to the data bus in a computer. These buffers have three output states: HIGH (1), LOW (0), and HIGH-Z (open). Tristate outputs are indicated on logic symbols by a small inverted triangle (∇), as shown in Figure 10–10, and are used for compatibility with bus structures such as those found in microprocessor-based systems.

Physically, a **bus** is a set of conductive paths that serve to interconnect two or more functional components of a system or several diverse systems. Electrically, a bus is a collection of specified voltage levels and/or current levels and signals that allow the various devices connected to the bus to communicate and work properly together.

For example, a microprocessor is connected to memories and input/output devices by certain bus structures. An address bus allows the microprocessor to address the memories, and the data bus provides for transfer of data between the microprocessor, the memories, and the input/output devices such as monitors, printers, keyboards, and modems. The control bus allows the microprocessor to control data transfers and timing for the various components.

Memory Array   SRAM chips can be organized in single bits, nibbles (4 bits), bytes (8 bits), or multiple bytes (16, 24, 32 bits, etc.).

Figure 10–11 shows the organization of a typical 32k × 8 SRAM. The memory cell array is arranged in 256 rows and 128 columns, each with 8 bits, as shown in part (a). There are actually $2^{15} = 32,768$ addresses and each address contains 8 bits. The capacity of this example memory is 32,768 bytes (typically expressed as 32 kB).

The SRAM in Figure 10–11(b) works as follows. First, the chip select, $\overline{CS}$, must be LOW for the memory to operate. Eight of the fifteen address lines are decoded by the row decoder

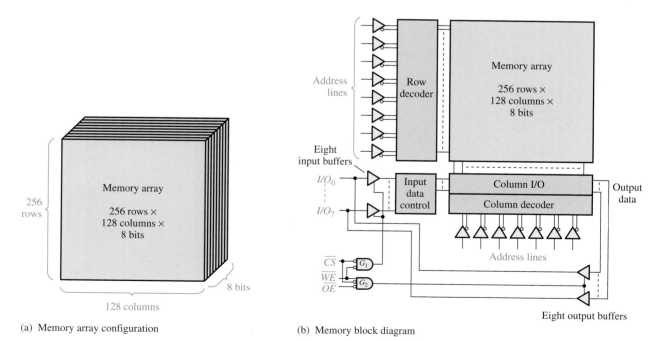

(a) Memory array configuration   (b) Memory block diagram

▲ **FIGURE 10–11**

Basic organization of an asynchronous 32k × 8 SRAM.

to select one of the 256 rows. Seven of the fifteen address lines are decoded by the column decoder to select one of the 128 8-bit columns.

Read　In the READ mode, the write enable input, $\overline{WE}$, is HIGH and the output enable, $\overline{OE}$, is LOW. The input tristate buffers are disabled by gate G_1, and the column output tristate buffers are enabled by gate G_2. Therefore, the eight data bits from the selected address are routed through the column I/O to the data lines (I/O_0 though I/O_7), which are acting as data output lines.

Write　In the WRITE mode, $\overline{WE}$ is LOW and $\overline{OE}$ is HIGH. The input buffers are enabled by gate G_1, and the output buffers are disabled by gate G_2. Therefore, the eight input data bits on the data lines are routed through the input data control and the column I/O to the selected address and stored.

Read and Write Cycles　Figure 10–12 shows typical timing diagrams for a memory read cycle and a write cycle. For the read cycle shown in part (a), a valid address code is applied to the address lines for a specified time interval called the *read cycle time, t_{RC}*. Next, the chip select ($\overline{CS}$) and the output enable ($\overline{OE}$) inputs go LOW. One time interval after the $\overline{OE}$ input goes LOW, a valid data byte from the selected address appears on the data lines. This time interval is called the *output enable access time, t_{GQ}*. Two other access times for the read cycle are the *address access time, t_{AQ}*, measured from the beginning of a valid address to the appearance of valid data on the data lines and the *chip enable access time, t_{EQ}*, measured from the HIGH-to-LOW transition of $\overline{CS}$ to the appearance of valid data on the data lines.

During each read cycle, one unit of data, a byte in this case, is read from the memory.

▶ FIGURE 10–12

Basic read and write cycle timing for the SRAM in Figure 10–11.

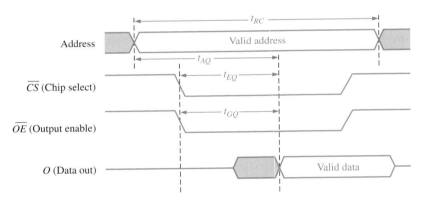

(a) Read cycle ($\overline{WE}$ HIGH)

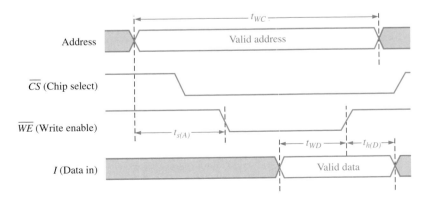

(b) Write cycle ($\overline{WE}$ LOW)

For the write cycle shown in Figure 10–12(b), a valid address code is applied to the address lines for a specified time interval called the *write cycle time*, t_{WC}. Next, the chip select $(\overline{CS})$ and the write enable $(\overline{WE})$ inputs go LOW. The required time interval from the beginning of a valid address until the $\overline{WE}$ input goes LOW is called the *address setup time*, $t_{s(A)}$. The time that the $\overline{WE}$ input must be LOW is the write pulse width. The time that the input $\overline{WE}$ must remain LOW after valid data are applied to the data inputs is designated t_{WD}; the time that the valid input data must remain on the data lines after the $\overline{WE}$ input goes HIGH is the *data hold time*, $t_{h(D)}$.

During each write cycle, one unit of data is written into the memory.

Basic Synchronous SRAM with Burst Feature

Unlike the asynchronous SRAM, a synchronous SRAM is synchronized with the system clock. For example, in a computer system, the synchronous SRAM operates with the same clock signal that operates the microprocessor so that the microprocessor and memory are synchronized for faster operation.

The fundamental concept of the synchronous feature of a SRAM can be shown with Figure 10–13, which is a simplified block diagram of a 32k × 8 memory for purposes of illustration. The synchronous SRAM is similar to the asynchronous SRAM in terms of the memory array, address decoder, and read/write and enable inputs. The basic difference is that the synchronous SRAM uses clocked registers to synchronize all inputs with the system clock. The address, the read/write input, the chip enable, and the input data are all latched into their respective registers on an active clock pulse edge. Once this information is latched, the memory operation is in sync with the clock.

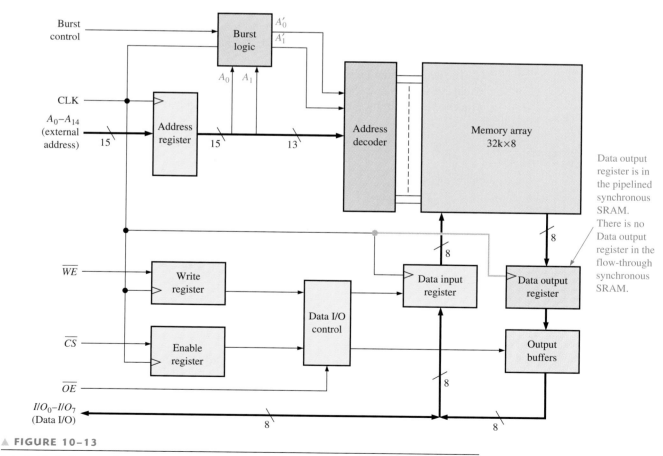

▲ **FIGURE 10–13**

A basic block diagram of a synchronous SRAM with burst feature.

For the purpose of simplification, a notation for multiple parallel lines or bus lines is introduced in Figure 10–13, as an alternative to drawing each line separately. A set of parallel lines can be indicated by a single heavy line with a slash and the number of separate lines in the set. For example, the following notation represents a set of 8 parallel lines:

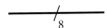

The address bits A_0 through A_{14} are latched into the Address register on the positive edge of a clock pulse. On the same clock pulse, the state of the write enable $(\overline{WE})$ line and chip select $(\overline{CS})$ are latched into the Write register and the Enable register respectively. These are one-bit registers or simply flip-flops. Also, on the same clock pulse the input data are latched into the Data input register for a Write operation, and data in a selected memory address are latched into the Data output register for a Read operation, as determined by the Data I/O control based on inputs from the Write register, Enable register, and the Output enable $(\overline{OE})$.

Two basic types of synchronous SRAM are the *flow-through* and the *pipelined*. The flow-through synchronous SRAM does not have a Data output register, so the output data flow asynchronously to the data I/O lines through the output buffers. The **pipelined** synchronous SRAM has a Data output register, as shown in Figure 10–13, so the output data are synchronously placed on the data I/O lines.

The Burst Feature   As shown in Figure 10–13, synchronous SRAMs normally have an address burst feature, which allows the memory to read or write at up to four locations using a single address. When an external address is latched in the address register, the two lowest-order address bits, A_0 and A_1, are applied to the burst logic. This produces a sequence of four internal addresses by adding 00, 01, 10, and 11 to the two lowest-order address bits on successive clock pulses. The sequence always begins with the base address, which is the external address held in the address register.

The address burst logic in a typical synchronous SRAM consists of a binary counter and exclusive-OR gates, as shown in Figure 10–14. For 2-bit burst logic, the internal burst address sequence is formed by the base address bits $A_2 - A_{14}$ plus the two burst address bits A_1' and A_0'.

▶ FIGURE 10–14

Address burst logic.

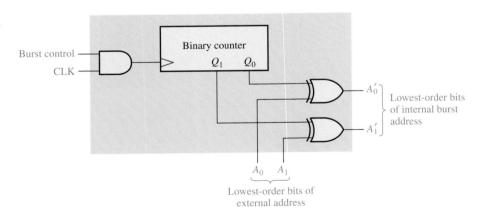

To begin the burst sequence, the counter is in its 00 state and the two lowest-order address bits are applied to the inputs of the XOR gates. Assuming that A_0 and A_1 are both 0, the internal address sequence in terms of its two lowest-order bits is 00, 01, 10, and 11.

Cache Memory

One of the major applications of SRAMs is in cache memories in computers. **Cache memory** is a relatively small, high-speed memory that stores the most recently used instructions or data from the larger but slower main memory. Cache memory can also use dynamic RAM (DRAM), which is covered next. Typically, SRAM is several times faster than DRAM. Overall, a cache memory gets stored information to the microprocessor much faster than if only high-capacity DRAM is used. Cache memory is basically a cost-effective method of improving system performance without having to resort to the expense of making all of the memory faster.

The concept of cache memory is based on the idea that computer programs tend to get instructions or data from one area of main memory before moving to another area. Basically, the cache controller "guesses" which area of the slow dynamic memory the CPU (central-processing unit) will need next and moves it to the cache memory so that it is ready when needed. If the cache controller guesses right, the data are immediately available to the microprocessor. If the cache controller guesses wrong, the CPU must go to the main memory and wait much longer for the correct instructions or data. Fortunately, the cache controller is right most of the time.

Cache Analogy There are many analogies that can be used to describe a cache memory, but comparing it to a home refrigerator is perhaps the most effective. A home refrigerator can be thought of as a "cache" for certain food items while the supermarket is the main memory where all foods are kept. Each time you want something to eat or drink, you can go to the refrigerator (cache) first to see if the item you want is there. If it is, you save a lot of time. If it is not there, then you have to spend extra time to get it from the supermarket (main memory).

L1 and L2 Caches A first-level cache (L1 cache) is usually integrated into the processor chip and has a very limited storage capacity. L1 cache is also known as *primary* cache. A second-level cache (L2 cache) is a separate memory chip or set of chips external to the processor and usually has a larger storage capacity than an L1 cache. L2 cache is also known as *secondary* cache. Some systems may have higher-level caches (L3, L4, etc.), but L1 and L2 are the most common. Also, some systems use a disk cache to enhance the performance of the hard disk because DRAM, although much slower than SRAM, is much faster than the hard disk drive. Figure 10–15 illustrates L1 and L2 cache memories in a computer system.

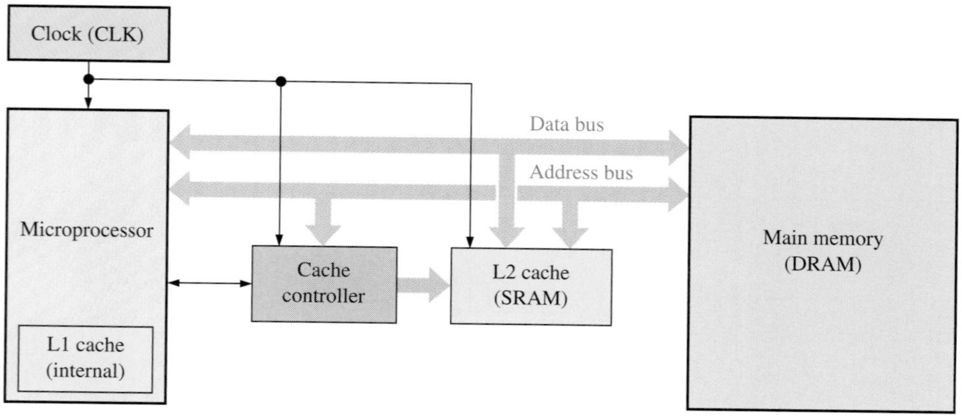

FIGURE 10–15
Block diagram showing L1 and L2 cache memories in a computer system.

Dynamic RAM (DRAM) Memory Cells

Dynamic memory cells store a data bit in a small capacitor rather than in a latch. The advantage of this type of cell is that it is very simple, thus allowing very large memory arrays to be constructed on a chip at a lower cost per bit. The disadvantage is that the storage capacitor cannot hold its charge over an extended period of time and will lose the stored data bit unless its charge is refreshed periodically. To refresh requires additional memory circuitry and complicates the operation of the **DRAM.** Figure 10–16 shows a typical DRAM cell consisting of a single MOS transistor (MOSFET) and a capacitor.

In this type of cell, the transistor acts as a switch. The basic simplified operation is illustrated in Figure 10–17 and is as follows. A LOW on the $R/\overline{W}$ line (WRITE mode) enables the tristate input buffer and disables the output buffer. For a 1 to be written into the cell, the D_{IN} line must be HIGH, and the transistor must be turned on by a HIGH on the row

▶ FIGURE 10–16

A MOS DRAM cell.

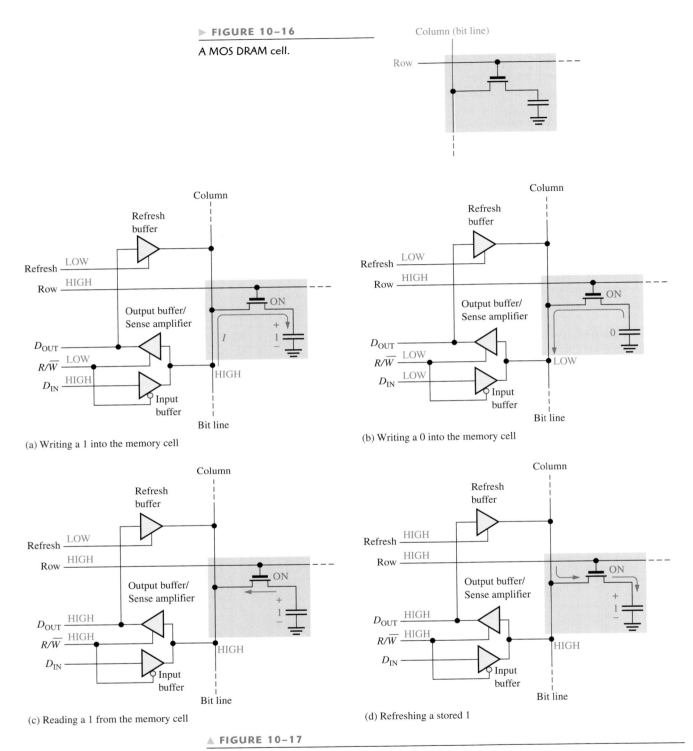

(a) Writing a 1 into the memory cell

(b) Writing a 0 into the memory cell

(c) Reading a 1 from the memory cell

(d) Refreshing a stored 1

▲ FIGURE 10–17

Basic operation of a DRAM cell.

line. The transistor acts as a closed switch connecting the capacitor to the bit line. This connection allows the capacitor to charge to a positive voltage, as shown in Figure 10–17(a). When a 0 is to be stored, a LOW is applied to the D_{IN} line. If the capacitor is storing a 0, it remains uncharged, or if it is storing a 1, it discharges as indicated in Figure 10–17(b). When the row line is taken back LOW, the transistor turns off and disconnects the capacitor from the bit line, thus "trapping" the charge (1 or 0) on the capacitor.

To read from the cell, the $R/\overline{W}$ (Read/Write) line is HIGH, enabling the output buffer and disabling the input buffer. When the row line is taken HIGH, the transistor turns on and connects the capacitor to the bit line and thus to the output buffer (sense amplifier), so the data bit appears on the data-output line (D_{OUT}). This process is illustrated in Figure 10–17(c).

For refreshing the memory cell, the $R/\overline{W}$ line is HIGH, the row line is HIGH, and the refresh line is HIGH. The transistor turns on, connecting the capacitor to the bit line. The output buffer is enabled, and the stored data bit is applied to the input of the refresh buffer, which is enabled by the HIGH on the refresh input. This produces a voltage on the bit line corresponding to the stored bit, thus replenishing the capacitor. This is illustrated in Figure 10–17(d).

Basic DRAM Organization

The major application of DRAMs is in the main memory of computers. The difference between DRAMs and SRAMs is the type of memory cell. As you have seen, the DRAM memory cell consists of one transistor and a capacitor and is much simpler than the SRAM cell. This allows much greater densities in DRAMs and results in greater bit capacities for a given chip area, although much slower access time.

Again, because charge stored in a capacitor will leak off, the DRAM cell requires a frequent refresh operation to preserve the stored data bit. This requirement results in more complex circuitry than in a SRAM. Several features common to most DRAMs are now discussed using a generic 1M × 1 bit DRAM as an example.

Address Multiplexing DRAMs use a technique called *address multiplexing* to reduce the number of address lines. Figure 10–18 shows the block diagram of a 1,048,576-bit (1 Mbit)

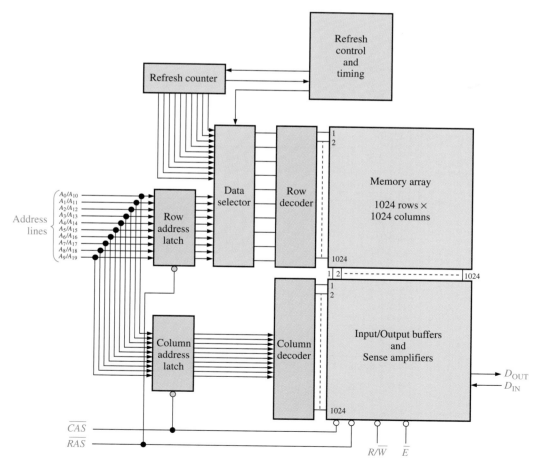

▲ **FIGURE 10–18**

Simplified block diagram of a 1M × 1 DRAM.

DRAM with a 1M × 1 organization. We will focus on the blue blocks to illustrate address multiplexing. The green blocks represent the refresh logic.

The ten address lines are time multiplexed at the beginning of a memory cycle by the row address select ($\overline{RAS}$) and the column address select ($\overline{CAS}$) into two separate 10-bit address fields. First, the 10-bit row address is latched into the row address latch. Next, the 10-bit column address is latched into the column address latch. The row address and the column address are decoded to select one of the 1,048,576 addresses ($2^{20} = 1,048,576$) in the memory array. The basic timing for the address multiplexing operation is shown in Figure 10–19.

▶ **FIGURE 10–19**

Basic timing for address multiplexing.

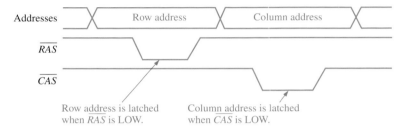

Read and Write Cycles   At the beginning of each read or write memory cycle, $\overline{RAS}$ and $\overline{CAS}$ go active (LOW) to multiplex the row and column addresses into the latches and decoders. For a read cycle, the $R/\overline{W}$ input is HIGH. For a write cycle, the $R/\overline{W}$ input is LOW. This is illustrated in Figure 10–20.

▶ **FIGURE 10–20**

Normal read and write cycle timing.

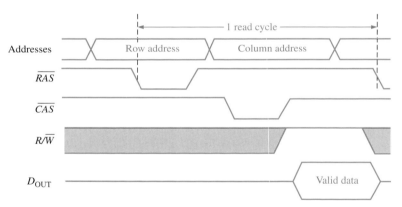

(a) Read cycle

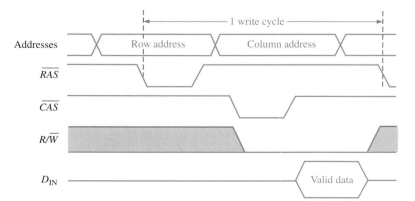

(b) Write cycle

Fast Page Mode   In the normal read or write cycle described previously, the row address for a particular memory location is first loaded by an active-LOW $\overline{RAS}$ and then the column address for that location is loaded by an active-LOW $\overline{CAS}$. The next location is selected by another $\overline{RAS}$ followed by a $\overline{CAS}$, and so on.

A "page" is a section of memory available at a single row address and consists of all the columns in a row. Fast page mode allows fast successive read or write operations at each column address in a selected row. A row address is first loaded by $\overline{RAS}$ going LOW and remaining LOW while $\overline{CAS}$ is toggled between HIGH and LOW. A single row address is selected and remains selected while $\overline{RAS}$ is active. Each successive $\overline{CAS}$ selects another column in the selected row. So, after a fast page mode cycle, all of the addresses in the selected row have been read from or written into, depending on $R/\overline{W}$. For example, a fast page mode cycle for the DRAM in Figure 10–18 requires $\overline{CAS}$ to go active 1024 times for each row selected by $\overline{RAS}$.

Fast page mode operation for read is illustrated by the timing diagram in Figure 10–21. When $\overline{CAS}$ goes to its nonasserted state (HIGH), it disables the data outputs. Therefore, the transition of $\overline{CAS}$ to HIGH must occur only after valid data are latched by the external system.

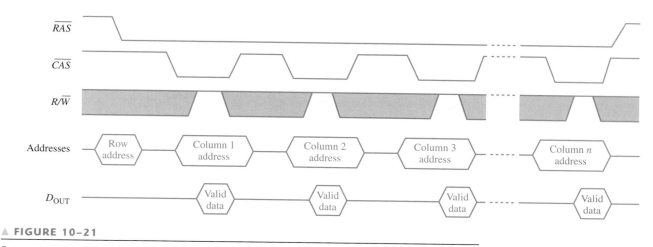

▲ **FIGURE 10–21**

Fast page mode timing for a read operation.

Refresh Cycles   As you know, DRAMs are based on capacitor charge storage for each bit in the memory array. This charge degrades (leaks off) with time and temperature, so each bit must be periodically refreshed (recharged) to maintain the correct bit state. Typically, a DRAM must be refreshed every 8 ms to 16 ms, although for some devices the refresh period can exceed 100 ms.

A read operation automatically refreshes all the addresses in the selected row. However, in typical applications, you cannot always predict how often there will be a read cycle and so you cannot depend on a read cycle to occur frequently enough to prevent data loss. Therefore, special refresh cycles must be implemented in DRAM systems.

Burst refresh and *distributed refresh* are the two basic refresh modes for refresh operations. In burst refresh, all rows in the memory array are refreshed consecutively each refresh period. For a memory with a refresh period of 8 ms, a burst refresh of all rows occurs once every 8 ms. The normal read and write operations are suspended during a burst refresh cycle. In distributed refresh, each row is refreshed at intervals interspersed between normal read or write cycles. For example, the memory in Figure 10–18 has 1024 rows. As an example, for an 8 ms refresh period, each row must be refreshed every 8 ms/1024 = 7.8 μs when distributed refresh is used.

The two types of refresh operations are $\overline{RAS}$-only refresh and $\overline{CAS}$ before $\overline{RAS}$ refresh. $\overline{RAS}$-only refresh consists of a $\overline{RAS}$ transition to the LOW (active) state, which latches the address of the row to be refreshed while $\overline{CAS}$ remains HIGH (inactive) throughout the cycle. An external counter is used to provide the row addresses for this type of operation.

The $\overline{CAS}$ before $\overline{RAS}$ refresh is initiated by $\overline{CAS}$ going LOW before $\overline{RAS}$ goes LOW. This sequence activates an internal refresh counter that generates the row address to be refreshed. This address is switched by the data selector into the row decoder.

Types of DRAMs

Now that you have learned the basic concept of a DRAM, let's briefly look at the major types. These are the *Fast Page Mode (FPM) DRAM,* the *Extended Data Output (EDO) DRAM,* the *Burst Extended Data Output (BEDO) DRAM,* and the *Synchronous (S) DRAM.*

FPM DRAM Fast page mode operation was described earlier. This type of DRAM traditionally has been the most common and has been the type used in computers until the development of the EDO DRAM. Recall that a page in memory is all of the column addresses contained within one row address.

The basic idea of the **FPM DRAM** is based on the probability that the next several memory addresses to be accessed are in the same row (on the same page). Fortunately, this happens a large percentage of the time. FPM saves time over pure random accessing because in FPM the row address is specified only once for access to several successive column addresses whereas for pure random accessing, a row address is specified for each column address.

Recall that in a fast page mode read operation, the $\overline{CAS}$ signal has to wait until the valid data from a given address are accepted (latched) by the external system (CPU) before it can go to its nonasserted state. When $\overline{CAS}$ goes to its nonasserted state, the data outputs are disabled. This means that the next column address cannot occur until after the data from the current column address are transferred to the CPU. This limits the rate at which the columns within a page can be addressed.

EDO DRAM The Extended Data Output DRAM, sometimes called *hyper page mode DRAM,* is similar to the FPM DRAM. The key difference is that the $\overline{CAS}$ signal in the **EDO DRAM** does not disable the output data when it goes to its nonasserted state because the valid data from the current address can be held until $\overline{CAS}$ is asserted again. This means that the next column address can be accessed before the external system accepts the current valid data. The idea is to speed up the access time.

BEDO DRAM The Burst Extended Data Output DRAM is an EDO DRAM with address burst capability. Recall from the discussion of the synchronous burst SRAM that the address burst feature allows up to four addresses to be internally generated from a single external address, which saves some access time. This same concept applies to the **BEDO DRAM.**

SDRAM Faster DRAMs are needed to keep up with the ever-increasing speed of microprocessors. The Synchronous DRAM is one way to accomplish this. Like the synchronous static RAM discussed earlier, the operation of the **SDRAM** is synchronized with the system clock, which also runs the microprocessor in a computer system. The same basic ideas described in relation to the synchronous burst SRAM, also apply to the SDRAM.

This synchronized operation makes the SDRAM totally different from the other asynchronous DRAM types. With asynchronous memories, the microprocessor must wait for the DRAM to complete its internal operations. However, with synchronous operation, the DRAM latches addresses, data, and control information from the processor under control of the system clock. This allows the processor to handle other tasks while the memory read or write operations are in progress, rather than having to wait for the memory to do its thing as is the case in asynchronous systems.

SECTION 10-2
REVIEW

1. List two types of SRAM.
2. What is a cache?
3. Explain how SRAMs and DRAMs differ.
4. Describe the refresh operation in a DRAM.
5. List four types of DRAM.

10-3 READ-ONLY MEMORIES (ROMs)

A ROM contains permanently or semipermanently stored data, which can be read from the memory but either cannot be changed at all or cannot be changed without specialized equipment. A ROM stores data that are used repeatedly in system applications, such as tables, conversions, or programmed instructions for system initialization and operation. ROMs retain stored data when the power is off and are therefore nonvolatile memories.

After completing this section, you should be able to

■ List the types of ROMs ■ Describe a basic mask ROM storage cell ■ Explain how data are read from a ROM ■ Discuss internal organization of a typical ROM ■ Discuss some ROM applications

The ROM Family

Figure 10–22 shows how semiconductor ROMs are categorized. The mask ROM is the type in which the data are permanently stored in the memory during the manufacturing process. The **PROM,** or programmable ROM, is the type in which the data are electrically stored by the user with the aid of specialized equipment. Both the mask ROM and the PROM can be of either MOS or bipolar technology. The **EPROM,** or erasable PROM, is strictly a MOS device. The **UV EPROM** is electrically programmable by the user, but the stored data must be erased by exposure to ultraviolet light over a period of several minutes. The electrically erasable PROM (**EEPROM** or E²PROM) can be erased in a few milliseconds.

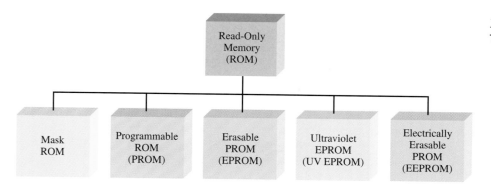

◀ **FIGURE 10-22**

The ROM family.

The Mask ROM

The mask ROM is usually referred to simply as a ROM. It is permanently programmed during the manufacturing process to provide widely used standard functions, such as popular conversions, or to provide user-specified functions. Once the memory is programmed, it cannot be changed. Most IC ROMs utilize the presence or absence of a transistor connection at a row/column junction to represent a 1 or a 0.

Figure 10–23 shows MOS ROM cells. The presence of a connection from a row line to the gate of a transistor represents a 1 at that location because when the row line is taken HIGH, all transistors with a gate connection to that row line turn on and connect the HIGH (1) to the associated column lines. At row/column junctions where there are no gate connections, the column lines remain LOW (0) when the row is addressed.

▶ **FIGURE 10–23**

ROM cells.

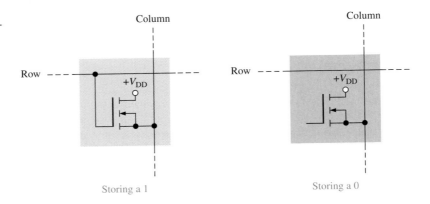

A Simple ROM

To illustrate the ROM concept, Figure 10–24 shows a small, simplified ROM array. The blue squares represent stored 1s, and the gray squares represent stored 0s. The basic read operation is as follows: When a binary address code is applied to the address input lines,

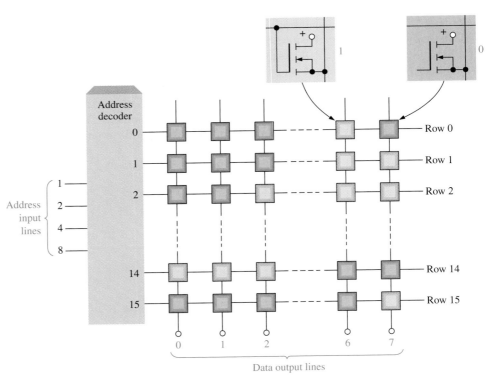

▲ **FIGURE 10–24**

A representation of a 16 × 8-bit ROM array.

the corresponding row line goes HIGH. This HIGH is connected to the column lines through the transistors at each junction (cell) where a 1 is stored. At each cell where a 0 is stored, the column line stays LOW because of the terminating resistor. The column lines form the data output. The eight data bits stored in the selected row appear on the output lines.

As you can see, the example ROM in Figure 10–24 is organized into 16 addresses, each of which stores 8 data bits. Thus, it is a 16×8 (16-by-8) ROM, and its total capacity is 128 bits or 16 bytes. ROMs can be used as look-up tables (LUTs) for code conversions and logic function generation.

EXAMPLE 10–1

Show a basic ROM, similar to the one in Figure 10–24, programmed for a 4-bit binary-to-Gray conversion.

Solution   Review Chapter 2 for the Gray code. Table 10–1 is developed for use in programming the ROM.

▼ TABLE 10–1

BINARY				GRAY			
B_3	B_2	B_1	B_0	G_3	G_2	G_1	G_0
0	0	0	0	0	0	0	0
0	0	0	1	0	0	0	1
0	0	1	0	0	0	1	1
0	0	1	1	0	0	1	0
0	1	0	0	0	1	1	0
0	1	0	1	0	1	1	1
0	1	1	0	0	1	0	1
0	1	1	1	0	1	0	0
1	0	0	0	1	1	0	0
1	0	0	1	1	1	0	1
1	0	1	0	1	1	1	1
1	0	1	1	1	1	1	0
1	1	0	0	1	0	1	0
1	1	0	1	1	0	1	1
1	1	1	0	1	0	0	1
1	1	1	1	1	0	0	0

The resulting 16×4 ROM array is shown in Figure 10–25. You can see that a binary code on the address input lines produces the corresponding Gray code on the output lines (columns). For example, when the binary number 0110 is applied to the address input lines, address 6, which stores the Gray code 0101, is selected.

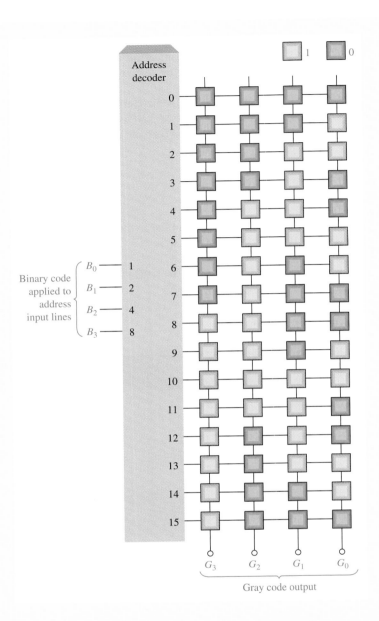

► FIGURE 10–25

Representation of a ROM programmed as a binary-to-Gray code converter.

Related Problem* Using Figure 10–25, determine the Gray code output when a binary code of 1011 is applied to the address input lines.

*Answers are at the end of the chapter.

Internal ROM Organization

Most IC ROMs have a more complex internal organization than that in the basic simplified example just presented. To illustrate how an IC ROM is structured, let's use a 1024-bit device with a 256 × 4 organization. The logic symbol is shown in Figure 10–26. When any one of 256 binary codes (eight bits) is applied to the address lines, four data bits appear on the outputs if the chip enable inputs are LOW. (256 addresses require eight address lines.)

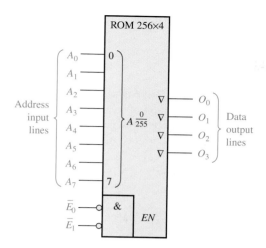

◄ **FIGURE 10–26**

A 256 × 4 ROM logic symbol. The A_{255}^{0} designator means that the 8-bit address code selects addresses 0 through 255.

Although the 256 × 4 organization of this device implies that there are 256 rows and 4 columns in the memory array, this is not actually the case. The memory cell array is actually a 32 × 32 matrix (32 rows and 32 columns), as shown in the block diagram in Figure 10–27.

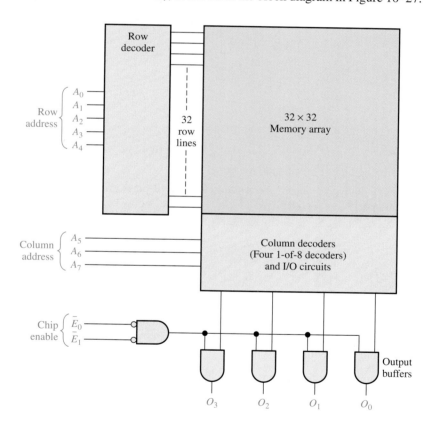

◄ **FIGURE 10–27**

A 1024-bit ROM with a 256 × 4 organization based on a 32 × 32 array.

The ROM in Figure 10–27 works as follows: Five of the eight address lines (A_0 through A_4) are decoded by the row decoder (often called the *Y* decoder) to select one of the 32 rows. Three of the eight address lines (A_5 through A_7) are decoded by the column decoder (often called the *X* decoder) to select four of the 32 columns. Actually, the column decoder consists of four 1-of-8 decoders (data selectors), as shown in Figure 10–27.

The result of this structure is that when an 8-bit address code (A_0 through A_7) is applied, a 4-bit data word appears on the data outputs when the chip enable lines ($\overline{E_0}$ and $\overline{E_1}$) are

LOW to enable the output buffers. This type of internal organization (architecture) is typical of IC ROMs of various capacities.

ROM Access Time

A typical timing diagram that illustrates ROM access time is shown in Figure 10–28. The **access time,** t_a, of a ROM is the time from the application of a valid address code on the input lines until the appearance of valid output data. Access time can also be measured from the activation of the chip enable ($\bar{E}$) input to the occurrence of valid output data when a valid address is already on the input lines.

▶ **FIGURE 10–28**

ROM access time (t_a) from address change to data output with chip enable already active.

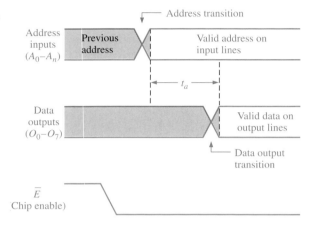

SECTION 10–3 REVIEW

1. What is the bit storage capacity of a ROM with a 512×8 organization?
2. List the types of read-only memories.
3. How many address bits are required for a 2048-bit memory organized as a 256×8 memory?

10–4 PROGRAMMABLE ROMs (PROMs AND EPROMs)

PROMs are basically the same as mask ROMs, once they have been programmed. As you have learned, ROMs are a type of programmable logic device. The difference is that PROMs come from the manufacturer unprogrammed and are custom programmed in the field to meet the user's needs.

After completing this section, you should be able to

▪ Distinguish between a mask ROM and a PROM ▪ Describe a basic PROM memory cell ▪ Discuss EPROMs including UV EPROMs and EEPROMs ▪ Analyze an EPROM programming cycle

PROMs

A **PROM** uses some type of fusing process to store bits, in which a memory *link* is burned open or left intact to represent a 0 or a 1. The fusing process is irreversible; once a PROM is programmed, it cannot be changed.

Figure 10–29 illustrates a MOS PROM array with fusible links. The fusible links are manufactured into the PROM between the source of each cell's transistor and its column line. In the programming process, a sufficient current is injected through the fusible link to burn it open to create a stored 0. The link is left intact for a stored 1.

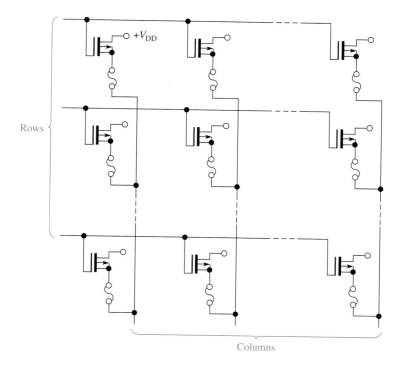

◀ FIGURE 10–29

MOS PROM array with fusible links. (All drains are commonly connected to V_{DD}.)

Three basic fuse technologies used in PROMs are metal links, silicon links, and *pn* junctions. A brief description of each of these follows.

1. Metal links are made of a material such as nichrome. Each bit in the memory array is represented by a separate link. During programming, the link is either "blown" open or left intact. This is done basically by first addressing a given cell and then forcing a sufficient amount of current through the link to cause it to open.

2. Silicon links are formed by narrow, notched strips of polycrystalline silicon. Programming of these fuses requires melting of the links by passing a sufficient amount of current through them. This amount of current causes a high temperature at the fuse location that oxidizes the silicon and forms an insulation around the now-open link.

3. Shorted junction, or avalanche-induced migration, technology consists basically of two *pn* junctions arranged back-to-back. During programming, one of the diode junctions is avalanched, and the resulting voltage and heat cause aluminum ions to migrate and short the junction. The remaining junction is then used as a forward-biased diode to represent a data bit.

EPROMs

An **EPROM** is an erasable PROM. Unlike an ordinary PROM, an EPROM can be reprogrammed if an existing program in the memory array is erased first.

An EPROM uses an NMOSFET array with an isolated-gate structure. The isolated transistor gate has no electrical connections and can store an electrical charge for indefinite periods of time. The data bits in this type of array are represented by the presence or absence of a stored gate charge. Erasure of a data bit is a process that removes the gate charge.

Two basic types of erasable PROMs are the ultraviolet erasable PROM (UV EPROM) and the electrically erasable PROM (EEPROM).

UV EPROMs   You can recognize the UV EPROM device by the transparent quartz lid on the package, as shown in Figure 10–30. The isolated gate in the **FET** of an ultraviolet EPROM is "floating" within an oxide insulating material. The programming process causes electrons to be removed from the floating gate. Erasure is done by exposure of the memory array chip to high-intensity ultraviolet radiation through the quartz window on top of the package. The positive charge stored on the gate is neutralized after several minutes to an hour of exposure time.

▶ **FIGURE 10–30**

Ultraviolet erasable PROM package.

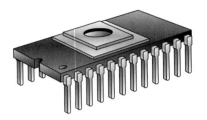

A typical UV EPROM is represented in Figure 10–31 by a logic diagram. Its operation is representative of that of other typical UV EPROMs of various sizes. As the logic symbol shows, this device has 2048 addresses ($2^{11} = 2048$), each with eight bits. Notice that the eight outputs are tristate (∇).

▶ **FIGURE 10–31**

The logic symbol for a 2048 × 8 UV EPROM.

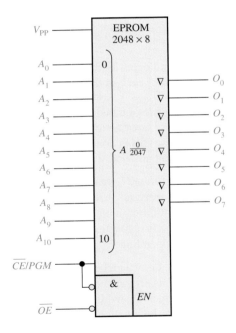

To read from the memory, the output enable input ($\overline{OE}$) must be LOW and the power-down/program ($\overline{CE}/PGM$) input LOW. To erase the stored data, the device is exposed to high-intensity ultraviolet light through the transparent lid. A typical UV lamp will erase the data in about 20 to 25 minutes. As in most UV EPROMs after erasure, all bits are 1s. Normal ambient light contains the correct wavelength of UV light for erasure over a period of time. Therefore, the transparent lid on the package must be kept covered.

To program the device, a high dc voltage is applied to V_{PP} and $\overline{OE}$ is HIGH. The eight data bits to be programmed into a given address are applied to the outputs (O_0 through O_7), and the address is selected on inputs A_0 through A_{10}. Next, a HIGH level pulse is applied to the $\overline{CE}/PGM$ input. The addresses can be programmed in any order.

A timing diagram for the programming is shown in Figure 10–32. These signals are normally produced by an EPROM programmer.

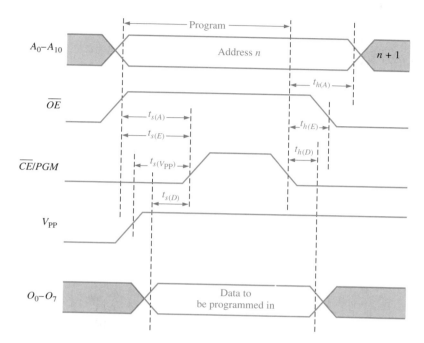

▶ **FIGURE 10–32**

Timing diagram for a 2048 × 8 UV EPROM programming cycle, with critical setup times (t_s) and hold times (t_h) indicated.

EEPROMs An electrically erasable PROM can be both erased and programmed with electrical pulses. Since it can be both electrically written into and electrically erased, the EEPROM can be rapidly programmed and erased in-circuit for reprogramming.

Two types of EEPROMs are the floating-gate MOS and the metal nitride-oxide silicon (MNOS). The application of a voltage on the control gate in the floating-gate structure permits the storage and removal of charge from the floating gate.

**SECTION 10–4
REVIEW**

1. How do PROMs differ from ROMs?
2. After erasure, all bits are (1s, 0s) in a typical EPROM.
3. What is the normal mode of operation for a PROM?

10–5 FLASH MEMORIES

The ideal memory has high storage capacity, nonvolatility, in-system read and write capability, comparatively fast operation, and cost effectiveness. The traditional memory technologies such as ROM, PROM, EPROM, EEPROM, SRAM, and DRAM individually exhibit one or more of these characteristics, but none of these technologies has all of them except the flash memory.

After completing this section, you should be able to

■ Discuss the basic characteristics of a flash memory ■ Describe the basic operation of a flash memory cell ■ Compare flash memories with other types of memories

Flash memories are high-density read/write memories (high-density translates into large bit storage capacity) that are nonvolatile, which means that data can be stored indefinitely without power. They are sometimes used in place of floppy or small-capacity hard disk drives in portable computers.

High-density means that a large number of cells can be packed into a given surface area on a chip; that is, the higher the density, the more bits that can be stored on a given size chip. This high density is achieved in flash memories with a storage cell that consists of a single floating-gate MOS transistor. A data bit is stored as charge or the absence of charge on the floating gate depending if a 0 or a 1 is stored.

Flash Memory Cell

A single-transistor cell in a flash memory is represented in Figure 10–33. The stacked gate MOS transistor consists of a control gate and a floating gate in addition to the drain and source. The floating gate stores electrons (charge) as a result of a sufficient voltage applied to the control gate. A *0 is stored when there is more charge* and a *1 is stored when there is less or no charge*. The amount of charge present on the floating gate determines if the transistor will turn on and conduct current from the drain to the source when a control voltage is applied during a read operation.

▶ **FIGURE 10–33**

The storage cell in a flash memory.

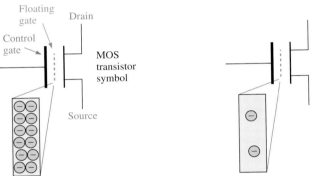

Many electrons = more charge = stored 0.   Few electrons = less charge = stored 1.

Basic Flash Memory Operation

There are three major operations in a flash memory: the *programming* operation, the *read* operation, and the *erase* operation.

Programming   Initially, all cells are at the 1 state because charge was removed from each cell in a previous erase operation. The programming operation adds electrons (charge) to the floating gate of those cells that are to store a 0. No charge is added to those cells that are to store a 1. Application of a sufficient positive voltage to the control gate with respect to the source during programming attracts electrons to the floating gate, as indicated in Figure 10–34. Once programmed, a cell can retain the charge for up to 100 years without any external power.

Read   During a read operation, a positive voltage is applied to the control gate. The amount of charge present on the floating gate of a cell determines whether or not the voltage applied to the control gate will turn on the transistor. If a 1 is stored, the control gate voltage is sufficient to turn the transistor on. If a 0 is stored, the transistor will not turn on because the control gate voltage is not sufficient to overcome the negative charge stored in the floating gate. Think of the charge on the floating gate as a voltage source that opposes the voltage applied to the control gate during a read operation. So the floating gate charge associated with a stored 0 prevents the control gate voltage from reaching the turn-on

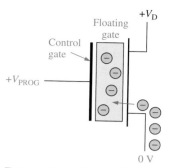

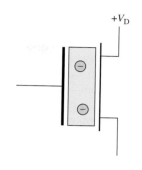

To store a 0, a sufficient positive voltage is applied to the control gate with respect to the source to add charge to the floating gate during programming.

To store a 1, no charge is added and the cell is left in the erased condition.

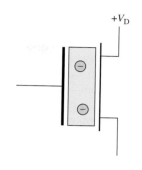

◀ **FIGURE 10–34**

Simplified illustration of storing a 0 or a 1 in a flash cell during the programming operation.

threshold, whereas the small or zero charge associated with a stored 1 allows the control gate voltage to exceed the turn-on threshold.

When the transistor turns on, there is current from the drain to the source of the cell transistor. The presence of this current is sensed to indicate a 1, and the absence of this current is sensed to indicate a 0. This basic idea is illustrated in Figure 10–35.

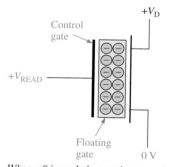

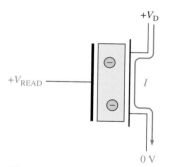

When a 0 is read, the transistor remains off because the charge on the floating gate prevents the read voltage from exceeding the turn-on threshold.

When a 1 is read, the transistor turns on because the absence of charge on the floating gate allows the read voltage to exceed the turn-on threshold.

◀ **FIGURE 10–35**

The read operation of a flash cell in an array.

Erase During an erase operation, charge is removed from all the memory cells. A sufficient positive voltage is applied to the transistor source with respect to the control gate. This is opposite in polarity to that used in programming. This voltage attracts electrons from the floating gate and depletes it of charge, as illustrated in Figure 10–36. A flash memory is always erased prior to being reprogrammed.

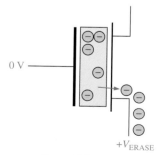

To erase a cell, a sufficient positive voltage is applied to the source with respect to the control gate to remove charge from the floating gate during the erase operation.

◀ **FIGURE 10–36**

Simplified illustration of removing charge from a cell during erase.

Basic Flash Memory Array

A simplified array of flash memory cells is shown in Figure 10–37. Only one row line is accessed at a time. When a cell in a given bit line turns on (stored 1) during a read operation, there is current through the bit line, which produces a voltage drop across the active load. This voltage drop is compared to a reference voltage with a comparator circuit and an output level indicating a 1 is produced. If a 0 is stored, then there is no current or little current in the bit line and an opposite level is produced on the comparator output.

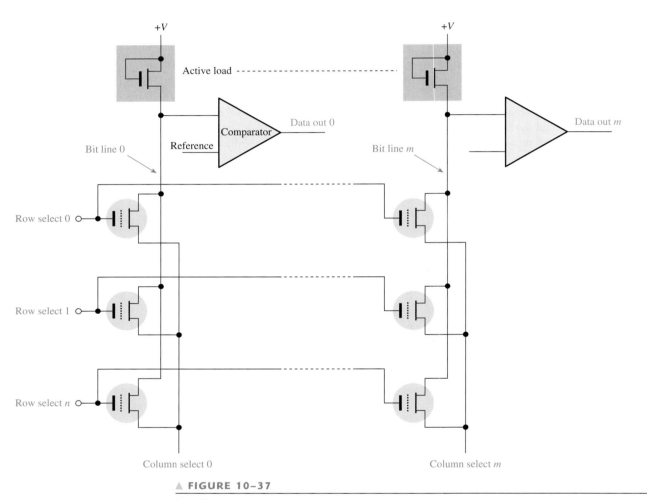

▲ **FIGURE 10–37**

Basic flash memory array.

The memory stick is a storage medium that uses flash memory technology in a physical configuration smaller than a stick of chewing gum. Memory sticks are typically available in 4 MB, 8 MB, 16 MB, 32 MB, 64 MB, and 128 MB capacities and as a kit with a PC card adaptor. Because of its compact design, it is ideal for use in small digital electronics products, such as laptop computers and digital cameras.

Comparison of Flash Memories with Other Memories

Let's compare flash memories with other types of memories with which you are already familiar.

Flash vs. ROM, EPROM, and EEPROM Read-only memories are high-density, non-volatile devices. However, once programmed the contents of a ROM can never be altered. Also, the initial programming is a time-consuming and costly process.

Although the EPROM is a high-density, nonvolatile memory, it can be erased only by removing it from the system and using ultraviolet light. It can be reprogrammed only with specialized equipment.

The EEPROM has a more complex cell structure than either the ROM or EPROM and so the density is not as high, although it can be reprogrammed without being removed from the system. Because of its lower density, the cost/bit is higher than ROMs or EPROMs.

A flash memory can be reprogrammed easily in the system because it is essentially a READ/WRITE device. The density of a flash memory compares with the ROM and EPROM because both have single transistor cells. A flash memory (like a ROM, EPROM, or EEPROM) is nonvolatile, which allows data to be stored indefinitely with power off.

Flash vs. SRAM As you have learned, static random-access memories are volatile READ/WRITE devices. A SRAM requires constant power to retain the stored data. In many applications, a battery backup is used to prevent data loss if the main power source is turned off. However, since battery failure is always a possibility, indefinite retention of the stored data in a SRAM cannot be guaranteed. Because the memory cell in a SRAM is basically a flip-flop consisting of several transistors, the density is relatively low.

A flash memory is also a READ/WRITE memory, but unlike the SRAM it is nonvolatile. Also, a flash memory has a much higher density than a SRAM.

Flash vs. DRAM Dynamic random-access memories are volatile high-density READ/WRITE devices. DRAMs require not only constant power to retain data but also that the stored data must be refreshed frequently. In many applications, backup storage such as hard disk must be used with a DRAM.

Flash memories exhibit higher densities than DRAMs because a flash memory cell consists of one transistor and does not need refreshing, whereas a DRAM cell is one transistor plus a capacitor that has to be refreshed. Typically, a flash memory consumes much less power than an equivalent DRAM and can be used as a hard disk replacement in many applications.

Table 10–2 provides a summary of the comparison of the memory technologies.

▼ TABLE 10–2

Comparison of types of memories.

MEMORY TYPE	NONVOLATILE	HIGH-DENSITY	ONE-TRANSISTOR CELL	IN-SYSTEM WRITABILITY
Flash	Yes	Yes	Yes	Yes
SRAM	No	No	No	Yes
DRAM	No	Yes	Yes	Yes
ROM	Yes	Yes	Yes	No
EPROM	Yes	Yes	Yes	No
EEPROM	Yes	No	No	Yes

10–6 MEMORY EXPANSION

Available memory can be expanded to increase the word length (number of bits in each address) or the word capacity (number of different addresses) or both. Memory expansion is accomplished by adding an appropriate number of memory chips to the address, data, and control buses. SIMMs, DIMMs, and RIMMs, which are types of memory expansion modules, are introduced.

After completing this section, you should be able to

- Define *word-length expansion* ■ Show how to expand the word length of a memory
- Define *word-capacity expansion* ■ Show how to expand the word capacity of a memory

Word-Length Expansion

To increase the **word length** of a memory, the number of bits in the data bus must be increased. For example, an 8-bit word length can be achieved by using two memories, each with 4-bit words as illustrated in Figure 10–38(a). As you can see in part (b), the 16-bit address bus is commonly connected to both memories so that the combination memory still has the same number of addresses ($2^{16} = 65,536$) as each individual memory. The 4-bit data buses from the two memories are combined to form an 8-bit data bus. Now when an address is selected, eight bits are produced on the data bus—four from each memory. Example 12–2 shows the details of $65,536 \times 4$ to $65,536 \times 8$ expansion.

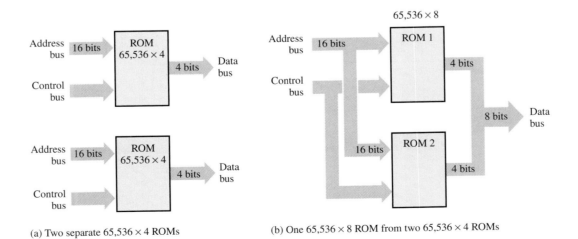

(a) Two separate $65,536 \times 4$ ROMs

(b) One $65,536 \times 8$ ROM from two $65,536 \times 4$ ROMs

▲ **FIGURE 10–38**

Expansion of two $65,536 \times 4$ ROMs into a $65,536 \times 8$ ROM to illustrate word–length expansion.

EXAMPLE 10–2

Expand the 65,536 × 4 ROM (64k × 4) in Figure 10–39 to form a 64k × 8 ROM. Note that "64k" is the accepted shorthand for 65,536. Why not "65k"? Maybe it's because 64 is also a power-of-two.

▶ **FIGURE 10–39**

A 64k × 4 ROM.

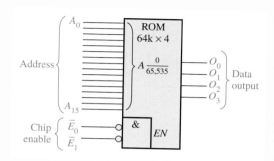

Solution   Two 64k × 4 ROMs are connected as shown in Figure 10–40. Notice that a specific address is accessed in ROM 1 and ROM 2 at the same time. The four bits from a selected address in ROM 1 and the four bits from the corresponding address in ROM 2 go out in parallel to form an 8-bit word on the data bus. Also notice that a LOW on the chip enable line, $\overline{E}$, which forms a simple control bus, enables *both* memories.

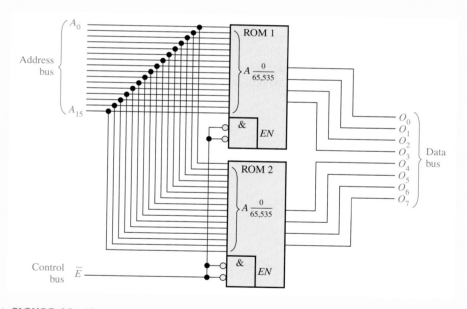

▲ **FIGURE 10–40**

Related Problem   Describe how you would expand a 64k × 1 ROM to a 64k × 8 ROM.

EXAMPLE 10–3

Use the memories in Example 10–2 to form a 64k × 16 ROM.

Solution   In this case you need a memory that stores 65,536 16-bit words. Four 64k × 4 ROMs are required to do the job, as shown in Figure 10–41.

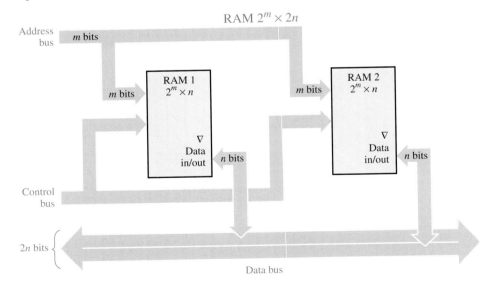

▲ FIGURE 10–41

Related Problem   How many 64k × 1 ROMs would be required to implement the memory shown in Figure 10–41?

A ROM has only data outputs, but a RAM has both data inputs and data outputs. For word-length expansion in a RAM (SRAM or DRAM), the data inputs *and* data outputs form the data bus. Because the same lines are used for data input and data output, tristate buffers are required. Most RAMs provide internal tristate circuitry. Figure 10–42 illustrates RAM expansion to increase the data word length.

▶ FIGURE 10–42

Illustration of word-length expansion with two $2^m \times n$ RAMs forming a $2^m \times 2n$ RAM.

EXAMPLE 10–4

Use 1M × 4 SRAMs to create a 1M × 8 SRAM.

Solution   Two 1M × 4 SRAMs are connected as shown in the simplified block diagram of Figure 10–43.

Related Problem   Use 1M × 8 SRAMs to create a 1M × 16 SRAM.

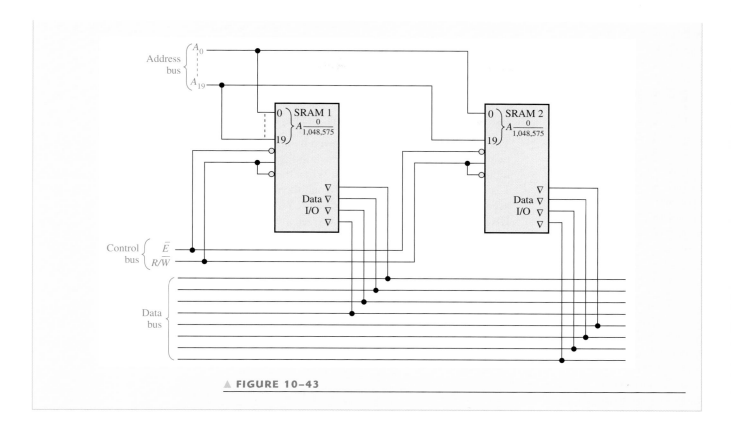

▲ FIGURE 10–43

Word-Capacity Expansion

When memories are expanded to increase the **word capacity,** the *number of addresses is increased.* To achieve this increase, the number of address bits must be increased, as illustrated in Figure 10–44, (where two $1M \times 8$ RAMs are expanded to form a $2M \times 8$ memory).

Each individual memory has 20 address bits to select its 1,048,576 addresses, as shown in part (a). The expanded memory has 2,097,152 addresses and therefore requires 21 address

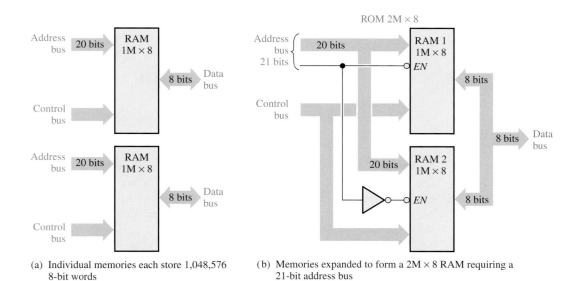

(a) Individual memories each store 1,048,576 8-bit words

(b) Memories expanded to form a $2M \times 8$ RAM requiring a 21-bit address bus

▲ FIGURE 10–44

Illustration of word-capacity expansion.

bits, as shown in part (b). The twenty-first address bit is used to enable the appropriate memory chip. The data bus for the expanded memory remains eight bits wide. Details of this expansion are illustrated in Example 10–5.

EXAMPLE 10–5

Use $512k \times 4$ RAMs to implement a $1M \times 4$ memory.

Solution   The expanded addressing is achieved by connecting the chip enable $(\overline{E}_0)$ input to the twentieth address bit (A_{19}), as shown in Figure 10–45. Input $\overline{E}_1$ is used as an enable input common to both memories. When the twentieth address bit (A_{19}) is LOW, RAM 1 is selected (RAM 2 is disabled), and the nineteen lower-order address bits $(A_0 - A_{18})$ access each of the addresses in RAM 1. When the twentieth address bit (A_{19}) is HIGH, RAM 2 is enabled by a LOW on the inverter output (RAM 1 is disabled), and the nineteen lower-order address bits $(A_0 - A_{18})$ access each of the RAM 2 addresses.

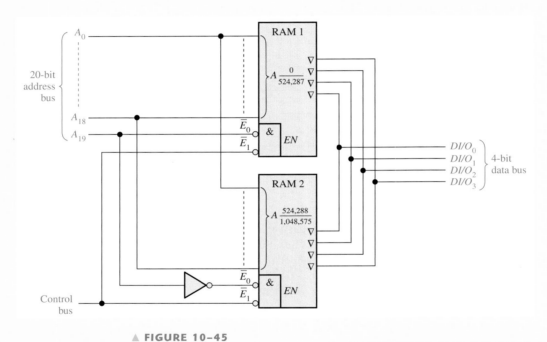

▲ FIGURE 10–45

Related Problem   What are the ranges of addresses in RAM 1 and in RAM 2 in Figure 10–45?

Memory Modules

RAMs are commonly supplied as single in-line memory modules (**SIMMs**) or as dual in-line memory modules (**DIMMs**). SIMMs and DIMMs are small circuit boards on which memory chips (ICs) are mounted with the inputs and outputs connected to an edge connector on the bottom of the board. DIMMs are generally faster, but they can only be installed in machines that are designed for them.

Two classifications of SIMMs are the 30-pin and the 72-pin. These are illustrated in Figure 10–46. Although memory capacities for SIMMs vary anywhere from 256 kB to 32 MB, the key difference in the two pin configurations is the size of the data path. Generally, 30-pin SIMMs are designed for 8-bit data buses, and more SIMMs are required for handling more data bits. The 72-pin SIMMs can accommodate a 32-bit data bus, so for 64-bit data buses a pair of SIMMs is required.

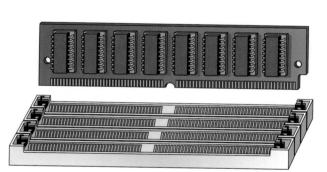

◀ **FIGURE 10-46**

30-pin and 72-pin SIMMs.

DIMMs look similar to SIMMs but provide an increase in memory density with only a relatively slight increase in physical size. The key difference is that DIMMs distribute the input and output pins on both sides of the PC board, whereas SIMMs use only one side. Common DIMM configurations are 72-pin, 100-pin, 144-pin, and 168-pin that accommodate both 32-bit and 64-bit data paths. Generally, DIMM capacities range from 4 MB to 512 MB.

SIMMs and DIMMs plug into sockets on a system board such as those illustrated in Figure 10–47 where several sockets are generally available for memory expansion. The sockets for SIMMs and DIMMs, of course, are different and not interchangeable.

Another standard memory module, similar to the DIMM but with a higher speed bus, is the RIMM (rambus in-line memory module). Also, many laptop computers use a variation of the DIMM called the SODIMM, which is smaller in size, has 144 pins, and has up to a 256 MB capacity.

◀ **FIGURE 10-47**

A SIMM/DIMM is inserted into a socket on a system board.

Memory components are extremely sensitive to static electricity. Use the following precautions when handling memory chips or modules such as SIMMs and DIMMs:

- Before handling, discharge your body's static charge by touching a grounded surface or wear a grounding wrist strap containing a high-value resistor if available. A convenient, reliable ground is the ac outlet ground.
- Do not remove components from their antistatic bags until you are ready to install them.
- Do not lay parts on the antistatic bags because only the inside is antistatic.
- When handling SIMMs or DIMMs, hold by the edges or the metal mounting bracket. Do not touch components on the boards or the edge connector pins.
- Never slide any part over any type of surface.
- Avoid plastic, vinyl, styrofoam, and nylon in the work area.

When installing SIMMs or DIMMs, follow these steps:

1. Line up the notches on the SIMM or DIMM board with the notches in the memory socket.
2. Push firmly on the module until it is securely seated in the socket.
3. Generally, the latches on both sides of the socket will snap into place when the module is completely inserted. These latches also release the module, so it can be removed from the socket.

**SECTION 10-6
REVIEW**

1. How many 16k × 1 RAMs are required to achieve a memory with a word capacity of 16k and a word length of eight bits?
2. To expand the 16k × 8 memory in question 1 to a 32k × 8 organization, how many more 16k × 1 RAMs are required?
3. What does SIMM stand for?
4. What does DIMM stand for?
5. What does the term RIMM stand for?

10-7 SPECIAL TYPES OF MEMORIES

In this section, the first in–first out (FIFO) memory, the last in–first out (LIFO) memory, the memory stack, and the charge-coupled device memory are covered.

After completing this section, you should be able to

■ Describe a FIFO memory ■ Describe a LIFO memory ■ Discuss memory stacks ■ Explain how to use a portion of RAM as a memory stack ■ Describe a basic CCD memory

First In–First Out (FIFO) Memories

This type of memory is formed by an arrangement of shift registers. The term **FIFO** refers to the basic operation of this type of memory, in which the first data bit written into the memory is the first to be read out.

One important difference between a conventional shift register and a FIFO register is illustrated in Figure 10–48. In a conventional register, a data bit moves through the register only as new data bits are entered; in a FIFO register, a data bit immediately goes through the register to the right-most bit location that is empty.

Conventional shift register					
Input	X	X	X	X	Output
0	0	X	X	X	→
1	1	0	X	X	→
1	1	1	0	X	→
0	0	1	1	1	→

FIFO shift register					
Input	—	—	—	—	Output
0	—	—	—	0	→
1	—	—	1	0	→
1	—	1	1	0	→
0	0	1	1	0	→

X = unknown data bits.
In a conventional shift register, data stay to the left until "forced" through by additional data.

— = empty positions.
In a FIFO shift register, data "fall" through (go right).

▲ **FIGURE 10–48**

Comparison of conventional and FIFO register operation.

Figure 10–49 is a block diagram of a FIFO serial memory. This particular memory has four serial 64-bit data registers and a 64-bit control register (marker register). When data are entered by a shift-in pulse, they move automatically under control of the marker register to the empty location closest to the output. Data cannot advance into occupied positions. However, when a data bit is shifted out by a shift-out pulse, the data bits remaining in the

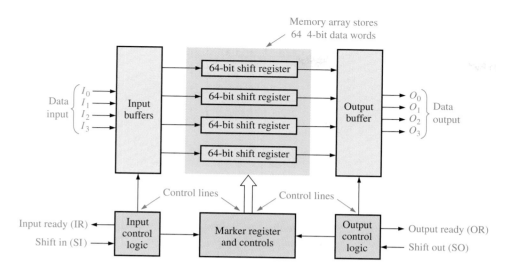

registers automatically move to the next position toward the output. In an asynchronous FIFO, data are shifted out independent of data entry, with the use of two separate clocks.

FIFO Applications

One important application area for the FIFO register is the case in which two systems of differing data rates must communicate. Data can be entered into a FIFO register at one rate and taken out at another rate. Figure 10–50 illustrates how a FIFO register might be used in these situations.

(a) Irregular telemetry data can be stored and retransmitted at a constant rate.

◀ FIGURE 10–50

Examples of the FIFO register in data-rate buffering applications.

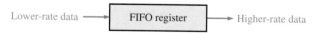

(b) Data input at a slow keyboard rate can be stored and then transferred at a higher rate for processing.

(c) Data input at a constant rate can be stored and then output in bursts.

(d) Data in bursts can be stored and reformatted into a constant-rate output.

Last In–First Out (LIFO) Memories

The **LIFO** (last in–first out) memory is found in applications involving microprocessors and other computing systems. It allows data to be stored and then recalled in reverse order; that is, the last data byte to be stored is the first data byte to be retrieved.

Register Stacks   A LIFO memory is commonly referred to as a push-down stack. In some systems, it is implemented with a group of registers as shown in Figure 10–51. A stack can consist of any number of registers, but the register at the top is called the *top-of-stack.*

▶ **FIGURE 10-51**

Register stack.

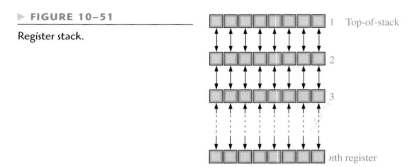

To illustrate the principle, a byte of data is loaded in parallel onto the top of the stack. Each successive byte pushes the previous one down into the next register. This process is illustrated in Figure 10–52. Notice that the new data byte is always loaded into the top register and the previously stored bytes are pushed deeper into the stack. The name *push-down stack* comes from this characteristic.

▶ **FIGURE 10-52**

Simplified illustration of pushing data onto the stack.

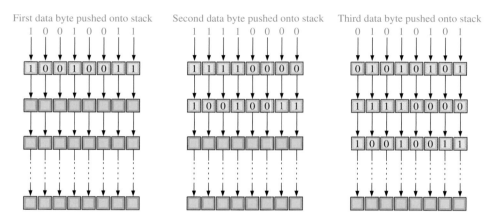

Data bytes are retrieved in the reverse order. The last byte entered is always at the top of the stack, so when it is pulled from the stack, the other bytes pop up into the next higher locations. This process is illustrated in Figure 10–53.

▶ **FIGURE 10-53**

Simplified illustration of pulling data from the stack.

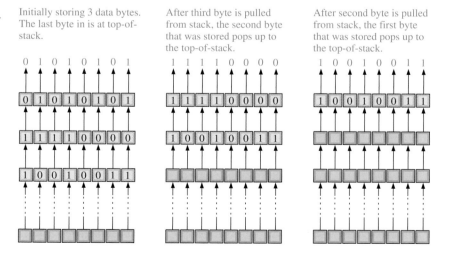

RAM Stack Another approach to LIFO memory used in microprocessor-based systems is the allocation of a section of RAM as the stack rather than the use of a dedicated set of registers. As you have seen, for a register stack the data moves up or down from one location to the next. In a RAM stack, the data itself does not move but the top-of-stack moves under control of a register called the stack pointer.

Consider a random-access memory that is byte organized—that is, one in which each address contains eight bits—as illustrated in Figure 10–54. The binary address 0000000000001111, for example, can be written as 000F in hexadecimal. A 16-bit address can have a *minimum* hexadecimal value of 0000_{16} and a *maximum* value of $FFFF_{16}$. With this notation, a 64 kB memory array can be represented as shown in Figure 10–54. The lowest memory address is 0000_{16} and the highest memory address is $FFFF_{16}$.

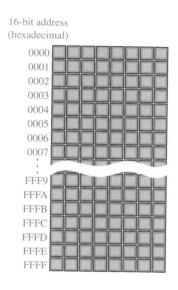

16-bit address
(hexadecimal)

0000
0001
0002
0003
0004
0005
0006
0007
FFF9
FFFA
FFFB
FFFC
FFFD
FFFE
FFFF

◀ **FIGURE 10–54**

Representation of a 64 kB memory with the 16-bit addresses expressed in hexadecimal.

Now, consider a section of RAM set aside for use as a stack. A special separate register, the stack pointer, contains the address of the top of the stack, as illustrated in Figure 10–55. A 4-digit hexadecimal representation is used for the binary addresses. In the figure, the addresses are chosen for purposes of illustration.

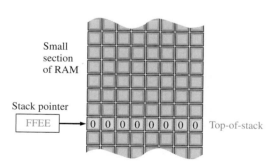

(a) The stack pointer is initially at FFEE before the data word 0001001000110100 (1234) is pushed onto the stack.

(b) The stack pointer is decremented by two and the data word 0001001000110100 is placed in the two locations prior to the original stack pointer location.

▲ **FIGURE 10–55**

Illustration of the PUSH operation for a RAM stack.

Now let's see how data are pushed onto the stack. The stack pointer is initially at address $FFEE_{16}$, which is the top of the stack as shown in Figure 10–55(a). The stack pointer is then decremented (decreased) by two to $FFEC_{16}$. This moves the top of the stack to a lower memory address, as shown in Figure 10–55(b). Notice that the top of the stack is not stationary as in the fixed register stack but moves downward (to lower addresses) in the RAM as data words are stored. Figure 10–55(b) shows that two bytes (one data word) are then pushed onto the stack. After the data word is stored, the top of the stack is at $FFEC_{16}$.

Figure 10–56 illustrates the POP operation for the RAM stack. The last data word stored in the stack is read first. The stack pointer that is at FFEC is incremented (increased) by two to address $FFEE_{16}$ and a POP operation is performed as shown in part (b). Keep in mind that RAMs are nondestructive when read, so the data word still remains in the memory after a POP operation. A data word is destroyed only when a new word is written over it.

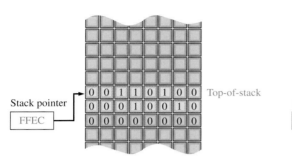

(a) The stack pointer is at FFEC before the data word is copied (popped) from the stack.

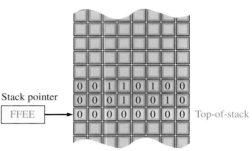

(b) The stack pointer is incremented by two and the last data word stored is copied (popped) from the stack.

▲ **FIGURE 10–56**

Illustration of the POP operation for the RAM stack.

A RAM stack can be of any depth, depending on the number of continuous memory addresses assigned for that purpose.

CCD Memories

The **CCD** (charge-coupled device) memory stores data as charges on capacitors. Unlike the DRAM, however, the storage cell does not include a transistor. High density is the main advantage of CCDs.

The CCD memory consists of long rows of semiconductor capacitors, called *channels*. Data are entered into a channel serially by depositing a small charge for a 0 and a large charge for a 1 on the capacitors. These charge packets are then shifted along the channel by clock signals as more data are entered.

As with the DRAM, the charges must be refreshed periodically. This process is done by shifting the charge packets serially through a refresh circuit. Figure 10–57 shows the basic concept of a CCD channel. Because data are shifted serially through the channels, the CCD memory has a relatively long access time. CCD arrays are used in some modern cameras to capture video images in the form of light-induced charge.

▷ **FIGURE 10–57**

A CCD (charge-coupled device) channel.

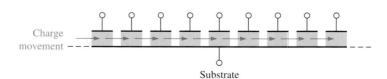

Charge movement

Substrate

SECTION 10–7 REVIEW

1. What is a FIFO memory?
2. What is a LIFO memory?
3. Explain the PUSH operation in a memory stack.
4. Explain the POP operation in a memory stack.
5. What does the term *CCD* stand for?

10–8 MAGNETIC AND OPTICAL STORAGE

In this section, the basics of magnetic disks, magnetic tape, magneto-optical disks, and optical disks are introduced. These storage media are very important, particularly in computer applications, where they are used for mass nonvolatile storage of data and programs.

After completing this section, you should be able to

■ Describe a magnetic hard disk ■ Describe a floppy disk ■ Discuss removable hard disks ■ Explain the principle of magneto-optical disks ■ Discuss the CD-ROM, CD-R, and CD-RW disks ■ Describe the WORM ■ Discuss the DVD-ROM

Magnetic Storage

Magnetic Hard Disks　Computers use hard disks as the internal mass storage media. **Hard disks** are rigid "platters" made of aluminum alloy or a mixture of glass and ceramic covered with a magnetic coating. Hard disk drives mainly come in two diameter sizes, 5.25 in. and 3.5 in. although 2.5 in. and 1.75 in. are also available. A hard disk drive is hermetically sealed to keep the disks dust-free.

Typically, two or more platters are stacked on top of each other on a common shaft or spindle that turns the assembly at several thousand rpm. A separation between each disk allows for a magnetic read/write head that is mounted on the end of an actuator arm, as shown in Figure 10–58. There is a read/write head for both sides of each disk since data are recorded on both sides of the disk surface. The drive actuator arm synchronizes all the read/write heads to keep them in perfect alignment as they "fly" across the disk surface with a separation of only a fraction of a millimeter from the disk. A small dust particle could cause a head to "crash," causing damage to the disk surface.

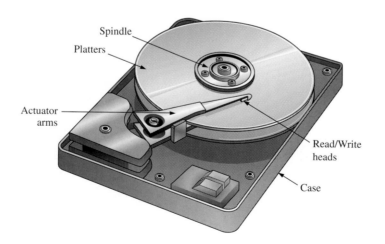

Spindle

Platters

Actuator arms

Read/Write heads

Case

◀ FIGURE 10–58

A hard disk drive.

Basic Read/Write Head Principles　The hard drive is a random-access device because it can retrieve stored data anywhere on the disk in any order. A simplified diagram of the magnetic surface read/write operation is shown in Figure 10–59. The direction or polarization of the magnetic domains on the disk surface is controlled by the direction of the magnetic flux lines (magnetic field) produced by the write head according to the direction of a current pulse in the winding. This magnetic flux magnetizes a small spot on the disk surface in the direction of the magnetic field. A magnetized spot of one polarity represents a binary 1, and one of the opposite polarity represents a binary 0. Once a spot on the disk surface is magnetized, it remains until written over with an opposite magnetic field.

▶ FIGURE 10-59

Simplified read/write head operation.

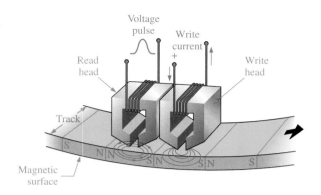

COMPUTER NOTE

Data are stored on a hard drive in the form of files. Keeping track of the location of files is the job of the device driver that manages the hard drive (sometimes referred to as hard drive BIOS). The device driver and the computer's operating system can access two tables to keep track of files and file names. The first table is called the FAT (File Allocation Table). The FAT shows what is assigned to specific files and keeps a record of open sectors and bad sectors. The second table is the Root Directory which has file names, type of file, time and date of creation, starting cluster number, and other information about the file.

When the magnetic surface passes a read head, the magnetized spots produce magnetic fields in the read head, which induce voltage pulses in the winding. The polarity of these pulses depends on the direction of the magnetized spot and indicates whether the stored bit is a 1 or a 0. The read and write heads are usually combined in a single unit.

Hard Disk Format   A hard disk is organized or formatted into tracks and sectors, as shown in Figure 10–60(a). Each track is divided into a number of sectors, and each track and sector has a physical address that is used by the operating system to locate a particular data record. Hard disks typically have from a few hundred to thousands of tracks. As you can see in the figure, there is a constant number of tracks/sector, with outer sectors using more surface area than the inner sectors. The arrangement of tracks and sectors on a disk is known as the *format.*

A hard disk stack is illustrated in Figure 10–60(b). Hard disk drives differ in the number of platters in a stack, but there is always a minimum of two. All of the same corresponding tracks on each platter are collectively known as a cylinder, as indicated.

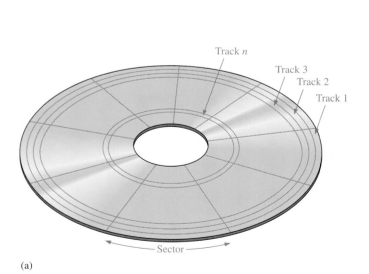

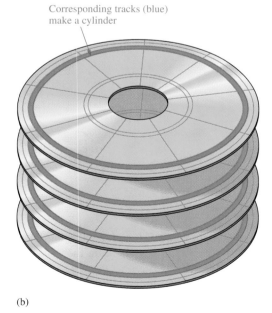

Corresponding tracks (blue) make a cylinder

Track *n*

Track 3
Track 2
Track 1

Sector

(a)

(b)

▲ FIGURE 10-60

Hard disk organization and formatting.

Hard Disk Performance Several basic parameters determine the performance of a given hard disk drive. A *seek* operation is the movement of the read/write head to the desired track. The **seek time** is the average time for this operation to be performed. Typically, hard disk drives have an average seek time of several milliseconds, depending on the particular drive.

The **latency period** is the time it takes for the desired sector to spin under the head once the head is positioned over the desired track. A worst case is when the desired sector is just past the head position and spinning away from it. The sector must rotate almost a full revolution back to the head position. *Average latency period* assumes that the disk must make half of a revolution. Obviously, the latency period depends on the constant rotational speed of the disk. Disk rotation speeds are different for different disk drives but typically are 3600 rpm, 4500 rpm, 5400 rpm, and 7200 rpm. Some disk drives rotate at 10,033 rpm and have an average latency period of less than 3 ms.

The sum of the average seek time and the average latency period is the *access time* for the disk drive.

Floppy Disks The floppy disk is an older technology and derives its name because it is made of a flexible polyester material with a magnetic coating on both sides. The early floppy disks were 5.25 inches in diameter and were packaged in a semiflexible jacket. Current **floppy disks** or diskettes are 3.5 inches in diameter and are encased in a rigid plastic jacket, as shown in Figure 10–61. A spring-loaded shutter covers the access window and remains closed until the disk is inserted into a disk drive. A metal hub has one hole to center the disk and another for spinning it within the protective jacket. Obviously, floppy disks are removable disks, whereas hard disks are not. Floppy disks are formatted into tracks and sectors similar to hard disks except for the number of tracks and sectors. The high-density 1.44 MB floppies have 80 tracks per side with 18 sectors.

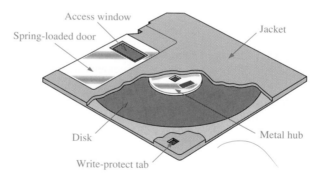

Access window
Spring-loaded door
Jacket
Disk
Metal hub
Write-protect tab

◄ **FIGURE 10–61**

The 3.5 inch floppy disk (diskette).

*Zip*TM The Zip drive is one type of removable magnetic storage device that has replaced the limited-capacity floppy. Like the floppy disk, the **Zip disk** cartridge is a flexible disk housed in a rigid case about the same size as that of the floppy disk but thicker. The typical Zip drive is much faster than the floppy drive because it has a 3000 rpm spin rate compared to the floppy's 300 rpm. The Zip drive has a storage capacity of up to 250 MB, over 173 times more than the 1.44 MB floppy.

*Jaz*TM Another type of removable magnetic storage device is the Jaz drive, which is similar to a hard disk drive except that two platters are housed in a removable cartridge protected by a dust-proof shutter. The **Jaz cartridges** are available with storage capacities of 1 or 2 GB.

Removable Hard Disk In addition to the popular Zip and Jaz removable drives, a removable hard disk drive with capacities of from 80 GB to 250 GB is available. Keep in mind that the technology is changing so rapidly that there most likely will be further advancements at the time you are reading this.

Magnetic Tape   Tape is used for backup data from mass storage devices and typically is slower than disks because data on tape is accessed serially rather than randomly. There are several types that are available, including QIC, DAT, 8 mm, and DLT.

QIC is an abbreviation for quarter-inch cartridge and looks much like audio tape cassettes with two reels inside. Various QIC standards have from 36 to 72 tracks that can store from 80 MB to 1.2 GB. More recent innovations under the Travan standard have lengthened the tape and increased its width allowing storage capacities up to 4 GB. QIC tape drives use read/write heads that have a single write head with a read head on each side. This allows the tape drive to verify data just written when the tape is running in either direction. In the record mode, the tape moves past the read/write heads at approximately 100 inches/second, as indicated in Figure 10–62.

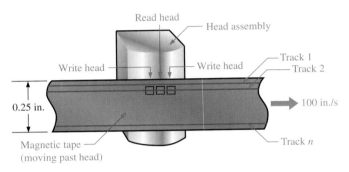

▲ **FIGURE 10–62**

QIC tape.

DAT, which is an abbreviation for digital audio tape, uses a technique called helical scan recording. DATs offer storage capacities ranging up to 12 GB but is more expensive than QIC.

A third type of tape format, the 8 mm tape, was originally designed for the video industry but has been adopted by the computer industry as a reliable way to store large amounts of computer data. 8 mm is similar to DAT but offers storage capacities up to 25 GB.

DLT is an abbreviation for digital linear tape. DLT is a half-inch wide tape, which is 60% wider than 8 mm and, of course, twice as wide as standard QIC. Basically, DLT differs in the way the tape-drive mechanism works to minimize tape wear compared to other systems. DLT offers the highest storage capacity of all the tape formats with capacities ranging up to 35 GB.

Magneto-Optical Storage

As the name implies, magneto-optical (MO) storage devices use a combination of magnetic and optical (laser) technologies. A **magneto-optical disk** is formatted into tracks and sectors similar to magnetic disks.

The basic difference between a purely magnetic disk and an MO disk is that the magnetic coating used on the MO disk requires heat to alter the magnetic polarization. Therefore, the MO is extremely stable at ambient temperature, making data unchangeable. To write a data bit, a high-power laser beam is focused on a tiny spot on the disk, and the temperature of that tiny spot is raised above a temperature level called the Curie point (about 200°C). Once heated, the magnetic particles at that spot can easily have their direction (polarization) changed by a magnetic field generated by the write head. Information is read from the disk with a less-powerful laser than used for writing, making use of the Kerr effect where the polarity of the reflected laser light is altered depending on the orientation of the magnetic particles. Spots of one polarity represent 0s and spots of the opposite polarity represent 1s. Basic MO operation is shown in Figure 10–63, which represents a small cross-sectional area of a disk.

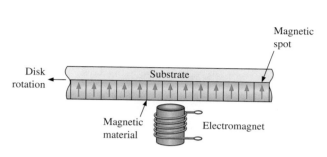

(a) Unrecorded disk

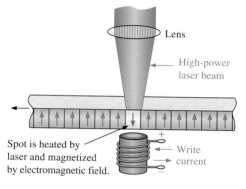

(b) Writing: A high-power laser beam heats the spot, causing the magnetic particles to align with the electromagnetic field.

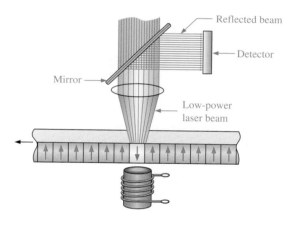

(c) Reading: A low-power laser beam reflects off of the reversed-polarity magnetic particles and its polarization shifts. If the particles are not reversed, the polarization of the reflected beam is unchanged.

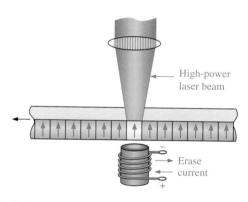

(d) Erasing: The electromagnetic field is reversed as the high-power laser beam heats the spot, causing the magnetic particles to be restored to the original polarity.

▲ FIGURE 10–63

Basic principle of a magneto-optical disk.

Optical Storage

CD-ROM   The basic Compact Disk–Read-Only Memory is a 120 mm diameter disk with a sandwich of three coatings: a polycarbonate plastic on the bottom, a thin aluminum sheet for reflectivity, and a top coating of lacquer for protection. The **CD-ROM** disk is formatted in a single spiral track with sequential 2 kB sectors and has a capacity of 680 MB. Data are prerecorded at the factory in the form of minute indentations called *pits* and the flat area surrounding the pits called *lands*. The pits are stamped into the plastic layer and cannot be erased.

A CD player reads data from the spiral track with a low-power infrared laser, as illustrated in Figure 10–64. The data are in the form of pits and lands as shown. Laser light reflected from a pit is 180° out-of-phase with the light reflected from the lands. As the disk rotates, the narrow laser beam strikes the series of pits and lands of varying lengths, and a photodiode detects the difference in the reflected light. The result is a series of 1s and 0s corresponding to the configuration of pits and lands along the track.

▶ FIGURE 10–64

Basic operation of reading data from a CD-ROM.

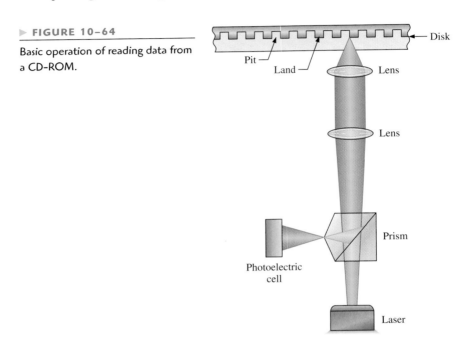

WORM   Write Once/Read Many (**WORM**) is a type of optical storage that can be written onto one time after which the data cannot be erased but can be read many times. To write data, a low-power laser is used to burn microscopic pits on the disk surface. 1s and 0s are represented by the burned and nonburned areas.

CD-R   This is essentially a type of WORM. The difference is that the CD-Recordable allows multiple write sessions to different areas of the disk. The **CD-R** disk has a spiral track like the CD-ROM, but instead of mechanically pressing indentations on the disk to represent data, the CD-R uses a laser to burn microscopic spots into an organic dye surface. When heated beyond a critical temperature with a laser during read, the burned spots change color and reflect less light than the nonburned areas. Therefore, 1s and 0s are represented on a CD-R by burned and nonburned areas, whereas on a CD-ROM they are represented by pits and lands. Like the CD-ROM, the data cannot be erased once it is written.

CD-RW   The CD-Rewritable disk can be used to read and write data. Instead of the dye-based recording layer in the CD-R, the **CD-RW** commonly uses a crystalline compound with a special property. When it is heated to a certain temperature, it becomes crystalline when it cools; but if it is heated to a certain higher temperature, it melts and becomes amorphous when it cools. To write data, the focused laser beam heats the material to the melting temperature resulting in an amorphous state. The resulting amorphous areas reflect less light than the crystalline areas, allowing the read operation to detect 1s and 0s. The data can be erased or overwritten by heating the amorphous areas to a temperature above the crystallization temperature but lower than the melting temperature that causes the amorphous material to revert to a crystalline state.

DVD-ROM Originally DVD was an abbreviation for Digital Video Disk but eventually came to represent *Digital Versatile Disk*. Like the CD-ROM, **DVD-ROM** data are pre-stored on the disk. However, the pit size is smaller than for the CD-ROM, allowing more data to be stored on a track. The major difference between CD-ROM and DVD-ROM is that the CD is single-sided, while the DVD has data on both sides. Also, in addition to double-sided DVD disks, there are also multiple-layer disks that use semitransparent data layers placed over the main data layers, providing storage capacities of tens of gigabytes. To access all the layers, the laser beam requires refocusing going from one layer to the other.

**SECTION 10–8
REVIEW**

1. List the major types of magnetic storage.
2. What is the storage capacity of floppy disks?
3. Generally, how is a magnetic disk organized?
4. How are data written on and read from a magneto-optical disk?
5. List the types of optical storage.

10–9 TROUBLESHOOTING

Because memories can contain large numbers of storage cells, testing each cell can be a lengthy and frustrating process. Fortunately, memory testing is usually an automated process performed with a programmable test instrument or with the aid of software for in-system testing. Most microprocessor-based systems provide automatic memory testing as part of their system software.

After completing this section, you should be able to

■ Discuss the checksum method of testing ROMs ■ Discuss the checkerboard pattern method of testing RAMs

ROM Testing

Since ROMs contain known data, they can be checked for the correctness of the stored data by reading each data word from the memory and comparing it with a data word that is known to be correct. One way of doing this is illustrated in Figure 10–65. This process

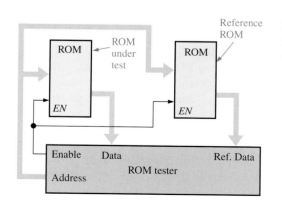

◄ **FIGURE 10–65**

Block diagram for a complete contents check of a ROM.

requires a reference ROM that contains the same data as the ROM to be tested. A special test instrument is programmed to read each address in both ROMs simultaneously and to compare the contents. A flowchart in Figure 10–66 illustrates the basic sequence.

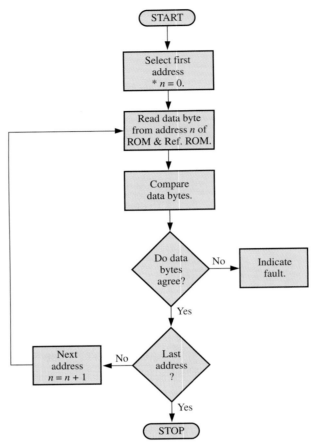

* n is the address number.

▲ FIGURE 10–66

Flowchart for a complete contents check of a ROM.

Checksum Method   Although the previous method checks each ROM address for correct data, it has the disadvantage of requiring a reference ROM for each different ROM to be tested. Also, a failure in the reference ROM can produce a fault indication.

In the checksum method a number, the sum of the contents of all the ROM addresses, is stored in a designated ROM address when the ROM is programmed. To test the ROM, the contents of all the addresses except the checksum are added, and the result is compared with the checksum stored in the ROM. If there is a difference, there is definitely a fault. If the checksums agree, the ROM is most likely good. However, there is a remote possibility that a combination of bad memory cells could cause the checksums to agree.

This process is illustrated in Figure 10–67 with a simple example. The checksum in this case is produced by taking the sum of each column of data bits and discarding the carries. This is actually an XOR sum of each column. The flowchart in Figure 10–68 illustrates the basic checksum test.

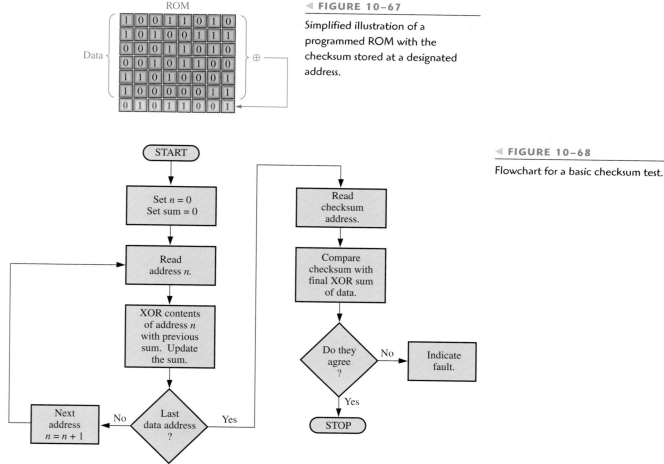

◀ **FIGURE 10–67**

Simplified illustration of a programmed ROM with the checksum stored at a designated address.

◀ **FIGURE 10–68**

Flowchart for a basic checksum test.

The checksum test can be implemented with a special test instrument, or it can be incorporated as a test routine in the built-in (system) software or microprocessor-based systems. In that case, the ROM test routine is automatically run on system start-up.

RAM Testing

To test a RAM for its ability to store both 0s and 1s in each cell, first 0s are written into all the cells in each address and then read out and checked. Next, 1s are written into all the cells in each address and then read out and checked. This basic test will detect a cell that is stuck in either a 1 state or a 0 state.

Some memory faults cannot be detected with the all-0s–all-1s test. For example, if two adjacent memory cells are shorted, they will always be in the same state, both 0s or both 1s. Also, the all-0s–all-1s test is ineffective if there are internal noise problems such that the contents of one or more addresses are altered by a change in the contents of another address.

The Checkerboard Pattern Test   One way to more fully test a RAM is by using a checkerboard pattern of 1s and 0s, as illustrated in Figure 10–69. Notice that all adjacent cells have opposite bits. This pattern checks for a short between two adjacent cells; if there is a short, both cells will be in the same state.

After the RAM is checked with the pattern in Figure 10–69(a), the pattern is reversed, as shown in part (b). This reversal checks the ability of all cells to store both 1s and 0s.

▶ FIGURE 10–69

The RAM checkerboard test pattern.

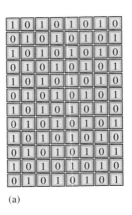

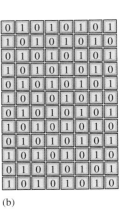

(a)   (b)

A further test is to alternate the pattern one address at a time and check all the other addresses for the proper pattern. This test will catch a problem in which the contents of an address are dynamically altered when the contents of another address change.

A basic procedure for the checkerboard test is illustrated by the flowchart in Figure 10–70. The procedure can be implemented with the system software in microprocessor-

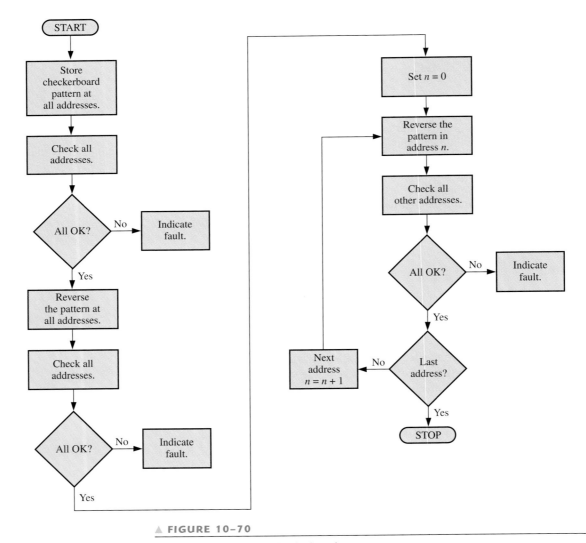

▲ FIGURE 10–70

Flowchart for basic RAM checkerboard test.

based systems so that either the tests are automatic when the system is powered up or they can be initiated from the keyboard.

DIGITAL SYSTEM APPLICATION

In this application, the memory logic for the security system that was introduced in Chapter 9 is developed. The security code logic that was completed in Chapter 9 will be combined with the memory logic to form the complete system.

General Operation

The block diagram of the complete security system is shown in Figure 10–71. The memory logic stores a 4-digit security code in BCD format. In the *disarm* mode, four digits are entered into the memory from the keypad. Once stored in the memory, the four BCD digits become the permanent security code for entry. If it is necessary to change the code, the memory is reprogrammed with a new one.

The memory is programmed by first putting the system into the *disarm* mode and using the store switch and keypad to enter the desired 4-digit code. This is a memory *write* operation. Once the memory is programmed with the security code, the *Arm/Disarm* switch is changed to *arm* mode, which sets the memory up for

read operations. A block diagram of the memory logic is shown in Figure 10–72.

The Memory Cell

The memory requires sixteen cells to store the four BCD digits of the security code. One possible design for a memory cell is shown in Figure 10–73. A J-K flip-flop is used as the basic storage element and can be operated in two modes (*read* and *write*). In the *write* mode, *AddSel* (address select) is HIGH and the $R/\overline{W}$ (read/write) input is LOW. Gates G1 and G2 are enabled, the input bit is applied to the *J* input, and its complement is applied to the *K* input. The input bit is then stored on the positive edge of the clock pulse. In the *read* mode, *AddSel* is HIGH and $R/\overline{W}$ is HIGH enabling G3. The stored bit on the *Q* output

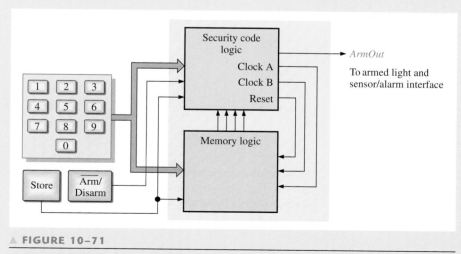

▲ **FIGURE 10–71**

Block diagram of the complete security system.

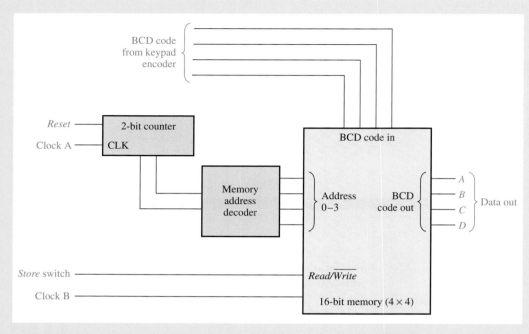

▲ **FIGURE 10–72**

Block diagram of the memory logic.

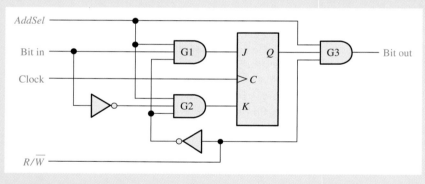

▲ **FIGURE 10–73**

Memory cell logic.

of the flip-flop appears on the output of G3 (Bit out).

The Memory Address Decoder

The memory address decoder logic is shown in Figure 10–74. A 2-bit binary sequence is applied to the select inputs (S_0, S_1) to select each of the four memory addresses using the *AddSel* lines.

The Memory Array

The memory has sixteen cells as shown in Figure 10–75. When one of the rows in the memory is selected by the address decoder and the read/write input is LOW, the 4-bit BCD input code is clocked into the four selected cells. The inputs to the address decoder are sequenced through each of four states (00, 01, 10, and 11) to sequentially select each row in the memory.

Figure 10–76 illustrates the programming of the memory as the security code 4739 is sequentially entered.

▶ **FIGURE 10–74**

The memory address decoder.

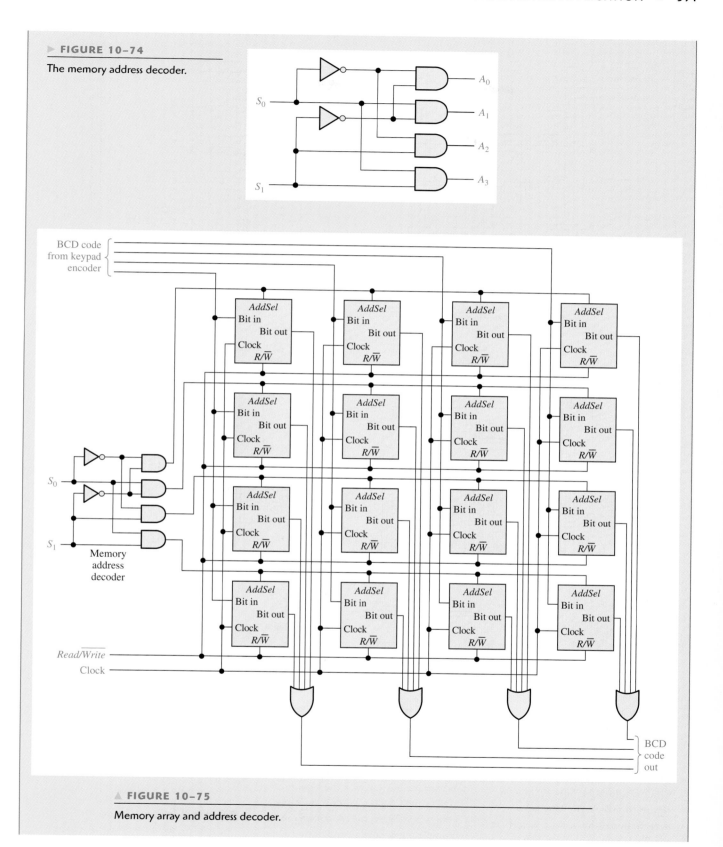

▲ **FIGURE 10–75**

Memory array and address decoder.

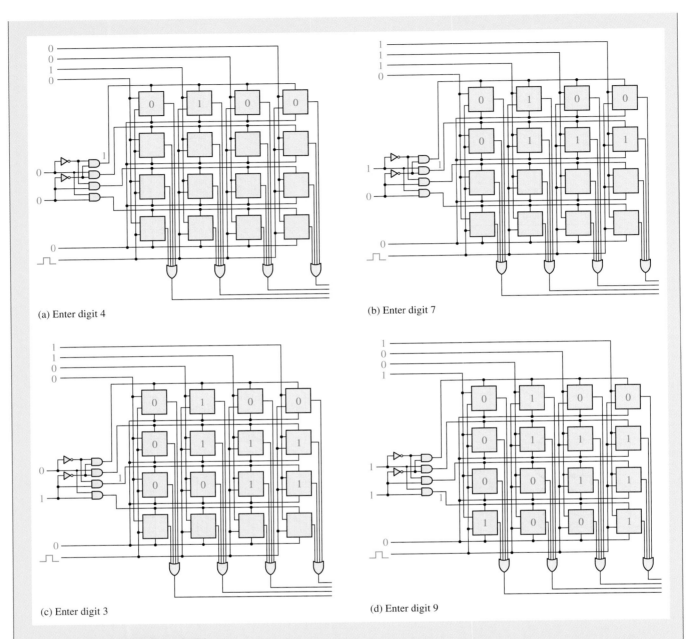

(a) Enter digit 4

(b) Enter digit 7

(c) Enter digit 3

(d) Enter digit 9

▲ FIGURE 10–76

Illustration of entering a security code (4739) into the memory.

The Complete Memory Logic

A keypad encoder is necessary for converting a key stroke to a BCD code, and a 2-bit counter is used to produce the sequence for selecting the memory addresses. This is shown in Figure 10–77. At the beginning of programming, the counter is reset to the 0 state by a reset input from the code entry logic and is advanced through its sequence by each key entry.

The Complete Security System

Now that the memory logic is complete, we can combine it with the security code logic from Chapter 9 shown as a block diagram in Figure 10–78 to form the complete security system shown as a block diagram in Figure 10–79.

System Assignment

- *Activity 1* Explain what the keypad encoder does in the memory.

- *Activity 2* Explain what the 2-bit counter does in the memory.

- *Optional Activity* Construct the complete security entry system using standard 74XX devices and other required components. Test the system.

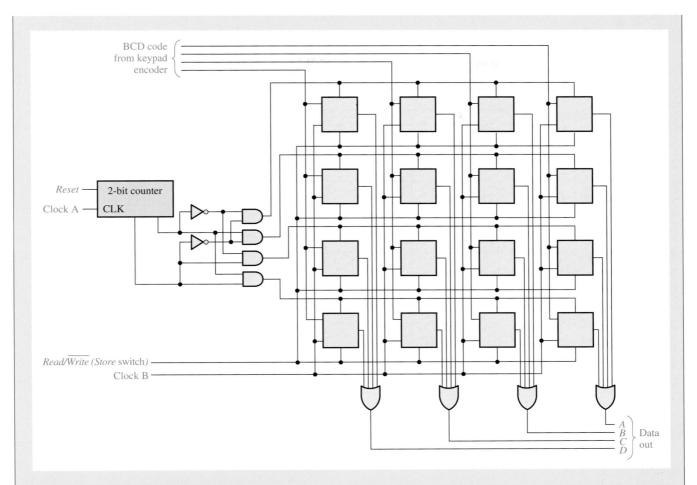

▲ **FIGURE 10–77**

The complete memory logic.

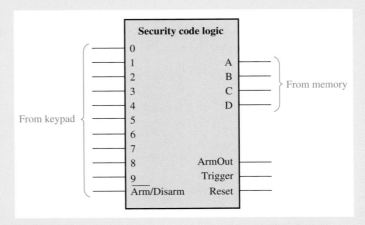

▲ **FIGURE 10–78**

The security code logic (from Chapter 11).

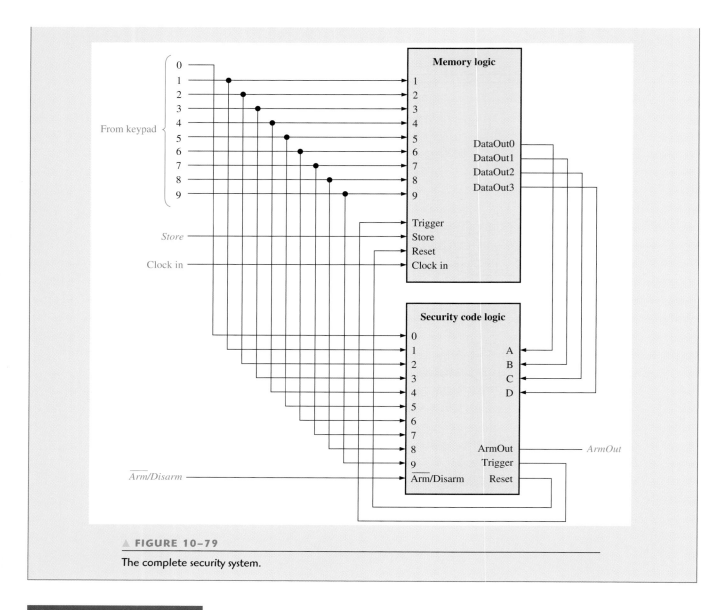

▲ **FIGURE 10-79**

The complete security system.

SUMMARY

■ Types of semiconductor memories:

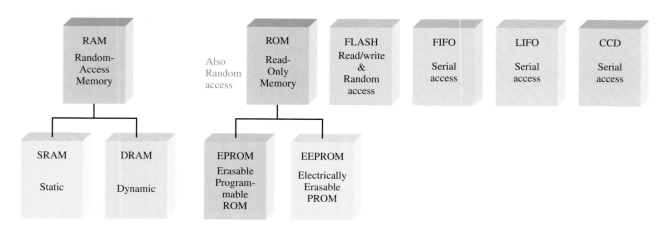

■ Types of SRAMs (Static RAMs) and DRAMs (Dynamic RAMs):

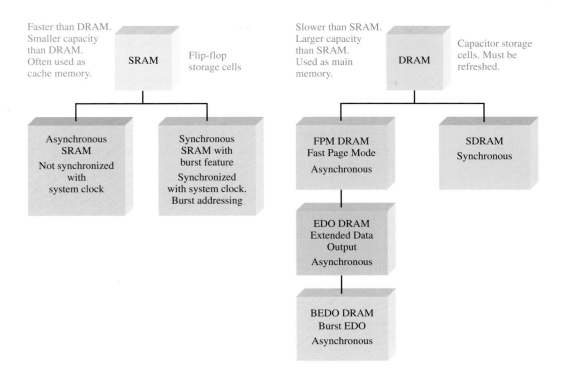

■ Types of magnetic storage:

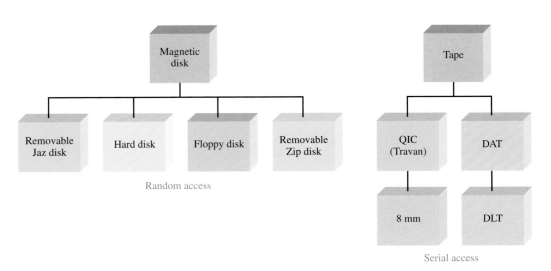

■ Types of optical (laser) storage:

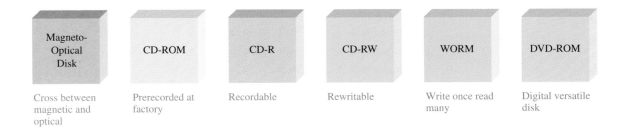

Key terms and other bold terms in the chapter are defined in the end-of-book glossary.

Address The location of a given storage cell or group of cells in a memory.

Bus A set of interconnections that interface one or more devices based on a standardized specification.

Byte A group of eight bits.

Capacity The total number of data units (bits, nibbles, bytes, words) that a memory can store.

Cell A single storage element in a memory.

DRAM Dynamic random-access memory; a type of semiconductor memory that uses capacitors as the storage elements and is a volatile, read/write memory.

EPROM Erasable programmable read-only memory; a type of semiconductor memory device that typically uses ultraviolet light to erase data.

FIFO First in–first out memory.

Flash memory A nonvolatile read/write random-access semiconductor memory in which data are stored as charge on the floating gate of a certain type of FET.

Hard disk A magnetic storage device; typically, a stack of two or more rigid disks enclosed in a sealed housing.

LIFO Last in–first out memory; a memory stack.

PROM Programmable read-only memory; a type of semiconductor memory.

RAM Random-access memory; a volatile read/write semiconductor memory.

Read The process of retrieving data from a memory.

ROM Read-only memory; a nonvolatile random-access semiconductor memory.

SRAM Static random-access memory; a type of volatile read/write semiconductor memory.

Word A complete unit of binary data.

Write The process of storing data in a memory.

Answers are at the end of the chapter.

1. The bit capacity of a memory that has 1024 addresses and can store 8 bits at each address is
   (a) 1024   (b) 8192   (c) 8   (d) 4096

2. A 32-bit data word consists of
   (a) 2 bytes   (b) 4 nibbles
   (c) 4 bytes   (d) 3 bytes and 1 nibble

3. Data are stored in a random-access memory (RAM) during the
   (a) read operation   (b) enable operation
   (c) write operation   (d) addressing operation

4. Data that are stored at a given address in a random-access memory (RAM) is lost when
   (a) power goes off
   (b) the data are read from the address
   (c) new data are written at the address
   (d) answers (a) and (c)

5. A ROM is a
   (a) nonvolatile memory   (b) volatile memory
   (c) read/write memory   (d) byte-organized memory

6. A memory with 256 addresses has
   (a) 256 address lines   (b) 6 address lines
   (c) 1 address line   (d) 8 address lines

7. A byte-organized memory has
   (a) 1 data output line (b) 4 data output lines
   (c) 8 data output lines (d) 16 data output lines

8. The storage cell in a SRAM is
   (a) a flip-flop (b) a capacitor
   (c) a fuse (d) a magnetic domain

9. A DRAM must be
   (a) replaced periodically (b) refreshed periodically
   (c) always enabled (d) programmed before each use

10. A flash memory is
    (a) volatile (b) a read-only memory
    (c) a read/write memory (d) nonvolatile
    (e) answers (a) and (c) (f) answers (c) and (d)

11. Hard disk, floppy disk, Zip disk, and Jaz disk are all
    (a) magneto-optical storage devices
    (b) semiconductor storage devices
    (c) magnetic storage devices
    (d) optical storage devices

12. Optical storage devices employ
    (a) ultraviolet light (b) electromagnetic fields
    (c) optical couplers (d) lasers

PROBLEMS

Answers to odd-numbered problems are at the end of the book.

SECTION 10–1 **Basics of Semiconductor Memory**

1. Identify the ROM and the RAM in Figure 10–80.

▶ FIGURE 10–80

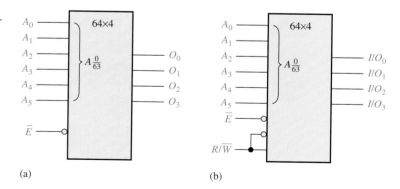

(a)

(b)

2. Explain why RAMs and ROMs are both random-access memories.

3. Explain the purposes of the address bus and the data bus.

4. What memory address (0 through 256) is represented by each of the following hexadecimal numbers:
   (a) $0A_{16}$ (b) $3F_{16}$ (c) CD_{16}

SECTION 10–2 **Random-Access Memories (RAMs)**

5. A static memory array with four rows similar to the one in Figure 10–9 is initially storing all 0s. What is its content after the following conditions? Assume a 1 selects a row.

   Row 0 = 1, Data in (Bit 0) = 1

   Row 1 = 0, Data in (Bit 1) = 1

   Row 2 = 1, Data in (Bit 2) = 1

   Row 3 = 0, Data in (Bit 3) = 0

6. Draw a basic logic diagram for a 512 × 8-bit static RAM, showing all the inputs and outputs.

7. Assuming that a 64k × 8 SRAM has a structure similar to that of the SRAM in Figure 10–11, determine the number of rows and 8-bit columns in its memory cell array.

8. Redraw the block diagram in Figure 10–11 for a 64k × 8 memory.

9. Explain the difference between a SRAM and a DRAM.

10. What is the capacity of a DRAM that has twelve address lines?

SECTION 10–3 **Read-Only Memories (ROMs)**

11. For the ROM array in Figure 10–81, determine the outputs for all possible input combinations, and summarize them in tabular form (Blue cell is a 1, gray cell is a 0).

▶ FIGURE 10–81

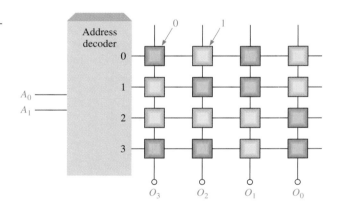

12. Determine the truth table for the ROM in Figure 10–82.

▶ FIGURE 10–82

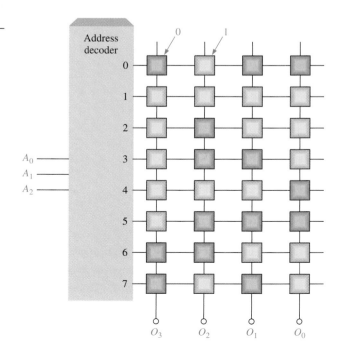

13. Using a procedure similar to that in Example 10–1, design a ROM for conversion of single-digit BCD to excess-3 code.

14. What is the total *bit* capacity of a ROM that has 14 address lines and 8 data outputs?

SECTION 10–4 **Programmable ROMs (PROMs and EPROMs)**

15. Assuming that the PROM matrix in Figure 10–83 is programmed by blowing a fuse link to create a 0, indicate the links to be blown to program an X^3 look-up table, where X is a number from 0 through 7.

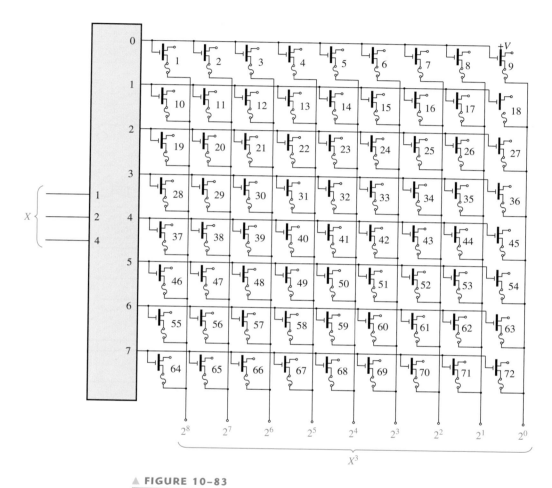

▲ FIGURE 10–83

16. Determine the addresses that are programmed and the contents of each address after the programming sequence in Figure 10–84 has been applied to an EPROM like the one shown in Figure 10–31.

SECTION 10–6 **Memory Expansion**

17. Use 16k × 4 DRAMs to build a 64k × 8 DRAM. Show the logic diagram.

18. Using a block diagram, show how 64k × 1 dynamic RAMs can be expanded to build a 256k × 4 RAM.

19. What is the word length and the word capacity of the memory of Problem 17? Problem 18?

SECTION 10–7 **Special Types of Memories**

20. Complete the timing diagram in Figure 10–85 by showing the output waveforms that are initially all LOW for a FIFO serial memory like that shown in Figure 10–49.

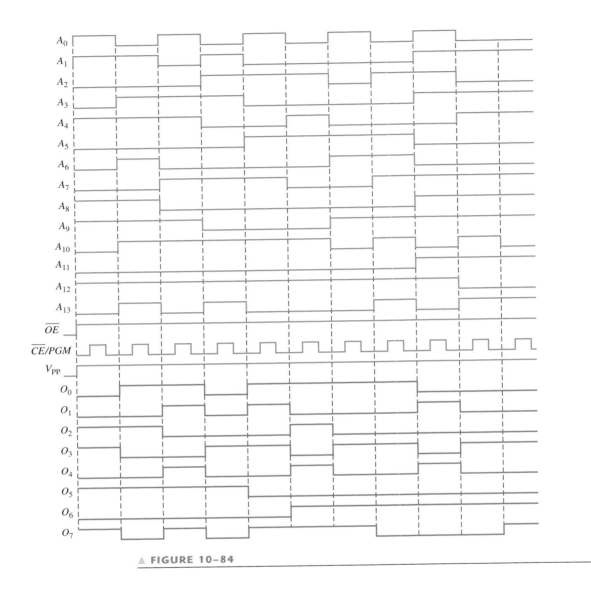

▲ **FIGURE 10–84**

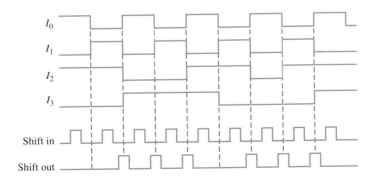

21. Consider a 4096×8 RAM in which the last 64 addresses are used as a LIFO stack. If the first address in the RAM is 000_{16}, designate the 64 addresses used for the stack.

22. In the memory of Problem 21, sixteen bytes are pushed into the stack. At what address is the first byte in located? At what address is the last byte in located?

SECTION 10–8   Magnetic and Optical Storage

23. Describe the general format of a hard disk.

24. Explain seek time and latency period in a hard disk drive.

25. Why does magnetic tape require a much longer access time than does a disk?

26. Explain the differences in a magneto-optical disk, a CD-ROM, and a WORM.

SECTION 10–9   Troubleshooting

27. Determine if the contents of the ROM in Figure 10–86 are correct.

▶ **FIGURE 10–86**

```
              ROM
         ┌─────────┐
         │1 0 1 1 1│
         │1 1 1 1 0│
         │1 1 0 1 1│
         │1 0 1 1 0│
         │1 1 1 0 1│
         │1 1 1 0 0│
         │0 0 0 0 1│
Checksum │0 1 1 0 0│
         └─────────┘
```

28. A 128×8 ROM is implemented as shown in Figure 10–87. The decoder decodes the two most significant address bits to enable the ROMs one at a time, depending on the address selected.

    (a) Express the lowest address and the highest address of each ROM as hexadecimal numbers.

    (b) Assume that a single checksum is used for the entire memory and it is stored at the highest address. Develop a flowchart for testing the complete memory system.

    (c) Assume that each ROM has a checksum stored at its highest address. Modify the flowchart developed in part (b) to accommodate this change.

    (d) What is the disadvantage of using a single checksum for the entire memory rather than a checksum for each individual ROM?

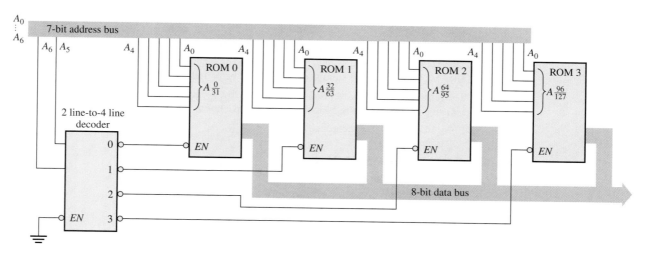

▲ **FIGURE 10–87**

29. Suppose that a checksum test is run on the memory in Figure 10–87 and each individual ROM has a checksum at its highest address. What IC or ICs will you replace for each of the following error messages that appear on the system's video monitor?

    (a) ADDRESSES 40–5F FAULTY

    (b) ADDRESSES 20–3F FAULTY

    (c) ADDRESSES 00–7F FAULTY

Digital System Application

30. Develop a timing diagram for the basic memory logic in Figure 10–72 to illustrate the entry of the digits 4321 into the SRAM. Include all the inputs and outputs of each device.

31. When programming the security system, what is the state of the counter in Figure 10–77 after two code digits have been entered?

32. What is the purpose of the memory logic?

33. Discuss the advantages and the disadvantages of using a PROM external to the CPLD instead of the memory on the CPLD chip in the memory logic.

Special Design Problems

34. Modify the design of the memory logic of the security entry system to accommodate a 5-digit entry code.

35. Make the appropriate modifications to the security code logic of the security entry system for a 5-digit entry code. Refer to the system application in Chapter 9.

ANSWERS

SECTION REVIEWS

SECTION 10–1   Basics of Semiconductor Memory

1. Bit is the smallest unit of data.
2. 256 bytes is 2048 bits.
3. A write operation stores data in memory.
4. A read operation takes a copy of data from memory.
5. A unit of data is located by its address.
6. A RAM is volatile and has read/write capability. A ROM is nonvolatile and has only read capability.

SECTION 10–2   Random-Access Memories (RAMs)

1. Asynchronous and synchronous with burst feature
2. A small fast memory between the CPU and main memory
3. SRAMs have latch storage cells that can retain data indefinitely while power is applied. DRAMs have capacitive storage cells that must be periodically refreshed.
4. The refresh operation prevents data from being lost because of capacitive discharge. A stored bit is restored periodically by recharging the capacitor to its nominal level.
5. FPM, EDO, BEDO, Synchronous

SECTION 10–3   Read-Only Memories (ROMs)

1. 512×8 equals 4096 bits.
2. Mask ROM, PROM, EPROM, UV EPROM, EEPROM
3. Eight bits of address are required for 256 byte locations ($2^8 = 256$).

SECTION 10–4   Programmable ROMs (PROMs and EPROMs)

1. PROMs are field-programmable; ROMs are not.
2. 1s are left after EPROM erasure.
3. Read is the normal mode of operation for a PROM.

SECTION 10–5   Flash Memories

1. Flash, ROM, EPROM, and EEPROM are nonvolatile.
2. Flash is nonvolatile; SRAM and DRAM are volatile.
3. Programming, read, erase

SECTION 10–6 **Memory Expansion**

1. Eight RAMs
2. Eight RAMs
3. SIMM; Single in-line memory module
4. DIMM: Dual in-line memory module
5. RIMM: Rambus in-line memory module

SECTION 10–7 **Special Types of Memories**

1. In a FIFO memory the *first* bit (or word) *in* is the *first* bit (or word) *out*.
2. In a LIFO memory the *last* bit (or word) *in* is the *first* bit (or word) *out*. A stack is a LIFO.
3. The operation or instruction that adds data to the memory stack
4. The operation or instruction that removes data from the memory stack
5. CCD is a charge-coupled device.

SECTION 10–8 **Magnetic and Optical Storage**

1. Magnetic storage: floppy disk, hard disk, tape, and magneto-optical disk.
2. Floppy disk storage capacity: 1.44 MB
3. A magnetic disk is organized in tracks and sectors.
4. A magneto-optical disk uses a laser beam and an electromagnet.
5. Optical storage: CD-ROM, CD-R, CD-RW, DVD-ROM, WORM

SECTION 10–9 **Troubleshooting**

1. The contents of the ROM are added and compared with a prestored checksum.
2. Checksum cannot be used because the contents of a RAM are not fixed.
3. (1) a short between adjacent cells; (2) an inability of some cells to store both 1s and 0s;
   (3) dynamic altering of the contents of one address when the contents of another address change.

RELATED PROBLEMS FOR EXAMPLES

10–1 $G_3G_2G_1G_0 = 1110$

10–2 Connect eight $64k \times 1$ ROMs in parallel to form a $64k \times 8$ ROM.

10–3 Sixteen $64k \times 1$ ROMs

10–4 See Figure 10–88.

▶ FIGURE 10–88

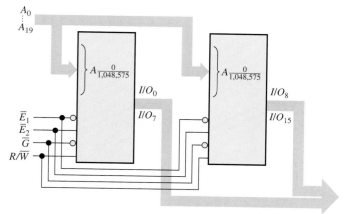

10–5 ROM 1: 0 to 524,287; ROM 2: 524,288 to 1,048,575

SELF-TEST

1. (b)	2. (c)	3. (c)	4. (d)	5. (a)	6. (d)
7. (c)	8. (a)	9. (b)	10. (f)	11. (c)	12. (d)

11

PROGRAMMABLE LOGIC AND SOFTWARE

CHAPTER OBJECTIVES

- Discuss the types of programmable logic, SPLDs and CPLDs, and explain their basic structure

- Describe the basic architecture of two types of SPLDs, the PAL and the GAL

- Describe the architecture of the Altera MAX 7000 family of CPLDs

- Describe the architecture of the Altera MAX II CPLD

- Explain the basic structure of a programmable logic array (PLA)

- Describe the architecture of the Xilinx CoolRunner II family of CPLDs
- Discuss the operation of macrocells
- Distinguish between CPLDs and FPGAs
- Explain the basic operation of a look-up table (LUT)
- Define *intellectual property* and *platform FPGA*
- Describe the architecture of the Altera Stratix FPGA family
- Discuss embedded functions
- Describe the architecture of the Xilinx Virtex FPGA family
- Show a basic software design flow for a programmable device
- Explain the design flow elements of design entry, functional simulation, synthesis, implementation, timing simulation, and downloading
- Discuss several methods of testing a programmable logic device, including boundary scan logic

KEY TERMS

- PAL
- GAL
- Macrocell
- Registered
- CPLD
- LAB
- LUT
- FPGA
- CLB
- Intellectual property
- Design flow
- Target device

- Schematic entry
- Text entry
- Compiler
- Downloading
- Bed-of-nails
- Flying probe
- Boundary scan
- Primitive
- Fitter tool
- Functional simulation
- Timing simulation

INTRODUCTION

In this chapter, the basic architecture (internal structure and organization) of SPLDs, CPLDs, and FPGAs is discussed. Several specific CPLDs are introduced, including the Altera MAX 7000, MAX II, and the Xilinx CoolRunner II. FPGAs that are introduced are the Altera Stratix and the Xilinx Virtex.

A discussion of software development tools covers the generic design flow for programming a device, including design entry, functional simulation, synthesis, implementation, timing simulation, and downloading. Also, there is a section on in-circuit troubleshooting of a circuit board once the programmable device is operating. Test methods include bed-of-nails, flying probe, and boundary scan.

■■■ DIGITAL SYSTEM APPLICATION PREVIEW

The Digital System Application illustrates a generic design flow process for programming the logic for driving a 7-segment display. The logic for each segment was developed in the Digital System Application in Chapter 4 and VHDL programs were written for each one. You could enter the VHDL programs using a text entry software tool. However, we will illustrate the design flow with a generic schematic entry approach.

www. **VISIT THE COMPANION WEBSITE**
Study aids for this chapter are available at
http://www.prenhall.com/floyd

11–1 PROGRAMMABLE LOGIC: SPLDs AND CPLDs

Two major types of simple programmable logic devices (SPLDs) are the PAL and the GAL. *PAL* stands for programmable array logic, and *GAL* stands for generic array logic. Generally, a PAL is one-time programmable (OTP), and a GAL is a type of PAL that is reprogrammable; however, because some reprogrammable SPLDs are still called PALs, the line between PALs and GALs is a little vague in common usage. The term *GAL* is a designation originally used by Lattice Semiconductor and later licensed to other manufacturers. The basic structure of both PALs and GALs is a programmable AND array and a fixed OR array, which is a basic sum-of-products architecture. The complex programmable logic device (CPLD) is basically a single device with multiple SPLDs that provides more capacity for larger logic designs.

After completing this section, you should be able to

■ Describe SPLD operation ■ Show how a sum-of-products expression is implemented in a PAL or GAL ■ Explain simplified PAL/GAL logic diagrams ■ Describe a basic PAL/GAL macrocell ■ Discuss the PAL16V8 and the GAL22V10 ■ Describe a basic CPLD

SPLD: The PAL

A **PAL** (programmable array logic) consists of a programmable array of AND gates that connects to a fixed array of OR gates. Generally, PALs are implemented with fuse process technology and are, therefore, one-time programmable (OTP).

The PAL structure allows any sum-of-products (SOP) logic expression with a defined number of variables to be implemented. As you have learned, any combinational logic function can be expressed in SOP form. A simple PAL structure is shown in Figure 11–1 for two input variables and one output; most PALs have many inputs and many outputs. As you learned in Chapter 3, a programmable array is essentially a grid or matrix of conductors that form rows and columns with a programmable link at each cross point. Each programmable link, which is a fuse in the case of a PAL, is called a *cell*. Each row is connected to the input of an AND gate, and each column is connected to an input variable or its complement. By programming the presence or absence of a fuse connection, any combination of input variables or complements can be applied to an AND gate to form any desired product term. The AND gates are connected to an OR gate, creating a sum-of-products (SOP) output.

▶ **FIGURE 11–1**

Basic AND/OR structure of a PAL.

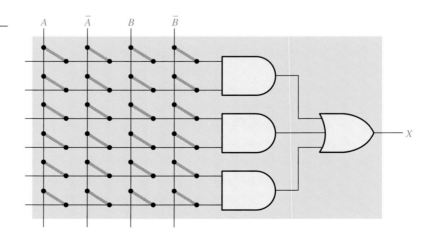

Implementing a Sum-of-Products Expression   An example of a simple PAL is programmed as shown in Figure 11–2 so that the product term AB is produced by the top AND gate, $A\overline{B}$ is produced by the middle AND gate, and $\overline{A}\,\overline{B}$ is produced by the bottom AND gate. As you can see, the fuses are left intact to connect the desired variables or their complements to the appropriate AND gate inputs. The fuses are opened where a variable or its complement is not used in a given product term. The final output from the OR gate is the SOP expression,

$$X = AB + A\overline{B} + \overline{A}\,\overline{B}$$

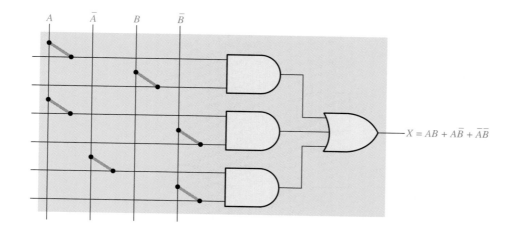

◀ **FIGURE 11–2**

PAL implementation of a sum-of-products expression.

$$X = AB + A\overline{B} + \overline{A}B$$

SPLD: The GAL

The **GAL** is essentially a PAL that can be reprogrammed. It has the same type of AND/OR organization that the PAL does. The basic difference is that a GAL uses a reprogrammable process technology, such as EEPROM (E^2CMOS), instead of fuses, as shown in Figure 11–3.

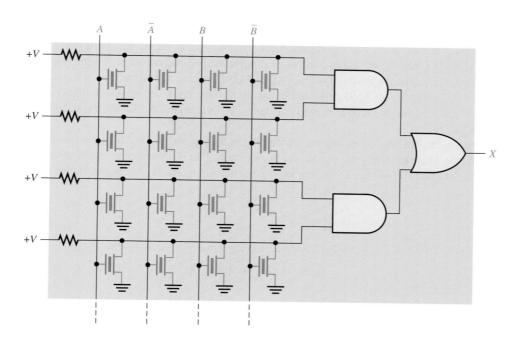

◀ **FIGURE 11–3**

Simplified GAL array.

Simplified Notation for PAL/GAL Diagrams

Actual PAL and GAL devices have many AND and OR gates in addition to other elements and are capable of handling many variables and their complements. Most PAL and GAL diagrams that you may see on a data sheet use simplified notation, as illustrated in Figure 11–4, to keep the schematic from being too complicated.

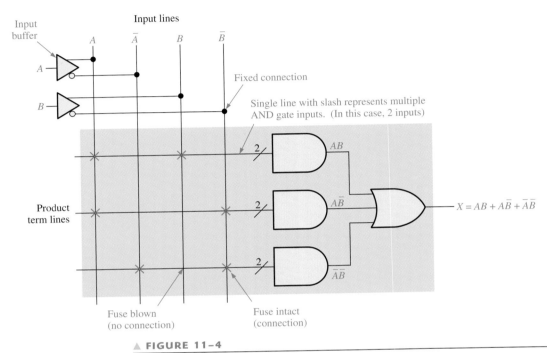

▲ **FIGURE 11–4**

A portion of a programmed PAL/GAL.

The input variables to a PAL or GAL are usually buffered to prevent loading by a large number of AND gate inputs to which they are connected. On the diagram, the triangle symbol represents a buffer that produces both the variable and its complement. The fixed connections of the input variables and buffers are shown using standard dot notation.

PALs and GALs have a large number of programmable interconnection lines, and each AND gate has multiple inputs. Typical PAL and GAL logic diagrams represent a multiple-input AND gate with an AND gate symbol having a single input line with a slash and a digit representing the actual number of inputs. Figure 11–4 illustrates this for the case of 2-input AND gates.

Programmable links in an array are indicated in a diagram by a red X at the cross point for an intact fuse or other type of link and the absence of an X for an open fuse or other type of link. In Figure 11–4, the 2-variable logic function $AB + A\bar{B} + \bar{A}\,\bar{B}$ is programmed.

EXAMPLE 11–1

Show how a PAL is programmed for the following 3-variable logic function:

$$X = A\bar{B}C + \bar{A}B\bar{C} + \bar{A}\,\bar{B} + AC$$

Solution   The programmed array is shown in Figure 11–5. The intact fusible links are indicated by small red Xs. The absence of an X means that the fuse is open.

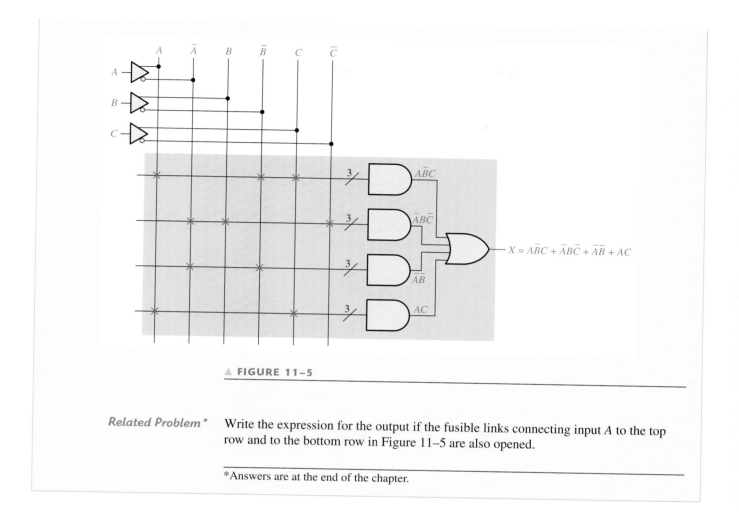

▲ **FIGURE 11–5**

Related Problem*   Write the expression for the output if the fusible links connecting input A to the top row and to the bottom row in Figure 11–5 are also opened.

*Answers are at the end of the chapter.

PAL/GAL General Block Diagram

A block diagram of a PAL or GAL is shown in Figure 11–6. Remember, the basic difference is that a GAL has a reprogrammable array and the PAL is one-time programmable. The programmable AND array outputs go to fixed OR gates that are connected to additional output logic. An OR gate combined with its associated output logic is typically called a *macrocell*. The complexity of the macrocell depends on the particular device, and in GALs it is often reprogrammable.

Macrocells

A **macrocell** generally consists of one OR gate and some associated output logic. The macrocells vary in complexity, depending on the particular type of PAL or GAL. A macrocell can be configured for combinational logic, registered logic, or a combination of both. **Registered** logic means that there is a flip-flop in the macrocell to provide for sequential logic functions. The registered operation of macrocells is covered in Section 11–4.

Figure 11–7 illustrates three basic types of macrocells with combinational logic. Part (a) shows a simple macrocell with the OR gate and an inverter with a tristate control that can make the inverter like an open circuit to completely disconnect the output. The output of the tristate inverter can be either LOW, HIGH, or disconnected. Part (b) is a macrocell that can be either

▶ **FIGURE 11-6**

General block diagram of a PAL or GAL.

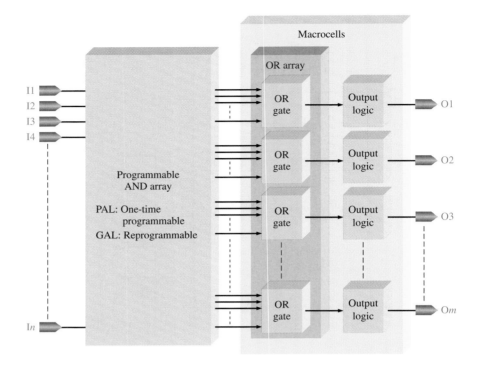

an input or an output. When the output is used as an input, the tristate inverter is disconnected, and the input goes to the buffer that is connected to the AND array. Part (c) is a macrocell that can be programmed to have either an active-HIGH or an active-LOW output, or it can be used as an input. One input to the exclusive-OR (XOR) gate can be programmed to be either HIGH

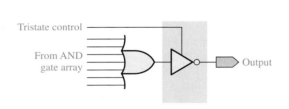

(a) Combinational output (active-LOW). An active-HIGH output would be shown without the bubble on the tristate gate symbol.

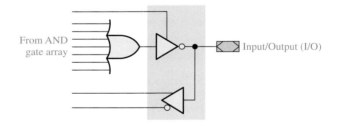

(b) Combinational input/output (active-LOW)

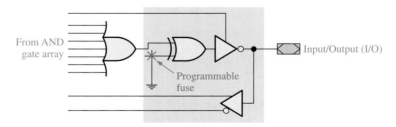

(c) Programmable polarity output

▲ **FIGURE 11-7**

Basic types of PAL/GAL macrocells for combinational logic.

or LOW. When the programmable XOR input is HIGH, the OR gate output is inverted because $0 \oplus 1 = 1$ and $1 \oplus 1 = 0$. Similarly, when the programmable XOR input is LOW, the OR gate output is not inverted because $0 \oplus 0 = 0$ and $1 \oplus 0 = 1$.

Specific SPLDs

Generally, SPLD package configurations range from 20 pins to 28 pins. Two factors that you can use to help determine whether a certain PAL or GAL is adequate for a given logic design are the number of inputs and outputs and the number of equivalent gates or density. Other parameters to consider are the maximum operating frequency, delay times, and dc supply voltage. Lattice, Actel, Atmel, and Cypress are among several companies that produce SPLDs. Various SPLD manufacturers may have different ways of defining density, so you have to use the specified number of equivalent gates with this in mind.

The 16V8 and the 22V10 are common types of PALs and GALs. The device designation indicates the number of inputs, the number of outputs, and the type of output logic. For example, 16V8 means that the device has sixteen inputs, eight outputs, and the outputs are variable (V). The letter L or H means the output is active-LOW or active-HIGH, respectively. The block diagram for a PAL16V8 and a typical SPLD package are shown in Figure 11–8.

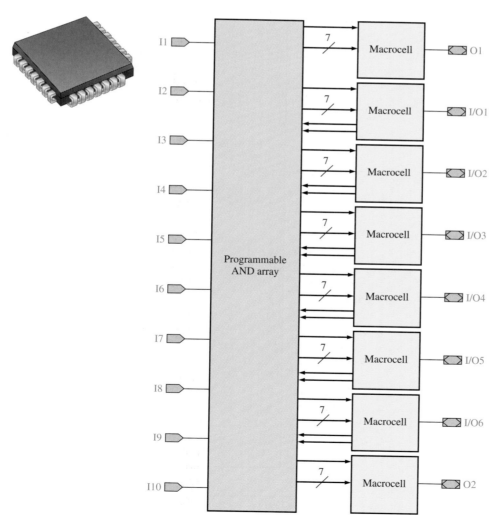

◀ **FIGURE 11–8**

Logic block diagram of a PAL16V8 and typical SPLD package.

Each macrocell has eight inputs from the AND gate array, so you can have up to eight product terms for each output. There are ten dedicated inputs (I), two dedicated outputs (O), and six pins that can be used as either inputs or outputs (I/O). Each output is active-LOW. The PAL16V8 has a density of approximately 300 equivalent gates.

A block diagram for a GAL22V10 and a typical SPLD package are shown in Figure 11–9. This device has twelve dedicated inputs and ten pins that can be either inputs or outputs. The macrocells have inputs from the AND array that vary from eight to sixteen, as indicated by the simplified notation. The GAL22V10 has a density of approximately 500 equivalent gates.

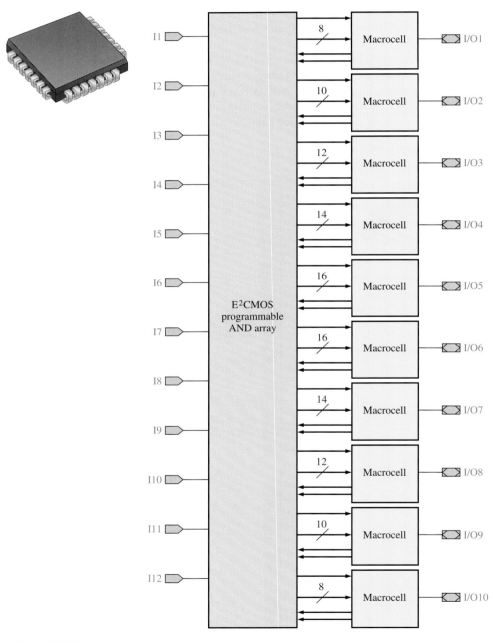

▲ **FIGURE 11–9**

Block diagram of the GAL22V10 and typical SPLD package.

The CPLD

A **CPLD** (complex programmable logic device) consists basically of multiple SPLD arrays with programmable interconnections, as illustrated in Figure 11–10. Although the way CPLDs are internally organized varies with the manufacturer, Figure 11–10 represents a generic CPLD. We will refer to each SPLD array in a CPLD as a **LAB** (logic array block). Other designations are sometimes used, such as *function block, logic block,* or *generic block*. The programmable interconnections are generally called the *PIA* (programmable interconnect array) although some manufacturers, such as Xilinx, use the term *AIM* (advanced interconnect matrix) or a similar designation. The LABs and the interconnections between LABs are programmed using software. A CPLD can be programmed for complex logic functions based on the SOP structure of the individual LABs (actually SPLDs). Inputs can be connected to any of the LABs, and their outputs can be interconnected to any other LABs via the PIA.

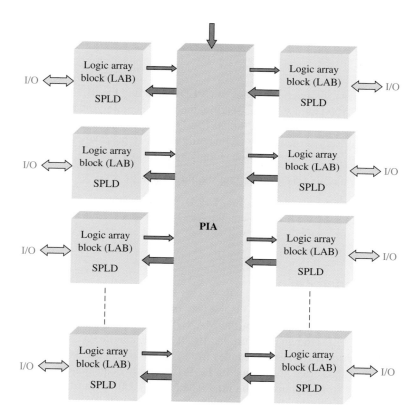

◀ **FIGURE 11–10**

Basic block diagram of a generic CPLD.

Most programmable logic manufacturers make a series of CPLDs that range in density, process technology, power consumption, supply voltage, and speed. Manufacturers usually specify CPLD density in terms of macrocells or logic array blocks. Densities can range from tens of macrocells to over 2000 macrocells in packages with up to several hundred pins. As PLDs become more complex, maximum densities will increase. Most CPLDs are reprogrammable and use EEPROM or SRAM process technology for the programmable links. Power consumption can range from a few milliwatts to a few hundred milliwatts. DC supply voltages are typically from 2.5 V to 5 V, depending on the specific device.

Several manufacturers, (for example, Altera, Xilinx, Lattice, and Cypress) produce CPLDs. In this chapter, we will focus on Altera and Xilinx products because they are two of the major companies in the market. The other companies offer similar devices and software, and you can easily make a transition to other products once you are familiar with one or two. As you will learn, CPLDs and other programmable logic devices are really a combination of hardware and software.

SECTION 11–1 REVIEW

Answers are at the end of the chapter.

1. What does PAL stand for?
2. What does GAL stand for?
3. What is the difference between a PAL and a GAL?
4. Basically, what does a macrocell contain?
5. What is a CPLD?

11–2   ALTERA CPLDs

Altera produces several families of CPLDs including the MAX II, the MAX 3000, and the MAX 7000 family. In this section, the focus is mainly on the MAX 7000 to illustrate the concepts of traditional CPLD architecture, keeping in mind that other series may vary somewhat in architecture and/or in parameters such as density, process technology, power consumption, supply voltage, and speed.

After completing this section, you should be able to

■ Describe a typical MAX family CPLD ■ Discuss the basic architecture of the MAX 7000 and the MAX II CPLDs ■ Explain how product terms are generated in CPLDs

MAX 7000 CPLD

The **architecture** of a CPLD is the way in which the internal elements are organized and arranged. The architecture of the MAX 7000 family is similar to the block diagram of a generic CPLD (shown in Figure 11–10). It has the classic PAL/GAL structure that produces SOP functions. The density ranges from 2 LABs to 16 LABs, depending on the particular device in the series. Remember, a LAB is roughly equivalent to one SPLD, and package sizes vary from 44 pins to 208 pins. The MAX 7000 series of CPLDs uses the EEPROM-based process technology. In-system programmable (ISP) versions use the JTAG standard interface.

Figure 11–11 shows a general block diagram of the Altera MAX 7000 series CPLD. Four LABs are shown, but there can be up to sixteen, depending on the particular device in the series. Each of the four LABs consists of sixteen macrocells, and multiple LABs are linked together via the PIA, which is a programmable global (goes to all LABs) bus structure to which the general-purpose inputs, the I/Os, and the macrocells are connected.

The Macrocell   A simplified diagram of a MAX 7000 series macrocell is shown in Figure 11–12. The macrocell contains a small programmable AND array with five AND gates, an OR gate, a product-term selection matrix for connecting the AND gate outputs to the OR gate, and associated logic that can be programmed for input, combinational logic output, or registered output. This macrocell is covered in more detail in Section 11–4.

Although based on the same concept, this macrocell differs somewhat from the macrocell discussed in Section 11–1 in relation to SPLDs because it contains a portion of the

General-purpose inputs

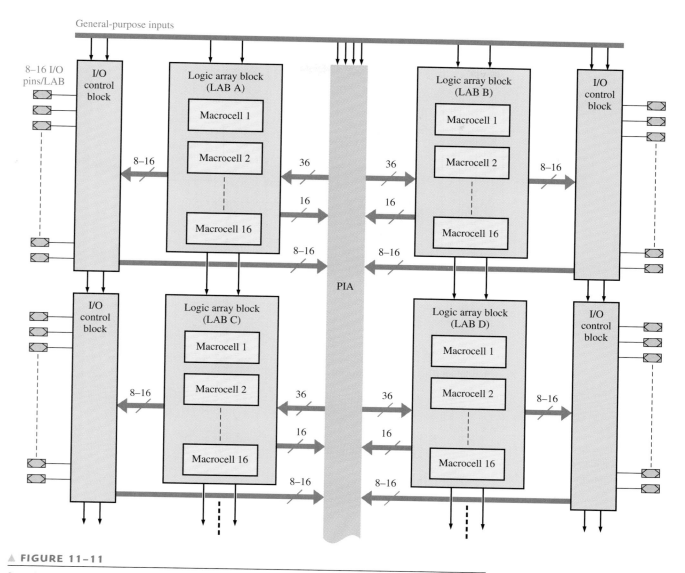

▲ FIGURE 11–11

Basic block diagram of the Altera MAX 7000 series CPLD.

programmable AND array and a product-term selection matrix. As shown in Figure 11–12, five AND gates feed product terms from the PIA into the product-term selection matrix. The product term from the bottom AND gate can be fed back inverted into the programmable array as a shared expander for use by other macrocells. The parallel expander inputs allow borrowing of unused product terms from other macrocells to expand an SOP expression. The product-term selection matrix is an array of programmable connections that is used to connect selected outputs from the AND array and from the expander inputs to the OR gate.

Shared Expanders A complemented product term that can be used to increase the number of product terms in an SOP expression is available from each macrocell in a LAB. Figure 11–13 illustrates how a shared expander term from another macrocell can be used to create additional product terms. In this case, each of the five AND gates in a macrocell array is limited to four inputs and, therefore, can produce up to a 4-variable product term, as illustrated in part (a). Figure 11–13(b) shows the expansion to two product terms.

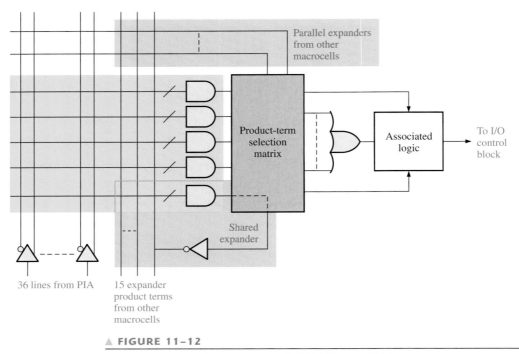

▲ **FIGURE 11–12**

Simplified diagram of a macrocell in a MAX 7000 series CPLD.

▶ **FIGURE 11–13**

Example of how a shared expander can be used in a macrocell to increase the number of product terms.

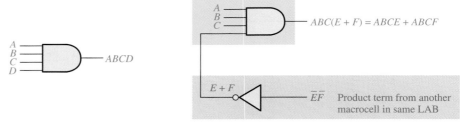

(a) A 4-input AND array gate can produce one 4-variable product term.

(b) AND gate is expanded to produce two product terms.

Each MAX 7000 macrocell can produce up to five product terms generated from its AND array. If a macrocell needs more than five product terms for its SOP output, it can use an expander term from another macrocell. Suppose that a design requires an SOP expression that contains six product terms. Figure 11–14 shows how a product term from another macrocell can be used to increase an SOP output. Macrocell 2, which is underutilized, generates a shared expander term $(E + F)$ that connects to the fifth AND gate in macrocell 1 to produce an SOP expression with six product terms. The red Xs and lines represent the connections produced in the hardware by the software compiler running the programmed design.

Parallel Expanders   Another way to increase the number of product terms for a macrocell is by using parallel expanders in which additional product terms are ORed with the terms generated by a macrocell instead of being combined in the AND array, as in the shared expander. A given macrocell can borrow unused product terms from neighboring macrocells (up to five product terms from three other macrocells for the MAX 7000). The basic concept is illustrated in Figure 11–15 where a simplified circuit that can produce two product terms borrows three additional product terms.

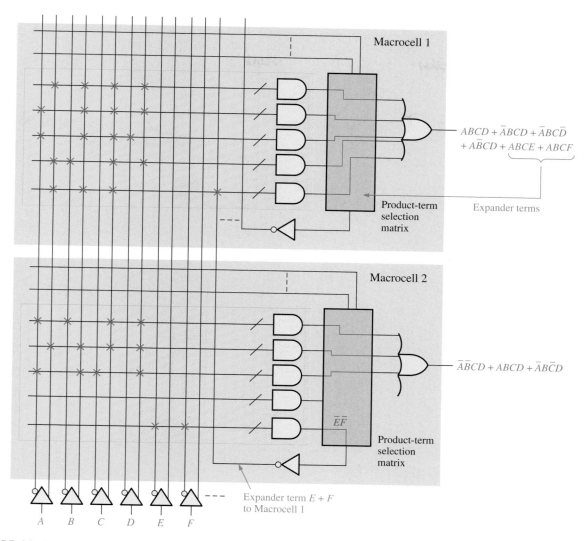

$$ABCD + \bar{A}BCD + \bar{A}BC\bar{D}$$
$$+ A\bar{B}CD + ABCE + ABCF$$

Expander terms

$$\bar{A}\bar{B}CD + ABCD + \bar{A}BC\bar{D}$$

Expander term $E + F$
to Macrocell 1

▲ **FIGURE 11–14**

Simplified illustration of using a shared expander term from another macrocell to increase an SOP expression.

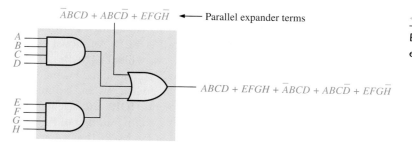

$$\bar{A}BCD + ABC\bar{D} + EFG\bar{H} \longleftarrow \text{Parallel expander terms}$$

$$ABCD + EFGH + \bar{A}BCD + ABC\bar{D} + EFG\bar{H}$$

◀ **FIGURE 11–15**

Basic concept of the parallel expander.

Figure 11–16 shows how one macrocell can borrow parallel expander terms from another macrocell to increase the SOP output. Macrocell 2 uses three product terms from macrocell 1 to produce an eight-term SOP expression. The red Xs and lines represent the connections produced in the hardware by the software compiler running the programmed design.

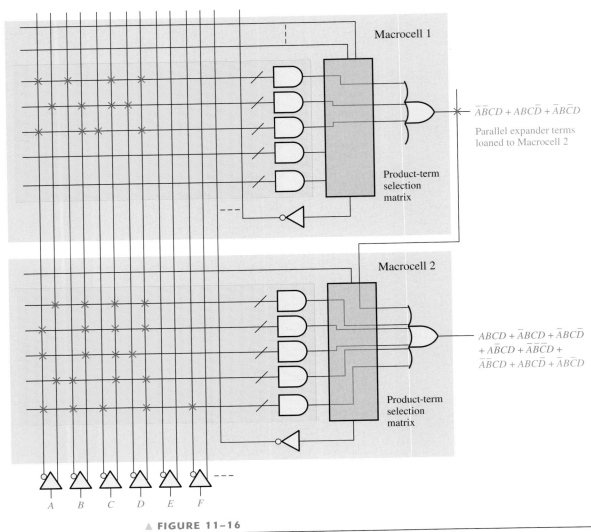

$\overline{A}BCD + ABC\overline{D} + \overline{A}B\overline{C}D$

Parallel expander terms
loaned to Macrocell 2

$ABCD + \overline{A}BCD + \overline{A}BC\overline{D}$
$+ A\overline{B}CD + \overline{A}B\overline{C}D +$
$\overline{A}\overline{B}CD + ABC\overline{D} + \overline{A}B\overline{C}\overline{D}$

▲ **FIGURE 11–16**

Simplified illustration of using parallel expander terms from another macrocell to increase an SOP expression.

The MAX II CPLD

The architecture of the MAX II CPLD differs dramatically from the MAX 7000 family and is what Altera calls a "post-macrocell" CPLD. As shown by the block diagram in Figure 11–17, this device contains logic array blocks (LABs) each with multiple logic elements (LEs). An LE is the basic logic design unit and is analogous to the macrocell. The programmable interconnects are arranged in a row and column arrangement running between the LABs, and input/output elements (IOEs) are oriented around the perimeter. The architecture of this family of CPLDs is similar to that of FPGAs, which we discuss in Section 11–5. In fact, you could think of the MAX II as a low-density FPGA.

A main difference between the MAX II CPLD and the classic SPLD-based CPLD is the way in which a logic function is developed. The MAX II uses look-up tables (LUT) instead of AND/OR arrays. An **LUT** is basically a type of memory that can be programmed to produce SOP functions (discussed in more detail in Section 11–5). These two approaches are contrasted in Figure 11–18.

As mentioned, the MAX II CPLD has a row/column arrangement of interconnects instead of the channel-type interconnects found in most classic CPLDs. These two approaches are contrasted in Figure 11–19 and can be understood by comparing Figure 11–11 and Figure 11–17.

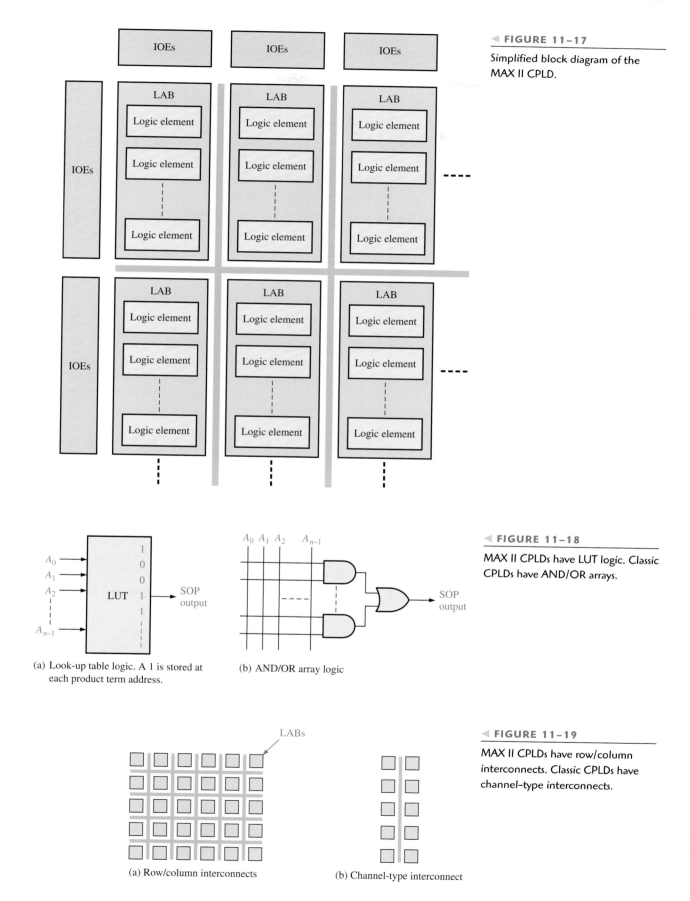

Simplified block diagram of the MAX II CPLD.

◀ FIGURE 11–18

MAX II CPLDs have LUT logic. Classic CPLDs have AND/OR arrays.

(a) Look-up table logic. A 1 is stored at each product term address.

(b) AND/OR array logic

◀ FIGURE 11–19

MAX II CPLDs have row/column interconnects. Classic CPLDs have channel-type interconnects.

(a) Row/column interconnects

(b) Channel-type interconnect

Most CPLDs use a nonvolatile process technology for the programmable links. MAX II, however, uses a SRAM-based process technology that is **volatile**—all programmed logic is lost when power is turned off. The memory embedded on the chip stores the program data using nonvolatile memory technology and reconfigures the CPLD on power up.

SECTION 11–2 REVIEW

1. What does LAB stand for?
2. Describe a LAB in the MAX 7000 CPLD.
3. What is the purpose of a shared expander?
4. What is the purpose of a parallel expander?
5. How does the MAX II differ from the MAX 7000?

11–3   XILINX CPLDs

Xilinx, like Altera, makes a series of CPLDs that range in density, process technology, power consumption, supply voltage, and speed. Xilinx produces several families of CPLDs including CoolRunner II, CoolRunner XPLA3, and the XC9500. The XC9500 is similar in architecture to the Altera MAX 7000 CPLD family and exhibits the classic PAL/GAL structure. In this section, we will focus on only the CoolRunner II to illustrate the concepts, keeping in mind that other series may vary somewhat in architecture, and/or in the parameters previously mentioned. This family of CPLDs is in-system programmable and JTAG compliant.

After completing this section, you should be able to

■ Describe a PLA and compare to a PAL ■ Discuss the architecture of the CoolRunner II CPLD ■ Describe a functional block

PLA (Programmable Logic Array)

As you have learned, the architecture of a CPLD is the way in which the internal elements are organized and arranged. The architecture of the Xilinx CoolRunner II family is based on a PLA (programmable logic array) structure rather than on a PAL (programmable array logic) structure. Figure 11–20 compares a simple PAL structure with a simple PLA struc-

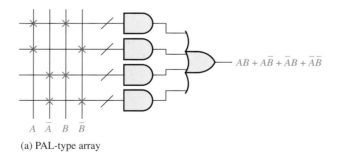

$AB + A\bar{B} + \bar{A}B + \bar{A}\bar{B}$

(a) PAL-type array

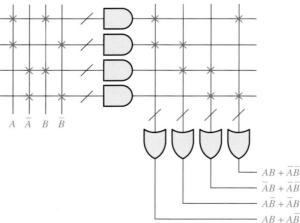

$AB + \bar{A}\bar{B}$
$\bar{A}B + A\bar{B}$
$A\bar{B} + \bar{A}B$
$AB + A\bar{B}$

(b) PLA-type array

▲ FIGURE 11–20

Comparison of a basic PLA to a basic PAL.

ture. As you know, the PAL has a programmable AND array followed by a fixed OR array and produces an SOP expression, as shown by the example in Figure 11–20(a). The **PLA** has a programmable AND array followed by a programmable OR array, as shown by the example in Figure 11–20(b).

CoolRunner II

The CoolRunner II CPLD uses a PLA type structure. This device has multiple function blocks (FBs) that are analogous to the LABs in the Altera MAX 7000 (Figure 11–11). Each function block contains sixteen macrocells, just as the LAB does. The function blocks are interconnected by an advanced interconnect matrix (AIM) that is analogous to the PIA in the MAX 7000. A basic architectural block diagram for the CoolRunner II is shown in Figure 11–21. As you can see, from a basic block diagram point of view,

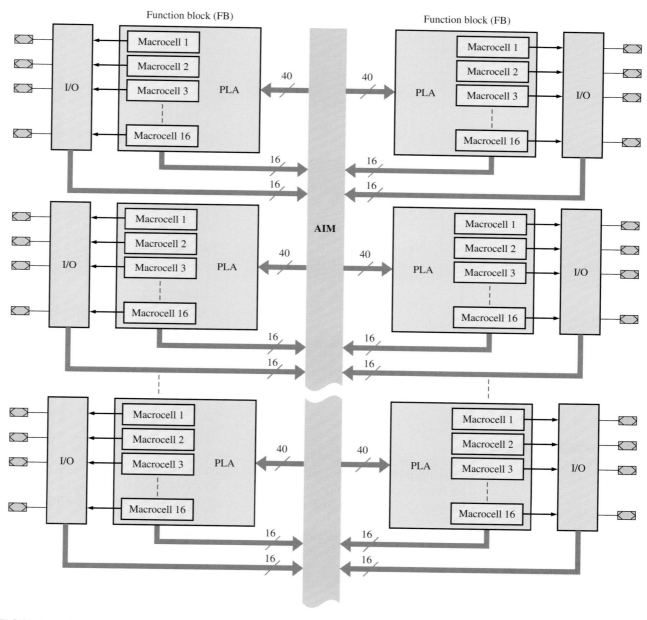

▲ FIGURE 11–21

Basic architectural block diagram of the Xilinx CoolRunner II CPLD.

there is not much difference between the Xilinx CPLD and the Altera CPLD; however, internally there are differences.

The CoolRunner II series of CPLDs contains from 32 macrocells to 512 macrocells. Since there are sixteen macrocells per function block, the number of function blocks range from 2 to 32. A greatly simplified diagram of a function block (FB) is shown in Figure 11–22. The programmable AND array has 56 AND gates, and the programmable OR array has 16 OR gates. With the PLA structure, any product term can be connected to any OR gate to create an SOP output. With maximum utilization, each FB can produce 16 SOP outputs each with 56 product terms. This macrocell is covered in detail in Section 11–4.

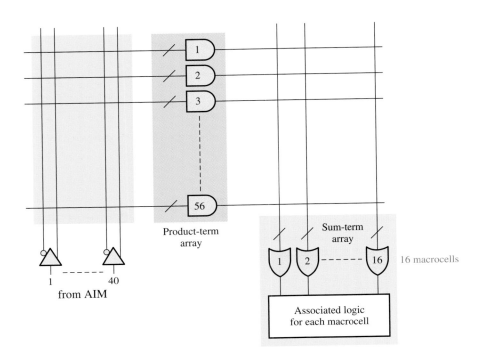

▲ FIGURE 11–22

Simplified diagram of a CoolRunner II function block (FB) with a PLA structure.

EXAMPLE 11–2

Show the programmed connections in the simplified FB of Figure 11–22 to produce the following SOP function from macrocell 1: $ABCD + A\overline{B}\overline{C}D + AB\overline{C}D$ and the following SOP function from macrocell 2: $\overline{A}BC\overline{D} + \overline{A}BCD + A\overline{B}CD + ABCD$.

Solution The red Xs in Figure 11–23 indicate programmed connections in the AND and OR arrays.

Related Problem How many SOP functions can be generated by the FB in Figure 11–23?

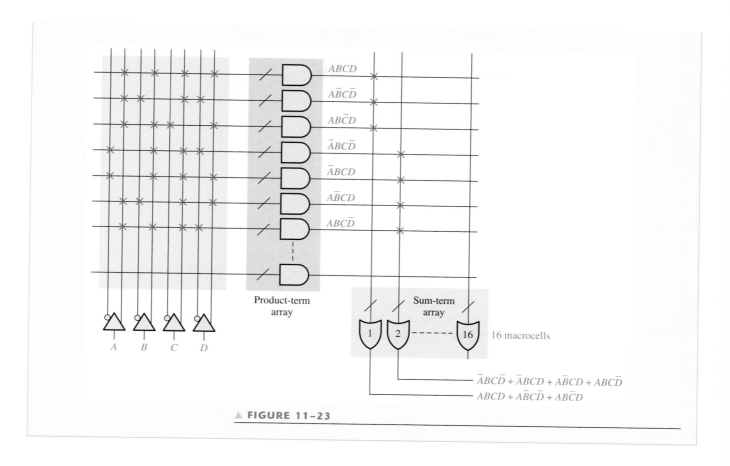

▲ **FIGURE 11–23**

1. What is the main difference between the Altera and the Xilinx CPLDs?
2. Describe a PLA.
3. How does a PLA differ from a PAL?
4. What does FB stand for?

11–4   MACROCELLS

CPLD macrocells were introduced in the previous sections for both Altera and Xilinx devices. Recall that a macrocell can be configured for combinational logic or registered logic outputs and inputs by programming. The term *registered* refers to the use of flip-flops. In this section, you will learn about the typical macrocell, including the combinational and the registered modes of operation. Although macrocell architecture varies among different CPLDs, representative devices are used for illustration.

After completing this section, you should be able to

■ Describe the operation of an Altera MAX 7000 CPLD macrocell ■ Describe the operation of a Xilinx CoolRunner II CPLD macrocell

Logic diagrams often use the symbol shown in Figure 11–24 to represent a multiplexer. In this case, the multiplexer has two data inputs and a select input that provides for programmable selection; the select input is usually not shown on a logic diagram.

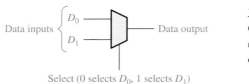

Data inputs $\left\{ \begin{array}{l} D_0 \\ D_1 \end{array} \right.$ ————— Data output

Select (0 selects D_0, 1 selects D_1)

◀ **FIGURE 11–24**

Commonly used symbol for a multiplexer. It can have any number of inputs.

The Altera MAX 7000 Macrocell

Figure 11–25 shows the complete macrocell including the flip-flop (register). The XOR gate provides for complementing the SOP function from the OR gate to produce a function in POS form. A 1 on the top input of the XOR gate complements the OR output, and a 0 lets the OR output pass uncomplemented (in SOP form). MUX 1 provides for selection of either the XOR output or an input from the I/O. MUX 2 can be programmed to select either the global clock or a clock signal based on a product term. MUX 3 can be programmed to select either a HIGH (V_{CC}) or a product-term enable for the flip-flop. MUX 4 can select the global clear or a product-term clear. MUX 5 is used to bypass the flip-flop and connect the combinational logic output to the I/O or to connect the registered output to the I/O. The flip-flop can be programmed as a D, T (toggle), J-K, or S-R flip-flop.

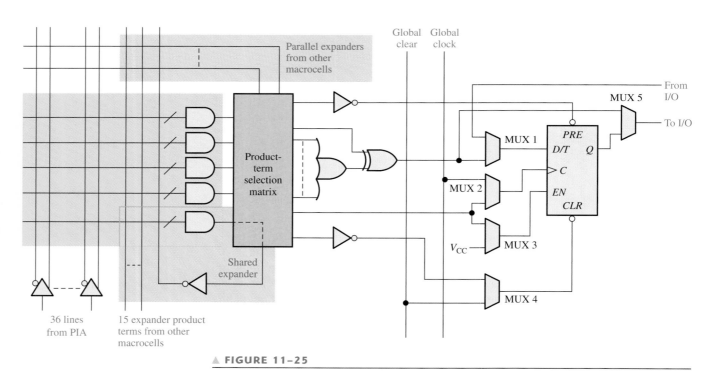

▲ **FIGURE 11–25**

A macrocell in the Altera MAX 7000 family of CPLDs.

The Combinational Mode When a macrocell is programmed to produce an SOP combinational logic function, the logic elements in the data path are as shown in red in Figure 11–26. As you can see, only one mux is used and the register (flip-flop) is bypassed.

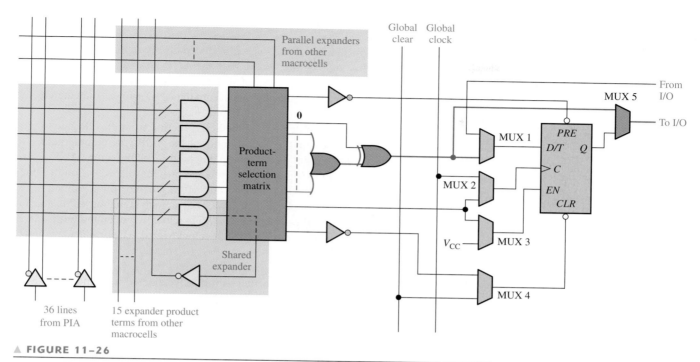

▲ FIGURE 11–26

A macrocell configured for generation of an SOP logic function. Red indicates data path.

The Registered Mode When a macrocell is programmed for the registered mode with the SOP combinational logic output providing the data input to the register and clocked by the global clock, the elements in the data path are as shown in red in Figure 11–27. As you can see, four muxes are used and the register (flip-flop) is active.

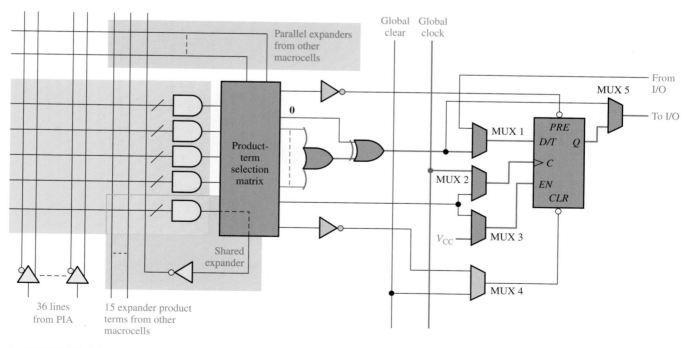

▲ FIGURE 11–27

A macrocell configured for generation of a registered logic function. Red indicates data path.

The Xilinx CoolRunner II Macrocell

The macrocell in the CoolRunner II CPLD was introduced briefly in Section 11–3. Recall that this device uses a PLA architecture, where both the AND array and the OR array are programmable. Figure 11–28 shows the complete logic for this macrocell, including the flip-flop (register). The OR gate has multiple inputs from the AND array as indicated by the slash through the input line.

The XOR gate provides for complementing the SOP function from the OR gate to produce a function in POS form. A 1 on the bottom input of the XOR gate complements the OR output and a 0 lets the OR output pass uncomplemented (in SOP form). MUX 1 provides for selection of SOP or POS logic outputs. MUX 2 provides for selection of either the XOR output or an input from the I/O. MUX 3 and MUX 4 can be programmed to select one of the global clocks (GCK0, GCK1, or GCK2) or a clock signal based on a product term (*CTC* or *PTC*). *CTC* is a shared term and *PTC* is a locally generated term. MUX 5 can be programmed to provide either polarity of the clock signal. The product term *PTC* is used to provide a clock enable to the flip-flop. MUX 6 can select one of four signals to set the flip-flop. These are *PTA* (locally generated product term), *CTS* (shared product term), *GSR* (global set/reset), and GND, which is normally selected when no active SET is required. MUX 7 provides the same functions to clear or reset the flip-flop as MUX 6. MUX 8 is used to bypass the flip-flop and connect the combinational logic output to the I/O or to connect the registered output to the I/O. The flip-flop can be programmed as a D, T (toggle), or as a latch.

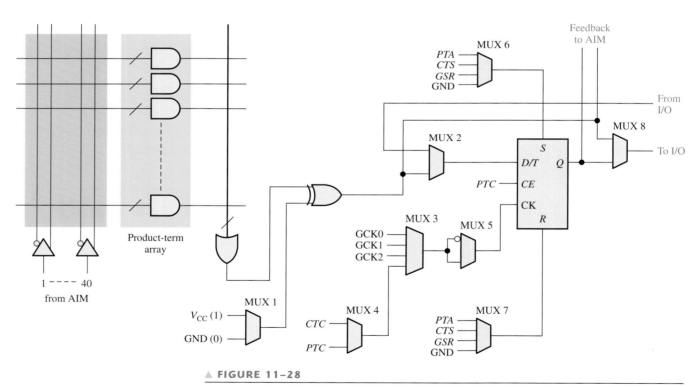

▲ **FIGURE 11–28**

A macrocell in the Xilinx CoolRunner II CPLD.

The Combinational Mode   When a macrocell is programmed to produce an SOP combinational logic function, the logic elements in the data path are as shown in red in Figure 11–29. As you can see, only two muxes are used and the register (flip-flop) is bypassed.

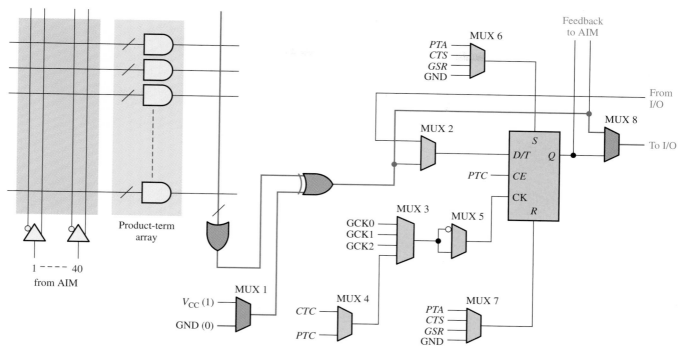

A macrocell configured for generation of an SOP logic function. Red indicates data path.

The Registered Mode When a macrocell is programmed for the registered mode with the SOP combinational logic output providing the data input to the register and clocked by one of the global clocks, the elements in the data path are as shown in red in Figure 11–30. As you can see, five muxes are used and the register (flip-flop) is active.

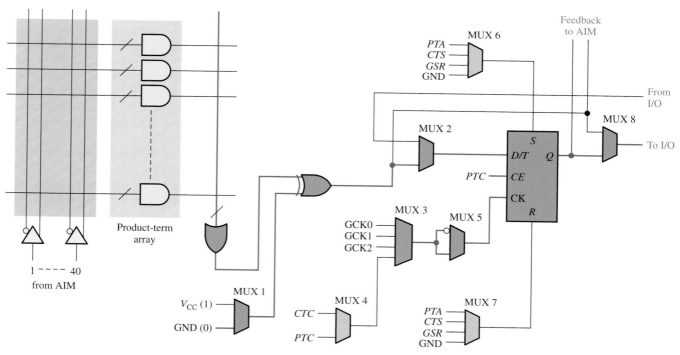

▲ **FIGURE 11–30**

A macrocell configured for generation of a registered logic function. Red indicates data path.

11–5   PROGRAMMABLE LOGIC: FPGAs

As you have learned, the classic CPLD architecture consists of PAL/GAL or PLA-type logic blocks with programmable interconnections. Basically, the FPGA (field-programmable gate array) differs in architecture, does not use PAL/PLA type arrays, and has much greater densities than CPLDs. A typical FPGA has many times more equivalent gates than a typical CPLD. The logic-producing elements in FPGAs are generally much smaller than in CPLDs, and there are many more of them. Also, the programmable interconnections are generally organized in a row and column arrangement in FPGAs.

After completing this section, you should be able to

■ Describe the basic structure of an FPGA   ■ Compare an FPGA to a CPLD   ■ Discuss LUTs   ■ Discuss the SRAM-based FPGA   ■ Define the FPGA core

The basic concept of an FPGA was introduced in Chapter 1. The three basic elements in an **FPGA** are the configurable logic block (CLB), the interconnections, and the input/output (I/O) blocks, as illustrated in Figure 11–31. The configurable logic blocks (CLBs) in an FPGA are not as complex as the LABs or FBs in a CPLD, but generally there are many more of them. When the CLBs are relatively simple, the FPGA architecture is called *fine grained*. When the CLBs are larger and more complex, the architecture is called *coarse grained*. The I/O blocks around the perimeter of the structure provide individually selectable input, output, or bidirectional access to the outside world. The distributed matrix of programmable interconnections provide for interconnection of the CLBs and connection to inputs and outputs. Large FPGAs can have tens of thousands of CLBs in addition to memory and other resources.

Most programmable logic manufacturers make a series of FPGAs that range in density, power consumption, supply voltage, speed, and to some degree vary in architecture. FPGAs are reprogrammable and use SRAM or antifuse process technology for the programmable links. Densities can range from hundreds of logic modules to approximately 180,000 logic modules in packages with up to over 1,000 pins. DC supply voltages are typically 1.2 V to 2.5 V, depending on the specific device.

Configurable Logic Blocks

Typically, an FPGA logic block consists of several smaller logic modules that are the basic building units, somewhat analogous to macrocells in a CPLD. Figure 11–32 shows the fundamental configurable logic blocks (CLBs) within the global row/column programmable interconnects that are used to connect logic blocks. Each **CLB** is made up of multiple smaller logic modules and a local programmable interconnect that is used to connect logic modules within the CLB.

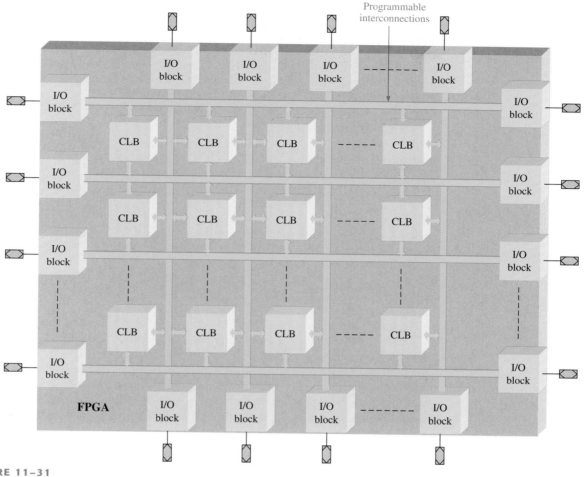

▲ FIGURE 11–31

Basic structure of an FPGA. CLB is configurable logic block.

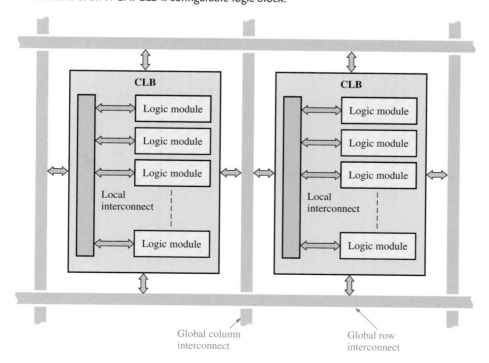

◀ FIGURE 11–32

Basic configurable logic blocks (CLBs) within the global row/column programmable interconnects.

Logic Modules   A logic module in an FPGA logic block can be configured for combinational logic, registered logic, or a combination of both. A flip-flop is part of the associated logic and is used for registered logic. (Flip-flops were covered in Chapter 7.) A block diagram of a typical LUT-based logic module is shown in Figure 11–33. As you know, an LUT (look-up table) is a type of memory that is programmable and used to generate SOP combinational logic functions. The LUT essentially does the same job as the PAL or PLA does.

▶ **FIGURE 11–33**

Basic block diagram of a logic module in an FPGA.

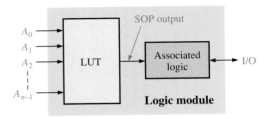

Generally, the organization of an LUT consists of a number of memory cells equal to 2^n, where n is the number of input variables. For example, three inputs can select up to eight memory cells, so an LUT with three input variables can produce an SOP expression with up to eight product terms. A pattern of 1s and 0s can be programmed into the LUT memory cells, as illustrated in Figure 11–34 for a specified SOP function. Each 1 means the associated product term appears in the SOP output, and each 0 means that the associated product term does not appear in the SOP output. The resulting SOP output expression is

$$\overline{A_2}\,\overline{A_1}\,\overline{A_0} + \overline{A_2}A_1A_0 + A_2\overline{A_1}A_0 + A_2A_1A_0$$

▶ **FIGURE 11–34**

The basic concept of an LUT programmed for a particular SOP output.

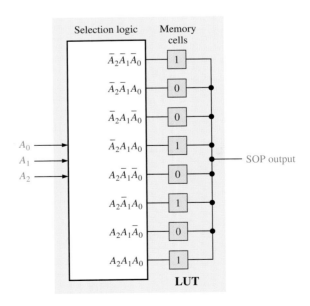

EXAMPLE 11–3

Show a basic 3-variable LUT programmed to produce the following SOP function:

$$A_2 A_1 \overline{A}_0 + A_2 \overline{A}_1 \overline{A}_0 + \overline{A}_2 A_1 A_0 + A_2 \overline{A}_1 A_0 + \overline{A}_2 \overline{A}_1 A_0$$

Solution   A 1 is stored for each product term in the SOP expression, as shown in Figure 11–35.

▶ **FIGURE 11–35**

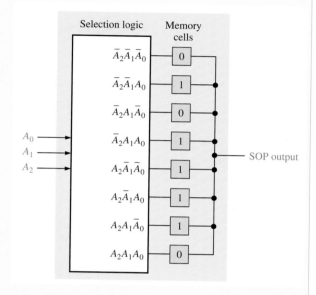

Related Problem   How many memory cells would be in an LUT with four input variables? What would be the maximum possible number of product terms in the SOP output?

SRAM-Based FPGAs

FPGAs are either nonvolatile because they are based on antifuse technology or they are volatile because they are based on SRAM technology. The term *volatile* means that all the data programmed into the configurable logic blocks are lost when power is turned off. Therefore, SRAM-based FPGAs include either a nonvolatile configuration memory embedded on the chip to store the program data and reconfigure the device each time power is turned back on or they use an external memory with data transfer controlled by a host processor. The concept of on-the-chip memory is illustrated in Figure 11–36(a). The concept of the host processor configuration is shown in part (b).

FPGA Cores

FPGAs, as we have discussed, are essentially like "blank slates" that the end user can program for any logic design. FPGAs are available that also contain hard-core logic. A **hard core** is a portion of logic in an FPGA that is put in by the manufacturer to provide a specific function and that cannot be reprogrammed. For example, if a customer needs a small microprocessor as part of a system design, it can be programmed into the FPGA by the customer or it can be provided as hard core by the manufacturer. If the embedded function has some programmable features, it is known as a **soft-core** function. An advantage of the hard-core approach is that

▶ FIGURE 11–36

Basic concepts of volatile FPGA configurations.

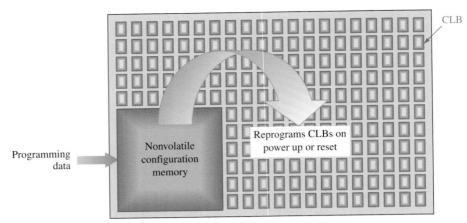

(a) Volatile FPGA with on-the-chip nonvolatile configuration memory

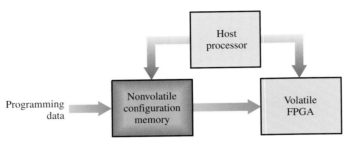

(b) Volatile FPGA with on-board memory and host processor

the same design can be implemented using much less of the available capacity of the FPGA than if the user programmed it in the field, resulting in less space on the chip ("real estate") and less development time for the user. Also, hard-core functions have been thoroughly tested. The disadvantage of the hard core is that the specifications are fixed during manufacturing and the customer must be able to use the hard-core logic "as is." It cannot be changed later.

Hard cores are generally available for functions that are commonly used in digital systems, such as a microprocessor, standard input/output interfaces, and digital signal processors. More than one hard-core function can be programmed in an FPGA. Figure 11–37 illustrates the concept of a hard core surrounded by configurable logic programmed by the user. This is a basic embedded system because the hard-core function is embedded in the user-programmed logic.

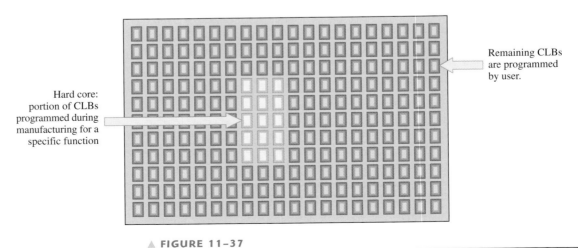

▲ FIGURE 11–37

Basic idea of a hard-core function embedded in an FPGA.

Hard core designs are generally developed by and are the property of the FPGA manufacturer. Designs owned by the manufacturer are termed **intellectual property** (IP). A company usually lists the types of intellectual property that are available on its website. Some intellectual properties are a mix of hard core and soft core. A processor that has some flexibility in the selection and adjustment of certain parameters by the user is an example.

Those FPGAs containing either or both hard-core and soft-core embedded processors and other functions are known as **platform FPGAs** because they can be used to implement an entire system without the need for external support devices.

**SECTION 11–5
REVIEW**

1. How does an FPGA differ from a CPLD?
2. What does CLB stand for?
3. Describe an LUT and discuss its purpose.
4. What is the difference between a local interconnect and a global interconnect in an FPGA?
5. What is an FPGA core?
6. Define the term *intellectual property* in relation to an FPGA manufacturer.

11–6   ALTERA FPGAs

Altera produces several families of FPGAs including the Stratix II, the Stratix, Cyclone, and the ACEX family. In this section, we will focus on only the Stratix II to illustrate the concepts, keeping in mind that other devices in the family may differ basically in certain aspects of their architecture, and/or in the parameters such as densities, speed, and power.

After completing this section, you should be able to

■ Discuss the basic architecture of a typical Stratix II family FPGA ■ Explain how product terms are generated in FPGAs ■ Discuss embedded functions

The Logic Array Block (LAB)

The block diagram of a generic FPGA was shown in Figure 11–31; the architecture of the Stratix II family and other Altera families is similar. They have the classic LUT structure for the logic modules, called adaptive logic modules (ALMs) by Altera, that produce SOP functions. Altera also calls the configurable logic blocks, shown in the generic device, logic array blocks (LABs). The density ranges from almost 2000 LABs to over 22,000 LABs, depending on the particular device in the family; and each LAB has eight ALMs. Package sizes vary from 341 pins to 1,173 pins. Devices requiring dc supply voltages of 1.2 V, 1.5 V, and 2.5 V are typically available. The Stratix II family of FPGAs uses the SRAM-based process technology.

Figure 11–38 is a simplified diagram of the Stratix II LAB structure. Each LAB consists of eight ALMs; multiple LABs are linked together via the global row and column interconnects. The local interconnect links the ALMs within each LAB.

▶ FIGURE 11–38

Simplified diagram of the Stratix II FPGA LAB (logic array block) structure. ALMs are adaptive logic modules.

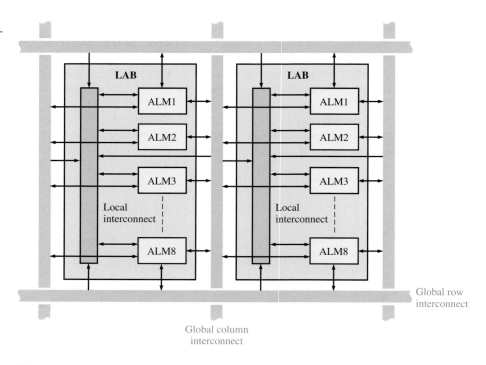

The Adaptive Logic Module (ALM)

The ALM is the basic design unit in the Stratix II FPGA. Each ALM contains an LUT-based combinational logic section and associated logic that can be programmed for two combinational logic outputs or registered outputs. Also, the ALM has adder logic, flip-flops, and other logic that allows for the implementation of arithmetic, counter, and shift register functions. A simplified diagram of a Stratix II ALM is shown in Figure 11–39.

Operating Modes of an ALM   An ALM can be programmed for the following modes of operation:

- Normal mode
- Extended LUT mode
- Arithmetic mode
- Shared arithmetic mode

▶ FIGURE 11–39

Simplified diagram of a Stratix II adaptive logic module (ALM).

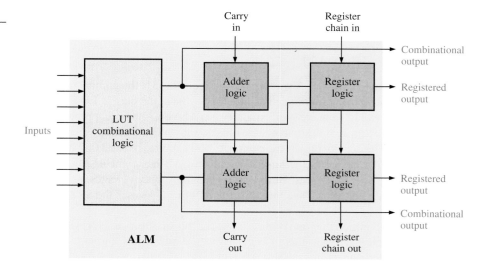

In addition to these four modes, an ALM can be utilized as a register chain to create counters and shift registers. In this section, we will discuss the normal mode and the extended LUT mode.

The *normal mode* is used primarily for generating combinational logic functions. An ALM can implement one or two combinational output functions with its two LUTs. Examples of four LUT configurations are illustrated in Figure 11–40. Two SOP functions, each with four variables or less, can be implemented in an ALM without sharing inputs. For example, you can have two 4-variable functions, one 4-variable function and one 3-variable function or two 3-variable functions. By sharing inputs, you can have any combination of a total of eight inputs up to a maximum of six inputs for each LUT. In the normal mode, you are limited to 6-variable SOP functions.

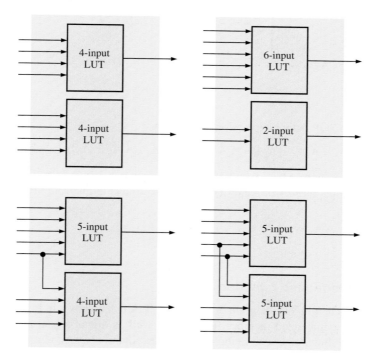

◀ FIGURE 11–40

Examples of possible LUT configurations in an adaptive logic module (ALM) in the normal mode.

The *extended LUT mode* allows expansion to a 7-variable function, as illustrated in Figure 11–41. The multiplexer formed by the AND-OR circuit with a complemented input is part of the dedicated logic in an ALM.

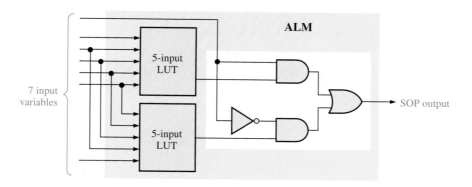

◀ FIGURE 11–41

Expansion of an ALM to produce a 7-variable SOP function in the extended LUT mode.

EXAMPLE 11–4

An ALM in a Stratix II FPGA is configured in the extended LUT mode, as shown in Figure 11–42. For the specific LUT outputs shown, determine the final SOP output.

▶ **FIGURE 11–42**

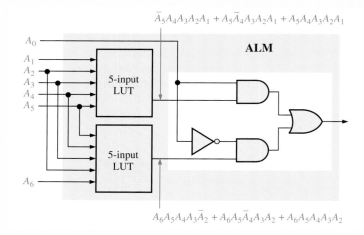

Solution   The SOP output expression is as follows:

$$\overline{A}_5 A_4 A_3 A_2 A_1 A_0 + A_5 \overline{A}_4 A_3 A_2 A_1 A_0 + A_5 A_4 A_3 A_2 A_1 A_0 + A_6 A_5 A_4 A_3 \overline{A}_2 \overline{A}_0 + A_6 A_5 \overline{A}_4 A_3 A_2 \overline{A}_0 + A_6 A_5 A_4 A_3 A_2 \overline{A}_0$$

Related Problem   Show an ALM configured in the normal mode to produce one SOP function with five product terms from one LUT and three product terms from the other LUT.

Embedded Functions

A general block diagram of the Stratix II FPGA is shown in Figure 11–43. The FPGA contains embedded memory functions as well as digital signal processing (DSP) functions. DSP functions, such as digital filters, are commonly used in many systems. As you can see in the block diagram, the embedded blocks are arranged throughout the FPGA interconnection matrix and input/output elements (IOEs) are placed around the FPGA perimeter.

SECTION 11–6 REVIEW

1. What is the basic logic design unit in the Stratix II FPGA?
2. How many ALMs are there in a LAB?
3. What produces combinational logic functions in an ALM?
4. How many SOP functions can one ALM produce?
5. Name the two types of embedded functions in the Stratix II.

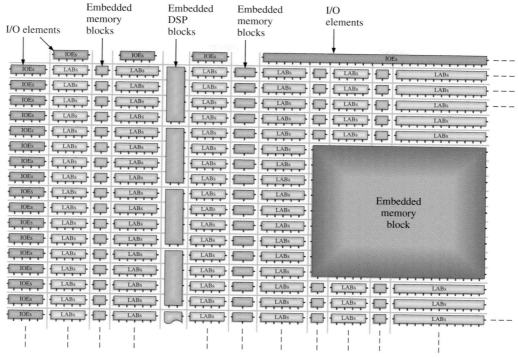

▲ FIGURE 11–43

Stratix II block diagram.

11–7 XILINX FPGAs

Xilinx has two major lines of FPGAs, the Spartan and the Virtex, and there are
different families within each line. Examples are the Spartan 3 and Spartan IIE, Virtex-4,
Virtex II, Virtex II Pro, and Virtex II Pro X. Xilinx designates the Virtex II, Virtex II
Pro, and Virtex II Pro X as platform FPGAs because they have embedded functions,
such as memories, processors, transceivers, and other hard and soft IP cores. The
PGA families differ generally in density and performance parameters. Most of the
Xilinx devices have a traditional FPGA architecture; however, the Virtex II Pro X has
what is called Application Specific Modular Block, ASMBL™ (pronounced *assemble*)
architecture with over a billion transistors in a single device.

After completing this section, you should be able to

- Describe a typical Virtex family FPGA ■ Discuss the basic Virtex architecture
- Explain how product terms are generated by an FPGA ■ Describe the ASMBL
architecture

Configurable Logic Blocks

The configurable logic area (called FPGA fabric) of most Xilinx FPGAs is divided into configurable logic blocks (CLBs) with each CLB containing multiple basic logic units called logic cells (LCs). Each logic cell is based on traditional 4-input LUT logic plus additional logic and a flip-flop. A 4-input LUT can produce from one product term to an SOP function with sixteen product terms. Two identical logic cells are called a *slice*. Figure 11–44 illustrates the levels of configurable logic from the logic cell to the CLB. Densities range from around 2000 to over 74,000 logic cells in a single Virtex device.

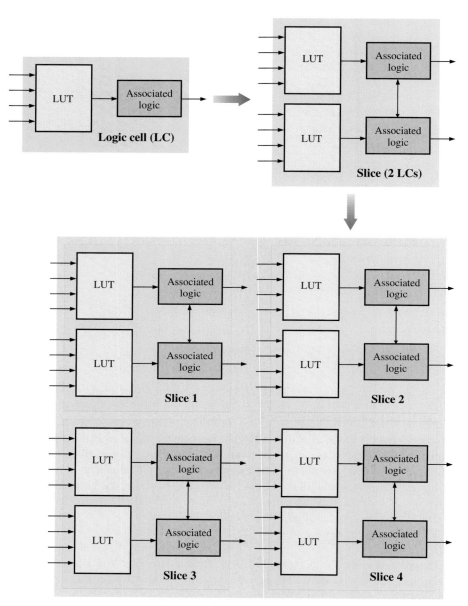

▲ **FIGURE 11–44**

Simplified CLB in a Virtex FPGA.

SOP Cascade Chains

A simplified slice (two logic cells) with cascade chain logic is shown in Figure 11–45. There is a dedicated multiplexer (MUX) within the associated logic of each LC that can be used in the cascade chain and a dedicated OR gate within each slice. Figure 11–45(a) shows an example of how one slice in a CLB can be configured as an AND gate to produce an 8-variable product term. Two slices can be configured to produce an SOP function with two 8-variable product terms, as shown in part (b). An entire CLB of four slices can be configured in a cascade chain to produce an SOP function with four 8-variable product terms, as shown in part (c) on the next page. Further SOP expansion can be done using additional CLBs.

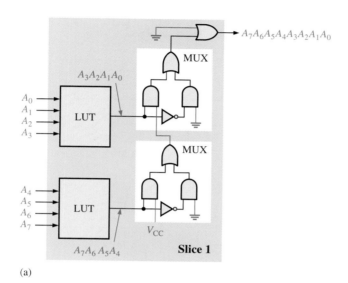

(a)

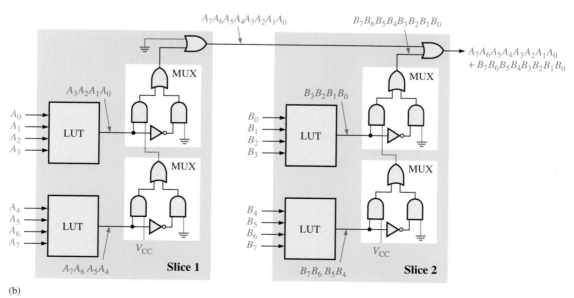

(b)

▲ **FIGURE 11–45**

Example of using cascade chains for expansion of an SOP function.

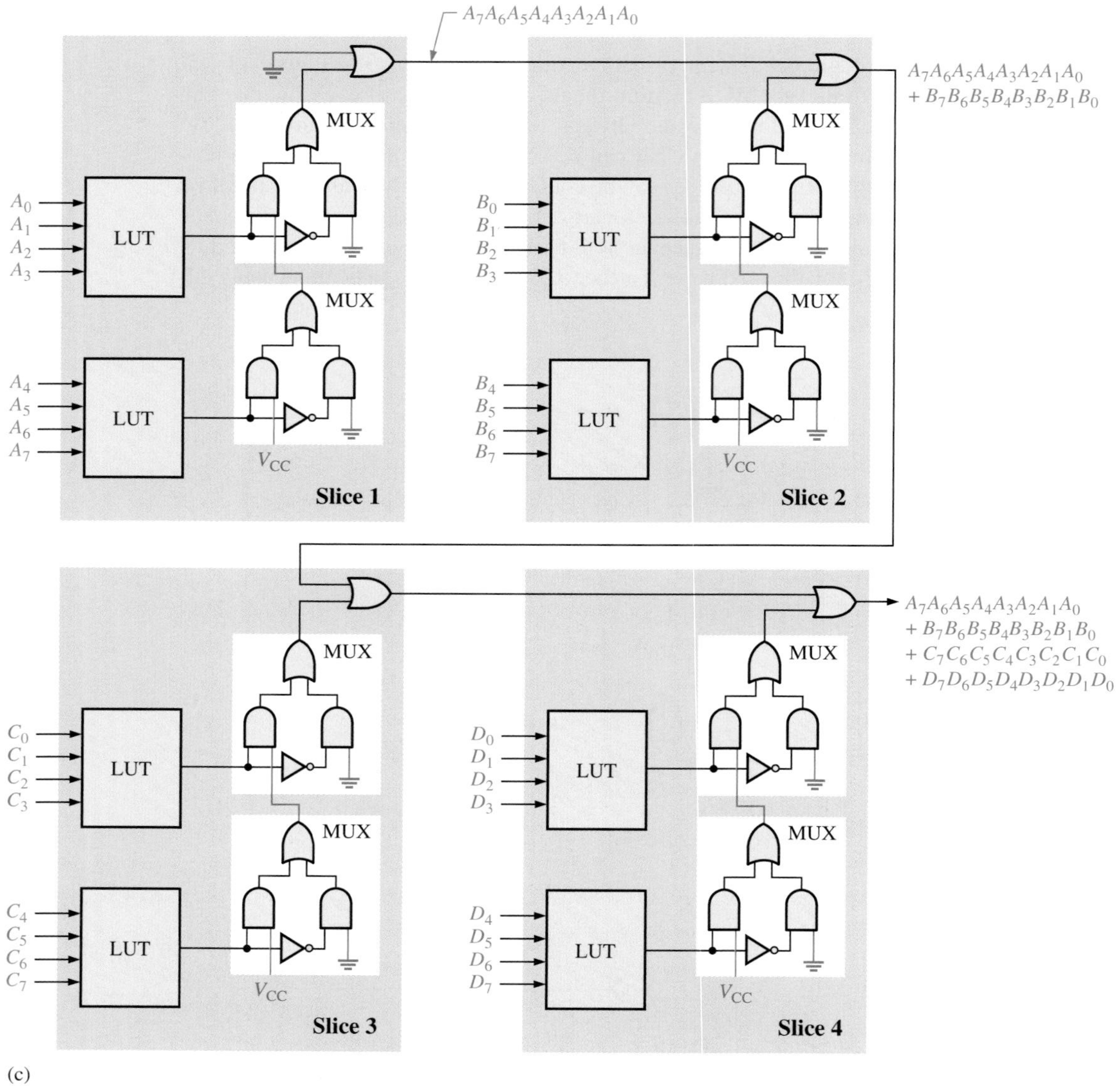

$A_7A_6A_5A_4A_3A_2A_1A_0$

$A_7A_6A_5A_4A_3A_2A_1A_0$
$+ B_7B_6B_5B_4B_3B_2B_1B_0$

$A_7A_6A_5A_4A_3A_2A_1A_0$
$+ B_7B_6B_5B_4B_3B_2B_1B_0$
$+ C_7C_6C_5C_4C_3C_2C_1C_0$
$+ D_7D_6D_5D_4D_3D_2D_1D_0$

(c)

▲ **FIGURE 11–45**

Continued

EXAMPLE 11–5

Show how a 16-input AND gate can be implemented in a CLB.

Solution   Two slices configured as shown in Figure 11–46 result in a 16-input AND gate.

Related Problem   Show how the two slices in Figure 11–46 could be configured to produce the SOP function, $A_7A_6A_5A_4 + A_3A_2A_1A_0 + B_7B_6B_5B_4 + B_3B_2B_1B_0$.

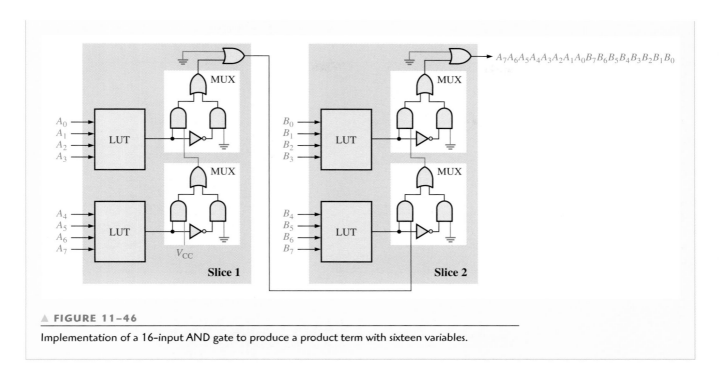

$A_7A_6A_5A_4A_3A_2A_1A_0B_7B_6B_5B_4B_3B_2B_1B_0$

▲ **FIGURE 11–46**

Implementation of a 16-input AND gate to produce a product term with sixteen variables.

Traditional FPGA Architecture vs. ASMBL Architecture

As you have learned, the traditional FPGA architecture appears as an array of logic blocks (CLBs or LABs) surrounded by configurable input/output cells. The amount of configurable logic (CLBs) in an FPGA depends on the number of I/O elements that can be physically placed around the perimeter. When IP cores such as DSP and embedded memory are required, the amount of configurable logic must be sacrificed and at some point additional I/Os may be required. As more IP cores are added, the physical size of the FPGA must be increased to maintain the necessary configurable logic and to increase the number of I/Os. This general concept is illustrated in Figure 11–47.

The more complex the logic on an FPGA, the more I/Os are required. The dependent relationship between logic and I/Os will result in an increase in chip size and cost. Also, another problem with platform FPGAs is that when additional embedded IP core functions are required, a major redesign or partial redesign in chip layout may be required, which is a very costly process.

The ASMBL Architecture Xilinx created a flexible approach to platform FPGAs in the Virtex II Pro X devices in order to overcome some of the limitations of the traditional architecture. The Application Specific Modular Block (ASMBL) architecture is a column-based structure instead of the row/column structure. The I/Os are interspersed throughout rather than positioned around the perimeter, so their number can be increased without increasing chip size. Each column is essentially a strip of logic that can be replaced by another type of logic strip without redesigning the chip layout. Examples of the types of logic strips are configurable logic blocks (CLBs), I/O blocks (IOBs), memory, and hard and soft IP cores such as DSP and processor.

Various numbers of each type of logic strip can be mixed to meet specific application requirements. For example, in the simplest configuration, you could have a mix of CLB strips and I/O block strips, as illustrated in Figure 11–48(a). More or fewer of either could be used depending on the requirements. If you require more memory, one or more CLB strips could be replaced, as indicated in part (b). If your specific area of application is digital signal processing, you could add DSP IP cores with a mix of memory, as shown in part (c). Part (d) shows the addition of processor cores.

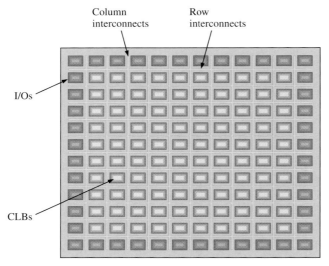

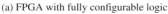

(a) FPGA with fully configurable logic

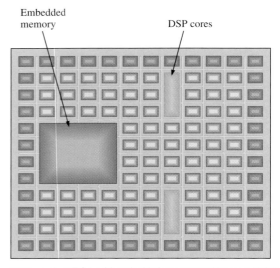

(b) Same size FPGA with embedded memory and IP cores (DSP) results in fewer CLBs and is limited by the perimeter I/Os.

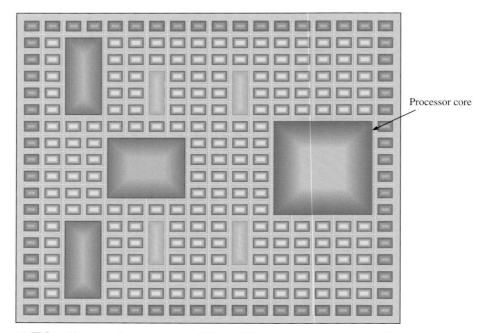

(c) FPGA with more embedded memory, additional DSP cores, and processor core will require a larger physical size at some point.

▲ **FIGURE 11–47**

Embedded IP functions (memory, DSP, and processor) result in less configurable logic and/or a larger physical chip size due to increased I/Os.

SECTION 11–7 REVIEW

1. What does CLB in a Xilinx FPGA consist of?
2. What does an LC consist of?
3. Describe a slice in a Xilinx FPGA.
4. What is an SOP cascade chain?
5. What does ASMBL stand for?

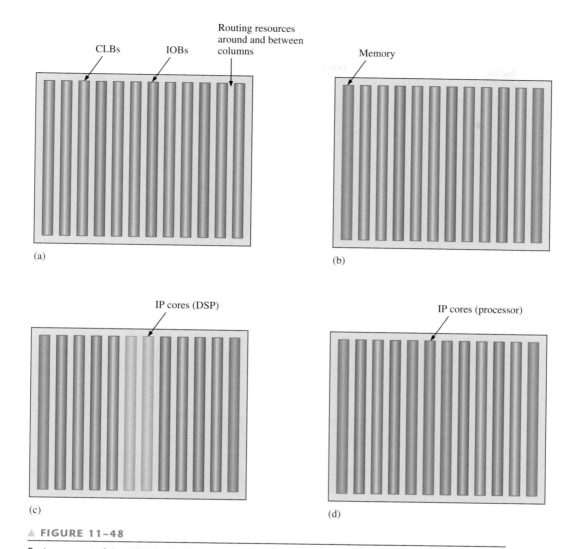

CLBs   IOBs   Routing resources around and between columns   Memory

(a)   (b)

IP cores (DSP)   IP cores (processor)

(c)   (d)

▲ FIGURE 11–48

Basic concept of the ASMBL platform–FPGA architecture.

11–8 PROGRAMMABLE LOGIC SOFTWARE

In order to be useful, programmable logic must have both hardware and software components combined into a functional unit. All manufacturers of SPLDs, CPLDs, and FPGAs provide software support for each hardware device. These software packages are in a category of software known as computer aided design (CAD). PLD programming was introduced in Chapter 1 and covered further in Chapter 3. In this section, programmable logic software is presented in a generic way.

After completing this section, you should be able to

■ Explain the programming process in terms of design flow ■ Describe the design entry phase ■ Describe the functional simulation phase ■ Describe the synthesis phase ■ Describe the implementation phase ■ Describe the timing simulation phase ■ Describe the download phase

The programming process is generally referred to as **design flow.** A basic design flow diagram for implementing a logic design in a programmable device is shown in Figure 11–49. Most specific software packages incorporate these elements in one form or another and process them automatically. The device being programmed is usually referred to as the **target device.**

▶ FIGURE 11–49

General design flow diagram for programming a SPLD, CPLD, or FPGA.

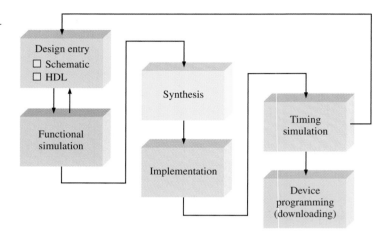

You must have four things to get started programming a device: a computer, development software, a programmable logic device (SPLD, CPLD, or FPGA), and a way to connect the device to the computer. These essentials are illustrated in Figure 11–50. Part (a)

(a) Computer

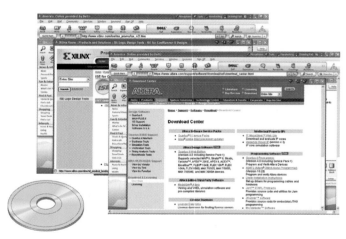

(b) Software (CD or Website download)

(c) Device

(d) Programming hardware (programming fixture or development board with cable for connection to computer port)

▲ FIGURE 11–50

Essential elements for programming an SPLD, CPLD, or FPGA.

shows a computer that meets the system requirements for the particular software you are using. Part (b) shows the software acquired either on a CD from the device manufacturer or downloaded from the device manufacturer's website. Most manufacturers provide free software that can be downloaded and used for a limited time. Part (c) shows a programmable logic device. Part (d) illustrates two means of physically connecting the device to the computer via cable by using either the programming fixture into which the device is inserted or the development board on which the device is mounted. After the software has been installed on your computer, you must become familiar with the particular software tools before attempting to connect and program a device. This learning process will require considerable effort and time.

Design Entry

Assume that you have a logic circuit design that you wish to implement in a programmable device. You can enter the design on your computer in either of two basic ways: **schematic entry** or **text entry.** In order to use text entry, you must be familiar with an HDL such as VHDL, Verilog, ABEL, or AHDL. Most programmable logic manufacturers provide software packages that support VHDL and Verilog because they are standard HDLs. Some also support ABEL, AHDL, or other proprietary HDLs. Schematic entry basically allows you to place symbols of logic gates and other logic functions from a library on the screen and connect them as required by your design. A knowlege of an HDL is not required for schematic entry. Figure 11–51 illustrates these two types of entry generically for a simple AND-OR logic circuit.

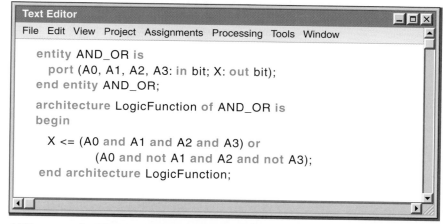

(a) Text entry using VHDL to describe an AND-OR logic circuit

◄ **FIGURE 11–51**

Examples of text and schematic entry screens.

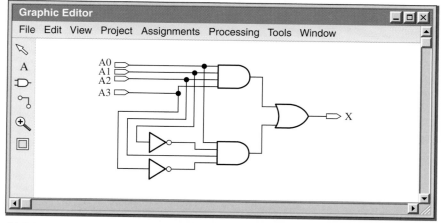

(b) Schematic entry of the same AND-OR logic circuit entered in (a)

Building a Logic Schematic   When you enter a complete logic circuit on the screen, it is called a "flat" schematic. More complex logic circuits may be hard to fit onto the screen. You can enter a logic circuit in segments, save each segment as a block symbol, and then connect the block symbols to form the complete circuit. This is called a hierarchical approach.

As an example, let's assume that you need a circuit that will produce the following SOP expression:

$$Z = (A_3A_2A_1A_0 + \overline{A}_3A_2\overline{A}_1A_0) + (A_3\overline{A}_2\overline{A}_1A_0 + A_3\overline{A}_2A_1\overline{A}_0 + \overline{A}_3A_2A_1\overline{A}_0)$$

Let's use the hierarchical approach and create the logic for each of the two parenthetical terms in the above equation; reduce each logic circuit to a single graphic block symbol; and then, when both circuits are complete, place them on the screen and connect their outputs to an OR gate. This is illustrated in the five parts of Figure 11–52. The entire circuit could be entered on the screen at one time, but the hierarchical approach is useful when the logic circuit is larger and must be broken down into parts.

In part (e) of Figure 11–52, the logic could be reduced to another block symbol and used in an even larger logic design; or it could be saved and reused in other designs, as illustrated in Figure 11–53.

After a logic circuit has been entered as a schematic, a program application called a **compiler** controls the various CAD tools that process the schematic and produces an implementation for the target device.

▶ **FIGURE 11–52**

Example of creating logic in segments and then combining the segments.

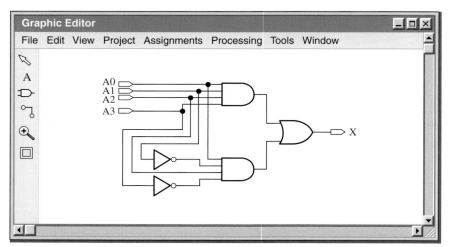

(a) Enter the two-product term AND-OR logic.

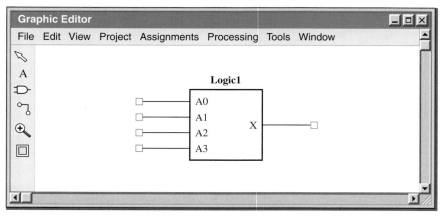

(b) Reduce the AND-OR logic to a block symbol defined as Logic1.

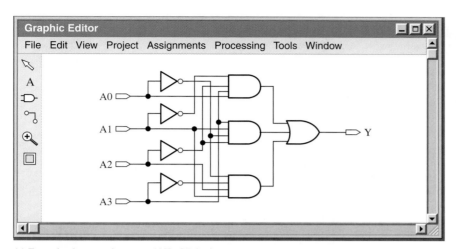

(c) Enter the three-product term AND-OR logic.

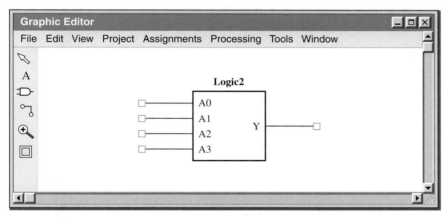

(d) Reduce to a block symbol defined as Logic2.

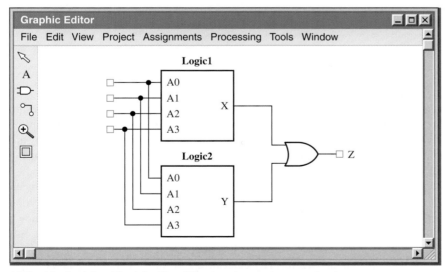

(e) Combine Logic1 and Logic2 with an OR gate and connect common inputs.

▶ **FIGURE 11–53**

The logic circuit with two logic blocks and an OR gate is reduced to another logic block, Logic3.

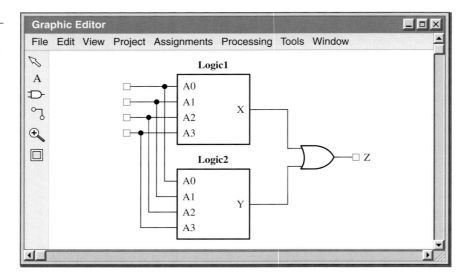

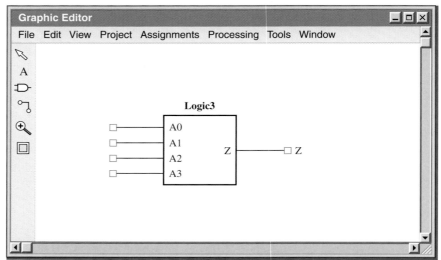

Functional Simulation

The purpose of the functional simulation in the design flow is to make sure that the design you entered works as it should, in terms of its logic operation, before synthesizing into a hardware design. Basically, after a logic circuit is compiled, it then can be simulated by applying input waveforms and checking the output for all possible input combinations using a waveform editor.

The waveform editor allows you to select the nodes (inputs and outputs) that you want to test. The selected input and output names appear on the Waveform Editor screen along with a symbol or other designation that identifies each as an input or an output, as shown in Figure 11–54. Initially, all four inputs default to 0, and the crosshatch indicates the output is unknown. You can select the time intervals for the display.

Next, you create each input waveform by entering a 1 or 0 for each time interval (interval between dashed lines in Figure 11–55). This is usually accomplished by a point, click, and select process with your mouse, depending on the specific software. For this particular case, you would create the waveforms so that all 16 possible combinations of 4 inputs are represented. Figure 11–55 shows the input waveforms (A0, A1, A2, A3) as they have been specified.

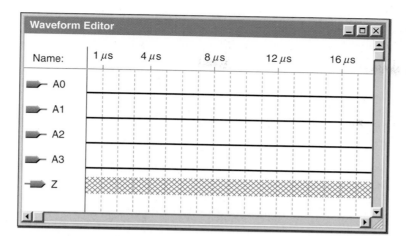

◀ FIGURE 11–54

Generic Waveform Editor screen with input and output names specified for the circuit that was entered in Figure 11–53.

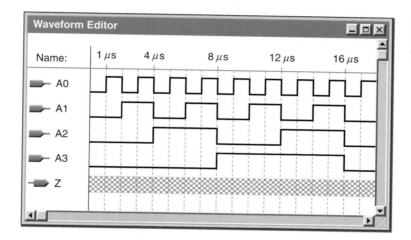

◀ FIGURE 11–55

Input waveforms are specified on the Waveform Editor screen.

After you have specified the input waveforms, generally a simulation control window opens, allowing you to set the start and end times for the simulation and specify the time intervals to be displayed. When the simulation is started, the output waveform Z will be displayed in the Waveform Editor, as shown in Figure 11–56. This allows you to verify that the design is good or that it is not working properly. In this case, the output waveform is correct for the selected input waveforms. An incorrect output waveform would indicate a flaw in the functionality of the logic; and you would have to go back, check the original design, and then re-enter a revised design.

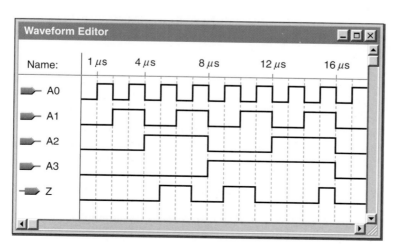

◀ FIGURE 11–56

After the functional simulation is run, the output waveform should indicate that the logic is functioning properly.

Synthesis

Once the design has been entered and functionally simulated to verify that its logic operation is correct, the compiler automatically goes through several phases to prepare the design to be downloaded to the target device. During the synthesis phase of the design flow, the design is optimized in terms of minimizing the number of gates, replacing logic elements with other logic elements that can perform the same function more efficiently, and eliminating any redundant logic. The final output from the synthesis phase is a netlist that describes the optimized version of the logic circuit.

Netlist   A **netlist** is basically a connectivity list that describes components and how they are connected together. Generally, a netlist contains references to descriptions of the components or elements used. Each time a component, such as a logic gate, is used in a netlist, it is called an *instance*. Each instance has a definition that lists the connections that can be made to that kind of component and some basic properties of that component. These connection points are called *ports* or *pins*. Usually, each instance will have a unique name; for example, if you have two instances of AND gates, one might be "and1" and the other "and2." Besides their names, they might otherwise be identical. Nets are the "wires" that connect things together in the circuit. Net-based netlists usually describe all the instances and their attributes, then describe each net, and specify which port they are connected to on each instance.

The AND-OR logic circuit that you entered in the design phase, shown in Figure 11–57(a), could result in the optimized circuit shown in Figure 11–57(b). In this illustration, the compiler removed the three OR gates and replaced them with one 5-input OR gate. Also, two redundant inverters were removed.

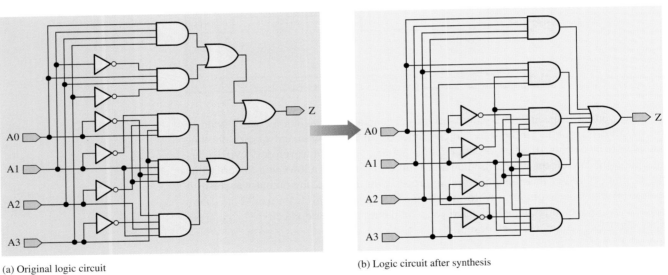

(a) Original logic circuit

(b) Logic circuit after synthesis

▲ **FIGURE 11–57**

Example of logic optimization during synthesis.

The synthesis software generates a netlist. To illustrate the concept of a generic netlist. Figure 11–58(a) shows net assignments, instance assignments, and I/O assignments in red. The netlist shown in Figure 11–58(b) does not necessarily resemble any specific netlist format or syntax. This hypothetical netlist simply indicates the type of information that would be necessary to describe a circuit. One format used for netlists is **EDIF** (Electronic Design Interchange Format).

Implementation (Software)

After the design has been synthesized, the compiler implements the design, which is basically a "mapping" of the design so that it will fit in the specific target device based on its

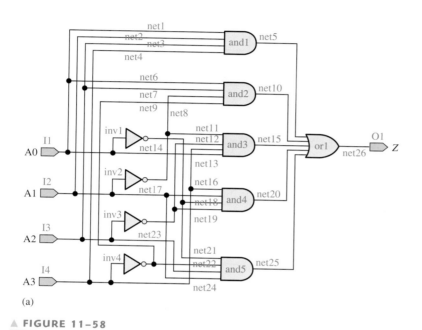

(a)

Netlist (Logic3)
net<name>: instance<name>, <from>; <to>;
instances: and1, and2, and3, and4, and5, or1, inv1, inv2,
inv3, inv4;
Input/outputs: I1, I2, I3, I4, O1;
net1: and1, inport1; I1;
net2: and1, inport2; I2;
net3: and1, inport3; I3;
net4: and1, inport4; I4;
net5: and1, outport1; or1, inport1;
net6: and2, inport1; I1;
net7: and2, inport2; I3;
net8: and2, inport3; inv2, outport1;
net9: and2, inport4; inv4, outport1;
net10: and2, outport1; or1, inport2;
net11: and3, inport1; inv2, outport1;
net12: and3, inport2; inv3, outport1;
net13: and3, inport3; I4;
net14: and3, inport4; I1;
net15: and3, outport1; or1, inport3;
net16: and4, inport1; I4;
net17: and4, inport2; I2;
net18: and4, inport3; inv1, outport1;
net19: and4, inport4; inv3, outport1;
net20: and4, outport1; or1, inport4;
net21: and5, inport1; inv1, outport1;
net22: and5, inport2; inv4, outport1;
net23: and5, inport3; I3;
net24: and5, inport4; I2;
net25: and5, outport1; or1, inport5;
net26: or1, outport1; O1;
end

(b)

▲ **FIGURE 11–58**

Synthesis produces a netlist for the optimized logic circuit.

architecture and pin configurations. This process is called *fitting* or *place and routing*. To accomplish the implementation phase of the design flow, the software must "know" about the specific device and have detailed pin information. Complete data on all potential target devices are generally stored in the software library.

Timing Simulation

This part of the design flow occurs after the implementation and before downloading to the target device. The timing simulation verifies that the circuit works at the design frequency and that there are no propagation delays or other timing problems that will affect the overall operation. Since a functional simulation has already been done, the circuit should work properly from a logic point of view. The development software uses information about the specific target device, such as propagation delays of the gates, to perform a timing simulation of the design. For the functional simulation, the specification of the target device was not required; but for the timing simulation, the target device must be chosen. The Waveform Editor can be used to view the result of the timing simulation just as with the functional simulation, as illustrated in Figure 11–59. If there are no problems with the timing, as shown in part (a), the design is ready to download. However, suppose that the timing simulation reveals a "glitch" due to propagation delay, as shown in Figure 11–59(b). A glitch is a very short duration spike in the waveform. In this event, you would need to carefully analyze the design for the cause, then re-enter the modified design, and repeat the design flow process. Remember, you have not committed the design to hardware at this point.

▶ **FIGURE 11–59**

Hypothetical examples of timing simulation results.

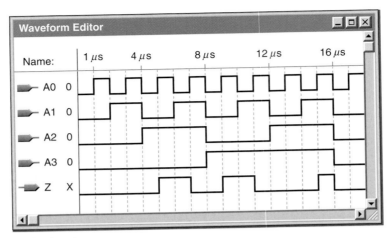

(a) Good result

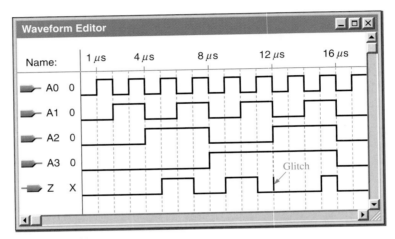

(b) Timing problem

Device Programming (Downloading)

Once the functional and timing simulations have verified that the design is working properly, you can initiate the download sequence. A **bitstream** is generated that represents the final design, and it is sent to the target device to automatically configure it. Upon completion, the design is actually in hardware and can be tested in-circuit. Figure 11–60 shows the basic concept of **downloading.**

A Real Estate Analogy

One way to think of the implementation and downloading processes is to use a real estate development analogy. The developer starts with a tract of land, surveys it, and divides it into lots (analogous to an unprogrammed device). A site plan of the development with all the lots, buildings, roads, and utilities is conceived and laid out (analogous to design entry). Next, it is necessary to verify that the number of buildings and infrastructure will fit on the tract of land and meet all local codes. The site plan also shows the placement of each building and shows the routing of each street and sidewalk (analogous to synthesis and implementation). Not until this mapping has occurred can the developer begin physically constructing the buildings, roads, and utilities (analogous to downloading). This real estate analogy is illustrated in Figure 11–61(a). The process for placing a logic design into a programmable device is depicted in part (b), where the mapping phase is analogous to the site plan and the downloading is analogous to placing physical structures on the site.

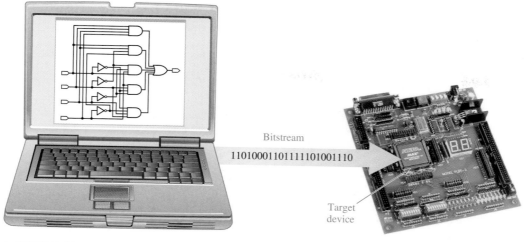

▲ **FIGURE 11-60**

Downloading a design to the target device.

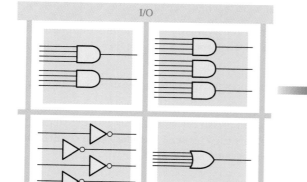

Site plan must be completed before the development is physically implemented. This is analogous to the implementation phase of the design flow for programmable logic.

(a)

Physical structures are placed only after the site plan verifies if and how everything will fit on the property. This is analogous to downloading the design into the target device.

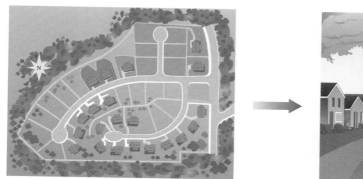

Fitting the software design to the "map" of the target device. (Analogous to site plan)

(b)

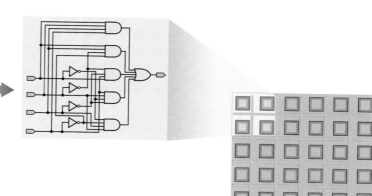

Downloading the design to the target device. (Analogous to placing physical structures)

▲ **FIGURE 11-61**

Analogy of real estate development to implementing (site plan) and downloading (construction) a logic design to target device.

11–9   BOUNDARY SCAN LOGIC

Boundary scan is used for both the testing and the programming of the internal logic of a programmable device. The JTAG standard for boundary scan logic is specified by IEEE Std. 1149.1. Most programmable logic devices are JTAG compliant. In this section, the basic architecture of a JTAG IEEE Std. 1149.1 device is introduced and discussed in terms of the details of its boundary scan register and control logic structure.

After completing this section, you should be able to

■ Describe the required elements of a JTAG compliant device ■ List the mandatory JTAG inputs and outputs ■ State the purpose of the boundary scan register ■ State the purpose of the instruction register ■ Explain what the bypass register is for

IEEE Std. 1149.1 Registers

All programmable logic devices that are compliant with IEEE Std. 1149.1 require the elements shown in the simplified diagram in Figure 11–62. These are the boundary scan register, the bypass register, the instruction register, and the TAP (test access port) logic. A fifth register, the identification register, is optional and not shown in the figure.

Boundary Scan (BS) Register   The interconnected BSCs (boundary scan cells) form the boundary scan register. The serial input to the register is the TDI (test data in), and the serial output is TDO (test data out). Data from the internal logic and the input and output pins of the device can also be parallel shifted into the BS register. The BS register is used to test connections between PLDs and the internal logic that has been programmed into the device.

Bypass (BP) Register   This required data register (typically only one flip-flop) optimizes the shifting process by shortening the path between the TDI and the TDO in case the BS register or other data register is not used.

Instruction Register   This required register stores instructions for the execution of various boundary scan operations.

Identification (ID) Register   An identification register is an optional data register that is not required by IEEE Std. 1149.1. However, it is used in some boundary scan architectures to store a code that identifies the particular programmable device.

IEEE Std. 1149.1 Boundary Scan Instructions

Several standard instructions are used to control the boundary scan logic. In addition to these, other optional instructions are available.

■ *BYPASS*   This instruction switches the BP register into the TDI/TDO path.

■ *EXTEST*   This instruction switches the BS register into the TDI/TDO path and allows external pin tests and interconnection tests between the output of one programmable logic device and the input of another.

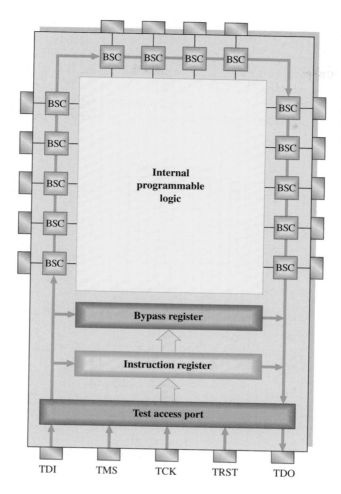

Greatly simplified diagram of a JTAG (IEEE Std. 1149.1) compliant programmable logic device (CPLD or FPGA). The BSCs (boundary scan cells) form the boundary scan register. Only a small number of BSCs are shown for illustration.

- *INTEST* This instruction switches the BS register into the TDI/TDO path and allows testing of the internal programmed logic.

- *SAMPLE/PRELOAD* This instruction is used to sample data at the device input pins and apply the data to the internal logic. Also, it is used to apply data (preload) from the internal logic to the device output pins.

- *IDCODE* This instruction switches the optional identification register into the TDI/TDO path so the ID code can be shifted out to the TDO.

IEEE Std. 1149.1 Test Access Port (TAP)

The Test Access Port (TAP) consists of control logic, four mandatory inputs and outputs, and one defined optional input, Test Reset (TRST).

- *Test Data In (TDI)* The TDI provides for serially shifting test and programming data as well as instructions into the boundary scan logic.

- *Test Data Out (TDO)* The TDO provides for serially shifting test and programming data as well as instructions out of the boundary scan logic.

- *Test Mode Select (TMS)* The TMS switches between the states of the TAP controller.

- *Test Clock (TCK)* The TCK provides timing for the TAP controller which generates control signals for the data registers and the instruction register.

A block diagram of the boundary scan logic is shown in Figure 11–63. Both instructions and data are shifted in on the TDI line. The TAP controller directs instructions into the instruction register or data into the appropriate data register. A decoded instruction from the instruction decoder selects which data register is to be accessed via MUX 1 and also if an instruction or data are to be shifted out on the TDO line via MUX 2. Also, a decoded instruction provides for setting up the boundary scan register in one of five basic modes. The boundary scan cell and its modes of operation are described next.

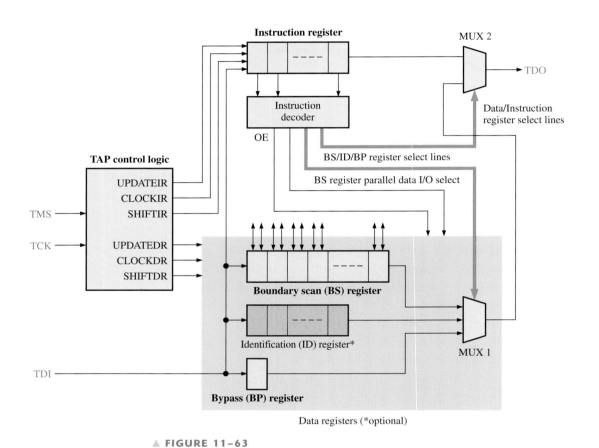

▲ FIGURE 11–63

Boundary scan logic diagram.

The Boundary Scan Cell (BSC)

The boundary scan register is made up of boundary scan cells. A block diagram of a basic BSC is shown in Figure 11–64. As indicated, data can be serially shifted in and out of the BSC. Also, data can be shifted into the BSC from the internal programmable logic, from a device input pin, or from the previous BSC. Additionally, data can be shifted out of the BSC to the internal programmable logic, to a device output pin, or to the next BSC.

The architecture of a generic boundary scan cell is shown in Figure 11–65. The cell consists of two identical logic circuits each containing two flip-flops and two multiplexers. Essentially, one circuit allows data to be shifted from the internal programmable logic or to a device output pin. The other circuit allows data to be shifted from a device input pin or to the internal programmable logic.

There are five modes in which the BSC can operate in terms of data flow. The first BSC mode allows data to flow serially from the previous BSC to the next BSC, as illustrated in

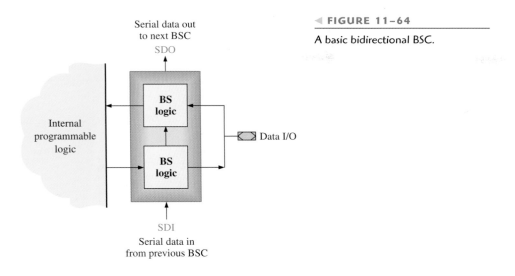

A basic bidirectional BSC.

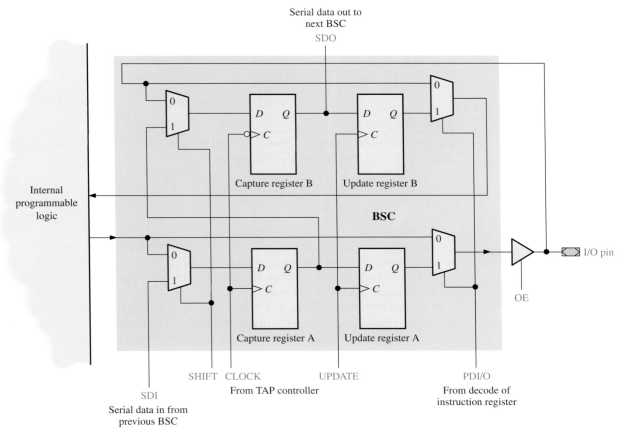

▲ FIGURE 11–65

Representative architecture of a typical boundary scan cell.

Figure 11–66. A 1 on the SHIFT input selects the SDI. The data on the SDI line are clocked into Capture register A on the positive edge of the CLOCK. The data are then clocked into Capture register B on the negative edge of the CLOCK and appear on the SDO line. This is equivalent to serially shifting data through the boundary scan register.

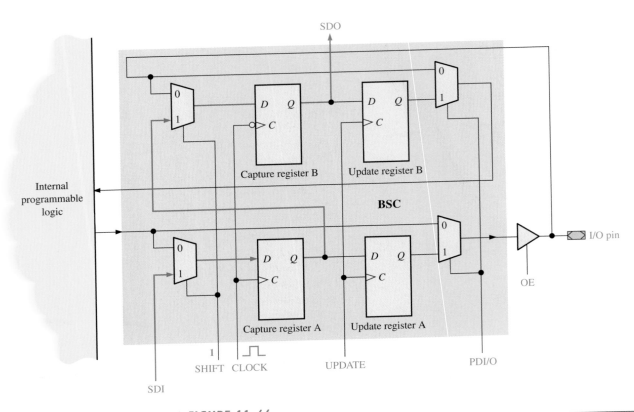

▲ **FIGURE 11–66**

Data path for serially shifting data from one BSC to the next. There is a 1 on the SHIFT input and a CLOCK pulse is applied. The red lines indicate data flow.

The second BSC mode allows data to flow directly from the internal programmable logic to a device output pin, as illustrated in Figure 11–67. The 0 on the PDI/O (parallel data I/O) control line selects the data from the internal programmable logic. The 1 on the OE (output enable) line enables the output buffer.

The third BSC mode allows data to flow directly from a device input pin to the internal programmable logic, as illustrated in Figure 11–68. The 0 on the PDI/O (parallel data I/O) control line selects the data from the input pin. The 0 on the OE (output enable) line disables the output buffer.

The fourth BSC mode allows data to flow from the SDI to the internal programmable logic, as illustrated in Figure 11–69. A 1 on the SHIFT input selects the SDI. The data on the SDI line are clocked into Capture register A on the positive edge of the CLOCK. The data are then clocked into Capture register B on the negative edge of the CLOCK and appear on the SDO line. A pulse on the UPDATE line clocks the data into Update register B. A 1 on the PDI/O line selects the output of Update register B and applies it to the internal programmable logic. The data also appear on the SDO line.

The fifth BSC mode allows data to flow from the SDI to a device output pin and to the SDO output, as illustrated in Figure 11–70. A 1 on the SHIFT input selects the SDI. The data on the SDI line are clocked into Capture register A on the positive edge of the CLOCK. The data are then clocked into Capture register B on the negative edge of the CLOCK and appear on the SDO line. A pulse on the UPDATE line clocks the data into Update register A. With a 1 on OE, a 1 on the PDI/O line selects the output of Update register A and applies it to the device output pin.

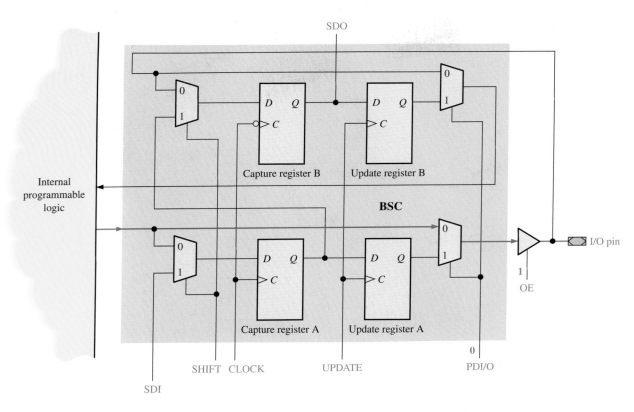

▲ FIGURE 11–67

Data path for transferring data from the internal programmable logic to a device output pin. There is a 0 on the PDI/O line and a 1 on the OE line.

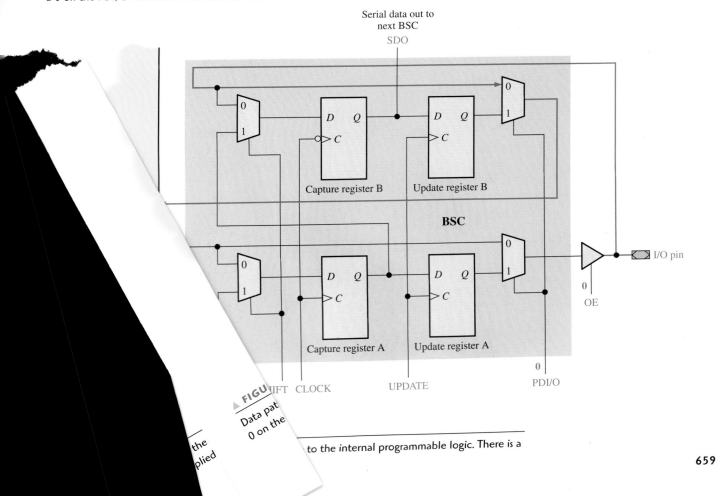

Serial data out to
next BSC

to the internal programmable logic. There is a

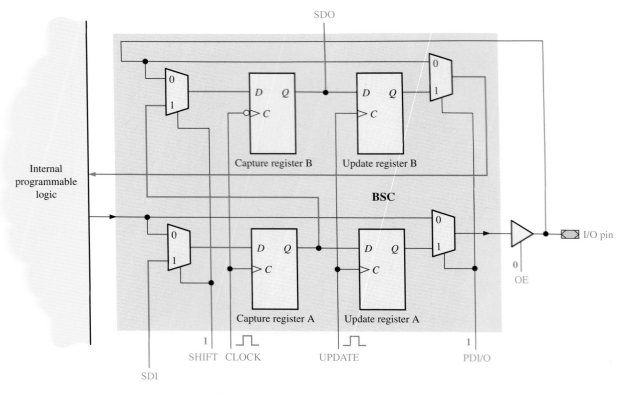

▲ FIGURE 11–69

Data path for transferring data from the SDI to the internal programmable logic and the SDO. There is a 1 on the SHIFT line, a 1 on the PDI/O line, and a 0 on the OE line. A pulse is applied to the CLOCK line followed by a pulse on the UPDATE line.

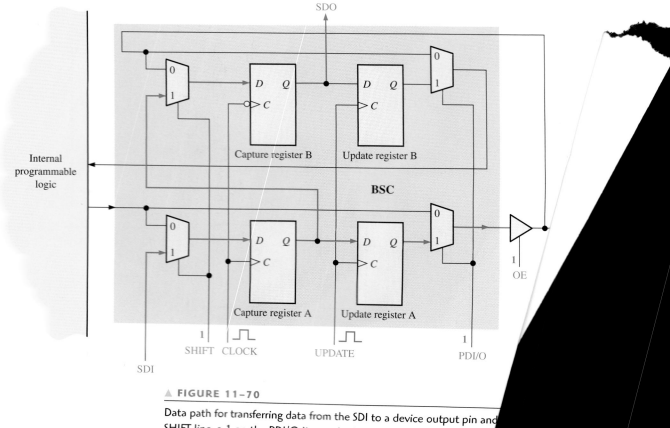

▲ FIGURE 11–70

Data path for transferring data from the SDI to a device output pin and
SHIFT line, a 1 on the PDI/O line and a 1 on the OE line. A pulse is a
followed by a pulse on the UPDATE line.

Boundary Scan Testing of Multiple Devices

Boundary scan testing can be applied to printed circuit boards on which multiple JTAG (IEEE Std. 1149.1) devices are mounted to check interconnections as well as internal logic. This concept is illustrated by tracing the path of data shown in red in Figure 11–71.

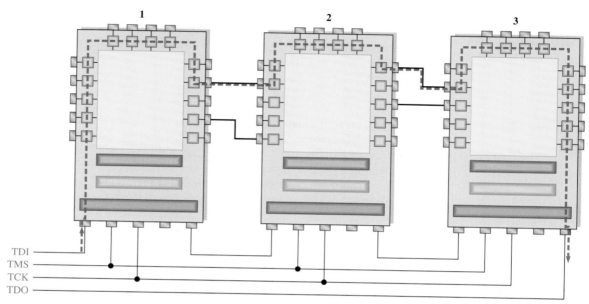

▲ **FIGURE 11–71**

Basic concept of boundary scan testing of multiple devices and interconnections. The test path is shown in red.

The bit is shifted into the TDI of device 1 and through the BS register of device 1 to a cell where the connection to be tested goes to device 2. The bit is shifted out to the device output pin and through the interconnection to the input pin of device 2. The bit continues through the BS register of device 2 to an output pin and through the interconnection to the input pin of device 3. It is then shifted through the BS register of device 3 to the TDO. If the bit coming out of the TDO is the same as the bit going into the TDI, the boundary scan cells through which it was shifted and the interconnections from device 1 to device 2 and from device 2 to device 3 are good.

SECTION 11–9 REVIEW

1. List the boundary scan inputs and outputs required by IEEE Std.1149.1.
2. What is the TAP?
3. Name the mandatory registers in boundary scan logic.
4. Describe five modes in which a boundary scan cell can operate in terms of data flow.

11–10 TROUBLESHOOTING

Two basic ways to test a device that has been programmed with a logic design are traditional and automated. In the traditional method, common laboratory test instruments can be used to check the operation. In the automated method, three fundamental approaches can be used for testing: bed-of-nails, flying probe, and boundary scan.

After completing this section, you should be able to

■ Describe traditional testing ■ Describe bed-of-nails and flying probe testing and discuss their limitations ■ Discuss the JTAG standard ■ Describe the basic concept of boundary scan ■ Explain the modes of boundary scan testing and briefly discuss BSDL

After you commit a logic design to hardware, you can test the device on a PC board. For relatively simple designs, you can check the device using standard laboratory test instruments such as the oscilloscope or logic analyzer, signal generator, and dc power supply. You can apply input signals to input pins on the board and check the output pins for the proper waveforms. This traditional approach, illustrated in Figure 11–72, is practical for one-of-a kind evaluation boards and for the preproduction testing of prototype circuits.

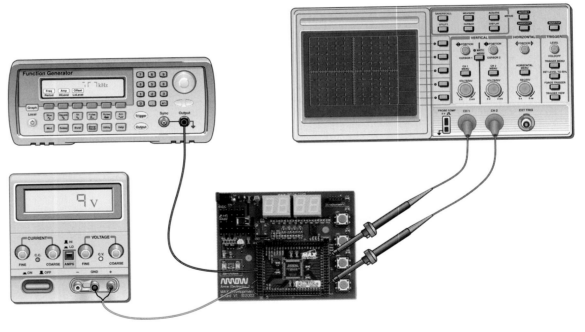

▲ FIGURE 11–72

Traditional testing using laboratory instruments.

Bed-of-Nails Testing

The testing of printed circuit boards at production levels must be done automatically. The **bed-of-nails** (BON) method was one of the first approaches to automated testing. The concept is illustrated in Figure 11–73 where the PC board is placed on a fixture with an array of small nail-like test probes that make contact with test pads on the board. The "nails" are arranged in an array that lines up with the test pad pattern on the board. With this method,

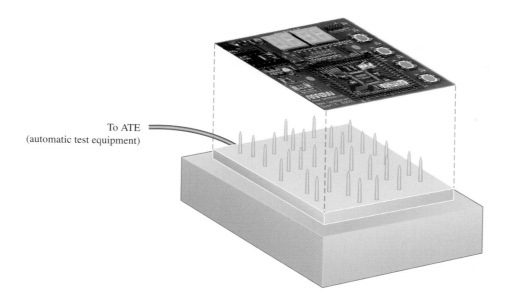

◀ **FIGURE 11–73**

Basic concept of bed–of–nails method for testing a circuit board.

To ATE (automatic test equipment)

the test points can be checked simultaneously with special automated test equipment. Basically, the purpose of automated production testing is to find any manufacturing flaws, such as open or shorted pins and wrong, missing, or misaligned components. This automated process does not primarily test for functionality of the logic. It is assumed that each component had been tested for functionality prior to installation on the circuit board and that the only flaws should be those created during manufacturing.

As integrated circuit devices became smaller and more complex, the trend to surface-mount technology increased, and circuit boards changed from double-sided to multilayer. The increased density and complexity of circuit boards and devices, with large numbers of very closely spaced pins, resulted in limited access to test points on the board using the bed-of-nails approach.

Flying Probe Testing

Another method for testing printed circuit boards is called the **flying probe** method. A typical flying probe and its basic operation are shown in Figure 11–74. A test probe is

(a) 3-axis movement

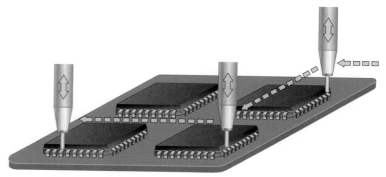

(b) Movement from point to point

▲ **FIGURE 11–74**

A flying probe testing a circuit board.

positioned above a circuit board that is to be tested. The probe can be automatically moved in three axes—along the *x*-axis, the *y*-axis, and the *z*-axis of the board—to make contact with any specified test points. The movement of the probe is controlled by software that uses the physical layout of the board to determine the coordinates. Many flying probe testers have multiple probes for one board.

The flying probe method of testing overcomes some of the limitations of the bed-of-nails. First, the BON method requires a different fixture for each type of circuit board, but the flying probe method requires no fixture. Also, the flying probe can access more points on a board because the probe can be moved to any position and it can access the top of the board where the components are. A drawback of the flying probe method is that it is slower than the BON and so is generally limited to testing prototypes and small production quantities.

Boundary Scan Testing

Limited access to test points led to the concept of placing the test points within the integrated circuit devices themselves. Most CPLDs and FPGAs include boundary scan logic as part of their internal structure independent of the functionality of the logic programmed into the device. These devices are JTAG compliant.

A circuit, known as a boundary scan cell, is placed between the programmable logic and each input and output pin of the device, as shown in Figure 11–75. The cells are basically memory cells that store a 1 or a 0. The cells connected to the programmable logic inputs are called input cells, and those connected to the programmable logic outputs are called output cells. **Boundary scan** testing is based on the JTAG standard (IEEE Std. 1149.1). The four JTAG inputs and outputs—TDI (test data in), TDO (test data out), TCK (test clock), and TMS (test mode select)—are known as the test access ports (TAP).

▶ **FIGURE 11–75**

Basic concept of boundary scan logic in a programmable device.

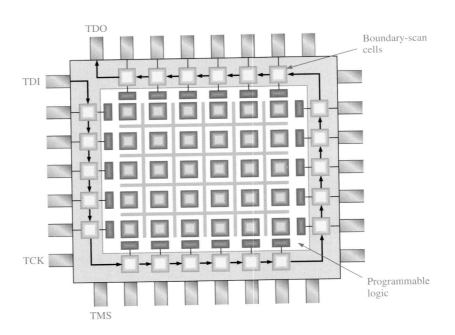

Intest When boundary scan cells are used to test the internal functionality of the device, the test mode is called Intest. The basic concept of boundary scan using Intest is as follows: A software-driven pattern of 1s and 0s is shifted in via the TDI pin and is placed on the programmable logic inputs. As a result of these applied input bits, the logic will produce output bit(s) in response. The resulting output bit(s) is (are) then shifted out on the TDO pin

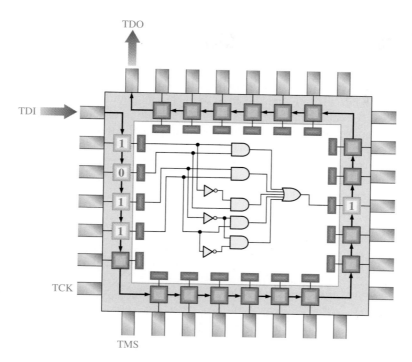

◀ FIGURE 11–76

Example of a bit pattern in the boundary scan Intest for the internal logic.

and checked for errors. An incorrect output, of course, indicates a fault in the programmed logic, I/O cells, or boundary scan cells.

Figure 11–76 shows a boundary scan Intest pattern 1011 for an AND-OR logic circuit that has been programmed into a device. Sixteen combinations of four TDI bits would test the circuit in all possible states according to the list in Table 11–1. The 4-bit combinations are serially shifted into the boundary scan cells, and the corresponding output is shifted out on TDO for checking. This process is controlled by boundary scan test software.

TDI	TDO
0000	1
0001	1
0010	0
0011	1
0100	1
0101	1
0110	1
0111	1
1000	1
1001	1
1010	0
1011	1
1100	1
1101	1
1110	1
1111	1

◀ TABLE 11–1

Boundary scan test bit pattern for the programmed device in Figure 11–76.

Extest   When boundary scan cells are used to test the external connections to the device in addition to some internal functionality, the test mode is called Extest. The basic concept of boundary scan using Extest is as follows: A software-driven pattern of 1s and 0s is applied to the input pins of the device and entered into the input cells. As a result of these applied input bits, the logic will produce output bit(s) in response. The resulting output bit(s) is (are) then taken from the output pin of the device and checked for errors. An incorrect output, of course, indicates a fault in the input or output pin connections or interconnections, an incorrect device, or improperly installed device. Obviously, some internal faults can also be detected in the Extest mode. For example, faults in the boundary scan cells, I/O cells or certain faults in the programmed logic will produce an incorrect output. Figure 11–77 shows an example of a boundary scan Extest that tests the four inputs and the output of the logic circuit.

▶ **FIGURE 11–77**

Example of a bit pattern in the boundary scan Extest for external faults.

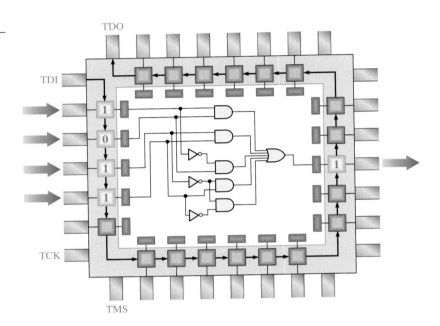

If a fault is detected in the Extest mode, it can be either external (a bad pin connection) or internal (a faulty connection, boundary scan cell, or logic element) to the device. Therefore, in order to isolate an Extest detected fault, an Intest should be run following the Extest. If both tests show a fault, then it is internal to the device.

In the Extest mode, it is necessary to probe contacts to the input and output pins of the device. These pins have to be available at a connector to the circuit board or on test pads so they can be checked by the automatic test equipment. Pins that are not brought through a JTAG board connector can be probed at a test pad using the bed-of-nails method and/or the flying probe method. These combined approaches are illustrated in Figure 11–78.

Boundary Scan Description Language (BSDL)   This test software is part of the JTAG standard IEEE 1149.1 and uses VHDL to describe how the boundary scan logic is implemented in a specific device and how it operates. BSDL provides a standard data format for describing how IEEE 1149.1 is implemented in a JTAG-compliant device. When you use

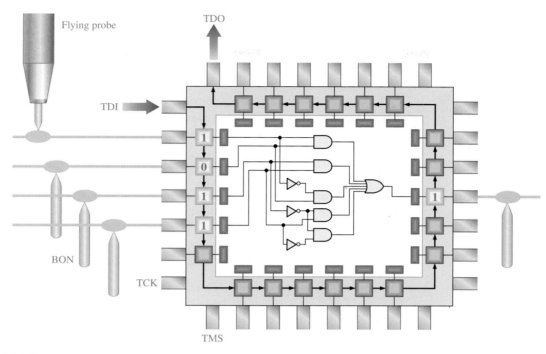

▲ **FIGURE 11–78**

A combination of boundary scan, bed-of-nails, and flying probe testing.

boundary scan test software tools that support BSDL, you can usually obtain BSDL from the device manufacturer.

Each device that contains dedicated boundary scan logic is supported by a BSDL file that describes that particular device. Certain things that are described in the BSDL file are the device type and port descriptions that name the I/O pins and test access port (TAP) pins and denote their nature such as input, output, or bidirectional. BSDL also provides a mapping of logical signals onto the physical pins and a description of the boundary scan logic architecture contained in the device. A bit test pattern for testing the device can be defined using BSDL.

SECTION 11–10
REVIEW

1. Describe the basic concept of bed-of-nails board testing.
2. What limits the bed-of-nails method?
3. How does the flying probe test method differ from BON?
4. Explain the basic concept of boundary scan.
5. What are the two modes of boundary scan test?
6. Name four JTAG signals used with boundary scan.
7. What is BSDL?

DIGITAL SYSTEM APPLICATION

In this system application, the generic software development tools using the schematic entry procedure that you learned in this chapter are applied to the BCD-to-7-segment decoder logic that was developed in Chapter 4. Only the major steps are shown here in a generic approach to illustrate the basic concept.

Figure 11–79 shows the individual logic circuits for each of the seven segments. Each segment logic circuit is entered as a separate file and then converted to block symbol form. After all seven segment logic files have been entered, they are combined in a single block that is the full decoder. All of the seven logic circuits could be entered at one time to create what is known as a "flat" schematic. However, we will use the hierarchical approach of design entry to keep the amount of logic on the Graphic Editor screen at one time more manageable. This approach is preferred when the schematic is fairly complex and can be broken down

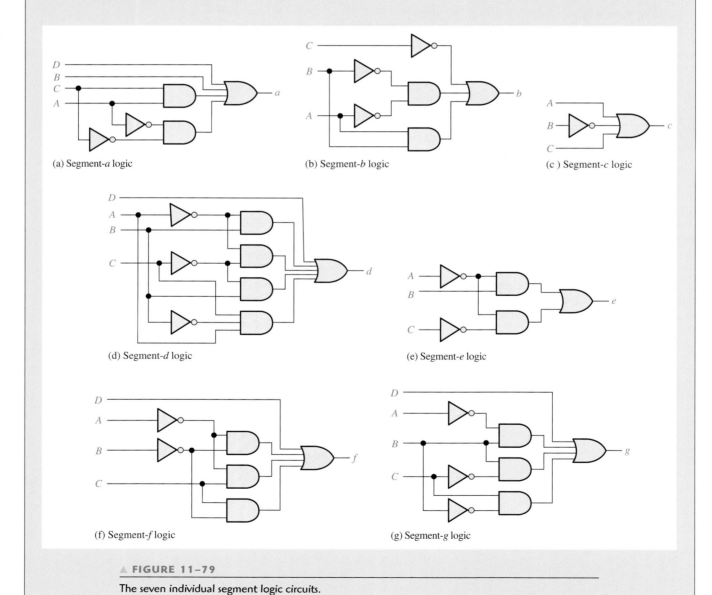

(a) Segment-*a* logic

(b) Segment-*b* logic

(c) Segment-*c* logic

(d) Segment-*d* logic

(e) Segment-*e* logic

(f) Segment-*f* logic

(g) Segment-*g* logic

▲ **FIGURE 11–79**

The seven individual segment logic circuits.

into several parts. Also, in situations where several people are working on circuits that will later be combined into a larger circuit or system, the hierarchical approach is essential.

Each of the seven logic circuits is entered individually, converted to a block symbol, and saved. When all seven circuits have been entered, each block symbol will be placed on the graphic entry screen. All the block symbols will be then connected to the inputs and outputs. Keep in mind that this is a generic description but illustrates the concept of some of the major tools

found in most software, such as Altera Quartus II and Xilinx ISE.

Design Entry of Segment-a Logic

After opening the software, we will set up a project for the 7-segment logic and open the Graphic Editor screen. To place the logic gate symbols on the screen, click on the gate icon as indicated in Figure 11–80. A Symbol screen appears to allow you to select the gates that you want from the software library. The logic gates are called **primitives** and can be selected from a list when you go to the Primitives heading.

For segment *a*, two instances of the 2-input AND gate are selected and placed on the screen as shown. Next, a 4-input OR gate is selected and one instance is placed on the screen. Finally, the inverter (NOT) is selected, and two instances of it are placed on the screen.

Next, in the Symbol window, select Pins from the library. Either an input pin or an output pin can be specified, as illustrated in Figure 11–81. For this particular circuit, four input pins and one output pin have been placed in the Graphic Editor screen, as shown in Figure 11–82.

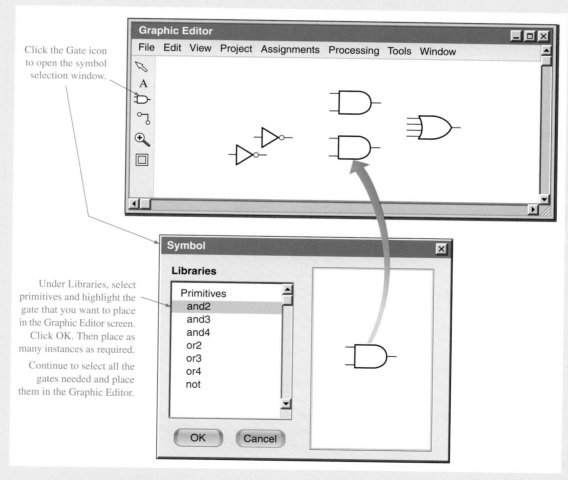

Click the Gate icon to open the symbol selection window.

Under Libraries, select primitives and highlight the gate that you want to place in the Graphic Editor screen. Click OK. Then place as many instances as required.

Continue to select all the gates needed and place them in the Graphic Editor.

▲ **FIGURE 11–80**

Illustration of selection and placement of logic symbols in the Graphic Editor.

▶ FIGURE 11–81

Selection of input and output pins in the Symbol window.

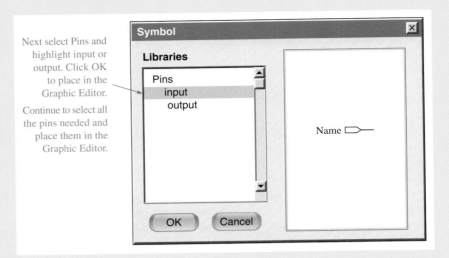

▲ FIGURE 11–82

Placement of pins and making circuit connections in the Graphic Editor.

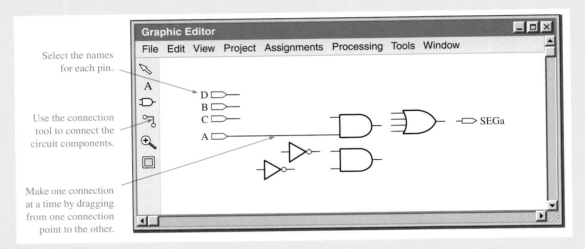

The complete schematic for the segment-*a* logic is shown in Figure 11–83.

Compile the Design

After the design has been entered, the next step is to compile it. The compiler is a software tool that manages the design flow process. The **fitter tool** in the compiler selects the optimum interconnections, pin assignments, and logic cell assignments to fit a design in the selected target device. Generally, a compiler dialog box appears; and when it is started, it indicates the progress being made as a bar graph and percentage of completion, as shown in Figure 11–84. An indication will appear to show if the compilation is successful or not.

Functional Simulation

Most software packages have at least two types of simulation tools: functional simulation and timing simulation. The **functional simulation** verifies functionality of the logic circuit and should be done as soon as the circuit has been successfully compiled.

The first step in setting up the simulation is to specify a signal as an input or output and assign a name. Next, create one waveform at a time by selecting the desired time interval and specifying the level as a HIGH (1) or a LOW (0). You move from interval to interval until the entire waveform is complete. This is repeated for all the other input waveforms, as indicated in Figure 11–85.

When all the input waveforms have been specified, the simulation is initiated and an output waveform is generated, as shown in Figure 11–85 for the segment-*a* logic.

Create Block Symbol

A successful simulation means that the logic circuit works as expected from a functional point of view; that is, the logic is correct.

▶ **FIGURE 11–83**

Complete schematic of segment-*a* logic.

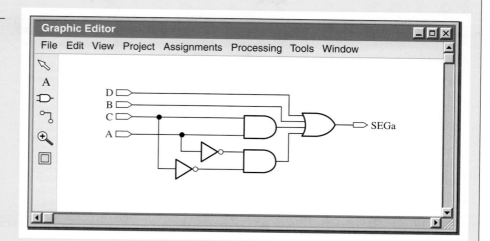

▶ **FIGURE 11–84**

Compiler dialog box.

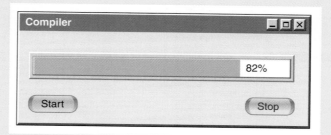

Pins and waveform names are assigned to match the schematic.

To create a waveform, select each time interval and specify a 0 or 1 for that time interval, one interval at a time.

You specify the input waveforms. The simulation tool produces the output waveform.

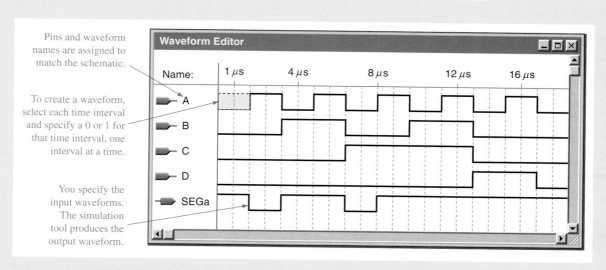

▲ **FIGURE 11–85**

Input waveforms and the resulting output waveform for segment-*a* logic.

The next step is to convert the logic schematic into a block symbol, as illustrated in Figure 11–86, and save it for use later. Once the block symbol has been saved, it can be accessed for use in the final design.

Design Entry for Segments *b* Through *g*

The same general procedure that was used to enter, simulate, and save as a block symbol the segment-*a* logic is repeated for each of the other six segment logic circuits. Figure 11–87 shows the screens for segment-*b* logic, segment-*c* logic, and segment-*d* logic.

Figure 11–88 shows the screens for segment-*e* logic, segment-*f* logic, and

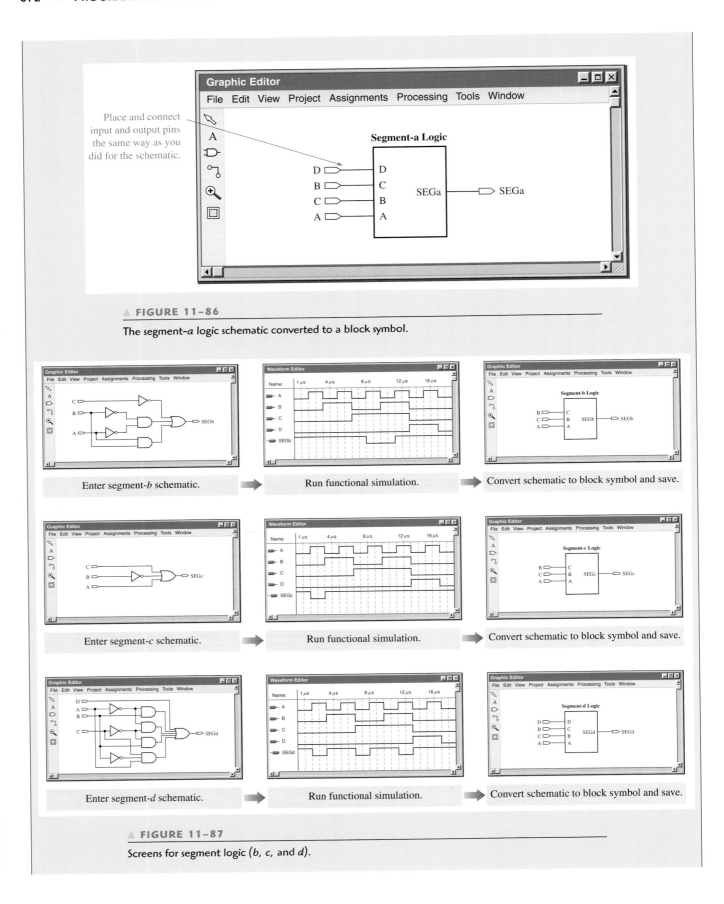

▲ FIGURE 11–86

The segment-*a* logic schematic converted to a block symbol.

Enter segment-*b* schematic. → Run functional simulation. → Convert schematic to block symbol and save.

Enter segment-*c* schematic. → Run functional simulation. → Convert schematic to block symbol and save.

Enter segment-*d* schematic. → Run functional simulation. → Convert schematic to block symbol and save.

▲ FIGURE 11–87

Screens for segment logic (*b*, *c*, and *d*).

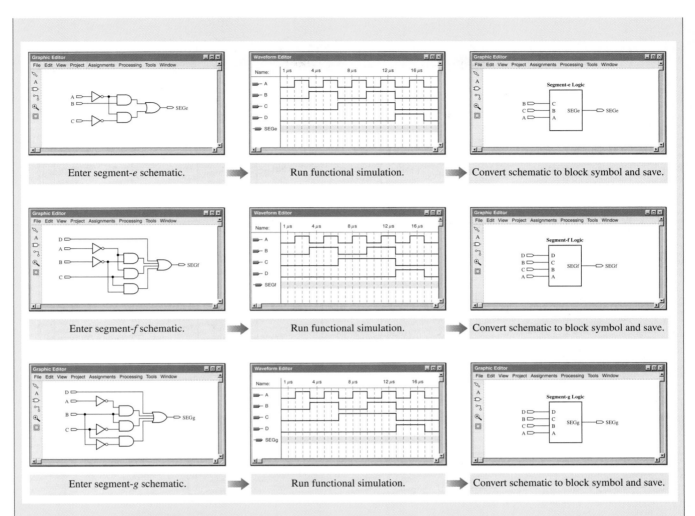

▲ **FIGURE 11–88**

Screens for segment logic (e, f, and g). The output waveforms in each case are to be completed as an activity.

segment-g logic. The functional simulations for segments e, f, and g are left as an activity.

Final Block Diagram

All of the block symbols for the segment logic have been saved and now can be recalled for use in the complete 7-segment logic. The symbols are accessed using the Symbol window, just as was done for the selection of logic gates. The file they are stored in is opened and the list appears, as shown in Figure 11–89. The input and output pins are added and all of the block symbols are connected to form the complete segment logic.

Timing Simulation

After the complete circuit has been entered in the Graphic Editor, it must be compiled in the same way that each of the segment logic circuits was compiled. After it has been successfully compiled, a timing simulation should be run. The timing simulation takes into account the propagation delays of each gate in the design as well as the functionality of the circuit. All the input waveforms are specified just as was done for the functional simulation. A timing simulation with no apparent problems, such as glitches, appears in the Waveform Editor, as shown in Figure 11–90. If there are any glitches, you would go back to the design and try to correct the condition if it could be a potential problem.

A successful timing simulation means that the logic circuit works as expected in terms of both functionality and timing. The next step is to convert the multiple-block diagram into a single block symbol, as illustrated in Figure 11–91, for possible later use.

Programming the Target Device

Assuming the timing simulation indicates no problems, the next step is to program the target device with the 7-segment logic. Start the programming sequence and select

Click the Gate icon to open the symbol selection window.

Select the file where you saved the block symbols for the segment logic. Click OK.

Continue to select all the block symbols and place them in th Graphic Editor.

Next select input and output pins and then interconnect all of the block symbols to the inputs and each block to its output.

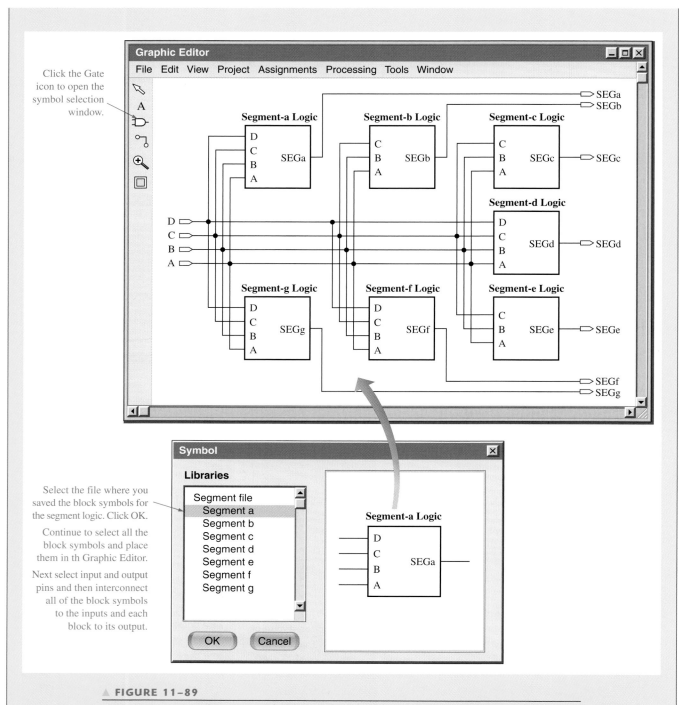

▲ FIGURE 11–89

Selection, placement, and interconnection of the segment logic in the Graphic Editor.

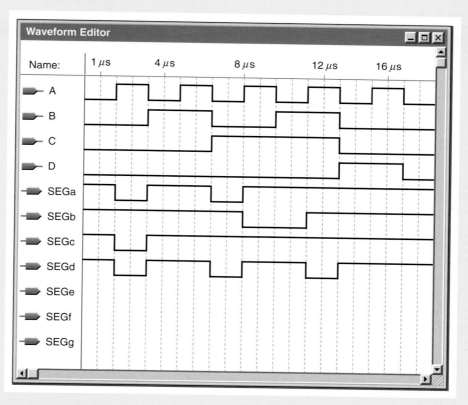

▲ FIGURE 11–90

Ideal timing simulation for the complete 7-segment logic in Figure 11–89. The segment waveforms for e, f, and g are to be completed as an activity.

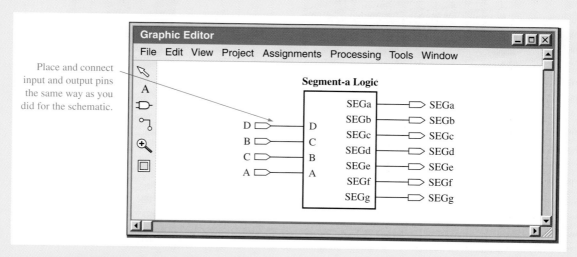

▲ FIGURE 11–91

The complete BCD-to-7-segment decoder as a single block symbol.

▶ FIGURE 11–92

Selection of the target device.

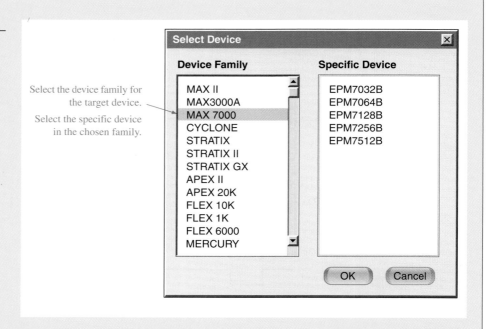

Select the device family for the target device.

Select the specific device in the chosen family.

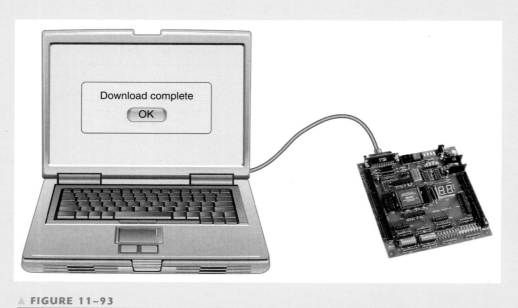

▲ FIGURE 11–93

The design has been downloaded to the target device on a development board.

the interface, such as JTAG. Next, select the target device from the Select Device window, as shown in Figure 11–92. Keep in mind that a very small percentage of the capacity of a typical programmable device will be used for a circuit the size of our example. Thousands of circuits with a similar number of gates can be programmed into a single PLD. Often the limitation is the number of inputs and outputs that are available on the target device package. Software packages generally allow you to assign device pin numbers for the inputs and outputs. However, if you choose not to, most software will automatically assign them for you and make available a listing of the assignments.

When the device selection is complete, the design is downloaded, as indicated by the Download Complete message in Figure 11–93.

In-Circuit Testing

After the design has been downloaded to the target device, hardware testing is usually the next step. Prior to this, only software simulation has been done to verify operation. Now, actual instruments are connected to the circuit on the development board, input signals are applied, and output signals are observed. The method of approach depends on the type and complexity of the logic that has been downloaded to the device, but generally a signal source such as a function generator or a pattern generator to supply the inputs and an oscilloscope or a logic analyzer for observing the outputs may be required.

In the case of the 7-segment decoder logic that has been downloaded to the target device, a simple test can be done using the resources available on the development board. The inputs can be connected to the switches on the board and the outputs to the 7-segment display on the board. After a power supply is connected, you can sequence through the BCD code with the switches and observe the output on the display. Most development boards have these features, as indicated in Figure 11–94.

System Assignment

- **Activity 1** Verify that the SEGa output waveform in Figure 11–85 is correct.

- **Activity 2** Verify that the output waveforms for segments b, c, and d in Figure 11–87 are correct.

- **Activity 3** Determine the correct output waveforms for segments e, f, and g in Figure 11–88 and Figure 11–90.

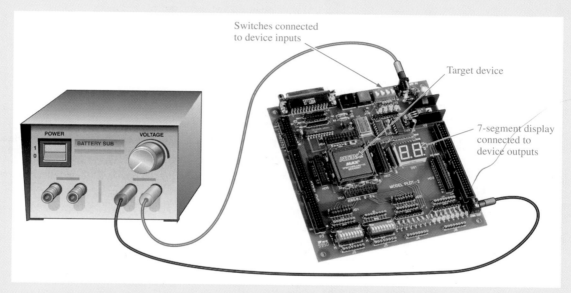

Switches connected to device inputs

Target device

7-segment display connected to device outputs

▲ FIGURE 11–94

Example of the target device mounted on a development board being tested with on board switches and 7-segment display.

SUMMARY

- A PAL is a one-time programmable (OTP) SPLD consisting of a programmable array of AND gates that connects to a fixed array of OR gates.
- The PAL structure allows any sum-of-products (SOP) logic expression with a defined number of variables to be implemented.
- The GAL is essentially a PAL that can be reprogrammed.
- In a PAL or GAL, a macrocell generally consists of one OR gate and some associated output logic.
- The PAL16V8 is a common type of programmable array logic device.
- The GAL22V10 is a common type of generic array logic device.

- A CPLD is a complex programmable logic device that consists basically of multiple SPLD arrays with programmable interconnections.
- Each SPLD array in a CPLD is called a logic array block (LAB).
- The MAX 7000 is an Altera family of CPLDs.
- In the MAX 7000 CPLD family, density ranges from 2 LABs to 16 LABs, depending on the particular device in the series, and each LAB has sixteen macrocells.
- The Altera MAX II CPLD differs dramatically from the MAX 7000 family and is known as a "post-macrocell" CPLD.
- The MAX II CPLD uses look-up tables (LUT) instead of AND/OR arrays.
- The architecture of the Xilinx CoolRunner II CPLD family is based on a PLA structure rather than on a PAL structure.
- The CoolRunner II family contains CPLDs ranging from 32 macrocells to 512 macrocells.
- A macrocell can be configured for either of two modes: the combinational mode or the registered mode.
- An FPGA (field-programmable gate array) differs in architecture, does not use PAL/PLA type arrays, and has much greater densities than typical CPLDs.
- Most FPGAs use either antifuse or SRAM-based process technology.
- Each configurable logic block (CLB) in an FPGA is made up of multiple smaller logic modules and a local programmable interconnect that is used to connect logic modules within the CLB.
- FPGAs are based on LUT architecture.
- LUT stands for *look-up table,* which is a type of memory that is programmable and used to generate SOP combinational logic functions.
- A hard core is a portion of logic embedded in an FPGA that is put in by the manufacturer to provide a specific function and which cannot be reprogrammed.
- A soft-core is a portion of logic embedded in an FPGA that has some programmable features.
- Designs owned by the manufacturer are termed *intellectual property* (IP).
- Altera produces several families of FPGAs including the Stratix II, the Stratix, Cyclone, and the ACEX family.
- Xilinx has two major lines of FPGAs, the Spartan and the Virtex, and there are different families within each line.
- The programming process is generally referred to as design flow.
- The device being programmed is usually referred to as the target device.
- In software packages for programmable logic, the operations are controlled by an application program called the compiler.
- During downloading, a bitstream is generated that represents the final design, and it is sent to the target device to automatically configure it.
- The bed-of-nails (BON) method was one of the first approaches to automated circuit board testing.
- Another method for testing printed circuit boards is called the flying probe method.
- A method of internally testing a programmable device is called boundary scan, which is based on the JTAG standard (IEEE Std. 1149.1).
- The boundary scan logic in a CPLD consists of a boundary scan register, a bypass register, an instruction register, and a test access port (TAP).

KEY TERMS

Key terms and other bold terms in the chapter are defined in the end-of-book glossary.

Bed-of-nails A method for the automated testing of printed circuit boards in which the board is mounted on a fixture that resembles a bed of nails that makes contact with test points.

Boundary scan A method for internally testing a PLD based on the JTAG standard (IEEE Std. 1149.1).

CLB Configurable logic block; a unit of logic in an FPGA that is made up of multiple smaller logic modules and a local programmable interconnect that is used to connect logic modules within the CLB.

Compiler An application program in development software packages that controls the operation of the software.

CPLD A complex programmable logic device that consists basically of multiple SPLD arrays with programmable interconnections.

Design flow The process or sequence of operations carried out to program a target device.

Downloading The final step in a design flow in which the logic design is implemented in the target device.

Fitter tool A compiler software tool that selects the optimum interconnections, pin assignments, and logic cell assignments to fit a design into the selected target device.

Flying probe A method for the automated testing of printed circuit boards, in which a probe or probes move from place to place to contact test points.

FPGA Field programmable gate array; a programmable logic device that uses the LUT as the basic logic element and generally employs either antifuse or SRAM-based process technology.

Functional simulation A software process that tests the logical or functional operation of a design.

GAL A reprogrammable type of SPLD that is similar to a PAL except that it uses a reprogrammable process technology, such as EEPROM (E^2CMOS), instead of fuses.

Intellectual property (IP) Designs owned by a manufacturer of programmable logic devices.

LAB Logic array block; an SPLD array in a CPLD.

LUT Look-up table; a type of memory that can be programmed to produce SOP functions.

Macrocell Part of a PAL, GAL, or CPLD that generally consists of one OR gate and some associated output logic.

PAL A type of one-time programmable SPLD that consists of a programmable array of AND gates that connects to a fixed array of OR gates.

Primitive A basic logic element such as a gate or flip-flop, input/output pins, ground, and V_{CC}.

Registered A macrocell operational mode that uses a flip–flop.

Schematic entry A method of placing a logic design into software using schematic symbols.

Target device The programmable logic device that is being programmed.

Text entry A method of placing a logic design into software using a hardware description language (HDL).

Timing simulation A software process that uses information on propagation delays and netlist data to test both the logical operation and the worst-case timing of a design.

SELF-TEST

Answers are at the end of the chapter.

1. Two types of SPLDs are
   (a) CPLD and PAL   (b) PAL and FPGA
   (c) PAL and GAL   (d) GAL and SRAM

2. A PAL consists of a
   (a) programmable AND array and a programmable OR array
   (b) programmable AND array and a fixed OR array
   (c) fixed AND array and a programmable OR array
   (d) fixed AND/OR array

3. A macrocell consists of a
   (a) fixed OR gate and other associated logic
   (b) programmable OR array and other associated logic
   (c) fixed AND gate and other associated logic
   (d) fixed AND/OR array with a flip-flop

4. The 16V8 is a type of
   (a) CPLD   (b) GAL
   (c) PAL   (d) FPGA

5. The basic AND/OR structure of SPLDs and CPLDs produces types of Boolean expressions known as

   (a) POS   (b) SOP   (c) product of complements   (d) sum of complements

6. The term *LAB* stands for

   (a) logic AND block   (b) logic array block

   (c) last asserted bit   (d) logic assembly block

7. The MAX 7000 is a

   (a) family of CPLDs   (b) family of SPLDs

   (c) family of FPGAs   (d) type of software

8. The CoolRunner II is a

   (a) family of CPLDs   (b) family of SPLDs

   (c) family of FPGAs   (d) type of software

9. Two modes of macrocell operation are

   (a) input and output   (b) registered and sequential

   (c) combinational and registered   (d) parallel and shared

10. When a macrocell is configured to produce an SOP function, it is in the

    (a) combinational mode   (b) parallel mode

    (c) registered mode   (d) shared mode

11. A typical macrocell consists of

    (a) gates, multiplexers, and a flip-flop   (b) gates and a shift register

    (c) a Gray code counter   (d) a fixed logic array

12. Based on the complexity of its configurable logic blocks (CLBs), an FPGA can be classified as either

    (a) volatile or nonvolatile

    (b) programmable or reprogrammable

    (c) fine grained or coarse grained

    (d) platform or embedded

13. Nonvolatile FPGAs are generally based on

    (a) fuse technology   (b) antifuse technology

    (c) EEPROM technology   (d) SRAM technology

14. An FPGA with an embedded logic function that cannot be programmed is said to be

    (a) nonvolatile   (b) platform

    (c) hard core   (d) soft core

15. Hard core designs are generally developed by and are the property of the FPGA manufacturer. These designs are called

    (a) intellectual property   (b) proprietary logic

    (c) custom designs   (d) IEEE standards

16. For text entry of a logic design,

    (a) logic symbols must be used   (b) an HDL must be used

    (c) only Boolean algebra is used   (d) a special code must be used

17. In a functional simulation, the user must specify the

    (a) specific target device   (b) output waveform

    (c) input waveforms   (d) HDL

18. The final output of the synthesis phase of a design flow is the

    (a) netlist   (b) bitstream

    (c) timing simulation   (d) device pin numbers

19. EDIF stands for

    (a) electronic device interchange format

    (b) electrical design integrated fixture

    (c) electrically destructive input function

    (d) electronic design interchange format

20. The boundary scan TAP stands for

    (a) test access point   **(b)** test array port

    (c) test access port   **(d)** terminal access path

21. A typical boundary scan cell contains

    (a) flip–flops only   **(b)** flip–flops and multiplexer logic

    (c) latches and flip–flops   **(d)** latches and an encoder

22. An automated printed circuit board test method that uses a fixture with many fixed contacts to the board test point is called

    (a) traditional   **(b)** flying probe

    (c) bed-of-nails   **(d)** boundary scan

23. An automated printed circuit board test method that uses a moving test point contact is called

    (a) traditional   **(b)** flying probe

    (c) bed-of-nails   **(d)** boundary scan

24. The JTAG standard has the following inputs and outputs

    (a) Intest, extest, TDI, TDO   **(b)** TDI, TDO, TCK, TMS

    (c) ENT, CLK, SHF, CLR   **(d)** TCK, TMS, TMO, TLF

25. The acronym BSDL stands for

    (a) board standard digital logic   **(b)** boundary scan down load

    (c) bistable digital latch   **(d)** boundary scan description language

PROBLEMS

Answers to odd-numbered problems are at the end of the book.

SECTION 11–1

Programmable Logic: SPLDs and CPLDs

1. Determine the Boolean output expression for the simple PAL array shown in Figure 11–95. The Xs represent connected links.

▶ **FIGURE 11–95**

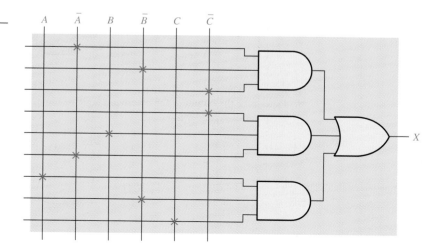

2. Show how the PAL-type array in Figure 11–96 should be programmed to implement each of the following SOP expressions. Use an X to indicate a connected link. Simplify the expressions, if necessary, to fit the PAL-type array shown.

(a) $Y = A\overline{B}C + \overline{A}B\overline{C} + ABC$

(b) $Y = A\overline{B}C + \overline{A}\,\overline{B}C + A\overline{B}\,\overline{C} + \overline{A}BC$

▶ FIGURE 11–96

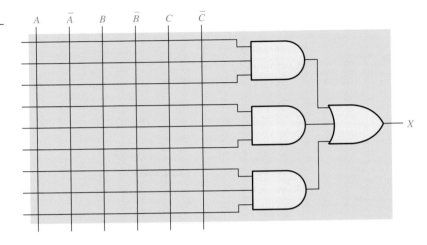

3. Interpret each of the PAL device numbers.

(a) PAL16L2   (b) PAL12H6

4. Explain how a programmed polarity output in a PAL works.

5. Describe how a CPLD differs from an SPLD.

SECTION 11–2 Altera CPLDs

6. Refer to the MAX 7000 block diagram in Figure 11–11 and determine the number of

(a) inputs from the PIA to a LAB   (b) outputs from a LAB to the PIA

(c) inputs from an I/O control block to the PIA   (d) outputs from a LAB to an I/O control block

7. Determine the product term for the AND gate in a CPLD array shown in Figure 11–97(a). If the AND gate is expanded, as shown in Figure 11–97(b), determine the SOP output.

▶ FIGURE 11–97

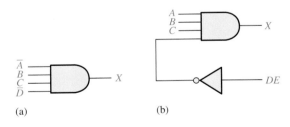

(a)   (b)

8. Determine the output of the macrocell logic in Figure 11–98 if $AB\overline{C}D + \overline{A}BCD$ is applied to the parallel expander input.

▶ FIGURE 11–98

Parallel expander input

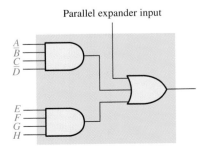

SECTION 11–3 Xilinx CPLDs

9. Determine the output of the PLA in Figure 11–99. The Xs represent connected links.

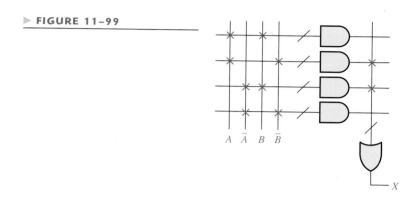

▶ **FIGURE 11–99**

10. Refer to the CoolRunner II CPLD block diagram in Figure 11–21 and determine the number of

  (a) inputs from the AIM to an FB

  (b) outputs from an FB to the AIM

  (c) inputs from an I/O block to the AIM

  (d) outputs from an FB to an I/O block

11. Determine the output expressions for X_1 and X_2 from macrocells 1 and 2 in Figure 11–100.

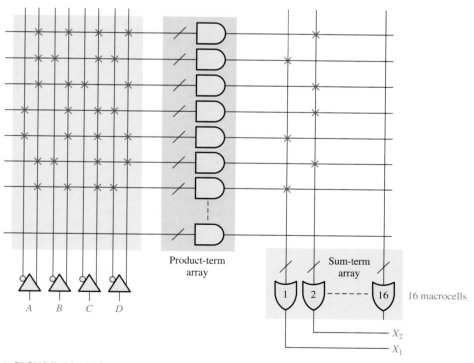

▲ **FIGURE 11–100**

D_0 ──────┐
             ├──── Data output
D_1 ──────┘

Select

SECTION 11–4   **Macrocells**

12. Determine the data output for the multiplexer in Figure 11–101 for each of the following conditions:

   (a) $D_0 = 1$, $D_1 = 0$, Select $= 0$ **(b)** $D_0 = 1$, $D_1 = 0$, Select $= 1$

13. Determine how the macrocell in Figure 11–102 is configured (combinational or registered) and the data bit that is on the output (to I/O) for each of the following conditions. The flip-flop is a D type. Refer to Figure 11–101 for MUX data input arrangement.

   (a) XOR output $= 1$, flip-flop Q output $= 1$, from I/O input $= 1$, MUX 1 select $= 1$, MUX 2 select $= 0$, MUX 3 select $= 0$, MUX 4 select $= 0$, and MUX 5 select $= 0$.

   (b) XOR output $= 0$, flip-flop Q output $= 0$, from I/O input $= 1$, MUX 1 select $= 1$, MUX 2 select $= 0$, MUX 3 select $= 1$, MUX 4 select $= 0$, and MUX 5 select $= 1$.

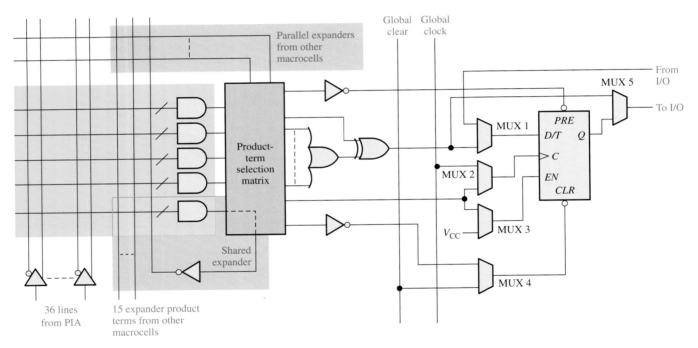

▲ FIGURE 11–102

14. For the CPLD macrocell in Figure 11–103, the following conditions are programmed: MUX 1 select $= 1$, MUX 2 select $= 1$, MUX 3 selects $= 01$, MUX 4 select $= 0$, MUX 5 select $= 1$, MUX 6 selects $= 11$, MUX 7 selects $= 11$, MUX 8 select $= 1$, and the OR output $= 1$. The flip-flop is a D type and the MUX inputs are from D_0 at the top to D_n at the bottom.

   (a) Is the macrocell configured for combinational or registered logic?

   (b) Which clock is applied to the flip-flop?

   (c) What is the data bit on the D input to the flip-flop?

   (d) What is the output of MUX 8?

15. Repeat Problem 14 for MUX 1 select $= 0$.

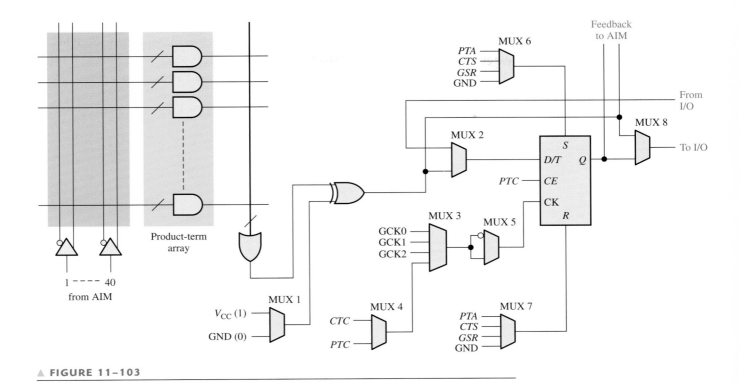

▲ FIGURE 11–103

SECTION 11–5 Programmable Logic: FPGAs

16. Generally, what elements make up a configurable logic block (CLB) in an FPGA? What elements make up a logic module?

17. Determine the output expression of the LUT for the internal conditions shown in Figure 11–104.

18. Show how to reprogram the LUT in Figure 11–104 to produce the following SOP output:

$$\overline{A}B\overline{C} + A\overline{B}\,\overline{C} + ABC$$

▶ FIGURE 11–104

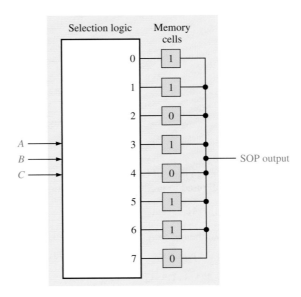

SECTION 11–6 **Altera FPGAs**

19. Name the basic elements that make up an adaptive logic module (ALM) in the Stratix II FPGA.

20. List the modes of operation for an ALM.

21. Show an ALM configured in the normal mode to produce one 4-variable SOP function and one 2-variable SOP function.

22. Determine the final SOP output function for the ALM shown in Figure 11–105.

▶ **FIGURE 11–105**

$A_4A_3\overline{A}_2A_1 + \overline{A}_4\overline{A}_3\overline{A}_2A_1$

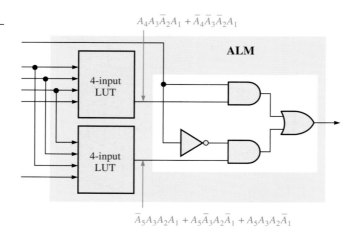

$\overline{A}_5A_3A_2A_1 + A_5\overline{A}_3A_2\overline{A}_1 + A_5A_3A_2\overline{A}_1$

SECTION 11–7 **Xilinx FPGAs**

23. Use one or more of the slices in Figure 11–106 to produce the SOP function:

$$A_7A_6A_5A_4A_3A_2A_1A_0 + B_7B_6B_5B_4B_3B_2B_1B_0$$

▶ **FIGURE 11–106**

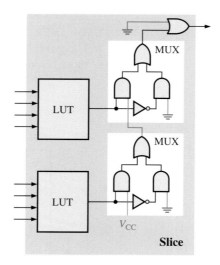

24. A slice from a Virtex FPGA is shown in Figure 11–106. Show how one or more of these slices can be configured to produce the SOP function:

$$A_7\overline{A}_6A_5A_4 + \overline{A}_3A_2A_1\overline{A}_0 + \overline{B}_7\overline{B}_6B_5B_4 + B_3\overline{B}_2\overline{B}_1B_0$$

Assume that the red elements as well as the LUTs are reconfigurable.

25. Determine the number of slices (Figure 11–106) required to generate the expression:

$$A_7A_6A_5A_4A_3A_2A_1A_0$$

26. Determine the number of slices required to generate the expression:

$$A_7A_6A_5A_4A_3A_2A_1A_0 + B_7B_6B_5B_4B_3B_2B_1B_0 + C_7C_6C_5C_4C_3C_2C_1C_0$$

SECTION 11–8   **Programmable Logic Software**

27. Show the logic diagram that you would enter in the Graphic Editor for the circuit described by each of the VHDL programs.

   a. **entity** AND_OR **is**
       port (A0, A1, A2, A3: **in** bit; X: **out** bit);
       end entity AND_OR;
       architecture LogicFunction **of** AND_OR **is**
       begin
           X <= (A0 **and** A1) **or** (A2 **and not** A3);
       end architecture LogicFunction;

   b. **entity** LogicCircuit **is**
       port (A, B, C, D: **in** bit; X: **out** bit);
       end entity LogicCircuit;
       architecture Function **of** LogicCircuit **is**
       begin
           X <= (A **and** B) **or** (C **and** D) **and**
                   (A **and not** B) **and** (**not** C **and not** D);
       end architecture Function;

28. Show the logic circuit that you would enter in the Graphic Editor for the following Boolean expression. Simplify before entering, if possible.

$$X = \overline{A}BCD + A\overline{B}CD + AB\overline{C}D + ABC\overline{D} + ABCD + \overline{A}\,\overline{B}\,\overline{C}D$$

29. The input waveforms for the logic circuit described in Problem 28 are as shown in the Waveform Editor of Figure 11–107. Determine the output waveform that is produced after running a simulation.

▶ **FIGURE 11–107**

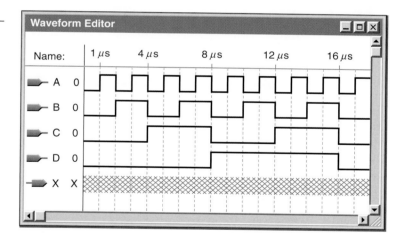

30. Repeat Problem 29 for the following Boolean expression:

$$X = \overline{A}BC\overline{D} + A\overline{B}\,\overline{C}D + ABCD + A\overline{B}C\overline{D} + \overline{A}B\overline{C}D$$

SECTION 11–9   **Boundary Scan Logic**

31. In a given boundary scan cell, assume that data flow serially from the previous BCS to the next BSC. Describe what happens as the data pass through the given BCS.

32. Describe the conditions and what happens in a given BCS when data flow directly from the internal programmable logic to a device output pin.

33. Describe the conditions and what happens in a given BCD when data flow from a device input pin to the internal programmable logic.

34. Describe the data path for transferring data from the SDI to the internal programmable logic.

SECTION 11–10 Troubleshooting

35. Develop a boundary scan test bit pattern to test the logic that is programmed into the device shown in Figure 11–108 for all possible input combinations.

► **FIGURE 11–108**

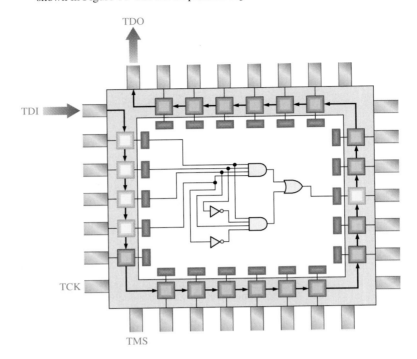

Digital System Application

36. If the logic for the seven segments shown in Figure 11–79 are entered in the Graphic Editor as a flat schematic, how many and which elements can be eliminated?

37. A simulation for the 7-segment logic is shown in the Waveform Editor in Figure 11–109. Determine what the problem may be with the simulated circuit.

► **FIGURE 11–109**

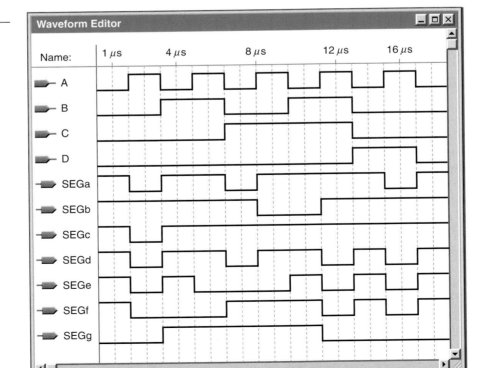

ANSWERS

SECTION 11-1   Programmable Logic: SPLDs and CPLDs

1. PAL: Programmable Array Logic
2. GAL: Generic Array Logic
3. A GAL is reprogrammable. A PAL is one-time programmable.
4. Basically, a macrocell consists of an OR gate and associated output logic including a flip-flop.
5. CPLD: Complex Programmable Logic Device

SECTION 11-2   Altera CPLDs

1. LAB: Logic Array Block
2. A LAB consists of 16 macrocells in the MAX 7000 family.
3. A shared expander is used to increase the number of product terms from a macrocell by ANDing additional sum terms (complemented product terms) from other macrocells.
4. A parallel expander is used to increase the number of product terms from a macrocell by ORing unused product terms from other macrocells in a LAB.
5. The MAX II is organized in a row/column architecture and uses LUTs in its macrocells. The MAX 7000 is organized in a traditional column architecture and uses SOP logic in its macrocells.

SECTION 11-3   Xilinx CPLDs

1. Altera uses PAL architecture. Xilinx uses PLA architecture.
2. A PLA has a programmable AND array and a programmable OR array.
3. A PAL has a fixed OR array.
4. FB: Function block

SECTION 11-4   Macrocells

1. The XOR gate is used as a programmable inverter for the data. It can be programmed to invert or not invert.
2. Combinational and registered
3. Registered refers to the use of a flip-flop.
4. Multiplexer

SECTION 11-5   Programmable Logic: FPGAs

1. Generally, an FPGA is organized with a row/column interconnect structure and uses LUTs rather than AND/OR logic for generating combinational logic functions.
2. CLB: Configurable Logic Block
3. LUT: Look-Up Table. A programmable type of memory that is used to store and generate combinational logic functions.
4. A local interconnect is used to connect logic modules within a CLB. A global interconnect is used to connect a CLB with other CLBs.
5. A core is a portion of logic embedded in an FPGA to provide a specific function.
6. *Intellectual property* refers to the hard-core designs that are developed and owned by the FPGA manufacturer.

SECTION 11-6   Altera FPGAs

1. The LAB (Logic Array Block) is the basic design unit in the Stratix II.
2. Typically, there are eight ALMs in a LAB.
3. An LUT produces combinational logic functions in an ALM.
4. Two
5. Memory and DSP (digital signal processing)

SECTION 11–7 **Xilinx FPGAs**

1. A CLB consists of eight logic cells or four slices.

2. A LC (logic cell) consists of an LUT and associated logic.

3. A slice consists of two logic cells (LCs).

4. A cascade chain is two or more slices connected to expand an SOP expression.

5. ASMBL: Application Specific Modular Block

SECTION 11–8 **Programmable Logic Software**

1. Design entry, functional simulation, synthesis, implementation, timing simulation, downloading

2. Computer running PLD development software, a programming fixture or a development board, and an interface cable

3. A netlist provides information necessary to describe a circuit.

4. The functional simulation comes before the timing simulation.

SECTION 11–9 **Boundary Scan Logic**

1. TDI, TMS, TCK, TRST, TDO

2. TAP: Test access port

3. Boundary scan register, bypass register, instruction register, and TAP

4. Transfer of data from SDI to SDO, transfer of data from internal programmable logic to device output pin, transfer of data from device input pin to internal programmable logic, transfer of data from SDI to internal programmable logic, and transfer of data from SDI to device output pin.

SECTION 11–10 **Troubleshooting**

1. Bed-of-nails testing utilizes a fixture consisting of a fixed array of test probes (resembling nails) onto which a circuit board to be tested is placed. Each test probe makes contact with a test point on the circuit board so that measurements can be made.

2. The bed-of-nails (BON) method is limited by the density of programmable logic devices, which makes many contacts on a device inaccessible.

3. In a flying probe fixture, one or more probes moves from one test point to another on a PC board in a controlled pattern.

4. Boundary scan enables the internal testing and programming of a programmable logic device and testing of interconnections between two or more devices. It is based on the JTAG IEEE Std. 1149.1. Boundary scan uses specific logic internal to the device for testing.

5. Intest and Extest

6. TDI, TDO, TCK, TMS

7. BSDL: Boundary Scan Description Language

RELATED PROBLEMS FOR EXAMPLES

6–1 $X = \overline{B}C + \overline{A}B\overline{C} + \overline{A}\,\overline{B} + C$ **6–2** Sixteen   **6–3** Sixteen; sixteen

6–4 See Figure 11–110.   **6–5** See Figure 11–111.

▶ FIGURE 11–110

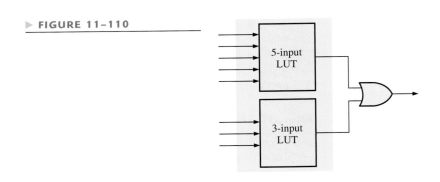

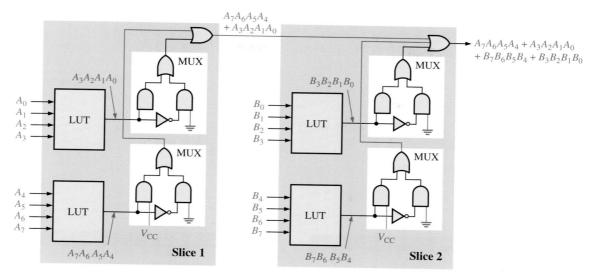

▲ FIGURE 11–111

SELF-TEST

1. (c)   **2.** (b)   **3.** (a)   **4.** (c)   **5.** (b)   **6.** (b)   **7.** (a)   **8.** (a)   **9.** (c)

10. (a)   **11.** (b)   **12.** (c)   **13.** (b)   **14.** (b)   **15.** (a)   **16.** (a)   **17.** (c)   **18.** (a)

19. (d)   **20.** (c)   **21.** (b)   **22.** (c)   **23.** (b)   **24.** (b)   **25.** (d)

12

INTRODUCTION TO COMPUTERS

CHAPTER OBJECTIVES

- Name the basic units of a computer
- Name the basic elements of a microprocessor
- Explain the basic operation of an Intel CPU
- Explain the basic architecture of the Intel microprocessor
- Explain the multiplexed bus operation of the Intel microprocessor
- Discuss the software model of the Intel Pentium processors

- Describe a simple assembly language program

- Describe the seven instruction groups for the Intel processors

- Distinguish between assembly language and machine language

- Compare polled I/O, interrupt-driven I/O, and software interrupts

- Describe the functions of PIC and PPI devices

- Define and explain the advantage of DMA

- Explain how functions are interfaced by the use of bus systems

- Define the basic characteristics and applications of the PCI and ISA internal bus standards

- Define the basic characteristics and applications of the RS-232C, IEEE 1394 (FireWire), USB, IEEE 488 (GPIB), and SCSI external bus standards

KEY TERMS

- Port
- Program
- CPU
- Interrupts
- Peripherals
- Microprocessor
- Address bus
- Data bus
- Control bus

- Machine language
- Assembly language
- High-level language
- Tristate
- Modem
- FireWire
- USB
- GPIB
- SCSI

INTRODUCTION

This chapter provides a brief introduction to computers, microprocessors, and buses. Naturally, a single chapter must be limited because one or more chapters could easily be devoted to each of the section topics. Keep in mind, however, that the purpose here is to give you a basic introduction. Thorough coverage of computers and microprocessors must wait until a later course. For further information on microprocessors, including data sheets, go to the Intel website at www.intel.com.

The Intel microprocessor families are briefly discussed. The Intel 8086/8088 processor is used as a "model" to illustrate basic microprocessor concepts with recent enhancements up through the Pentium described in further detail. The 8086/8088 was the first generation of the Intel 80X86 family. Although the Pentium is more powerful and contains advanced features, it is related in architecture and basic functions such as the register structure.

WWW. **VISIT THE COMPANION WEBSITE**
Study aids for this chapter are available at
http://www.prenhall.com/floyd

12–1 THE BASIC COMPUTER

Special-purpose computers control various functions in automobiles or appliances, control manufacturing processes in industry, provide games for entertainment, and are used in navigation systems such as GPS (Global Positioning System), to name a few areas. However, the most familiar type of computer is the general-purpose computer that can be programmed to do many different types of things.

After completing this section, you should be able to

■ Describe the basic elements in a computer ■ Discuss what each part of a computer does ■ Explain what a peripheral device is

All computers consist of basic functional blocks that include a *central processing unit* (CPU), *memory,* and *input/output ports*. These functional blocks are connected together with three internal buses, as shown in the block diagram of Figure 12–1. The three buses are the *data bus,* the *address bus,* and the *control bus*. Input and output devices are connected through the input/output ports. A **port** is a physical interface on a computer through which data are passed to and from peripherals.

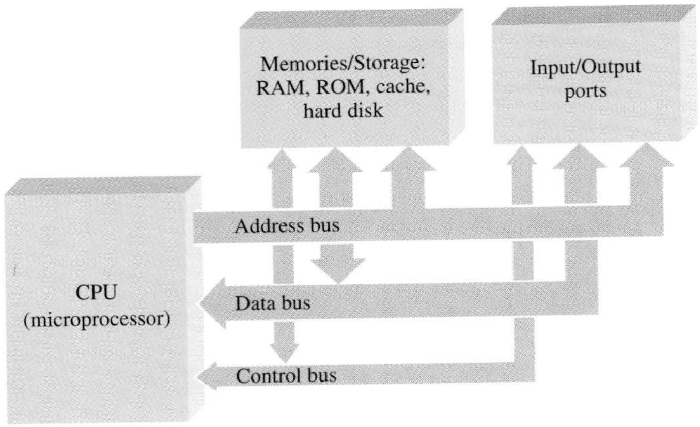

► FIGURE 12–1

Basic computer block diagram.

Instructions and data are stored in memory in specific locations determined by the **program,** a list of instructions designed to solve a specific problem. Each location has a unique address associated with it. Instructions are obtained by the CPU by placing an address on the address bus. Instructions are transferred via the data bus as they are requested by the CPU. The CPU executes the instructions sequentially; frequently, the instructions modify data stored in memory or obtained from an input device. Processed data may be stored back in memory or sent to an output device via the data bus. Signals on the control bus are generated by the CPU to coordinate all of these operations.

Central Processing Unit (CPU)

The **CPU** is the "brain" of the computer; it oversees everything that the computer does. The CPU is a microprocessor with associated circuits that control the running of the computer software programs. Basically, the CPU obtains (fetches) each program instruction from memory and carries out (executes) the instruction.

After completing one instruction, the CPU moves on to the next one and in most cases can operate on more than one instruction at the same time. This "fetch and execute" process is repeated until all of the instructions in a specific program have been executed. For ex-

COMPUTER NOTE

Grace Hopper, a mathematician and pioneer programmer, developed considerable troubleshooting skills as a naval officer working with the Harvard Mark I computer in the 1940s. She found and documented in the Mark I's log the first real computer bug. It was a moth that had been trapped in one of the electromechanical relays inside the machine, causing the computer to malfunction. From then on, when asked if anything was being accomplished, those working on the computer would reply that they were "debugging" the system. The term stuck, and finding problems in a computer (or other electronic device), particularly the software, would always be known as debugging.

ample, an application program may require the sum of a series of numbers. The instructions to add the numbers are stored in the form of binary codes that direct the CPU to fetch a series of numbers from memory, add them, and store the sum back in memory.

Memories and Storage

Several types of memories are used in a typical computer. The *RAM* (random-access memory) stores binary data and programs temporarily during processing. Data are numbers and other information, and programs are lists of instructions. Data can be written into and read out of a RAM at any time. The RAM is volatile, meaning that the information is lost if power is turned off or fails. Therefore, any data or program that needs to be saved should be moved to nonvolatile memory (such as a CD or hard disk) before power is removed.

The *ROM* (read-only memory) stores a permanent system program called the **BIOS** (Basic Input/Output System) and certain locations of system programs in memory. The ROM is nonvolatile, which means it retains what is stored, even when the power is *off*. As the name implies, the programs and data in ROM cannot be altered. Sometimes it is referred to as "firmware" because it is permanent software for a given system.

The BIOS is the lowest level of the computer's operating system. It contains instructions that tell the CPU what to do when power is first applied; the first instruction executed is in the BIOS. It controls the computer's basic start-up functions that include a self-test and a disk self-loader to bring up the rest of the operating system. In addition, the BIOS stores locations of system programs that handle certain requests from peripherals called **interrupts,** which cause the current processing to be temporarily stopped.

The *cache* memory is a small RAM that is used to store a limited amount of frequently used data that can be accessed much faster than the main RAM. The cache stores "close at hand" information that will be used again instead of having to retrieve it from farther away in the main memory. Most microprocessors have internal cache memory called level-1, or simply L1. External cache memory is in a separate memory chip and is referred to as level-2, or L2.

The *hard disk* is the major storage medium in a computer because it can store large amounts of data and is nonvolatile. The high-level operating systems as well as applications software and data files are all stored on the hard disk.

Removable storage is part of most computer systems. The most common types of removable storage media are the CDs, floppy disks, and Zip disks (magnetic storage media). Floppy disks have limited storage capability of about 1.4 MB (megabyte). CDs are available as CD-ROMs (Compact Disk–Read-Only Memory) and as CD-RWs (Rewritable) and can store huge amounts of data (typically 650 MB). Zip drives typically store 250 MB.

Input/Output Ports

Generally, the computer sends data to a peripheral device through an output port and receives information through an input port. Ports can be configured in software to be either an input or output port. The keyboard, mouse, video monitor, printer, and other peripherals communicate to the CPU through individual ports. Ports are generally classified as either serial ports, with a single data line, or parallel ports, with multiple data lines.

Buses

Peripherals are connected to the computer ports with standard interface buses. A bus can be thought of as a highway for digital signals that consists of a set of physical connections, as well as electrical specifications for the signals. Examples of serial buses are FireWire and USB (Universal Serial Bus). The most common parallel bus is simply called the *parallel bus,* which connects to a port commonly referred to as the printer port (although this port can be used by other peripherals.) Another example of a parallel bus, for connecting lab instruments to a computer, is called the General Purpose Interface Bus (GPIB).

The three basic types of internal buses that interconnect the CPU with memory and storage and with input and output ports are the address bus, data bus, and control bus. These buses are usually lumped into what is called the *local bus*. The address bus is used by the CPU to specify memory locations or addresses and to select ports. The data bus is used to transfer program instructions and data between the CPU, memories, and ports. The control bus is used for transferring control signals to and from the CPU.

Computer Software

In addition to the hardware, another major aspect of a computer is the software. The software makes the hardware perform. The two major categories of software used in computers are system software and applications software.

System Software   The system software is called the operating system of a computer and allows the user to interface with the computer. The most common operating systems used in desktop and laptop computers are Windows, MacOS, and UNIX. Many other operating systems are used in special-purpose computers and in mainframe computers.

System software performs two basic functions. It manages all the hardware and software in a computer. For example, the operating system manages and allots space on the hard disk. It also provides a consistent interface between applications software and hardware. This allows an applications program to work on various computers that may differ in hardware details.

The operating system on your computer allows you to have several programs running at the same time. This is called multitasking. For example, you can be using the word processor while downloading something from the Internet and printing an e-mail message.

Applications Software   You use applications software to accomplish a specific job or task. Table 12–1 lists several types of applications software.

▶ **TABLE 12–1**

Applications software.

APPLICATION	FUNCTION	EXAMPLES
Word processing	Prepare text documents and letters	Microsoft Word, WordPerfect
Drawing	Prepare technical drawings and pictures	CorelDraw, Freehand, Illustrator
Spreadsheet	Manipulate numbers and words in an array	Excel, Lotus 123
Desktop publishing	Prepare newsletters, flyers, books, and other printed material	Quark XPress, Pagemaker
Photography	Manipulate digital pictures, add special effects to pictures	Photoshop, Image Expert
Accounting	Tax preparation, bookkeeping	Quickbooks, Turbotax, MYOB
Presentations	Prepare slide shows and technical presentations	PowerPoint, Harvard Graphics
Data management	Manipulate large databases	Filemaker, Access
Multimedia	Digital video editing, produce moving images in presentations	Premier, Dreamweaver, After Effects
Speech recognition	Converts speech to text	NaturallySpeaking
Website preparation	Tools to create web pages and websites on the Internet	FrontPage, Acrobat
Circuit simulation	Create and test electronic circuits	Multisim

Sequence of Operation   When you first turn on your computer, this is what happens:

1. BIOS from ROM is loaded into RAM and a self-test is performed to check all major components and memory. Also, the BIOS provides information about storage, boot sequence, and the like.

2. The operating system (such as Windows) on the hard disk is loaded into RAM.

3. Application programs (such as Microsoft Word) are stored on the hard disk. When you select one, it is loaded into RAM. Sometimes, only portions are loaded as needed.

4. Files required by the application are loaded from the hard disk into RAM.

5. When a file is saved and the application is closed, the file is written back to the hard disk and both the application and the file are removed from RAM.

The Computer System

The block diagram in Figure 12–2 shows the main elements in a typical computer system and how they are interconnected. For a computer to accomplish a given task, it must communicate with the "outside world" by interfacing with people, sensing devices, or devices to be controlled in some way. To do this, there is a keyboard for data entry, a mouse, a video monitor, a printer, a modem, and a CD drive in most basic systems. These are called **peripherals**.

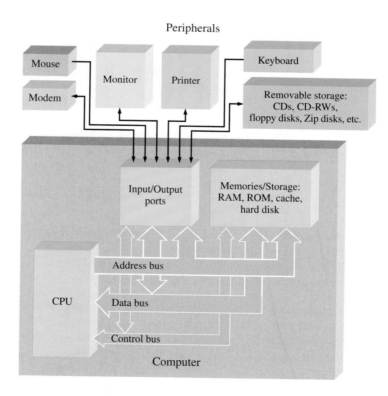

◀ FIGURE 12–2

Basic block diagram of a typical computer system including common peripherals. The computer itself is shown within the gray block.

SECTION 12–1 REVIEW

Answers are at the end of the chapter.

1. What are the major elements or blocks in a computer?
2. What is the difference between RAM and ROM?
3. What are peripherals?
4. What is the difference between computer hardware and computer software?

12–2 MICROPROCESSORS

The **microprocessor** is a digital integrated circuit that can be programmed with a series of instructions to perform various operations on data. A microprocessor is the CPU of a computer. It can do arithmetic and logic operations, move data from one place to another, and make decisions based on certain instructions.

After completing this section, you should be able to

- Describe the basic elements of a microprocessor ■ Discuss microprocessor buses
- Discuss a microprocessor instruction set

Basic Elements

A microprocessor consists of several units, each designed for a specific job. The specific units, their design and organization, are called the architecture (do not confuse the term with the VHDL element). The architecture determines the instruction set and the process for executing those instructions. Four basic units that are common to all microprocessors are the arithmetic logic unit (ALU), the instruction decoder, the register array, and the control unit, as shown in Figure 12–3.

▶ FIGURE 12–3

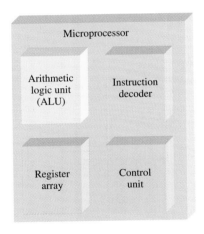

Arithmetic Logic Unit   The **ALU** is the key processing element of the microprocessor. It is directed by the control unit to perform arithmetic operations (addition, subtraction, multiplication, and division) and logic operations (NOT, AND, OR, and exclusive-OR), as well as many other types of operations. Data for the ALU are obtained from the register array.

Instruction Decoder   The instruction decoder can be considered as part of the ALU, although we are treating it as a separate function in this discussion because the instructions and the decoding of them are key to a microprocessor's operation. The microprocessor accomplishes a given task as directed by programs that consist of lists of instructions stored in memory. The instruction decoder takes each binary instruction in the order in which it appears in memory and decodes it.

Register Array   The **register array** is a collection of registers that are contained within the microprocessor. During the execution of a program, data and memory addresses are temporarily stored in registers that make up this array. The ALU can access the registers very quickly, making the program run more efficiently. Some registers are classed as general-purpose, meaning they can be used for any purpose dictated by the program. Other registers have specific capabilities and functions and cannot be used as general-purpose registers. Still others are called program invisible registers, used only by the microprocessor and not available to the programmer.

Control Unit The **control unit** is "in charge" of the processing of instructions once they are decoded. It provides the timing and control signals for getting data into and out of the microprocessor and for synchronizing the execution of instructions.

Microprocessor Buses

The three buses mentioned earlier are the connections for microprocessors to allow data, addresses, and instructions to be moved.

The Address Bus The **address bus** is a "one-way street" over which the microprocessor sends an address code to a memory or other external device. The size or width of the address bus is specified by the number of conductive paths or bits. Early microprocessors had sixteen address lines that could select 65,536 (2^{16}) unique locations in memory. The more bits there are in the address, the higher the number of memory locations that can be accessed. The number of address bits has advanced to the point where the Pentium 4 has 36 address bits and can access over 68 G (68,000,000,000) memory locations.

The Data Bus The **data bus** is a "two-way street" on which data or instruction codes are transferred into the microprocessor or the result of an operation or computation is sent out. The original microprocessors had 8-bit data buses. Today's microprocessors have up to 64-bit data buses.

The Control Bus The **control bus** is used by the microprocessor to coordinate its operations and to communicate with external devices. The control bus has signals that enable either a memory or an input/output operation at the proper time to read or write data. Control bus lines are also used to insert special wait states for slower devices and prevent bus contention, a condition that can occur if two or more devices try to communicate at the same time.

Microprocessor Programming

All microprocessors work with an instruction set that implements the basic operations. The Pentium, for example, has hundreds of variations of its instruction set divided into seven basic groups.

- Data transfer
- Arithmetic and logic
- Bit manipulation
- Loops and jumps
- Strings
- Subroutines and interrupts
- Control

Each instruction consists of a group of bits (1s and 0s) that is decoded by the microprocessor before being executed. These binary code instructions are called machine language and are all that the microprocessor recognizes. The first computers were programmed by actually writing instructions in binary code, which was a tedious job and prone to error. This primitive method of programming in binary code has evolved to a higher form where coded instructions are represented by English-like words to form what is known as assembly language. This will be discussed further in Section 12–4.

Technological Progress

The first microprocessor, the Intel 4004, was introduced in 1971. Basically, all it could do was add and subtract only 4 bits at a time. In 1974, the Intel 8080 became the first microprocessor to be used as the CPU in a computer. The 8080 chip had 6,000 transistors, an 8-bit data bus, and it ran at a clock frequency of 2 MHz. The 8080 could perform about 0.64 million instructions per second (MIPS). The Intel family has evolved from the 8080 through several different processors to the Pentium 4. This latest microprocessor (it may not be at the time you are reading this) has about 42,000,000 transistors on the chip and a 64-bit data bus. It runs at clock frequencies of up to over 3 GHz and it can do approximately 1,700 MIPS. The instruction sets have also changed drastically, but the Pentium 4 can execute any instruction code that ran on the 8086, the 1979 device that came after the 8080.

The number of transistors available has a tremendous impact on the performance and the types of things that a microprocessor can do. For example, the large number of transistors on a chip has made a technology called *pipelining* possible. Basically, pipelining allows more than one instruction to be in the process of execution at one time. Also, modern microprocessors have multiple instruction decoders, each with its own pipeline. This allows several streams of instructions to be processed simultaneously.

SECTION 12–2 REVIEW

1. What are the four basic elements in a microprocessor?
2. What are the three types of buses in a microprocessor?
3. What function does a microprocessor perform in a computer?
4. What are the three basic operations that a microprocessor performs?
5. What is pipelining?

12–3 A SPECIFIC MICROPROCESSOR FAMILY

The original Intel microprocessor family has undergone a tremendous change over the years from the 8086/8088 to the Pentium family, both in speed and in complexity. However, the basic register set and other features of the 8086/8088 have been retained (and expanded) throughout the evolutionary process so that all of the newer Intel processors respond to the same instructions (as well as a number of new instructions) as the original devices. This section starts with a limited introduction of basic concepts of microprocessor architecture, operation, and programming. The section ends with a brief overview of the principal changes to the register structure that forms the software model of the newer processors. The approach is to show a basic processor and discuss the enhancements to the Intel line as it has evolved.

After completing this section, you should be able to

■ Discuss the basic microprocessor operation ■ Describe the bus interface unit ■ State the purpose of the segment registers ■ State the purpose of the instruction pointer ■ Describe the execution unit ■ Describe the general set of registers ■ State the purpose of the flag register ■ Discuss the software model of the Pentium processor

Basic Operation

A microprocessor executes a program by repeatedly cycling through the following three steps:

1. Fetch an instruction from memory and place it in the CPU.

2. Decode the instruction; if other information is required by the instruction, fetch the other information. In the decode step, the program counter is updated to point to the next instruction.

3. Execute the instruction (do what the instruction says). Results are returned to registers and memory during this step.

The architecture of the 8086/8088 microprocessor provided for two separate internal units: the execution unit (**EU**), which executes instructions, and the bus interface unit (**BIU**), which interfaces with the system buses and fetches instructions, reads operands, and writes results. These units are shown in Figure 12–4.

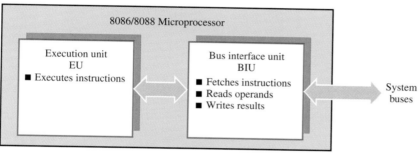

▲ **FIGURE 12–4**

The 8086/8088 has two separate internal units, the EU and the BIU.

The BIU performs all the bus operations for the EU, such as data transfers from memory or I/O. While the EU is executing instructions, the BIU "looks ahead" and fetches more instructions from memory. This action is called **prefetching** or *pipelining*. The concept of prefetching is to allow the processor to execute instructions at the same time as the next instruction was being fetched, eliminating idle time. The prefetched instructions are stored in an internal high-speed memory called the instruction **queue** (pronounced "Q"). The queue allows the BIU to keep the EU supplied with instructions. The EU does not have to wait for the next instruction to be fetched from memory; but instead it retrieves the next instruction directly from the queue in much less time. In the Pentium, this process is taken a step further. Two complete execution units enable two instructions to execute at the same time provided they are independent. Certain compilers are designed to take advantage of the two execution units by a process known as **instruction pairing** to remove dependencies.

Basic 8086/8088 Architecture

Figure 12–5 is a block diagram of the architecture (internal organization) of an 8088 microprocessor. Externally, the 8088 had 20 address bits that could address 1 MB (1,048,576 bytes) of memory and used an 8-bit data bus. Internally, the 8088 had a 16-bit data bus and a 4-byte queue. The 8086 was identical except that it had an external 16-bit data bus and a 6-byte instruction queue.

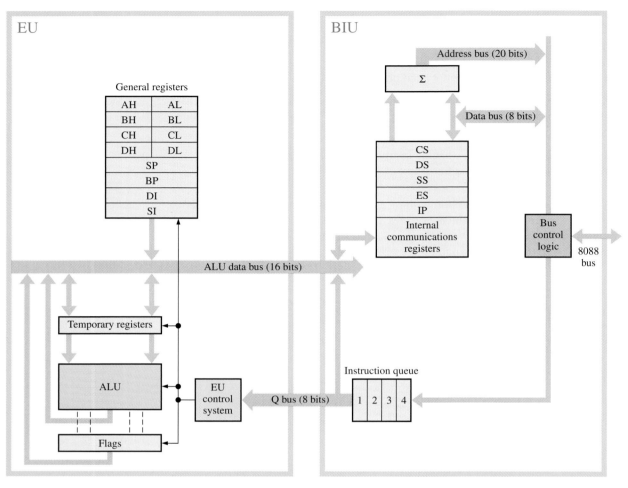

▲ FIGURE 12–5

The internal organization of the 8088 microprocessor.

The Bus Interface Unit (BIU)

The major parts of the BIU are the 4-byte instruction queue, the segment registers (CS, DS, SS, and ES), the instruction pointer (IP), and the address summing block (Σ). The 16-bit internal data buses and the Q bus interconnect the BIU and the EU.

Instruction Queue   The instruction queue increases the average speed with which a program is executed (called the **throughput**) by storing up to four bytes (six in the 8086). As described earlier, this technique allowed the 8088 essentially to do two things, fetch and execute, at one time. This feature has been expanded in subsequent processors to include much larger and faster queues.

Segment Registers   The 8086/8088 processors had four segment registers (CS, DS, SS, and ES) that were all 16-bit registers used in the process of forming a 20-bit address. A **segment** is a 64 kB block of memory and can begin at any point in the 1 MB (1,048,576 bytes) of memory space, provided it begins on a 16-byte boundary (evenly divisible by 16).

In designing the 8086/8088 and subsequent processors, Intel chose a unique method of generating the required 20-bit physical address using two 16-bit registers. One of the registers that formed the physical address (or actual) was always a segment register; the other register was a 16-bit general register containing address information. The principal advantage of the method selected was to allow codes to be easily relocatable. A **relocatable code** can be moved anywhere within the memory space without changing the basic code.

Each of the four segments identify the starting address of a 64 kB (65,536-byte) block representing a "window" in the entire 1 MB (20-bit) memory space. The starting address of a segment is represented by the 16-bit number in the segment register plus an implied 4 bits appended to the right that are always assumed to be zero. In other words, the segment registers contain the most significant 16 bits that represent the physical starting address of the segment.

The four segment registers (CS, DS, SS, and ES) can be changed by the program to point to other 64 kB blocks if necessary. (For small codes, it is normally not necessary to change the segments.) The four segments can be separate locations within the memory space or can overlap, depending on the size and requirements of the particular code. They can even be defined as the same 64 kB block. In the 8086/8088, currently addressable memory segments were those defined by the segment address contained in the CS (code segment) register, the DS (data segment) register, the SS (stack segment) register, and the ES (extra segment) register. In later processors, other segment registers were added.

As mentioned, within each segment are 64 kB of memory. To find a given memory location, a segment address is combined with an offset address. The segment address represents the most significant sixteen bits (four hex digits) of the physical address which represent the beginning address of a segment. The offset address is sixteen additional bits that represent the distance from the start of the segment to the physical address within the segment. Figure 12–6 illustrates how the memory is divided into segments and shows examples of nonoverlapping and overlapping segments.

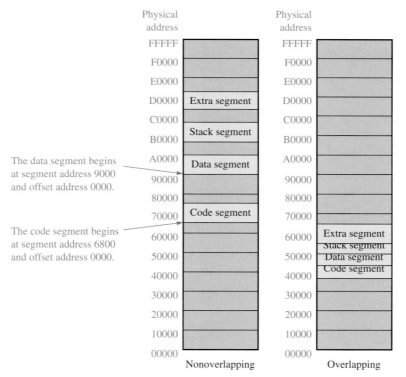

◀ **FIGURE 12–6**

Nonoverlapping and overlapping segments in the first 1 MB of memory. Each segment represents 64 kB.

Instruction Pointer (IP) and Address Summing Block The 16-bit **IP** (instruction pointer) points to the offset of the next instruction to be executed in memory. The IP always references the CS (code segment) register; thus, the physical address of the next instruction is formed by combining the code segment and the instruction pointer. The IP always contains the offset address of the next instruction, and the CS register always contains the segment address. This address is shown in assembly language as CS:IP.

To form the 20-bit physical address of the next instruction, the 16-bit offset address in the IP is added to the segment address contained in the CS register, which has been shifted four bits to the left, as indicated in Figure 12–7. As mentioned earlier, an assumed binary 0000 is the least significant position. The addition is then done by the address summing block.

▶ **FIGURE 12–7**

Formation of the 20-bit physical address from the segment base address and the offset address.

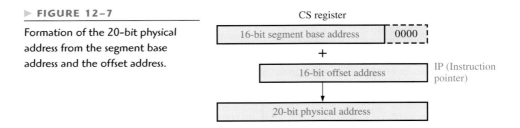

Figure 12–8 illustrates the addressing of a location in memory by the segment: offset method. In this figure, $A000_{16}$ is in the segment register and $A0B0_{16}$ is in the IP. When the CS register is shifted and added to the IP, we get $A0000_{16} + A0B0_{16} = AA0B0_{16}$ for the physical address.

▶ **FIGURE 12–8**

Illustration of the segmented addressing method.

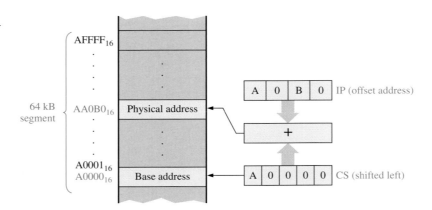

EXAMPLE 12–1

The hexadecimal contents of the CS register and the IP are shown in Figure 12–9. Determine the physical address in memory of the next instruction.

▶ **FIGURE 12–9**

Solution   Shifting the CS base address left four bits (one hex digit) effectively places a 0_{16} in the LSD position, as shown in Figure 12–10. The shifted base address and the offset address are added to produce the 20-bit physical address.

▶ FIGURE 12–10

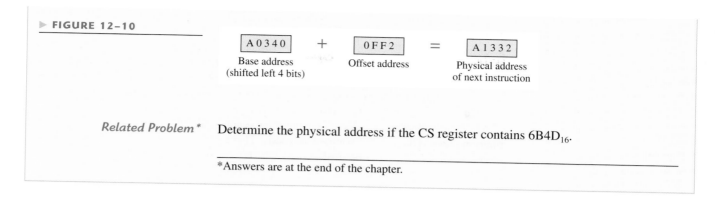

*Related Problem** Determine the physical address if the CS register contains $6B4D_{16}$.

_____

*Answers are at the end of the chapter.

It is important to understand how the segment:offset method is used to form the physical address; however, in programming work, it isn't usually necessary for the programmer to specify actual physical addresses. This job is done by the assembler program using labels supplied by the programmer. When a physical address is required, the programmer generally specifies it with the segment:offset method. Thus, the address for Example 12–1 would be given as simply A034:0FF2.

The Execution Unit (EU)

The EU decodes instructions fetched by the BIU, generates appropriate control signals, and executes the instructions. The main parts of the EU are the arithmetic logic unit (ALU), the general registers, and the flags.

The ALU   This unit does all the arithmetic and logic operations, working with either 8-bit or 16-bit operands.

The General Registers   This set of 16-bit registers is divided into two sets of four registers each, as shown in Figure 12–11. One set consists of the data registers, and the other set consists of the pointer and index registers. The **pointer** and index registers are generally used to keep offset addresses (as used here, a pointer refers to a specific memory location). In the case of the stack pointer (SP) and the base pointer (BP), the default reference to form a physical address is the stack segment (SS). The index pointers (SI and DI) and the base register (BX) generally default to the data segment (DS) register (an exception is made for certain instructions to this general rule).

	15	8 7	0	
Data set	AH	AL		Accumulator
	BH	BL		Base index
	CH	CL		Count
	DH	DL		Data

	15	0	
Pointer & index set	SP		Stack pointer
	BP		Base pointer
	DI		Destination index
	SI		Source index

▲ FIGURE 12–11

The general register set.

Each of the 16-bit data registers (AX, BX, CX, DX) has two separately accessible 8-bit sections. Depending on the program, they can be used either as a 16-bit register or as two 8-bit registers. The low-order bytes of the data registers are designated as *AL, BL, CL,* and *DL.* The high-order bytes are designated as *AH, BH, CH,* and *DH.* These registers can be used in most arithmetic and logic operations in any manner specified by the programmer for storing data prior to and after processing. Also, some of these registers are used specifically by certain program instructions.

The pointer and index registers are the stack pointer (SP), the base pointer (BP), the destination index (DI), and the source index (SI). These registers are used in various forms of memory addressing under control of the EU.

The Flags   The flag register contains nine independent status and control bits (**flags**), as shown in Figure 12–12. A status flag is a one-bit indicator used to reflect a certain condition after an arithmetic or logic operation by the ALU, such as a carry (CF), a zero result (ZF), or the sign of a result (SF), among others. The control flags are used to alter processor operations in certain situations.

▶ **FIGURE 12–12**

The status and control flags.

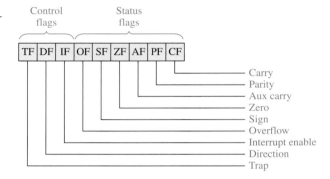

Software Model of the Pentium Family of Processors

As Intel introduced newer microprocessors, capabilities and speed increased dramatically. With the Pentium processor, the earlier pipeline concept introduced in the 8086/8088 was increased to two-integer pipelines. The external coprocessor was incorporated within the microprocessor, and address and data buses were greatly expanded. Other improvements (such as clock speed, reduced instruction clock cycles, branch prediction capability, and an integral floating-point unit) made the Pentium a significantly better processor than its predecessors. In addition to processor improvements, many improvements to other parts of computers occurred (such as bus protocols and size, speed, memory size, and cost). Despite all of these changes, the designers of the newer processors maintained compatibility with earlier software; that is, the newest Pentium could still run the software for any of the processors that preceded it. This was done by maintaining the basic software model (register structure) of the original 8086/8088 microprocessor.

The registers described previously for the 8086/8088 microprocessor are a subset of the registers in the Pentium family of processors. Beginning with the 80386 processor, the register set was expanded to include 32-bit registers. The 32-bit registers kept the original names but an E (for Extended) was added as a prefix to the register names; thus, the 32-bit designation for the AX register is the EAX. In addition, two new segment registers were added. The extended registers are shown in Figure 12–13. The gray areas represent the registers only available on the 386 and above.

In addition to the extended registers, the addressable space in memory was increased dramatically with the introduction of newer processors. To keep the upward compatibility, Intel reserved the first 1 MB of memory for codes running in real mode. **Real mode** is any

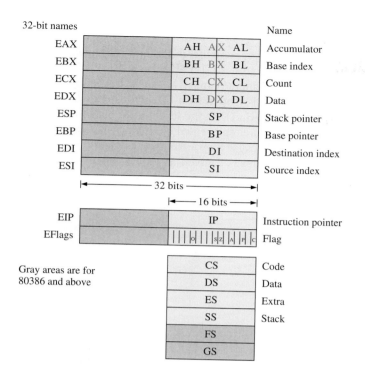

Registers for the Intel processors from 8086/8088 through Pentium.

operation that allows the processor to only access the first 1 MB of memory to simulate the 8086/8088. Code written for an earlier processor can run in real mode on a newer processor (although the reverse is not strictly true). Code written in real mode is generally compatible (with some exceptions) with all of the Intel processors from the 8086/8088 upward.

**SECTION 12–3
REVIEW**

1. Name the general-purpose registers in the Intel microprocessor.
2. What is the purpose of the BIU?
3. Does the EU interface with the system buses?
4. What is the function of the instruction queue?
5. What is the advantage of the segment:offset method of forming addresses?
6. What is instruction pairing?

12–4   COMPUTER PROGRAMMING

Assembly language is a way to express machine language in English-like terms, so there is a one-to-one correspondence. Assembly language has limited applications and is not portable from one processor to another, so most computer programs are written in high-level languages such as C, C++, JAVA, BASIC, COBOL, and FORTRAN. High-level languages are portable and therefore can be used in different computers. High-level languages must be converted to the machine language for a specific microprocessor by a process called *compiling*.

After completing this section, you should be able to

■ Describe some programming concepts ■ Discuss the levels of programming languages

Levels of Programming Languages

A hierarchy diagram of computer programming languages relative to the computer hardware is shown in Figure 12–14. At the lowest level is the computer hardware (CPU, memory, disk drive, input/output). Next is the **machine language** that the hardware understands because it is written with 1s and 0s (remember, a logic gate can recognize only a LOW (0) or a HIGH (1). At the level above machine language is **assembly language** where the 1s and 0s are represented by English-like words. Assembly languages are considered low-level because they are closely related to machine language and are machine dependent, which means a given assembly language can only be used on a specific microprocessor.

▶ **FIGURE 12–14**

Hierarchy of programming languages relative to computer hardware.

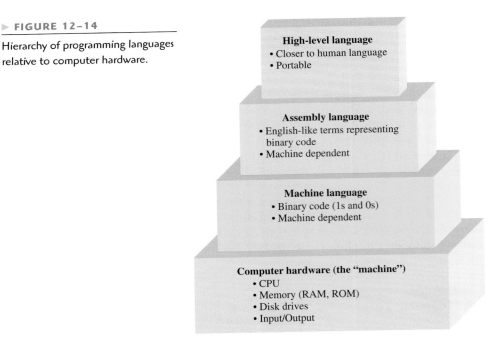

At the level above assembly language is **high-level language,** which is closer to human language and further from machine language. An advantage of high-level language over assembly language is that it is portable, which means that a program can run on a variety of computers. Also, high-level language is easier to read, write, and maintain than assembly language. Most system software (e.g., Windows and Unix), and applications software (e.g., word processors and spreadsheets) are written with high-level languages.

Assembly Language

To avoid having to write out long strings of 1s and 0s to represent microprocessor instructions, English-like terms called mnemonics or **op-codes** are used. Each type of microprocessor has its own set of mnemonic instructions that represent binary codes for the instructions. All of the mnemonic instructions for a given microprocessor are called the instruction set. Assembly language uses the instruction set to create programs for the microprocessor; and because an assembly language is directly related to the machine language (binary code instructions), it is classified as a low-level language. Assembly language is one step removed from machine language.

Assembly language and the corresponding machine language that it represents is specific to the type of microprocessor or microprocessor family. Assembly language is not portable; that is, you cannot run an assembly language program written for one type of microprocessor on another type of microprocessor. For example, an assembly program for the Motorola processors will not work on the Intel processors. Even within a given family different microprocessors may have different instruction sets.

An **assembler** is a program that converts an assembly language program to machine language that is recognized by the microprocessor. Also, programs called **cross-assemblers** translate an assembly language program for one type of microprocessor to an assembly language for another type of microprocessor.

Assembly language is rarely used to create large application programs. However, assembly language is often used in a subroutine (a small program within a larger program) that can be called from a high-level language program. Assembly language is useful in subroutine applications because it usually runs faster and has none of the restrictions of a high-level language. Assembly language is also used in machine control, such as for industrial processes. Another area for assembly language is in video game programming.

Conversion of a Program to Machine Language

All programs written in either an assembly language or a high-level language must be converted into machine language in order for a particular computer to recognize the program instructions.

Assemblers   An assembler translates and converts a program written in assembly language into machine code, as indicated in Figure 12–15. The term **source program** is often used to refer to a program written in either assembly or high-level language. The term **object program** refers to a machine language translation of a source program.

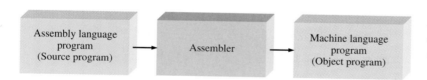

◀ FIGURE 12–15

Assembly to machine conversion using an assembler.

Compilers   A *compiler* is a program that compiles or translates a program written in a high-level language and converts it into machine code, as shown in Figure 12–16. The compiler examines the entire source program and collects and reorganizes the instructions. Every high-level language comes with a specific compiler for a specific computer, making the high-level language independent of the computer on which it is used. Some high-level languages are translated using what is called an *interpreter* that translates each line of program code to machine language.

◀ FIGURE 12–16

High-level to machine conversion with a compiler.

All high-level languages, such as C, C++, FORTRAN, and COBOL, will run on any computer. A given high-level language is valid for any computer, but the compiler that goes with it is specific to a particular type of CPU. This is illustrated in Figure 12–17, where the same high-level language program (written in C++ in this case) is converted by different machine-specific compilers.

Example of an Assembly Language Program   For a simple assembly language program, let's say that we want the computer to add a list of numbers from the memory and place the

▶ FIGURE 12–17

Machine independence of a
program written in a high-level
language.

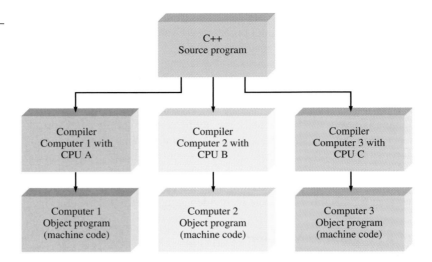

sum of the numbers back into the memory. A zero is used as the last number in the list to indicate the end of the list of numbers. The steps required to accomplish this task are as follows:

1. Clear a register (in the microprocessor) for the total or sum of the numbers.

2. Point to the first number in the memory (RAM).

3. Check to see if the number is zero. If it is zero, all the numbers have been added.

4. If the number is not zero, add the number in the memory to the total in the register.

5. Point to the next number in the memory.

6. Repeat steps 3, 4, and 5.

A flowchart is often used to diagram the sequence of steps in a computer program.
Figure 12–18 shows the flowchart for the program represented by the six steps listed above.

The assembly language program implements the addition problem shown in the flowchart in Figure 12–18. Two of the registers in the microprocessor are named ax and bx. The
comments preceded by a semicolon are not recognized by the microprocessor; they are for
explanation only.

```
        mov ax,0              ;Replaces the contents of the ax register with zero.
                              ;Register ax will store the total of the addition.
        mov bx,50H            ;Places memory address hexadecimal 50 into the bx register.
next: cmp word ptr [bx],0     ;Compares the number stored in the memory location pointed to by
                              ;the bx register to zero.
        jz done              ;If the number in the memory location is zero, jump to "done".
        add ax,[bx]          ;Add the number in the memory location pointed to by the bx register to
                              ;the number in the ax register and place the sum into the ax register.
        add bx,02            ;2 is added to the address in the bx register. Two addresses are
                              ;required to store each number which is two bytes long.
        jmp next             ;Loop back to "next" and repeat the process.
done: mov [bx],ax            ;Replace the zero last number in the memory location pointed to by the
                              ;bx register with the total in the ax register.
        nop                  ;No operation, this indicates the end of the program.
```

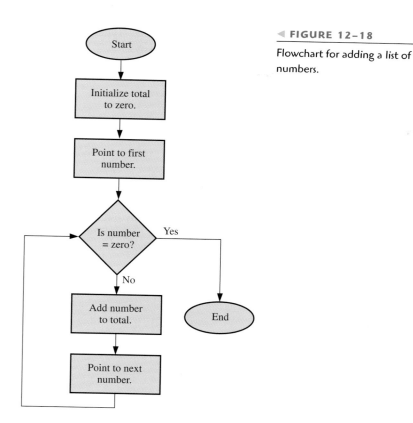

◀ **FIGURE 12–18**

Flowchart for adding a list of numbers.

Depending on the assembler, most programs in assembly language will have a number of assembler directives that are used by the assembler to do a variety of tasks. These tasks include setting up segments, using the appropriate instruction set, describing data sizes, and performing many other "housekeeping" functions. To simplify the explanation, only one directive (required) was shown in the preceding program. The directive was **word ptr,** which is used to indicate the size of the data pointed to by the BX register.

The Debug Assembler

With a few small changes, you can run the preceding program, if you choose, by using a built-in assembler, present in DOS-based PCs. You will be able to observe it execute step by step. All DOS-based PCs have a program called *Debug* that includes a primitive assembler. To use Debug, go into DOS and type **Debug<cr>** at the DOS prompt. (**<cr>** stands for "carriage return," which means to press the ENTER key.) You should see a minus sign, which is the Debug prompt. Debug has a number of commands to observe or enter data or programs. The complete list of Debug commands will be shown if you type **?** at the Debug prompt. Before writing and executing an assembly program, you can enter some data by typing the information shown in red:

```
-a 50
```

This tells Debug to start assembling instructions at the current data segment at an offset of 50H. Although these are data that are being entered, it is simpler to enter a 16-bit word this way. (Keep in mind that all data in Debug is entered in HEX.) Debug responds with a segment address, which will undoubtedly be different than that shown (20D8), but it doesn't matter. The offset address (50H) will be the same. Type the information shown in red (remember each <cr> means press the ⎡ENTER⎤ key).

```
20D8:0050 dw 30 <cr>
20D8:0052 dw 15 <cr>
20D8:0054 dw a0 <cr>
20D8:0056 dw 0c <cr>
20D8:0058 dw 00 <cr>
<cr>
```

The **dw** is an assembler directive. It is not stored itself; it merely informs the assembler that each data point is two bytes long. In the data shown, only one byte is used, but the program will save each point in two locations with a zero in the high-order position.

You can now enter the program at location 100 as follows:

```
          -a 100 <cr>
20D8:0100 mov ax,0 <cr>
20D8:0103 mov bx,50 <cr>
20D8:0106 cmp word ptr [bx],0 <cr>
20D8:0109 jz 112 <cr>
20D8:010B add ax,[bx] <cr>
20D8:010D add bx,2 <cr>
20D8:0110 jmp 106 <cr>
20D8:0112 mov [bx],ax <cr>
20D8:0114 nop <cr>
20D8:0115 <cr>
```

To confirm that the program has been entered correctly, you can type **u 100 114** at the Debug prompt, and the code you typed will be shown on the screen. (It will be shown in capital letters). Now type **r** after the Debug prompt, and a list of the 16-bit registers and condition of the flags will appear. Notice that the IP should have 100, the starting address of the code. Underneath the list of registers, the first instruction (MOV AX, 0000) will be shown.

You can cause Debug to execute this instruction with the **t** (trace) command. This will bring onto the screen the latest condition of all of the registers and show the next instruction (MOV BX, 0050). Executing this with a **t** command will show that the number 0050 has been moved into the BX register. The steps up to this point are shown in Figure 12–19. Notice that the first data point is shown in the lower right.

Continuing in this way, you can execute the entire code and observe the changes to the registers as the microprocessor follows the instructions, as shown in Figure 12–20. This program has a common programming structure called a *loop*. A loop is a repetitive group of instructions that are executed until some condition is met; in this case, the condition is finding a zero in the data. After the zero has been found, the last instruction will be executed and the sum will be stored (in this case, 00F1 is the hex sum) in place of the zero that indicated the last data point. You can observe this by pressing **d 0050 005F** (display between addresses 0050 and 005F) at the Debug prompt when you reach the last instruction (NOP), as shown in Figure 12–20. The result appears as the 9th and 10th bytes (the 5th word) on the line following the display instruction. Notice that the least significant part of the answer is shown first. When executed in "real time" by the microprocessor, this program actually uses only about 1 μs to do this entire process. If you choose to repeat the process, you will need to reload the zero at location 0058 because it has been replaced with the sum. (Recall that the program uses the zero as a "last data point" sensor.)

```
—u 100 114
20D8:0100 B80000          MOV      AX,0000
20D8:0103 BB5000          MOV      BX,0050
20D8:0106 833F00          CMP      WORD PTR [BX],+00
20D8:0109 7407            JZ       0112
20D8:010B 0307            ADD      AX,[BX]
20D8:010D 83C302          ADD      BX,+02
20D8:0110 EBF4            JMP      0106
20D8:0112 8907            MOV      [BX],AX
20D8:0114 90              NOP
—r
AX=0000   BX=0000   CX=0000   DX=0000   SP=FFEE   BP=0000   SI=0000   DI=0000
DS=20D8   ES=20D8   SS=20D8   CS=20D8   IP=0100      NV UP EI PL ZR NA PE NC
20D8:0100 B8000           MOV      AX,0000
—t

AX=0000   BX=0000   CX=0000   DX=0000   SP=FFEE   BP=0000   SI=0000   DI=0000
DS=20D8   ES=20D8   SS=20D8   CS=20D8   IP=0103      NV UP EI PL ZR NA PE NC
20D8:0103 BB5000          MOV      BX,0050
—t

AX=0000   BX=0050   CX=0000   DX=0000   SP=FFEE   BP=0000   SI=0000   DI=0000
DS=20D8   ES=20D8   SS=20D8   CS=20D8   IP=0106      NV UP EI PL ZR NA PE NC
20D8:0106 833F00          CMP      WORD PTR [BX],+00                    DS:0050=0030
_
```

▲ **FIGURE 12–19**

Steps in beginning to execute the addition program with Debug.

```
DS=20D8   ES=20D8   SS=20D8   CS=20D8   IP=0110      NV UP EI PL NZ NA PO NC
20D8:0110 EBF4                 JMP      0106
—t

AX=00F1   BX=0058   CX=0000   DX=0000   SP=FFEE   BP=0000   SI=0000   DI=0000
DS=20D8   ES=20D8   SS=20D8   CS=20D8   IP=0106      NV UP EI PL NZ NA PO NC
20D8:0106 833F00              CMP      WORD PTR [BX],+00                DS:0058=0000
—t

AX=00F1   BX=0058   CX=0000   DX=0000   SP=FFEE   BP=0000   SI=0000   DI=0000
DS=20D8   ES=20D8   SS=20D8   CS=20D8   IP=0109      NV UP EI PL ZR NA PE NC
20D8:0109 7407                JZ       0112
—t

AX=00F1   BX=0058   CX=0000   DX=0000   SP=FFEE   BP=0000   SI=0000   DI=0000
DS=20D8   ES=20D8   SS=20D8   CS=20D8   IP=0112      NV UP EI PL ZR NA PE NC
20D8:0112 8907                MOV      [BX],AX                         DS:0058=0000
—t

AX=00F1   BX=0058   CX=0000   DX=0000   SP=FFEE   BP=0000   SI=0000   DI=0000
DS=20D8   ES=20D8   SS=20D8   CS=20D8   IP=0114      NV UP EI PL ZR NA PE NC
20D8:0114 90                  NOP
—d 0050 005f
20D8:0050    30 00 15 00 A0 00 0C 00-F1 00 00 00 00 20 20 20    0............
_
                     └──────────┬──────────┘ └──┬──┘
                          Original data          Sum
```

▲ **FIGURE 12–20**

Last portion of tracing the addition program. The sum 00F1 is shown in blue with the low part (F1) given first.

EXAMPLE 12–2

Write the instructions for an assembly language program that will find the largest unsigned number in the data and place it in the last position. Assume the last data point is signaled with a zero.

Solution The flowchart is shown in Figure 12–21.

▶ **FIGURE 12–21**

Flowchart. The variable *BIG* represents the largest value.

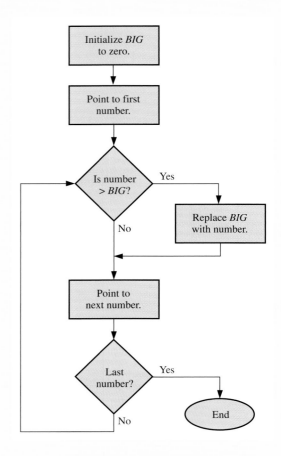

The data is assumed to be the same as before. The program listing (with comments) is as follows:

```
        mov ax,0000          ; initial value of BIG is in the ax register
        mov bx,0050          ; point to a location in memory (50H) where the data starts
repeat: cmp [bx],ax          ; is the data point larger than BIG?
        jbe check            ; if the data point is smaller, go to "check"
        mov ax,[bx]          ; otherwise, put new largest data point in ax
 check: add bx,02            ; point to the next number in memory (two bytes per word)
        cmp word ptr [bx],0  ; test for last data point
        jnz repeat           ; continue if the data point is not a zero
        mov [bx],ax          ; save BIG in memory
        nop                  ; no operation
```

The Debug listing of this program is shown in Figure 12–22. Data are entered in the same way as before starting at location 0050. In this case the same data are used, but you may choose new data if you prefer. It is important that a zero be entered as the last data point because the program continues until it finds this point. The zero is replaced each time the program is run with the largest data point. The program is entered by starting the assembly at location 100 by entering the program after the **a 100** command is issued.

▶ **FIGURE 12–22**

Listing of Debug portion of program.

```
—a 100
20D8:0100 mov ax,0
20D8:0103 mov bx,50
20D8:0106 cmp [bx],ax
20D8:0108 jbe 10c
20D8:010A mov ax, [bx]
20D8:010C add bx,2
20D8:010F cmp word ptr [bx],0
20D8:0112 jnz 106
20D8:0114 mov [bx],ax
20D8:0116 nop
20D8:0117
```

The program can be traced to watch the execution, one step at a time. Alternatively, you can enter **g = 100 116** to "go" between address 100 and 116. Figure 12–23 shows the data before and after execution. Note that each data point is stored in two bytes (although the data is only one byte long) because the data was defined as words. Also, note that the low byte preceded the high byte in memory. After the program is run, the last data point (formerly zero) is seen to be equal to the largest value (A0 in this example). This value is seen to be in both the AX register and in memory.

```
—a 100
20D8:0100 mov ax,0
20D8:0103 mov bx,50
20D8:0106 cmp [bx],ax
20D8:0108 jbe 10c
20D8:010A mov ax, [bx]
20D8:010C add bx,2
20D8:010F cmp word ptr [bx],0
20D8:0112 jnz 106
20D8:0114 mov [bx],ax
20D8:0116 nop
20D8:0117                                      End of data signal
—d 50 5f
20D8:0050   30 00 15 00 A0 00 0C 00-00 00 00 00 00 20 20 20   0...........
—g= 100 116
                Data before executing program

AX=00A0  BX=0058  CX=0000  DX=0000  SP=FFEE  BP=0000  SI=0000  DI=0000
DS=20D8  ES=20D8  SS=20D8  CS=20D8  IP=0116        NV UP EI PL ZR NA PE NC
20D8:0116 90                NOP
—d 50 5f
20D8:0050   30 00 15 00 A0 00 0C 00-A0 00 00 00 00 20 20 20   0...........
—
—
—
—
                Data are unchanged.         End of data now has
                                            largest data point.
```

▲ **FIGURE 12–23**

Data before and after a run.

Related Problem   Explain how you could change the flowchart to find the smallest number in the list instead of the largest.

Types of Instructions

The programs in this section only show a few of the hundreds of variations of instructions available to programmers. To simplify learning the Intel instruction set, instructions are divided into seven categories. These categories are described here.

Data Transfer   The most basic data transfer instruction MOV was introduced in the example programs. The MOV instruction, for example, can be used in several ways to copy a byte, a word (16 bits), or a double word (32 bits) between various sources and destinations such as registers, memory, and I/O ports. (A better mnemonic for MOV might have been "COPY" because this is what the instruction actually does.) Other data transfer instructions include IN (get data from a port), OUT (send data to a port), PUSH (copy data onto the stack, a separate area of memory), POP (copy data from the stack), and XCHG (exchange).

Arithmetic   There are a number of instructions and variations of these instructions for addition, subtraction, multiplication, and division. The ADD instruction was used in both example programs. Other arithmetic instructions include INC (increment), DEC (decrement), CMP (compare), SUB (subtract), MUL (multiply), and DIV (divide). Variations of these instructions allow for carry operations and for signed or unsigned arithmetic. These instructions allow for specification of operands located in memory, registers, and I/O ports.

Bit Manipulation   This group of instructions includes those used for three classes of operations: logical (Boolean) operations, shifts, and rotations. The logical instructions are NOT, AND, OR, XOR, and TEST. An example of a shift instruction is SAR (shift arithmetic right). An example of a rotate instruction is ROL (rotate left). When bits are shifted out of an operand, they are lost; but when bits are rotated out of an operand, they are looped back into the other end. These logical, shift, and rotate instructions can operate on bytes or words in registers or memory.

Loops and Jumps   These instructions are designed to alter the normal (one after the other) sequence of instructions. Most of these instructions test the processor's flags to determine which instruction should be processed next. In Example 12–2, the instructions JBE and JNZ were used to alter the path. Other instructions in this group include JMP (unconditional jump), JA (jump above), JO (jump overflow), LOOP (decrement the CX register and repeat if not zero) and many others.

Strings   A **string** is a **contiguous** (one after the other) sequence of bytes or words. Strings are common in computer programs. A simple example is a sentence that the programmer wishes to display on the screen. There are five basic string instructions that are designed to copy, load, store, compare, or scan a string—either as a byte at a time or a word at a time. Examples of string instructions are MOVSB (copy a string, one byte at a time) and MOVSW (copy a string, one word at a time).

Subroutine and Interrupts   A **subroutine** is a miniprogram that can be used repeatedly but programmed only once. For example, if a programmer needs to convert ASCII numbers from a keyboard to a BCD format, a simple programming structure is to make the required instructions a separate process and "call" the process whenever necessary. Instructions in this group include CALL (begin the subroutine) and RET (return to the main program).

Processor Control   This is a small group of instructions that allow direct control of some of the processor's flags and other miscellaneous tasks. An example is the STC (set carry flag) instruction.

High-Level Programming

The basic steps to take when you write a high-level computer program, regardless of the particular programming language that you use, are as follows:

1. Determine and specify the problem that is to be solved or task that is to be done.

2. Create an algorithm; that is, develop a series of steps to accomplish the task.

3. Express the steps using a particular programming language and enter them on the software text editor.

4. Compile (or assemble) and run the program.

A simple program will show an example of high-level programming. The following C++ program implements the same addition problem defined by the flowchart in Figure 12–18 and implemented using assembly language.

```
int total = 0;              //Initialize the total to 0.

while (*number ! = 0X00)    //Loop while the value is not found. The
                            //asterisk preceding the pointer identifier
                            //number says that the contents of the
                            //memory location pointed to by the
                            //identifier number are being evaluated.
{
total = total + *number;    //Accumulative summation of total.

number++;                   //Increment pointer to next number in memory.
}
```

This C++ program is equivalent to the assembly program that adds a series of numbers and produces a total value.

```
in total = 0;                                    mov ax, 0

while (*number ! = 0X00)                         mov bx, 50H

{                                          next:  cmp word ptr [bx], 0

    total = total + *number;                      jz done

    number++;                      ➡           add ax, [bx]

}                               Equivalent        add bx, 02

                                                  jmp next

        C++                               done:  mov [bx], ax

                                                  nop


                                                  Assembly
```

1. Define *program.*
2. What is an op-code?
3. What is a string?

12–5 INTERRUPTS

In this section, the establishment of communications between a peripheral and the CPU is presented. Three methods are discussed: polled I/O, interrupt-driven I/O, and software interrupts.

After completing this section, you should be able to

■ Discuss the need for interrupts in a computer system ■ Describe the basic concept of a polled I/O ■ Describe the basic concept of an interrupt-driven I/O ■ Discuss a software interrupt

In microprocessor-based systems such as the personal computer, peripheral devices require periodic service from the CPU. The term *service* generally means sending data to or taking data from the device or performing some updating process. There are three ways that a service routine can be started: *polled I/O, interrupt driven I/O,* or *software interrupts.* Recall that an interrupt is a signal or instruction that causes the current process to be temporarily stopped while a service routine is run.

In general, peripheral devices are very slow compared with the CPU. A printer may average only a few characters per second (one character is represented by eight bits), depending on the type of material being printed and the type of printer. A keyboard input rate may be one or two characters per second, depending on the speed of the operator. So, in between the times that the CPU is required to service a peripheral, it can do a lot of processing. In most systems, this processing time must be maximized by using an efficient method of servicing the peripherals.

Polled I/O

One method of servicing the peripherals is called **polling.** In this method, the CPU must test each peripheral device in sequence at certain intervals to see if it needs or is ready for servicing. Figure 12–24 illustrates the basic polled I/O method.

The CPU sequentially selects each peripheral device via the multiplexer to see if it needs service by checking the state of its ready line. Certain peripherals may need service at irregular and unpredictable intervals, that is, more frequently on some occasions than on others. Nevertheless, the CPU must poll the device at the highest rate. For example, let's say that a certain peripheral occasionally needs service every 1000 μs but most of the time requires service only once every 100 ms. As you can see, precious processing time is wasted if the CPU polls the device, as it must, at its maximum rate (every 1000 μs) because most of the time the device will not need service when it is polled.

Each time the CPU polls a device, it must stop the program that it is currently processing, go through the polling sequence, provide service if needed, and then return to the point where it left off in its current program.

Another problem with the sequentially polled I/O approach is that if two or more devices need service at the same time, the first one polled will be serviced first; the other devices will have to wait although they may need servicing much more urgently than the first device polled. As you can see, polling is suitable only for devices that can be ser-

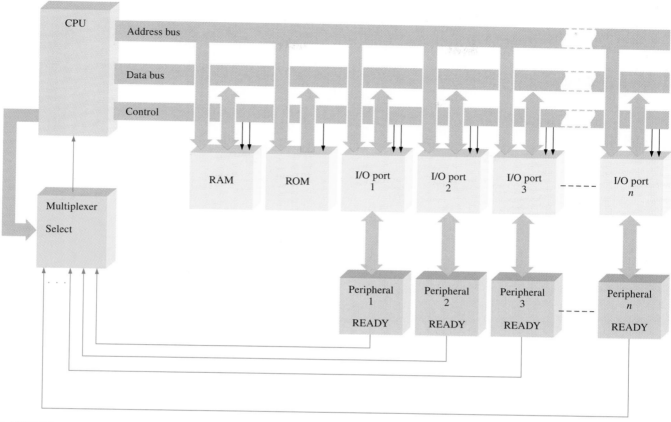

▲ FIGURE 12–24

The basic polled I/O configuration.

viced at regular and predictable intervals and only in situations in which there are no priority considerations.

Interrupt-Driven I/O

This approach overcomes the disadvantages of the polling method. In the interrupt-driven method, the CPU responds to a need for service only when service is requested by a peripheral device. Thus, the CPU can concentrate on running the current program without having to break away unnecessarily to see if a device needs service.

When the CPU receives an I/O interrupt signal, it temporarily stops its current program, acknowledges the interrupt, and fetches a special program (service routine) from memory for the particular device that has issued the interrupt. When the service routine is complete, the CPU returns to where it left off.

A device called a programmable interrupt controller (**PIC**) handles the interrupts on a priority basis. It accepts service requests from the peripherals. If two or more devices request service at the same time, the one assigned the highest priority is serviced first, then the one with the next highest priority, and so on. After issuing an interrupt (*INTR*) signal to the CPU, the PIC provides the CPU with information that "points" the CPU to the beginning memory address of the appropriate service routine. This process is called *vectoring*. Figure 12–25 shows a basic interrupt-driven I/O configuration.

Software Interrupts

Another type of interrupt is called a **software interrupt.** Software interrupts are program instructions that can invoke the same service routines described previously. The difference

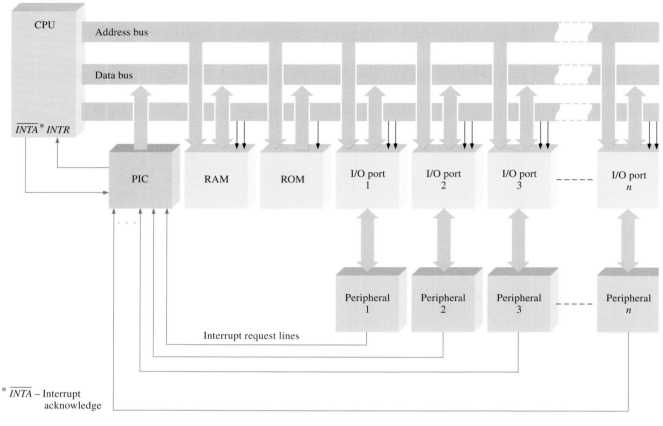

▲ **FIGURE 12–25**

A basic interrupt-driven I/O configuration.

is they are invoked from software rather than from external hardware. When invoked, the interrupt service routine executes exactly as if a hardware interrupt had occurred. The first five interrupts are defined by Intel. Others are defined by the BIOS and by DOS to perform many of the I/O operations, such as reading and writing data to the disk, writing data to the display, and reading data from the keyboard.

SECTION 12–5 REVIEW	
	1. How does an interrupt-driven I/O differ from a polled I/O?
	2. What is the main advantage of an interrupt-driven I/O?
	3. What is a software interrupt?

12–6 DIRECT MEMORY ACCESS (DMA)

In this short section, the technique of data transfer called direct memory access (DMA) is defined. A comparison of a CPU-handled transfer and a DMA transfer is presented.

After completing this section, you should be able to

■ Define the term *DMA* ■ Compare a memory I/O data transfer handled by the CPU to a DMA transfer

All I/O data transfers discussed so far have passed through the CPU. For example, when data are to be transferred from RAM to a peripheral device, the CPU reads the first data byte from the memory and loads it into an internal register within the microprocessor. Then the CPU writes the data byte to the appropriate I/O port. This read/write operation is repeated for each byte in the group of data to be transferred. Figure 12–26 illustrates this process.

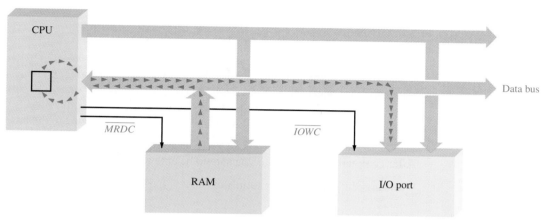

▲ FIGURE 12-26

A memory I/O transfer handled by the CPU.

For large blocks of data, intermediate stops by the microprocessor consume a lot of time. For this reason, many systems use a technique called **DMA** (direct memory access) to speed up data transfers between RAM and certain peripheral devices. Basically, DMA bypasses the CPU for certain types of data transfers, thus eliminating the time consumed by the normal fetch and execute cycles required for each read or write operation.

For direct memory transfers, a device called the DMA controller takes control of the system buses and allows data to flow directly between RAM and the peripheral device, as indicated in Figure 12–27. Transfers between the disk drive and RAM are particularly suited for DMA because of the large amounts of data involved and the serial nature of the transfers. The DMA controller can handle data transfers several times faster than the CPU.

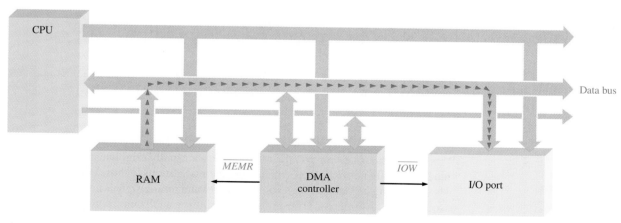

▲ FIGURE 12-27

A DMA transfer.

12–7 INTERNAL INTERFACING

As you have seen, all the components in a computer are interconnected by buses, which serve as communication paths. Physically, a bus is a set of conductive paths that serves to interconnect two or more functional components of a system or several diverse systems. Electrically, a bus is a collection of specified voltage levels and/or current levels and signals that allow the various devices connected to the bus to work properly together.

After completing this section, you should be able to

■ Discuss the concept of a multiplexed bus ■ Explain the reason for tristate outputs

Basic Multiplexed Buses

In computers the microprocessor controls and communicates with the memories and the input/output (I/O) devices via the *internal bus structure,* as indicated in Figure 12–28. A bus is multiplexed so that any of the devices connected to it can either send or receive data to or from one of the other devices. A sending device is often called a **source,** and a receiving device is often called an **acceptor.** At any given time, there is only one source active. For example, the RAM may be sending data to the input/output (I/O) interface under control of the microprocessor.

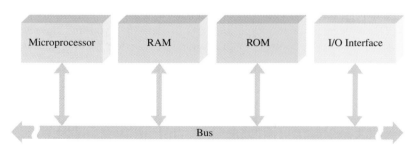

▲ **FIGURE 12–28**

The interconnection of microprocessor-based system components by a bidirectional, multiplexed bus.

Bus Signals

With synchronous bus control, the microprocessor usually originates all control and timing signals. The other devices then synchronize their operations to those control and timing signals. With asynchronous bus control, the control and timing signals are generated jointly by a source and an acceptor. The process of jointly establishing communication is called **handshaking.** A simple example of a handshaking sequence is given in Figure 12–29.

An important control function is called **bus arbitration.** Arbitration prevents two sources from trying to use the bus at the same time.

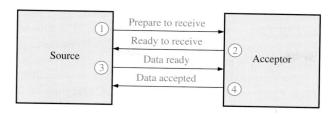

An example of a handshaking sequence.

Connecting Devices to a Bus

Tristate buffers are normally used to interface the outputs of a source device to a bus. Usually more than one source is connected to a bus, but only one can have access at any given time. All the other sources must be disconnected from the bus to prevent **bus contention.**

Tristate circuits are used to connect a source to a bus or disconnect it from a bus, as illustrated in Figure 12–30(a) for the case of two sources. The select input is used to connect

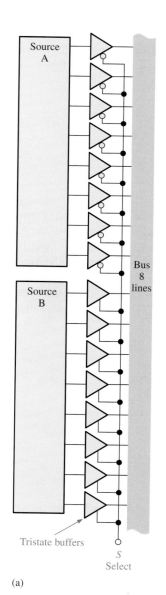

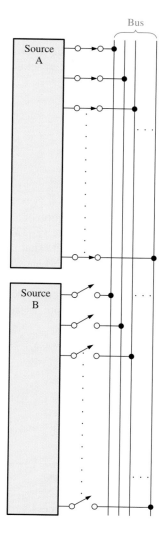

◀ **FIGURE 12–30**

Tristate buffer interface to a bus.

(a)   (b)

either source A or source B but not both at the same time to the bus. When the select input is LOW, source A is connected and source B is disconnected. When the select input is HIGH, source B is connected and source A is disconnected. A switch equivalent of this action is shown in part (b) of the figure.

When the enable input of a tristate circuit is not active, the device is in a high-impedance (**high-Z**) state and acts like an open switch. Many digital ICs provide internal tristate buffers for the output lines. A tristate output is indicated by a ∇ symbol as shown in Figure 12–31.

▶ **FIGURE 12–31**

Method of indicating tristate outputs on an IC device.

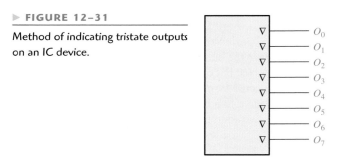

Tristate Buffer Operation Figure 12–32(a) shows the logic symbol for a noninverting tristate buffer with an active-HIGH enable. Part (b) of the figure shows one with an active-LOW enable.

▶ **FIGURE 12–32**

Tristate buffer symbols.

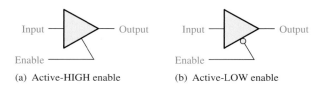

(a) Active-HIGH enable   (b) Active-LOW enable

The basic operation of a tristate buffer can be understood in terms of switching action as illustrated in Figure 12–33. When the enable input is active, the gate operates as a normal noninverting circuit. That is, the output is HIGH when the input is HIGH and LOW when the input is LOW, as shown in parts (a) and (b) respectively. The HIGH and LOW levels represent two of the states. The buffer operates in its third state when the enable input is not active. In this state, the circuit acts as an open switch, and the output is completely disconnected from the input, as shown in part (c). This is sometimes called the *high-impedance* or *high-Z* state.

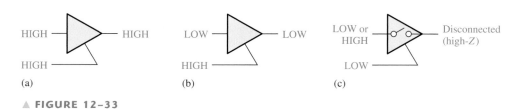

(a)   (b)   (c)

▲ **FIGURE 12–33**

Tristate buffer operation.

Many microprocessors, memories, and other integrated circuit functions have tristate buffers that serve to interface with the buses. Such buffers are necessary when two or more devices are connected to a common bus. To prevent the devices from interfering with each other, the tristate buffers are used to disconnect all devices except the ones that are communicating at any given time.

Bus Contention

Bus contention occurs when two or more devices try to output opposite logic levels on the same common bus line. The most common form of bus contention is when one device has not completely turned off before another device connected to the bus line is turned on. This generally occurs in memory systems when switching from the READ mode to the WRITE mode or vice versa and is the result of a timing problem.

Multiplexed I/Os

Some devices that send and receive data have combined input and output lines, called I/O ports, that must be multiplexed onto the data bus. Bidirectional tristate buffers interface this type of device with the bus, as illustrated in Figure 12–34(a).

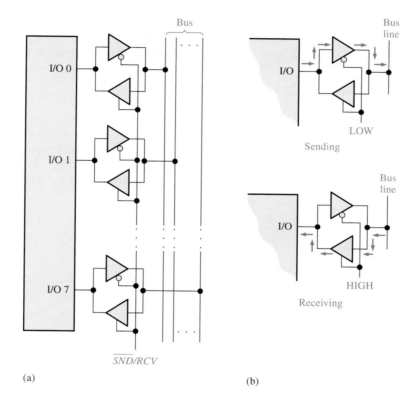

◀ FIGURE 12–34

Multiplexed I/O operation.

(a)

(b)

Each I/O port has a pair of tristate buffers. When the $\overline{SND}/RCV(\overline{\text{Send}}/\text{Receive})$ line is LOW, the upper tristate buffer in each pair is enabled and the lower one disabled. In this state, the device is acting as a source and sending data to the bus. When the $\overline{SND}/RCV$ line is HIGH, the lower tristate buffer in each pair is enabled so that the device is acting as an acceptor and receiving data from the bus. This operation is illustrated in Figure 12–34(b). Some devices provide for multiplexed I/O operation with internal circuitry.

SECTION 12–7 REVIEW

1. Why are tristate buffers required to interface digital devices to a bus?
2. What is the purpose of a bus system?

12–8 STANDARD BUSES

A bus can be thought of as a "highway" for digital signals; it consists of a set of physical connections (printed circuit traces or wires) over which the data and other information are moved from one place to another. A bus also consists of a standard set of specifications that designate the characteristics and types of signals that can travel along its pathway. Internal buses interconnect the various components within a computer system: the processor, memory, disk drive, controller, and interface cards. External or I/O buses provide for transfer of digital signals between a computer and the "outside world" and interface the computer with peripheral equipment (a video monitor, keyboard, mouse, and printer) or with other equipment that is to be controlled by a computer, such as test and measurement instruments.

After completing this section, you should be able to

- ▪ Discuss various internal and external serial and parallel buses ▪ Define local bus
- ▪ Describe PCI and ISA internal bus standards ▪ Describe the RS-232C bus
- ▪ Describe the FireWire ▪ Discuss the USB ▪ Explain the GPIB ▪ Discuss SCSI

Internal Buses

Internal buses in a computer carry addresses, data, and control signals between the microprocessor, cache memory, SRAM, DRAM, disk drives, expansion slots, and other internal devices. Personal computers consist of three types of internal buses: the *local bus,* the *PCI bus,* and the *ISA bus.* Figure 12–35 shows the basic arrangement of a bus system.

▶ **FIGURE 12–35**

Simplified illustration of the basic bus system in a typical personal computer.

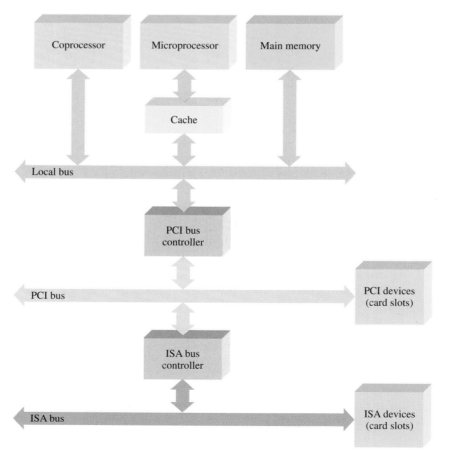

Local Bus   This bus directly connects the microprocessor to the cache memory, the main memory, the coprocessor, and the PCI bus controller. The **local bus** is the only internal bus that connects directly to the microprocessor. Generally, this bus includes the data bus, the address bus, and the control bus that allows the microprocessor to communicate with the other devices. The local bus can be considered as the *primary* bus in a computer system. For example, the Pentium local bus consists of the address bus containing 32 memory address lines, the data bus containing 64 data lines, and the control bus containing numerous control lines.

PCI (Peripheral Control Interconnect) Bus   This bus is for interfacing the microprocessor with external devices via expansion slots (connectors). The **PCI bus** was developed by Intel; and since it was first introduced in 1993, it has become the standard personal computer interface bus, replacing several earlier bus standards. PCI is a 64-bit bus, although it is often implemented as a 32-bit bus in which the address and data buses are multiplexed. It can operate at clock speeds of 33 MHz or 66 MHz.

The PCI bus is isolated from the local bus by a bus controller unit that acts as a "bridge" between the two buses. PCI is considered a *secondary* bus and is clocked independently of the microprocessor. The PCI can connect the microprocessor to peripheral devices, such as a hard drive, via expansion slots with adapter cards.

PCI supports "plug-and-play," the ability of a computer to automatically configure expansion boards and other devices. This allows a device to be connected to a computer without concern about setting switches, changing jumper wires, or dealing with any other configuration elements. This is accomplished with a 256-byte memory that allows the computer to interrogate the PCI interface.

ISA (Industry Standard Architecture) Bus   This expansion bus was developed by IBM for its AT personal computer and is the standard bus into which virtually all printed circuit cards made before 1993 plug. The **ISA** is currently incorporated into most modern personal computers, as a companion to the PCI bus, for purposes of backward compatibility.

The ISA has either an 8-bit or a 16-bit data bus and can operate at 8.33 MHz. An expanded version called the EISA provided a 32-bit data bus, but it has largely been discontinued due to its slow speed and replaced with the PCI bus.

External Buses

External devices are connected to a computer via an input/output (I/O) interface called a *port*. There are two basic types of computer ports, the *serial port* and the *parallel port,* and most computers have a parallel port and at least one serial port for connecting modems, printers, mice, and other peripheral devices.

A serial port is used for serial data communication, where only 1 bit is transferred at a time. Modems and mice are examples of typical serial devices. Also, serial ports are sometimes used for interfacing test and measurement equipment with a computer. A parallel port is used for parallel data communication, where at least 1 byte (8 bits) is transferred at a time. There are several bus standards currently in use for both serial and parallel ports. The most prominent ones are described next.

Serial I/O Interface Buses

RS-232C   This is one of the oldest and most common standards for serial interface approved by the Electronic Industries Association (EIA). The RS-232C is also referred to as the EIA-232. Most **modems** (*mo*dulator/*dem*odulator) conform to the EIA-232 standard, and most personal computers have an RS-232C port. The mouse, some display screens, and serial printers—in addition to modems—are designed to connect to the RS-232C port. The RS-232C is commonly used for interfacing data terminal equipment (**DTE**) with data communications equipment (**DCE**). For example, a computer is classified as DTE and a modem is classified as DCE.

The EIA-232 standard specifies twenty-five lines between DTE and DCE requiring a twenty-five pin plug (DB-25), as shown in Figure 12–36. In personal computer applications, not all of the RS-232C signals are required. A minimum of three and a maximum of eleven are typically used. For this reason, a 9-pin connection (DB-9) was defined by IBM for its serial interface.

▶ **FIGURE 12–36**

The RS-232C 25-pin connector plug.

Figure 12–37(a) lists the signals and pin assignments for a 25-pin RS-232C connector, and part (b) lists the signals and pin assignments for a 9-pin connector. The eleven pins and signals shown in blue in part (a) indicate the signals typically used for personal computer applications, and the three minimum signals are indicated by an asterisk (pins 2, 3, 7).

The specified maximum cable length for the RS-232C is 50 feet with a rate of data transfer of 20 kbaud. If a shorter cable is used, the baud rate can be higher. The specification of data transfer rate in baud and bits per second (b/s) are not necessarily always equal. This is because the baud rate is modem terminology and is defined as the number of signal changes per second, which is called the modulation rate. In modems, one signal change sometimes transfers several bits of data. At lower rates the baud is equal to bits per second, but at higher rates the baud may be less than the bits per second.

To overcome the limitations of the RS-232C, two other standards, the RS-422 and RS-423 were developed. These newer standards specify much longer cable lengths and higher data transfer rates under certain conditions. For example, both the RS-422 and RS-423 specify a maximum cable length of 4000 feet. The maximum RS-422 data transfer rate is 10 Mbaud for 40 feet of cable and 100 kbaud for 4000 feet. For the RS-423, the data transfer rate is 100 kbaud for 30 feet and 1 kbaud for 4000 feet. The RS-232C still remains the most common.

IEEE 1394   This external serial bus standard supports data transfer rates of up to 400 Mb/s and is typically used for, although not limited to, interfacing with graphics and video peripherals, such as digital cameras. The **IEEE 1394** standard is often called **FireWire,** a name trademarked by Apple computer who first developed it. Other companies use other names to describe their IEEE 1394 products. IEEE stands for *Institute of Electrical and Electronics Engineers.*

Up to 63 devices can be connected to FireWire based on a daisy-chain arrangement. The FireWire cable consists of six wires, two twisted pair for data and two for power. Also, this standard allows "hot plugging," the ability to add and remove devices connected to a computer while the computer is running.

USB (Universal Serial Bus)   The **USB** supports two data transfer rates, a high-speed rate of 12 Mb/s and a low-speed rate of 1.5 Mb/s. A USB port can be used to connect up to 127 peripheral devices and permits both plug-and-play and hot plugging. The USB cable has four wires, two for data and two for power, and connects the computer to USB peripheral

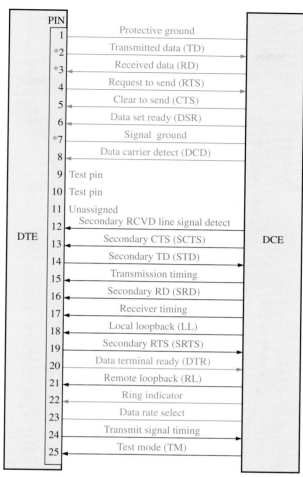

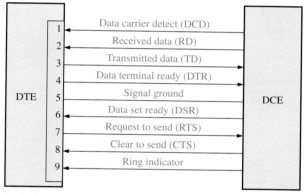

(a) Full RS-232C 25-pin interface with typical personal computer configuration indicated by blue and three minimum signals marked by an asterisk (pins 2, 3, 7).

(b) 9-pin RS-232C interface

▲ FIGURE 12–37

The RS-232C pin assignments and signals for both connector versions.

devices, any of which can also act as "hubs" for connecting to other USB peripheral devices. Figure 12–38 illustrates a computer system with USB interfacing.

Parallel I/O Interface Buses

IEEE 488   This bus standard has been around a long time and is also known as the General-Purpose Interface Bus (**GPIB**). Widely used in test and measurement applications, it was developed by Hewlett-Packard in the 1960s. The **IEEE 488** specifies 24 lines that are used to transfer eight parallel data bits at a time and provide eight control signals that include three handshake lines and five bus-management lines. Also included are eight ground lines used for shielding and ground returns. The maximum data transfer rate for the IEEE 488 standard is 1 MB/s. A superset of this standard, called the HS488, has a maximum data rate of 8 MB/s.

To connect test equipment to a computer using the IEEE 488 bus, an interface card is installed in the computer, which turns the computer into a system **controller.** In a typical GPIB

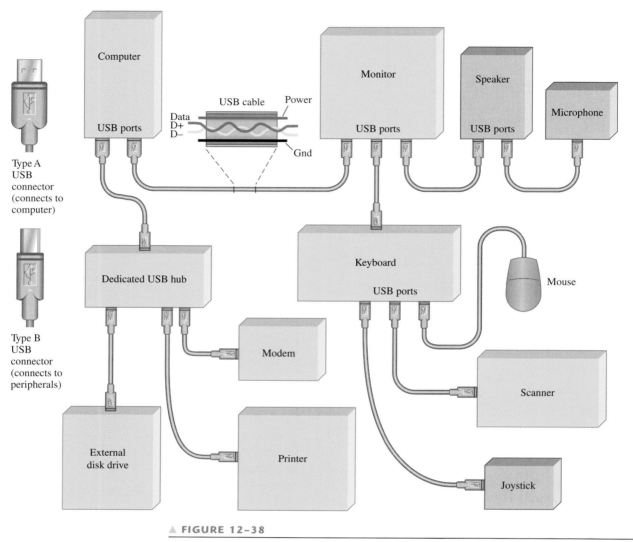

▲ FIGURE 12–38

Example of a computer system with USB interfacing.

setup, up to 14 controlled devices (test and measurement instruments) can be connected to the system controller. When the system controller issues a command for a controlled device to perform a specified operation, such as a frequency measurement, it is said that the controller "talks" and the controlled device "listens."

A **listener** is an instrument capable of receiving data over the GPIB when it is addressed by the system controller (computer). Examples of listeners are printers, monitors, programmable power supplies, and programmable signal generators. A **talker** is an instrument capable of sending data over the GPIB. Examples are DMMs and frequency counters that can output bus-compatible data. Some instruments can send and receive data and are called talker/listeners; examples are computers, modems, and certain measurement instruments. The system controller can specify each of the other instruments on the bus as either a talker or a listener for the purpose of data transfer. The controller is usually a talker/listener.

A typical GPIB arrangement is shown in Figure 12–39 as an example. The three basic bus signal groupings are shown as the *data bus, data transfer control bus,* and *interface management bus.*

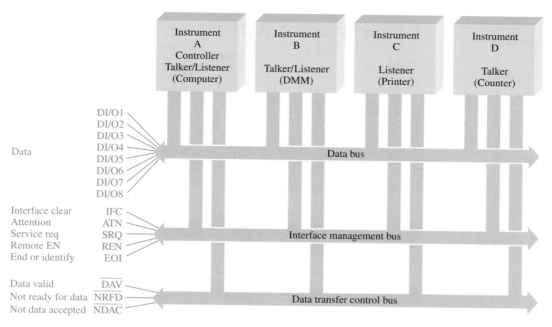

▲ FIGURE 12–39

A typical IEEE 488 (GPIB) connection.

The parallel data lines are designated DI/O1 through DI/O8 (data input/output). One byte of data is transferred on this bidirectional part of the bus. Every byte that is transferred undergoes a handshaking operation via the data transfer control. The three active-LOW handshaking lines indicate if data are valid ($\overline{\text{DAV}}$), if the addressed instrument is not ready for data ($\overline{\text{NRFD}}$), or if the data are not accepted ($\overline{\text{NDAC}}$). More than one instrument can accept data at the same time, and the slowest instrument sets the rate of transfer. Figure 12–40 shows the timing diagram for the GPIB handshaking sequence, and Table 12–2 describes the handshaking signals.

The five signals of the interface management bus control the orderly flow of data. The ATN (attention) line is monitored by all instruments on the bus. When ATN is active, the system controller selects the specific interface operation, designates the talkers and the listeners, and provides specific addressing for the listeners. Each GPIB instrument has a

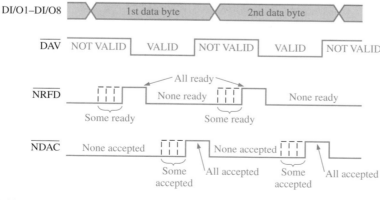

▲ FIGURE 12–40

Timing diagram for the GPIB handshaking sequence.

▶ TABLE 12-2

The GPIB handshaking signals.

NAME	DESCRIPTION
$\overline{DAV}$	**Data Valid:** After the talker detects a HIGH on the $\overline{NRFD}$ line, a LOW is placed on this line by the talker when the data on its I/O are settled and valid.
$\overline{NRFD}$	**Not Ready for Data:** The listener places a LOW on this line to indicate that it is not ready for data. A HIGH indicates that it is ready. The $\overline{NRFD}$ line will not go HIGH until all addressed listeners are ready to accept data.
$\overline{NDAC}$	**Not Data Accepted:** The listener places a LOW on this line to indicate that it has not accepted data. When it accepts data from its I/O, it releases its $\overline{NDAC}$ line. The $\overline{NDAC}$ line to the talker does not go HIGH until the last listener has accepted data.

▶ TABLE 12-3

The GPIB management lines.

NAME	DESCRIPTION
ATN	**Attention:** Causes all the devices on the bus to interpret data, as a controller command or address and activates the handshaking function.
IFC	**Interface Clear:** Initializes the bus.
SRQ	**Service Request:** Alerts the controller that a device needs to communicate.
REN	**Remote Enable:** Enables devices to respond to remote program control.
EOI	**End or Identity:** Indicates the last byte of data to be transferred.

specific identifying address that is used by the system controller. Table 12–3 describes the GPIB interface management lines and their functions.

The GPIB is limited to a maximum cable length of 15 meters, and there can be no more than one instrument per meter with a maximum capacitive loading of 50 pF each. The cable length limitation can be overcome by the use of bus extenders and modems. A bus extender provides for cable-interfacing of instruments that are separated by a distance greater than allowed by the GPIB specifications or for communicating over greater distances via modem-interfaced telephone lines. The use of bus extenders and/or modems is illustrated in Figure 12–41.

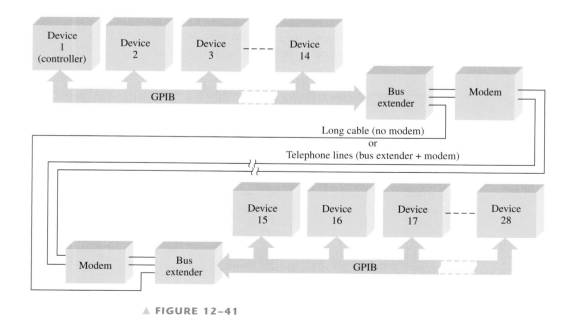

▲ FIGURE 12-41

A bus extender and modem can be used for interfacing remote GPIB systems.

SCSI (Small Computer System Interface)   Pronounced *scuzzy,* this is a widely used standard for interfacing personal computers and peripherals. Although **SCSI** is an ANSI (American National Standards Institute) standard, there are several variations and connector types originating from a variety of manufacturers. One type of SCSI may not be compatible with another type. SCSI-1 is the 25-pin connector version that provides an 8-bit data bus and supports data transfer rates of 4 MB/s. Some other versions of the SCSI bus standard are listed as follows:

■ *SCSI-2*   This version is the same as SCSI-1, but uses a 50-pin connector and supports multiple devices.

■ *Wide SCSI*   This uses a wider connector than SCSI-2 to support 16-bit data transfers.

■ *Fast SCSI*   This provides for 8-bit data transfer but supports data transfer rates of 10 MB/s.

■ *Fast Wide SCSI*   This version allows 16-bit data transfer at 20 MB/s.

■ *Ultra SCSI*   This version transfers 8 bits of data at 20 MB/s.

■ *SCSI-3*   This version has 16 data lines and runs at 40 MB/s.

■ *Ultra SCSI-2*   This version transfers 8 bits at 40 MB/s.

■ *Wide Ultra SCSI-2*   This version provides for 16-bite data transfer and operates at 80 MB/s.

The signal descriptions for a SCSI 25-pin connector are given in Table 12–4, and the pin configuration is shown in Figure 12–42.

▼ TABLE 12–4

SCSI signals.

PIN NUMBER	SIGNAL NAME	SIGNAL DESCRIPTION	PIN NUMBER	SIGNAL NAME	SIGNAL DESCRIPTION
1	REQ/	Request	14	GND	Signal ground
2	MSG/	Message	15	C/D/	Command/Data
3	I/O/	Input/Output	16	GND	Signal ground
4	RST/	SCSI bus reset	17	ATN/	Attention
5	ACK/	Acknowledge	18	GND	Signal ground
6	BSY/	Busy	19	SEL/	Select
7	GND	Signal ground	20	DBP/	Data parity
8	DB0/	Data bit 0	21	DB1/	Data bit 1
9	GND	Signal ground	22	DB2/	Data bit 2
10	DB3/	Data bit 3	23	DB4/	Data bit 4
11	DB5/	Data bit 5	24	GND	Signal ground
12	DB6/	Data bit 6	25	TPWR	Terminator power
13	DB7/	Data bit 7			

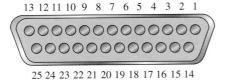

13 12 11 10 9 8 7 6 5 4 3 2 1

25 24 23 22 21 20 19 18 17 16 15 14

◀ FIGURE 12–42

SCSI 25-pin connector.

1. Name the two major bus categories in terms of the way data are transferred.
2. Classify each of the following buses as serial or parallel:
   (a) SCSI   (b) RS-232C   (c) USB   (d) GPIB
3. Explain the basic difference between a serial bus and a parallel bus.
4. How many devices can be connected to the USB?
5. Is the FireWire a faster bus than the USB in terms of data transfer?

SUMMARY

- Basic units of a computer are shown in Figure 12–43.

▶ **FIGURE 12–43**

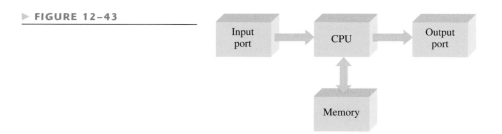

- The three basic computer buses are the *address bus, data bus,* and the *control bus.* The size of any bus is specified by the number of separate conductive paths.
- Typical peripheral devices include the keyboard, external disk drives, mouse, printer, modem, and scanner.
- The number of address lines increased from 20 for the 8086/8088 to 32 for the Pentium family. The data bus was originally 16 bits for the 8086 and is 64 bits for the Pentium family.
- General registers are a subset of those in all of the Intel processors. They include

  Accumulator (AX, which includes AH and AL)
  Base (BX, which includes BH and BL)
  Count (CX, which includes CH and CL)
  Data (DX, which includes DH and DL)
  Stack pointer (SP)
  Base pointer (BP)
  Destination index (DI)
  Source index (SI)

  Beginning with the 80386 processor, this basic set was expanded to the extended register set.
- The basic segment registers are a subset of those in all of the Intel processors. The segment registers are

  Code segment (CS)
  Data segment (DS)
  Extra segment (ES)
  Stack segment (SS)

  Beginning with the 80386 processor, two new segment registers were added.
- The flag registers are a subset of those in all of the Intel processors. They include

  Trap (TF)
  Direction (DF)
  Interrupt enable (IF)
  Overflow (OF)

Sign (SF)
Zero (ZF)
Auxiliary carry (AF)
Parity (PF)
Carry (CF)

■ The basic "language" of a computer is called machine code in which instructions are given as a series of binary codes.

■ In assembly language, machine instructions are replaced with a short alphabetic English mnemonic that has a one-to-one correspondence to machine code. Assembly language also uses directives to allow the programmer to specify other parameters that are not translated directly into machine code.

■ Ports are an interface to external devices. They can be set up as input, output, or a combination of both. They can be accessed as dedicated or as memory-mapped and can be serviced by polling, interrupt-driven, or software.

■ Table 12–5 compares standard buses.

▼ TABLE 12–5

	INTERNAL BUSES		EXTERNAL BUSES				
	PCI	ISA	RS-232C	IEEE 1394	USB	IEEE 488	SCSI
Type	Parallel	Parallel	Serial	Serial	Serial	Parallel	Parallel
Data lines	32/64	8/16	—	—	—	8	8/16
Data rate	33/66 MHz	8.33 MHz	20 kbaud	400 Mb/s	1.5/12 Mb/s	1 Mb/s	4 Mb/s (1)
							10 Mb/s (Fast)
							20 Mb/s (Ultra)
							40 Mb/s (3)
							80 Mb/s (Ultrawide 2)
Number of devices	—	—	1	63	127	14	16

KEY TERMS

Key terms and other bold terms in the chapter are defined in the end-of-book glossary.

Address bus A one-way group of conductors from the microprocessor to a memory, or other external device, on which the address code is sent.

Assembly language A programming language that uses English-like words and has a one-to-one correspondence to machine language.

Control bus A set of conductive paths that connects the CPU to other parts of the computer to coordinate its operations and to communicate with external devices.

CPU Central processing unit; the "brain" of a computer that processes the program instructions.

Data bus A bidirectional set of conductive paths on which data or instruction codes are transferred into a microprocessor or on which the result of an operation is sent out from a microprocessor.

FireWire The IEEE-1394 standard serial bus.

GPIB General-purpose interface bus based on the IEEE-488 standard.

High-level language A type of computer language closest to human language that is a level above assembly language.

Interrupt A computer signal or instruction that causes the current process to be temporarily stopped while a service routine is run.

Machine language Computer instructions written in binary code that are understood by a computer; the lowest level of programming language.

Microprocessor A large-scale digital integrated circuit that can be programmed to perform arithmetic, logic, or other operations; the CPU of a computer.

Modem A modulator/demodulator for interfacing digital devices to analog transmission systems such as telephone lines.

Peripheral A device such as a printer or modem that provides communication with a computer.

Port A physical interface on a computer through which data are passed to or from a peripheral.

Program A list of instructions that a computer follows in order to achieve a specified result.

SCSI Small computer systems interface; an external parallel bus standard.

Tristate A type of output on logic circuits that exhibits three states: HIGH, LOW, and high Z; used to interface the outputs of a source device to a bus.

USB Universal serial bus; an external serial bus standard.

SELF-TEST

Answers are at the end of the chapter.

1. A basic computer does not include
   - (a) an arithmetic logic unit
   - (b) a control unit
   - (c) peripheral units
   - (d) a memory unit

2. A 20-bit address bus supports
   - (a) 100,000 memory addresses
   - (b) 1,048,576 memory addresses
   - (c) 2,097,152 memory addresses
   - (d) 20,000 memory addresses

3. The number of bits on the data bus in the Pentium processors is
   - (a) 16
   - (b) 24
   - (c) 32
   - (d) 64

4. A bus that is used to transfer information both to and from the microprocessor is the
   - (a) address bus
   - (b) data bus
   - (c) both of the above
   - (d) none of the above

5. An example of a peripheral unit is
   - (a) the address register
   - (b) the MPU
   - (c) the video monitor
   - (d) the interface adapter

6. Two types of memory transfers handled by the CPU are
   - (a) direct and interrupt
   - (b) read and write
   - (c) bussed and multiplexed
   - (d) input and output

7. In the Intel family, the maximum number of 8-bit I/O devices is
   - (a) 64
   - (b) 1000
   - (c) 64,000
   - (d) 1 million
   - (e) unlimited

8. Polling is a method used for
   - (a) determining the state of the microprocessor
   - (b) establishing communications between the CPU and a peripheral
   - (c) establishing a priority for communication with several peripherals
   - (d) determining the next instruction

9. Of the following, which is a 8-bit register?
   - (a) AH
   - (b) BX
   - (c) SS
   - (d) IP

10. Essentially, a mnemonic is a(n)
    - (a) flowchart
    - (b) operand
    - (c) machine code
    - (d) instruction

11. DMA stands for
    - (a) digital microprocessor address
    - (b) direct memory access
    - (c) data multiplexed access
    - (d) direct memory addressing

12. A computer program is a list of
    (a) memory addresses that contain data to be used in an operation
    (b) addresses that contain instructions to be used in an operation
    (c) instructions arranged to achieve a specific result

13. A type of assembly language that alters the course of the program is called a
    (a) loop   (b) jump
    (c) both of the above (d) none of the above

14. A type of interrupt that is invoked from within a program is called a
    (a) software interrupt (b) polled interrupt
    (c) direct interrupt   (d) I/O interrupt

15. Most devices are interfaced to a bus with
    (a) totem-pole outputs (b) tristate buffers
    (c) *pnp* transistors (d) resistors

16. The PCI bus consists of
    (a) 8 or 16 data lines (b) 32 or 64 data lines   (c) 1 serial data line

17. The devices operating on a GPIB are called
    (a) source and load   (b) talker and listener
    (c) transmitter and receiver (d) donor and acceptor

18. The RS-232C is
    (a) a standard interface for parallel data   (b) a standard interface for serial data
    (c) an enhancement of the IEEE 488 interface (d) the same as SCSI

19. The FireWire bus is the same as the
    (a) IEEE 488 bus (b) USB   (c) IEEE 1394
    (d) RS-422   (e) RS-423

20. The USB can support up to
    (a) 63 devices (b) 14 devices
    (c) 100 devices (d) 127 devices

PROBLEMS

Answers to odd-numbered problems are at the end of the book.

SECTION 12–1 The Basic Computer

1. Name the basic elements of a computer.
2. Name two categories of computer software.
3. What is a bus?
4. What is a port?

SECTION 12–2 Microprocessors

5. Name the basic elements of a microprocessor.
6. List three operations that a microprocessor performs.
7. List the three microprocessor buses.
8. What are the seven basic groups of the Pentium instruction set?

SECTION 12–3 A Specific Microprocessor Family

9. What are the three basic steps a processor repeatedly cycles through?
10. What is meant by "pipelining"?
11. Name the six segment registers.

12. Assume the code segment register contains the hex number 0F05 and the instruction pointer register contains the number 0100. What is the physical address of the next instruction to be executed?

13. Explain the difference between the AH, the AL, the AX, and the EAX registers.

14. **(a)** What is a flag?

    (b) What two purposes are flags used for?

15. Explain the advantage of instruction pairing in the Pentium processor.

SECTION 12–4   **Computer Programming**

16. What is an assembler?

17. Draw a flowchart for a program that adds the numbers from one to 10 and saves the result in a memory location named TOTAL.

18. Draw a flowchart showing how you could count the number of bytes in a string and place the count in a location in memory called COUNT. Assume the string starts at a location named START and has a 20H (ASCII for a space) to signal the end. You should not count the space character.

19. Explain what happens when the instruction **mov ax,[bx]** is executed.

20. What is a compiler?

SECTION 12–5   **Interrupts**

21. Compare polled I/O to interrupt-driven I/O.

22. What is meant by the term *vectoring?*

23. What is meant by a software interrupt?

SECTION 12–6   **Direct Memory Access (DMA)**

24. Explain what happens in a DMA operation.

25. How is the CPU used in DMA?

SECTION 12–7   **Internal Interfacing**

26. In a simple serial transfer of eight data bits from a source device to an acceptor device, the handshaking sequence in Figure 12–44 is observed on the four generic bus lines. By analyzing the time relationships, identify the function of each signal, and indicate if it originates at the source or at the acceptor.

▶ FIGURE 12–44

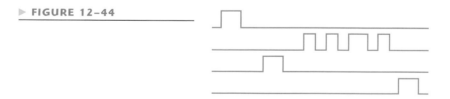

27. Determine the signal on the bus line in Figure 12–45 for the data-input and enable waveforms shown.

▶ FIGURE 12–45

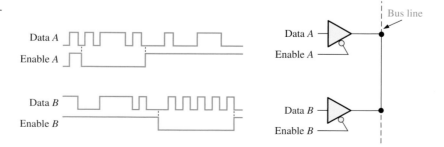

28. In Figure 12–46(a), data from the two sources are being placed on the data bus under control of the select line. The select waveform is shown in Figure 12–46(b). Determine the data-bus waveforms for the device output codes indicated.

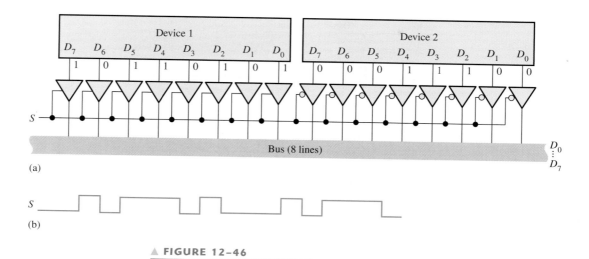

(a)

(b)

▲ **FIGURE 12–46**

SECTION 12–8 **Standard Buses**

29. Explain the basic difference between a local bus and the PCI bus.

30. Define "plug-and-play."

31. How does the PCI bus differ from the ISA bus?

32. If a shorter RS-232C is used, can data be transmitted at a faster rate?

33. DCE and DTE are part of which bus specification? Explain the acronyms, DCE and DTE.

34. List the wires in a USB cable.

35. Eight GPIB-compatible instruments are connected to the bus. How many more can be added without exceeding the specifications?

36. Consider the GPIB interface between a talker and a listener as shown in Figure 12–47(a). From the handshaking timing diagram in part (b), determine how many data bytes are actually transferred to the listening device.

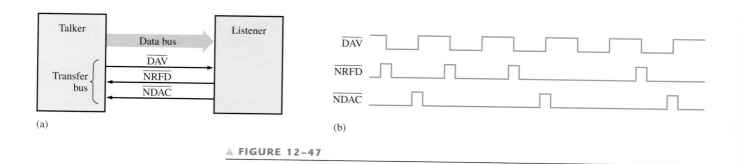

(a)

(b)

▲ **FIGURE 12–47**

37. Describe the operations depicted in the GPIB timing diagram of Figure 12–48. Develop a basic block diagram of the system involved in this operation.

38. A talker sends a data byte to a listener in a GPIB system. Simultaneously, a DTE sends a data byte to a DCE on an RS-232C interface. Which system will receive the complete data byte first? Why?

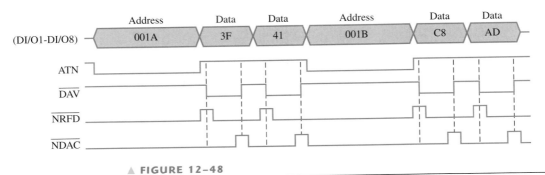

▲ FIGURE 12–48

ANSWERS

SECTION REVIEWS

SECTION 12–1 The Basic Computer

1. Basic elements of a computer are CPU, memories, input/output ports, buses.
2. RAM is random access memory, and ROM is read-only memory.
3. Peripherals are devices external to the computer.
4. Hardware is the microprocessor, memory, hard disk, etc. Software is the program that runs the computer.

SECTION 12–2 Microprocessors

1. Elements of a microprocessor are ALU, instruction decoder, register array, and control unit.
2. Microprocessor buses are address, data, and control.
3. A microprocessor functions as the CPU.
4. Arithmetic/logic operations, moves data, and makes decisions.
5. Pipelining is the process of executing more than one instruction at the same time.

SECTION 12–3 A Specific Microprocessor Family

1. The general-purpose registers are

   Accumulator (AX: AH, AL)   Stack pointer (SP)
   Base index (BX: BH, BL)   Base pointer (BP)
   Count (CX: CH, CL)   Destination index (DI)
   Data (DX: DH, DL)   Source index (SI)

2. The BIU provides addressing and data interface.
3. No, the EU does not interface with the buses.
4. The instruction queue stores prefetched instructions for the EU to increase throughput.
5. Codes can be easily relocated within memory.
6. Instruction pairing is the process of combining independent instructions so that they can be executed simultaneously by the two execution units in the Pentium.

SECTION 12–4 Computer Programming

1. A program is a list of computer instructions arranged to achieve a specific result.
2. An op-code is the code for an instruction.
3. A string is a contiguous sequence of bytes or words.

SECTION 12–5   **Interrupts**

1. For an interrupt-driven I/O, the CPU provides service to a peripheral only when requested to do so by the peripheral; for a polled I/O, the CPU periodically checks a peripheral to see if it needs service.

2. Interrupt-driven I/Os save CPU time.

3. A software interrupt is an instruction that invokes an interrupt service routine.

SECTION 12–6   **Direct Memory Access (DMA)**

1. DMA is direct memory access.

2. A DMA transfer of data from memory to I/O or vice versa saves CPU time. Direct memory access is often used in transferring data between RAM and a disk drive.

SECTION 12–7   **Internal Interfacing**

1. Tristate buffers allow devices to be completely disconnected from the bus when not in use, thus preventing interference with other devices.

2. A bus interconnects all the devices in a system and makes communication between devices possible.

SECTION 12–8   **Standard Buses**

1. Serial and parallel data transfer

2. **(a)** parallel   **(b)** serial   **(c)** serial   **(d)** parallel

3. Serial—one bit at a time; Parallel—8 or more bits at a time.

4. 127 USB devices

5. FireWire is faster than USB.

RELATED PROBLEMS FOR EXAMPLES

12–1 $6C4C2_{16}$

12–2 Change first block (initialization block) to "BIG = FFFF"; this is the largest possible unsigned number. Change first question to "Is number < BIG?"

SELF-TEST

1. (c)   **2.** (b)   **3.** (d)   **4.** (b)   **5.** (c)   **6.** (b)   **7.** (c)   **8.** (b)

9. (a)   **10.** (d)   **11.** (b)   **12.** (c)   **13.** (c)   **14.** (a)   **15.** (b)   **16.** (b)

17. (b)   **18.** (b)   **19.** (c)   **20.** (d)

13

INTRODUCTION TO DIGITAL SIGNAL PROCESSING

CHAPTER OBJECTIVES

- List the essential elements in a digital signal processing system
- Explain how analog signals are converted to digital form
- Discuss the purpose of filtering
- Describe the sampling process
- State the purpose of analog-to-digital conversion
- Explain how several types of ADCs operate
- Explain the basic concepts of a digital signal processor (DSP)
- Describe the basic architecture of a DSP

- Name some of the functions that a DSP performs

- State the purpose of digital-to-analog conversion

- Explain how DACs operate

INTRODUCTION

Digital signal processing is a powerful technology that is widely used in many applications, such as automotive, consumer, graphics/imaging, industrial, instrumentation, medical, military, telecommunications, and voice/speech applications. Digital signal processing incorporates mathematics, software programming, and processing hardware to manipulate analog signals. For example, digital signal processing can be used to enhance images, compress data for efficient transmission and storage, recognize and generate speech, and clean up noisy or deteriorated audio.

This chapter provides a brief look at digital signal processing. To completely cover the topic in the depth necessary to have a detailed understanding would take much more than a single chapter. Entire books are available on the subject; a list of references is available at the end of the chapter. Much information, including data sheets, on the TMS320 family of DSPs is available at the Texas Instruments website (www.ti.com). Information about other DSPs can be found on the Motorola website (www.motorola.com) and the Analog Devices website (www.analogdevices.com).

FIXED-FUNCTION LOGIC

ADC0804

DIGITAL SIGNAL PROCESSORS

TMS320C62xx   TMS320C64xx   TMS320C67xx

WWW. VISIT THE COMPANION WEBSITE

Study aids for this chapter are available at

http://www.prenhall.com/floyd

13–1 DIGITAL SIGNAL PROCESSING BASICS

Digital signal processing converts signals that naturally occur in analog form, such as sound, video, and information from sensors, to digital form and uses digital techniques to enhance and modify analog signal data for various applications.

After completing this section, you should be able to

■ Define *ADC* ■ Define *DSP* ■ Define *DAC* ■ Draw a basic block diagram of a digital signal processing system

A digital signal processing system first translates a continuously varying analog signal into a series of discrete levels. This series of levels follows the variations of the analog signal and resembles a staircase, as illustrated for the case of a sine wave in Figure 13–1. The process of changing the original analog signal to a "stairstep" approximation is accomplished by a sample-and-hold circuit.

▶ FIGURE 13–1

An original analog signal (sine wave) and its "stairstep" approximation.

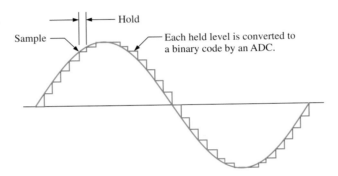

Next, the "stairstep" approximation is quantized into binary codes that represent each discrete step on the "stairsteps" by a process called analog-to-digital (A/D) conversion. The circuit that performs A/D conversion is an **analog-to-digital converter (ADC)**.

Once the analog signal has been converted to a binary coded form, it is applied to a **DSP** (digital signal processor). The DSP can perform various operations on the incoming data, such as removing unwanted interference, increasing the amplitude of some signal frequencies and reducing others, encoding the data for secure transmissions, and detecting and correcting errors in transmitted codes. DSPs make possible, among many other things, the cleanup of sound recordings, the removal of echos from communications lines, the enhancement of images from CT scans for better medical diagnosis, and the scrambling of cellular phone conversations for privacy.

After a DSP processes a signal, the signal can be converted back to a much improved version of the original analog signal. This is accomplished by a **digital-to-analog converter (DAC)**. Figure 13–2 shows a basic block diagram of a typical digital signal processing system.

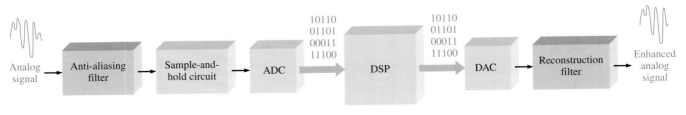

▲ FIGURE 13–2

Basic block diagram of a typical digital signal processing system.

DSPs are actually a specialized type of microprocessor but are different from general-purpose microprocessors in a couple of significant ways. Typically, microprocessors are designed for general-purpose functions and operate with large software packages. DSPs are used for special-purpose applications; they are very fast number crunchers that must work in real time by processing information as it happens using specialized algorithms (programs). The analog-to-digital converter (ADC) in a system must take samples of the incoming analog data often enough to catch all the relevant fluctuations in the signal amplitude, and the DSP must keep pace with the sampling rate of the ADC by doing its calculations as fast as the sampled data are received. Once the digital data are processed by the DSP, they go to the digital-to-analog converter (DAC) for conversion back to analog form.

SECTION 13–1 REVIEW	
Answers are at the end of the chapter.	1. What does DSP stand for?
	2. What does ADC stand for?
	3. What does DAC stand for?
	4. An analog signal is changed to a binary coded form by what circuit?
	5. A binary coded signal is changed to analog form by what circuit?

13–2 CONVERTING ANALOG SIGNALS TO DIGITAL

In order to process signals using digital techniques, the incoming analog signal must be converted into digital form.

After completing this section, you should be able to

■ Explain the basic process of converting an analog signal to digital ■ Describe the purpose of the sample-and-hold function ■ Define the Nyquist frequency ■ Define the reason for *aliasing* and discuss how it is eliminated ■ Describe the purpose of an ADC

Sampling and Filtering

The first two blocks in the system diagram of Figure 13–2 are the anti-aliasing filter and the sample-and-hold circuit. The sample-and-hold function does two operations, the first of which is sampling. **Sampling** is the process of taking a sufficient number of discrete values at points on a waveform that will define the shape of waveform. The more samples you take, the more accurately you can define a waveform. Sampling converts an analog signal into a series of impulses, each representing the amplitude of the signal at a given instant in time. Figure 13–3 illustrates the process of sampling.

When an analog signal is to be sampled, there are certain criteria that must be met in order to accurately represent the original signal. All analog signals (except a pure sine wave) contain a spectrum of component frequencies called *harmonics*. The harmonics of an analog signal are sine waves of different frequencies and amplitudes. When the harmonics of a given periodic waveform are added, the result is the original signal. Before a signal can be sampled, it must be passed through a low-pass filter (anti-aliasing filter) to eliminate harmonic frequencies above a certain value as determined by the Nyquist frequency.

The Sampling Theorem Notice in Figure 13–3 that there are two input waveforms. One is the analog signal and the other is the sampling pulse waveform. The sampling theorem states that, in order to represent an analog signal, the sampling frequency, f_{sample}, must be at

▶ **FIGURE 13–3**

Illustration of the sampling process.

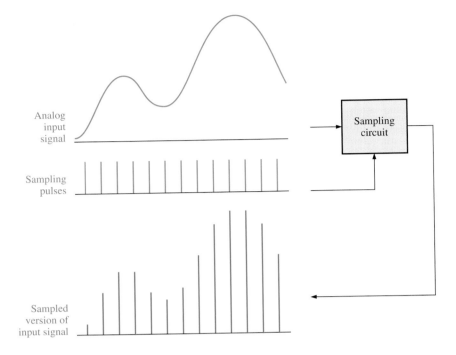

least twice the highest frequency component $f_{a(max)}$ of the analog signal. Another way to say this is that the highest analog frequency can be no greater than one-half the sampling frequency. The frequency $f_{a(max)}$ is known as the **Nyquist frequency** and is expressed in Equation 13–1. In practice, the sampling frequency should be more than twice the highest analog frequency.

Equation 13–1

$$f_{sample} \geq 2f_{a(max)}$$

To intuitively understand the sampling theorem, a simple "bouncing-ball" analogy may be helpful. Although it is not a perfect representation of the sampling of electrical signals, it does serve to illustrate the basic idea. If a ball is photographed (sampled) at one instant during a single bounce, as illustrated in Figure 13–4(a), you cannot tell anything about the path of the ball except that it is off the floor. You can't tell whether it is going up or down or the distance of its bounce. If you take photos at two equally-spaced instants during one bounce, as shown in part (b), you can obtain only a minimum amount of information about its movement and nothing about the distance of the bounce. In this particular case, you know only that the ball has been in the air at the times the two photos were taken and that the maximum height of the bounce is at least equal to the height shown in each photo. If you take four photos, as shown in part (c), then the path that the ball follows during a bounce

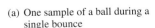

(a) One sample of a ball during a single bounce

(b) Two samples of a ball during a single bounce. This is the absolute minimum required to tell anything about its movement, but generally insufficient to describe its path.

(c) Four samples of a ball during a single bounce form a rough picture of the path of the ball.

▲ **FIGURE 13–4**

Bouncing ball analogy of sampling theory.

begins to emerge. The more photos (samples) that you take, the more accurately you can determine the path of the ball as it bounces.

The Need for Filtering   Low-pass filtering is necessary to remove all frequency components (harmonics) of the analog signal that exceed the Nyquist frequency. If there are any frequency components in the analog signal that exceed the Nyquist frequency, an unwanted condition known as **aliasing** will occur. An alias is a signal produced when the sampling frequency is not at least twice the signal frequency. An alias signal has a frequency that is less than the highest frequency in the analog signal being sampled and therefore falls within the spectrum or frequency band of the input analog signal causing distortion. Such a signal is actually "posing" as part of the analog signal when it really isn't, thus the term *alias*.

Another way to view aliasing is by considering that the sampling pulses produce a spectrum of harmonic frequencies above and below the sample frequency, as shown in Figure 13–5. If the analog signal contains frequencies above the Nyquist frequency, these frequencies overlap into the spectrum of the sample waveform as shown and interference occurs. The lower frequency components of the sampling waveform become mixed in with the frequency spectra of the analog waveform, resulting in an aliasing error.

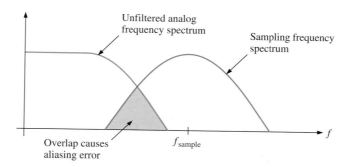

◀ **FIGURE 13–5**

A basic illustration of the condition $f_{sample} < 2f_{a(max)}$.

A low-pass anti-aliasing filter must be used to limit the frequency spectrum of the analog signal for a given sample frequency. To avoid an aliasing error, the filter must at least eliminate all analog frequencies above the minimum frequency in the sampling spectrum, as illustrated in Figure 13–6. Aliasing can also be avoided by sufficiently increasing the sampling frequency. However, the maximum sampling frequency is usually limited by the performance of the analog-to-digital converter (ADC) that follows it.

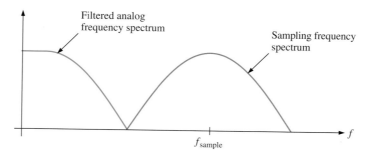

◀ **FIGURE 13–6**

After low-pass filtering, the frequency spectra of the analog and the sampling signals do not overlap, thus eliminating aliasing error.

An Application   An example of the application of sampling is in digital audio equipment. The sampling rates used are 32 kHz, 44.1 kHz, or 48 kHz (the number of samples per second). The 48 kHz rate is the most common, but the 44.1 kHz rate is used for audio CDs and prerecorded tapes. According to the Nyquist rate, the sampling frequency must be at least twice the audio signal. Therefore, the CD sampling rate of 44.1 kHz captures frequencies up to about 22 kHz, which exceeds the 20 kHz specification that is common for most audio equipment.

Many applications do not require a wide frequency range to obtain reproduced sound that is acceptable. For example, human speech contains some frequencies near 10 kHz and, therefore, requires a sampling rate of at least 20 kHz. However, if only frequencies up to 4 kHz (ideally requiring an 8 kHz minimum sampling rate) are reproduced, voice is very understandable. On the other hand, if a sound signal is not sampled at a high enough rate, the effect of aliasing will become noticeable with background noise and distortion.

Holding the Sampled Value

The holding operation is part of the sample-and-hold block shown in Figure 13–2. After filtering and sampling, the sampled level must be held constant until the next sample occurs. This is necessary for the ADC to have time to process the sampled value. This sample-and-hold operation results in a "stairstep" waveform that approximates the analog input waveform, as shown in Figure 13–7.

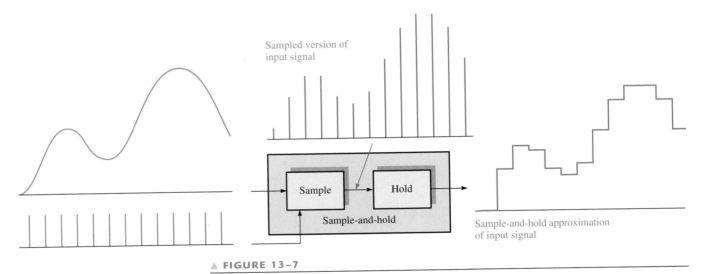

▲ **FIGURE 13–7**

Illustration of a sample–and–hold operation.

Analog-to-Digital Conversion

Analog-to-digital conversion is the process of converting the output of the sample-and-hold circuit to a series of binary codes that represent the amplitude of the analog input at each of the sample times. The sample-and-hold process keeps the amplitude of the analog input signal constant between sample pulses; therefore, the analog-to-digital conversion can be done using a constant value rather than having the analog signal change during a conversion interval, which is the time between sample pulses. Figure 13–8 il-

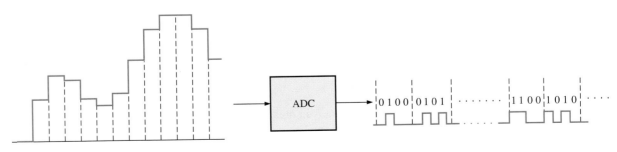

▲ **FIGURE 13–8**

Basic function of an analog-to-digital (ADC) converter (The binary codes and number of bits are arbitrarily chosen for illustration only). The ADC output waveform that represents the binary codes is also shown.

lustrates the basic function of an analog-to-digital (ADC) converter. The sample intervals are indicated by dashed lines.

Quantization   The process of converting an analog value to a code is called **quantization.** During the quantization process, the ADC converts each sampled value of the analog signal to a binary code. The more bits that are used to represent a sampled value, the more accurate is the representation.

To illustrate, let's quantize a reproduction of the analog waveform into four levels (0–3). As shown in Figure 13–9, two bits are required. Note that each quantization level is represented by a 2-bit code on the vertical axis, and each sample interval is numbered along the horizontal axis. The quantization process is summarized in Table 13–1.

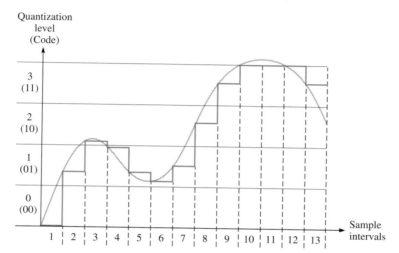

◄ **FIGURE 13–9**

Sample-and-hold output waveform with four quantization levels. The original analog waveform is shown in light gray for reference.

◄ **TABLE 13–1**

Two-bit quantization for the waveform in Figure 13–9.

SAMPLE INTERVAL	QUANTIZATION LEVEL	CODE
1	0	00
2	1	01
3	2	10
4	1	01
5	1	01
6	1	01
7	1	01
8	2	10
9	3	11
10	3	11
11	3	11
12	3	11
13	3	11

If the resulting 2-bit digital codes are used to reconstruct the original waveform, which is done by digital-to-analog converters (DAC), you would get the waveform shown in Figure 13–10. As you can see, quite a bit of accuracy is lost using only two bits to represent the sampled values.

▶ **FIGURE 13–10**

The reconstructed waveform in Figure 13–9 using four quantization levels (2 bits). The original analog waveform is shown in light gray for reference.

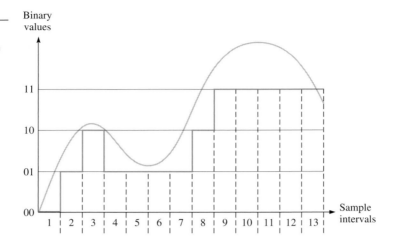

Now, let's see how more bits will improve the accuracy. Figure 13–11 shows the same waveform with sixteen quantization levels (4 bits). The 4-bit quantization process is summarized in Table 13–2.

▶ **FIGURE 13–11**

Sample-and-hold output waveform with sixteen quantization levels. The original analog waveform is shown in light gray for reference.

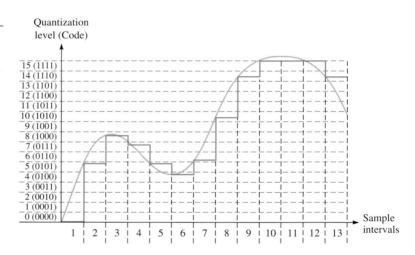

▶ **TABLE 13–2**

Four-bit quantization for the waveform in Figure 13–11.

SAMPLE INTERVAL	QUANTIZATION LEVEL	CODE
1	0	0000
2	5	0101
3	8	1000
4	7	0111
5	5	0101
6	4	0100
7	6	0110
8	10	1010
9	14	1110
10	15	1111
11	15	1111
12	15	1111
13	14	1110

If the resulting 4-bit digital codes are used to reconstruct the original waveform, you would get the waveform shown in Figure 13–12. As you can see, the result is much more like the original waveform than for the case of four quantization levels in Figure 13–10. This shows that greater accuracy is achieved with more quantization bits. Most integrated circuit ADCs use from 8 to 24 bits, and the sample-and-hold function is sometimes contained on the ADC chip. Several types of ADCs are introduced in the next section.

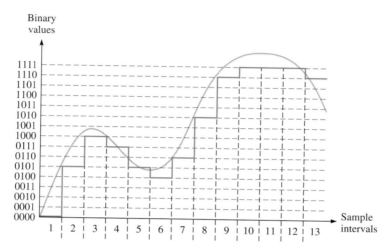

◀ **FIGURE 13–12**

The reconstructed waveform in Figure 13–11 using sixteen quantization levels (4 bits). The original analog waveform is shown in light gray for reference.

SECTION 13–2 REVIEW

1. What does sampling mean?
2. Why must you hold a sampled value?
3. If the highest frequency component in an analog signal is 20 kHz, what is the minimum sample frequency?
4. What does quantization mean?
5. What determines the accuracy of the quantization process?

13–3 ANALOG-TO-DIGITAL CONVERSION METHODS

As you have seen, analog-to-digital conversion is the process by which an analog quantity is converted to digital form. It is necessary when measured quantities must be in digital form for processing or for display or storage. Some common types of analog-to-digital converters (ADCs) are now examined. Two important ADC parameters are *resolution,* which is the number of bits, and *throughput,* which is the sampling rate an ADC can handle in units of samples per second (sps).

After completing this section, you should be able to

■ Explain basically what an operational amplifier is ■ Show how the op-amp can be used as an inverting amplifier or a comparator ■ Explain how a flash ADC works ■ Discuss dual-slope ADCs ■ Describe the operation of a successive-approximation ADC ■ Describe a delta-sigma ADC ■ Discuss testing ADCs for a missing code, incorrect code and offset

A Quick Look at an Operational Amplifier

Before getting into analog-to-digital converters (ADCs), let's look briefly at an element that is common to most types of ADCs and digital-to-analog converters (DACs). This element is the operational amplifier, or op-amp for short. This is an abbreviated coverage of the op-amp.

An **op-amp** is a linear amplifier that has two inputs (inverting and noninverting) and one output. It has a very high voltage gain and a very high input impedance, as well as a very low output impedance. The op-amp symbol is shown in Figure 13–13(a). When used as an inverting amplifier, the op-amp is configured as shown in part (b). The feedback resistor, R_f, and the input resistor, R_i, control the voltage gain according to the formula in Equation 13–2, where V_{out}/V_{in} is the closed-loop voltage gain (closed loop refers to the feedback from output to input provided by R_f). The negative sign indicates inversion.

Equation 13–2

$$\frac{V_{out}}{V_{in}} = -\frac{R_f}{R_i}$$

In the inverting amplifier configuration, the inverting input of the op-amp is approximately at ground potential (0 V) because feedback and the extremely high open-loop gain make the differential voltage between the two inputs extremely small. Since the noninverting input is grounded, the inverting input is at approximately 0 V, which is called *virtual ground*.

When the op-amp is used as a comparator, as shown in Figure 13–13(c), two voltages are applied to the inputs. When these input voltages differ by a very small amount, the op-amp is driven into one of its two saturated output states, either HIGH or LOW, depending on which input voltage is greater.

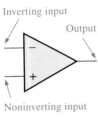

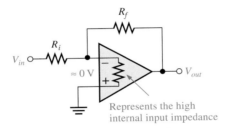

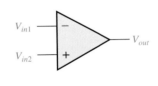

(a) Op-amp symbol

(b) Op-amp as an inverting amplifier with gain of R_f/R_i

(c) Op-amp as a comparator

▲ FIGURE 13–13

The operational amplifier (op-amp).

Flash (Simultaneous) Analog-to-Digital Converter

The flash method utilizes comparators that compare reference voltages with the analog input voltage. When the input voltage exceeds the reference voltage for a given comparator, a HIGH is generated. Figure 13–14 shows a 3-bit converter that uses seven comparator circuits; a comparator is not needed for the all-0s condition. A 4-bit converter of this type requires fifteen comparators. In general, $2^n - 1$ comparators are required for conversion to an *n*-bit binary code. The number of bits used in an ADC is its **resolution.** The large number of comparators necessary for a reasonable-sized binary number is one of the disadvantages of the **flash ADC.** Its chief advantage is that it provides a fast conversion time because of a high *throughput,* measured in samples per second (sps).

The reference voltage for each comparator is set by the resistive voltage-divider circuit. The output of each comparator is connected to an input of the priority encoder. The encoder is enabled by a pulse on the *EN* input, and a 3-bit code representing the value of the input appears on the encoder's outputs. The binary code is determined by the highest-order input having a HIGH level.

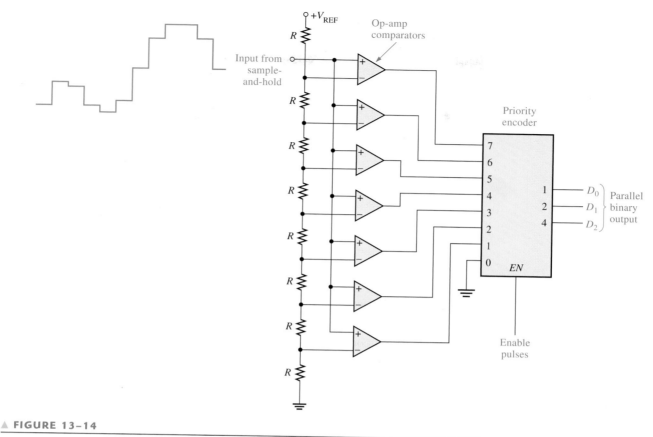

▲ FIGURE 13–14

A 3-bit flash ADC.

The frequency of the enable pulses and the number of bits in the binary code determine the accuracy with which the sequence of binary codes represents the input of the ADC. There should be one enable pulse for each sampled level of the input signal.

EXAMPLE 13–1

Determine the binary code output of the 3-bit flash ADC in Figure 13–14 for the input signal in Figure 13–15 and the encoder enable pulses shown. For this example, $V_{REF} = +8$ V.

▶ FIGURE 13–15

Sampling of values on a waveform for conversion to binary code.

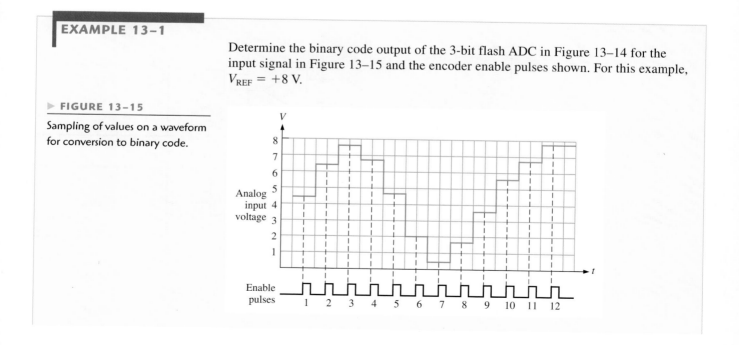

Solution   The resulting digital output sequence is listed as follows and shown in the waveform diagram of Figure 13–16 in relation to the enable pulses:

100, 110, 111, 110, 100, 010, 000, 001, 011, 101, 110, 111

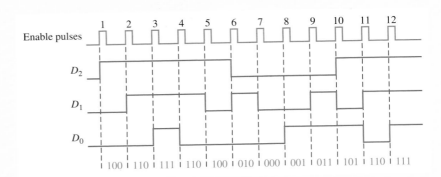

▲ FIGURE 13–16

Resulting digital outputs for sample-and-hold values. Output D_0 is the LSB of the 3-bit binary code.

*Related Problem**   If the enable pulse frequency in Figure 13–15 were halved, determine the binary numbers represented by the resulting digital output sequence for 6 pulses. Is any information lost?

*Answers are at the end of the chapter.

Dual-Slope Analog-to-Digital Converter

A dual-slope ADC is common in digital voltmeters and other types of measurement instruments. A ramp generator (integrator) is used to produce the dual-slope characteristic. A block diagram of a dual-slope ADC is shown in Figure 13–17.

▶ FIGURE 13–17

Basic dual-slope ADC.

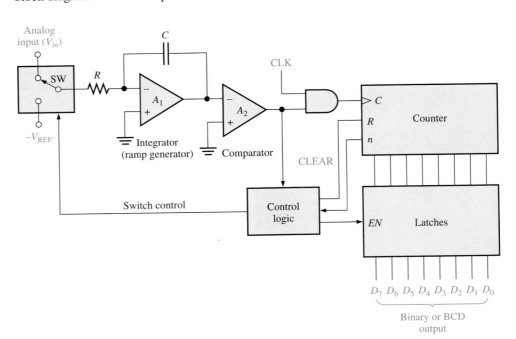

Figure 13–18 illustrates dual-slope conversion. Start by assuming that the counter is reset and the output of the integrator is zero. Now assume that a positive input voltage is applied to the input through the switch (SW) as selected by the control logic. Since the

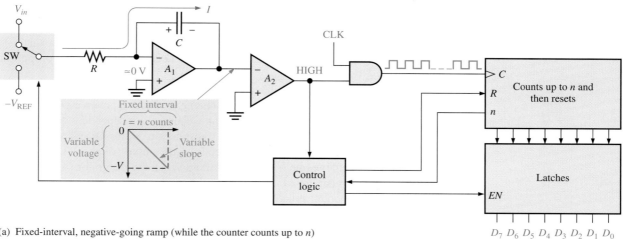

(a) Fixed-interval, negative-going ramp (while the counter counts up to n)

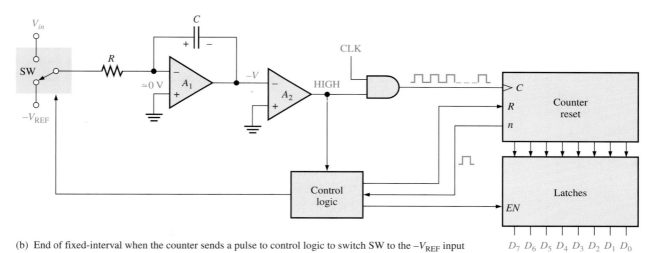

(b) End of fixed-interval when the counter sends a pulse to control logic to switch SW to the $-V_{REF}$ input

(c) Fixed-slope, positive-going ramp while the counter counts up again. When the ramp reaches 0 V, the counter stops, and the counter output is loaded into latches.

▲ FIGURE 13–18

Illustration of dual-slope conversion.

inverting input of A_1 is at virtual ground, and assuming that V_{in} is constant for a period of time, there will be constant current through the input resistor R and therefore through the capacitor C. Capacitor C will charge linearly because the current is constant, and as a result, there will be a negative-going linear voltage ramp on the output of A_1, as illustrated in Figure 13–18(a).

When the counter reaches a specified count, it will be reset, and the control logic will switch the negative reference voltage $(-V_{REF})$ to the input of A_1, as shown in Figure 13–18(b). At this point the capacitor is charged to a negative voltage $(-V)$ proportional to the input analog voltage.

Now the capacitor discharges linearly because of the constant current from the $-V_{REF}$, as shown in Figure 13–18(c). This linear discharge produces a positive-going ramp on the A_1 output, starting at $-V$ and having a constant slope that is independent of the charge voltage. As the capacitor discharges, the counter advances from its RESET state. The time it takes the capacitor to discharge to zero depends on the initial voltage $-V$ (proportional to V_{in}) because the discharge rate (slope) is constant. When the integrator (A_1) output voltage reaches zero, the comparator (A_2) switches to the LOW state and disables the clock to the counter. The binary count is latched, thus completing one conversion cycle. The binary count is proportional to V_{in} because the time it takes the capacitor to discharge depends only on $-V$, and the counter records this interval of time.

Successive-Approximation Analog-to-Digital Converter

One of the most widely used methods of analog-to-digital conversion is successive-approximation. It has a much faster conversion time than the dual-slope conversion, but it is slower than the flash method. It also has a fixed conversion time that is the same for any value of the analog input.

Figure 13–19 shows a basic block diagram of a 4-bit successive approximation ADC. It consists of a DAC (DACs are covered in Section 13–5), a successive-approximation register (SAR), and a comparator. The basic operation is as follows: The input bits of the DAC are enabled (made equal to a 1) one at a time, starting with the most significant bit (MSB). As each bit is enabled, the comparator produces an output that indicates whether the input signal voltage is greater or less than the output of the DAC. If the DAC output is greater than the input signal, the comparator's output is LOW, causing the bit in the register to reset. If the output is less than the input signal, the 1 bit is retained in the register. The system does this with the MSB first, then the next most significant bit, then

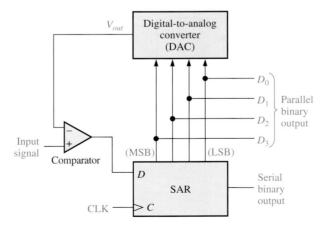

▲ **FIGURE 13–19**

Successive-approximation ADC.

the next, and so on. After all the bits of the DAC have been tried, the conversion cycle is complete.

In order to better understand the operation of the successive-approximation ADC, let's take a specific example of a 4-bit conversion. Figure 13–20 illustrates the step-by-step conversion of a constant input voltage (5.1 V in this case). Let's assume that the DAC has the following output characteristic: $V_{out} = 8$ V for the 2^3 bit (MSB), $V_{out} = 4$ V for the 2^2 bit, $V_{out} = 2$ V for the 2^1 bit, and $V_{out} = 1$ V for the 2^0 bit (LSB).

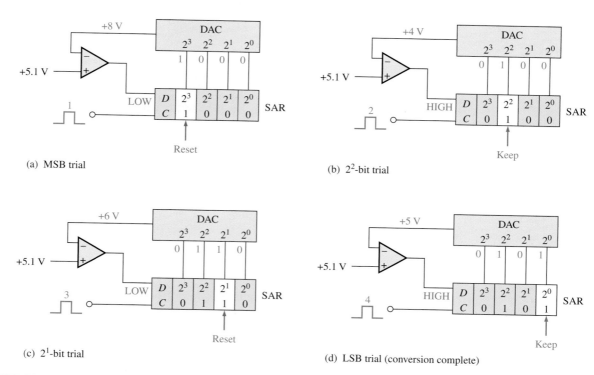

(a) MSB trial

(b) 2^2-bit trial

(c) 2^1-bit trial

(d) LSB trial (conversion complete)

▲ **FIGURE 13–20**

Illustration of the successive-approximation conversion process.

Figure 13–20(a) shows the first step in the conversion cycle with the MSB = 1. The output of the DAC is 8 V. Since this is greater than the input of 5.1 V, the output of the comparator is LOW, causing the MSB in the SAR to be reset to a 0.

Figure 13–20(b) shows the second step in the conversion cycle with the 2^2 bit equal to a 1. The output of the DAC is 4 V. Since this is less than the input of 5.1 V, the output of the comparator switches to a HIGH, causing this bit to be retained in the SAR.

Figure 13–20(c) shows the third step in the conversion cycle with the 2^1 bit equal to a 1. The output of the DAC is 6 V because there is a 1 on the 2^2 bit input and on the 2^1 bit input; 4 V + 2 V = 6 V. Since this is greater than the input of 5.1 V, the output of the comparator switches to a LOW, causing this bit to be reset to a 0.

Figure 13–20(d) shows the fourth and final step in the conversion cycle with the 2^0 bit equal to a 1. The output of the DAC is 5 V because there is a 1 on the 2^2 bit input and on the 2^0 bit input; 4 V + 1 V = 5 V.

The four bits have all been tried, thus completing the conversion cycle. At this point the binary code in the register is 0101, which is approximately the binary value of the input of 5.1 V. Additional bits will produce an even more accurate result. Another conversion cycle now begins, and the basic process is repeated. The SAR is cleared at the beginning of each cycle.

THE ADC0804 ANALOG-TO-DIGITAL CONVERTER

The ADC0804 is an example of a successive-approximation ADC. A block diagram is shown in Figure 13–21. This device operates from a +5 V supply and has a resolution of eight bits with a conversion time of 100 μs. Also, it has an on-chip clock generator. The data outputs are tristate, so they can be interfaced with a microprocessor bus system.

▶ **FIGURE 13–21**

The ADC0804 analog-to-digital converter.

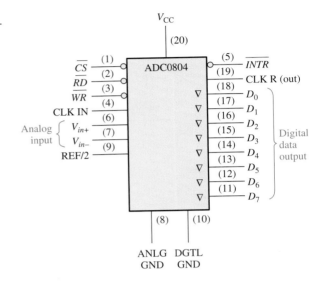

The basic operation of the device is as follows: The ADC0804 contains the equivalent of a 256-resistor DAC network. The successive-approximation logic sequences the network to match the analog differential input voltage ($V_{in+} - V_{in-}$) with an output from the resistive network. The MSB is tested first. After eight comparisons (sixty-four clock periods), an 8-bit binary code is transferred to output latches, and the interrupt ($\overline{INTR}$) output goes LOW. The device can be operated in a free-running mode by connecting the $\overline{INTR}$ output to the write ($\overline{WR}$) input and holding the conversion start ($\overline{CS}$) LOW. To ensure startup under all conditions, a LOW $\overline{WR}$ input is required during the power-up cycle. Taking $\overline{CS}$ low anytime after that will interrupt the conversion process.

When the $\overline{WR}$ input goes LOW, the internal successive-approximation register (SAR) and the 8-bit shift register are reset. As long as both $\overline{CS}$ and $\overline{WR}$ remain LOW, the ADC remains in a RESET state. Conversion starts one to eight clock periods after $\overline{CS}$ or $\overline{WR}$ makes a LOW-to-HIGH transition.

When a LOW is at both the $\overline{CS}$ and $\overline{RD}$ inputs, the tristate output latch is enabled and the output code is applied to the D_0-D_7 lines. When either the $\overline{CS}$ or the $\overline{RD}$ input returns to a HIGH, the D_0-D_7 outputs are disabled.

Sigma-Delta Analog-to-Digital Converter

Sigma-delta is a widely used method of analog-to-digital conversion, particularly in telecommunications using audio signals. The method is based on **delta modulation** where the difference between two successive samples (increase or decrease) is quantized; other ADC methods were based on the absolute value of a sample. Delta modulation is a 1-bit quantization method.

The output of a delta modulator is a single-bit data stream where the relative number of 1s and 0s indicates the level or amplitude of the input signal. The number of 1s over a given number of clock cycles establishes the signal amplitude during that interval. A maximum number of 1s corresponds to the maximum positive input voltage. A number of 1s equal to one-half the maximum corresponds to an input voltage of zero. No 1s (all 0s) corresponds to the maximum negative input voltage. This is illustrated in a simplified way in Figure 13–22. For example, assume that 4096 1s occur during the interval when the input signal is a positive maximum. Since zero is the midpoint of the dynamic range of the input signal, 2048 1s occur during the interval when the input signal is zero. There are no 1s during the interval when the input signal is a negative maximum. For signal levels in between, the number of 1s is proportional to the level.

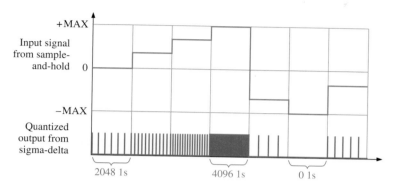

◀ FIGURE 13–22

A simplified illustration of sigma-delta analog-to-digital conversion.

The Sigma-Delta ADC Functional Block Diagram   The basic block diagram in Figure 13–23 accomplishes the conversion illustrated in Figure 13–22. The analog input signal and the analog signal from the converted quantized bit stream from the DAC in the feedback loop are applied to the summation (Σ) point. The difference (Δ) signal out of the Σ is integrated, and the 1-bit ADC increases or decreases the number of 1s depending on the difference signal. This action attempts to keep the quantized signal that is fed back equal to the incoming analog signal. The 1-bit quantizer is essentially a comparator followed by a latch.

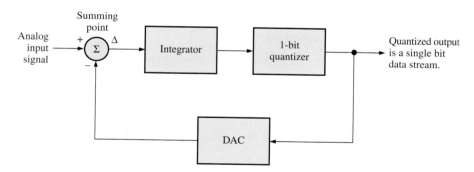

◀ FIGURE 13–23

Partial functional block diagram of a sigma–delta ADC.

To complete the sigma-delta conversion process using one particular approach, the single bit data stream is converted to a series of binary codes, as shown in Figure 13–24. The counter counts the 1s in the quantized data stream for successive intervals. The code in the

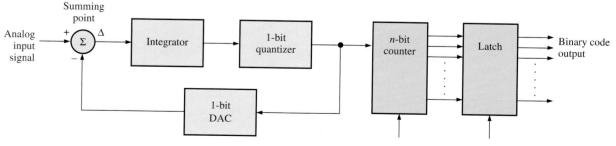

▲ FIGURE 13–24

One type of sigma–delta ADC.

counter then represents the amplitude of the analog input signal for each interval. These codes are shifted out into the latch for temporary storage. What comes out of the latch is a series of *n*-bit codes, which completely represent the analog signal.

Another approach uses a digital decimation filter to produce the output instead of the counter and latch. This subject is beyond the scope of our coverage.

Testing Analog-to-Digital Converters

One method for testing ADCs is shown in Figure 13–25. A DAC is used as part of the test setup to convert the ADC output back to analog form for comparison with the test input.

A test input in the form of a linear ramp is applied to the input of the ADC. The resulting binary output sequence is then applied to the DAC test unit and converted to a stairstep ramp. The input and output ramps are compared for any deviation.

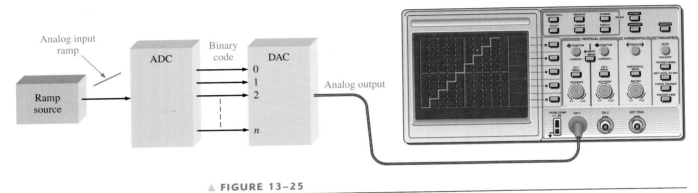

▲ **FIGURE 13–25**

A method for testing ADCs.

Analog-to-Digital Conversion Errors

Again, a 4-bit conversion is used to illustrate the principles. Let's assume that the test input is an ideal linear ramp.

Missing Code   The stairstep output in Figure 13–26(a) indicates that the binary code 1001 does not appear on the output of the ADC. Notice that the 1000 value stays for two intervals and then the output jumps to the 1010 value.

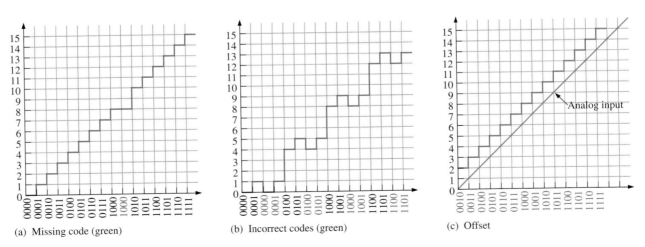

(a) Missing code (green)   (b) Incorrect codes (green)   (c) Offset

▲ **FIGURE 13–26**

Illustrations of analog-to-digital conversion errors.

In a flash ADC, for example, a failure of one of the op-amp comparators can cause a missing-code error.

Incorrect Codes   The stairstep output in Figure 13–26(b) indicates that several of the binary code words coming out of the ADC are incorrect. Analysis indicates that the 2^1-bit line is stuck in the LOW (0) state in this particular case.

Offset   Offset conditions are shown in 13–26(c). In this situation the ADC interprets the analog input voltage as greater than its actual value.

EXAMPLE 13–2

A 4-bit flash ADC is shown in Figure 13–27(a). It is tested with a setup like the one in Figure 13–25. The resulting reconstructed analog output is shown in Figure 13–27(b). Identify the problem and the most probable fault.

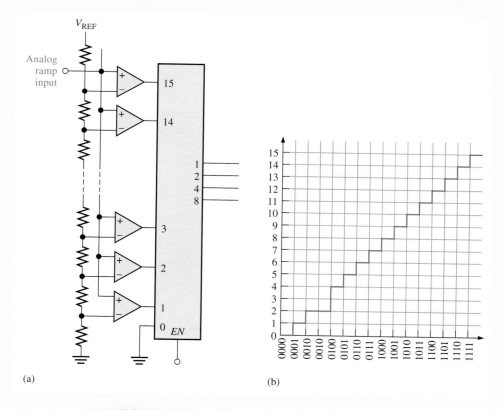

(a)   (b)

▲ FIGURE 13–27

Solution   The binary code 0011 is missing from the ADC output, as indicated by the missing step. Most likely, the output of comparator 3 is stuck in its inactive state (LOW).

Related Problem   Reconstruct the analog output in a test setup like in Figure 13–25 if the ADC in Figure 13–27(a) has comparator 8 stuck in the HIGH output state.

**SECTION 13–3
REVIEW**

1. What is the fastest method of analog-to-digital conversion?
2. Which analog-to-digital conversion method produces a single-bit data stream?
3. Does the successive-approximation converter have a fixed conversion time?
4. Name two types of output errors in an ADC.

13–4 THE DIGITAL SIGNAL PROCESSOR (DSP)

Essentially, a digital signal processor (DSP) is a special type of microprocessor that processes data in real time. Its applications focus on the processing of digital data that represents analog signals. A DSP, like a microprocessor, has a central processing unit (CPU) and memory units in addition to many interfacing functions. Every time you use your cellular telephone, you are using a DSP, and this is only one example of its many applications.

After completing this chapter, you should be able to

■ Explain the basic concepts of a DSP ■ List some of the applications of DSPs
■ Describe the basic functions of a DSP in a cell phone ■ Discuss the TMS320C6000 series DSP

The digital signal processor (DSP) is the heart of a digital signal processing system. It takes its input from an ADC and produces an output that goes to a DAC, as shown in Figure 13–28. As you have learned, the ADC changes an analog waveform into data in the form of a series of binary codes that are then applied to the DSP for processing. After being processed by the DSP, the data go to a DAC for conversion back to analog form.

▷ **FIGURE 13–28**

The DSP has a digital input and produces a digital output.

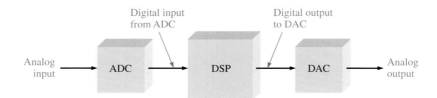

DSP Programming

DSPs are typically programmed in either assembly language or in C. Because programs written in assembly language can usually execute faster and because speed is critical in most DSP applications, assembly language is used much more in DSPs than in general-purpose microprocessors. Also, DSP programs are usually much shorter than traditional microprocessor programs because of their very specialized applications where much redundancy is used. In general, the instruction sets for DSPs tend to be smaller than for microprocessors.

DSP Applications

The DSP, unlike the general-purpose microprocessor, must typically process data in *real time;* that is, as it happens. Many applications in which DSPs are used cannot tolerate any noticeable delays, requiring the DSP to be extremely fast. In addition to cell phones, digital signal processors (DSPs) are used in multimedia computers, video recorders, CD players, hard disk drives, digital radio modems, and other applications to improve the signal quality. Also, DSPs are becoming more common in television applications.

An important application of DSPs is in signal compression and decompression. In CD systems, for example, the music on the CD is in a compressed form so that it doesn't use as much storage space. It must be decompressed in order to be reproduced. Also signal compression is used in cell phones to allow a greater number of calls to be handled simultaneously in a local cell. Some other areas where it has had a major impact are as follows.

Telecommunications   The field of telecommunications involves transferring all types of information from one location to another, including telephone conversations, television signals, and digital data. Among other functions, the DSP facilitates multiplexing many signals onto one transmission channel because information in digital form is relatively easy to multiplex and demultiplex.

At the transmitting end of a telecommunications system, DSPs are used to compress digitized voice signals for conservation of bandwidth. Compression is the process of reducing the data rate. Generally, a voice signal is converted to digital form at 8000 samples per second (sps), based on a Nyquist frequency of 4 kHz. If 8 bits are used to encode each sample, the data rate is 64 kbits/s. In general, reducing (compressing) the data rate from 64 kbits/s to 32 kbits/s results in no loss of sound quality. When the data are compressed to 8 kbits/s, the sound quality is reduced noticeably. When compressed to the minimum of 2 kbits/s, the sound is greatly distorted but still usable for some applications where only word recognition and not quality is important. At the receiving end of a telecommunications system, the DSP decompresses the data to restore the signal to its original form.

Echoes, a problem in many long distance telephone connections, occur when a portion of a voice signal is returned with a delay. For shorter distances, this delay is barely noticeable; but as the distance between the transmitter and the receiver increases, so does the delay time of the echo. DSPs are used to effectively cancel the annoying echo, which results in a clear, undisturbed voice signal.

Music Processing   The DSP is used in the music industry to provide filtering, signal addition and subtraction, and signal editing in music preparation and recording. Also, another application of the DSP is to add artificial echo and reverberation, which are usually minimized by the acoustics of a sound studio, in order to simulate ideal listening environments from concert halls to small rooms.

Speech Generation and Recognition   DSPs are used in speech generation and recognition to enhance the quality of man/machine communication. The most common method used to produce computer-generated speech is digital recording. In digital recording, the human voice is digitized and stored, usually in a compressed form. During playback the stored voice data are uncompressed and converted back into the original analog form. Approximately an hour of speech can be stored using about 3 MB of memory.

Speech recognition is much more difficult to accomplish than speech generation. Even with today's computers, speech recognition is very limited and, with a few exceptions, the results are only moderately successful. The DSP is used to isolate and analyze each word in the incoming voice signal. Certain parameters are identified in each word and compared with previous examples of the spoken word to create the closest match. Most systems are limited to a few hundred words at best. Also, significant pauses between words are usually required and the system must be "trained" for a given individual's voice. Speech recognition is an area of tremendous research effort and will eventually be applied in many commercial applications.

Radar   In *ra*dio *d*etection *an*d *r*anging (radar) applications, DSPs provide more accurate determination of distance using data compression techniques, decrease noise using filtering techniques, thereby increasing the range, and optimize the ability of the radar system to identity specific types of targets. DSPs are also used in similar ways in sonar systems.

Image Processing   The DSP is used in image-processing applications such as the computed tomography (CT) and magnetic resonance imaging (MRI), which are widely used in the medical field for looking inside the human body. In CT, X-rays are passed through a

COMPUTER NOTE

Sound cards used in computers use an ADC to convert sound from a microphone, audio CD player, or other source into a digital signal. The ADC sends the digital signal to a digital signal processor (DSP). Based on instructions from a ROM, one function of the DSP is to compress the digital signal so it uses less storage space. The DSP then sends the compressed data to the computer's processor which, in turn, sends the data to a hard drive or CD ROM for storage. To play a recorded sound, the stored data is retrieved by the processor and sent to the DSP where it is decompressed and sent to a DAC. The output of the DAC, which is a reproduction of the original sound signal, is applied to the speakers.

section of the body from many directions. The resulting signals are converted to digital form and stored. This stored information is used to produce calculated images that appear to be slices through the human body that show great detail and permit better diagnosis.

Instead of X-rays, MRI uses magnetic fields in conjunction with radio waves to probe inside the human body. MRI produces images, just as CT, and provides excellent discrimination between different types of tissue as well as information such as blood flow through arteries. MRI depends entirely on digital signal processing methods.

In applications such as video telephones, digital television, and other media that provide moving pictures, the DSP uses image compression to reduce the number of bits needed, making these systems commercially feasible.

Filtering   DSPs are commonly used to implement digital filters for the purposes of separating signals that have been combined with other signals or with interference and noise and for restoring signals that are distorted. Although analog filters are quite adequate for some applications, the digital filter is generally much superior in terms of the performance that can be achieved. One drawback to digital filters is that the execute time required produces a delay from the time the analog signal is applied until the time the output appears. Analog filters present no delay problems because as soon as the input occurs, the response appears on the output. Analog filters are also less expensive than digital filters. Regardless of this, the overall performance of the digital filter is far superior in many applications.

The DSP in a Cellular Telephone

The digital cellular telephone is an example of how a DSP can be used. Figure 13–29 shows a simplified block diagram of a digital cell phone. The voice **codec** (codec is the abbreviation for coder/decoder) contains, among other functions, the ADC and DAC necessary to convert between the analog voice signal and a digital voice format. Sigma-delta conversion is typically used in most cell phone applications. For transmission, the voice signal from the microphone is converted to digital form by the ADC in the codec and then it goes to the DSP for processing. From the DSP, the digital signal goes to the rf (radio frequency) section where it is modulated and changed to the radio frequency for transmission. An incoming rf signal containing voice data is picked up by the antenna, demodulated, and changed to a digital signal. It is then applied to the DSP for processing, after which the digital signal goes to the codec for conversion back to the original voice signal by the DAC. It is then amplified and applied to the speaker.

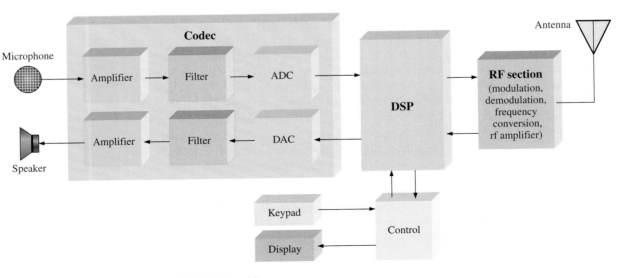

▲ **FIGURE 13–29**

Simplified block diagram of a digital cellular phone.

Functions Performed by the DSP   In a cellular phone application, the DSP performs many functions to improve and facilitate the reception and transmission of a voice signal. Some of these DSP functions are as follows:

- *Speech compression.*   The rate of the digital voice signal is reduced significantly for transmission in order to meet the bandwidth requirements.

- *Speech decompression.*   The rate of the received digital voice signal is returned to its original rate in order to properly reproduce the analog voice signal.

- *Protocol handling.*   The cell phone communicates with the nearest base in order to establish the location of the cell phone, allocates time and frequency slots, and arranges handover to another base station as the phone moves into another cell.

- *Error detection and correction.*   During transmission, error detection and correction codes are generated and, during reception, detect and correct errors induced in the rf channel by noise or interference.

- *Encryption.*   Converts the digital voice signal to a form for secure transmission and converts it back to original form during reception.

Basic DSP Architecture

As mentioned before, a DSP is basically a specialized microprocessor optimized for speed in order to process data in real time. Many DSPs are based on what is known as the *Harvard architecture,* which consists of a central processing unit (CPU) and two memories, one for data and the other for the program, as shown by the block diagram in Figure 13–30.

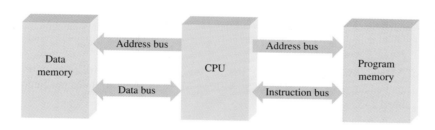

◀ **FIGURE 13–30**

Many DSPs use the Harvard architecture (two memories).

Specific DSPs—The TMS320C6000 Series

DSPs are manufactured by several companies including Texas Instruments, Motorola, and Analog Devices. DSPs are available for both fixed-point and floating-point processing. Recall from Chapter 2 that these two methods differ in the way numbers are stored and manipulated. All floating-point DSPs can also handle numbers in fixed-point format. Fixed-point DSPs are less expensive than the floating-point versions and, generally, can operate faster. The details of DSP architecture can vary significantly, even within the same family. Let's look briefly at one particular DSP series as an example of how a DSP is generally organized.

Examples of DSPs available in the TMS320C6000 series include the TMS320C62xx, the TMS320C64xx, and the TMS320C67xx, which are part of Texas Instrument's TMS320 family of devices. A general block diagram for these devices is shown in Figure 13–31.

The DSPs have a central processing unit (CPU), also known as the **DSP core,** that contains 64 general-purpose 32-bit registers in the C64xx and 32 general-purpose 32-bit registers in the C62xx and the C67xx. The C67xx can handle floating-point operations, whereas the C62xx and C64xx are fixed-point devices.

Each DSP has eight functional units that contain two 16-bit multipliers and six arithmetic logic units (ALUs). The performance of the three DSPs in the C6000 series in terms

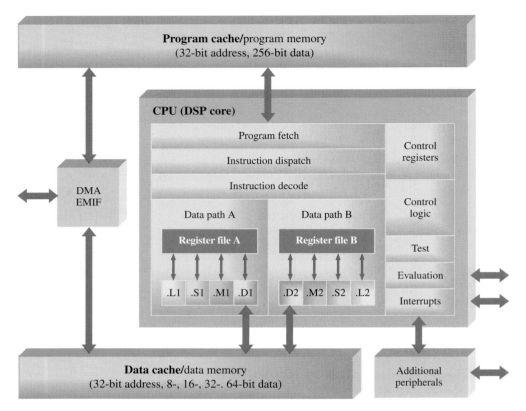

▲ FIGURE 13–31

General block diagram of the TMS320C6000 series DSP.

of **MIPS** (Million Instructions Per Second), **MFLOPS** (Million Floating-point Operations Per Second), and **MMACS** (Million Multiply/Accumulates per Second) is shown in Table 13–3.

▶ TABLE 13–3

TMS320C6000 series DSP data processing performance.

DSP	TYPE	APPLICATION	PROCESSING SPEED	MULTIPLY/ ACCUMULATE SPEED
C62xx	Fixed-point	General-purpose	1200–2400 MIPS	300–600 MMACS
C64xx	Fixed-point	Special-purpose	3200–4800 MIPS	1600–2400 MMACS
C67xx	Floating-point	General-purpose	600–1000 MFLOPS	200–333 MMACS

Data Paths in the CPU   In the CPU, the program fetch, instruction dispatch, and instruction decode sections can provide eight 32-bit instructions to the functional units during every clock cycle. The CPU is split into two data paths, and instruction processing occurs in both data paths A and B. Each data path contains half of the general-purpose registers (16 in the C62xx and C67xx or 32 in the C64xx) and four functional units. The control register and logic are used to configure and control the various processor operations.

Functional Units   Each data path has four functional units. The M units (labeled .M1 and .M2 in Figure 13–31) are dedicated multipliers. The L units (labeled .L1 and .L2) perform arithmetic, logic, and miscellaneous operations. The S units (labeled .S1 and .S2) perform compare, shift, and miscellaneous arithmetic operations. The D units (labeled .D1 and .D2) perform load, store, and miscellaneous operations.

Pipeline   A **pipeline** allows multiple instructions to be processed simultaneously. A pipeline operation consists of three stages through which all instructions flow: *fetch, decode, execute.* Eight instructions at a time are first fetched from the program memory; they are then decoded, and finally they are executed.

During **fetch,** the eight instructions (called a packet) are taken from memory in four *phases,* as shown in Figure 13–32.

- ■ *Program address generate (PG).*   The program address is generated by the CPU.
- ■ *Program address send (PS).*   The program address is sent to the memory.
- ■ *Program access ready wait (PW).*   A memory read operation occurs.
- ■ *Program fetch packet receive (PR).*   The CPU receives the packet of instructions.

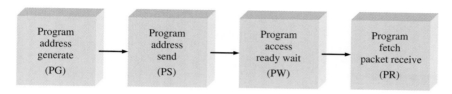

◀ **FIGURE 13–32**

The four fetch phases of the pipeline operation.

Two phases make up the instruction **decode** stage of pipeline operation, as shown in Figure 13–33. The instruction dispatch (DP) phase is where the instruction packets are split into execute packets and assigned to the appropriate functional units. The instruction decode (DC) phase is where the instructions are decoded.

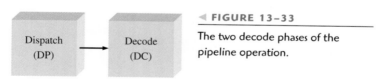

◀ **FIGURE 13–33**

The two decode phases of the pipeline operation.

The **execute** stage of the pipeline operation is where the instructions from the decode stage are carried out. The execute stage has a maximum of five phases (E1 through E5), as shown in Figure 13–34. All instructions do not use all five phases. The number of phases used during execution depends on the type of instruction. Part of the execution of an instruction requires getting data from the data memory.

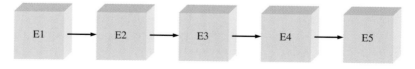

◀ **FIGURE 13–34**

The five execute phases of pipeline operation.

Internal DSP Memory and Interfaces   As you can see in Figure 13–31, there are two internal memories, one for data and one for program. The program memory is organized in 256 bit packets (eight 32-bit instructions) and there are 64 kB of capacity. The data memory also has a capacity of 64 kB and can be accessed in 8-, 16-, 32-, or 64-bit word lengths, depending on the specific device in the series. Both internal memories are accessed with a 32-bit address. The DMA (Direct Memory Access) is used to transfer data without going through the CPU. The EMIF (External Memory Interface) is used to support external memories when required in an application. Additional interface is provided for serial I/O ports and other external devices.

Timers   There are two general-purpose timers in the DSP that can be used for timed events, counting, pulse generation, CPU interrupts, and more.

Packaging   These particular processors are available in 352-pin ball grid array (BGA) packages, as shown in Figure 13–35, and are implemented with CMOS technology.

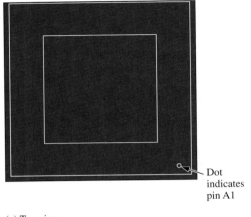

(a) Top view

Dot indicates pin A1

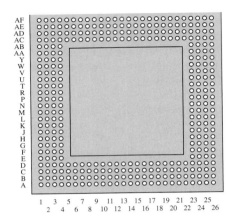

(b) Bottom view

(c) Side view

▲ FIGURE 13–35

A 352-pin BGA package.

SECTION 13–4 REVIEW

1. What is meant by the Harvard architecture?
2. What is a DSP core?
3. Name two categories of DSPs according to the type of numbers handled.
4. What are the two types of internal memory?
5. Define (a) MIPS   (b) MFLOPS   (c) MMACS.
6. Basically, what does pipelining accomplish?
7. Name the three stages of pipeline operation.
8. What happens during the fetch phase?

13–5   DIGITAL-TO-ANALOG CONVERSION METHODS

Digital-to-analog conversion is an important part of the digital processing system. Once the digital data have been processed by the DSP, they are converted back to analog form. In this section, we will examine the theory of operation of two basic types of digital-to-analog converters (DACs) and learn about their performance characteristics.

After completing this section, you should be able to

■ Explain the operation of a binary-weighted-input DAC ■ Explain the operation of an *R/2R* ladder DAC ■ Discuss resolution, accuracy, linearity, monotonicity, and settling time in a DAC ■ Discuss the testing of DACs for nonmonotonicity, differential nonlinearity, low or high gain, and offset error

Binary-Weighted-Input Digital-to-Analog Converter

One method of digital-to-analog conversion uses a resistor network with resistance values that represent the binary weights of the input bits of the digital code. Figure 13–36 shows a 4-bit DAC of this type. Each of the input resistors will either have current or have no current, depending on the input voltage level. If the input voltage is zero (binary 0), the current is also zero. If the input voltage is HIGH (binary 1), the amount of current depends on the input resistor value and is different for each input resistor, as indicated in the figure.

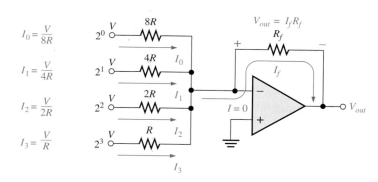

$$I_0 = \frac{V}{8R}$$

$$I_1 = \frac{V}{4R}$$

$$I_2 = \frac{V}{2R}$$

$$I_3 = \frac{V}{R}$$

Since there is practically no current into the op-amp inverting ($-$) input, all of the input currents sum together and go through R_f. Since the inverting input is at 0 V (virtual ground), the drop across R_f is equal to the output voltage, so $V_{out} = I_f R_f$.

The values of the input resistors are chosen to be inversely proportional to the binary weights of the corresponding input bits. The lowest-value resistor (R) corresponds to the highest binary-weighted input (2^3). The other resistors are multiples of R (that is, $2R$, $4R$, and $8R$) and correspond to the binary weights 2^2, 2^1, and 2^0, respectively. The input currents are also proportional to the binary weights. Thus, the output voltage is proportional to the sum of the binary weights because the sum of the input currents is through R_f.

Disadvantages of this type of DAC are the number of different resistor values and the fact that the voltage levels must be exactly the same for all inputs. For example, an 8-bit converter requires eight resistors, ranging from some value of R to $128R$ in binary-weighted steps. This range of resistors requires tolerances of one part in 255 (less than 0.5%) to accurately convert the input, making this type of DAC very difficult to mass-produce.

EXAMPLE 13–3

Determine the output of the DAC in Figure 13–37(a) if the waveforms representing a sequence of 4-bit numbers in Figure 13–37(b) are applied to the inputs. Input D_0 is the least significant bit (LSB).

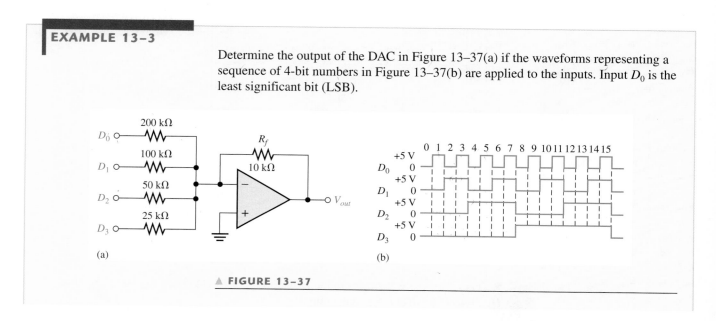

(a)

(b)

▲ FIGURE 13–37

Solution   First, determine the current for each of the weighted inputs. Since the inverting $(-)$ input of the op-amp is at 0 V (virtual ground) and a binary 1 corresponds to $+5$ V, the current through any of the input resistors is 5 V divided by the resistance value.

$$I_0 = \frac{5\ \text{V}}{200\ \text{k}\Omega} = 0.025\ \text{mA}$$

$$I_1 = \frac{5\ \text{V}}{100\ \text{k}\Omega} = 0.05\ \text{mA}$$

$$I_2 = \frac{5\ \text{V}}{50\ \text{k}\Omega} = 0.1\ \text{mA}$$

$$I_3 = \frac{5\ \text{V}}{25\ \text{k}\Omega} = 0.2\ \text{mA}$$

Almost no current goes into the inverting op-amp input because of its extremely high impedance. Therefore, assume that all of the current goes through the feedback resistor R_f. Since one end of R_f is at 0 V (virtual ground), the drop across R_f equals the output voltage, which is negative with respect to virtual ground.

$$V_{out(D0)} = (10\ \text{k}\Omega)(-0.025\ \text{mA}) = -0.25\ \text{V}$$
$$V_{out(D1)} = (10\ \text{k}\Omega)(-0.05\ \text{mA}) = -0.5\ \text{V}$$
$$V_{out(D2)} = (10\ \text{k}\Omega)(-0.1\ \text{mA}) = -1\ \text{V}$$
$$V_{out(D3)} = (10\ \text{k}\Omega)(-0.2\ \text{mA}) = -2\ \text{V}$$

From Figure 13–37(b), the first binary input code is 0000, which produces an output voltage of 0 V. The next input code is 0001, which produces an output voltage of -0.25 V. For this, the output voltage is -0.25 V. The next code is 0010, which produces an output voltage of -0.5 V. The next code is 0011, which produces an output voltage of -0.25 V $+$ -0.5 V $= -0.75$ V. Each successive binary code increases the output voltage by -0.25 V, so for this particular straight binary sequence on the inputs, the output is a stairstep waveform going from 0 V to -3.75 V in -0.25 V steps. This is shown in Figure 13–38.

▶ **FIGURE 13–38**

Output of the DAC in Figure 13–37.

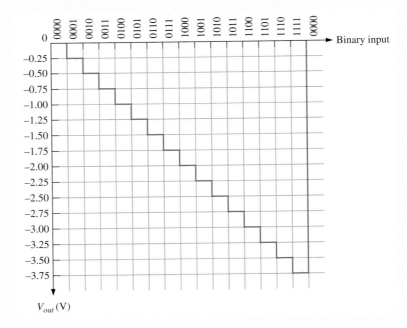

Related Problem   Reverse the input waveforms to the DAC in Figure 13–37 (D_3 to D_0, D_2 to D_1, D_1 to D_2, D_0 to D_3) and determine the output.

The R/2R Ladder Digital-to-Analog Converter

Another method of digital-to-analog conversion is the R/2R ladder, as shown in Figure 13–39 for four bits. It overcomes one of the problems in the binary-weighted-input DAC in that it requires only two resistor values.

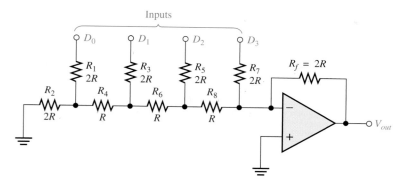

◀ FIGURE 13–39

An R/2R ladder DAC.

Start by assuming that the D_3 input is HIGH (+5 V) and the others are LOW (ground, 0 V). This condition represents the binary number 1000. A circuit analysis will show that this reduces to the equivalent form shown in Figure 13–40(a). Essentially no current goes through the 2R equivalent resistance because the inverting input is at virtual ground. Thus, all of the current ($I = 5$ V/2R) through R_7 also goes through R_f, and the output voltage is −5 V. The operational amplifier keeps the inverting (−) input near zero volts (≈0 V) because of negative feedback. Therefore, all current goes through R_f rather than into the inverting input.

Figure 13–40(b) shows the equivalent circuit when the D_2 input is at +5 V and the others are at ground. This condition represents 0100. If we thevenize* looking from R_8, we get 2.5 V in series with R, as shown. This results in a current through R_f of $I = 2.5$ V/2R, which gives an output voltage of −2.5 V. Keep in mind that there is no current into the op-amp inverting input and that there is no current through the equivalent resistance to ground because it has 0 V across it, due to the virtual ground.

Figure 13–40(c) shows the equivalent circuit when the D_1 input is at +5 V and the others are at ground. This condition represents 0010. Again thevenizing looking from R_8, you get 1.25 V in series with R as shown. This results in a current through R_f of $I = 1.25$ V/2R, which gives an output voltage of −1.25 V.

In part (d) of Figure 13–40, the equivalent circuit representing the case where D_0 is at +5 V and the other inputs are at ground is shown. This condition represents 0001. Thevenizing from R_8 gives an equivalent of 0.625 V in series with R as shown. The resulting current through R_f is $I = 0.625$ V/2R, which gives an output voltage of −0.625 V.

Notice that each successively lower-weighted input produces an output voltage that is halved, so that the output voltage is proportional to the binary weight of the input bits.

Performance Characteristics of Digital-to-Analog Converters

The performance characteristics of a DAC include resolution, accuracy, linearity, monotonicity, and settling time, each of which is discussed in the following list:

■ *Resolution.* The resolution of a DAC is the reciprocal of the number of discrete steps in the output. This, of course, is dependent on the number of input bits. For example, a 4-bit DAC has a resolution of one part in $2^4 − 1$ (one part in fifteen). Expressed as a percentage, this is $(1/15)100 = 6.67\%$. The total number of discrete steps equals $2^n − 1$, where n is the number of bits. Resolution can also be expressed as the number of bits that are converted.

---

*Thevenin's theorem states that any circuit can be reduced to an equivalent voltage source in series with an equivalent resistance.

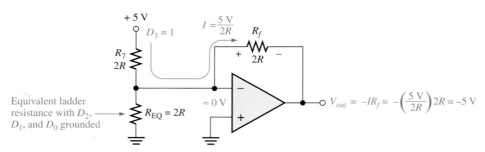

(a) Equivalent circuit for $D_3 = 1$, $D_2 = 0$, $D_1 = 0$, $D_0 = 0$

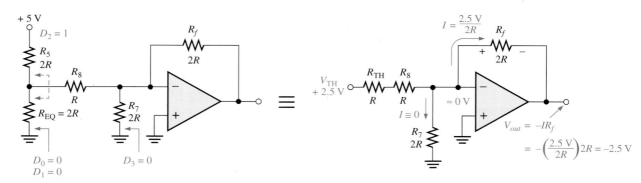

(b) Equivalent circuit for $D_3 = 0$, $D_2 = 1$, $D_1 = 0$, $D_0 = 0$

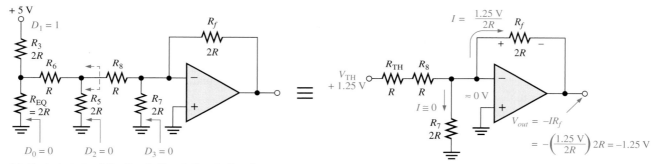

(c) Equivalent circuit for $D_3 = 0$, $D_2 = 0$, $D_1 = 1$, $D_0 = 0$

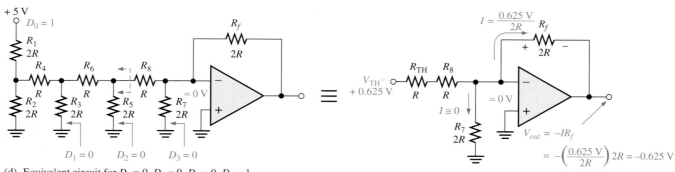

(d) Equivalent circuit for $D_3 = 0$, $D_2 = 0$, $D_1 = 0$, $D_0 = 1$

▲ FIGURE 13–40

Analysis of the *R/2R* ladder DAC.

- *Accuracy.* Accuracy is derived from a comparison of the actual output of a DAC with the expected output. It is expressed as a percentage of a full-scale, or maximum, output voltage. For example, if a converter has a full-scale output of 10 V and the accuracy is ±0.1%, then the maximum error for any output voltage is (10 V)(0.001) = 10 mV. Ideally, the accuracy should be no worse than ± 1/2 of a least significant bit. For an 8-bit converter, the least significant bit is 0.39% of full scale. The accuracy should be approximately ±0.2%.

- *Linearity.* A linear error is a deviation from the ideal straight-line output of a DAC. A special case is an offset error, which is the amount of output voltage when the input bits are all zeros.

- *Monotonicity.* A DAC is **monotonic** if it does not take any reverse steps when it is sequenced over its entire range of input bits.

- *Settling time.* Settling time is normally defined as the time it takes a DAC to settle within ± 1/2 LSB of its final value when a change occurs in the input code.

EXAMPLE 13–4

Determine the resolution, expressed as a percentage, of the following:

(a) an 8-bit DAC   **(b)** a 12-bit DAC

Solution **(a)** For the 8-bit converter,

$$\frac{1}{2^8 - 1} \times 100 = \frac{1}{255} \times 100 = \mathbf{0.392\%}$$

(b) For the 12-bit converter,

$$\frac{1}{2^{12} - 1} \times 100 = \frac{1}{4095} \times 100 = \mathbf{0.0244\%}$$

Related Problem Calculate the resolution for a 16-bit DAC.

Testing Digital-to-Analog Converters

The concept of DAC testing is illustrated in Figure 13–41. In this basic method, a sequence of binary codes is applied to the inputs, and the resulting output is observed. The binary code sequence extends over the full range of values from 0 to $2^n - 1$ in ascending order, where n is the number of bits.

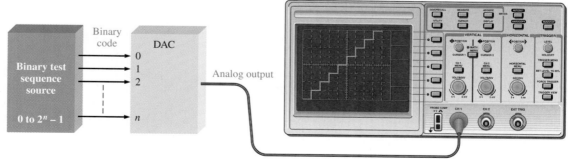

▲ **FIGURE 13–41**

Basic test setup for a DAC.

The ideal output is a straight-line stairstep as indicated. As the number of bits in the binary code is increased, the resolution is improved. That is, the number of discrete steps increases, and the output approaches a straight-line linear ramp.

Digital-to-Analog Conversion Errors

Several digital-to-analog conversion errors to be checked for are shown in Figure 13–42, which uses a 4-bit conversion for illustration purposes. A 4-bit conversion produces fifteen discrete steps. Each graph in the figure includes an ideal stairstep ramp for comparison with the faulty outputs.

Nonmonotonicity   The step reversals in Figure 13–42(a) indicate nonmonotonic performance, which is a form of nonlinearity. In this particular case, the error occurs because the 2^1 bit in the binary code is interpreted as a constant 0. That is, a short is causing the bit input line to be stuck LOW.

Differential Nonlinearity   Figure 13–42(b) illustrates differential nonlinearity in which the step amplitude is less than it should be for certain input codes. This particular output could be caused by the 2^2 bit having an insufficient weight, perhaps because of a faulty input resistor. We could also see steps with amplitudes greater than normal if a particular binary weight were greater than it should be.

▶ **FIGURE 13–42**

Illustrations of several digital-to-analog conversion errors.

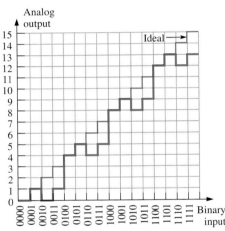

(a) Nonmonotonic output (green)

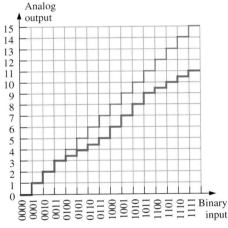

(b) Differential nonlinearity (green)

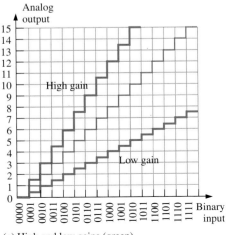

(c) High and low gains (green)

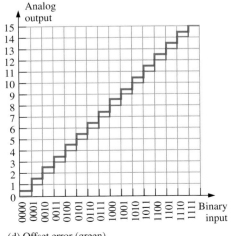

(d) Offset error (green)

Low or High Gain   Output errors caused by low or high gain are illustrated in Figure 13–42(c). In the case of low gain, all of the step amplitudes are less than ideal. In the case of high gain, all of the step amplitudes are greater than ideal. This situation may be caused by a faulty feedback resistor in the op-amp circuit.

Offset Error   An offset error is illustrated in Figure 13–42(d). Notice that when the binary input is 0000, the output voltage is nonzero and that this amount of offset is the same for all steps in the conversion. A faulty op-amp may be the culprit in this situation.

EXAMPLE 13–5

The DAC output in Figure 13–43 is observed when a straight 4-bit binary sequence is applied to the inputs. Identify the type of error, and suggest an approach to isolate the fault.

▶ **FIGURE 13–43**

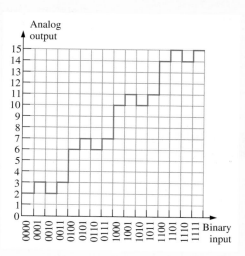

Solution   The DAC in this case is nonmonotonic. Analysis of the output reveals that the device is converting the following sequence, rather than the actual binary sequence applied to the inputs.

0010, 0011, 0010, 0011, 0110, 0111, 0110, 0111, 1010, 1011, 1010, 1011, 1110, 1111, 1110, 1111

Apparently, the 2^1 bit is stuck in the HIGH (1) state. To find the problem, first monitor the bit input pin to the device. If it is changing states, the fault is internal to the DAC and it should be replaced. If the external pin is not changing states and is always HIGH, check for an external short to $+V$ that may be caused by a solder bridge somewhere on the circuit board.

Related Problem   Determine the output of a DAC when a straight 4-bit binary sequence is applied to the inputs and the 2^0 bit is stuck HIGH.

The Reconstruction Filter

The output of the DAC is a "stairstep" approximation of the original analog signal after it has been processed by the DSP. The purpose of the low-pass reconstruction filter (sometimes called a postfilter) is to smooth out the DAC output by eliminating the higher frequency content that results from the fast transitions of the "stairsteps," as roughly illustrated in Figure 13–44.

The reconstruction filter smooths the output of the DAC.

SECTION 13–5 REVIEW	1. What is the disadvantage of the DAC with binary weighted inputs?
	2. What is the resolution of a 4-bit DAC?
	3. How do you detect nonmonotonic behavior in a DAC?
	4. What effect does low gain have on a DAC output?

SUMMARY

- Digital signal processing is the digital processing of analog signals, usually in real-time, for the purpose of modifying or enhancing the signal in some way.

- In general, a digital signal processing system consists of an anti-aliasing filter, a sample-and-hold circuit, an analog-to-digital converter, a DSP (digital signal processor), a digital-to-analog converter, and a reconstruction filter.

- Sampling converts an analog signal into a series of impulses, each representing the signal amplitude at a given instant in time.

- The sampling theorem states that the sampling frequency must be at least twice the highest sampled frequency (Nyquist frequency).

- Analog-to-digital conversion changes an analog signal into a series of digital codes.

- Four types of analog-to-digital converters (ADCs) are flash (simultaneous), dual-slope, successive-approximation, and sigma-delta.

- A DSP is a specialized microprocessor optimized for speed in order to process data as it occurs (real-time).

- Most DSPs are based on the Harvard architecture, which means that there is a data memory and a program memory.

- A pipeline operation consists of fetch, decode, and execute stages.

- Digital-to-analog conversion changes a series of digital codes that represent an analog signal back into the analog signal.

- Two types of digital-to-analog converters (DACs) are binary-weighted input and $R/2R$ ladder.

KEY TERMS

Key terms and other bold terms in the chapter are defined in the end-of-book glossary.

Aliasing The effect created when a signal is sampled at less than twice the signal frequency. Aliasing creates unwanted frequencies that interfere with the signal frequency.

Analog-to-digital converter (ADC) A circuit used to convert an analog signal to digital form.

Decode A stage of the DSP pipeline operation in which instructions are assigned to functional units and are decoded.

Digital-to-analog converter (DAC) A circuit used to convert the digital representation of an analog signal back to the analog signal.

DSP Digital signal processor; a special type of microprocessor that processes data in real time.

DSP core The central processing unit of a DSP.

Execute A stage of the DSP pipeline operation in which the decoded instructions are carried out.

Fetch A stage of the DSP pipeline operation in which an instruction is obtained from the program memory.

MFLOPS Million floating-point operations per second.

MIPS Million instructions per second.

MMACS Million multiply/accumulates per second.

Nyquist frequency The highest signal frequency that can be sampled at a specified sampling frequency; a frequency equal to or less than half the sampling frequency.

Pipeline Part of the DSP architecture that allows multiple instructions to be processed simultaneously.

Quantization The process whereby a binary code is assigned to each sampled value during analog-to-digital conversion.

Sampling The process of taking a sufficient number of discrete values at points on a waveform that will define the shape of the waveform.

SELF-TEST

Answers are at the end of the chapter.

1. An ADC is an
   (a) alphanumeric data coder
   (b) analog-to-digital converter
   (c) analog device carrier
   (d) analog-to-digital comparator

2. A DAC is a
   (a) digital-to-analog computer
   (b) digital analysis calculator
   (c) data accumulation converter
   (d) digital-to-analog converter

3. A digital signal processing system usually operates in
   (a) real time
   (b) imaginary time
   (c) compressed time
   (d) computer time

4. Sampling of an analog signal produces
   (a) a series of impulses that are proportional to the amplitude of the signal
   (b) a series of impulses that are proportional to the frequency of the signal
   (c) digital codes that represent the analog signal amplitude
   (d) digital codes that represent the time of each sample

5. According to the sampling theorem, the sampling frequency should be
   (a) less than half the highest signal frequency
   (b) greater than twice the highest signal frequency
   (c) less than half the lowest signal frequency
   (d) greater than the lowest signal frequency

6. A hold action occurs
   (a) before each sample
   (b) during each sample
   (c) after the analog-to-digital conversion
   (d) immediately after a sample

7. The quantization process
   (a) converts the sample-and-hold output to binary code
   (b) converts a sample impulse to a level
   (c) converts a sequence of binary codes to a reconstructed analog signal
   (d) filters out unwanted frequencies before sampling takes place

8. Generally, an analog signal can be reconstructed more accurately with
   (a) more quantization levels
   (b) fewer quantization levels
   (c) a higher sampling frequency
   (d) a lower sampling frequency
   (e) either answer (a) or (c)

9. A flash ADC uses
   (a) counters   (b) op-amps   (c) an integrator   (d) flip-flops
   (e) answers (a) and (c)

10. A dual-slope ADC uses
    (a) a counter   (b) op-amps   (c) an integrator   (d) a differentiator
    (e) answers (a) and (c)

11. The output of a sigma-delta ADC is
    (a) parallel binary codes   (b) multiple-bit data
    (c) single-bit data   (d) a difference voltage

12. The term *Harvard architecture* means
    (a) a CPU and a main memory
    (b) a CPU and two data memories
    (c) a CPU, a program memory, and a data memory
    (d) a CPU and two register files

13. The minimum number of general-purpose registers in the TMS320C6000 series DSPs is
    (a) 32   (b) 64   (c) 16   (d) 8

14. The two internal memories in the TMS320C6000 series each have a capacity of
    (a) 1 MB   (b) 512 kB   (c) 64 kB   (d) 32 kB

15. In the TMS320C6000 series pipeline operation, the number of instructions processed simultaneously is
    (a) eight   (b) four   (c) two   (d) one

16. The stage of the pipeline operation in which instructions are retrieved from the memory is called
    (a) execute   (b) accumulate
    (c) decode   (d) fetch

17. In a binary-weighted DAC, the resistors on the inputs
    (a) determine the amplitude of the analog signal
    (b) determine the weights of the digital inputs
    (c) limit the power consumption
    (d) prevent loading on the source

18. In an *R/2R* DAC, there are
    (a) four values of resistors
    (b) one resistor value
    (c) two resistor values
    (d) a number of resistor values equal to the number of inputs

PROBLEMS

Answers to odd–numbered problems are at the end of the book.

SECTION 13–1   Digital Signal Processing Basics

1. Explain the purpose of analog-to-digital conversion.

2. Fill in the appropriate functional names for the digital signal processing system block diagram in Figure 13–45.

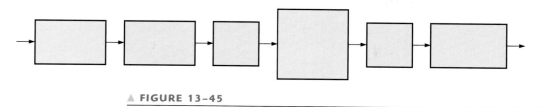

▲ **FIGURE 13–45**

3. Explain the purpose of digital-to-analog conversion.

SECTION 13–2 **Converting Analog Signals to Digital**

4. The waveform shown in Figure 13–46 is applied to a sampling circuit and is sampled every 3 ms. Show the output of the sampling circuit. Assume a one-to-one voltage correspondence between the input and output.

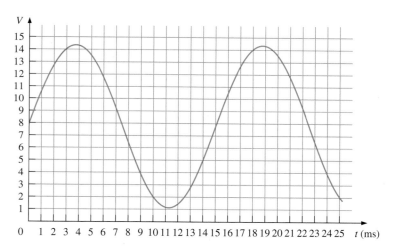

▲ **FIGURE 13–46**

5. The output of the sampling circuit in Problem 4 is applied to a hold circuit. Show the output of the hold circuit.

6. If the output of the hold circuit in Problem 5 is quantized using two bits, what is the resulting sequence of binary codes?

7. Repeat Problem 6 using 4-bit quantization.

8. (a) Reconstruct the analog signal from the 2-bit quantization in Problem 6.

   (b) Reconstruct the analog signal from the 4-bit quantization in Problem 7.

9. Graph the analog function represented by the following sequence of binary numbers:

   1111, 1110, 1101, 1100, 1010, 1001, 1000, 0111, 0110, 0101, 0100, 0101, 0110, 0111, 1000, 1001, 1010, 1011, 1100, 1100, 1100, 1011, 1010, 1001.

SECTION 13–3 **Analog-to-Digital Conversion Methods**

10. The input voltage to a certain op-amp inverting amplifier is 10 mV, and the output is 2 V. What is the closed-loop voltage gain?

11. To achieve a closed-loop voltage gain of 330 with an inverting amplifier, what value of feedback resistor do you use if $R_i = 1.0 \text{ k}\Omega$?

12. Determine the binary output code of a 3-bit flash ADC for the analog input signal in Figure 13–47.

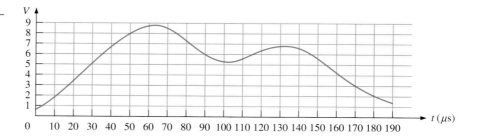

13. Repeat Problem 12 for the analog waveform in Figure 13–48.

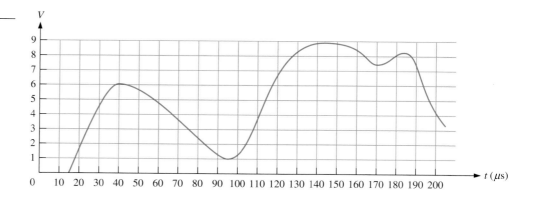

14. For a certain 2-bit successive-approximation ADC, the maximum ladder output is +8 V. If a constant +6 V is applied to the analog input, determine the sequence of binary states for the SAR.

15. Repeat Problem 14 for a 4-bit successive-approximation ADC.

16. An ADC produces the following sequence of binary numbers when an analog signal is applied to its input: 0000, 0001, 0010, 0011, 0100, 0101, 0110, 0111, 0110, 0101, 0100, 0011, 0010, 0001, 0000.

 (a) Reconstruct the input digitally.

 (b) If the ADC failed so that the code 0111 were missing, what would the reconstructed output look like?

SECTION 13–4 **The Digital Signal Processor (DSP)**

17. A TMS320C62xx DSP has 32-bit instructions and is operating at 2000 MIPS. How many bytes per second is the DSP processing?

18. If the clock rate of a TMS320C64xx DSP is 400 MHz, how many instructions can it provide to the CPU functional units in one second?

19. How many floating-point operations can a DSP do in one second if it is specified at 1000 MFLOPS?

20. List and describe the four phases of the fetch operation in a TMS320C6000 series DSP.

21. List and describe the two phases of the decode operation in a TMS320C6000 series DSP.

SECTION 13–5 **Digital-to-Analog Conversion Methods**

22. In the 4-bit DAC in Figure 13–36, the lowest-weighted resistor has a value of 10 kΩ. What should the values of the other input resistors be?

23. Determine the output of the DAC in Figure 13–49(a) if the sequence of 4-bit numbers in part (b) is applied to the inputs. The data inputs have a low value of 0 V and a high value of +5 V.

▶ FIGURE 13–49

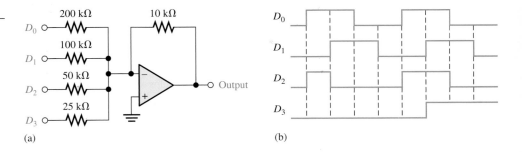

(a)

(b)

24. Repeat Problem 23 for the inputs in Figure 13–50.

▶ FIGURE 13–50

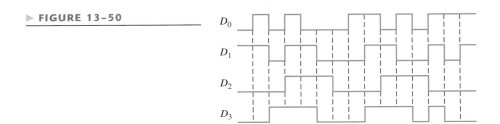

25. Determine the resolution expressed as a percentage, for each of the following DACs:

   (a) 3-bit   (b) 10-bit   (c) 18-bit

26. Develop a circuit for generating an 8-bit binary test sequence for the test setup in Figure 13–41.

27. A 4-bit DAC has failed in such a way that the MSB is stuck in the 0 state. Draw the analog output when a straight binary sequence is applied to the inputs.

28. A straight binary sequence is applied to a 4-bit DAC, and the output in Figure 13–51 is observed. What is the problem?

▶ FIGURE 13–51

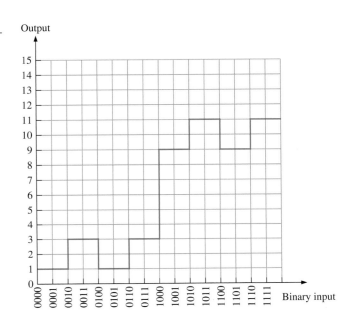

ANSWERS

SECTION REVIEWS

SECTION 13–1 **Digital Signal Processing Basics**

1. DSP stands for digital signal processor.
2. ADC stands for analog-to-digital converter.
3. DAC stands for digital-to-analog converter.
4. The ADC changes an analog signal to binary coded form.
5. The DAC changes a binary coded signal to analog form.

SECTION 13–2 **Converting Analog Signals to Digital**

1. Sampling is the process of converting an analog signal into a series of impulses, each representing the amplitude of the analog signal.
2. A sampled value is held to allow time to convert the value to a binary code.
3. The minimum sampling frequency is 40 kHz.
4. Quantization is the process of converting a sampled level to a binary code.
5. The number of bits determine quantization accuracy.

SECTION 13–3 **Analog-to-Digital Conversion Methods**

1. The simultaneous (flash) method is fastest.
2. The sigma-delta method produces a single-bit data stream.
3. Yes, successive approximation has a fixed conversion time.
4. Missing code, incorrect code, and offset are types of ADC output errors.

SECTION 13–4 **The Digital Signal Processor (DSP)**

1. Harvard architecture means that there is a CPU and two memories, one for data and one for programs.
2. The DSP core is the CPU.
3. DSPs can be fixed-point or floating-point.
4. Internal memory types are data and program.
5. (a) MIPS—million instructions per second
   (b) MFLOPS—million floating-point operations per second
   (c) MMACS—million multiply/accumulates per second
6. Pipelining provides for the processing of multiple instructions simultaneously.
7. The stages of pipeline operation are fetch, decode, and execute.
8. During fetch, instructions are retrieved from the program memory.

SECTION 13–5 **Digital-to-Analog Conversion Methods**

1. In a binary-weighted DAC, each resistor has a different value.
2. $(1/(2^4 - 1))100\% = 6.67\%$
3. A step reversal indicates nonmonotonic behavior in a DAC.
4. Step amplitudes in a DAC are less than ideal with low gain.

RELATED PROBLEMS FOR EXAMPLES

13–1 100, 111, 100, 000, 011, 110. Yes, information is lost.
13–2 See Figure 13–52.
13–3 See Figure 13–53.
13–4 $(1/(216 - 1))100\% = 0.00153\%$
13–5 See Figure 13–54.

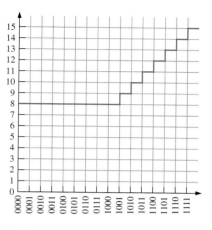

▲ FIGURE 13–52

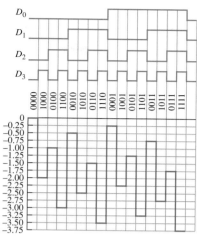

▲ FIGURE 13–53

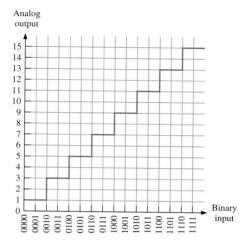

▲ FIGURE 13–54

SELF-TEST

1. (b)	**2.** (d)	**3.** (a)	**4.** (a)	**5.** (b)	**6.** (d)
7. (a)	**8.** (e)	**9.** (b)	**10.** (e)	**11.** (c)	**12.** (c)
13. (a)	**14.** (c)	**15.** (a)	**16.** (d)	**17.** (b)	**18.** (c)

References

Dahnoun, Naim. *Digital Signal Processing Implementation Using the TMS320C6000 DSP Platform.* Reading, Mass.: Addison-Wesley Longman. 2000.

Hayes, Monson. *Schaum's Outline of Digital Signal Processing.* New York: McGraw-Hill. 1998.

Kuo, Sen, and Bob Lee. *Real-Time Digital Signal Processing: Implementations, Applications, and Experiments with the TMS320C55x.* New York: John Wiley & Sons. 2001.

Lyons, Richard. *Understanding Digital Signal Processing.* Reading, Mass.: Addison-Wesley Longman. 1996.

Marven, Craig, and Gillian Ewers. *A Simple Approach to Digital Signal Processing.* New York: John Wiley & Sons. 1996.

Oppenheim, Alan, and Ronald Schafer. *Digital Signal Processing.* Englewood Cliffs, N.J.: Prentice-Hall. 1974.

Orfanidis, Sophocles. *Introduction to Signal Processing.* Upper Saddle River, N.J.: Prentice-Hall. 1996.

Proakis, John, and Dimitris Manolakis. *Digital Signal Processing: Principles, Algorithms, and Applications,* 3d ed. Upper Saddle River, N.J.: Prentice-Hall. 1996.

Steiglitz, Ken. *Digital Signal Processing Primer: With Applications to Digital Audio and Computer Music.* Reading, Mass.: Addison-Wesley Longman. 1996.

Williams, Douglas, and Vijay Madisetti. *Digital Signal Processing Handbook.* Boca Raton, Fl.: CRC Press. 1997.

14

INTEGRATED CIRCUIT TECHNOLOGIES

CHAPTER OBJECTIVES

- Determine the noise margin of a device from data sheet parameters
- Calculate the power dissipation of a device
- Explain how propagation delay time affects the frequency of operation or speed of a circuit
- Interpret the speed-power product as a measure of performance
- Use data sheets to obtain information about a specific device
- Explain what the fan-out of a gate means

- Describe how basic TTL and CMOS gates operate at the component level

- Recognize the difference between TTL totem-pole outputs and TTL open-collector outputs and understand the limitations and uses of each

- Connect circuits in a wired-AND configuration

- Describe the operation of tristate circuits

- Properly terminate unused gate inputs

- Compare the performance of TTL and CMOS families

- Handle CMOS devices without risk of damage due to electrostatic discharge

- State the advantages of ECL

- Describe the PMOS and NMOS circuits

- Describe an E^{2}CMOS cell

KEY TERMS

- TTL
- CMOS
- Noise immunity
- Noise margin
- Power dissipation
- Propagation delay time
- Fan-out
- Current sourcing

- Current sinking
- Unit load
- Pull-up resistor
- Tristate
- Totem pole
- Open-collector
- ECL
- E^{2}CMOS

INTRODUCTION

This chapter is intended to be used as a "floating" chapter. That is, all or portions of this chapter can be covered at any point throughout the book or completely omitted, depending on the course objectives. Section 3–8 should be covered before beginning this chapter.

In Chapter 3 (Section 3–8) you learned about basic integrated circuit logic gates. This chapter provides an introduction to the circuit technology used to implement those gates, as well as other types of IC devices.

Two major IC technologies, CMOS and TTL, are covered and their operating parameters are defined. Also, the operational characteristics of various families within these circuit technologies are compared. Other circuit technologies are also introduced. It is important to keep in mind that the particular circuit technology used to implement a logic gate has no effect on the logic operation of the gate. In terms of its truth table operation, a certain type of gate that is implemented with CMOS is the same as that type of gate implemented with TTL. The only differences in the gates are the electrical characteristics such as power dissipation, switching speed, and noise immunity.

WWW. VISIT THE COMPANION WEBSITE
Study aids for this chapter are available at
http://www.prenhall.com/floyd

14–1 BASIC OPERATIONAL CHARACTERISTICS AND PARAMETERS

When you work with digital ICs, you should be familiar not only with their logical operation but also with such operational properties as voltage levels, noise immunity, power dissipation, fan-out, and propagation delay time. In this section, the practical aspects of these properties are discussed.

After completing this section, you should be able to

■ Determine the power and ground connections ■ Describe the logic levels for CMOS and TTL ■ Discuss noise immunity ■ Determine the power dissipation of a logic circuit ■ Define the propagation delay time of a logic gate ■ Discuss speed-power product and explain its significance ■ Discuss loading and fan-out of TTL and CMOS

DC Supply Voltage

The nominal value of the dc supply voltage for **TTL** (transistor-transistor logic) devices is +5 V. TTL is also designated T^2L. **CMOS** (complementary metal-oxide semiconductor) devices are available in different supply voltage categories: +5 V, +3.3 V, 2.5 V, and 1.2 V. Although omitted from logic diagrams for simplicity, the dc supply voltage is connected to the V_{CC} pin of an IC package, and ground is connected to the GND pin. Both voltage and ground are distributed internally to all elements within the package, as illustrated in Figure 14–1 for a 14-pin package.

▶ **FIGURE 14–1**

Example of V_{CC} and ground connection and distribution in an IC package. Other pin connections are omitted for simplicity.

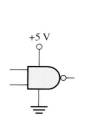

+5 V

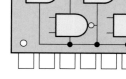

V_{CC}

(a) Single gate   (b) IC dual in-line package

CMOS Logic Levels

Logic levels were discussed briefly in Chapter 1. There are four different logic-level specifications: V_{IL}, V_{IH}, V_{OL}, and V_{OH}. For CMOS circuits, the ranges of input voltages (V_{IL}) that can represent a valid LOW (logic 0) are from 0 V to 1.5 V for the +5 V logic and 0 V to 0.8 V for the 3.3 V logic. The ranges of input voltages (V_{IH}) that can represent a valid HIGH (logic 1) are from 3.5 V to 5 V for the 5 V logic and 2 V to 3.3 V for the 3.3 V logic, as indicated in Figure 14–2. The ranges of values from 1.5 V to 3.5 V for 5 V logic and 0.8 V to 2 V for 3.3 V logic are regions of unpredictable performance, and values in these ranges are unallowed. When an input voltage is in one of these ranges, it can be interpreted as either a HIGH or a LOW by the logic circuit. Therefore, CMOS gates cannot be operated reliably when the input voltages are in these unallowed ranges.

The ranges of CMOS output voltages (V_{OL} and V_{OH}) for both 5 V and 3.3 V logic are also shown in Figure 14–2. Notice that the minimum HIGH output voltage, $V_{OH(min)}$, is greater than the minimum HIGH input voltage, $V_{IH(min)}$. Also, notice that the maximum LOW output voltage, $V_{OL(max)}$, is less than the maximum LOW input voltage, $V_{IL(max)}$.

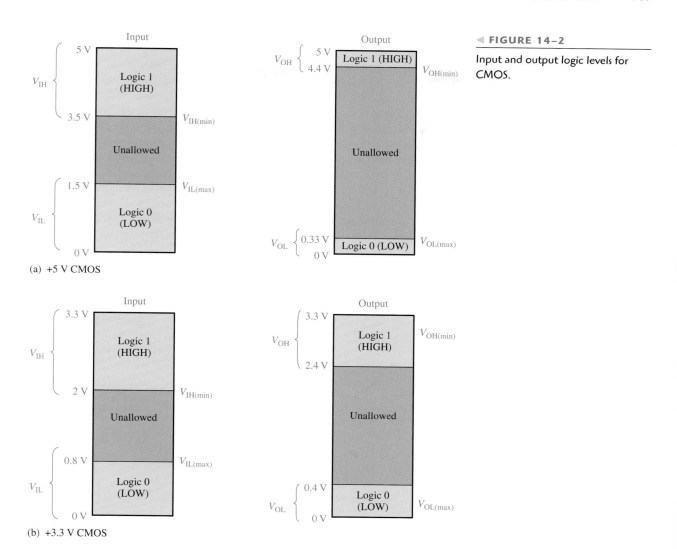

► FIGURE 14–2

Input and output logic levels for CMOS.

(a) +5 V CMOS

(b) +3.3 V CMOS

TTL Logic Levels

The input and output logic levels for TTL are given in Figure 14–3. Just as for CMOS, there are four different logic level specifications: V_{IL}, V_{IH}, V_{OL}, and V_{OH}.

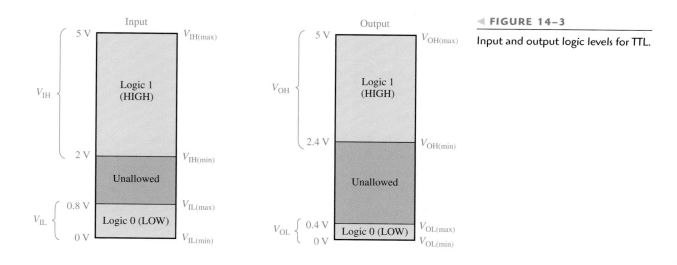

► FIGURE 14–3

Input and output logic levels for TTL.

Noise Immunity

Noise is unwanted voltage that is induced in electrical circuits and can present a threat to the proper operation of the circuit. Wires and other conductors within a system can pick up stray high-frequency electromagnetic radiation from adjacent conductors in which currents are changing rapidly or from many other sources external to the system. Also, power-line voltage fluctuation is a form of low-frequency noise.

In order not to be adversely affected by noise, a logic circuit must have a certain amount of **noise immunity.** This is the ability to tolerate a certain amount of unwanted voltage fluctuation on its inputs without changing its output state. For example, if noise voltage causes the input of a 5 V CMOS gate to drop below 3.5 V in the HIGH state, the input is in the unallowed region and operation is unpredictable (see Figure 14–2). Thus, the gate may interpret the fluctuation below 3.5 V as a LOW level, as illustrated in Figure 14–4(a). Similarly, if noise causes a gate input to go above 1.5 V in the LOW state, an uncertain condition is created, as illustrated in part (b).

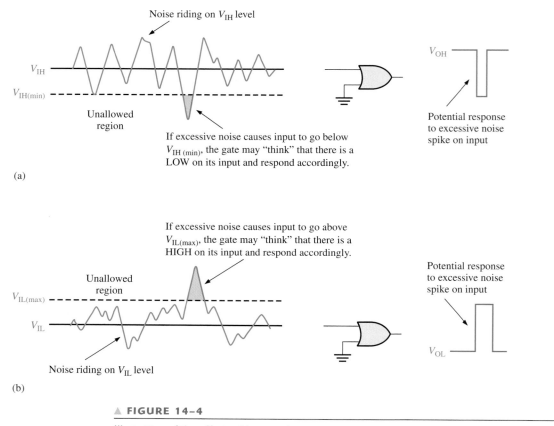

(a)

(b)

▲ **FIGURE 14–4**

Illustration of the effects of input noise on gate operation.

Noise Margin

A measure of a circuit's noise immunity is called the **noise margin,** which is expressed in volts. There are two values of noise margin specified for a given logic circuit: the HIGH-level noise margin (V_{NH}) and the LOW-level noise margin (V_{NL}). These parameters are defined by the following equations:

Equation 14–1

$$V_{NH} = V_{OH(min)} - V_{IH(min)}$$

Equation 14–2

$$V_{NL} = V_{IL(max)} - V_{OL(max)}$$

Sometimes you will see the noise margin expressed as a percentage of V_{CC}. From the equations, V_{NH} is the difference between the lowest possible HIGH output from a driving gate ($V_{OH(min)}$) and the lowest possible HIGH input that the load gate can tolerate ($V_{IH(min)}$). Noise margin, V_{NL}, is the difference between the maximum possible LOW input that a gate can tolerate ($V_{IL(max)}$) and the maximum possible LOW output of the driving gate ($V_{OL(max)}$). Noise margins are illustrated in Figure 14–5.

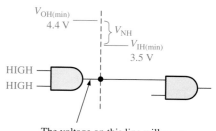

$V_{OH(min)}$
4.4 V
V_{NH}
$V_{IH(min)}$
3.5 V

HIGH
HIGH

The voltage on this line will never be less than 4.4 V unless noise or improper operation is introduced.

(a) HIGH-level noise margin

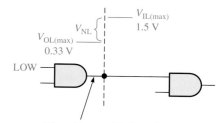

$V_{IL(max)}$
1.5 V
V_{NL}
$V_{OL(max)}$
0.33 V

LOW

The voltage on this line will never exceed 0.33 V unless noise or improper operation is introduced.

(b) LOW-level noise margin

▲ FIGURE 14–5

Illustration of noise margins. Values are for 5 V CMOS, but the principle applies to any logic family.

EXAMPLE 14–1

Determine the HIGH-level and LOW-level noise margins for CMOS and for TTL by using the information in Figures 14–2 and 14–3.

Solution For 5 V CMOS,

$$V_{IH(min)} = 3.5 \text{ V}$$
$$V_{IL(max)} = 1.5 \text{ V}$$
$$V_{OH(min)} = 4.4 \text{ V}$$
$$V_{OL(max)} = 0.33 \text{ V}$$
$$V_{NH} = V_{OH(min)} - V_{IH(min)} = 4.4 \text{ V} - 3.5 \text{ V} = \mathbf{0.9 \text{ V}}$$
$$V_{NL} = V_{IL(max)} - V_{OL(max)} = 1.5 \text{ V} - 0.33 \text{ V} = \mathbf{1.17 \text{ V}}$$

For TTL,

$$V_{IH(min)} = 2 \text{ V}$$
$$V_{IL(max)} = 0.8 \text{ V}$$
$$V_{OH(min)} = 2.4 \text{ V}$$
$$V_{OL(max)} = 0.4 \text{ V}$$
$$V_{NH} = V_{OH(min)} - V_{IH(min)} = 2.4 \text{ V} - 2 \text{ V} = \mathbf{0.4 \text{ V}}$$
$$V_{NL} = V_{IL(max)} - V_{OL(max)} = 0.8 \text{ V} - 0.4 \text{ V} = \mathbf{0.4 \text{ V}}$$

A TTL gate is immune to up to 0.4 V of noise for both the HIGH and LOW input states.

Related Problem* Based on the preceding noise margin calculations, which family of devices, 5 V CMOS or TTL, should be used in a high-noise environment?

---

*Answers are at the end of the chapter.

Power Dissipation

A logic gate draws current from the dc supply voltage source, as indicated in Figure 14–6. When the gate is in the HIGH output state, an amount of current designated by I_{CCH} is drawn; and in the LOW output state, a different amount of current, I_{CCL}, is drawn.

▶ **FIGURE 14–6**

Currents from the dc supply.
Conventional current direction is
shown. Electron flow notation is
opposite.

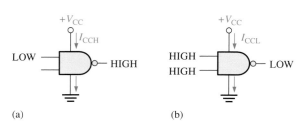

(a)   (b)

As an example, if I_{CCH} is specified as 1.5 mA when V_{CC} is 5 V and if the gate is in a static (nonchanging) HIGH output state, the **power dissipation** (P_D) of the gate is

$$P_D = V_{CC}I_{CCH} = (5 \text{ V})(1.5 \text{ mA}) = 7.5 \text{ mW}$$

When a gate is pulsed, its output switches back and forth between HIGH and LOW, and the amount of supply current varies between I_{CCH} and I_{CCL}. The average power dissipation depends on the duty cycle and is usually specified for a duty cycle of 50%. When the duty cycle is 50%, the output is HIGH half the time and LOW the other half. The average supply current is therefore

Equation 14–3
$$I_{CC} = \frac{I_{CCH} + I_{CCL}}{2}$$

The average power dissipation is

Equation 14–4
$$P_D = V_{CC}I_{CC}$$

EXAMPLE 14–2

A certain gate draws 2 μA when its output is HIGH and 3.6 μA when its output is LOW. What is its average power dissipation if V_{CC} is 5 V and the gate is operated on a 50% duty cycle?

Solution   The average I_{CC} is

$$I_{CC} = \frac{I_{CCH} + I_{CCL}}{2} = \frac{2.0 \,\mu\text{A} + 3.6 \,\mu\text{A}}{2} = 2.8 \,\mu\text{A}$$

The average power dissipation is

$$P_D = V_{CC}I_{CC} = (5 \text{ V})(2.8 \,\mu\text{A}) = \textbf{14} \,\mu\textbf{W}$$

Related Problem   A certain IC gate has an $I_{CCH} = 1.5 \,\mu$A and $I_{CCL} = 2.8 \,\mu$A. Determine the average power dissipation for 50% duty cycle operation if V_{CC} is 5 V.

Power dissipation in a TTL circuit is essentially constant over its range of operating frequencies. Power dissipation in CMOS, however, is frequency dependent. It is extremely low under static (dc) conditions and increases as the frequency increases. These characteristics are shown in the general curves of Figure 14–7. For example, the power dissipation of a low-power Schottky (LS) TTL gate is a constant 2.2 mW. The power dissipation of an HCMOS gate is 2.75 μW under static conditions and 170 μW at 100 kHz.

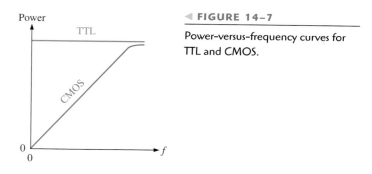

Power-versus-frequency curves for TTL and CMOS.

Propagation Delay Time

When a signal passes (propagates) through a logic circuit, it always experiences a time delay, as illustrated in Figure 14–8. A change in the output level always occurs a short time, called the **propagation delay time,** later than the change in the input level that caused it.

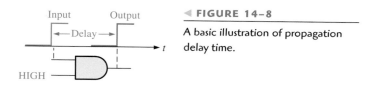

▸ FIGURE 14–8

A basic illustration of propagation delay time.

As mentioned in Chapter 3, there are two propagation delay times specified for logic gates:

- t_{PHL}: The time between a designated point on the input pulse and the corresponding point on the output pulse when the output is changing from HIGH to LOW.

- t_{PLH}: The time between a designated point on the input pulse and the corresponding point on the output pulse when the output is changing from LOW to HIGH.

These propagation delay times are illustrated in Figure 14–9, with the 50% points on the pulse edges used as references.

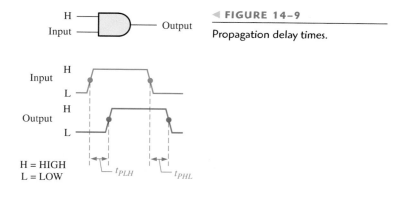

▸ FIGURE 14–9

Propagation delay times.

The propagation delay time of a gate limits the frequency at which it can be operated. The greater the propagation delay time, the lower the maximum frequency. Thus, a higher-speed circuit is one that has a smaller propagation delay time. For example, a gate with a delay of 3 ns is faster than one with a 10 ns delay.

Speed-Power Product

The speed-power product provides a basis for the comparison of logic circuits when *both* propagation delay time and power dissipation are important considerations in the selection of the type of logic to be used in a certain application. The lower the speed-power product, the better. The unit of speed-power product is the picojoule (pJ). For example, HCMOS has a speed-power product of 1.2 pJ at 100 kHz while LS TTL has a value of 22 pJ.

Loading and Fan-Out

When the output of a logic gate is connected to one or more inputs of other gates, a load on the driving gate is created, as shown in Figure 14–10. There is a limit to the number of load gate inputs that a given gate can drive. This limit is called the **fan-out** of the gate.

▶ **FIGURE 14–10**

Loading a gate output with gate inputs.

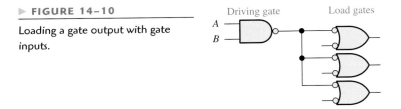

CMOS Loading   Loading in CMOS differs from that in TTL because the type of transistors used in CMOS logic present a predominantly capacitive load to the driving gate, as illustrated in Figure 14–11. In this case, the limitations are the charging and discharging times associated with the output resistance of the driving gate and the input capacitance of the load gates. When the output of the driving gate is HIGH, the input capacitance of the load gate is charging through the output resistance of the driving gate. When the output of the driving gate is LOW, the capacitance is discharging, as indicated in Figure 14–11.

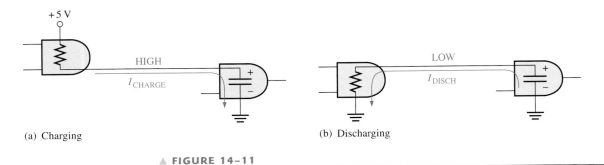

(a) Charging

(b) Discharging

▲ **FIGURE 14–11**

Capacitive loading of a CMOS gate.

When more load gate inputs are added to the driving gate output, the total capacitance increases because the input capacitances effectively appear in parallel. This increase in capacitance increases the charging and discharging times, thus reducing the maximum frequency at which the gate can be operated. Therefore, the fan-out of a CMOS gate depends on the frequency of operation. The fewer the load gate inputs, the greater the maximum frequency.

TTL Loading   A TTL driving gate sources current to a load gate input in the HIGH state (I_{IH}) and sinks current from the load gate in the LOW state (I_{IL}). **Current sourcing** and **current sinking** are illustrated in simplified form in Figure 14–12, where the resistors represent the internal input and output resistance of the gate for the two conditions.

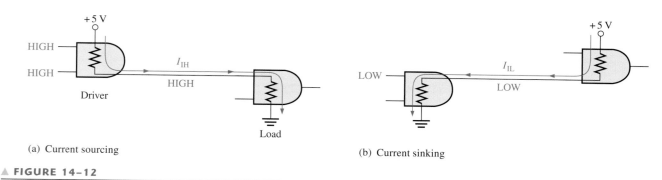

(a) Current sourcing

(b) Current sinking

▲ **FIGURE 14–12**

Basic illustration of current sourcing and current sinking in logic gates.

As more load gates are connected to the driving gate, the loading on the driving gate increases. The total source current increases with each load gate input that is added, as illustrated in Figure 14–13. As this current increases, the internal voltage drop of the driving gate increases, causing the output, V_{OH}, to decrease. If an excessive number of load gate inputs are connected, V_{OH} drops below $V_{OH(min)}$, and the HIGH-level noise margin is reduced, thus compromising the circuit operation. Also, as the total source current increases, the power dissipation of the driving gate increases.

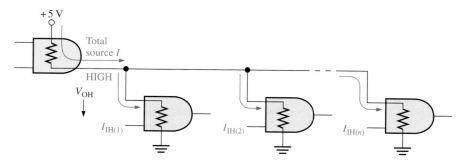

◀ **FIGURE 14–13**

HIGH-state TTL loading.

The fan-out is the maximum number of load gate inputs that can be connected without adversely affecting the specified operational characteristics of the gate. For example, low-power Schottky (LS) TTL has a fan-out of 20 unit loads. One input of the same logic family as the driving gate is called a **unit load.**

The total sink current also increases with each load gate input that is added, as shown in Figure 14–14. As this current increases, the internal voltage drop of the driving gate increases, causing V_{OL} to increase. If an excessive number of loads are added, V_{OL} exceeds $V_{OL(max)}$, and the LOW-level noise margin is reduced.

In TTL, the current-sinking capability (LOW output state) is the limiting factor in determining the fan-out.

LOW-stage TTL loading.

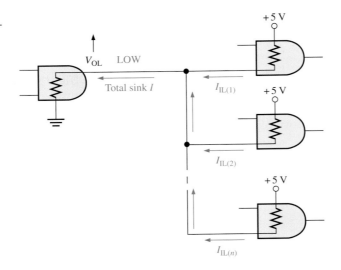

1. Define V_{IH}, V_{IL}, V_{OH}, and V_{OL}.
2. Is it better to have a lower value of noise margin or a higher value?
3. Gate A has a greater propagation delay time than gate B. Which gate can operate at a higher frequency?
4. How does excessive loading affect the noise margin of a gate?

14–2 CMOS CIRCUITS

Basic internal CMOS circuitry and its operation are discussed in this section. The abbreviation CMOS stands for complementary metal-oxide semiconductor. The term *complementary* refers to the use of two types of transistors in the output circuit. An *n*-channel MOSFET (MOS field-effect transistor) and a *p*-channel MOSFET are used.

After completing this section, you should be able to

- Identify a MOSFET by its symbol ■ Discuss the switching action of a MOSFET
- Describe the basic operation of a CMOS inverter circuit ■ Describe the basic operation of CMOS NAND and NOR gates ■ Explain the operation of a CMOS gate with an open-drain output ■ Discuss the operation of tristate CMOS gates ■ List the precautions required when handling CMOS devices

The MOSFET

Metal-oxide semiconductor field-effect transistors (**MOSFETs**) are the active switching elements in CMOS circuits. These devices differ greatly in construction and internal operation from bipolar junction transistors used in TTL circuits, but the switching action is basically the same: they function ideally as open or closed switches, depending on the input.

Figure 14–15(a) shows the symbols for both *n*-channel and *p*-channel MOSFETs. As indicated, the three terminals of a MOSFET are **gate, drain,** and **source.** When the gate voltage of an *n*-channel MOSFET is more positive than the source, the MOSFET is on (*saturation*), and there is, ideally, a closed switch between the drain and the source. When the gate-to-source voltage is zero, the MOSFET is off (*cutoff*), and there is, ideally, an open

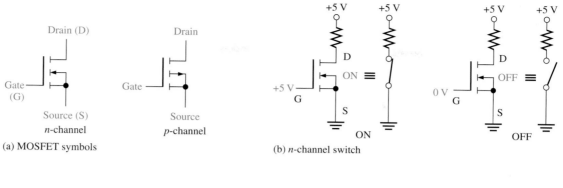

▲ **FIGURE 14–15**

Basic symbols and switching action of MOSFETs.

switch between the drain and the source. This operation is illustrated in Figure 14–15(b). The *p*-channel MOSFET operates with opposite voltage polarities, as shown in part (c).

Sometimes a simplified MOSFET symbol as shown in Figure 14–16 is used.

◄ **FIGURE 14–16**

Simplified MOSFET symbol.

CMOS Inverter

Complementary MOS (CMOS) logic uses the MOSFET in complementary pairs as its basic element. A complementary pair uses both *p*-channel and *n*-channel enhancement MOSFETs, as shown in the inverter circuit in Figure 14–17.

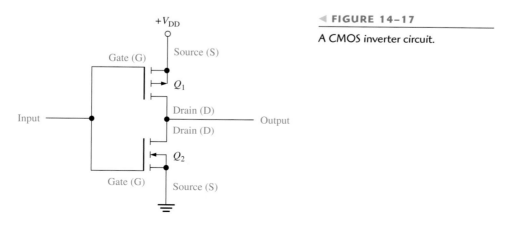

◄ **FIGURE 14–17**

A CMOS inverter circuit.

When a HIGH is applied to the input, as shown in Figure 14–18(a), the p-channel MOSFET Q_1 is off and the n-channel MOSFET Q_2 is on. This condition connects the output to ground through the *on* resistance of Q_2, resulting in a LOW output. When a LOW is applied to the input, as shown in Figure 14–18(b), Q_1 is on and Q_2 is off. This condition connects the output to $+V_{DD}$ (dc supply voltage) through the *on* resistance of Q_1, resulting in a HIGH output.

▶ **FIGURE 14–18**

Operation of a CMOS inverter.

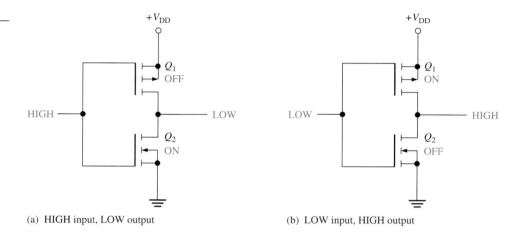

(a) HIGH input, LOW output

(b) LOW input, HIGH output

CMOS NAND Gate

Figure 14–19 shows a CMOS NAND gate with two inputs. Notice the arrangement of the complementary pairs (n-channel and p-channel MOSFETs).

▶ **FIGURE 14–19**

A CMOS NAND gate circuit.

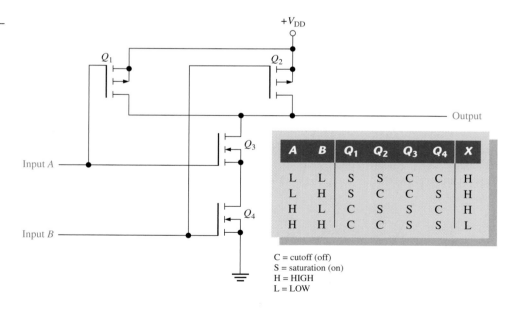

A	B	Q_1	Q_2	Q_3	Q_4	X
L	L	S	S	C	C	H
L	H	S	C	C	S	H
H	L	C	S	S	C	H
H	H	C	C	S	S	L

C = cutoff (off)
S = saturation (on)
H = HIGH
L = LOW

The operation of a CMOS NAND gate is as follows:

■ When both inputs are LOW, Q_1 and Q_2 are on, and Q_3 and Q_4 are off. The output is pulled HIGH through the *on* resistance of Q_1 and Q_2 in parallel.

- When input A is LOW and input B is HIGH, Q_1 and Q_4 are on, and Q_2 and Q_3 are off. The output is pulled HIGH through the low *on* resistance of Q_1.

- When input A is HIGH and input B is LOW, Q_1 and Q_4 are off, and Q_2 and Q_3 are on. The output is pulled HIGH through the low *on* resistance of Q_2.

- Finally, when both inputs are HIGH, Q_1 and Q_2 are off, and Q_3 and Q_4 are on. In this case, the output is pulled LOW through the *on* resistance of Q_3 and Q_4 in series to ground.

CMOS NOR Gate

Figure 14–20 shows a CMOS NOR gate with two inputs. Notice the arrangement of the complementary pairs.

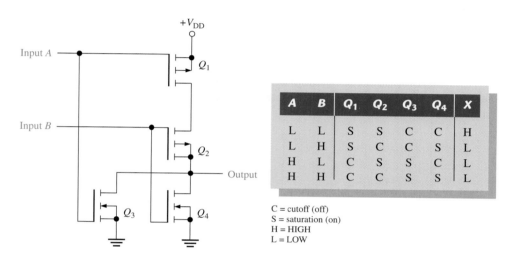

◀ FIGURE 14–20

A CMOS NOR gate circuit.

A	B	Q_1	Q_2	Q_3	Q_4	X
L	L	S	S	C	C	H
L	H	S	C	C	S	L
H	L	C	S	S	C	L
H	H	C	C	S	S	L

C = cutoff (off)
S = saturation (on)
H = HIGH
L = LOW

The operation of a CMOS NOR gate is as follows:

- When both inputs are LOW, Q_1 and Q_2 are on, and Q_3 and Q_4 are off. As a result, the output is pulled HIGH through the *on* resistance of Q_1 and Q_2 in series.

- When input A is LOW and input B is HIGH, Q_1 and Q_4 are on, and Q_2 and Q_3 are off. The output is pulled LOW through the low *on* resistance of Q_4 to ground.

- When input A is HIGH and input B is LOW, Q_1 and Q_4 are off, and Q_2 and Q_3 are on. The output is pulled LOW through the *on* resistance of Q_3 to ground.

- When both inputs are HIGH, Q_1 and Q_2 are off, and Q_3 and Q_4 are on. The output is pulled LOW through the *on* resistance of Q_3 and Q_4 in parallel to ground.

Open-Drain Gates

The term *open-drain* means that the drain terminal of the output transistor is unconnected and must be connected externally to V_{DD} through a load. An open-drain gate is the CMOS counterpart of an open-collector TTL gate (discussed in Section 14–3). An open-drain output circuit is a single *n*-channel MOSFET as shown in Figure 14–21(a). An external **pull-up resistor** must be used, as shown in part (b), to produce a HIGH output state. Also, open-drain outputs can be connected in a wired-AND configuration, a concept that is discussed in the next section in relation to TTL.

Open-drain CMOS gates.

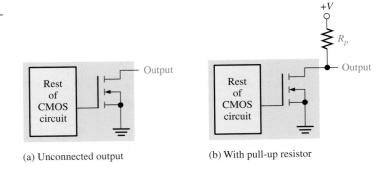

(a) Unconnected output   (b) With pull-up resistor

Tristate CMOS Gates

Tristate outputs are available in both CMOS and TTL logic. The **tristate** output combines the advantages of the totem-pole and open-collector circuits. As you recall, the three output states are HIGH, LOW, and high-impedance (**high-Z**). When selected for normal logic-level operation, as determined by the state of the enable input, a tristate circuit operates in the same way as a regular gate. When a tristate circuit is selected for high-Z operation, the output is effectively disconnected from the rest of the circuit by the internal circuitry. Figure 14–22 illustrates the operation of a tristate circuit. The inverted triangle (∇) designates a tristate output.

The three states of a tristate circuit.

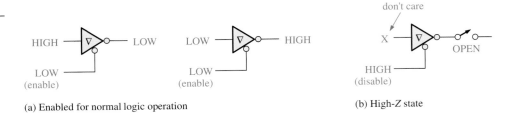

(a) Enabled for normal logic operation   (b) High-Z state

The circuitry in a tristate CMOS gate, as shown in Figure 14–23, allows each of the output transistors Q_1 and Q_2 to be turned off at the same time, thus disconnecting the output from the rest of the circuit.

A tristate CMOS inverter.

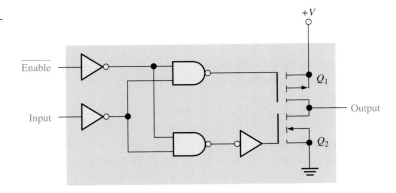

When the enable input is LOW, the device is enabled for normal logic operation. When the enable input is HIGH, both Q_1 and Q_2 are off and the circuit is in the high-Z state.

Precautions for Handling CMOS

As you have learned, all CMOS devices are subject to damage from electrostatic discharge (ESD). Therefore, they must be handled with special care. Review the following precautions:

1. All CMOS devices are shipped in conductive foam to prevent electrostatic charge buildup. When they are removed from the foam, the pins should not be touched.

2. The devices should be placed with pins down on a grounded surface, such as a metal plate, when removed from protective material. Do not place CMOS devices in polystyrene foam or plastic trays.

3. All tools, test equipment, and metal workbenches should be earth-grounded. A person working with CMOS devices should, in certain environments, have his or her wrist grounded with a length of cable and a large-value series resistor. The resistor prevents severe shock should the person come in contact with a voltage source.

4. Do not insert CMOS devices (or any other ICs) into sockets or PC boards with the power on.

5. All unused inputs should be connected to the supply voltage or ground as indicated in Figure 14–24. If left open, an input can acquire electrostatic charge and "float" to unpredicted levels.

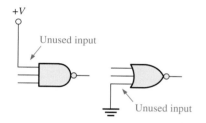

◀ **FIGURE 14–24**

Handling unused CMOS inputs.

6. After assembly on PC boards, protection should be provided by storing or shipping boards with their connectors in conductive foam. The CMOS input and output pins may also be protected with large-value resistors connected to ground.

SECTION 14–2 REVIEW

1. What type of transistor is used in CMOS logic?
2. What is meant by the term *complementary MOS?*
3. Why must CMOS devices be handled with care?

14–3 TTL CIRCUITS

The internal circuit operation of TTL logic gates with totem-pole outputs is covered in this section. Also, the operation of TTL gates with open-collector outputs and the operation of tristate gates are covered.

After completing this section, you should be able to

■ Identify a bipolar junction transistor (BJT) by its symbol ■ Describe the switching action of a BJT ■ Describe the basic operation of a TTL inverter circuit ■ Describe the basic operation of TTL, AND, NAND, OR, and NOR gate circuits ■ Explain what a totem-pole output is ■ Explain the operation and use a TTL gate with an open-collector output ■ Explain the operation of a gate with a tristate output

The Bipolar Junction Transistor

The **bipolar** junction transistor (**BJT**) is the active switching element used in all TTL circuits. Figure 14–25 shows the symbol for an *npn* BJT with its three terminals; **base, emitter,** and **collector.** A BJT has two **junctions,** the base-emitter junction and the base-collector junction.

▶ **FIGURE 14–25**

The symbol for a BJT.

Collector (C)

Base (B)

Emitter (E)

The basic switching operation is as follows: When the base is approximately 0.7 V more positive than the emitter and when sufficient current is provided into the base, the **transistor** turns on and goes into saturation. In saturation, the transistor ideally acts like a closed switch between the collector and the emitter, as illustrated in Figure 14–26(a). When the base is less than 0.7 V more positive than the emitter, the transistor turns off and becomes an open switch between the collector and the emitter, as shown in part (b). To summarize in general terms, a HIGH on the base turns the transistor on and makes it a closed switch. A LOW on the base turns the transistor off and makes it an open switch. In TTL, some BJTs have multiple emitters.

▶ **FIGURE 14–26**

The ideal switching action of the BJT. Conventional current direction is shown. Electron flow notation is opposite.

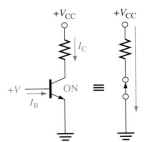

(a) Saturated (ON) transistor and ideal switch equivalent

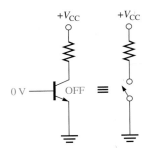

(b) OFF transistor and ideal switch equivalent

TTL Inverter

The logic function of an inverter or any type of gate is always the same, regardless of the type of circuit technology that is used. Figure 14–27 shows a standard TTL circuit for an inverter. In this figure Q_1 is the input coupling transistor, and D_1 is the input clamp **diode.** Transistor Q_2 is called a *phase splitter,* and the combination of Q_3 and Q_4 forms the output circuit often referred to as a **totem-pole** arrangement.

When the input is a HIGH, the base-emitter junction of Q_1 is **reverse biased,** and the base-collector junction is **forward biased.** This condition permits current through R_1 and the base-collector junction of Q_1 into the base of Q_2, thus driving Q_2 into saturation. As a result, Q_3 is turned on by Q_2, and its collector voltage, which is the output, is near ground potential. We therefore have a LOW output for a HIGH input. At the same time, the collector of Q_2 is at a sufficiently low voltage level to keep Q_4 off.

When the input is LOW, the base-emitter junction of Q_1 is forward biased, and the base-collector junction is reverse biased. There is current through R_1 and the base-emitter junc-

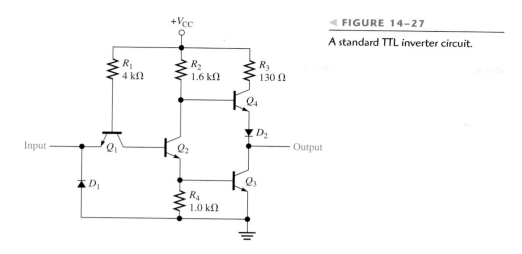

◀ FIGURE 14–27

A standard TTL inverter circuit.

tion of Q_1 to the LOW input. A LOW provides a path to ground for the current. There is no current into the base of Q_2, so it is off. The collector of Q_2 is HIGH, thus turning Q_4 on. A saturated Q_4 provides a low-resistance path from V_{CC} to the output; we therefore have a HIGH on the output for a LOW on the input. At the same time, the emitter of Q_2 is at ground potential, keeping Q_3 off.

Diode D_1 in the TTL circuit prevents negative spikes of voltage on the input from damaging Q_1. Diode D_2 ensures that Q_4 will turn off when Q_2 is on (HIGH input). In this condition, the collector voltage of Q_2 is equal to the base-to-emitter voltage, V_{BE}, of Q_3 plus the collector-to-emitter voltage, V_{CE}, of Q_2. Diode D_2 provides an additional V_{BE} equivalent drop in series with the base-emitter junction of Q_4 to ensure its turn-off when Q_2 is on.

The operation of the TTL inverter for the two input states is illustrated in Figure 14–28. In the circuit in part (a), the base of Q_1 is 2.1 V above ground, so Q_2 and Q_3 are on. In the circuit in part (b), the base of Q_1 is about 0.7 V above ground—not enough to turn Q_2 and Q_3 on.

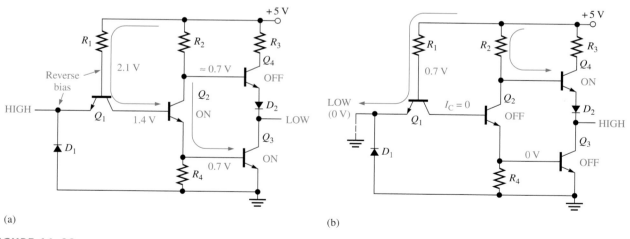

(a)   (b)

▲ FIGURE 14–28

Operation of a TTL inverter.

TTL NAND Gate

A 2-input TTL NAND gate is shown in Figure 14–29. Basically, it is the same as the inverter circuit except for the additional input emitter of Q_1. In TTL technology multiple-emitter

▶ FIGURE 14–29

A TTL NAND gate circuit.

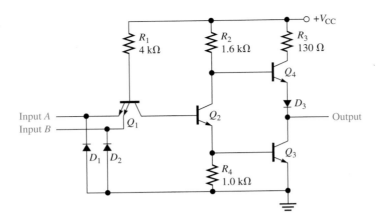

transistors are used for the input devices. These multiple-emitter transistors can be compared to the diode arrangement, as shown in Figure 14–30.

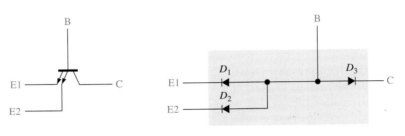

▲ FIGURE 14–30

Diode equivalent of a TTL multiple-emitter transistor.

Perhaps you can understand the operation of this circuit better by visualizing Q_1 in Figure 14–29 replaced by the diode arrangement in Figure 14–30. A LOW on either input A or input B forward-biases the respective diode and reverse-biases D_3 (Q_1 base-collector junction). This action keeps Q_2 off and results in a HIGH output in the same way as described for the TTL inverter. Of course, a LOW on both inputs will do the same thing.

A HIGH on both inputs reverse-biases both input diodes and forward-biases D_3 (Q_1 base-collector junction). This action turns Q_2 on and results in a LOW output in the same way as described for the TTL inverter. You should recognize this operation as that of the NAND function: The output is LOW only if all inputs are HIGH.

Open-Collector Gates

The TTL gates described in the previous sections all had the totem-pole output circuit. Another type of output available in TTL integrated circuits is the **open-collector** output. This is comparable to the open-drain output of CMOS. A standard TTL inverter with an open-collector is shown in Figure 14–31(a). The other types of gates are also available with open-collector outputs.

Notice that the output is the collector of transistor Q_3 with nothing connected to it, hence the name *open collector*. In order to get the proper HIGH and LOW logic levels out of the circuit, an external pull-up resistor must be connected to V_{CC} from the collector of Q_3, as shown in Figure 14–31(b). When Q_3 is off, the output is pulled up to V_{CC} through the external resistor. When Q_3 is on, the output is connected to near-ground through the saturated transistor.

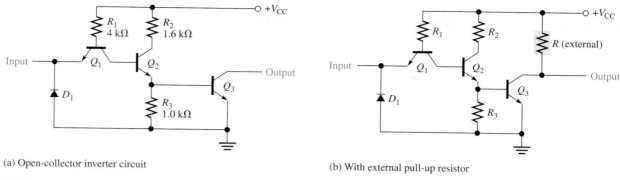

(a) Open-collector inverter circuit

(b) With external pull-up resistor

▲ **FIGURE 14–31**

TTL inverter with open-collector output.

◀ **FIGURE 14–32**

Open-collector symbol in an inverter.

The ANSI/IEEE standard symbol that designates an open-collector output is shown in Figure 14–32 for an inverter and is the same for an open-drain output.

Tristate TTL Gates

Figure 14–33 shows the basic circuit for a TTL tristate inverter. When the enable input is LOW, Q_2 is off, and the output circuit operates as a normal totem-pole configuration, in which the output state depends on the input state. When the enable input is HIGH, Q_2 is on. There is thus a LOW on the second emitter of Q_1, causing Q_3 and Q_5 to turn off, and diode D_1 is forward biased, causing Q_4 also to turn off. When both totem-pole transistors are off, they are effectively open, and the output is completely disconnected from the internal circuitry, as illustrated in Figure 14–34.

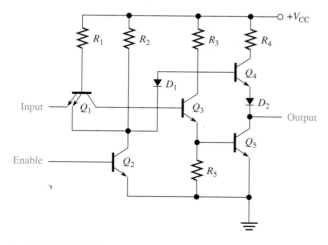

▲ **FIGURE 14–33**

Basic tristate inverter circuit.

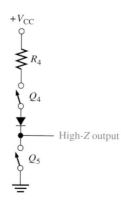

▲ **FIGURE 14–34**

An equivalent circuit for the tristate output in the high-Z state.

Schottky TTL

The basic or standard TTL NAND gate circuit was discussed earlier. It is a current-sinking type of logic that draws current from the load when in the LOW output state and sources negligible current to the load when in the HIGH output state. Most TTL logic used today is some form of Schottky TTL, which provides a faster switching time by incorporating *Schottky* diodes to prevent the transistors from going into saturation, thereby decreasing the time for a transistor to turn on or off. Figure 14–35 shows a Schottky gate circuit. Notice the symbols for the Schottky transistor and Schottky diodes. Schottky devices are designated by an *S* in their part number, such as 74S00. Other types of Schottky TTL are low-power Schottky designated by LS, advanced Schottky designated by AS, advanced low-power Schottky designated by ALS, and fast designated by F.

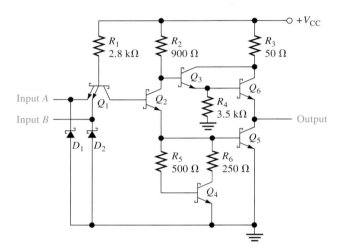

▲ **FIGURE 14–35**

Schottky TTL NAND gate.

**SECTION 14–3
REVIEW**

1. An *npn* BJT is on when the base is more negative than the emitter.   (T or F)
2. In terms of switching action, what do the *on* and *off* states of a BJT represent?
3. What are the two major types of output circuits in TTL?
4. Explain how tristate logic differs from normal, two-state logic.

14–4 PRACTICAL CONSIDERATIONS IN THE USE OF TTL

Although CMOS is the more predominant IC technology in industry and commercial applications, TTL is still used. In educational applications, TTL is usually preferred because it does not have the handling restrictions that CMOS does due to ESD. Because of this, several practical considerations in the use and application of TTL circuits will be covered using standard TTL for illustration.

After completing this section, you should be able to

■ Describe current sinking and current sourcing ■ Use an open-collector circuit for wired-AND operation ■ Describe the effects of connecting two or more totem-pole outputs ■ Use open-collector gates to drive LEDs and lamps ■ Explain what to do with unused TTL inputs

Current Sinking and Current Sourcing

The concepts of current sinking and current sourcing were introduced in Section 14–1. Now that you are familiar with the totem-pole-output circuit configuration used in TTL, let's look closer at the sinking and sourcing action.

Figure 14–36 shows a standard TTL inverter with a totem-pole output connected to the input of another TTL inverter. When the driving gate is in the HIGH output state, the driver

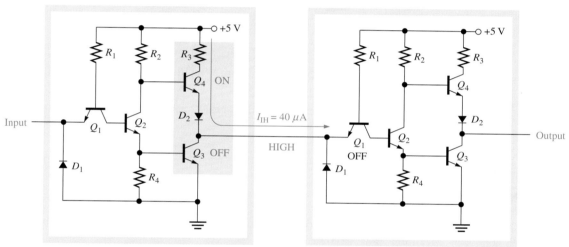

(a) Current sourcing (I_{IH} value is maximum)

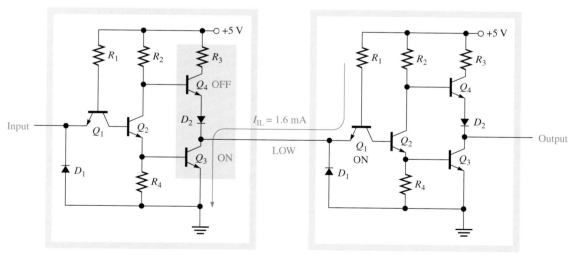

(b) Current sinking (I_{IL} value is maximum)

▲ FIGURE 14–36

Current sinking and sourcing action in TTL.

is sourcing current to the load, as shown in Figure 14–36(a). The input to the load gate is like a reverse-biased diode, so there is practically no current required by the load. Actually, since the input is nonideal, there is a maximum of 40 μA from the totem-pole output of the driver into the load gate input.

When the driving gate is in the LOW output state, the driver is sinking current from the load, as shown in Figure 14–36(b). This current is 1.6 mA maximum for standard TTL and is indicated on a **data sheet** with a negative value because it is *out* of the input.

EXAMPLE 14–3

When a TTL NAND gate drives five TTL inputs, how much current does the driver output source, and how much does it sink? (Refer to Figure 14–36.)

Solution Total source current (in HIGH output state):

$$I_{IH(max)} = 40 \ \mu\text{A per input}$$

$$I_{T(source)} = (5 \text{ inputs})(40 \ \mu\text{A/input}) = 5(40 \ \mu\text{A}) = \textbf{200} \ \boldsymbol{\mu}\textbf{A}$$

Total sink current (in LOW output state):

$$I_{IL(max)} = -1.6 \ \text{mA per input}$$

$$I_{T(sink)} = (5 \text{ inputs})(-1.6 \ \text{mA/input}) = 5(-1.6 \ \text{mA}) = \textbf{-8.0 mA}$$

Related Problem Repeat the calculations for an LS TTL NAND gate. Refer to a data sheet on the Texas Instruments CD-ROM.

EXAMPLE 14–4

Refer to the data sheet on the Texas Instruments CD-ROM, and determine the fan-out of the 7400 NAND gate.

Solution According to the data sheet, the current parameters are as follows:

$$I_{IH(max)} = 40 \ \mu\text{A} \qquad I_{OH(max)} = -400 \ \mu\text{A}$$
$$I_{IL(max)} = -1.6 \ \text{mA} \qquad I_{OL(max)} = 16 \ \text{mA}$$

Fan-out for the HIGH output state is calculated as follows: Current $I_{OH(max)}$ is the maximum current that the gate can source to a load. Each load input requires an $I_{IH(max)}$ of 40 μA. The HIGH-state fan-out is

$$\left| \frac{I_{OH(max)}}{I_{IH(max)}} \right| = \frac{400 \ \mu\text{A}}{40 \ \mu\text{A}} = 10$$

For the LOW output state, fan-out is calculated as follows: $I_{OL(max)}$ is the maximum current that the gate can sink. Each load input produces an $I_{IL(max)}$ of -1.6 mA. The LOW-state fan-out is

$$\left| \frac{I_{OL(max)}}{I_{IL(max)}} \right| = \frac{16 \ \text{mA}}{1.6 \ \text{mA}} = 10$$

In this case both the HIGH-state fan-out and the LOW-state fan-out are the same.

Related Problem Determine the fan-out for a 74LS00 NAND gate.

Using Open-Collector Gates for Wired-AND Operation

The outputs of open-collector gates can be wired together to form what is called a *wired-AND* configuration. Figure 14–37 illustrates how four inverters are connected to produce a 4-input negative-AND gate. A single external pull-up resistor, R_p, is required in all wired-AND circuits.

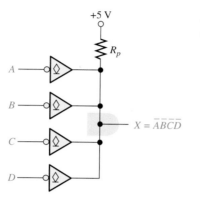

A wired-AND configuration of four inverters.

$$X = \overline{A}\,\overline{B}\,\overline{C}\,\overline{D}$$

When one (or more) of the inverter inputs is HIGH, the output X is pulled LOW because an output transistor is on and acts as a closed switch to ground, as illustrated in Figure 14–38(a). In this case only one inverter has a HIGH input, but this is sufficient to pull the output LOW through the saturated output transistor Q_1 as indicated.

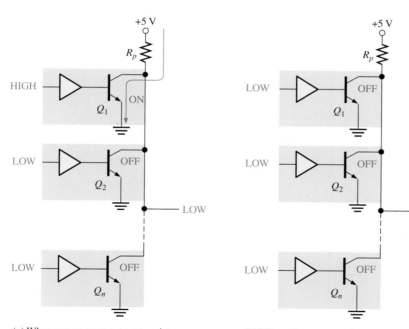

(a) When one or more output transistors are on, the output is LOW.

(b) When all output transistors are off, the output is HIGH.

Open-collector wired negative-AND operation with inverters.

For the output X to be HIGH, *all* inverter inputs must be LOW so that all the open-collector output transistors are off, as indicated in Figure 14–38(b). When this condition exists, the output X is pulled HIGH through the pull-up resistor. Thus, the output X is HIGH only when *all* the inputs are LOW. Therefore, we have a negative-AND function, as expressed in the following equation:

$$X = \overline{A}\,\overline{B}\,\overline{C}\,\overline{D}$$

EXAMPLE 14–5

Write the output expression for the wired-AND configuration of open-collector AND gates in Figure 14–39.

▶ **FIGURE 14–39**

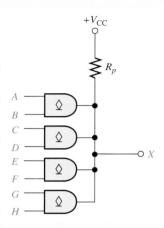

Solution The output expression is

$$X = ABCDEFGH$$

The wired-AND connection of the four 2-input AND gates creates an 8-input AND gate.

Related Problem Determine the output expression if NAND gates are used in Figure 14–39.

EXAMPLE 14–6

Three open-collector AND gates are connected in a wired-AND configuration as shown in Figure 14–40. Assume that the wired-AND circuit is driving four standard TTL inputs (-1.6 mA each).

(a) Write the logic expression for X.

(b) Determine the minimum value of R_p if $I_{OL(max)}$ for each gate is 30 mA and $V_{OL(max)}$ is 0.4 V.

▶ **FIGURE 14–40**

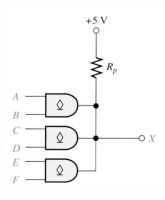

Solution **(a)** $X = ABCDEF$

(b) $4(1.6 \text{ mA}) = 6.4 \text{ mA}$

$$I_{R_P} = I_{\text{OL(max)}} - 6.4 \text{ mA} = 30 \text{ mA} - 6.4 \text{ mA} = 23.6 \text{ mA}$$

$$R_P = \frac{V_{\text{CC}} - V_{\text{OL(max)}}}{I_{R_P}} = \frac{5 \text{ V} - 0.4 \text{ V}}{23.6 \text{ mA}} = \textbf{195 } \boldsymbol{\Omega}$$

Related Problem Show the wired-AND circuit for a 10-input AND function using 74LS09 quad 2-input AND gates.

Connection of Totem-Pole Outputs

Totem-pole outputs cannot be connected together because such a connection might produce excessive current and result in damage to the devices. For example, in Figure 14–41, when Q_1 in device A and Q_2 in device B are both on, the output of device A is effectively shorted to ground through Q_2 of device B.

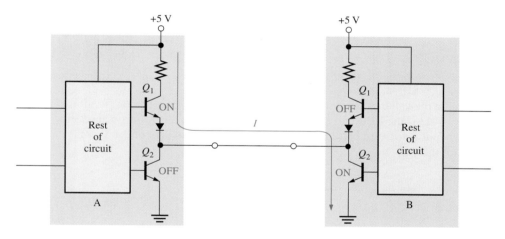

◄ **FIGURE 14–41**

Totem-pole outputs wired together. Such a connection may cause excessive current through Q_1 of device A and Q_2 of device B and should never be used.

Open-Collector Buffer/Drivers

A TTL circuit with a totem-pole output is limited in the amount of current that it can sink in the LOW state ($I_{\text{OL(max)}}$) to 16 mA for standard TTL and 8 mA for LS TTL. In many special applications, a gate must drive external devices, such as LEDs, lamps, or relays, that may require more current than that.

Because of their higher voltage and current-handling capability, circuits with open-collector outputs are generally used for driving LEDs, lamps, or relays. However, totem-pole outputs can be used, as long as the output current required by the external device does not exceed the amount that the TTL driver can sink.

With an open-collector TTL gate, the collector of the output transistor is connected to an **LED** or incandescent lamp, as illustrated in Figure 14–42. In part (a) the limiting resistor, R_L, is used to keep the current below maximum LED current. When the output of the gate is LOW, the output transistor is sinking current, and the LED is on. The LED is off when the output transistor is off and the output is HIGH. A typical open-collector buffer gate can sink up to 40 mA. In part (b) of the figure, the lamp requires no limiting resistor because the filament is resistive. Typically, up to +30 V can be used on the open collector, depending on the particular logic family.

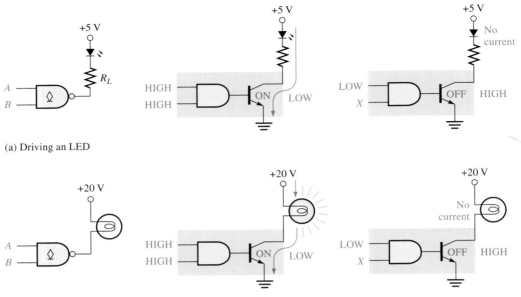

(a) Driving an LED

(b) Driving a low-current lamp

▲ **FIGURE 14–42**

Some applications of open–collector drivers.

EXAMPLE 14–7

Determine the value of the limiting resistor, R_L, in the open-collector circuit of Figure 14–43 if the LED current is to be 20 mA. Assume a 1.5 V drop across the LED when it is forward biased and a LOW-state output voltage of 0.1 V at the output of the gate.

▶ **FIGURE 14–43**

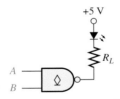

Solution

$$V_{R_L} = 5\ \text{V} - 1.5\ \text{V} - 0.1\ \text{V} = 3.4\ \text{V}$$

$$R_L = \frac{V_{R_L}}{I} = \frac{3.4\ \text{V}}{20\ \text{mA}} = \mathbf{170\ \Omega}$$

Related Problem   Determine the value of the limiting resistor, R_L, if the LED requires 35 mA.

Unused TTL Inputs

An unconnected input on a TTL gate acts as a HIGH because an open input results in a reverse-biased emitter junction on the input transistor, just as a HIGH level does. This effect is illustrated in Figure 14–44. However, because of noise sensitivity, it is best not to leave unused TTL inputs unconnected (open). There are several alternative ways to handle unused inputs.

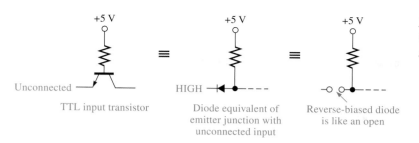

Comparison of an open TTL input and a HIGH-level input.

Tied-Together Inputs The most common method for handling unused gate inputs is to connect them to a used input of the same gate. For AND gates and NAND gates, all tied-together inputs count as one unit load in the LOW state; but for OR gates and NOR gates, each input tied to another input counts as a separate unit load in the LOW state. In the HIGH state, each tied-together input counts as a separate load for all types of TTL gates. In Figure 14-45(a) are two examples of the connection of two unused inputs to a used input.

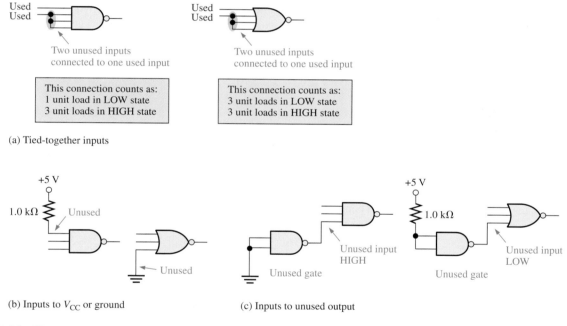

(a) Tied-together inputs

(b) Inputs to V_{CC} or ground

(c) Inputs to unused output

▲ **FIGURE 14-45**

Methods for handling unused TTL inputs.

The AND and NAND gates present only a single unit load no matter how many inputs are tied together, whereas OR and NOR gates present a unit load for each tied-together input. This is because the NAND gate uses a multiple-emitter input transistor; so no matter how many inputs are LOW, the total LOW-state current is limited to a fixed value. The NOR gate uses a separate transistor for each input; therefore, the LOW-state current is the sum of the currents from all the tied-together inputs.

Inputs to V_{CC} or Ground Unused inputs of AND and NAND gates can be connected to V_{CC} through a 1.0 kΩ resistor. This connection pulls the unused inputs to a HIGH level. Unused inputs of OR and NOR gates can be connected to ground. These methods are illustrated in Figure 14-45(b).

Inputs to Unused Output   A third method of terminating unused inputs may be appropriate in some cases when an unused gate or inverter is available. The unused gate output must be a constant HIGH for unused AND and NAND inputs and a constant LOW for unused OR and NOR inputs, as illustrated in Figure 14–45(c).

SECTION 14–4
REVIEW

1. In what output state does a TTL circuit sink current from a load?
2. Why does a TTL circuit source less current into a TTL load than it sinks?
3. Why can TTL circuits with totem-pole outputs not be connected together?
4. What type of TTL circuit must be used for a wired-AND configuration?
5. Why type of TTL circuit would you use to drive a lamp?
6. An unconnected TTL input acts as a LOW.   (T or F)

14–5   COMPARISON OF CMOS AND TTL PERFORMANCE

In this section, the main operational and performance characteristics of selected CMOS series are compared with those of the major TTL series and with BiCMOS.

After completing this section, you should be able to

■ Compare TTL (bipolar), BiMOS, and CMOS devices in terms of propagation delay, maximum clock frequency, power dissipation, and drive capability

In the past, the superior characteristic of TTL (bipolar) compared to CMOS was its relatively high speed and output current capability. Today, these advantages of TTL have diminished to the point where CMOS is often equal or superior in many areas and has become the dominant IC technology, although TTL is still available and in use, as you know. One family of IC logic devices, BiCMOS, combines CMOS logic with TTL output circuitry in an effort to combine the advantages of both.

Table 14–1 provides a comparison of the performance of several IC logic families.

▼ TABLE 14–1

Comparison of selected performance parameters of several 74XX IC families.

| | BIPOLAR (TTL) | | | BiCMOS | CMOS | | | | | |
| | | | | | 5 V | | | 3.3 V | | |
	F	LS	ALS	ABT	HC	AC	AHC	LV	LVC	ALVC
Speed										
Gate propagation delay, t_p (ns)	3.3	10	7	3.2	7	5	3.7	9	4.3	3
FF maximum clock freq. (MHz)	145	33	45	150	50	160	170	90	100	150
Power Dissipation Per Gate										
Bipolar: 50% dc (mW)	6	2.2	1.4							
CMOS: quiescent (μW)				17	2.75	0.55	2.75	1.6	0.8	0.8
Output Drive										
I_{OL} (mA)	20	8	8	64	4	24	8	12	24	24

1. What is a BiCMOS circuit?
2. In general, what is the main advantage of CMOS over bipolar (TTL)?

14–6 EMITTER-COUPLED LOGIC (ECL) CIRCUITS

Emitter-coupled logic, like TTL, is a bipolar technology. The typical ECL circuit consists of a different amplifier input circuit, a bias circuit, and emitter-follower outputs. ECL is much faster than TTL because the transistors do not operate in saturation and is used in more specialized high-speed applications.

After completing this section, you should be able to

■ Describe how ECL differs from TTL and CMOS ■ Explain the advantages and disadvantages of ECL

An **ECL** OR/NOR gate is shown in Figure 14–46(a). The emitter-follower outputs provide the OR logic function and its NOR complement, as indicated by Figure 14–46(b).

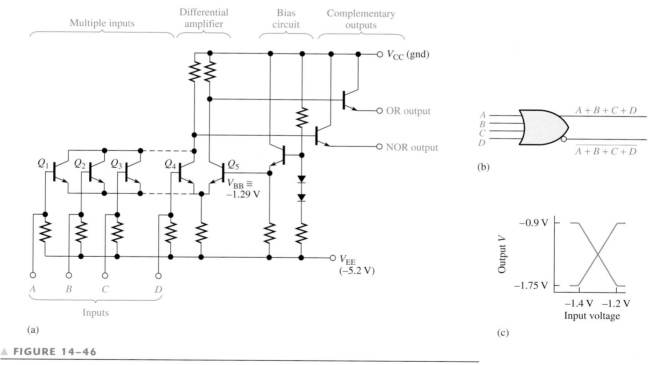

▲ **FIGURE 14–46**

An ECL OR/NOR gate circuit.

Because of the low output impedance of the emitter-follower and the high input impedance of the differential amplifier input, high fan-out operation is possible. In this type of circuit, saturation is not possible. The lack of saturation results in higher power consumption and limited voltage swing (less than 1 V), but it permits high-frequency switching.

The V_{CC} pin is normally connected to ground, and the V_{EE} pin is connected to -5.2 V from the power supply for best operation. Notice that in Figure 14–46(c) the output varies from a LOW level of -1.75 V to a HIGH level of -0.9 V with respect to ground. In positive logic a 1 is the HIGH level (less negative), and a 0 is the LOW level (more negative).

Noise Margin

As you have learned, the noise margin of a gate is the measure of its immunity to undesired voltage fluctuations (noise). Typical ECL circuits have noise margins from about 0.2 V to 0.25 V. These are less than for TTL and make ECL less suitable in high-noise environments.

Comparison of ECL with TTL and CMOS

Table 14–2 shows a comparison of key performance parameters for F, AHC, and ECL.

▶ **TABLE 14–2**

Comparison of ECL series performance parameters with F and AHC.

	BIPOLAR (TTL) F	CMOS AHC	BIPOLAR (ECL)
Speed			
Gate propagation delay, t_p (ns)	3.3	3.7	0.22–1
FF maximum clock freq. (MHz)	145	170	330–2800
Power Dissipation Per Gate			
Bipolar: 50% dc	8.9 mW		25 mW–73 mW
CMOS: quiescent		2.5 μW	

SECTION 14–6 REVIEW

1. What is the primary advantage of ECL over TTL?
2. Name two disadvantages of ECL compared with TTL.

14–7   PMOS, NMOS, AND E²CMOS

The PMOS and NMOS circuits are used largely in LSI functions, such as long shift registers, large memories, and microprocessor products. Such use is a result of the low power consumption and very small chip area required for MOS transistors. E²CMOS is used in reprogrammable PLDs.

After completing this section, you should be able to

■ Describe a basic PMOS gate   ■ Describe a basic NMOS gate   ■ Describe a basic E²CMOS cell

PMOS

One of the first high-density **MOS** circuit technologies to be produced was **PMOS.** It utilizes enhancement-mode p-channel MOS transistors to form the basic gate building blocks. Figure 14–47 shows a basic PMOS gate that produces the NOR function in positive logic.

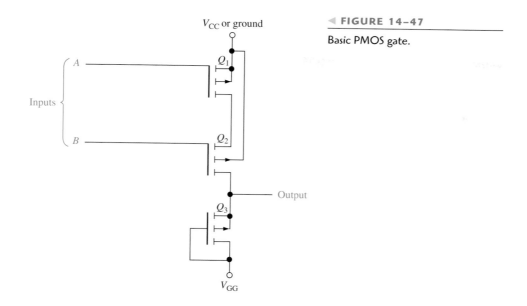

◀ **FIGURE 14–47**

Basic PMOS gate.

The operation of the PMOS gate is as follows: The supply voltage V_{GG} is a negative voltage, and V_{CC} is a positive voltage or ground (0 V). Transistor Q_3 is permanently biased to create a constant drain-to-source resistance. Its sole purpose is to function as a current-limiting resistor. If a HIGH (V_{CC}) is applied to input A or B, then Q_1 or Q_2 is off, and the output is pulled down to a voltage near V_{GG}, which represents a LOW. When a LOW voltage (V_{GG}) is applied to both input A and input B, both Q_1 and Q_2 are turned on. This causes the output to go to a HIGH level (near V_{CC}). Since a LOW output occurs when either or both inputs are HIGH, and a HIGH output occurs only when all inputs are LOW, we have a NOR gate.

NMOS

The NMOS devices were developed as processing technology improved. The n-channel MOS transistor is used in NMOS circuits, as shown in Figure 14–48 for a NAND gate and a NOR gate.

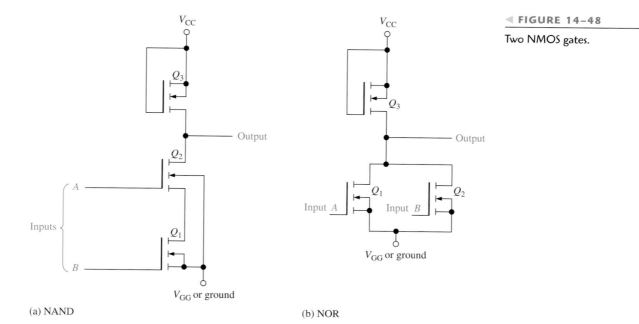

(a) NAND

(b) NOR

◀ **FIGURE 14–48**

Two NMOS gates.

In Figure 14–48(a), Q_3 acts as a resistor to limit current. When a LOW (V_{GG} or ground) is applied to one or both inputs, then at least one of the transistors (Q_1 or Q_2) is off, and the output is pulled up to a HIGH level near V_{CC}. When HIGHs (V_{CC}) are applied to both A and B, both Q_1 and Q_2 conduct, and the output is LOW. This action, of course, identifies this circuit as a NAND gate.

In Figure 14–48(b), Q_3 again acts as a resistor. A HIGH on either input turns Q_1 or Q_2 on, pulling the output LOW. When both inputs are LOW, both transistors are off, and the output is pulled up to a HIGH level.

E²CMOS

E²CMOS (electrically erasable CMOS) technology is based on a combination of CMOS and NMOS technologies and is used in programmable devices such as PROMs and CPLDs. An E²CMOS cell is built around a MOS transistor with a floating gate that is externally charged or discharged by a small programming current. A schematic of this type of cell is shown in Figure 14–49.

▶ **FIGURE 14–49**

An E²CMOS cell.

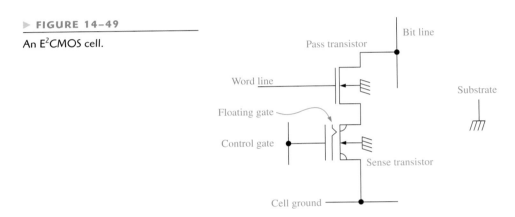

When the floating gate is charged to a positive potential by removing electrons, the sense transistor is turned on, storing a binary zero. When the floating gate is charged to a negative potential by placing electrons on it, the sense transistor is turned off, storing a binary 1. The control gate controls the potential of the floating gate. The pass transistor isolates the sense transistor from the array during read and write operations that use the word and bit lines.

The cell is programmed by applying a programming pulse to either the control gate or the bit line of a cell that has been selected by a voltage on the word line. During the programming cycle, the cell is first erased by applying a voltage to the control gate to make the floating gate negative. This leaves the sense transistor in the *off* state (storing a 1). A write pulse is applied to the bit line of a cell in which a 0 is to be stored. This will charge the floating gate to a point where the sense transistor is on (storing a 0). The bit stored in the cell is read by sensing presence or absence of a small cell current in the bit line. When a 1 is stored, there is no cell current because the sense transistor is off. When a 0 is stored, there is a small cell current because the sense transistor is on. Once a bit is stored in a cell, it will remain indefinitely unless the cell is erased or a new bit is written into the cell.

SECTION 14–7 REVIEW

1. What is the main feature of NMOS and PMOS technology in integrated circuits?
2. What is the mechanism for charge storage in an E²CMOS cell?

SUMMARY

- *Formulas:*

  14–1 $V_{NH} = V_{OH(min)} - V_{IH(min)}$ High-level noise margin

  14–2 $V_{NL} = V_{IL(max)} - V_{OL(max)}$ Low-level noise margin

  14–3 $I_{CC} = \dfrac{I_{CCH} + I_{CCL}}{2}$ Average dc supply current

  14–4 $P_D = V_{CC}I_{CC}$ Power dissipation

- Totem-pole outputs of TTL cannot be connected together.
- Open-collector and open-drain outputs can be connected for wired-AND.
- CMOS devices offer lower power dissipation than any of the TTL series.
- A TTL device is not as vulnerable to electrostatic discharge (ESD) as is a CMOS device.
- Because of ESD, CMOS devices must be handled with great care.
- ECL is the fastest type of logic circuit.
- E²CMOS is used in PROMs and other PLDs.

KEY TERMS

Key terms and other bold terms in the chapter are defined in the end-of-book glossary.

CMOS Complementary metal-oxide semiconductor; a type of integrated logic circuit that uses *n*- and *p*-channel MOSFETs (metal-oxide semiconductor field-effect transistors).

Current sinking The action of a logic circuit in which it accepts current into its output from a load.

Current sourcing The action of a logic circuit in which it sends current from its output to a load.

ECL Emitter-coupled logic; a class of integrated logic circuits that are implemented with nonsaturating bipolar junction transistors.

E²CMOS Electrically erasable CMOS; the IC technology used in programmable logic devices (PLDs).

Fanout The number of equivalent gate inputs of the same family series that a logic gate can drive.

Noise immunity The ability of a logic circuit to reject unwanted signals (noise).

Noise margin The difference between the maximum LOW output of a gate and the maximum acceptable LOW input of an equivalent gate; also, the difference between the minimum HIGH output of a gate and the minimum HIGH input of an equivalent gate. Noise margin is sometimes expressed as a percentage of the dc supply voltage.

Open-collector A type of output for a TTL circuit in which the collector of the output transistor is left internally disconnected and is available for connection to an external load that requires relatively high current or voltage.

Power dissipation The product of the dc supply voltage and the dc supply current in an electronic circuit.

Propagation delay time The time interval between the occurrence of an input transition and the occurrence of the corresponding output transition in a logic circuit.

Pull-up resistor A resistor with one end connected to the dc supply voltage used to keep a given point in a logic circuit HIGH when in the inactive state.

Totem pole A type of output in TTL circuits.

Tristate A type of output in logic circuits that exhibits three states: HIGH, LOW, and high *Z*.

TTL Transistor-transistor logic; a type of integrated circuit that uses bipolar junction transistors.

Unit load A measure of fan-out. One gate input represents a unit load to a driving gate.

1. When the frequency of the input signal to a CMOS gate is increased, the average power dissipation
   (a) decreases   (b) increases
   (c) does not change   (d) decreases exponentially

2. CMOS operates more reliably than TTL in a high-noise environment because of its
   (a) lower noise margin   (b) input capacitance
   (c) higher noise margin   (d) smaller power dissipation

3. Proper handling of a CMOS device is necessary because of its
   (a) fragile construction   (b) high-noise immunity
   (c) susceptibility to electrostatic discharge   (d) low power dissipation

4. Which of the following is not a TTL circuit?
   (a) 74F00   (b) 74AS00   (c) 74HC00   (d) 74ALS00

5. An open TTL NOR gate input
   (a) acts as a LOW   (b) acts as a HIGH
   (c) should be grounded   (d) should be connected to V_{CC} through a resistor
   (e) answers (b) and (c) (f) answers (a) and (c)

6. An LS TTL gate can drive a maximum of
   (a) 20 unit loads   (b) 10 unit loads
   (c) 40 unit loads   (d) unlimited unit loads

7. If two unused inputs of a LS TTL gate are connected to an input being driven by another LS TTL gate, the total number of remaining unit loads that can be driven by this gate is
   (a) seven   (b) eight   (c) seventeen   (d) unlimited

8. The main advantage of ECL over TTL or CMOS is
   (a) ECL is less expensive   (b) ECL consumes less power
   (c) ECL is available in a greater variety of circuit types   (d) ECL is faster

9. ECL cannot be used in
   (a) high-noise environments   (b) damp environments   (c) high-frequency applications

10. The basic mechanism for storing a data bit in an E^2CMOS cell is
    (a) control gate   (b) floating drain
    (c) floating gate   (d) cell current

SECTION 14–1   **Basic Operational Characteristics and Parameters**

1. A certain logic gate has a $V_{OH(min)} = 2.2$ V, and it is driving a gate with a $V_{IH(min)} = 2.5$ V. Are these gates compatible for HIGH-state operation? Why?

2. A certain logic gate has a $V_{OL(max)} = 0.45$ V, and it is driving a gate with a $V_{IL(max)} = 0.75$ V. Are these gates compatible for LOW-state operation? Why?

3. A TTL gate has the following actual voltage level values: $V_{IH(min)} = 2.25$ V, $V_{IL(max)} = 0.65$ V. Assuming it is being driven by a gate with $V_{OH(min)} = 2.4$ V and $V_{OL(max)} = 0.4$ V, what are the HIGH- and LOW-level noise margins?

4. What is the maximum amplitude of noise spikes that can be tolerated on the inputs in both the HIGH state and the LOW state for the gate in Problem 3?

5. Voltage specifications for three types of logic gates are given in Table 14–3. Select the gate that you would use in a high-noise industrial environment.

▶ TABLE 14–3

	$V_{OH(MIN)}$	$V_{OL(MAX)}$	$V_{IH(MIN)}$	$V_{IL(MAX)}$
Gate A	2.4 V	0.4 V	2 V	0.8 V
Gate B	3.5 V	0.2 V	2.5 V	0.6 V
Gate C	4.2 V	0.2 V	3.2 V	0.8 V

6. A certain gate draws a dc supply current from a +5 V source of 2 mA in the LOW state and 3.5 mA in the HIGH state. What is the power dissipation in the LOW state? What is the power dissipation in the HIGH state? Assuming a 50% duty cycle, what is the average power dissipation?

7. Each gate in the circuit of Figure 14–50 has a t_{PLH} and a t_{PHL} of 4 ns. If a positive-going pulse is applied to the input as indicated, how long will it take the output pulse to appear?

▶ FIGURE 14–50

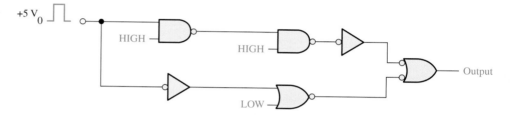

8. For a certain gate, t_{PLH} = 3 ns and t_{PHL} = 2 ns. What is the average propagation delay time?

9. Table 14–4 lists parameters for three types of gates. Basing your decision on the speed-power product, which one would you select for best performance?

▶ TABLE 14–4

	t_{PLH}	t_{PHL}	P_D
Gate A	1 ns	1.2 ns	15 mW
Gate B	5 ns	4 ns	8 mW
Gate C	10 ns	10 ns	0.5 mW

10. Which gate in Table 14–4 would you select if you wanted the gate to operate at the highest possible frequency?

11. A standard TTL gate has a fan-out of 10. Are any of the gates in Figure 14–51 overloaded? If so, which ones?

▶ FIGURE 14–51

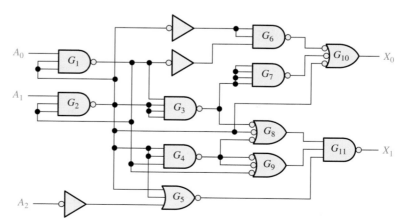

12. Which CMOS gate network in Figure 14–52 can operate at the highest frequency?

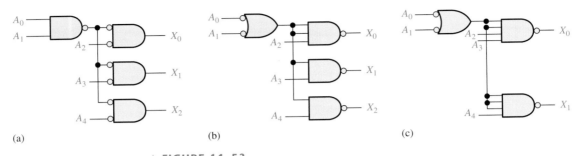

(a)   (b)   (c)

▲ **FIGURE 14–52**

SECTION 14–2   CMOS Circuits

13. Determine the state (on or off) of each MOSFET in Figure 14–53.

▶ **FIGURE 14–53**

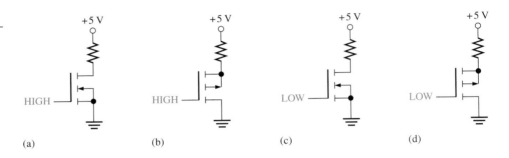

(a)   (b)   (c)   (d)

14. The CMOS gate network in Figure 14–54 is incomplete. Indicate the changes that should be made.

▶ **FIGURE 14–54**

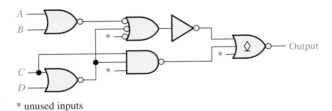

* unused inputs

15. Devise a circuit, using appropriate CMOS logic gates and/or inverters, with which signals from four different sources can be connected to a common line at different times without interfering with each other.

SECTION 14–3   TTL Circuits

16. Determine which BJTs in Figure 14–55 are off and which are on.

▶ **FIGURE 14–55**

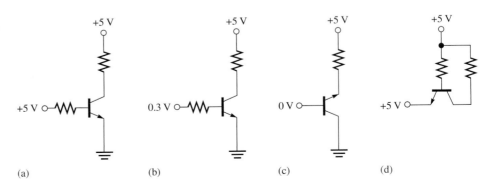

(a)   (b)   (c)   (d)

17. Determine the output state of each TTL gate in Figure 14–56.

▶ **FIGURE 14–56**

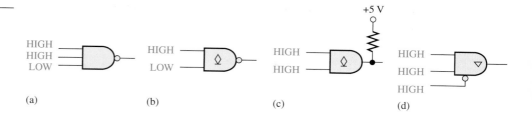

(a)　　　　　　　(b)　　　　　　　(c)　　　　　　　(d)

18. The TTL gate network in Figure 14–57 is incomplete. Indicate the changes that should be made.

▶ **FIGURE 14–57**

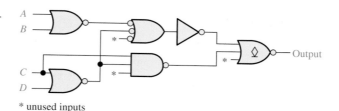

* unused inputs

SECTION 14–4 **Practical Considerations in the Use of TTL**

19. Determine the output level of each TTL gate in Figure 14–58.

▶ **FIGURE 14–58**

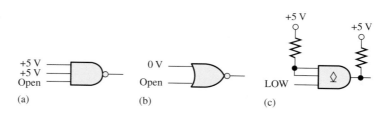

(a)　　　　　　　(b)　　　　　　　(c)

20. For each part of Figure 14–59, tell whether each driving gate is sourcing or sinking current. Specify the maximum current out of or into the output of the driving gate or gates in each case. All gates are standard TTL.

▶ **FIGURE 14–59**

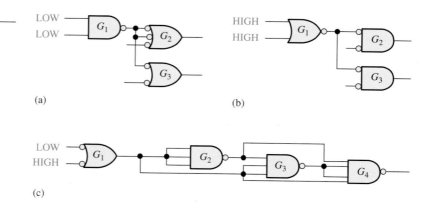

(a)　　　　　　　　　　　　　(b)

(c)

21. Use open-collector inverters to implement the following logic expressions:

　　(a) $X = \overline{A}\,\overline{B}\,\overline{C}$　　　**(b)** $X = A\overline{B}C\overline{D}$　　　**(c)** $X = ABC\overline{D}\,\overline{E}\,\overline{F}$

22. Write the logic expression for each of the circuits in Figure 14–60.

▶ **FIGURE 14–60**

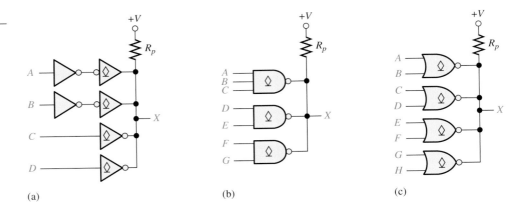

(a) (b) (c)

23. Determine the minimum value for the pull-up resistor in each circuit in Figure 14–60 if $I_{OL(max)} = 40$ mA and $V_{OL(max)} = 0.25$ V for each gate. Assume that 10 standard TTL unit loads are being driven from output X and the supply voltage is 5 V.

24. A certain relay requires 60 mA. Devise a way to use open-collector NAND gates with $I_{OL(max)} = 40$ mA to drive the relay.

SECTION 14–5   Comparison of CMOS and TTL Performance

25. Select the IC family with the best speed-power product in Table 14–1.

26. Determine from Table 14–1 the logic family that is most appropriate for each of the following requirements:

(a) shortest propagation delay time

(b) fastest flip-flop toggle rate

(c) lowest power dissipation

(d) best compromise between speed and power for a logic gate

27. Determine the total propagation delay from each input to each output for each circuit in Figure 14–61.

▶ **FIGURE 14–61**

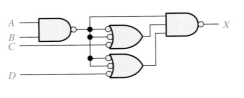

(a) 74FXX gates

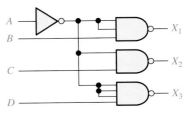

(b) 74HCXX gates

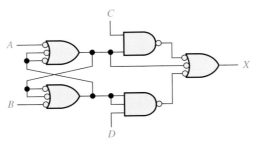

(c) 74AHCXX gates

28. One of the flip-flops in Figure 14–62 may have an erratic output. Which one is it if any and why?

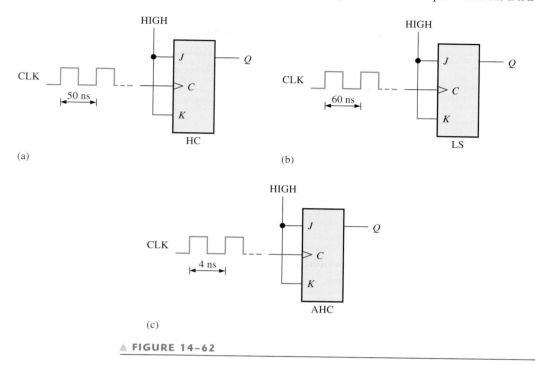

(a)

(b)

(c)

▲ FIGURE 14–62

SECTION 14–6   **Emitter-Coupled Logic (ECL) Circuits**

29. What is the basic difference between ECL circuitry and TTL circuitry?

30. Select ECL, HCMOS, or the appropriate TTL series for each of the following requirements:

(a) highest speed

(b) lowest power

(c) best compromise between high speed and low power (speed-power product)

ANSWERS

SECTION REVIEWS

SECTION 14–1   **Basic Operational Characteristics and Parameters**

1. V_{IH}: HIGH level input voltage: V_{IL}: LOW level input voltage; V_{OH}: HIGH level output voltage; V_{OL}: LOW level output voltage

2. A higher value of noise margin is better.

3. Gate B can operate at a higher frequency.

4. Excessive loading reduces the noise margin of a gate.

SECTION 14–2   **CMOS Circuits**

1. MOSFETs are used in CMOS logic.

2. A complementary output circuit consists of an n-channel and a p-channel MOSFET.

3. Because electrostatic discharge can damage CMOS devices

SECTION 14–3   **TTL Circuits**

1. False, the *npn* BJT is off.

2. The *on* state of a BJT is a closed switch; the *off* state is an open switch.

3. Totem-pole and open-collector are types of TTL outputs.

4. Tristate logic provides a high-impedance state, in which the output is disconnected from the rest of the circuit.

SECTION 14–4 **Practical Considerations in the Use of TTL**

1. Sink current occurs in a LOW output state.

2. Source current is less than sink current because a TTL load looks like a reverse-biased diode in the HIGH state.

3. The totem-pole transistors cannot handle the current when one output tries to go HIGH and the other is LOW.

4. Wired-AND must use open-collector.

5. Lamp driver must be open-collector.

6. False, an unconnected TTL input generally acts as a HIGH.

SECTION 14–5 **Comparison of CMOS and TTL Performance**

1. BiCMOS uses bipolar transistors for input and output circuitry and CMOS in between.

2. CMOS has lower power dissipation than bipolar.

SECTION 14–6 **Emitter-Coupled Logic (ECL) Circuits**

1. ECL is faster than TTL.

2. ECL has more power and less noise margin than TTL.

SECTION 14–7 **PMOS, NMOS, and E²CMOS**

1. NMOS and PMOS are high density.

2. The floating gate is the mechanism for storing charge in an E²CMOS cell.

RELATED PROBLEMS FOR EXAMPLES

14–1 CMOS **14–2** 10.75 μW

14–3 $I_{T(source)} = 5(20\ \mu A) = 100\ \mu A$

$I_{T(sink)} = 5(-0.4\ mA) = -2.0\ mA$

14–4 Fan-out = 20

14–5 $X = (\overline{AB})(\overline{CD})(\overline{EF})(\overline{GH}) = (\overline{A} + \overline{B})(\overline{C} + \overline{D})(\overline{E} + \overline{F})(\overline{G} + \overline{H})$

14–6 See Figure 14–63. **14–7** $R_L = 97\ \Omega$

▶ **FIGURE 14–63**

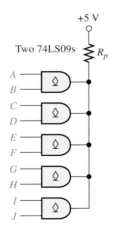

SELF-TEST

1. (b) **2.** (c) **3.** (c) **4.** (c) **5.** (e)

6. (a) **7.** (c) **8.** (d) **9.** (a) **10.** (c)

A: Conversions

DECIMAL	BCD(8421)	OCTAL	BINARY	DECIMAL	BCD(8421)	OCTAL	BINARY	DECIMAL	BCD(8421)	OCTAL	BINARY
0	0000	0	0	34	00110100	42	100010	68	01101000	104	1000100
1	0001	1	1	35	00110101	43	100011	69	01101001	105	1000101
2	0010	2	10	36	00110110	44	100100	70	01110000	106	1000110
3	0011	3	11	37	00110111	45	100101	71	01110001	107	1000111
4	0100	4	100	38	00111000	46	100110	72	01110010	110	1001000
5	0101	5	101	39	00111001	47	100111	73	01110011	111	1001001
6	0110	6	110	40	01000000	50	101000	74	01110100	112	1001010
7	0111	7	111	41	01000001	51	101001	75	01110101	113	1001011
8	1000	10	1000	42	01000010	52	101010	76	01110110	114	1001100
9	1001	11	1001	43	01000011	53	101011	77	01110111	115	1001101
10	00010000	12	1010	44	01000100	54	101100	78	01111000	116	1001110
11	00010001	13	1011	45	01000101	55	101101	79	01111001	117	1001111
12	00010010	14	1100	46	01000110	56	101110	80	10000000	120	1010000
13	00010011	15	1101	47	01000111	57	101111	81	10000001	121	1010001
14	00010100	16	1110	48	01001000	60	110000	82	10000010	122	1010010
15	00010101	17	1111	49	01001001	61	110001	83	10000011	123	1010011
16	00010110	20	10000	50	01010000	62	110010	84	10000100	124	1010100
17	00010111	21	10001	51	01010001	63	110011	85	10000101	125	1010101
18	00011000	22	10010	52	01010010	64	110100	86	10000110	126	1010110
19	00011001	23	10011	53	01010011	65	110101	87	10000111	127	1010111
20	00100000	24	10100	54	01010100	66	110110	88	10001000	130	1011000
21	00100001	25	10101	55	01010101	67	110111	89	10001001	131	1011001
22	00100010	26	10110	56	01010110	70	111000	90	10010000	132	1011010
23	00100011	27	10111	57	01010111	71	111001	91	10010001	133	1011011
24	00100100	30	11000	58	01011000	72	111010	92	10010010	134	1011100
25	00100101	31	11001	59	01011001	73	111011	93	10010011	135	1011101
26	00100110	32	11010	60	01100000	74	111100	94	10010100	136	1011110
27	00100111	33	11011	61	01100001	75	111101	95	10010101	137	1011111
28	00101000	34	11100	62	01100010	76	111110	96	10010110	140	1100000
29	00101001	35	11101	63	01100011	77	111111	97	10010111	141	1100001
30	00110000	36	11110	64	01100100	100	1000000	98	10011000	142	1100010
31	00110001	37	11111	65	01100101	101	1000001	99	10011001	143	1100011
32	00110010	40	100000	66	01100110	102	1000010				
33	00110011	41	100001	67	01100111	103	1000011				

Powers of Two

2^n	n	2^{-n}
1	0	1.0
2	1	0.5
4	2	0.25
8	3	0.125
16	4	0.062 5
32	5	0.031 25
64	6	0.015 625
128	7	0.007 812 5
256	8	0.003 906 25
512	9	0.001 953 125
1 024	10	0.000 976 562 5
2 048	11	0.000 488 281 25
4 096	12	0.000 244 140 625
8 192	13	0.000 122 070 312 5
16 384	14	0.000 061 035 156 25
32 768	15	0.000 030 517 578 125
65 536	16	0.000 015 258 789 062 5
131 072	17	0.000 007 629 394 531 25
262 144	18	0.000 003 814 697 265 625
524 288	19	0.000 001 907 348 632 812 5
1 048 576	20	0.000 000 953 674 316 406 25
2 097 152	21	0.000 000 476 837 158 203 125
4 194 304	22	0.000 000 238 418 579 101 562 5
8 388 608	23	0.000 000 119 209 289 550 781 25
16 777 216	24	0.000 000 059 604 644 775 390 625
33 554 432	25	0.000 000 029 802 322 387 695 312 5
67 108 864	26	0.000 000 014 901 161 193 847 656 25
134 217 728	27	0.000 000 007 450 580 596 923 828 125
268 435 456	28	0.000 000 003 725 290 298 461 914 062 5
536 870 912	29	0.000 000 001 862 645 149 230 957 031 25
1 073 741 824	30	0.000 000 000 931 322 574 615 478 515 625
2 147 483 648	31	0.000 000 000 465 661 287 307 739 257 812 5
4 294 967 296	32	0.000 000 000 232 830 643 653 869 628 906 25
8 589 934 592	33	0.000 000 000 116 415 321 826 934 814 453 125
17 179 869 184	34	0.000 000 000 058 207 660 913 467 407 226 562 5
34 359 738 368	35	0.000 000 000 029 103 830 456 733 703 613 281 25
68 719 476 736	36	0.000 000 000 014 551 915 228 366 851 806 640 625
137 438 953 472	37	0.000 000 000 007 275 957 614 183 425 903 320 312 5
274 877 906 944	38	0.000 000 000 003 637 978 807 091 712 951 660 156 25
549 755 813 888	39	0.000 000 000 001 818 989 403 545 856 475 830 078 125
1 099 511 627 776	40	0.000 000 000 000 909 494 701 772 928 237 915 039 062 5
2 199 023 255 552	41	0.000 000 000 000 454 747 350 886 464 118 957 519 531 25
4 398 046 511 104	42	0.000 000 000 000 227 373 675 443 232 059 478 759 765 625
8 796 093 022 208	43	0.000 000 000 000 113 686 837 721 616 029 739 379 882 812 5
17 592 186 044 416	44	0.000 000 000 000 056 843 418 860 808 014 869 689 941 406 25
35 184 372 088 832	45	0.000 000 000 000 028 421 709 430 404 007 434 844 970 703 125
70 368 744 177 664	46	0.000 000 000 000 014 210 854 715 202 003 717 422 485 351 562 5
140 737 488 355 328	47	0.000 000 000 000 007 105 427 357 601 001 858 711 242 675 781 25
281 474 976 710 656	48	0.000 000 000 000 003 552 713 678 800 500 929 355 621 337 890 625
562 949 953 421 312	49	0.000 000 000 000 001 776 356 839 400 250 464 677 810 668 945 312 5
1 125 899 906 842 624	50	0.000 000 000 000 000 888 178 419 700 125 232 338 905 334 472 656 25
2 251 799 813 685 248	51	0.000 000 000 000 000 444 089 209 850 062 616 169 452 667 236 328 125
4 503 599 627 370 496	52	0.000 000 000 000 000 222 044 604 925 031 308 084 726 333 618 164 062 5
9 007 199 254 740 992	53	0.000 000 000 000 000 111 022 302 462 515 654 042 363 166 809 082 031 25
18 014 398 509 481 984	54	0.000 000 000 000 000 055 511 151 231 257 827 021 181 583 404 541 015 625
36 028 797 018 963 968	55	0.000 000 000 000 000 027 755 575 615 628 913 510 590 791 702 270 507 812 5
72 057 594 037 927 936	56	0.000 000 000 000 000 013 877 787 807 814 456 755 295 395 851 135 253 906 25
144 115 188 075 855 872	57	0.000 000 000 000 000 006 938 893 903 907 228 377 647 697 925 567 626 953 125
288 230 376 151 711 744	58	0.000 000 000 000 000 003 469 446 951 953 614 188 823 848 962 783 813 476 562 5
576 460 752 303 423 488	59	0.000 000 000 000 000 001 734 723 475 976 807 094 411 924 481 391 906 738 281 25
1 152 921 504 606 846 976	60	0.000 000 000 000 000 000 867 361 737 988 403 547 205 962 240 695 953 369 140 625
2 305 843 009 213 693 952	61	0.000 000 000 000 000 000 433 680 868 994 201 773 602 981 120 347 976 684 570 312 5
4 611 686 018 427 387 904	62	0.000 000 000 000 000 000 216 840 434 497 100 886 801 490 560 173 988 342 285 156 25
9 223 372 036 854 775 808	63	0.000 000 000 000 000 000 108 420 217 248 550 443 400 745 280 086 994 171 142 578 125
18 446 744 073 709 551 616	64	0.000 000 000 000 000 000 054 210 108 624 275 221 700 372 640 043 497 085 571 289 062 5
36 893 488 147 419 103 232	65	0.000 000 000 000 000 000 027 105 054 312 137 610 850 186 320 021 748 542 785 644 531 25
73 786 976 294 838 206 464	66	0.000 000 000 000 000 000 013 552 527 156 088 805 425 093 160 010 874 271 392 822 265 625
147 573 952 589 676 412 928	67	0.000 000 000 000 000 000 006 776 263 578 034 402 712 546 580 005 437 135 696 411 132 812 5
295 147 905 179 352 825 856	68	0.000 000 000 000 000 000 003 388 131 789 017 201 356 273 290 002 718 567 848 205 566 406 25
590 295 810 358 705 651 712	69	0.000 000 000 000 000 000 001 694 065 894 508 600 678 136 645 001 359 283 924 102 783 203 125
1 180 591 620 717 411 303 424	70	0.000 000 000 000 000 000 000 847 032 947 254 300 339 068 322 500 679 641 962 051 391 601 562 5
2 361 183 241 434 822 606 848	71	0.000 000 000 000 000 000 000 423 516 473 627 150 169 534 161 250 339 820 981 025 695 800 781 25
4 722 366 482 869 645 213 696	72	0.000 000 000 000 000 000 000 211 758 236 813 575 084 767 080 625 169 910 490 512 847 900 390 625

B: Traffic Light Interface

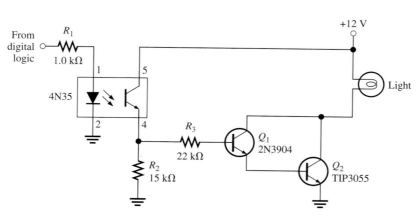

The development board with programmed CPLD and an interface board running model traffic lights in the lab. Courtesy of Dave Buchla.

▲ **FIGURE B–1**

Interface circuit used with model traffic lights. One circuit drives one light.

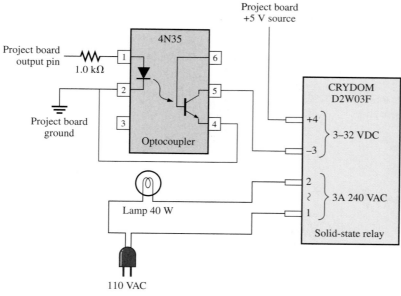

Student holding a development board with the programmed CPLD running real traffic lights in the lab. The interface circuits are in the metal box mounted on the light support. Courtesy of Doug Joksch.

▲ **FIGURE B–2**

Interface circuit used with actual traffic lights. One circuit drives one light.

Answers to Odd-Numbered Problems

Chapter 1

1. Digital can be transmitted and stored more efficiently and reliably.

3. (a) 11010001   (b) 000101010

5. (a) 550 ns   (b) 600 ns
   (c) 2.7 μs   (d) 10 V

7. 250 Hz

9. 50%

11. 8 μs; 1 μs

13. AND gate

15. (a) adder   (b) multiplier
    (c) multiplexer (d) comparator

17. 01010000

19. DIP pins go through holes in a circuit board. SMT pins connect to surface pads.

21. ABEL, CUPL

23. (a) Design entry: The step in a programmable logic design flow where a description of the circuit is entered in either schematic (graphic) form or in text form using an HDL.

   (b) Simulation: The step in a design flow where the entered design is simulated based on defined input waveforms.

   (c) Compilation: A program process that controls the design flow process and translates a design source code to object code for testing and downloading.

   (d) Download: The process in which the design is transferred from software to hardware.

25. 7 V

27. A collection of circuits interconnected to perform a specified function

29. Enter a new value on the keypad.

Chapter 2

1. (a) 1   (b) 100   (c) 100,000

3. (a) 400; 70; 1   (b) 9000; 300; 50; 6
   (c) 100,000; 20,000; 5000; 0; 0; 0

5. (a) 3   (b) 4   (c) 7   (d) 8   (e) 9
   (f) 12   (g) 11   (h) 15

7. (a) 51.75   (b) 42.25   (c) 65.875
   (d) 120.625   (e) 92.65625   (f) 113.0625
   (g) 90.625   (h) 127.96875

9. (a) 5 bits   (b) 6 bits   (c) 6 bits
   (d) 7 bits   (e) 7 bits   (f) 7 bits
   (g) 8 bits   (h) 8 bits

11. (a) 1010   (b) 10001   (c) 11000
    (d) 110000   (e) 111101   (f) 1011101
    (g) 1111101   (h) 10111010

13. (a) 1111   (b) 10101   (c) 11100
    (d) 100010   (e) 101000   (f) 111011
    (g) 1000001   (h) 1001001

15. (a) 100   (b) 100   (c) 1000
    (d) 1101   (e) 1110   (f) 11000

17. (a) 1001   (b) 1000   (c) 100011
    (d) 110110   (e) 10101001   (f) 10110110

19. (a) 010   (b) 001   (c) 0101
    (d) 00101000   (e) 0001010   (f) 11110

21. (a) 00011101   (b) 11010101
    (c) 01100100   (d) 11111011

23. (a) 00001100   (b) 10111100
    (c) 01100101   (d) 10000011

25. (a) -102 (b) $+116$ (c) -64

27. (a) 0 10001101 11110000101011000000000
    (b) 1 10001010 11000001100000000000000

29. (a) 00110000   (b) 00011101
    (c) 11101011   (d) 100111110

31. (a) 11000101   (b) 11000000

33. 100111001010

35. (a) 00111000
    (b) 01011001
    (c) 101000010100
    (d) 010111001000
    (e) 0100000100000000
    (f) 1111101100010111
    (g) 1000101010011101

37. (a) 35   (b) 146   (c) 26   (d) 141
    (e) 243   (f) 235   (g) 1474   (h) 1792

39. (a) 60_{16} (b) $10B_{16}$ (c) $1BA_{16}$

41. (a) 10   (b) 23   (c) 46   (d) 52   (e) 67
    (f) 367   (g) 115   (h) 532   (i) 4085

43. (a) 001011   (b) 101111
    (c) 001000001
    (d) 011010001
    (e) 101100000
    (f) 100110101011
    (g) 001011010111001
    (h) 100101110000000
    (i) 0010000000010001011

45. (a) 00010000   (b) 00010011

    (c) 00011000   (d) 00100001

    (e) 00100101   (f) 00110110

    (g) 01000100   (h) 01010111

    (i) 01101001   (j) 10011000

    (k) 000100100101   (l) 000101010110

47. (a) 000100000100   (b) 000100101000

    (c) 000100110010   (d) 000101010000

    (e) 000110000110   (f) 001000010000

    (g) 001101011001   (h) 010101000111

    (i) 0001000001010001

49. (a) 80   (b) 237   (c) 346   (d) 421

    (e) 754   (f) 800   (g) 978   (h) 1683

    (i) 9018   (j) 6667

51. (a) 00010100   (b) 00010010

    (c) 00010111   (d) 00010110

    (e) 01010010   (f) 000100001001

    (g) 000110010101   (h) 0001001001101001

53. The Gray code makes only one bit change at a time when going from one number in the sequence to the next.

55. (a) 1100   (b) 00011   (c) 10000011110

57. (a) CAN   (b) J   (c) =

    (d) #   (e) >   (f) B

59. 48 65 6C 6C 6F 2E 20 48 6F 77 20 61 72 65 20 79 6F 75 3F

61. (b) is incorrect.

63. (a) 110100100   (b) 000001001

    (c) 111111110

65. 001010001

67. (a) 110100010   (b) 100000101

Chapter 3

1. See Figure P–1.

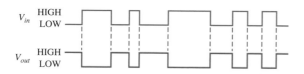

▲ **FIGURE P–1**

3. See Figure P–2.

▶ **FIGURE P–2**

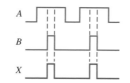

5. See Figure P–3.

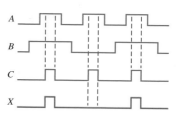

▲ **FIGURE P–3**

7. See Figure P–4.

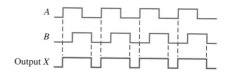

▲ **FIGURE P–4**

9. See Figure P–5.

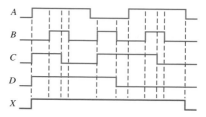

▲ **FIGURE P–5**

11. See Figure P–6.

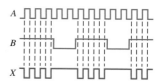

▲ **FIGURE P–6**

13. See Figure P–7.

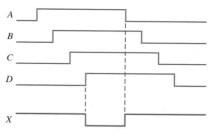

▲ **FIGURE P–7**

15. See Figure P–8.

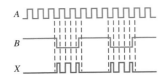

▲ **FIGURE P–8**

17. See Figure P–9.

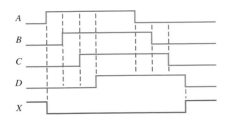

▲ **FIGURE P–9**

19. $XOR = A\overline{B} + \overline{A}B$; $OR = A + B$

21. See Figure P–10.

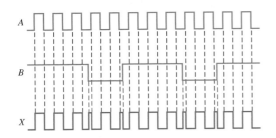

▲ **FIGURE P–10**

23. $X_1 = \overline{A}B, X_2 = \overline{A}\,\overline{B}, X_3 = A\overline{B}$

25. CMOS

27. $t_{PLH} = 4.3$ ns; $t_{PHL} = 10.5$ ns

29. 20 mW

31. The gates in parts (b), (c), (e) are faulty.

33. (a) defective output (stuck LOW or open)

   (b) Pin 4 input or pin 6 output internally open.

35. The seat belt input to the AND gate is open.

37. See Figure P–11.

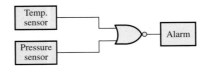

▲ **FIGURE P–11**

39. Add an inverter to the enable input line of the AND gate.

41. See Figure P–12.

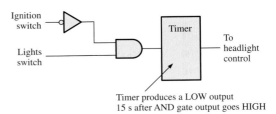

▲ **FIGURE P–12**

43. The inputs are now active-LOW. Change the OR gates to NAND gates (negative-OR) and add two inverters.

45. Gate inputs shorted together

47. Gate output open

Chapter 4

1. $X = A + B + C + D$

3. $X = \overline{A} + \overline{B} + \overline{C}$

5. (a) $AB = 1$ when $A = 1, B = 1$

   (b) $A\overline{B}C = 1$ when $A = 1, B = 0, C = 1$

   (c) $A + B = 0$ when $A = 0, B = 0$

   (d) $\overline{A} + B + \overline{C} = 0$ when $A = 1, B = 0, C = 1$

   (e) $\overline{A} + \overline{B} + C = 0$ when $A = 1, B = 1, C = 0$

   (f) $\overline{A} + B = 0$ when $A = 1, B = 0$

   (g) $A\overline{B}\,\overline{C} = 1$ when $A = 1, B = 0, C = 0$

7. (a) Commutative   (b) Commutative

   (c) Distributive

9. (a) $\overline{A}B$ (b) $A + \overline{B}$

   (c) $\overline{A}\,\overline{B}\,\overline{C}$ (d) $\overline{A} + \overline{B} + \overline{C}$

   (e) $\overline{A} + \overline{B}\,\overline{C}$ (f) $\overline{A} + \overline{B} + \overline{C} + \overline{D}$

   (g) $(\overline{A} + \overline{B})(\overline{C} + \overline{D})$ (h) $\overline{A}B + C\overline{D}$

11. (a) $(\overline{A} + \overline{B} + \overline{C})(\overline{E} + \overline{F} + \overline{G})(\overline{H} + \overline{I} + \overline{J})$
$(\overline{K} + \overline{L} + \overline{M})$

   (b) $\overline{A}B\overline{C} + BC$

   (c) $\overline{A}\,\overline{B}\,\overline{C}\,\overline{D}\,\overline{E}\,\overline{F}\,\overline{G}\,\overline{H}$

13. (a) $X = ABCD$ (b) $X = AB + C$

   (c) $X = \overline{\overline{A}B}$ (d) $X = (A + B)C$

15. See Figure P–13.

17. (a) A (b) AB (c) C

   (d) A (e) $\overline{A}C + \overline{B}C$

19. (a) $BD + BE + \overline{D}F$ (b) $\overline{A}\,BC + \overline{A}\,\overline{B}D$

   (c) B (d) $AB + CD$

   (e) ABC

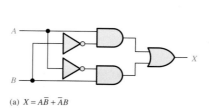

(a) $X = A\overline{B} + \overline{A}B$

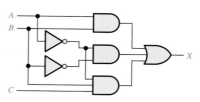

(b) $X = AB + \overline{A}\overline{B} + \overline{A}BC$

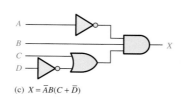

(c) $X = \overline{A}B(C + \overline{D})$

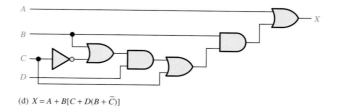

(d) $X = A + B[C + D(B + \overline{C})]$

▲ **FIGURE P–13**

---

21. (a) $A\overline{B} + AC + BC$ (b) $AC + \overline{B}C$ (c) $AB + AC$

23. (a) Domain: A, B, C
Standard SOP: $A\overline{B}C + A\overline{B}\,\overline{C} + ABC + \overline{A}BC$

   (b) Domain: A, B, C
Standard SOP: $ABC + A\overline{B}C + \overline{A}\,BC$

   (c) Domain: A, B, C
Standard SOP: $ABC + AB\overline{C} + A\overline{B}C$

25. (a) $101 + 100 + 111 + 011$ (b) $111 + 101 + 001$

   (c) $111 + 110 + 101$

27. (a) $(A + B + C)(A + B + \overline{C})(A + \overline{B} + C)$
$(\overline{A} + \overline{B} + C)$

   (b) $(A + B + C)(A + \overline{B} + C)(A + \overline{B} + \overline{C})$
$(\overline{A} + B + C)(\overline{A} + \overline{B} + C)$

   (c) $(A + B + C)(A + B + \overline{C})(A + \overline{B} + C)$
$(A + \overline{B} + \overline{C})(\overline{A} + B + C)$

29. (a) See Table P–1.

29. (b) See Table P–2.

▼ **TABLE P–2**

X	Y	Z	Q
0	0	0	1
0	0	1	1
0	1	0	0
0	1	1	1
1	0	0	0
1	0	1	1
1	1	0	1
1	1	1	0

31. (a) See Table P–3.

▼ **TABLE P–1**

A	B	C	X
0	0	0	0
0	0	1	0
0	1	0	1
0	1	1	0
1	0	0	0
1	0	1	1
1	1	0	0
1	1	1	1

▼ **TABLE P–3**

A	B	C	X
0	0	0	1
0	0	1	0
0	1	0	1
0	1	1	1
1	0	0	0
1	0	1	1
1	1	0	1
1	1	1	0

31. (b) See Table P–4.

▼ **TABLE P–4**

W	X	Y	Z	Q
0	0	0	0	1
0	0	0	1	1
0	0	1	0	1
0	0	1	1	1
0	1	0	0	0
0	1	0	1	1
0	1	1	0	1
0	1	1	1	0
1	0	0	0	1
1	0	0	1	1
1	0	1	0	1
1	0	1	1	1
1	1	0	0	0
1	1	0	1	1
1	1	1	0	1
1	1	1	1	1

33. (a) See Table P–5.

▼ **TABLE P–5**

A	B	C	X
0	0	0	0
0	0	1	0
0	1	0	0
0	1	1	1
1	0	0	1
1	0	1	1
1	1	0	1
1	1	1	1

33. (b) See Table P–6.

▼ **TABLE P–6**

A	B	C	D	X
0	0	0	0	1
0	0	0	1	0
0	0	1	0	1
0	0	1	1	1
0	1	0	0	0
0	1	0	1	0
0	1	1	0	0
0	1	1	1	0
1	0	0	0	1
1	0	0	1	0
1	0	1	0	0
1	0	1	1	1
1	1	0	0	1
1	1	0	1	1
1	1	1	0	1
1	1	1	1	1

35. See Figure P–14.

▶ **FIGURE P–14**

AB \ C	0	1
00	000	001
01	010	011
11	110	111
10	100	101

37. See Figure P–15.

▶ **FIGURE P–15**

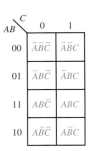

AB \ C	0	1
00	$\bar{A}\bar{B}\bar{C}$	$\bar{A}\bar{B}C$
01	$\bar{A}B\bar{C}$	$\bar{A}BC$
11	$AB\bar{C}$	ABC
10	$A\bar{B}\bar{C}$	$A\bar{B}C$

39. (a) No simplification **(b)** AC

(c) $\overline{D}\,\overline{F} + E\overline{F}$

41. (a) $AB + AC$

(b) $A + BC$

(c) $B\overline{C}D + A\overline{C}D + BC\overline{D} + AC\overline{D}$

(d) $A\overline{B} + CD$

43. $\overline{B} + C$

45. $\overline{A}\,\overline{B}\,\overline{C}D + C\overline{D} + BC + A\overline{D}$

47. (a) $(A + \overline{B} + C + \overline{D})(\overline{A} + B + \overline{C} + D)$
$(\overline{A} + \overline{B} + \overline{C} + \overline{D})$

(b) $(W + \overline{Z})(W + X)(\overline{Y} + \overline{Z})(X + \overline{Y})$

49. $(A + C + D)(A + \overline{B} + C)(\overline{A} + B + \overline{D})$
$(B + \overline{C} + \overline{D})(\overline{A} + \overline{B} + \overline{C} + D)$

51. $X = \overline{A}\,\overline{B}\,\overline{C}\,\overline{D}\,\overline{E} + \overline{A}BCDE + AB\overline{C}\,\overline{D}\,\overline{E} +$
$\overline{A}\,\overline{B}DE + \overline{A}BD\overline{E} + \overline{B}\,\overline{C}DE + AB\overline{C}D$

53. **entity** AND_OR **is**

port (A, B, C, D, E, F, G, H, I: **in** bit; X: **out** bit);

end entity AND_OR;

architecture Logic **of** AND_OR **is**

begin

X <= (A **and** B **and** C) **or** (D **and** E **and** F) **or**
(G **and** H **and** I);

end architecture Logic;

55. LED. LEDs emit light, LCDs do not.

57. One less inverter and six fewer gates.

59. Add an inverter to the output of the OR gate in each of the segment logic circuits.

61. See Figure P–16.

63. Inverter output open

65. b-segment OR gate output open

▶ **FIGURE P–16**

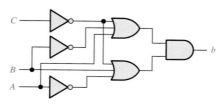

segment $b = (\overline{C} + B + \overline{A})(\overline{C} + \overline{B} + A)$

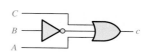

segment $c = C + \overline{B} + A$

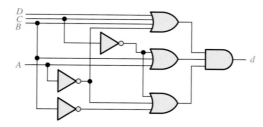

segment $d = (D + C + B + \overline{A})(\overline{C} + B + A)(\overline{C} + \overline{B} + \overline{A})$

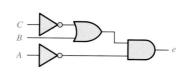

segment $e = \overline{A}(\overline{C} + B)$

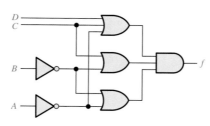

segment $f = (D + C + \overline{A})(C + \overline{B})(\overline{B} + \overline{A})$

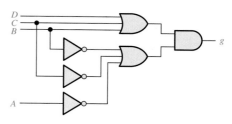

segment $g = (D + C + B)(\overline{C} + \overline{B} + \overline{A})$

Chapter 5

1. See Figure P–17.

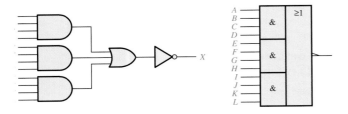

3. (a) $X = ABB$ (b) $X = AB + B$
  (c) $X = \overline{A} + B$ (d) $X = (A + B) + AB$
  (e) $X = \overline{\overline{ABC}}$ (f) $X = (A + B)(\overline{B} + C)$

5. (a)

A	B	X
0	0	0
0	1	0
1	0	0
1	1	1

5. (b)

A	B	X
0	0	0
0	1	1
1	0	0
1	1	1

5. (c)

A	B	X
0	0	1
0	1	1
1	0	0
1	1	1

5. (d)

A	B	X
0	0	0
0	1	1
1	0	1
1	1	1

5. (e)

A	B	C	X
0	0	0	1
0	0	1	1
0	1	0	1
0	1	1	0
1	0	0	1
1	0	1	1
1	1	0	1
1	1	1	1

5. (f)

A	B	C	X
0	0	0	0
0	0	1	0
0	1	0	0
0	1	1	1
1	0	0	1
1	0	1	1
1	1	0	0
1	1	1	1

7. $X = \overline{A\overline{B} + \overline{A}B} = (\overline{A} + B)(A + \overline{B})$

9. See Figure P–18.

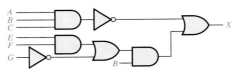

(a) $X = AB + \overline{B}C$

(b) $X = A(B + \overline{C})$

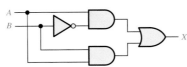

(c) $X = A\overline{B} + AB$

(d) $X = \overline{ABC} + B(EF + \overline{G})$

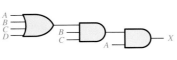

(e) $X = A[BC(A + B + C + D)]$

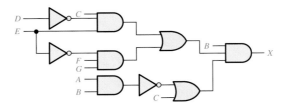

(f) $X = B(C\overline{D}E + \overline{E}FG)\,(\overline{AB} + C)$

11. See Figure P–19.

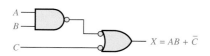

13. $X = AB$

15. **(a)** No simplification

    (b) No simplification

    (c) $X = A$

    (d) $X = \overline{A} + \overline{B} + \overline{C} + EF + \overline{G}$

    (e) $X = ABC$

    (f) $X = BC\overline{D}E + \overline{A}B\overline{E}FG + BC\overline{E}FG$

17. **(a)** $X = AC + AD + BC + BD$

    (b) $X = \overline{A}CD + \overline{B}CD$

    (c) $X = ABD + CD + E$

    (d) $X = \overline{A} + B + D$

    (e) $X = ABD + \overline{C}D + \overline{E}$

    (f) $X = \overline{A}\,\overline{C} + \overline{A}\,\overline{D} + \overline{B}\,\overline{C} + \overline{B}\,\overline{D} + \overline{E}\,\overline{G} + \overline{E}\,\overline{H} + \overline{F}\,\overline{G} + \overline{F}\,\overline{H}$

19. See Figure P–20.

21. See Figure P–21.

23. See Figure P–22.

25. See Figure P–23 on page 836.

27. $X = A + \overline{B}$; see Figure P–24.

29. $X = A\overline{B}\,\overline{C}$; see Figure P–25.

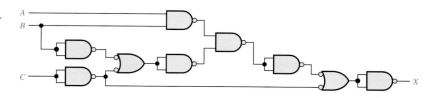

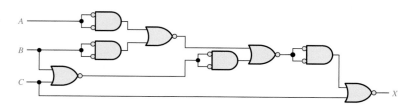

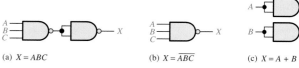

(a) $X = ABC$ (b) $X = \overline{ABC}$ (c) $X = A + B$ (d) $X = A + B + \overline{C}$

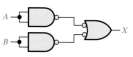

(e) $X = \overline{AB} + \overline{CD}$

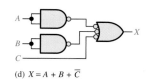

(f) $X = (A + B)(C + D)$

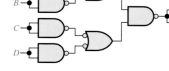

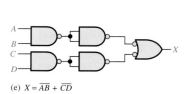

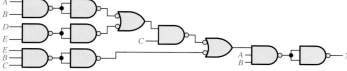

(g) $X = AB[C(\overline{DE} + \overline{AB}) + \overline{BCE}]$

▶ FIGURE P–23

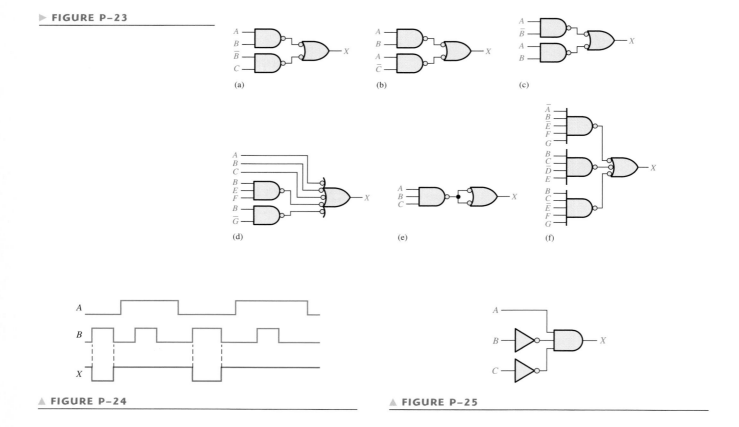

(a)　　　　　　(b)　　　　　　(c)

(d)　　　　　　(e)　　　　　　(f)

▲ FIGURE P–24

▲ FIGURE P–25

31. The output pulse width is greater than the specified minimum.

33. **(e)** **entity** Circuit 5_52e **is**

  port (A, B, C: **in** bit; X: **out** bit);

  end entity Circuit5_52e;

  architecture LogicFunction **of** Circuit5_52e **is**

  begin

  X <= (**not** A **and** B) **or** B **or** (B **and not** C) **or**
  (**not** A **and not** C) **or** (B **and not** C) **or not** C;

  end architecture LogicFunction;

  (f) **entity** Circuit5_52f **is**

  port (A, B, C: **in** bit; X: **out** bit);

  end entity Circuit5_52f;

  architecture LogicFunction **of** Circuit5_52f **is**

  begin

  X <= (A **or** B) **and** (**not** B **or** C);

  end architecture LogicFunction;

35. Number gates from top to bottom and left to right G1, G2, G3, etc. Relabel inputs IN1, IN2, IN3, etc. and output OUT.

  entity Circuit5_53f **is**

  port (IN1, IN2, IN3, IN4, IN5, IN6, IN7, IN8: **in** bit;
  OUT: **out** bit);

  end entity Circuit5_53f;

  architecture LogicFunction **of** Circuit5_53f **is**

  component NAND_gate **is**

  port (A, B: **in** bit; X: **out** bit);

  end component NAND_gate;

  signal G1OUT, G2OUT, G3OUT, G4OUT, G5OUT,
  G6OUT: bit;

  begin

  G1: NAND_gate **port map** (A => IN1, B => IN2, X =>
  G1OUT);

  G2: NAND_gate **port map** (A => IN3, B => IN4, X =>
  G2OUT);

  G3: NAND_gate **port map** (A => IN5, B => IN6, X =>
  G3OUT);

  G4: NAND_gate **port map** (A => IN7, B => IN8, X =>
  G4OUT);

  G5: NAND_gate **port map** (A => G1OUT, B =>
  G2OUT, X => G5OUT);

  G6: NAND_gate **port map** (A => G3OUT, B =>
  G4OUT, X => G6OUT);

  G7: NAND_gate **port map** (A => G5OUT, B =>
  G6OUT, X => OUT);

  end architecture LogicFunction;

37. --Data flow approach

    entity Fig5_64 **is**

        port (A, B, C, D, E: **in** bit; X: **out** bit);

    end entity Fig5_64;

    architecture DataFlow **of** Fig5_64 **is**

    begin

        X <= (A **and** B **and** C) **or** (D **and not** E);

    end architecture DataFlow;

    --Structural approach

    entity Fig5_64 **is**

        port (IN1, IN2, IN3, IN4, IN5: **in** bit; OUT: **out** bit);

    end entity Fig5_64;

    architecture Structure **of** Fig5_64 **is**

    component AND_gate **is**

        port (A, B: **in** bit; X: **out** bit);

    end component AND_gate;

    component OR_gate **is**

        port (A, B: **in** bit; X: **out** bit);

    end component OR_gate;

    component Inverter **is**

        port (A: **in** bit; X: **out** bit);

    end component Inverter;

        signal G1OUT, G2OUT, G3OUT, INVOUT: bit;

    begin

        G1: AND_gate **port map** (A => IN1, B => IN2, X => G1OUT);

        G2: AND_gate **port map** (A => G1OUT, B => IN3, X => G2OUT);

        INV: Inverter **port map** (A => IN5, X => INVOUT);

        G3: AND_gate **port map** (A => IN4, B => INVOUT, X => G3OUT);

        G4: OR_gate **port map** (A => G2OUT, B => G3OUT, X => OUT);

    end architecture Structure;

39. See Table P–7.

41. The AND gates are numbered top to bottom G1, G2, G3, G4. The OR gate is G5 and the inverters are, top to bottom. G6 and G7. Change A_1, A_2, B_1, B_2 to IN1, IN2, IN3, IN4 respectively. Change X to OUT.

    entity Circuit5_62 **is**

        port (IN1, IN2, IN3, IN4: **in** bit; OUT: **out** bit);

    end entity Circuit5_62;

    architecture Logic **of** Circuit5_62 **is**

    component AND_gate **is**

        port (A, B: **in** bit; X: **out** bit);

    end component AND_gate;

    component OR_gate **is**

        port (A, B, C, D: **in** bit; X: **out** bit);

▼ **TABLE P–7**

INPUTS				OUTPUT
A	B	C	D	X
0	0	0	0	0
1	0	0	0	0
0	1	0	0	0
1	1	0	0	0
0	0	1	0	0
1	0	1	0	0
0	1	1	0	0
1	1	1	0	0
0	0	0	1	0
1	0	0	1	0
0	1	0	1	0
1	1	0	1	1
0	0	1	1	0
1	0	1	1	1
0	1	1	1	1
1	1	1	1	1

    end component OR_gate;

    component Inverter **is**

        port (A: **in** bit; X: **out** bit);

    end component Inverter;

        signal G1OUT, G2OUT, G3OUT, G4OUT, G5OUT, G6OUT, G7OUT: bit;

    begin

        G1: AND_gate **port map** (A => IN1, B => IN2, X => G1OUT);

        G2: AND_gate **port map** (A => IN2, B => G6OUT, X => G2OUT);

        G3: AND_gate **port map** (A => G6OUT, B => G7OUT, X => G3OUT);

        G4: AND_gate **port map** (A => G7OUT, B => IN1, X => G4OUT);

        G5: OR_gate **port map** (A => G1OUT, B => G2OUT, C => G3OUT, D => G4OUT, X => OUT);

        G6: Inverter **port map** (A => IN3, X => G6OUT);

        G7: Inverter **port map** (A => IN4, X => G7OUT);

    end architecture Logic;

43. $X = ABC + D\overline{E}$. Since X is the same as the G_3 output, either G_1 or G_2 has failed, with its output stuck LOW.

45. See Figure P–26 on page 838.

47. **(a)** See Figure P–27.   **(b)** $X = E$

    (c) $X = E$

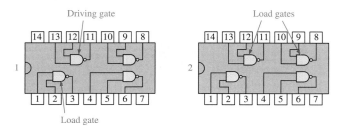

▲ FIGURE P–26

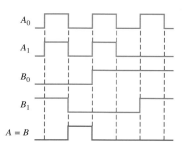

▲ FIGURE P–27

49. See Figure P–28.

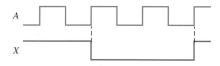

▲ FIGURE P–28

51. See Figure P–29.

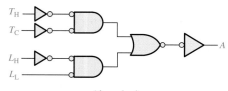

Heater logic

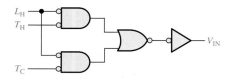

Alarm logic

▲ FIGURE P–29

53. X = lamp on, A = front door switch on, B = back door switch on. See Figure P–30.

▶ FIGURE P–30

55. See Figure P–31. Inverters (not shown) are used to convert each HIGH key closure to LOW.

57. Pin C of OR gate open.

59. No fault

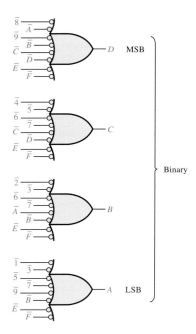

▲ FIGURE P–31

Chapter 6

1. (a) $A \oplus B = 0$, $\Sigma = 1$, $(A \oplus B)C_{in} = 0$, $AB = 1$, $C_{out} = 1$
   (b) $A \oplus B = 1$, $\Sigma = 0$, $(A \oplus B)C_{in} = 1$, $AB = 0$, $C_{out} = 1$
   (c) $A \oplus B = 1$, $\Sigma = 1$, $(A \oplus B)C_{in} = 0$, $AB = 0$, $C_{out} = 0$

3. (a) $\Sigma = 1$, $C_{out} = 0$;
   (b) $\Sigma = 1$, $C_{out} = 0$;
   (c) $\Sigma = 0$, $C_{out} = 1$;
   (d) $\Sigma = 1$, $C_{out} = 1$

5. 11100

7. $\Sigma_1 = 0110$; $\Sigma_2 = 1011$; $\Sigma_3 = 0110$; $\Sigma_4 = 0001$; $\Sigma_5 = 1000$

9. 225 ns

11. $A = B$ is HIGH when $A_0 = B_0$ and $A_1 = B_1$; see Figure P–32.

13. (a) $A > B = 1$; $A = B = 0$; $A < B = 0$
   (b) $A < B = 1$; $A = B = 0$; $A > B = 0$
   (c) $A = B = 1$; $A < B = 0$; $A > B = 0$

15. See Figure P–33.

17. $X = A_3 A_2 \bar{A}_1 \bar{A}_0 + \bar{A}_3 \bar{A}_2 \bar{A}_1 A_0 + A_3 \bar{A}_2 A_1$

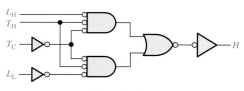

▲ FIGURE P–32

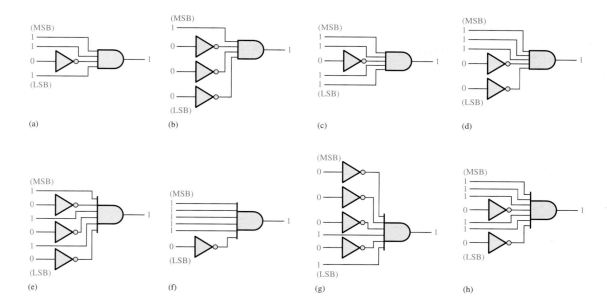

▲ **FIGURE P–33**

19. See Figure P–34.

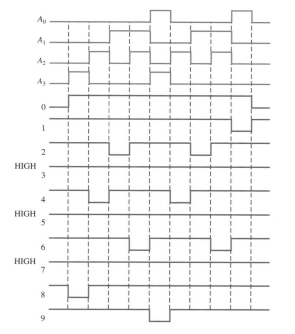

▲ **FIGURE P–34**

21. $A_3A_2A_1A_0 = 1011$, invalid BCD

23. (a) $2 = 0010 = 0010_2$

    (b) $8 = 1000 = 1000_2$

    (c) $13 = 00010011 = 1101_2$

    (d) $26 = 00100110 = 11010_2$

    (e) $33 = 00110011 = 100001_2$

25. (a) 1010000000 Gray → 1100000000 binary

    (b) 0011001100 Gray → 0010001000 binary

    (c) 1111000111 Gray → 1010000101 binary

    (d) 0000000001 Gray → 0000000001 binary

    See Figure P–35.

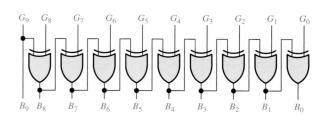

▲ **FIGURE P–35**

27. See Figure P–36.

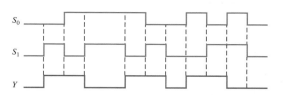

▲ **FIGURE P–36**

29. See Figure P–37 on page 840.

31. See Figure P–38.

33. (a) OK

    (b) segment g burned out; output G open

    (c) Segment b output stuck LOW

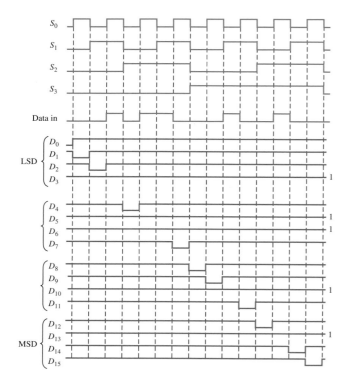

▲ FIGURE P–37

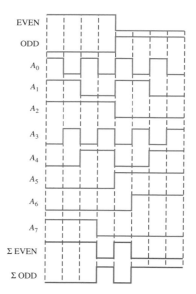

▲ FIGURE P–38

35. (a) The A_1 input of the top adder is open: All binary values corresponding to a BCD number having a value of 0, 1, 4, 5, 8, or 9 will be off by 2. This will first be seen for a BCD value of 0000 0000.

(b) The carry out of the top adder is open: All values not normally involving an output carry will be off by 32. This will first be seen for a BCD value of 0000 0000.

(c) The Σ_4 output of the top adder is shorted to ground: Same binary values above 15 will be short by 16. The first BCD value to indicate this will be 0001 1000.

(d) The Σ_3 output of the bottom adder is shorted to ground: Every other set of 16 values starting with 16 will be short 16. The first BCD value to indicate this will be 0001 0110.

37. 1. Place a LOW on pin 7 (Enable).

2. Apply a HIGH to D_0 and a LOW to D_1 through D_7.

3. Go through the binary sequence on the select inputs and check Y and $\overline{Y}$ according to Table P–8.

▼ **TABLE P–8**

S_2	S_1	S_0	Y	$\overline{Y}$
0	0	0	1	0
0	0	1	0	1
0	1	0	0	1
0	1	1	0	1
1	0	0	0	1
1	0	1	0	1
1	1	0	0	1
1	1	1	0	1

4. Repeat the binary sequence of select inputs for each set of data inputs listed in Table P–9. A HIGH on the Y output should occur only for the corresponding combinations of select inputs shown.

39. Apply a HIGH in turn to each Data input, D_0 through D_7 with LOWs on all the other inputs. For each HIGH applied to a data input, sequence through all eight binary combinations of select inputs ($S_2 S_1 S_0$) and check for HIGH on the corresponding data output and LOWs on all the other data outputs.

41. See Figure P–39.

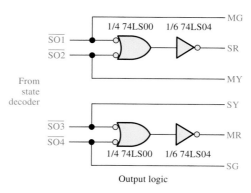

Output logic

▲ FIGURE P–39

▼ **TABLE P–9**

D_0	D_1	D_2	D_3	D_4	D_5	D_6	D_7	Y	$\overline{Y}$	S_2	S_1	S_0
L	H	L	L	L	L	L	L	1	0	0	0	1
L	L	H	L	L	L	L	L	1	0	0	1	0
L	L	L	H	L	L	L	L	1	0	0	1	1
L	L	L	L	H	L	L	L	1	0	1	0	0
L	L	L	L	L	H	L	L	1	0	1	0	1
L	L	L	L	L	L	H	L	1	0	1	1	0
L	L	L	L	L	L	L	H	1	0	1	1	1

43. $\Sigma = \overline{A}\,\overline{B}C_{in} + \overline{A}B\overline{C}_{in} + A\overline{B}\,\overline{C}_{in} + ABC_{in}$

$C_{out} = \overline{A}BC_{in} + A\overline{B}C_{in} + AB\overline{C}_{in} + ABC_{in}$

See Figure P–40.

45. See the block diagram in Figure P–41 on page 842.

47. See Figure P–42.

49. See Figure P–43.

51. LSB adder carry out open

53. Pin 12 of upper 74148 open

▶ **FIGURE P–40**

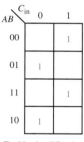
$\Sigma = $ No simplification

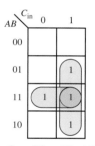
$C_{out} = BC_{in} + AB + AC_{in}$

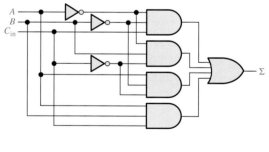

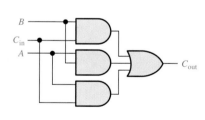

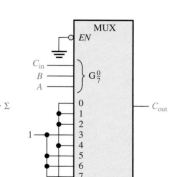

▶ FIGURE P–41

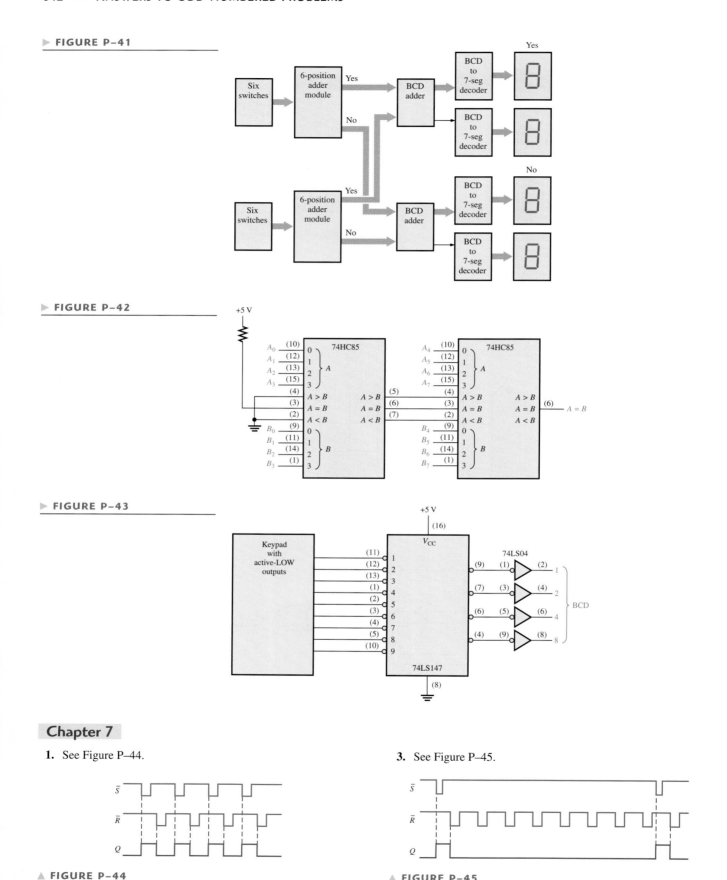

▶ FIGURE P–42

▶ FIGURE P–43

Chapter 7

1. See Figure P–44.

3. See Figure P–45.

▲ FIGURE P–44

▲ FIGURE P–45

5. See Figure P–46.

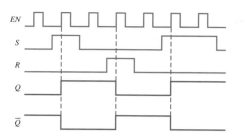

▲ **FIGURE P–46**

7. See Figure P–47.

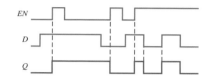

▲ **FIGURE P–47**

9. See Figure P–48.

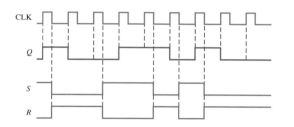

▲ **FIGURE P–48**

11. See Figure P–49.

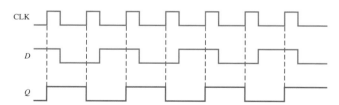

▲ **FIGURE P–49**

13. See Figure P–50.

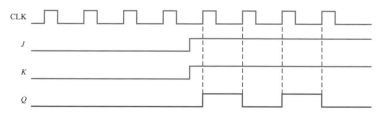

▲ **FIGURE P–50**

15. See Figure P–51.

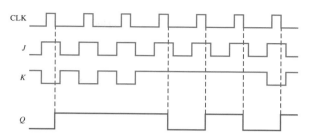

▲ **FIGURE P–51**

17. See Figure P–52.

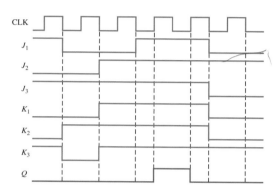

▲ **FIGURE P–52**

19. Direct current and dc supply voltage

21. 14.9 MHz

23. 150 mA, 750 mW

25. divide-by-2; see Figure P–53.

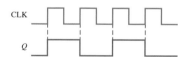

▲ **FIGURE P–53**

27. 4.62 μs

29. $C_1 = 1\ \mu$F, $R_1 = 227$ kΩ (use 220 kΩ). See Figure P–54.

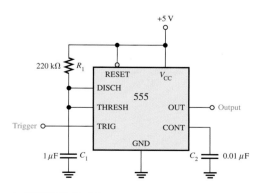

▲ **FIGURE P–54**

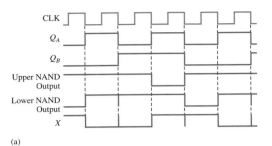

(a)

(b) Same as (a)

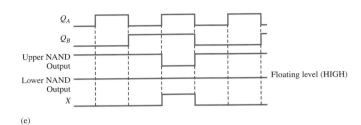

(e)

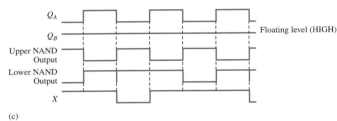

(c)

(d) $X = $ LOW if $Q_B = 1$; $X = \overline{Q}_A$ if $Q_B = 0$

▲ FIGURE P–55

31. $R_1 = 18 \text{ k}\Omega$, $R_2 = 9.1 \text{ k}\Omega$.

33. The wire from pin 6 to pin 10 and the ground wire are reversed on the protoboard.

35. $\overline{CLR}$ shorted to ground.

37. See Figure P–55. Delays not shown.

39. See Figure P–56.

   4 s: $C_1 = 1 \ \mu\text{F}$, $R_1 = 3.63 \text{ M}\Omega$ (use 3.9 MΩ)

   25 s: $C_1 = 2.2 \ \mu\text{F}$, $R_1 = 10.3 \text{ M}\Omega$ (use 10 MΩ)

41. See Figure P–57.

43. $\overline{Q}$ output of U1 open.

45. $\overline{SET}$ input of U1 open.

47. K input of U2 open.

Chapter 8

1. See Figure P–58.

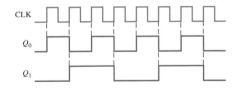

▲ FIGURE P–58

3. Worst-case delay is 24 ns; it occurs when all flip-flops change state from 011 to 100 or from 111 to 000.

5. 8 ns

7. Initially, each flip-flop is reset.

   At CLK1:

   $J_0 = K_0 = 1$ Therefore Q_0 goes to a 1.

   $J_1 = K_1 = 0$ Therefore Q_1 remains a 0.

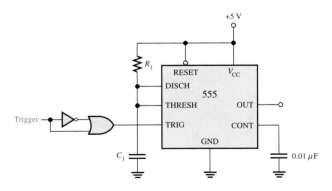

▲ FIGURE P–56

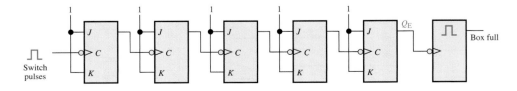

▲ FIGURE P–57

$J_2 = K_2 = 0$ Therefore Q_2 remains a 0.

$J_3 = K_3 = 0$ Therefore Q_3 remains a 0.

At CLK2:

$J_0 = K_0 = 1$ Therefore Q_0 goes to a 0.

$J_1 = K_1 = 1$ Therefore Q_1 goes to a 1.

$J_2 = K_2 = 0$ Therefore Q_2 remains a 0.

$J_3 = K_3 = 0$ Therefore Q_3 remains a 0.

At CLK3:

$J_0 = K_0 = 1$ Therefore Q_0 goes to a 1.

$J_1 = K_1 = 0$ Therefore Q_1 remains a 1.

$J_2 = K_2 = 0$ Therefore Q_2 remains a 0.

$J_3 = K_3 = 0$ Therefore Q_3 remains a 0.

A continuation of this procedure for the next seven clock pulses will show that the counter progresses through the BCD sequence.

9. See Figure P–59.

11. See Figure P–60.

13. See Figure P–61.

15. The sequence is 0000, 1111, 1110, 1101, 1010, 0101. The counter "locks up" in the 1010 and 0101 states and alternates between them.

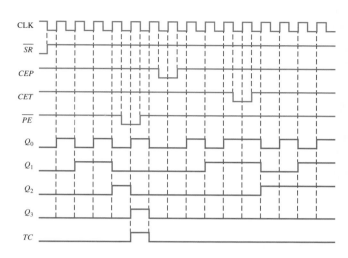

▲ FIGURE P–60

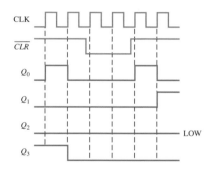

▲ FIGURE P–59

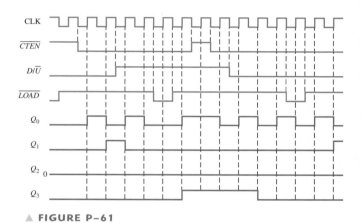

▲ FIGURE P–61

17. See Figure P–62.

19. See Figure P–63.

▶ FIGURE P–62

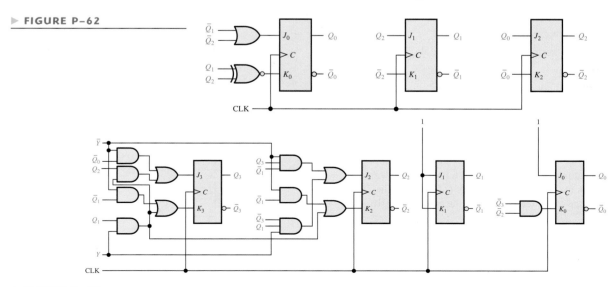

▲ FIGURE P–63

▶ FIGURE P-64

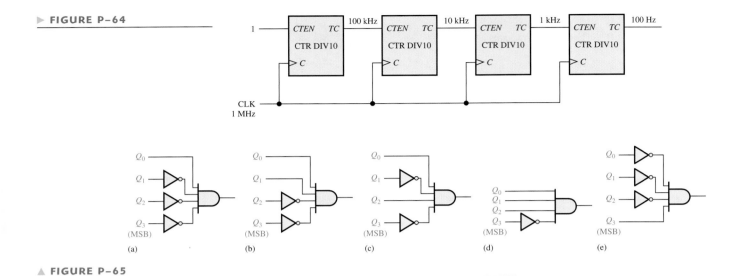

▲ FIGURE P-65

21. See Figure P–64 for divide-by-10,000. Add one more DIV10 counter to create a divide-by-100,000.

23. See Figure P–65.

25. CLK2, output 0; CLK4, outputs 2, 0; CLK6, output 4; CLK8, outputs 6, 4, 0; CLK10, output 8; CLK12, outputs 10, 8; CLK14, output 12; CLK16, outputs 14, 12, 8

27. A glitch of the AND gate output occurs on the 111 to 000 transition. Eliminate by ANDing $\overline{CLK}$ with counter outputs (strobe) or use Gray code.

29. Hours tens: 0001

   Hours units: 0010

   Minutes tens: 0000

   Minutes units: 0001

   Seconds tens: 0000

   Seconds units: 0010

31. 64

33. **(a)** Q_0 and Q_1 will not change from their initial state.

   (b) normal operation except Q_0 floating

   (c) Q_0 waveform is normal; Q_1 remains in initial state.

   (d) normal operation

   (e) The counter will not change from its initial state.

35. The K input of FF1 must be connected to ground rather than to the J input. Check for a wiring error.

37. Q_0 input to AND gate open and acting as a HIGH

39. See Table P–10.

41. The decode 6 gate interprets count 4 as a 6 (0110) and clears the counter back to 0 (actually 0010 since Q_1 is open). The apparent sequence of the tens portion of the counter is 0010, 0011, 0010, 0011, 0110.

43. See Figure P–66.

45. Increase the $R_{EXT}C_{EXT}$ time constant of the 25 s one-shot by 2.4 times.

47. See Figure P–67.

▼ TABLE P-10

STAGE	OPEN	LOADED COUNT	f_{OUT}
1	0	63C1	250.006 Hz
1	1	63C2	250.012 Hz
1	2	63C4	250.025 Hz
1	3	63C8	250.050 Hz
2	0	63D0	250.100 Hz
2	1	63E0	250.200 Hz
2	2	63C0	250 Hz
2	3	63C0	250 Hz
3	0	63C0	250 Hz
3	1	63C0	250 Hz
3	2	67C0	256.568 Hz
3	3	6BC0	263.491 Hz
4	0	73C0	278.520 Hz
4	1	63C0	250 Hz
4	2	63C0	250 Hz
4	3	E3C0	1.383 kHz

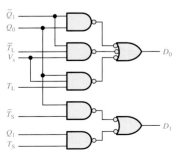

▲ FIGURE P-66

▶ **FIGURE P–67**

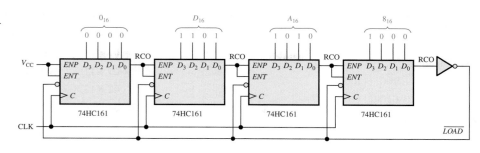

49. See Figure P–68.

51. See Figure P–69.

53. See Figure P–70.

55. Q output of U3 open

57. Pin A of G3 open

59. Pin 9 open

▶ **FIGURE P–68**

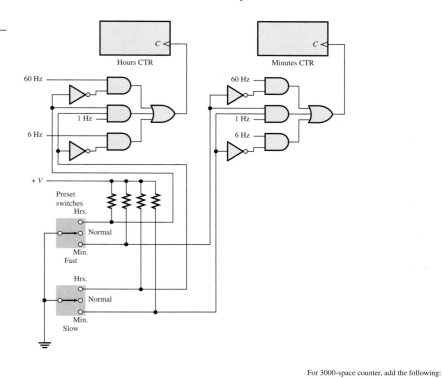

▲ **FIGURE P–69**

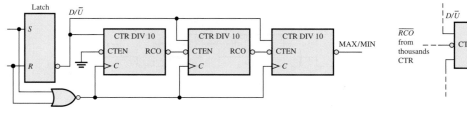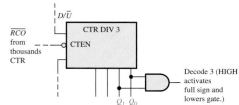

▶ **FIGURE P–70**

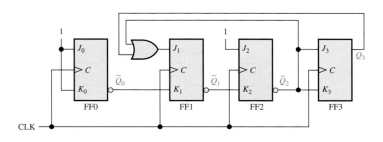

Chapter 9

1. Shift registers store binary data.
3. See Figure P–71.
5. Initially: 101001111000
   CLK1:   010100111100
   CLK2:   001010011110
   CLK3:   000101001111
   CLK4:   000010100111
   CLK5:   100001010011
   CLK6:   110000101001
   CLK7:   111000010100

CLK8:   011100001010
CLK9:   001110000101
CLK10:   000111000010
CLK11:   100011100001
CLK12:   110001110000

7. See Figure P–72.
9. See Figure P–73.
11. See Figure P–74.
13. See Figure P–75.
15. See Figure P–76.
17. See Figure P–77.

▶ FIGURE P–71

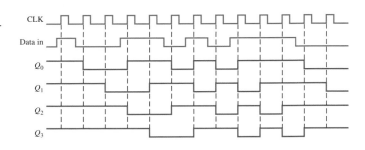

▶ FIGURE P–72

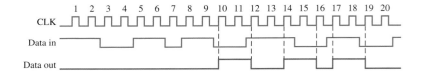

▶ FIGURE P–73

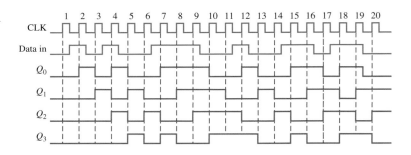

▶ FIGURE P–74

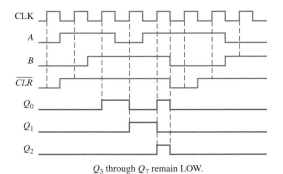

Q_3 through Q_7 remain LOW.

► **FIGURE P–75**

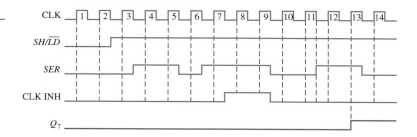

► **FIGURE P–76**

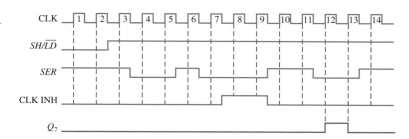

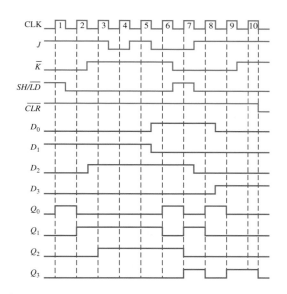

▲ **FIGURE P–77**

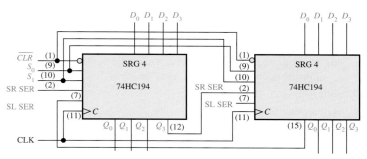

▲ **FIGURE P–78**

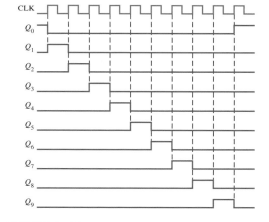

▲ **FIGURE P–79**

19.

Initially (76):	01001100	
CLK1:	10011000	left
CLK2:	01001100	right
CLK3:	00100110	right
CLK4:	00010011	right
CLK5:	00100110	left
CLK6:	01001100	left
CLK7:	00100110	right
CLK8:	01001100	left
CLK9:	00100110	right
CLK10:	01001100	left
CLK11:	10011000	left

21. See Figure P–78.

23. **(a)** 3   **(b)** 5   **(c)** 7   **(d)** 8

25. See Figure P–79.

27. See Figure P–80 on page 850.

29. An incorrect code may be produced.

31. D_3 input open

33. **(a)** No clock at switch closure because of faulty NAND (negative-OR) gate or one-shot; open clock (C) input

▲ FIGURE P–80

to key code register; open $SH/\overline{LD}$ input to key code register

(b) Diode in third row open; Q_2 output of ring counter open

(c) The NAND (negative-OR) gate input connected to the first column is open or shorted.

(d) The "2" input to the column encoder is open.

35. (a) Contents of data output register remain constant.

(b) Contents of both registers do not change.

(c) Third stage output of data output register remains HIGH.

(d) Clock generator is disabled after each pulse by the flip-flop being continuously SET and then RESET.

37. shift register A: 1001

shift register C: 00000100

39. Control flip-flop: 7476

Clock generator: 555

Counter: 74LS163

Data input register: 74LS164

Data output register: 74LS199

One-shot: 74121

41. See Figure P–81.

43. See Figure P–82.

45. CLK input of U3 open

47. Pin 14 open

49. CLK input of U6 open

▶ FIGURE P–81

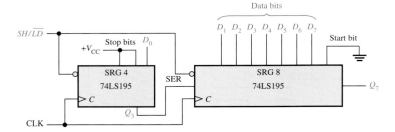

▶ FIGURE P–82

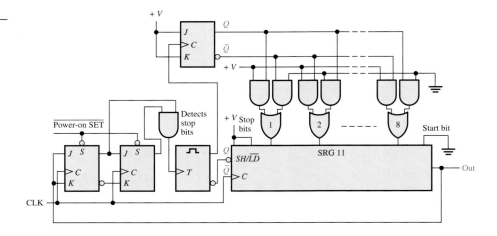

Chapter 10

1. (a) ROM (b) RAM

3. *Address bus* provides for transfer of address code to memory for accessing any memory location in any order for a read or write operation. *Data bus* provides for transfer of data between the microprocessor and the memory or I/O.

5.

	Bit 0	Bit 1	Bit 2	Bit 3
Row 0	1	0	0	0
Row 1	0	0	0	0
Row 2	0	0	1	0
Row 3	0	0	0	0

7. 512 rows × 128 8-bit columns

9. A SRAM stores bits in flip-flops indefinitely as long as power is applied. A DRAM stores bits in capacitors that must be refreshed periodically to retain the data.

11. See Table P–11.

▼ TABLE P–11

INPUTS		OUTPUTS			
A_1	A_0	O_3	O_2	O_1	O_0
0	0	0	1	0	1
0	1	1	0	0	1
1	0	1	1	1	0
1	1	0	0	1	0

13. See Figure P–83.

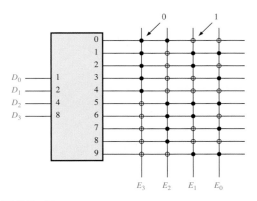

▲ FIGURE P–83

15. Blown links: 1–17, 19–23, 25–31, 34, 37, 38, 40–47, 53, 55, 58, 61, 62, 63, 65, 67, 69

17. Use eight 16k × 4 DRAMs with sixteen address lines. Two of the address lines are decoded to enable the selected memory chips. Four data lines go to each chip.

19. 8 bits, 64k words; 4 bits, 256k words

21. lowest address: $FC0_{16}$

   highest address: FFF_{16}

23. A hard disk is formatted into tracks and sectors. Each track is divided into a number of sectors with each sector of a track having a physical address. Hard disks typically have from a few hundred to a few thousand tracks.

25. Magnetic tape has a longer access time than disk because data must be accessed sequentially rather than randomly.

27. Checksum content is in error.

29. (a) ROM 2 (b) ROM 1 (c) All ROMs

31. 10

33. PROM will retain code when power is off. The code in PROM cannot be readily changed unless it is an EEPROM.

35. To accommodate a 5-bit entry code, shift register C must be loaded with five 0s instead of four. The HIGH (1) must be moved left one place on the parallel inputs.

Chapter 11

1. $X = \overline{A}\,\overline{B}\,\overline{C} + \overline{A}B\overline{C} + AB\overline{C}$

3. (a) PAL16L2 is a programmable array logic device with 16 inputs and two active-LOW outputs.

   (b) PAL12H6 is a programmable array logic device with 12 inputs and 6 active-HIGH outputs.

5. A CPLD basically consists of multiple SPLDs that can be connected with a programmable interconnect array.

7. (a) $\overline{A}BC\overline{D}$ (b) $ABC(\overline{D} + \overline{E}) = ABC\overline{D} + ABC\overline{E}$

9. $X = A\overline{B} + \overline{A}B$

11. $X_1 = \overline{A}\overline{B}C\overline{D} + \overline{A}BCD + ABC\overline{D}$;
   $X_2 = ABCD + AB\overline{C}D + \overline{A}BC\overline{D} + \overline{A}\overline{B}CD$

13. (a) Combinational; 1 (b) Registered; 0

15. (a) Registered (b) GCK1 (c) 0 (d) 0

17. SOP output $= \overline{A}\,\overline{B}\,\overline{C} + \overline{A}\,\overline{B}C + \overline{A}B\overline{C} + A\overline{B}C + AB\overline{C}$

19. LUT for combinational logic, adder logic, and register logic

21. See Figure P–84.

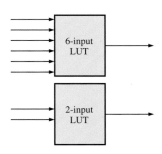

▲ FIGURE P–84

23. See Figure P–85 on page 852.

25. One slice

27. See Figure P–86.

▶ FIGURE P–85

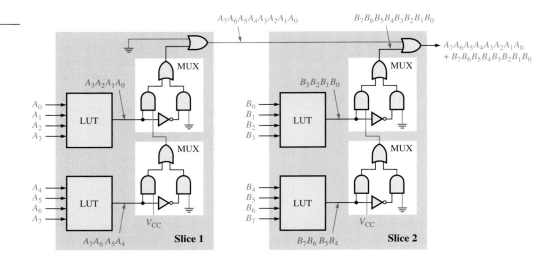

$A_7A_6A_5A_4A_3A_2A_1A_0$ $B_7B_6B_5B_4B_3B_2B_1B_0$

$A_7A_6A_5A_4A_3A_2A_1A_0$
$+ B_7B_6B_5B_4B_3B_2B_1B_0$

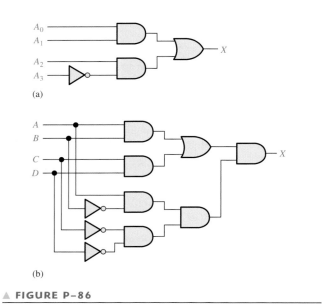

(a)

(b)

▲ FIGURE P–86

29. See Figure P–87.

31. Shift input = 1, data are applied to SDI, go through the MUX, and are clocked into Capture register A on the leading edge of the clock pulse. From the output of Capture register A, the data go through the upper MUX and are clocked into Capture register B on the trailing edge of the clock pulse.

33. PDI/O = 0 and OE = 0. The data are applied to the input pin and go through the selected MUX to the internal programmable logic.

35. 000011001010001111011

0	000011001010001111011
1	000011001010001111011
3	000011001010001111011
6	000011001010001111011
12	000011001010001111011
9	000011001010001111011
2	000011001010001111011
5	000011001010001111011
10	000011001010001111011
4	000011001010001111011

▶ FIGURE P–87

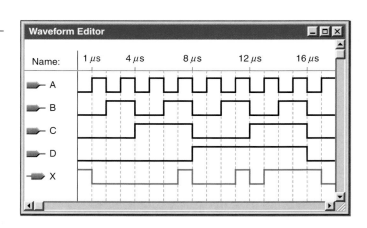

8	00001100101010001111011
1	00001100101010001111011
3	00001100101010001111011
7	00001100101010001111011
15	00001100101010001111011
14	00001100101010001111011
13	00001100101010001111011
11	00001100101010001111011

37. The D input to the logic is faulty or not connected.

Chapter 12

1. CPU, memory, I/O ports, buses

3. A bus is a set of connections and electrical specifications for moving information in a computer.

5. ALU, instruction decoder, register array, and control unit

7. Address bus, data bus, and control bus

9. Fetch, decode, execute

11. CS, DS, SS, ES, FS, GS

13. AH and AL are 8-bit registers and represent the high and low part of the 16-bit AX register. The EAX is a 32-bit register which includes the AX register as the lower 16 bits.

15. Pairing allows two instructions to execute at the same time.

17. See Figure P–88.

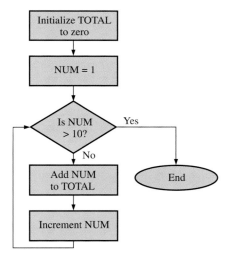

▲ **FIGURE P–88**

19. When the instruction **mov ax, [bx]** is executed, the word in memory pointed to by the bx register is copied to the ax register.

21. In a polled I/O, the CPU polls each device in turn to see if it needs service; in an interrupt-driven system, the peripheral device signals the CPU when it requires service.

23. A program instruction that invokes an interrupt service routine.

25. The CPU is bypassed in DMA.

27. See Figure P–89.

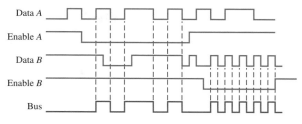

▲ **FIGURE P–89**

29. The local bus is the collection of buses interfacing directly with the processor. The PCI bus is used for expansion devices and is connected to the local bus through a bus controller.

31. The PCI bus is a 33- or 66 MHz, 32- or 66-bit expansion bus. The ISA bus is an 8.33 MHz, 8- or 16-bit expansion bus.

33. DCE stands for data communications equipment, such as a modem. DTE stands for data terminal equipment, such as a computer. Both acronyms are associated with the RS-232/EIA-232 standard.

35. six

37. A controller is sending data to two listeners. The first two bytes of data (3F and 41) go the listener with address 001A. The second two bytes go to the listener with address 001B. The handshaking signals ($\overline{DAV}$, $\overline{NRFD}$, and $\overline{NDAC}$) indicate the data transfer is successful. See Figure P–90.

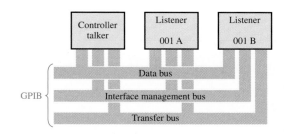

▲ **FIGURE P–90**

Chapter 13

1. An analog-to-digital converter converts an analog signal to a digital code.

3. A digital-to-analog converter changes a digital code to the corresponding analog signal.

5. See Figure P–91 on page 854.

7. 1000, 1110, 1011, 0100, 0001, 0111, 1110, 1011, 0100.

9. See Figure P–92.

11. 330 kΩ

13. 000, 001, 100, 110, 101, 100, 011, 010, 001, 001, 011, 110, 111, 111, 111, 111, 111, 111, 111, 100

15. See Table P–12.

17. 8000 MB/s

▶ **FIGURE P–91**

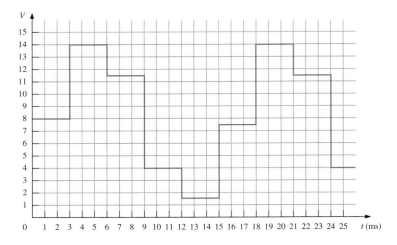

▶ **FIGURE P–92**

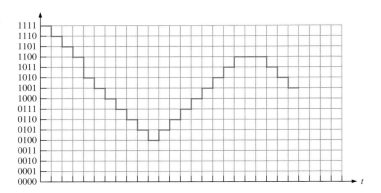

▼ **TABLE P–12**

SAR	COMMENT
1000	Greater than V_{in}, reset MSB
0100	Less than V_{in}, keep the I
0110	Equal to V_{in}, keep the (final state)

19. 1,000,000,000

21. Instruction dispatch (DP): Instruction packets are split into execute packets and assigned to functional units; Instruction decode (DC): Instructions are decoded.

23. See Figure P–93.

25. (a) 14.3% (b) 0.098% (c) 0.00038%

27. See Figure P–94.

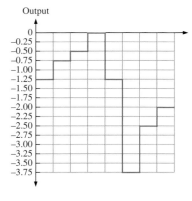

▲ **FIGURE P–93**

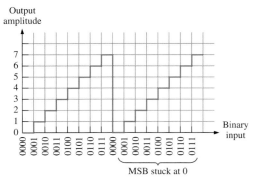

▲ **FIGURE P–94**

Chapter 14

1. No; $V_{OH(min)} < V_{IH(min)}$

3. 0.15 V in HIGH state; 0.25 V in LOW state.

5. Gate C **7.** 12 ns

9. Gate C **11.** Yes, G_2

13. (a) on (b) off
(c) off (d) on

15. See Figure P–95 for one possible circuit.

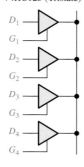

74HC125 (Tristate)

▲ FIGURE P–95

▶ FIGURE P–96

17. (a) HIGH (b) Floating
(c) HIGH (d) High-Z

19. (a) LOW (b) LOW (c) LOW

21. See Figure P–96.

23. (a) $R_p = 198\ \Omega$
(b) $R_p = 198\ \Omega$
(c) $R_p = 198\ \Omega$

25. ALVC

27. (a) A, B to X: 9.9 ns
C, D to X: 6.6 ns
(b) A to X_1, X_2, X_3: 14 ns
B to X_1: 7 ns
C to X_2: 7 ns
D to X_3; 7 ns
(c) A to X: 11.1 ns
B to X: 11.1 ns
C to X: 7.4 ns
D to X: 7.4 ns

29. ECL operates with nonsaturated BJTs.

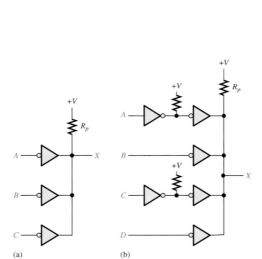

(a) (b)

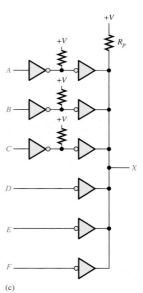

(c)

Glossary

acceptor A receiving device on a bus.

access time The time from the application of a valid memory address to the appearance of valid output data.

addend In addition, the number that is added to another number called the augend.

adder A logic circuit used to add two binary numbers.

address The location of a given storage cell or group of cells in a memory; a unique memory location containing one byte.

address bus A one-way group of conductors from the microprocessor to a memory, or other external device, on which the address code is sent.

adjacency Characteristic of cells in a Karnaugh map in which there is a single-variable change from one cell to another cell next to it on any of its four sides.

aliasing The effect created when a signal is sampled at less than twice the signal frequency. Aliasing creates unwanted frequencies that interfere with the signal frequency.

alphanumeric Consisting of numerals, letters, and other characters.

ALU Arithmetic logic unit; the key processing element of a microprocessor that performs arithmetic and logic operations.

amplitude In a pulse waveform, the height or maximum value of the pulse as measured from its low level.

analog Being continuous or having continuous values, as opposed to having a set of discrete values.

analog-to-digital (A/D) conversion The process of converting an analog signal to digital form.

analog-to-digital converter (ADC) A device used to convert an analog signal to a sequence of digital codes.

AND A basic logic operation in which a true (HIGH) output occurs only when all the input conditions are true (HIGH).

AND array An array of AND gates consisting of a matrix of programmable interconnections.

AND gate A logic gate that produces a HIGH output only when all of the inputs are HIGH.

ANSI American National Standards Institute.

antifuse A type of PLD nonvolatile programmable link that can be left open or can be shorted once as directed by the program.

architecture The VHDL unit that describes the internal operation of a logic function; the internal functional arrangement of the elements that give a device its particular operating characteristics.

array In a PLD, a matrix formed by rows of product-term lines and columns of input lines with a programmable cell at each junction. In VHDL, an array is an ordered set of individual items called elements with a single identifier name.

ASCII American Standard Code for Information Interchange; the most widely used alphanumeric code.

assembler A program that converts English-like mnemonics into machine code.

assembly language A programming language that uses English-like words and has a one-to-one correspondence to machine language.

associative law In addition (ORing) and multiplication (ANDing) of three or more variables, the order in which the variables are grouped makes no difference.

astable Having no stable state. An astable multivibrator oscillates between two quasi-stable states.

asynchronous Having no fixed time relationship; not occurring at the same time.

asynchronous counter A type of counter in which each stage is clocked from the output of the preceding stage.

augend In addition, the number to which the addend is added.

base One of the three regions in a bipolar junction transistor.

base address The beginning address of a segment of memory.

BCD Binary coded decimal; a digital code in which each of the decimal digits, 0 through 9, is represented by a group of four bits.

BEDO DRAM Burst extended data output dynamic random-access memory.

bed-of-nails A method for the automated testing of printed circuit boards in which the board is mounted on a fixture that resembles a bed of nails that makes contact with test points.

bidirectional Having two directions. In a bidirectional shift register, the stored data can be shifted right or left.

binary Having two values or states; describes a number system that has a base of two and utilizes 1 and 0 as its digits.

BIOS Basic input/output system; a set of programs in ROM that interfaces the I/O devices in a computer system.

bipolar Having two opposite charge carriers within the transistor structure.

bistable Having two stable states. Flip-flops and latches are bistable multivibrators.

bit A binary digit, which can be either a 1 or 0.

bitstream A series of bits describing a final design that is sent to the target device during programming.

bit time The interval of time occupied by a single bit in a sequence of bits; the period of the clock.

BIU Bus interface unit; the portion of the CPU that interfaces with the system buses and fetches instructions, reads operands, and writes results.

BJT Bipolar junction transistor; a semiconductor device used for switching or amplification. A BJT has two junctions, the base-emitter junction and the base-collector junction.

Boolean addition In Boolean algebra, the OR operation.

Boolean algebra The mathematics of logic circuits.

Boolean expression An expression of variables and operators used to express the operation of a logic circuit.

Boolean multiplication In Boolean algebra, the AND operation.

boundary scan A method for internally testing a PLD based on the JTAG standard (IEEE Std. 1149.1).

buffer A circuit that prevents loading of an input or output.

bus A set of interconnections that interface one or more devices based on a standardized specification.

bus arbitration The process that prevents two sources from using a bus at the same time.

bus contention An adverse condition that could occur if two or more devices try to communicate at the same time on a bus.

byte A group of eight bits.

cache memory A relatively small, high-speed memory that stores the most recently used instructions or data from the larger but slower main memory.

capacity The total number of data units (bits, nibbles, bytes, words) that a memory can store.

carry The digit generated when the sum of two binary digits exceeds 1.

carry generation The process of producing an output carry in a full-adder when both input bits are 1s.

carry propagation The process of rippling an input carry to become the output carry in a full-adder when either or both of the input bits are 1s and the input carry is a 1.

cascade To connect "end-to-end" as when several counters are connected from the terminal count output of one counter to the enable input of the next counter.

cascading Connecting the output of one device to the input of a similar device, allowing one device to drive another in order to expand the operational capability.

CCD Charge-coupled device; a type of semiconductor memory that stores data in the form of charge packets and is serially accessed.

CD-R CD-Recordable; an optical disk storage device on which data can be stored once.

CD-ROM An optical disk storage device on which data are prestored and can only be read.

CD-RW CD-Rewritable; an optical disk storage on which data can be written and overwritten many times.

cell An area on a Karnaugh map that represents a unique combination of variables in product form; a single storage element in a memory; a fused cross point of a row and column in a PLD; a single storage element in a memory.

character A symbol, letter, or numeral.

circuit An arrangement of electrical and/or electronic components interconnected in such a way as to perform a specified function.

CLB Configurable logic block; a unit of logic in an FPGA that is made up of multiple smaller logic modules and a local programmable interconntect that is used to connect logic modules within the CLB.

clear An asynchronous input used to reset a flip-flop (make the Q output 0); to place a register or counter in the state in which it contains all 0s.

clock The basic timing signal in a digital system; a periodic waveform in which the interval between pulses equals the time for one bit; the triggering input of a flip-flop.

CMOS Complementary metal oxide semiconductor; a class of integrated logic circuits that is implemented with a type of field-effect transistor.

code A set of bits arranged in a unique pattern and used to represent such information as numbers, letters, and other symbols; in VHDL, program statements.

codec A combined coder and decoder.

collector One of the three regions in a bipolar transistor.

combinational logic A combination of logic gates interconnected to produce a specified Boolean function with no storage or memory capability; sometimes called *combinatorial logic*.

commutative law In addition (ORing) and multiplication (ANDing) of two variables, the order in which the variables are ORed or ANDed makes no difference.

comparator A digital circuit that compares the magnitudes of two quantities and produces an output indicating the relationship of the quantities.

compiler An application program in development software packages that controls the design flow process and translates source code into object code in a format that can be logically tested or downloaded to a target device.

complement The inverse or opposite of a number; in Boolean algebra, the inverse function, expressed with a bar over the variable. The complement of a 1 is a 0, and vice versa.

component A VHDL feature that can be used to predefine the logic function for multiple use throughout a program or programs.

contiguous Joined together.

control bus A set of conductive paths that connects the CPU to other parts of the computer to coordinate its operations and to communicate with external devices.

controller An instrument that can specify each of the other instruments on the bus as either a talker or a listener for the purpose of data transfer.

control unit The portion within the microprocessor that provides the timing and control signals for getting data into and out of the microprocessor and for synchronizing the execution of instructions.

counter A digital circuit capable of counting electronic events, such as pulses, by progressing through a sequence of binary states.

CPLD A complex programmable logic device that consists basically of multiple SPLD arrays with programmable interconnections.

CPU Central processing unit; the main part of a computer responsible for control and processing of data; the core of a DSP that processes the program instructions.

cross-assembler A program that translates an assembly language program for one type of microprocessor to an assembly language for another type of microprocessor.

current sinking The action of a circuit in which it accepts current into its output from a load.

current sourcing The action of a circuit in which it sends current out of its output and into a load.

DAT Digital audio tape; a type of magnetic tape format.

data Information in numeric, alphabetic, or other form.

data bus A bidirectional set of conductive paths on which data or instruction codes are transferred into a microprocessor or on which the result of an operation is sent out from the microprocessor.

data selector A circuit that selects data from several inputs one at a time in a sequence and places them on the output; also called a multiplexer.

data sheet A document that specifies parameter values and operating conditions for an integrated circuit or other device.

DCE Data communications equipment.

Debug A code within DOS that allows various operations on files and includes a primitive assembler; to eliminate a problem in hardware or software.

decade Characterized by ten states or values.

decade counter A digital counter having ten states.

decimal Describes a number system with a base of ten.

decode A stage of the DSP pipeline operation in which instructions are assigned to functional units and are decoded.

decoder A digital circuit (device) that converts coded information into another (familiar) or noncoded form.

decrement To decrease the binary state of a counter by one.

delta modulation A method of analog-to-digital conversion using a 1-bit quantization process.

design flow The process or sequence of operations carried out to program a target device.

D flip-flop A type of bistable multivibrator in which the output assumes the state of the D input on the triggering edge of a clock pulse.

demultiplexer (demux) A circuit (digital device) that switches digital data from one input line to several output lines in a specified time sequence.

dependency notation A notational system for logic symbols that specifies input and output relationships, thus fully defining a given function; an integral part of ANSI/IEEE Std. 91-1984.

difference The result of a subtraction.

digit A symbol used to express a quantity.

digital Related to digits or discrete quantities; having a set of discrete values as opposed to continuous values.

digital-to-analog (D/A) conversion The process of converting a sequence of digital codes to an analog form.

digital-to-analog converter (DAC) A device in which information in digital form is converted to analog form.

DIMM Dual in-line memory module.

diode A semiconductor device that conducts current in only one direction.

DIP Dual in-line package; a type of IC package whose leads must pass through holes to the other side of a PC board.

distributive law The law that states that ORing several variables and then ANDing the result with a single variable is equivalent to ANDing the single variable with each of the several variables and then ORing the product.

dividend In a division operation, the quantity that is being divided.

divisor In a division operation, the quantity that is divided into the dividend.

DLT Digital linear tape; a type of magnetic tape format.

DMA Direct memory access; a method to directly interface a peripheral device to memory without using the CPU for control.

domain All of the variables in a Boolean expression.

"Don't care" A combination of input literals that cannot occur and can be used as a 1 or a 0 on a Karnaugh map for simplification.

downloading A design flow process in which the logic design is transferred from software to hardware.

drain One of the terminals of a field-effect transistor.

DRAM Dynamic random-access memory; a type of semiconductor memory that uses capacitors as the storage elements and is a volatile, read/write memory.

DSP Digital signal processor; a special type of microprocessor that processes data in real time.

DSP core The central processing unit of a digital system processor.

DTE Data terminal equipment.

duty cycle The ratio of pulse width to period expressed as a percentage.

DVD-ROM Digital versatile disk-ROM; also known as digital video disk-ROM; a type of optical storage device on which data is prestored with a much higher capacity than a CD-ROM.

dynamic memory A type of semiconductor memory having capacitive storage cells that lose stored data over a period of time and, therefore, must be refreshed.

ECL Emitter-coupled logic; a class of integrated logic circuits that are implemented with nonsaturating bipolar junction transistors.

E²CMOS Electrically erasable CMOS (EECMOS); the circuit technology used for the reprogrammable cells in a PLD.

edge-triggered flip-flop A type of flip-flop in which the data are entered and appear on the output on the same clock edge.

EDIF Electronic design interchange format; a standard form of netlist.

EDO DRAM Extended data output dynamic random-access memory.

EEPROM Electrically erasable programmable read-only memory; a type of nonvolatile PLD programmable link based on electrically-erasable programmable read-only memory cells and can be turned on or off repeatedly by programming.

emitter One of the three regions in a bipolar junction transistor.

enable To activate or put into an operational mode; an input on a logic circuit that enables its operation.

encoder A digital circuit (device) that converts information to a coded form.

entity The VHDL unit that describes the inputs and outputs of a logic function.

EPROM Erasable programmable read-only memory; A type of PLD nonvolatile programmable link based on electrically programmable read-only memory cells and can be turned either on or off once with programming.

error detection The process of detecting bit errors in a digital code.

EU Execution unit; the portion of a CPU that executes instructions; it contains the arithmetic logic unit (ALU), the general registers, and the flags.

even parity The condition of having an even number of 1s in every group of bits.

exclusive-NOR (XNOR)gate A logic gate that produces a LOW only when the two inputs are at opposite levels.

exclusive-OR (XOR) A basic logic operation in which a HIGH occurs when the two inputs are at opposite levels.

exclusive-OR (XOR) gate A logic gate that produces a HIGH only when the two inputs are at opposite levels.

execute A CPU process in which an instruction is carried out; a stage of the DSP pipeline operation in which the decoded instructions are carried out.

exponent The part of a floating-point number that represents the number of places that the decimal point (or binary point) is to be moved.

fall time The time interval between the 90% point and the 10% point on the negative-going edge of a pulse.

fan-out The number of equivalent gate inputs of the same family series that a logic gate can drive.

feedback The output voltage or a portion of it that is connected back to the input of a circuit.

FET Field-effect transistor.

fetch A CPU process in which an instruction is obtained from the memory; a stage of the DSP pipeline operation in which an instruction is obtained from the program memory.

fifo First in—first out memory.

FireWire The IEEE-1394 standard serial bus.

fitter tool A compiler software tool that selects the optimum interconnections, pin assignments, and logic cell assignments to fit a design into the selected target device.

flag A bit that indicates the result of an arithmetic or logic operation or is used to alter an operation.

flash ADC A simultaneous analog-to-digital converter.

flash memory A nonvolatile read/write random-access semiconductor memory in which data is stored as charge on the floating gate of a certain FET.

flip-flop A basic storage circuit that can store only one bit at a time; a synchronous bistable device.

floating-point number A number representation based on scientific notation in which the number consists of an exponent and a mantissa.

floppy disk A magnetic storage device; a flexible disk with a diameter of 3.5 inches and a storage capacity of 1.44 MB encased in a rigid plastic housing.

flying probe A method for the automated testing of printed circuit boards, in which a probe or probes move from place to place to contact test points.

forward bias A voltage polarity condition that allows a semiconductor *pn* junction in a transistor or diode to conduct current.

FPGA Field programmable gate array; a programmable logic device that uses the LUT as the basic logic elements and generally employs either antifuse or SRAM-based process technology.

FPM DRAM Fast page mode dynamic random-access memory.

frequency (*f*) The number of pulses in one second for a periodic waveform. The unit of frequency is the hertz.

full-adder A digital circuit that adds two bits and an input carry to produce a sum and an output carry.

functional simulation A software process that tests the logical or functional operation of a design.

fuse A type of PLD nonvolatile programmable link that can be left shorted or can be opened once as directed by the program; also called a fusible link.

GAL Generic array logic; a reprogrammable type of SPLD that is similar to a PAL except that it uses a reprogrammable process technology, such as EEPROM (E^2 CMOS), instead of fuses.

gate A logic circuit that performs a specified logic operation, such as AND or OR; one of the three terminals of a field-effect transistor.

glitch A voltage or current spike of short duration, usually unintentionally produced and unwanted.

graphic (schematic) entry A method of entering a logic design into software by graphically creating a logic diagram (schematic) on a design screen.

GPIB General-purpose interface bus based on the IEEE 488 standard.

Gray code An unweighted digital code characterized by a single bit change between adjacent code numbers in a sequence.

half-adder A digital circuit that adds two bits and produces a sum and an output carry. It cannot handle input carries.

Hamming code A type of error-correction code.

handshaking The process of signal interchange by which two digital devices or systems jointly establish communication.

hard core A fixed portion of logic in an FPGA that is put in by the manufacturer to provide a specific function.

hard disk A magnetic disk storage device; typically, a stack of two or more rigid disks enclosed in a sealed housing.

hardware The circuitry and physical components of a computer system (as opposed to the directions called software).

HDL Hardware description language; a language used for describing a logic design using software.

hexadecimal Describes a number system with a base of 16.

high-level language A type of computer language closest to human language that is a level above assembly language.

high-Z The high-impedance state of a tristate circuit in which the output is effectively disconnected from the rest of the circuit.

hold time The time interval required for the control levels to remain on the inputs to a flip-flop after the triggering edge of the clock in order to reliably activate the device.

HPIB Hewlett-Packard interface bus; same as GPIB (general-purpose interface bus).

hysteresis A characteristic of a threshold-triggered circuit, such as the Schmitt trigger, where the device turns on and off at different input levels.

IEEE Institute of Electrical and Electronics Engineers.

IEEE 488 bus Same as GPIB (general-purpose interface bus); a standard parallel bus used widely for test and measurement interfacing.

IEEE 1394 A serial bus for high-speed data transfer; also known as FireWire.

I²L Integrated injection logic; an IC technology.

Implementation The software process where the logic structures described by the netlist are mapped into the structure of the target device.

increment To increase the binary state of a counter by one.

input The signal or line going into a circuit; a signal that controls the operation of a circuit.

input/output (I/O) A terminal of a device that can be used as either an input or as an output.

instruction One step in a computer program; a unit of information that tells the CPU what to do.

instruction pairing The process of combining certain independent instructions so that they can be executed simultaneously by two separate execution units.

integer A whole number.

integrated circuit (IC) A type of circuit in which all of the components are integrated on a single chip of semiconductive material of very small size.

intellectual property (IP) Designs owned by the manufacturer of programmable logic devices.

interfacing The process of making two or more electronic devices or systems operationally compatible with each other so that they function properly together.

interrupt A computer signal or instruction that causes the current process to be temporarily stopped while a service routine is run.

inversion The conversion of a HIGH level to a LOW level or vice versa; also called complementation.

inverter A NOT circuit; a circuit that changes a HIGH to a LOW or vice versa.

I/O port Input/output port; the interface between an internal bus and a peripheral.

IP Instruction pointer; a special register within the CPU that holds the offset address of the next instruction to be executed.

ISA bus Industry standard architecture bus; an internal parallel bus standard.

ISP In-system programming; a method for programming SPLDs after they are installed on a printed circuit board and operating in a system.

Jaz cartridge A magnetic storage device; hard disks encased in a rigid plastic cartridge with storage capacities of 1 GB or 2 GB.

J-K flip-flop A type of flip-flop that can operate in the SET, RESET, no-change, and toggle modes.

Johnson counter A type of register in which a specific prestored pattern of 1s and 0s is shifted through the stages, creating a unique sequence of bit patterns.

JTAG Joint test action group; the IEEE Std. 1149.1 standard interface for in-system programming.

junction The boundary between an n region and a p region in a BJT.

Karnaugh map An arrangement of cells representing the combinations of literals in a Boolean expression and used for a systematic simplification of the expression.

LAB Logic array block; an SPLD array in a CPLD.

latch A bistable digital circuit used for storing a bit.

latency period The time it takes for the desired sector to spin under the head once the head is positioned over the desired track of a magnetic hard disk.

LCCC Leadless ceramic chip carrier; an SMT package that has metallic contacts molded into its body.

LCD Liquid crystal display.

leading edge The first transition of a pulse.

least significant bit (LSB) Generally, the right-most bit in a binary whole number or code.

LED Light-emitting diode.

LIFO Last in—first out memory, memory stack.

listener An instrument capable of receiving data on a GPIB (general-purpose interface bus).

literal A variable or the complement of a variable.

load To enter data into a shift register.

local bus An internal bus that connects the microprocessor to the cache memory, the main memory, the coprocessor, and the PCI bus controller.

local interconnect A set of lines that allows interconnections among the eight logic elements in a logic array block without using the row and column interconnects.

logic In digital electronics, the decision-making capability of gate circuits, in which a HIGH represents a true statement and a LOW represents a false one.

logic array block (LAB) A group of macrocells that can be interconnected with other LABs or to other I/Os using a programmable interconnect array; also called a function block.

logic element The smallest section of logic in an FPGA that typically contains an LUT, associated logic, and a flip-flop.

look-ahead carry A method of binary addition whereby carries from preceding adder stages are anticipated, thus eliminating carry propagation delays.

LSI Large-scale integration; a level of fixed-function IC complexity in which there are from more than 100 to 10,000 equivalent gates per chip.

LUT Look-up table; a type of memory that can be programmed to produce SOP functions.

machine code The basic binary instructions understood by the processor.

machine language Computer instructions written in binary code that are understood by a computer; the lowest level of programming language.

macrocell An SOP logic array with combinational and registered outputs; part of a PAL or GAL that generally consists of one OR gate and some associated output logic. Multiple interconnected macrocells form a CPLD.

magneto-optical disk A storage device that uses electromagnetism and a laser beam to read and write data.

magnitude The size or value of a quantity.

mantissa The magnitude of a floating-point number.

memory array An array of memory cells arranged in rows and columns.

MFLOPS Million floating-point operations per second.

microprocessor A large-scale digital integrated circuit device that can be programmed with a series of instructions to perform specified functions on data.

minimization The process that results in an SOP or POS Boolean expression that contains the fewest possible terms with the fewest possible literals per term.

minuend The number from which another number is subtracted.

MIPS Million instructions per second.

MMACS Million multiply/accumulates per second.

mnemonic An English-like instruction that is converted by an assembler into a machine code for use by a processor.

modem A modulator/demodulator for interfacing digital devices to analog transmission systems such as telephone lines.

modulus The number of unique states through which a counter will sequence.

monostable Having only one stable state. A monostable multivibrator, commonly called a one-shot, produces a single pulse in response to a triggering input.

monotonicity The characteristic of a DAC defined by the absence of any incorrect step reversals; one type of digital-to-analog linearity.

MOS Metal-oxide semiconductor; a type of transistor technology.

MOSFET Metal-oxide semiconductor field-effect transistor.

most significant bit (MSB) The left-most bit in a binary whole number or code.

MSI Medium-scale integration; a level of fixed-function IC complexity in which there are from 10 to 100 equivalent gates per chip.

multiplexer (mux) A circuit (digital device) that switches digital data from several input lines onto a single output line in a specified time sequence.

multiplicand The number that is being multiplied by another number.

multiplier The number that multiplies the multiplicand.

multivibrator A class of digital circuits in which the output is connected back to the input (an arrangement called feedback) to produce either two stable states, one stable state, or no stable states, depending on the configuration.

NAND gate A logic circuit in which a LOW output occurs only if all the inputs are HIGH.

negative-AND An equivalent NOR gate operation in which the HIGH is the active input when all inputs are LOW.

negative-OR An equivalent NAND gate operation in which the HIGH is the active input when one or more of the inputs are LOW.

netlist A detailed listing of information necessary to describe a circuit, such as types of elements, inputs, and outputs, and all interconnections.

nibble A group of four bits.

NMOS An *n*-channel metal-oxide semiconductor.

node A common connection point in a circuit in which a gate output is connected to one or more gate inputs.

noise immunity The ability of a circuit to reject unwanted signals.

noise margin The difference between the maximum LOW output of a gate and the maximum acceptable LOW input of an equivalent gate; also, the difference between the minimum HIGH output of a gate and the minimum HIGH input of an equivalent gate.

nonvolatile A term that describes a memory that can retain stored data when the power is removed.

NOR gate A logic gate in which the output is LOW when any or all of the inputs are HIGH.

NOT A basic logic operation that performs inversions.

numeric Related to numbers.

Nyquist frequency The highest signal frequency that can be sampled at a specified sampling frequency; a frequency equal to or less than half the sampling frequency.

object program A machine language translation of a high-level source program.

octal Describes a number system with a base of eight.

odd parity The condition of having an odd number of 1s in every group of bits.

offset address The distance in number of bytes of a physical address from the base address.

OLMC Output logic macrocell; the part of a GAL that can be programmed for either combinational or registered outputs; a block of logic in a GAL that contains a fixed OR gate and other logic for handling inputs and/or outputs.

one-shot A monostable multivibrator.

op code Operation code; the code representing a particular microprocessor instruction; a mnemonic.

open-collector A type of output in a logic circuit in which the collector of the output transistor is left disconnected from any internal circuitry and is available for external connection; normally used for driving higher-current or higher-voltage loads.

operational amplifier (op-amp) A device with two differential inputs that has very high gain, very high input impedance, and very low output impedance.

OR A basic logic operation in which a true (HIGH) output occurs when one or more of the input conditions are true (HIGH).

OR gate A logic gate that produces a HIGH output when one or more inputs are HIGH.

oscillator An electronic circuit that is based on the principle of regenerative feedback and produces a repetitive output waveform; a signal source.

OTP One-time programmable.

output The signal or line coming out of a circuit.

overflow The condition that occurs when the number of bits in a sum exceeds the number of bits in each of the numbers added.

PAL Programmable array logic; a type of one-programmable SPLD that consists of a programmable array of AND gates that connects to a fixed array of OR gates.

parallel In digital systems, data occurring simultaneously on several lines; the transfer or processing of several bits simultaneously.

parity In relation to binary codes, the condition of evenness or oddness of the number of 1s in a code group.

parity bit A bit attached to each group of information bits to make the total number of 1s odd or even for every group of bits.

PCI bus Peripheral control interconnect bus; an internal parallel bus standard.

period (T) The time required for a periodic waveform to repeat itself.

periodic Describes a waveform that repeats itself at a fixed interval.

peripheral A device or instrument that provides communication with a computer or provides auxiliary services or functions for the computer.

physical address The actual location of a data unit in memory.

PIC Programable interrupt controller; handles the interrupts on a priority basis.

pipeline As applied to memories, an implementation that allows a read or write operation to be initiated before the previous operation is completed; part of the DSP architecture that allows multiple instructions to be processed simultaneously.

PLA Programmable logic array; an SPLD with programmable AND and OR arrays.

platform FPGA An FPGA that contains either or both hard core and soft core embedded processors and other functions.

PLCC Plastic leaded chip carrier; an SMT package whose leads are turned up under its body in a J-type shape.

PLD Programmable logic device; an integrated circuit that can be programmed with any specified logic function.

PMOS A *p*-channel metal-oxide semiconductor.

pointer The contents of a register (or registers) that contain an address.

polling The process of checking a series of peripheral devices to determine if any require service from the CPU.

port A physical interface on a computer through which data are passed to or from peripherals.

positive logic The system of representing a binary 1 with a HIGH and a binary 0 with a LOW.

power dissipation The product of the dc supply voltage and the dc supply current in an electronic circuit; the amount of power required by a circuit.

prefetching The process of executing instructions at the same time as other instructions are "fetched," eliminating idle time; also called pipelining.

preset An asynchronous input used to set a flip-flop (make the Q output 1).

primitive A basic logic element such as a gate or flip-flop, input/output pins, ground, and V_{CC}.

priority encoder An encoder in which only the highest value input digit is encoded and any other active input is ignored.

probe An accessory used to connect a voltage to the input of an oscilloscope or other instrument.

product The result of a multiplication.

product-of-sums (POS) A form of Boolean expression that is basically the ANDing of ORed terms.

product term The Boolean product of two or more literals equivalent to an AND operation.

program A list of computer instructions arranged to achieve a specific result; software.

programmable interconnect array (PIA) An array consisting of conductors that run throughout the CPLD chip and to which connections from the macrocells in each LAB can be made.

PROM Programmable read-only semiconductor memory; an SPLD with a fixed AND array and programmable OR array; used as a memory device and normally not as a logic circuit device.

propagation delay time The time interval between the occurrence of an input transition and the occurrence of the corresponding output transition in a logic circuit.

pseudo-operation An instruction to the assembler (as opposed to a processor).

pull-up resistor A resistor with one end connected to the dc supply voltage used to keep a given point in a circuit HIGH when in the inactive state.

pulse A sudden change from one level to another, followed after a time, called the pulse width, by a sudden change back to the original level.

pulse width (t_w) The time interval between the 50% points of the leading and trailing edges of the pulse; the duration of the pulse.

QIC Quarter-inch cassette; a type of magnetic tape.

quantization The process whereby a binary code is assigned to each sampled value during analog-to-digital conversion.

queue A high-speed memory that stores instructions or data.

quotient The result of a division.

race A condition in a logic network in which the difference in propagation times through two or more signal paths in the network can produce an erroneous output.

RAM Random-access memory; a volatile read/write semiconductor memory.

read The process of retrieving data from a memory.

real mode Operation of an Intel processor in a manner to emulate the 8086's 1 MB of memory.

recycle To undergo transition (as in a counter) from the final or terminal state back to the initial state.

refresh To renew the contents of a dynamic memory by recharging the capacitor storage cells.

register A digital circuit capable of storing and shifting binary information; typically used as a temporary storage device.

register array A set of temporary storage locations within the microprocessor for keeping data and addresses that need to be accessed quickly by the program.

registered A CPLD macrocell output configuration where the output comes from a flip-flop.

relocatable code A program that can be moved anywhere within the memory space without changing the basic code.

remainder The amount left over after a division.

RESET The state of a flip-flop or latch when the output is 0; the action of producing a RESET state.

resolution The number of bits used in an ADC.

reverse bias A voltage polarity condition that prevents a *pn* junction of a transistor or diode from conducting current.

ring counter A register in which a certain pattern of 1s and 0s is continuously recirculated.

ripple carry A method of binary addition in which the output carry from each adder becomes the input carry of the next higher-order adder.

ripple counter An asynchronous counter.

rise time The time required for the positive-going edge of a pulse to go from 10% of its full value to 90% of its full value.

ROM Read-only semiconductor memory, accessed randomly; also referred to as mask-ROM.

sampling The process of taking a sufficient number of discrete values at points on a waveform that will define the shape of the waveform.

schematic (graphic) entry A method of placing a logic design into software using schematic symbols.

Schottky A specific type of transistor-transistor logic circuit technology.

SCSI Small computer system interface; an external parallel bus standard.

SDRAM Synchronous dynamic random-access memory.

seek time The time for the read/write head in a hard drive to position itself over the desired track for a read operation.

segment A 64k block of memory.

sequential circuit A digital circuit whose logic states follow a specified time sequence.

serial Having one element following another, as in a serial transfer of bits; occurring, as pulses, in sequence rather than simultaneously.

SET The state of a flip-flop or latch when the output is 1; the action of producing a SET state.

set-up time The time interval required for the control levels to be on the inputs to a digital circuit, such as a flip-flop, prior to the triggering edge of clock pulse.

shift To move binary data from stage to stage within a shift register or other storage device or to move binary data into or out of the device.

signal A type of VHDL object that holds data.

signal tracing A troubleshooting technique in which waveforms are observed in a step-by-step manner beginning at the input and working toward the output or vice versa. At each point the observed waveform is compared with the correct signal for that point.

sign bit The left-most bit of a binary number that designates whether the number is positive (0) or negative (1).

SIMM Single-in-line memory module.

SMT Surface-mount technology; an IC package technique in which the packages are smaller than DIPs and are mounted on the printed surface of the PC board.

soft core A portion of logic in an FPGA; similar to hard core except it has some programmable features.

software Computer programs; programs that instruct a computer what to do in order to carry out a given set of tasks.

software interrupt An instruction that invokes an interrupt service routine.

SOIC Small-outline integrated circuit; an SMT package that resembles a small DIP but has its leads bent out in a "gull-wing" shape.

source A sending device of a bus; one of the terminals of a field-effect transistor.

source program A program written in either assembly or high-level language.

speed-power product A performance parameter that is the product of the propagation delay time and the power dissipation in a digital circuit.

SPLD Simple programmable logic device; an array of AND gates and OR gates that can be programmed to achieve specified logic functions. Four types are PROM, PLA, PAL, and GAL.

SRAM Static random-access memory; a type of PLD volatile programmable link based on static random-access memory cells and can be turned on or off repeatedly with programming.

S-R flip-flop A SET-RESET flip-flop.

SSI Small-scale integration; a level of fixed-function IC complexity in which there are up to 10 equivalent gates per chip.

SSOP Shrink small-outline package.

stage One storage element (flip-flop) in a register.

state diagram A graphic depiction of a sequence of states or values.

state machine A logic system exhibiting a sequence of states conditioned by internal logic and external inputs; any sequential circuit exhibiting a specified sequence of states.

static memory A volatile semiconductor memory that uses flip-flops as the storage cells and is capable of retaining data without refreshing.

storage The capability of a digital device to retain bits; the process of retaining digital data for later use.

string A contiguous sequence of bytes or words.

strobing A process of using a pulse to sample the occurrence of an event at a specified time in relation to the event.

subroutine A series of instructions that can be assembled together and used repeatedly by a program but programmed only once.

subtracter A logic circuit used to subtract two binary numbers.

subtrahend The number that is being subtracted from the minuend.

sum The result when two or more numbers are added together.

sum-of-products (SOP) A form of Boolean expression that is basically the ORing of ANDed terms.

sum term The Boolean sum of two or more literals equivalent to an OR operation.

synchronous Having a fixed time relationship; occurring at the same time.

synchronous counter A type of counter in which each stage is clocked by the same pulse.

synthesis The software process where the design is translated into a netlist.

talker An instrument capable of transmitting data on a GPIB (general-purpose interface bus).

target device A PLD mounted on a programming fixture or development board into which a software logic design is to be downloaded; the programmable logic device that is being programmed.

terminal count The final state in a counter's sequence.

text entry A method of entering a logic design into software using a hardware description language (HDL).

throughput The average speed with which a program is executed.

timer A circuit that can be used as a one-shot or as an oscillator; a circuit that produces a fixed time interval output.

timing diagram A graph of digital waveforms showing the proper time relationship of two or more waveforms and how each waveform changes in relation to the others.

timing simulation A software process that uses information on propagation delays and netlist data to test both the logical operation and the worst-case timing of a design.

toggle The action of a flip-flop when it changes state on each clock pulse.

totem-pole A type of output in TTL circuits.

trailing edge The second transition of a pulse.

transistor A semiconductor device exhibiting current and/or voltage gain. When used as a switching device, it approximates an open or closed switch.

trigger A pulse used to initiate a change in the state of a logic circuit.

tristate A type of output in logic circuits that exhibits three states: HIGH, LOW, and high-Z; also known as 3-state.

troubleshooting The technique of systematically identifying, isolating, and correcting a fault in a circuit or system.

truth table A table showing the inputs and corresponding output level of a logic circuit.

TSSOP Thin shrink small-outline package.

TTL Transistor-transistor logic; a class of integrated logic circuit that uses bipolar junction transistors.

TVSOP Thin very small-outline package.

ULSI Ultra large-scale integration; a level of IC complexity in which there are more than 100,000 equivalent gates per chip.

unit load A measure of fan-out. One gate input represents a unit load to the output of a gate within the same IC family.

universal gate Either a NAND gate or a NOR gate. The term *universal* refers to the property of a gate that permits any logic function to be implemented by that gate or by a combination of gates of that kind.

universal shift register A register that has both serial and parallel input and output capability.

up/down counter A counter that can progress in either direction through a certain sequence.

USB Universal serial bus; an external serial bus standard.

UV EPROM Ultraviolet erasable programmable ROM.

variable symbol used to represent a logical quantity that can have a value of 1 or 0, usually designated by an italic letter.

VHDL A standard hardware description language; IEEE Std. 1076-1993.

VLSI Very large-scale integration; a level of IC complexity in which there are from more than 10,000 to 100,000 equivalent gates per chip.

volatile The characteristic of a programmable logic device that loses programmed data when power is turned off.

weight The value of a digit in a number based on its position in the number.

word A complete unit of binary data.

word capacity The number of words that a memory can store.

word length The number of bits in a word.

WORM Write once-read many; a type of optical storage device.

write The process of storing data in a memory.

zero suppression The process of blanking out leading or trailing zeros in a digital display.

Zip disk A type of magnetic storage; a flexible disk with a capacity of 100 MB housed in a rigid plastic cartridge about the size of a floppy.

Index